Light Scattering Reviews 8

Alexander A. Kokhanovsky (Editor)

Light Scattering Reviews 8

Springer

Published in association with
Praxis Publishing
Chichester, UK

PRAXIS

Editor
Dr. Alexander A. Kokhanovsky
Institute of Environmental Physics
University of Bremen
Bremen
Germany

SPRINGER–PRAXIS BOOKS IN ENVIRONMENTAL SCIENCES (*LIGHT SCATTERING SUB-SERIES*)
EDITORIAL *ADVISORY BOARD MEMBER*: Dr. Alexander A. Kokhanovsky, Ph.D., Institute of Environmental Physics, University of Bremen, Bremen, Germany

ISBN 978-3-642-44030-4 ISBN 978-3-642-32106-1 (eBook)
DOI 10.1007/978-3-642-32106-1
Springer Heidelberg New York Dordrecht London

Cover design: Jim Wilkie
Project copy editor: Mike Shardlow
Author-generated LaTex, processed by EDV-Beratung Herweg, Germany

Printed on acid-free paper

Springer is part of Springer Science+Business Media (www.springer.com)

Contents

List of Contributors

Victor P. Afanas'ev
Department of General Physics and Fusion
Moscow Power Engineering Institute
Krasnokazarmennaya Str. 14
Moscow 111250
Russia
v.af@mail.ru

Anthony Baran
Met Office
FitzRoy Road
Exeter, Devon, EX1 3PB
United Kingdom
anthony.baran@metoffice.gov.uk

Yves Bérubé-Lauzière
Département de Génie Électrique et de Génie Informatique
Université de Sherbrooke
2500 Boul. Université
Sherbrooke J1K 2R1
Canada
Yves.Berube-Lauziere@usherbrooke.ca

Lei Bi
Texas A&M University
O&M Building
Department of Atmospheric Sciences
MS 3150
College Station, Texas 77843
USA
bilei@neo.tamu.edu

Anatoli G. Borovoi
V. E. Zuev Institute of Atmospheric Optics
Academician Zuev square 1
634021 Tomsk
Russia
borovoi@iao.ru

Jorge Bouza Domínguez
Département de Génie Électrique et de Génie Informatique
Université de Sherbrooke
2500 Boul. Université
Sherbrooke J1K 2R1
Canada
Jorge.Bouza.Dominguez@USherbrooke.ca

Dmitry S. Efremenko
Remote Sensing Technology Institute
German Aerospace Center
Munchner Str. 20
Wessling 82234
Germany
dmitry.efremenko@dlr.de

Victor Farafonov
Department of Applied Mathematics
St. Petersburg University of Aerospace Instrumentation
Bol. Morskaya Street 67
St. Petersburg 190000
Russia
far@aanet.ru

Evgenii E. Gorodnichev
Department of Theoretical Physics
Moscow Engineering Physics Institute
Kashirskoe Shosse 31
Moscow 115409
Russia
gorodn@theor.mephi.ru

Sergei V. Ivliev
Department of Theoretical Physics
Moscow Engineering Physics Institute
Kashirskoe Shosse 31
Moscow 115409
Russia
ivliev@theor.mephi.ru

Alexander I. Kuzovlev
Department of Theoretical Physics
Moscow Engineering Physics Institute
Kashirskoe Shosse 31
Moscow 115409
Russia
aikuzov@yandex.ru

Christian Lanconelli
Institute of Atmospheric Sciences and Climate (ISAC),
National Council of Research (CNR),
Via Piero Gobetti 101
40129 Bologna
Italy
c.lanconelli@isac.cnr.it

Chao Liu
Texas A&M University
O&M Building
Department of Atmospheric Sciences
MS 3150
College Station, Texas 77843
USA
chao_liu@geos.tamu.edu

Alexander V. Lubenchenko
Department of General Physics and Fusion
Moscow Power Engineering institute
Krasnokazarmennaya Str. 14
Moscow 111250
Russia
lem_a@mail.ru

Angelo Lupi
Institute of Atmospheric Sciences and Climate (ISAC)
National Council of Research (CNR)
Via Piero Gobetti 101
40129 Bologna
Italy
a.lupi@isac.cnr.it

Mauro Mazzola
Institute of Atmospheric Sciences and Climate (ISAC)
National Council of Research (CNR)
Via Piero Gobetti 101
40129 Bologna
Italy
m.mazzola@isac.cnr.it

Vijay Natraj
Earth Atmospheric Science, M/S 233-200
Jet Propulsion Laboratory
California Institute of Technology
4800 Oak Grove Drive
Pasadena, CA 91109
USA
Vijay.Natraj@jpl.nasa.gov

Arthur G. Nikoghossian
V. A. Ambartsumian Byurakan Astrophysical Observatory
Aragatsotn region, Byurakan
Armenia
nikoghoss@yahoo.com

R. Lee Panetta
Texas A&M University
Room 906B, O&M Building
Department of Atmospheric Sciences
MS 3150
College Station, Texas 77843
USA
r-panetta@tamu.edu

Dmitrii B. Rogozkin
Department of Theoretical Physics
Moscow Engineering Physics Institute
Kashirskoe Shosse 31
Moscow 115409
Russia
rogozkin@theor.mephi.ru

Claudio Tomasi
Institute of Atmospheric Sciences and Climate (ISAC)
National Council of Research (CNR)
Via Piero Gobetti 101
40129 Bologna
Italy
c.tomasi@isac.cnr.it

Ping Yang
Texas A&M University
O&M Building
Department of Atmospheric Sciences
MS 3150
College Station, Texas 77843
USA
pyang@tamu.edu

Notes on the contributors

Viktor P. Afanas'ev graduated from the Moscow Power Engineering Institute, Russia, in 1970. His PhD work was devoted to the diffusion of polarized radiation in a medium with a magnetic field. He obtained his PhD in Physics and Mathematics in 1978 at the Moscow Institute of Physics and Engineering and his habilitation title at Moscow State University, Russia, in 2003. His research interests involve radiative and particle transfer. He is developing new electron and ion spectroscopy methods for surface state analysis. He is the author and co-author of about a hundred papers in peer-reviewed journals. His hobby activities are Nordic skiing, canoeing and fishing.

Anthony J. Baran joined the Met Office, England, in 1990. He gained his first degree in Physics with Astrophysics from the University of London, and PhD also from the University of London, and has a total of four degrees. Before joining the Met Office, Dr Baran was engaged in a variety of other fields, including astrophysics, plasma physics and turbulence applied to wind turbine engineering. Since joining the Met Office, Dr Baran has published over sixty-five peer-reviewed papers, and given many invited talks at international conferences and visited a number of universities in France, Germany, and the USA. His research interests cover the areas of electromagnetic and light scattering by nonspherical particles, radiative transfer, physically consistent parameterization of ice optics in climate and high-resolution numerical weather prediction models, physically robust cirrus and aerosol remote sensing methods, including polarimetric measurements. More recently, the nature of scattering by volcanic aerosol and Saharan dust have also been of interest. Dr Baran has also been the principal investigator of airborne cirrus campaigns, and been involved in a number of international projects.

Yves Bérubé-Lauzière (BSc, 1991, and MSc, 1993, in Mathematics/Physics, Université de Montréal, Canada) obtained his PhD (2002) in Electrical/Computer Engineering from McGill University, Canada, for work in collaboration with the Institut National d'Optique (INO), Québec City, on automatic road sign recognition using computer vision techniques. During his PhD, he also worked as full-time researcher at INO in computer vision and biomedical optics, where he developed an optical tartar detection instrument for dentistry, and a small animal planar molecular imager. He joined the Université de Sherbrooke in 2003 as full-time professor. His research interests are in diffuse optical tomography instrumentation and reconstruction for biomedical applications.

Lei Bi received his BS degree from Anhui Normal University in 2003 and his MS degree in Nuclear and Particle Physics from Beijing Normal University in 2006. In 2011, he obtained his PhD degree in Physics from Texas A&M University. Dr Bi is currently a postdoctoral research associate in the Department of Atmospheric Sciences, Texas A&M University. His main research interests are in developing computational programs to solve light scattering by small particles of arbitrary shape and chemical composition and in developing the optical property databases of ice crystals and aerosols for application in atmospheric radiative transfer simulations.

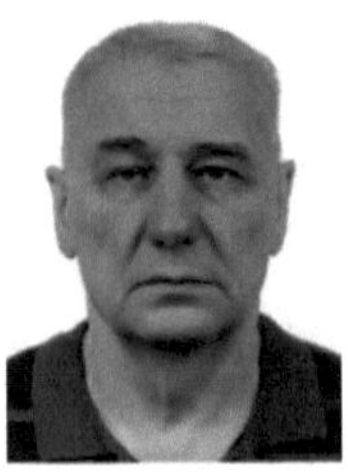

Anatoli G. Borovoi graduated from the Tomsk State University in 1963. He gained his PhD (1967) and Doctor of Sciences (1983) degrees from the same university. His scientific interests include theories of single and multiple scattering of waves in particulate media, wave propagation in random media, speckles and remote sensing. Working at the Institute of Atmospheric Optics (Tomsk) from 1969 until now, he has headed theoretical and experimental works concerning the propagation of laser beams through the turbulent atmosphere with precipitations, speckle-optics, and the development of methods for optical diagnostics of scattering media. He has published more than one hundred papers. His recent papers have been devoted to light scattering by ice crystals of cirrus clouds.

Jorge Bouza Domínguez (Ph.D., 2012, Electrical/Computer engineering, Université de Sherbrooke, Québec, Canada. B.Sc., 2001, Nuclear Physics, University of Havana, Cuba). During his Ph.D., he worked on light propagation modeling in tissues and optical tomographic reconstruction algorithms. He has worked and taught at undergraduate and postgraduate levels for over 10 years in mathematics, applied physics and, specially, in medical imaging. In the field of medical imaging, he has contributed to the development of diverse imaging techniques related to neurosciences and clinical cancer imaging. He joined Bishop's University in 2012 as adjunct professor. His research interests are in medical imaging and biomedical optics, with emphasis in biomedical instrumentation, optical tomography and molecular imaging.

Dmitry S. Efremenko graduated from the Moscow Power Engineering Institute, Russia, in 2009. He received his PhD in Physics and Mathematics from Moscow State University, Russia, in 2011. Currently, he is a research scientist at the German Aerospace Center (DLR) and works in the field of inverse problems for atmospheric remote sensing and radiative transfer.

Victor Farafonov graduated from St Petersburg State University, Russia, in 1976. He received a PhD in Mathematical Physics and the Doctor of Sciences degree in Physics and Mathematics from the St Petersburg State University in 1981 and 1991, respectively.

He is a head of the Applied Mathematics Chair at St Petersburg University of Aerospace Instrumentation. His research interests are related to electromagnetic scattering and special functions, in particular spheroidal wave functions. He has published about 120 papers in the field of the light scattering theory.

Evgenii E. Gorodnichev graduated from the Theoretical Physics Department of the Moscow Engineering Physics Institute in 1985. He received his PhD in Theoretical and Mathematical Physics and Doctor of Science both from the Moscow Engineering Physics Institute in 1989 and 2012, respectively. His PhD work was devoted to coherent phenomena in wave propagation through disordered media; his DSc thesis was devoted to polarization effects in multiple scattering of light. His principal scientific interests are concerned with multiple scattering of polarized light in random media. He has published over sixty works on light scattering theory. He is currently an associate professor at the Department of Theoretical Physics of the Moscow Engineering Physics Institute.

Sergei V. Ivliev graduated from the Theoretical Physics Department of the Moscow Engineering Physics Institute in 1983. He received his PhD in Theoretical and Mathematical Physics from the Moscow Engineering Physics Institute in 1991. His PhD work was devoted to the exchange and correlation effects in a system of interacting electrons. His principal scientific interests are concerned with electron transport and the interaction of waves with complex media. Ivliev has about forty scientific publications. He is currently an associate professor at the Department of Theoretical Physics of the Moscow Engineering Physics Institute.

Alexander I. Kuzovlev graduated from the Theoretical Physics Department of the Moscow Engineering Physics Institute in 1983. He received his PhD in theoretical and mathematical physics from the Moscow Engineering Physics Institute in 1988. His PhD work was devoted to the albedo problem of the radiative transfer theory. His principal scientific interests are concerned with radiative transfer. He has published over seventy papers on scattering theory. He is currently an associate professor at the Department of Theoretical Physics of the Moscow Engineering Physics Institute.

Christian Lanconelli graduated in Physics in July 2002 at the University of Bologna, discussing a thesis on measurements and models of the reflectivity of land surfaces. In 2007 he obtained his PhD in Physics at the University of Ferrara, carrying out the experimental and modeling activities at the ISAC-CNR Institute of Bologna. His principal investigation field is radiative transfer in the atmosphere, with particular attention to the parameterization of surface reflectance and aerosol and cloud radiative forcing, also including studies concerning the cloud effects on the SW and LW radiation budget at the Earth's surface. He has gained laboratory and field experience with spectrometric measurements of reflectivity, solar photometry, nephelometric and absorption photometric techniques, and measurements of radiation fluxes at the ground. Since his PhD thesis, he has participated in various field campaigns planned as parts of the ISAC-CNR activities in Antarctica and in the Arctic Svalbard region.

Chao Liu received a bachelor's degree in Physics at Tangji University. He is currently a PhD student in Atmospheric Sciences at Texas A&M University. His research interests are the single-scattering properties of the atmospheric particles and the development of numerical algorithms to compute these properties.

Alexander V. Lubenchenko, born 1966, in Kaskelen, Kazakhstan. He is a professor of Moscow Power Engineering Institute. He obtained his Doctor of Technics degree in 2006. His research focus is on new methods for the solution of the transfer equation for radiation and particles in media with an anisotropic scattering law based on invariant embedding. He has over 80 publications.

Angelo Lupi is currently research assistant at ISAC-CNR in Bologna, having obtained an MSc in Physics at the University of Bologna, and a PhD in Polar Science at the University of Siena. He has gained a 10-year professional experience in atmospheric physics, focusing his studies on radiative transfer processes occurring in the atmosphere, analysis and modeling of aerosol optical and physical properties, and analysis and modeling of surface reflectance properties. Since 2003, he has participated to four Antarctic expeditions. He has published more than 20 papers in international journals.

Mauro Mazzola, MSc (2002, University of Milan) and PhD (2006, University of Ferrara) in Physics, has been a fellow of the ISAC-CNR Institute since 2007. His research topics concern measurements and modeling of aerosol optical and physical properties aimed at evaluating the aerosol radiative effects, and analysing ground-based remote sensing measurements, *in situ* optical data and satellite observations. He has gained experimental field experience at mid-latitude and polar sites, through the participation to numerous international measurement campaigns. He is the author of 13 publications in international journals.

Vijay Natraj received bachelor and master degrees in Chemical Engineering from the National University of Singapore in 1998 and 2002, respectively, and a PhD degree in Chemical Engineering from the California Institute of Technology in 2008. His PhD dissertation was on radiative transfer modeling for CO_2 retrievals from space. As part of his graduate work, he developed a fast and accurate polarized radiative transfer model to account for polarization by the atmosphere and the surface in the near-infrared. This model will be used operationally for the Orbiting Carbon Observatory-2 (OCO-2) mission. After completing his doctoral degree, Dr Natraj continued to work as a researcher in the Department of Planetary Sciences at Caltech, where he still holds a visiting position. He also holds a visiting assistant researcher position at the University of California at Los Angeles (UCLA) Joint Institute for Regional Earth System Science and Engineering (JIFRESSE). Dr. Natraj joined the Jet Propulsion Laboratory (JPL) as a Scientist in February 2010. He leads the retrieval algorithm group at JPL for the Geostationary Coastal and Air Pollution Events (GEO-CAPE) and Panchromatic Fourier Transform Spectrometer (PanFTS) projects. He is also the principal investigator of a project to retrieve aerosol vertical profiles from O_2 A band and O_2–O_2 absorption measurements. His research interests are in the areas of scattering, polarization, aerosol and cloud modeling, fast radiative transfer computations, and information theoretical analysis.

Arthur Nikoghossian (1965, graduated with honours from the astrophysical department of Yerevan St. University, Armenia) supported his Candidate thesis (1969) under the leadership of Ambartsumian in non-linear theory of radiative transfer. It was one of the pioneering works in the field. The Phys.-Math.Sci. Doctor degree was obtained in 1986 (Leningrad St. University) for development in the theory of the spectral lines formation for partial redistribution of the radiation over frequencies and directions. From 2006 he was the principal researcher of Byurakan Astrophysical Observatory. The long-term collaboration with Observatoir de Paris-Meudon and Institut d?Astrophysique in France was devoted to solar physics (quiescent prominences and corona) and radiation transfer theory. The research interests are in tne theory of stellar atmospheres, radiation transfer, non-stationary stars and solar physics. He is the author of about hundred papers in peer-reviewed journals. More than three decades he tought the course of theoretical astrophysics and other astrophysical and mathematical courses in Yerevan St. University. He is a member of the IAU (Commission 36, Stellar Atmospheres) and the European and the Euro-Asian Astronomical Societies.

R. Lee Panetta received a BS degree in Mathematics from McGill University and a PhD in mathematics at the University of Wisconsin-Madison. He is currently Professor in both the Atmospheric Sciences and Mathematics Departments at Texas A&M University. He has previously held a faculty position at Occidental College and visiting positions at the Space Science and Engineering Center at the University of Wisconsin, the Geophysical Fluid Dynamics Laboratory at Princeton, and the Joint Institute for the Study of the Atmosphere and Oceans at the University of Washington. His research interests include computational methods in electromagnetic scattering, pattern-forming partial differential equations, mathematical models of turbulent geophysical flows, and the general circulation of Earth's atmosphere.

Dmitrii B. Rogozkin graduated from the Theoretical Physics Department of the Moscow Engineering Physics Institute in 1979. He received his PhD, in Theoretical and Mathematical Physics, and Doctor of Science from the Moscow Engineering Physics Institute in 1984 and 1998, respectively. His PhD work was devoted to analytical methods of solving and radiative transfer equation under conditions of anisotropic scattering. His DSc thesis was devoted to interference phenomena in multiple scattering from disordered media. Currently his attention is focused on the study of coherent and polarization phenomena in multiple wave scattering. He has about eighty publications. He is currently Professor at the Department of Theoretical Physics of the Moscow Engineering Physics Institute.

Claudio Tomasi worked as Researcher at the National Council of Research (CNR) from 1970 to 1991, and as director of research from 1991 to 2006. He retired in April 2006, but continues his research activity as Associate Researcher at the Institute of Atmospheric Sciences and Climate (ISAC-CNR). He is author of more than 140 papers published in international peer-reviewed journals and international conference proceedings, and of more than 150 articles published in national journals, technical reports, scientific memos, and books. He is currently P. I. of two research projects (AEROCLOUDS, POLAR-AOD) devoted to study direct aerosol-induced radiative forcing in the Po valley (Italy) and polar aerosol radiative parameters, respectively, and is partner of the CLIMSLIP project, paying particular attention in evaluating changes in the radiation budget of the surface–atmosphere system induced by variations in the surface albedo.

Ping Yang received the BS (Theoretical Physics) and MS (Atmospheric Physics) degrees from Lanzhou University and Lanzhou Institute of Plateau Atmospheric Physics, Chinese Academy of Sciences, Lanzhou, China, in 1985 and 1988, respectively, and the PhD degree in meteorology from the University of Utah, Salt Lake City, USA, in 1995. He is currently a professor and the holder of the David Bullock Harris Chair in Geosciences, the Department of Atmospheric Sciences, Texas A&M University, College Station, Texas, USA. His research interests cover the areas of remote sensing and radiative transfer. He has been actively conducting research in the modeling of the optical and radiative properties of clouds and aerosols and their applications to space-borne and ground-based remote sensing. He received a best paper award from the Climate and Radiation Branch, NASA Goddard Space Center in 2000, the U.S. National Science Foundation CAREER grant in 2003, and the Dean's Distinguished Achievement Award for Faculty Research, College of Geosciences, Texas A&M University in 2004. He is a fellow of the Optical Society of America (OSA). He currently serves as an associate editor for the Journal of Atmospheric Sciences, the *Journal of Quantitative Spectroscopy & Radiative Transfer*, and the *Journal of Applied Meteorology and Climatology, and is also on the Editorial Board for Theoretical and Applied Climatology.* Dr Yang has published one textbook and more than 190 peer-reviewed papers.

Preface

This eighth volume of *Light Scattering Reviews* is aimed at the discussion of recent trends and results in radiative transfer and light scattering theories. Radiative transfer theory is based on the phenomenological radiative transfer equation whereas the main equations of light scattering theory are derived on the basis of the Maxwell theory. The first part of the book discusses recent results in single light scattering theory. Aerosol and cloud particles composed of liquids (mainly water) are generally spherical in shape although they can contain nonspherical inclusions. Solids suspended in the atmosphere (road dust, ice crystals, etc.) are irregularly shaped particles. The methods of calculation of optical properties of spherical particles are well developed. This is not the case for irregularly shaped particles, where the main tool used for the computations is the physical and geometrical optics approximations. Interestingly enough, due to large sizes of scatterers and weak absorption in the visible, various optical models of irregularly shaped particles produce similar light scattering patterns, for both intensity and degree of polarization of scattered light. These patterns are usually featureless and differ considerably from those for Mie scatterers.

The first chapter of this volume, prepared by A. Baran, discusses various models used to represent single light scattering by crystalline clouds and nonspherical aerosols. Bi and Yang describe in detail the physical-optics hybrid methods for computing the scattering and absorption properties of ice crystals and dust aerosols. A. Borovoi considers the physical optics approximation and the shadow-forming field, which is a useful concept used for understanding light scattering properties of large scatterers. R. L. Panetta et al. introduce a pseudo-spectral time domain method, valid also for small macroscopic particles, where physical optics cannot be used. Farafonov in the last chapter of Part I discusses the application of non-orthogonal bases in the theory of light scattering by spheroidal particles. The last two chapters of Part I section are based on the direct solution of Maxwell equations.

Part II of the volume is aimed at the discussion of diverse radiative transfer problems, such as optical imaging in biological media (Dominguez and Bérubé-Lauzière), transillumination of turbid media by polarized light (Gorodnichev et al.), and surface state analysis (Efremenko et al.). The astrophysical applications are presented by A. Nikoghossian. V. Natraj gives a comprehensive account of fast radiative transfer methods. In the last chapter of the volume, C. Tomasi et al. study the aerosol direct effect on climate using extensive radiative transfer calculations.

Bremen, Germany
October, 2012

Alexander A. Kokhanovsky

Part I

Single Light Scattering

1 Light scattering by irregular particles in the Earth's atmosphere

Anthony J. Baran

1.1 Introduction

On planet Earth, we are fortunate to have an atmosphere, which sustains life, along with the radiation radiated from our nearest star, the Sun. The interaction between radiation emitted by the Sun and the Earth's atmosphere, not only helps to sustain life, but also gives rise to the observed display of colours in the Earth's atmosphere. The interaction between electromagnetic radiation emitted by the Sun and the Earth's atmosphere is responsible for the depth of blue in the sky. The red sky at sunset. The yellowness of the Sun, rainbows, the whiteness of clouds, the appearance of haloes around the Sun and moon, and the vivid array of colours that might appear in the sky after volcanic eruptions. All these manifestations of colour arise from the basic interaction between electromagnetic radiation and matter. On the surface of the Earth, we observe these manifestations of colour, through the process of single-scattering or multiple scattering. Indeed, if we were observers in space, the Sun would appear a different colour, in fact white, simply because there is no atmosphere in space.

Through the process of scattering, we are indeed fortunate to enjoy the rich tapestry of colour that nature can provide, which has inspired great painters and poets, such as Turner and Wordsworth. Indeed, Keats once commented that the rainbow appears so beautiful that science by describing its causes will destroy its beauty. However, this is a comment that few would agree with, since science has enhanced its beauty through the unification of electricity and magnetism, so elegantly coupled together, through the Maxwell equations, achieved in 1861. It is this set of equations, which can predict all the scattering that we observe on Earth, and indeed on any other planet or in interstellar space. The Maxwell equations are universal; far from destroying the beauty of the rainbow, they enhance its beauty, through four simple universal equations.

Given that the Earth's atmosphere is composed of matter, which can take on a variety of forms, such as molecules of different gases, aerosol particles, water droplets and ice crystals. These different states of matter exist in different sizes, materials, densities, and spatial and temporal distributions, and over a spectrum of size. Fundamentally, it is the size and material composition of the particles that determine how incident electromagnetic radiation will interact with atmospheric particles. For instance, molecules being the smallest size will interact with incident electromagnetic radiation of particular wavelengths, i.e., the shorter wavelengths, such as blue light. This preferential wavelength selection manifests itself as scattered blue light around the sky. The intensity of scattered blue light also depends on the density profile of atmospheric molecules and the presence of ozone. Since, the atmosphere is composed of gases, which vary spatially and temporally, as the Sun sets near the horizon, most of the blue light has been scattered out of the line of sight, and this results in the longer wavelength, red light, being preferentially scattered into our eyes.

As the size of particles increase with respect to the wavelength of incident electromagnetic radiation, i.e., size $\gg$ incident wavelength, results in processes such as refraction into the particle, and internal reflection around the particle. These processes in water droplets form the rainbow and, in ice crystals, they can result in the formation of the 22° and 46° halo, observed around the Sun. Atmospheric particles such as aerosols may be composed of differing materials, and these different materials, can absorb visible light, which may result in scattered electromagnetic radiation, with a vivid array of colours. Aerosols can be injected into the Earth's upper atmosphere by meteor impacts, wind-blown desert dust storms, and volcanic eruptions.

Clouds also manifest themselves in the Earth's atmosphere, and these in the lower atmosphere, due to the warmer temperatures, are water clouds, which are composed of water droplets. In the upper atmosphere, usually at altitudes greater than about 6 km, at colder temperatures, the water molecules freeze into the simplest ice crystal form – hexagonal ice crystals. These ice crystal clouds are called cirrus, and appear as wispy tufts of hair, taking on a ghostly appearance. Given that the basic ice crystal formed is hexagonal in shape, this geometry, which is tumbling randomly in the Earth's atmosphere, is the reason why the 22° and 46° halo exist. Ice crystals, being essentially non-absorbing at visible wavelengths, refracts light into the crystal via its mantle surfaces, and light refracts back out of the mantle surface, and through its ends, resulting in the familiar 22° and 46° haloes. Therefore, optical phenomena observed in the sky depend on the phase, composition, shape and orientation of the particles.

The differences in scattering and absorption, between the different particle ensembles, will result in a net radiative effect that either cools or warms the surface of the Earth. The net radiative effect is defined as follows; it is the sum of the short-wave radiative effect and long-wave radiative effect. The short-wave radiative effect is the difference between the reflected short-wave flux and the clear-sky reflected short-wave flux, and the long-wave radiative effect is similarly defined. In general, the short-wave radiative effect is generally negative (i.e., cooling effect), whilst the long-wave radiative effect is generally positive (i.e., warming effect). The net radiative effect can therefore be positive, neutral or negative. Therefore, in order to determine the net radiative effect of different particle ensembles it is important to understand and to predict, how electromagnetic radiation interacts with atmospheric particulates, if the future climate state of planet Earth is to be reliably predicted.

However, as Figs. 1.1(a) and 1.1(b) demonstrate, the representation of particulate scattering and absorption in a climate model is far from understood. The figure

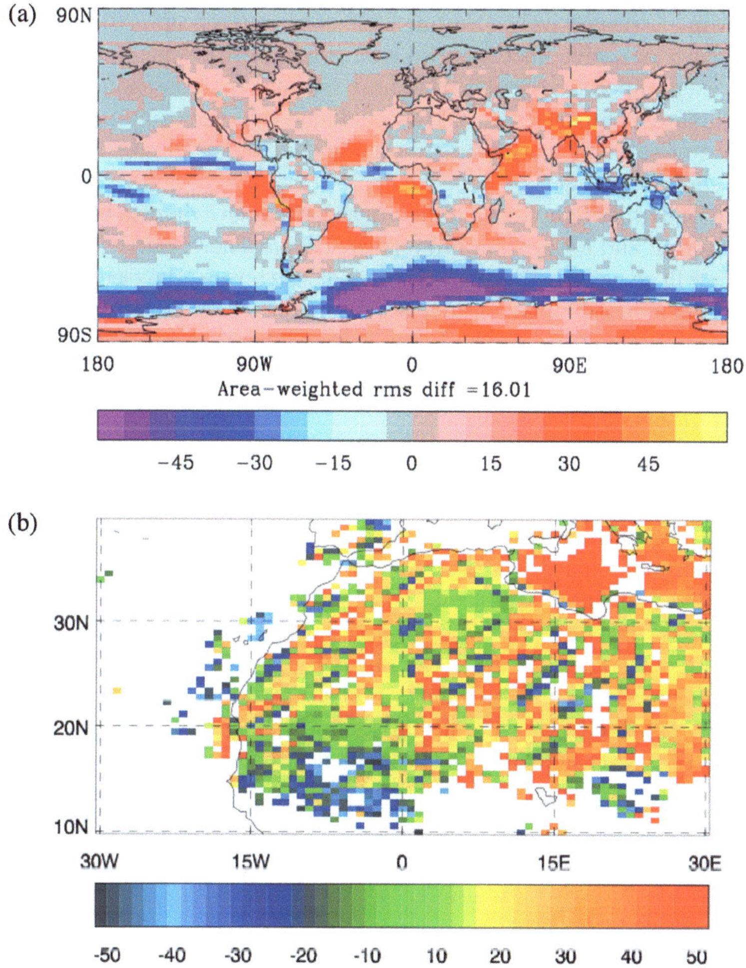

Fig. 1.1. Differences between GCM predicted top-of-atmosphere reflected short-wave flux (units of $\mathrm{W\,m^{-2}}$) and space-based measurements for June-July-August, averaged over 10 years, for (a) cloud and (b) mineral dust aerosol (b) from J. Mulcahy, personal communication).

compares space-based measurements of the Earth's top-of-atmosphere (TOA) reflected short-wave flux with a general circulation model (GCM) prediction of the TOA reflected short-wave flux. Figure 1.1(a) compares this difference for cloud, and Fig.1.1(b) compares the difference for mineral dust aerosol. In general, the figure shows that the GCM is too reflective, when compared against measurements, though in the tropics the cloud is too dark. However, how much incident solar radiation is reflected back to space depends not only on scattering, but also on how the particles are vertically distributed, in terms of their size and mass and their altitude, and on their spatial and temporal distribution. All these parameters are currently highly uncertain, and can lead to uncertainties in the instantaneous short-wave radiative effect of cirrus and mineral dust aerosol of about $\pm 30\,\mathrm{W\,m^{-2}}$

and $\pm 46\,\mathrm{W\,m^{-2}}$, respectively (Baran, 2009; Osborne et al., 2011). Not surprisingly, with such radiative uncertainties and differences between measurements and models, as exemplified by Fig. 1.1, the most recent fourth assessment report of the Panel on Climate Change (IPCC, 2007) concluded that the coupling between clouds and aerosol to the Earth's atmosphere remains one of the greatest uncertainties in predicting climate change.

Having briefly highlighted the difficulties that climate models have in representing the scattering and absorption properties of atmospheric particulates, the rest of this chapter will review the basic definitions of scattering, the electromagnetic and light scattering methods used to solve scattering problems, the myriad of sizes and shapes of particle that exist in the Earth's atmosphere, and the idealized irregular models that have been proposed to represent their light scattering properties.

1.2 Basic definitions of scattering

In this chapter, it is assumed that atmospheric particulates are randomly oriented in three-dimensional space, and that each particle possesses a plane of symmetry. Incident sunlight on this ensemble of nonspherical particles is unpolarized, in which case the incident Stokes vector ($I_{\mathrm{inc}}, Q_{\mathrm{inc}}, U_{\mathrm{inc}}, V_{\mathrm{inc}}$) is linearly related to the scattered Stokes vector ($I_{\mathrm{sca}}, Q_{sca}, U_{\mathrm{sca}}, V_{\mathrm{sca}}$) by a 4×4 scattering matrix. Each element of the scattering matrix is dependent on the scattering angle, θ, but not on the azimuth angle, ϕ, due to the simplifying assumptions previously stated. Thus, the scattered Stokes vector is related to the incident Stokes vector (van de Hulst, 1957), via the following 4×4 scattering matrix.

$$\begin{pmatrix} I_{\mathrm{sca}} \\ Q_{\mathrm{sca}} \\ U_{\mathrm{sca}} \\ V_{\mathrm{sca}} \end{pmatrix} = \frac{C_{\mathrm{sca}}}{4\pi r^2} \begin{pmatrix} P_{11} & P_{12} & 0 & 0 \\ P_{21} & P_{22} & 0 & 0 \\ 0 & 0 & P_{33} & P_{34} \\ 0 & 0 & P_{43} & P_{44} \end{pmatrix} \begin{pmatrix} I_{\mathrm{inc}} \\ Q_{\mathrm{inc}} \\ U_{\mathrm{inc}} \\ V_{\mathrm{inc}} \end{pmatrix} \tag{1.1}$$

where in Eq. (1.1) C_{sca} is the particle scattering cross-section (scattering efficiency multiplied by the particle orientation-averaged geometric cross-section) and r is the distance of the particle from the observer. The scattering efficiency, Q_{sca}, is a dimensionless quantity, and in the case of zero absorption and particle size $\gg$ incident wavelength (referred to as the limit of geometric optics), then $Q_{\mathrm{sca}} = Q_{\mathrm{ext}} \sim 2.0$, where Q_{ext} is the extinction efficiency (van de Hulst, 1957). The 4×4 matrix, given by Eq. (1.1), is called the 'scattering phase matrix', 'scattering matrix' or the 'Mueller' matrix. The term phase used here is unfortunate as Eq. (1.1) does not by itself contain any information on phase.

If phase were to be included, then the amplitude scattering matrix must be solved (Mishchenko, 2000). The amplitude scattering matrix more generally describes a scattering event, using electromagnetic theory. The linearity of the boundary conditions imposed by the Maxwell equations means that the electric or magnetic vector fields of the incident plane waves, denoted by i, can be related to the electric or magnetic vector fields of the scattered plane waves, denoted by s, by resolving them into their parallel ($\|$) and perpendicular ($\perp$) counterparts, defined with respect to the scattering plane. The scattered and incident electric fields can be related to each other via the following 2×2 amplitude scattering matrix (Bohren and Huffman, 1983):

$$\begin{pmatrix} E_{s\perp} \\ E_{s\|} \end{pmatrix} = \frac{e^{ik(r-z)}}{-ikr} \begin{pmatrix} S_2 & S_3 \\ S_4 & S_1 \end{pmatrix} \begin{pmatrix} E_{i\perp} \\ E_{i\|} \end{pmatrix} \tag{1.2}$$

where in Eq. (1.2), k is the wavenumber ($2\pi/\lambda$, and λ is the incident wavelength) of a plane wave propagating along the z-axis. Equation (1.2) describes the complete particle scattering pattern, inclusive of interference. In general, the matrix elements of Eq. (1.2) could also be functions of θ and ϕ to obtain the complete scattering pattern.

Since Eq. (1.1) considers only randomly oriented particles, each possessing a plane of symmetry, then out of the 8 elements shown in Eq. (1.1), only 6 are independent, due to $P_{21} = P_{12}$ and $P_{43} = -P_{34}$ (van de Hulst, 1957). Likewise, the scattering matrix elements of Eq. (1.2) could also be simplified by considering simple symmetric shapes such as the sphere. Given that incident sunlight is unpolarized, then the incident Stokes vector is $[1, 0, 0, 0]^{\mathrm{T}}$, which means that from Eq. (1.1), $I_{\mathrm{sca}} \propto P_{11} I_{\mathrm{inc}}$. The P_{11} element of the scattering phase matrix is called the scattering phase function and it is proportional to the scattered intensity, and is normalized to unity by the following equation:

$$\frac{1}{2} \int_0^{\pi} P_{11}(\theta) \sin\theta \, d\theta = 1 \tag{1.3}$$

Note also, from Eq. (1.1), that after multiplying out the scattering phase matrix by the incident unpolarized Stokes vector, the resulting scattered Stokes vector has become linearly polarized, since $Q_{\mathrm{sca}} \propto P_{12} I_{\mathrm{inc}}$. Incident unpolarized electromagnetic radiation on an ensemble of particles generally leads to scattered linearly polarized light , unless $P_{12} = 0$, which is true in the exact forward (i.e., $\theta_{\mathrm{s}} = 0$) and exact backscattering (i.e., $\theta_{\mathrm{s}} = 180°$) directions. It is therefore important to generally include polarization in radiative transfer calculations, as its omission can lead to serious errors in aerosol scattered radiance calculations (Mishchenko et al., 1994; Levy et al., 2004; Feng et al., 2009).

The degree to which unpolarized incident light becomes linearly polarized is called the degree of linear polarization (DLP), in terms of matrix elements from Eq. (1.1), it is defined by the following equation:

$$DLP = -\frac{P_{12}}{P_{11}} \tag{1.4}$$

The DLP is an important quantity, as it gives the proportion of scattered light that is horizontally (i.e., DLP < 0) and perpendicularly (i.e., DLP > 0) polarized, with respect to some horizontal plane. Thus, the DLP can take on both positive and negative values, and the DLP depends upon the size and shape of the particle, as well as scattering angle. Therefore, it is an important quantity to measure for the remote sensing of cirrus and aerosol particles.

In the exact backscattering direction the scattering matrix is diagonal (Mishchenko and Hovenier, 1995) and consists of two independent elements of the scattering matrix, which are P_{11} and P_{22}. It has been seen from above that incident un polarized light can become linearly polarized due to the process of single scattering, in turn, incident vertical or horizontal polarized light can become 'depolarized' due to the process of single scattering. Laser light can be a source of polarized light. If it is assumed that a laser beam is 100% linearly polarized parallel to a fixed plane such as the scattering plane, then the incident Stokes parameter is $[1, 1, 0, 0]^{\mathrm{T}}$, and in the case of single scattering in the exact backscattering direction Q_{s} differs from I_{s}, and this phenomenon is called linear depolarization. The degree to which polarized light becomes depolarized due to single scattering is called the linear depolarization ratio, δ_{L}, defined as the ratio of flux between the cross-polarized component of the backscattered light to the co-polarized component, and is given by the following equation (Mishchenko et al., 2002):

$$\delta_{\mathrm{L}} = \frac{P_{11}(180^\circ) - P_{22}(180^\circ)}{P_{11}(180^\circ) + P_{22}(180^\circ)} \tag{1.5}$$

In the case of perfectly symmetric particles such as spheres or if the incident light is aligned with the symmetry axis of an axially symmetric particle, then, $\delta_{\mathrm{L}} = 0$, since $P_{11} = P_{22}$. For the case of randomly oriented axially symmetric particles or particles lacking symmetry, then $\delta_{\mathrm{L}} > 0$. Therefore, δ_{L} is also an important quantity to measure for the remote sensing of clouds and aerosol and can be used to differentiate between spherical and nonspherical particles. However, 100% incident linearly polarized light may be linearly depolarized not only because of particle shape and size (Mishchenko and Sassen, 1998), but also due to multiple scattering. It is therefore important to take into account multiple scattering when considering atmospheric linear depolarization (Noel and Chepfer, 2004; Hu et al., 2007). The expression for δ_{L}, given by Eq. (1.5), can be simply re-arranged, to derive the ratio of $P_{22}(180^\circ)$ to $P_{11}(180^\circ)$, given by the following equation:

$$\frac{P_{22}(180^\circ)}{P_{11}(180^\circ)} = \frac{1 - \delta_{\mathrm{L}}}{1 + \delta_{\mathrm{L}}} \tag{1.6}$$

Equation (1.6) has been used to constrain scattering models of cirrus using a space-based LIght Detection And Ranging (LIDAR) instrument (Baum et al., 2011).

Equation (1.1) implicitly assumes that particles are randomly oriented in three-dimensional space. A question that naturally arises is how true is this assumption? Optical phenomena such as Sun dogs, caused by oriented hexagonal ice plates, are quite commonly observed. However, to obtain a global perspective of the commonality of oriented cirrus particles, space-based measurements must be used. There have been lidar observations of horizontally oriented hexagonal plates in cirrus (Chepfer et al., 1999; Noel and Sassen, 2005; Westbrook et al., 2009). However, more recent work by Yoshida et al. (2010) using global Cloud-Aerosol Lidar and Infrared Pathfinder Satellite Observations (CALIPSO) data, conclude that oriented plates occur at temperatures between about −10°C and −20°C, whilst randomly oriented hexagonal columns occur at temperatures colder than this. Similarly, Noel and Chepfer (2010) also, using global Calipso data, conclude that at temperatures colder than −30°C, oriented particles are very infrequent. At temperatures warmer than this, between about −10°C and −30°C, they find that orientation can be more common, with 30% and 50% of low and high latitude clouds being composed of oriented ice particles, respectively. However, the work of (Bréon and Dubrulle, 2004;

Noel and Chepfer, 2004) found that the actual fraction of horizontally oriented ice crystals is more likely to be about 10^{-2}. With such small fractions of oriented ice crystals, it would be surprising if oriented particles dominated the cirrus solar and thermal radiation fields, since cirrus temperatures are usually less than about $-40°$C (Guignard et al., 2012).

Resolving the orientation of mineral dust aerosol particles may be more problematic. It is known that electric fields can exist within dust plumes at some considerable distance from their source (Ulanowski et al., 2007; Harrison et al., 2010 and Nicoll et al., 2011). Moreover, if the aerosol particles are charged, then alignment may occur. This charging of mineral dust particles might help to explain why these aerosols are transported over such long distances (Ulanowski et al., 2007). However, the alignment of atmospheric particles is still an active area of research, and further intensity and polarized measurements are required before definitive conclusions can be made.

To predict the transfer of incident radiation through clouds of aerosol or ice crystals, the following optical properties are required, if polarization is neglected. The P_{11} element of the scattering phase matrix, volume extinction coefficient, K_{ext}, volume scattering coefficient, K_{sca}, volume absorption coefficient, K_{abs}, and the single-scattering albedo (the ratio of the scattered energy to the total amount of attenuated energy), ω_0. The volume extinction/scattering coefficient is defined by,

$$K_{\mathrm{ext}_\lambda,\mathrm{sca}_\lambda} = \int Q_{\mathrm{ext}_\lambda,\mathrm{sca}_\lambda}(\vec{q})\langle S(\vec{q})\rangle n(\vec{q})\,\mathrm{d}\vec{q} \tag{1.7}$$

where both the extinction and scattering efficiency factors are functions of the incident wavelength, λ. The term $\langle S(\vec{q})\rangle$ is the orientation averaged geometric cross-section, where the vector $\vec{q}$ represents the size and shape of particles and $n(\vec{q})$ is the size spectra of particles (PSD). From K_{ext} and K_{sca} the volume absorption coefficient, K_{abs}, can be found from.

$$K_{\mathrm{abs}} = K_{\mathrm{ext}} - K_{\mathrm{sca}} \tag{1.8}$$

Then, by definition, ω_0, is given by.

$$\omega_0 = K_{\mathrm{sca}}/(K_{\mathrm{abs}} + K_{\mathrm{sca}}) \tag{1.9}$$

The subscript λ has been dropped for reasons of clarity.

Equation (1.8) can be re-arranged, in terms of K_{ext}, and then integrated, to find the total optical depth, τ, of the layer of particles of some vertical geometric depth, dz. The total optical depth is given by.

$$\tau_{\mathrm{ext}} = \int_{z_1}^{z_2} (K_{\mathrm{sca}} + K_{\mathrm{abs}})\,\mathrm{d}z \tag{1.10}$$

To forward model the flux or irradiance in GCMs, a further parameter is required, and this parameter is called the asymmetry parameter. The asymmetry parameter, g, is a measure of how much asymmetry there is in the forward peak of the P_{11} element of the scattering phase matrix. The asymmetry parameter is then a parameterization of the scattering phase function, and is described by a single number. The formal mathematical definition of the asymmetry parameter is that it is the average cosine of the scattering angle and it is given by the following equation.

$$g = \langle \cos\theta \rangle = \frac{1}{2}\int_{-1}^{1} P_{11}(\cos\theta)\,\cos\theta\,\mathrm{d}(\cos\theta) \tag{1.11}$$

The asymmetry parameter can take on values, at least mathematically, between ± 1, depending on the size, shape, and complex refractive index of the particle. It is an important quantity in climate models, because it determines how much solar radiation is reflected back to space. For high g values, clouds appear dark, while for low g values clouds appear bright.

The optical properties defined by Eqs (1.7, 1.9 and 1.11) are often referred to as the 'scalar' or 'total' optical properties, as they are given by total integrals, and so have no angular dependence. It is the scalar optical properties that are required in GCMs, to forward model solar and infrared irradiance (flux). The next section of this chapter reviews the current and possibly future methods, which are or might be employed to solve for the general scattering properties of atmospheric particles.

1.3 Electromagnetic and light scattering methods

The fundamental problem to be solved is Eq. (1.2), for any particle ensemble of shapes or composition (complex refractive index), independent of incident wavelength or frequency. To date, there is no one electromagnetic or light scattering method available that can be arbitrarily applied to any particle of given size and complex refractive index. This is because atmospheric particles, as will be understood later in this chapter, can be very large and complex, rendering current numerical solutions to the Maxwell equations impossible to obtain. For this reason, approximations are still sought to solve Eq. (1.1) or Eq. (1.2). The range of applicability of current electromagnetic methods depends on the particle size, shape and complex refractive index. When the electromagnetic limit is reached, approximations such as geometric optics or physical optics are applied to solve Eq. (1.1) and Eq. (1.2), respectively. Currently, it is possible to obtain general scattering solutions over the size space observed in the Earth's atmosphere, by bridging the gap between small and large particle sizes, by combining electromagnetic and physical optics methods, respectively. However, this is still unsatisfactory, as physical optics is still an approximation, ideally the electromagnetic method should be applied to the entire size domain observed in the Earth's atmosphere, or solutions using physical optics methods are practically indistinguishable from electromagnetic solutions.

The electromagnetic or light scattering method used to solve Eqs (1.1), (1.2), in part, depends on the ratio of the particle size to incident wavelength. This ratio is called the size parameter, X, and is more formally defined as the ratio of the circumference of the equivalent sphere to incident wavelength. The equivalent sphere is usually defined as the equal volume or area (surface or geometric projected) sphere, of some radius r_v or r_a, respectively. The size parameter of aerosols and ice crystals, in the Earth's atmosphere, can range from less than unity to $\gg$1000s. Currently, small size parameter space, is defined as about $X = 60$ (electromagnetic methods are applied) and large size parameter space, is defined as $X \gg 60$ (geometric optics methods are applied). Though, for $20 < X \ll 60$, physical optics methods can be applied. The ultimate goal of this field is to render the partition between small X and large X as meaningless. However, since this one classical method, has so far eluded researchers, approximations are still usefully employed to solve Eqs (1.1), (1.2).

Small X is covered by electromagnetic methods; these methods usually fall into two categories, depending on how the Maxwell equations are solved. The so-called 'volume-based' and 'surface-based' methods. The volume-based methods require discretization of the whole volume of the scatterer, and therefore the computational resources required by these methods are high. However, volume-based methods can, in principle, be applied to any homogeneous or inhomogeneous arbitrary shape of any complex refractive index. Examples of volume-based methods, suitable for application to the problem of scattering by nonspherical atmospheric particles, are the Finite-Difference-Time-Domain (FDTD) (Yee, 1966; Yang et al., 2000; Sun et al., 1999) and the Discrete Dipole Approximation (DDA) (Purcell and Pennypacker, 1973; Draine and Flatau, 1994; Zubko et al., 2008, and references therein). A further improvement to the FDTD method is the PseudoSpectral Time Domain (PSTD) method, which allows full electromagnetic scattering solutions for larger X values relative to the FDTD method, for the sphere, size parameters up to $X = 80$ have been achieved (Chen et al., 2008). The Boundary Element method has been applied to the hexagonal ice column for certain orientations, up to $X \sim 50$, has so far been achieved by Mano (2000).

The most well-known surface-based method is the T-matrix method (Waterman, 1971; Wriedt and Doicu, 1998; Mishchenko and Travis, 1998; Havemann and Baran, 2001; Kahnert et al., 2002; Havemann et al., 2003; Petrov et al., 2008), which is sometimes referred to as the Extended Boundary Condition Method (EBCM) or null-field method. In this method, the linearity property of the Maxwell equations is used to simply relate the incident Stokes vector to the scattered Stokes vector via the so-called transition matrix. The great advantage of the T-matrix method is that the matrix depends only on the shape of the particle, its complex refractive index, size parameter and its relation to the coordinate system. It is independent of the directions of incidence or scattering, and so needs only to be computed once, if the T-matrix is known, then the averaged scattering properties of the particle or system of particles can easily be computed (Mischenko, 1991). The FDTD and T-matrix methods have been shown to be in good agreement for calculating the phase matrix elements of Eq. (1.1), assuming the randomly oriented hexagonal ice column of aspect ratio unity (i.e., ratio of diameter to length), for $X \sim 20$ (Baran et al., 2001a). The T-matrix method has been more recently applied to multiple scattering problems, which involve systems of many particles, which in principle, can vary in shape and thus the averaged-scattered field of the whole ensemble could, in principle, be solved using the method outlined by Ganesh and Hawkins (2010).

A method that includes the coupling of volume and surface-based electromagnetic approaches is the Discrete Dipole Method of Moments (Mackowski, 2002). A comprehensive review of all the current electromagnetic methods can be found in the following references (Mishchenko et al., 2002; Kahnert, 2003; and Wriedt, 2009).

For cases where $X \gg 50$, solutions to Eqs (1.1), (1.2) are sought, using approximations applied to nonspherical particles. Approximate light scattering methods that are currently being applied to intermediate size parameter space include the physical optics methods. In this method, wave interference and diffraction, are implicitly included, so that a complete solution to scattering in the far-field is obtained. In these methods, the incident wave is still treated as a ray or beam, so the size of the particle still needs to be larger than the incident wavelength, around $X \sim 20$ (Bi et al., 2011), for the ray or beam concept to have any physical meaning. The first physical optics approach, of relevance to atmospheric light scattering applied to general shapes, is the Modified Kirchhoff Approximation (MKA), developed by Muinonen (1989). Although the MKA approach can only be applied to particles in random orientation, it can however be applied to X as low as about 10. Later improvements to this approach, were introduced by Yang and Liou (1996) using the so-called Improved Geometric Optics (IGO) method. The improvement over the MKA method is that IGO calculates the surface field using Fresnel formulae, and by taking into account the phase and area illuminated by each individual wavelet. The surface fields are expressed in the form of tangential and electric and magnetic currents, which are then transformed to the far-field, using the rigorous electromagnetic equivalence principle.

More recently Bi et al. (2011), have introduced the Physical-Geometric Optics Hybrid method (PGOH), which uses the method of beam-tracing to compute the single-scattering properties of nonspherical particles of arbitrary orientation, complex refractive index (including high absorption), and in-principle shape, for size parameters much greater than 20. The nonspherical particle edge contributions to Q_{ext} and Q_{abs} are also approximated, and included into the solution for the total or scalar optical properties. Good agreement was found between PGOH and DDA in computing the P_{11} element of the scattering phase matrix, assuming the finite oriented hexagonal ice column, for $X = 50$ (X in terms of the length of the column) over a wide range of complex refractive index (Bi et al., 2011). Moreover, in random orientation excellent agreement, at the exact backscattering angle of 180°, was found between the PGOH and DDA methods (Bi et al., 2011). Due to the approach of Bi et al. (2011), the limiting factor of the PGOH method is no longer the size parameter but the shape of the nonspherical particle. The more complex the nonspherical particle becomes, a greater number of beams are required to trace around the particle in order to accurately represent scattering by such particles. Further details of the PGOH method with results, is described in Chapter 2 of this book.

A further method that has been shown to be useful in bridging the gap between small and large size parameter space, is the Ray Tracing Diffraction on Facets method (Hesse, 2008). This method has been demonstrated to be computationally very fast, since it is only a modification to the method of geometric optics, and can in principle be applied to any arbitrary dielectric three-dimensional nonspherical particle. As stated earlier, the method of RTDF is a modification to geometric optics, so the Fresnelian interactions are treated in the usual way, except that, as rays meet each facet of the particle, they are deflected to take account of internal diffraction, caused by each facet acting as an aperture. Due to diffraction being implicitly accounted for, within the nonspherical particle, this means that RTDF can be applied to bridge the gap between approximations and electromagnetic methods. Moreover, RTDF also has the advantage that at large values of X, the solutions to Eq. (1.1), should be more accurate than ordinary geometric optics or ray tracing (Clarke et al., 2006). In the same size parameter region, a further novel physical optics method has been proposed by Borovoi and Grishin (2003) for computing the amplitude scattering matrix, which includes interference information implicitly and assumes Fraunhofer diffraction. However, in this method, diffraction effects inside the crystal are not accounted for.

For large size parameters encountered in the Earth's atmosphere, the classical ray-tracing method (Wendling et al., 1979; Liou and Takano, 1994; Macke et al., 1996a) can be applied to any three-dimensional particle, as long as all of its dimensions are much greater than the incident wavelength. However, in the highly absorptive case, the simple ray-tracing approach is inaccurate (Yang and Liou, 1997). In Chapter 5 of this book the meaning of physical and geometric optics is more fundamentally discussed.

In principle, with the current methods, the problem of light scattering can be solved for any nonspherical particle with arbitrary complex refractive index, across the whole of X encountered in the atmospheric sciences. However, beyond $X \sim 50$ the methods outlined are still approximate. At visible wavelengths, for typical ice crystal sizes encountered in the Earth's atmosphere, the approximate methods still have to be applied. This is still problematic and unsatisfactory because a size parameter of 50 is about a size of 10 μm at visible wavelengths, which represents a very small fraction of typical size spectra encountered in cirrus.

The remaining fundamental problem is to develop new numerical methods that solve the scattering problem efficiently, such that the solution to Eq. (1.2), becomes independent of X (size or frequency). The traditional electromagnetic approaches outlined above are being developed slowly. The limiting factor, for electromagnetic methods, is ultimately determined by the size, irregularity and composition of the particle, size of the equation systems, or size of matrix inversion, numerical stability, convergence, or volume discretization within the particle and in the entire space domain. For sufficiently large particles or small particles at sufficiently short incident wavelengths, the computational demands are so great, that exact (within the computational accuracy of the numerical method), solutions are currently unobtainable. Therefore, it is timely to investigate different approaches to the scattering problem from other areas of mathematical physics.

One area that is currently receiving significant attention is high-frequency scalar wave scattering by impenetrable objects. The problems encountered in this area, are not unlike the problems encountered, when trying to solve the scattering equations using Maxwell's equations. However, the traditional approach, using electromagnetic methods, is to tend to higher size parameters, until the equation system can no longer be solved, due to computational limitations or the particle geometry is too complex. However, recent developments in scalar wave scattering have improved the computational cost of the numerical algorithms used to solve the Helmholtz equation, to such an extent that the cost grows as $\log(k)$, where k is the wave number as defined above, and as k tends to very large values, or high frequencies, the cost is still approximately $\log(k)$. The fundamental method used to solve the Helmholtz equation is Green's representation theorem. This theorem, allows finding the solution for the entire scattered field, by re-transforming the scattering problem, to an integral equation on the boundary of the particle (it is similar to the Kirchhoff diffraction integral), and therefore naturally includes interference and diffraction. This approach therefore reduces the dimensionality of the scattering problem, 2-D becomes 1-D, and so on. Although, Boundary Integral Equation (BIE) methods are well established, the novel aspect of more recent work in reducing the computational cost to $\log(k)$, is the inclusion of information from the high-frequency asymptotics into the approximation space (Chandler-Wilde et al., 2012).

Clearly, the approach used in high-frequency scalar wave scattering is potentially powerful, if it can be extended to higher-dimensional space, i.e., the vector nature of light, and general penetrable shapes, then the approach could be a solution to the general scattering problem. This method, already considers the equivalent to very large X. This is, the other way round to the traditional electromagnetic approach, and is therefore highly desirable. The generalization of high-frequency scalar wave scattering to the problem of particle transmission is currently a very active area of applied mathematical research, and represents an alternative and illuminating approach to the more traditional approaches, which tend to be computationally expensive.

To simulate the short-wave and long-wave irradiance of aerosol or cirrus over a period of decades, GCMs are required. To do these simulations over reasonable time scales, the scalar optical properties (K_{ext}, ω_0, and g) must be used. To compute the scalar optical properties, it is possible to employ any of the previously described electromagnetic or light scattering methods. However, other approximations that overcome limitations on particle size and shape have been proposed. Methods based on the anomalous diffraction approximation (ADT) (van de Hulst, 1957) have been applied to compute the extinction and absorption efficiencies of nonspherical particles; these include modifications to ADT proposed by (Mitchell et al., 1996; Mitchell et al., 2001; and Mitchell et al., 2006). These modifications include parameterizations of the edge effects and internal reflections, so that these physical processes are added to Q_{ext} and Q_{abs}. This method is called the Modified Anomalous Diffraction Approach (MADA). In the original ADT, the size of the sphere is assumed to be much greater than the incident wavelength, and the real refractive index is close to unity, so only the phase change of the incident wave with respect to the diffracted wave is considered, ignoring internal reflections or edge effects. Therefore, for real refractive indices larger than 1.0, methods based on ADT will be in error, and the error will increase as the real refractive index becomes larger than unity. Since there is no angle-dependence in ADT (except for $\theta \to 0°$), then neither the scattering phase matrix or the asymmetry parameter can be calculated. In the MADA approach, the empirically derived coefficients that are used to estimate the edge effects are based on exact methods. Not surprisingly therefore, comparisons between MADA, FDTD and T-matrix were generally in good agreement, in calculating Q_{ext} of randomly oriented hexagonal ice columns, between the wavelengths of 2.2 and 16.0 μm (Mitchell et al., 2006).

Other approximate methods, to compute the scalar optical properties of nonspherical particles, using the equivalent spherical volume-to-area ratio, are described in the following references (Grenfell and Warren, 1999; Neshya et al., 2003;

and Grenfell et al., 2005). Approximating the scalar optical properties of compact hexagonal aggregates, by an ensemble of equivalent volume-to-area cylinders, was proposed by Baran (2003). Using this approach, it was shown that an ensemble of symmetric shapes could approximate the scalar optical property FDTD solutions of non-symmetric shapes, to well within 4%, at wavelengths across the terrestrial window region. A similar method to Baran (2003) was applied by Weinman and Kim (2007), to compute the total optical properties of irregular particles, at frequencies in the microwave region, and, on comparing with exact methods, they found similar results to Baran (2003). In the paper by Lee et al. (2003), they investigated using circular cylinders as surrogates for pristine hexagonal ice crystals in the terrestrial window region. They show using T-matrix, FDTD and IGO, that randomly oriented finite circular ice cylinders can approximate the single-scattering properties of randomly oriented finite hexagonal ice columns, to within a few percent, at wavelengths between 8.0 and 12.0 μm. In the next section of this chapter, the shapes and sizes of mineral dust aerosol, volcanic dust aerosol and cirrus ice crystals are discussed.

1.4 A myriad of sizes and shapes

1.4.1 The sizes and shapes of mineral dust and volcanic ash particles in the Earth's atmosphere

It is important to measure the sizes and shapes of mineral dust and volcanic aerosol, because of their influence on the Earth–atmosphere radiation balance, and civil aviation aircraft, respectively. Mineral dust aerosol is transported into the Earth's atmosphere, by dust storms or winds; as a consequence, this aerosol type is one of the most common found in the Earth's atmosphere (Penner et al., 2001). It is also known that wind-blown dust storms transport mineral dust aerosol from Africa, over the Atlantic Ocean, to America (Prospero et al., 2010; Otto et al., 2011, and references therein). With such a spatial extent, mineral dust aerosol has a significant influence on the Earth–atmosphere radiation balance (Haywood et al., 2003; 2005; 2011; Highwood et al., 2003; Osborne et al., 2011; and Otto et al., 2011). Although, it is now well established that mineral dust aerosol is transported from Africa to America, it is also known that large sizes of aerosol, $> 62.5\,\mu$m, can be transported thousands of kilometres from their original source (Ulanowski et al., 2007).

The physical reason why such large mineral dust aerosol particles are transported over long distances could be due to the fine and coarse mineral dust aerosol being charged with opposite polarity, which may reduce the fall speeds of the coarse particles (Ulanowski et al., 2007). Alignment of mineral dust particles will clearly have important consequences for their radiative properties. For instance, Ulanowski et al. (2007) estimated that, due to the vertical alignment of the particles, the optical depth becomes anisotropic, and is reduced by as much as 10%, in that direction, for the case they examined, at the wavelength 0.780 μm. If vertical alignment of mineral dust is commonplace, then this will have important implications for modelling their radiative properties in climate models, and clearly, further research is required in this area, as this effect is currently neglected in climate simulations.

To study the possible alignment of mineral dust aerosol, the linear depolarization of mineral dust aerosol as a function of zenith angle, would be a useful measure (Asano, 1983).

Volcanic eruptions are also a source of dust in the atmosphere, the most dramatic recent example of this, which grounded civil aviation aircraft in Europe, was the eruption of the Icelandic Eyjafjallajökull volcano, during April 2010. The plume of this volcano, advected over northern European airports, due to the fine nature of the ash particles, and their potential concentrations, caused European airports to close. The closure of European airports proved to be very expensive for the airline industry and insurance companies. Due to the financial impact of this one event, interest in the nature, mass, composition, and sizes of volcanic particles, has increased considerably (Johnson et al., 2012; Turnbull et al., 2012). However, volcanic eruptions can also have significant impacts on the Earth's climate. The eruption of Mt. Pinatubo in the Philippines in June 1991 ejected an estimated 20 million tonnes of sulphur dioxide into the Earth's stratosphere (Bluth et al., 1992). Though, later estimates from a number of different authors, estimated the total mass to be 14–20 million tonnes of sulphur dioxide (McCormick and Veiga, 1992; Stowe et al., 1992; Lambert et al., 1993; Strong and Stowe, 1993; and Baran and Foot, 1994). The sulphur dioxide deposited in the stratosphere, in a few weeks, converts to sulphates, which circumnavigated the globe in about one month (McCormick and Veiga, 1992; Long and Stowe, 1994). The sulphate aerosol particles reflect solar radiation back to space, and absorb near-infrared solar radiation and long-wave terrestrial radiation, thereby cooling the surface of the Earth by about 0.5°C (Soden et al., 2002) and warming the lower tropical stratosphere by about 3 K (Ramachandran et al., 2000).

The radiative effects of aerosols, most critically depend on their size, concentration, composition, vertical distribution and shape. The size spectrum of aerosols can be measured using a number of microphysical probes, some are based on single particle scattering (Johnson et al., 2012). The size spectrum of aerosol, in the nominal size range $0.01\,\mu$m to $0.1\,\mu$m, is called the 'nucleation mode', and $0.1\,\mu$m to $0.6\,\mu$m is called the 'accumulation mode'. For nominal aerosol sizes greater than about $0.6\,\mu$m, the term 'coarse mode' is used. The size spectrum is measured as a function of concentration, and can be measured by an instrument called the Passive Cavity Aerosol Spectrometer Probe (PCASP). The principle of PCASP is based on scattering a $0.630\,\mu$m laser beam, by the fine mode aerosol, into a scattering angle range of between 35° to 145°. For aerosol size greater than $3.0\,\mu$m, an alternative single-particle scattering probe is used. This probe is called the Cloud and Aerosol Spectrometer (CAS), and this can measure aerosol size and concentration in the size range $0.6\,\mu$m to $50.0\,\mu$m. The CAS instrument works on the principle of scattering $0.630\,\mu$m laser light into the forward scattering directions of 4° to 12°, to size the particles. Particles in this size range are called the 'coarse-mode' aerosol. Therefore, aerosols have both a fine-mode and a coarse-mode, which must be fully measured to estimate their radiative effect. The PCASP and CAS, both bin the measured concentration as a function of nominal diameter. This binning of concentration as a function of diameter is called the 'particle size distribution', often abbreviated to PSD. A typical aerosol PSD is shown in Fig. 1.2, measured with the PCASP and CAS (Johnson et al., 2012) probes. The figure shows a fine- and coarse-mode, detected by PCASP and CAS, respectively. Note, that the aerosol mass distribution is dominated by the aerosol coarse-mode, with a peak at about 4.0 μm. This is why aerosol, with peak coarse-modes typically in terms of micrometre sizes, interacts with solar and terrestrial radiation (Haywood et al., 2003), to warm or cool the surface of the planet, depending on surface type (Haywood et al., 2011).

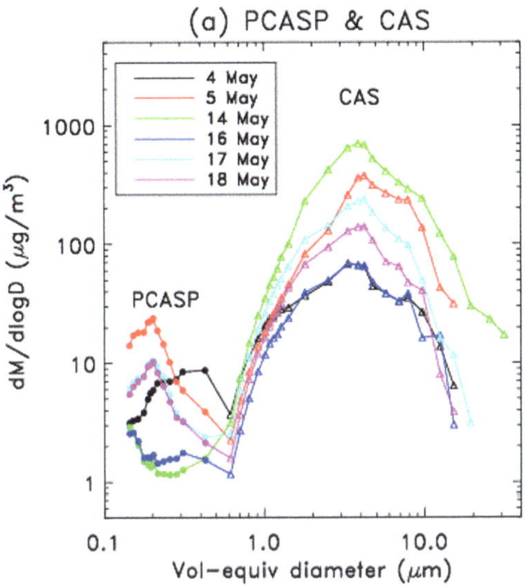

Fig. 1.2. Aerosol PSD measured by PCASP and CAS, during a number of flights during April 2010. The PSD is shown as a function of mass concentration ($dM/d\log D$) and volume equivalent diameter (after Johnson et al., 2012).

The aerosol PSD, shown in Fig. 1.2, is typically modelled using log-normal size distribution functions (Whitby, 1978; Tanre et al., 1996; Kokhanovsky, 1998; Zhang et al., 1998; Haywood et al., 2003; Nousiainen and Vermeulen, 2003; Bi et al., 2009; Johnson and Osborne 2011; Osborne et al., 2011; and Johnson et al., 2012), using either mono-modal or multi-modal PSDs. The lognormal PSD is given by the following equation (Kokhanovsky, 1998):

$$f(r) = \frac{1}{r \ln \sigma_g \sqrt{2\pi}} \exp\left[-\frac{\ln^2 r/r_g}{2\ln^2 \sigma_g}\right] \tag{1.12}$$

Where r is the aerosol radius, and r_g and σ_g are the geometric mean radius and geometric standard deviations of the PSD, respectively.

Note that, polydispersions of spherical particles of very different PSDs, with similar effective variance, $v_{ef} = \langle r^2(r - r_{ef})^2\rangle / \langle r^2\rangle r_{ef}^2$, and effective radius, $r_{ef} = \langle r^3\rangle / \langle r^2\rangle$, have similar scalar optical properties (Hansen and Travis, 1974), where the angle backets mean averaged quantities. The effective radius is weighted by the geometric cross-section of spherical particles, and the cross-section is related to the scattered intensity of incident light. For a given r_{ef} of spherical particles, the same extinction coefficient can be derived, as integrating each optical property, for each bin size, over the PSD. Therefore, the effective radius is a very important quantity in atmospheric optics (Hansen and Travis, 1974; Kokhanovsky, 1998). A typical lognormal fit, using Eq. (1.12), to the coarse-mode of aerosol, is shown in Fig. 1.3. The figure shows that lognormal fits are usually a good representation of the coarse-mode (Johnson et al., 2012).

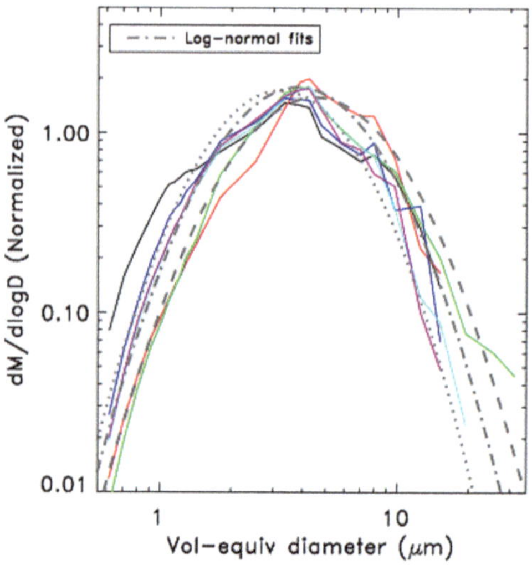

Fig. 1.3. The CAS normalized mass concentration as a function of volume equivalent diameter, showing the lognormal fits to the data shown in Fig. 1.2. Note that the peak of the coarse mode occurs at a volume equivalent diameter of about 4.0 μm (after Johnson et al., 2012).

An example of the shapes of particulate aerosol is shown in Fig. 1.4. The figure shows examples of Scanning Electron Microscope (SEM) images (R. A. Burgess, personal communication, and from Johnson et al., 2012) of ground-based Arizona dust, Saharan dust and volcanic ash aerosol. The images show that aerosol is irregular and not spherical or smooth spheroids, as many papers in the literature assume. Moreover, not only are the particles sharp-edged, but may also aggregate, and have roughened surfaces. The figure highlights the need for more sophisticated treatments of aerosol morphology assumed in electromagnetic and light scattering calculations. Idealized irregular models, proposed to represent irregular aerosols, such as those shown in Fig. 1.4, are reviewed in Section 1.5.

The composition of mineral dust and volcanic aerosol is also of importance in electromagnetic and light scattering calculations, as their material compositions determine their complex refractive index. The complex refractive index of mineral dust aerosol has been compiled by Balkanski et al. (2007), between the wavelengths 0.3 μm to 100 μm, for different internal mineralogical mixtures of 0.9%, 1.5% and 2.7% of haematite. The Balkanski et al. (2007) compilation is generally used to compute the scattering properties of mineral dust aerosol. The composition of volcanic aerosol can differ substantially, from similar properties to mineral dust aerosol

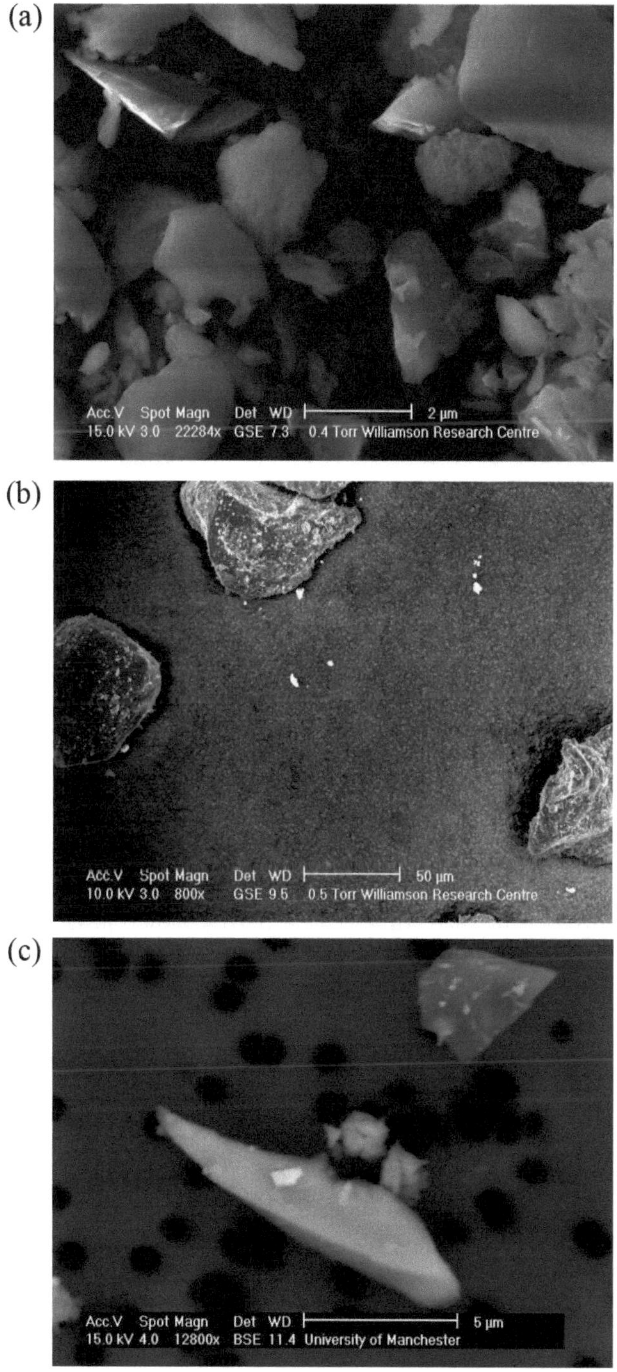

Fig. 1.4. Scanning electron microscope images of (a) desert dust from Arizona, (b) Saharan mineral dust aerosol, and (c) volcanic dust aerosol from the April 2010 eruption of Eyjafjallajökull (from R. A. Burgess, personal communication, and R. A. Burgess, and after Johnson et al., 2012).

(Schumann et al., 2011), to basaltic ash particles (Pollack et al., 1973). The complex refractive index for a number of volcanic aerosols has been compiled by Pollack et al. (1973), from wavelengths 0.21 μm to 50 μm. Since the publication of the Pollack et al. (1973) compilation, there have been no revisions to those refractive indices. Given the need to accurately estimate the mass concentration of volcanic aerosol using space-based or *in situ* measurements, then measuring new refractive indices of volcanic aerosol in the solar and infrared region should be a priority (Newman et al., 2012). There have been a few measurements of the complex refractive index of mineral dust and volcanic aerosol at low microwave frequencies, from 8.0 to 12.0 GHz and from 10.0 to 18.0 GHz, measured by Ghobrial and Sharief (1987) and Bredow et al. (1995), respectively. However, there have been no determinations of the complex refractive index of Earth-based atmospheric aerosols at submillimetre frequencies (i.e., >300 GHz) (Baran, 2012b). The submillimetre region, when combined with the microwave region, of the electromagnetic spectrum, may be useful in characterizing the mass concentration of aerosol plumes, close to their source (Baran 2012b). Therefore, it is important to extend measurements of the complex refractive index of aerosols into the submillimetre region.

1.4.2 The sizes and shapes of ice crystals in the Earth's atmosphere

Cirrus or ice crystal cloud, generally form at altitudes greater than about 6 km. Therefore, they are cold, pure ice crystal cloud, occurs at temperatures below about $-40°$C (Guignard et al., 2012). On the global scale, the spatial and temporal distribution of cirrus can only be determined using space-based measurements. These measurements show that cirrus covers about 30% of the mid-latitudes at any given time, whilst in the tropics; the coverage can be 60% to 80% at any given time (Wylie et al., 1999; Stubenrauch et al., 2006; Sassen et al., 2008; Nazaryan et al., 2008; Lee et al., 2009; Guignard et al., 2012). With this spatial and temporal distribution, it is clear that cirrus is an important contribution to the Earth's radiation budget and hydrological cycle. Indeed, the most recent fourth assessment report of the Intergovernmental Panel on Climate Change (IPCC, 2007) concluded that, understanding the coupling between all clouds and the Earth's atmosphere remains one of the largest uncertainties in climate prediction. Cirrus is one such cloud type, where the radiative coupling between it and the Earth's atmosphere is still highly uncertain. Therefore, determining the size and shapes of ice crystals, their vertical variability and extent in the Earth's atmosphere, are important quantities to determine, if their representation in climate models is to be further improved.

In the mid-latitudes, cirrus is usually generated synoptically, and this mode of generation forms layers of ice crystals, with the ice crystal size (number or mass weighted) increasing with cloud depth, measured relative to cloud-top. At cloud-top, ice crystal size can be less than 10 μm, increasing to many thousands of micrometres at cloud-base (Heymsfield and Miloshevich, 2003; Korolev and Isaac 2003; Field et al., 2007; Field et al., 2008; Korolev et al., 2011). The distance the ice crystal is from cloud-top, provides a proxy for time, the greater the vertical extent of the cloud layers, the deeper the ice crystal falls, due to gravitational settling. The longer the ice crystals take to fall, the more time there is for ice crystal growth, due to vapour deposition, and perhaps more importantly ice crystal aggregation (Heymsfield et al., 2002; Field et al., 2005; Field et al., 2008).

The overall PSD shape of ice crystals is bi-modal (Heymsfield and Miloshevich 1995; Ivanova et al., 2001; Field et al., 2005; Baker and Lawson 2006; Field et al., 2007, Baran et al., 2011a; Zhao et al., 2011), and the nature of the cirrus PSD is that both small and large ice crystals can coexist, but with the larger ice crystals appearing less frequently. In the mid-latitudes, the updraughts are not usually sufficient, to transport larger ice crystals toward cloud-top. However, in the tropics, where there are much greater updraughts, due to vigorous convection, larger ice crystals can be transported to cloud-top, near the convective core, and here larger ice crystals can appear more frequently in the PSD (Heymsfield and Miloshevich 2003; Yuan et al., 2011). As the cirrus flows out from the convective core, the ice crystals gravitationally settle, so that at cloud-top, the PSD is dominated by smaller ice crystals, and as the PSD evolves with depth from the cloud-top, ice crystal aggregation takes place and a broader PSD results, with sizes that could be several centimetres in size (Heymsfield, 2003). There have been fewer cirrus *in situ* measurements taking place in the Arctic where, not surprisingly, there are difficulties with accessing this particular region. However, there have been some studies, which suggest that ice crystal size in this region can be larger than about 40 μm, and up to about 1000 μm (Korolev et al., 1999; Liou 2002; Lawson et al., 2006).

Example images of ice crystal shapes are shown in Figs. 1.5, 1.6 (provided by Andrew Heymsfield, personal communication) and 1.7 (Baran et al., 2011a), as a function of altitude. The images were obtained using an instrument called, the Cloud Particle Imager (CPI), and is described in Lawson et al. (2006). The images shown in Fig. 1.5 are for ice crystal sizes greater than 100 μm (Heymsfield and Miloshevich, 2003). In Fig. 1.5, there is little evidence for pristine hexagonal ice columns or plates, the most common shapes appear to be rosettes or chains of rosettes, with the rosettes appearing spatial rather than compact. In Fig. 1.5, there is also evidence of air inclusions, both in single hexagonal ice columns and branches of the rosettes (Schmitt and Heymsfield, 2007). The occurrence of hollowness may be a common feature of ice crystals, as noted by other authors such as (Weickmann, 1947; Magono and Lee, 1966; Heymsfield and Platt, 1984). In Fig. 1.6, the most common shapes, larger than about 100 μm, seem to be bullet-rosettes or aggregates of rosettes. The shape of ice crystals, less than 100 μm in size, is currently unknown, due to the limited resolving power of the CPI. The CPI cannot resolves the shapes of ice crystals less than about 35 μm in size, due to its spatial resolution being 3 μm. However, the CPI images small ice crystals as quasi-spherical or spheroidal in shape, due to diffraction. However, these small ice particles may still be irregular (Ulanowski et al., 2006). Moreover, recent laboratory cloud chamber studies conducted by López et al. (2012), at temperatures between $-30°$C and $-40°$C, show that small ice particles, studied under a microscope, of average effective diameter 14 μm, were deformed, with sharp-edged long protrusions extending out of their surfaces. Characterizing the shapes and sizes of ice particles less than 100 μm in size, is very important, as they can have an important influence on the radiative properties of cirrus (Ivanova et al., 2001; Yang et al., 2003; Ulanowski et al., 2006; Ulanowski et al., 2010; Mauno et al., 2011; Um and McFarquhar, 2011; Thelen and Edwards, 2012).

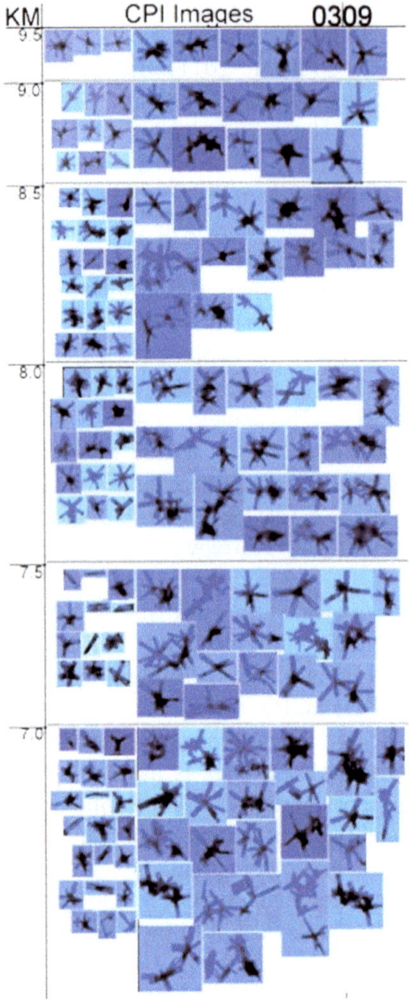

Fig. 1.5. A set of ice crystal images shown as a function of height. The images were obtained using the CPI instrument (from A. J. Heymsfield, personal communication).

Figure 1.7 shows examples of tropical ice crystal CPI images, which originate from a fresh tropical anvil. These images show that the most common ice crystal shape appears to be hexagonal ice plates and aggregates of plates, though there are other more indeterminate ice crystal shapes present. The hexagonal ice plate aggregates are large in size, and if they are sufficient in number, they may well dominate the radiative properties of fresh anvils (Um and McFarquhar, 2009). The occurrence of aggregates of hexagonal plates has also been noted by (Lawson et al., 2003; Connolly et al., 2005; Evans et al., 2005; Baran 2009; Xie et al., 2011; Gayet et al., 2012). Laboratory cloud chamber studies have shown that plate aggregates might form in the presence of electric fields (Saunders and Wahab, 1975; Wahab 1974). But, the hexagonal plate aggregates shown in Fig. 1.7, could also

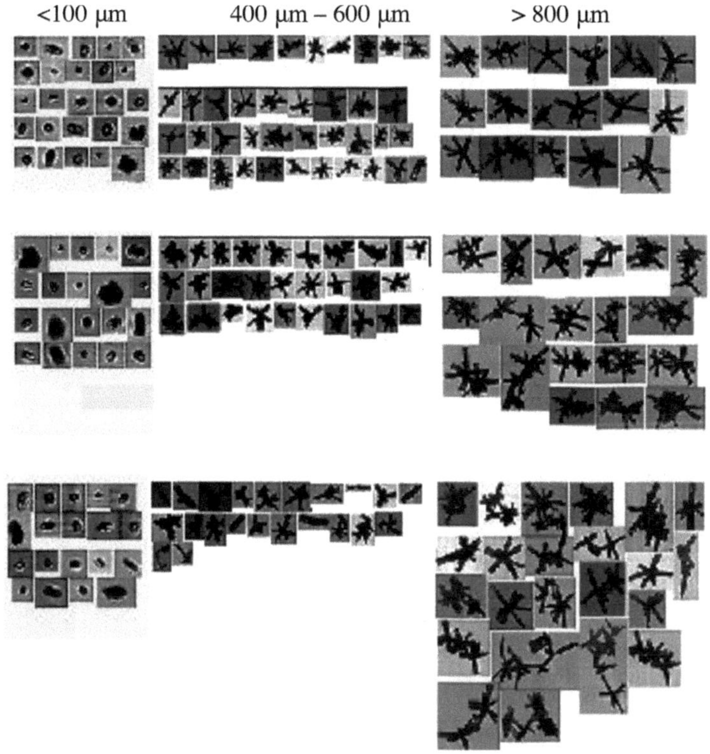

Fig. 1.6. Same as Fig. 1.5 but ice crystal size is shown on the top, along the x-axis. The altitude of the top row is 9.5 km, second row 8.0 km, and third row 7 km (from A. J. Heymsfield, personal communication).

form through the process of ice crystal aggregation (Westbrook et al., 2004; Baran 2009). However, not all convective cloud produces chains of aggregates (Lawson et al., 2003).

From the currently available evidence, it can be said that the most common types of ice crystal that inhabit cirrus are pristine or spatial ice crystals and non-symmetric aggregates of these (Korolev et al., 2000; Um and McFarquhar 2007; Baran et al., 2009; McFarquhar and Heymsfield, 1996; Lawson et al., 2003; Connolly et al., 2005; Evans et al., 2005; Baran et al., 2011a; and Xie et al., 2011). Although ice crystal aggregates may, themselves, be non-symmetric overall, the individual monomers that make up the aggregate may be symmetric (Stoelinga et al., 2007).

The CPI images of ice crystal sizes less than 100 μm, shown in Figs. 1.6 and 1.7, may also be a result of ice crystal shattering on the inlet of the probes. Instruments such as the CPI are closed instruments, that is, the ice crystals are funnelled through an inlet, into the instrument's sampling volume. The inlet has sharp edges, and this can cause large ice crystals to shatter into smaller ice crystals on impact with the probe. These smaller ice crystals, resulting from the shattering of larger ice crystals, artificially increase the measured concentrations of small ice crystals, less than about 250 μm in size (Korolev et al., 2011). Although shattering has been

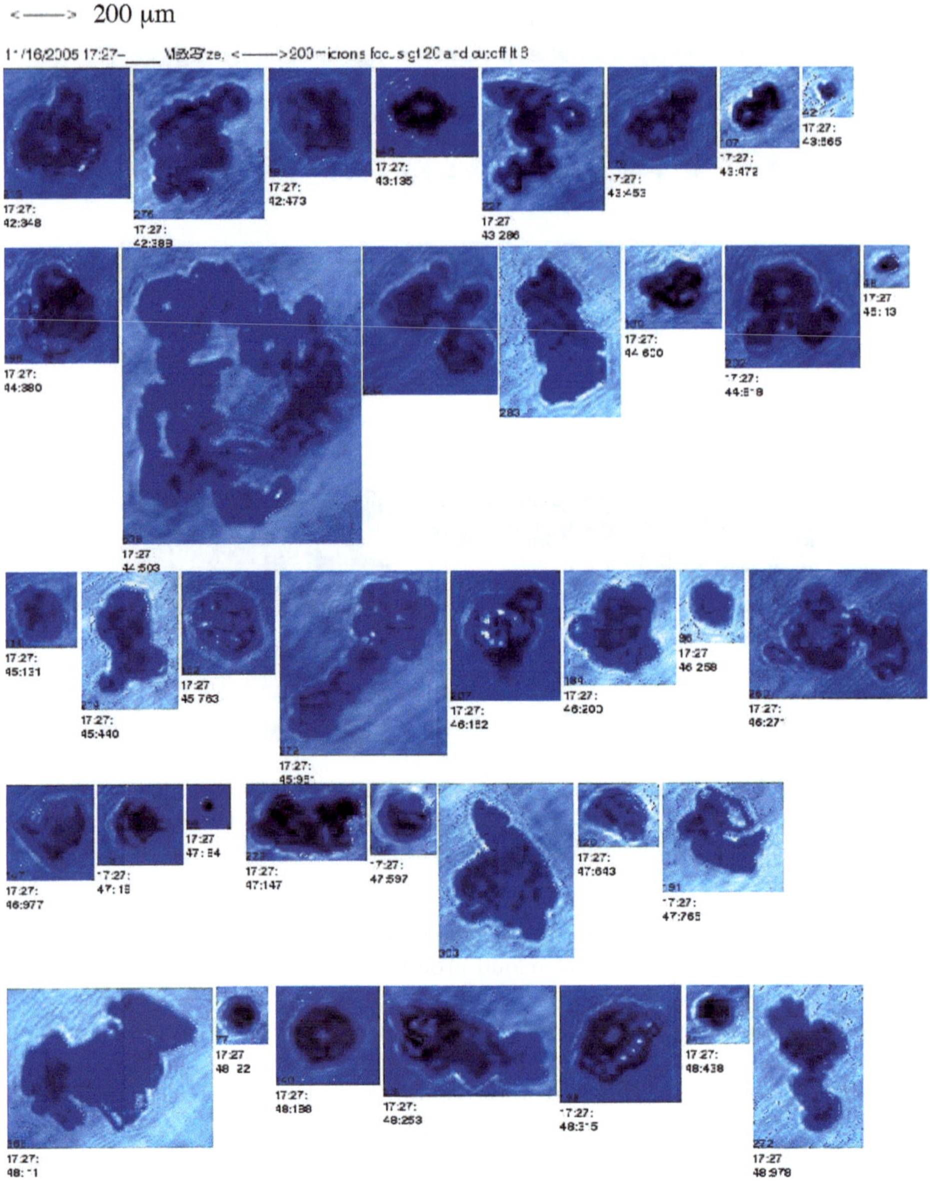

Fig. 1.7. CPI images of ice crystals that occurred in a fresh tropical anvil; note the appearance of hexagonal plate aggregates. The scale on the top left indicates the ice crystal size in each of the images (after Baran et al., 2011b).

known for some time (Cooper 1978), solutions to the ice crystal shattering problem have only been offered in recent times (Field et al., 2003; Field et al., 2006; Lawson et al., 2006; McFarquhar et al., 2007; Heymsfield 2007; Korolev et al., 2011; and Lawson 2011).

It is now generally accepted, that historical *in situ* measurements of cirrus PSDs, have been artificially skewed, for particle sizes less than about 250 μm in size (Korolev et al., 2011). To remove the shattered artifacts from the measured PSD, it is necessary to measure the inter-arrival time of each ice crystal (shattered ice crystals arrive in groups and so have short inter-arrival times), so that ice crystals with short inter-arrival times are filtered out from the measured PSD (Field et al., 2003). However, filtering itself may not be sufficient to remove all shattered ice crystals, and what is also required, is the placing of specially designed tips at the probe inlets (Korolev et al., 2011). However, in a more recent paper by Lawson (2011), it was found that only inter-arrival time was required, due to the dominance of small ice crystal sizes, in the data that was examined. Clearly, shattering has important consequences for measurements of ice crystal concentration and ice crystal size. However, for estimates of ice water content (IWC), using the measured size spectrum, ice crystal shattering is less of a problem, as shown by Field et al. (2006) and Korolev et al. (2011). Indeed, it was shown by Field et al. (2006), that ice crystal shattering does not affect IWC by more than 10%. Whilst Korolev et al. (2011) concluded that IWC and radar reflectivity can generally be derived to well within a factor 2% and 20%, respectively, when shattering is present. Clearly, to remove shattering from measured historical PSDs, the method of filtering must be applied, if inter-arrival times are available. Future *in situ* measurements of the PSD should be based on microphysical probes that are fitted with anti-shatter tips. Historical PSDs need not necessarily be abandoned, especially if the PSDs are used to estimate IWC and/or radar reflectivity. Historical PSDs must also be examined for evidence of shattering using ice crystal inter-arrival times, before applying these PSDs to calculate the bulk scattering properties of cirrus.

To characterize the shape and size of ice crystals much less than 100 μm in size requires instrumentation that is able to overcome the limitations of optical resolution. One such probe is called the Small Ice Detector (SID); a description of SID is given in Kaye et al. (2008). The latest SID probe is collectively known as SID-3. The SID series of instruments is based on single-scattering measurements of the 2-D light scattering pattern (i.e., both θ and ϕ are measured), and can size ice crystals in the range 1 μm to several hundred μm (Ulanowski et al., 2010; Ulanowski et al., 2011). The latest, SID-3, uses intensified charge coupled device cameras to measure high-resolution 2-D light scattering patterns, enabling the estimation of ice particle surface roughness, size, and concavity (Ulanowski et al., 2011).

A random example of 2-D light scattering patterns measured by SID-3 is shown in Fig. 1.8. These images were obtained in mid-latitude cirrus, around the UK, in the winter of 2010 (Ulanowski et al., 2010, personal communication). Unlike most earlier results from cloud chambers, where 2-D patterns had sharp, well-defined bright arcs and spots (Kaye et al., 2008), the majority of the cloud data, collected during this particular campaign, was characterized by much more random, 'speckled' 2-D scattering patterns. Laboratory experiments demonstrate that such pattern features are characteristic of particles with rough or irregular surfaces. Quantitative comparison of laboratory and cloud data has been done using pattern texture measures, originally developed for surface roughness analysis using laser speckle. The results were consistent with the presence of significant roughness in

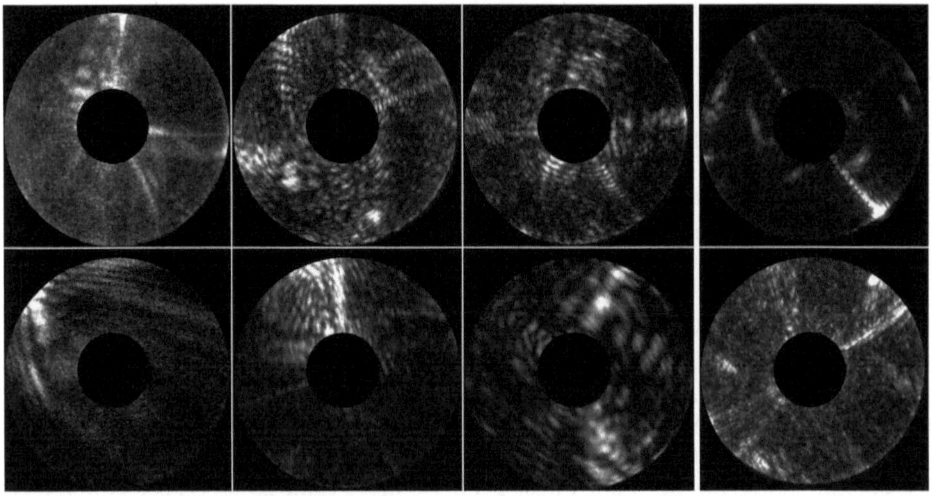

Fig. 1.8. Six randomly selected SID-3 scattering patterns from ice particles seen in mid-latitude cirrus (top row) and mixed phase (bottom row) flights, compared to patterns from ice-analogue rosettes with smooth (top right) and moderately rough surfaces (bottom right) (after Ulanowski et al., 2010).

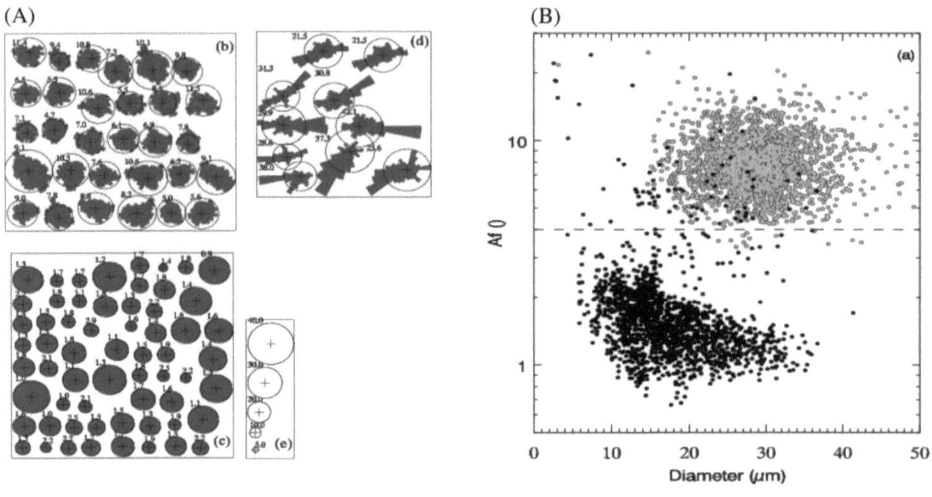

Fig. 1.9. SID-2 measurements obtained in mid-latitude cirrus in which supercooled liquid water drops co-existed with ice crystals, showing (A) 2-D scattering patterns of nonspherical ice crystals (b, d) and (c) 2-D scattering patterns of supercooled liquid drops, with the size of each drop shown in the panel labelled (e). (B) The SID-2 estimated asphericity parameter (AF) as a function of diameter for ice crystals (grey circles) and supercooled liquid water drops (black circles) (after Cotton et al., 2010).

the majority of cirrus and mixed phase cloud ice crystals, at levels comparable to those found in the rough ice analogues studied previously (Ulanowski et al., 2006; Ulanowski et al., 2010).

Another example of how single light scattering measurements can be used to characterize the phase and shapes of ice crystals is shown in Fig. 1.9(A) and Fig. 1.9(B) (Cotton et al., 2010), respectively. The figure shows SID-2 measurements, again obtained in mid-latitude cirrus, of the particle shape and phase of ice crystals encountered. Fig. 1.9(A) shows the SID-2 measured 2-D scattering pattern and the asphericity factor, Af, is shown in Fig. 1.9(B). Fig. 1.9(A) shows that, the sphere, bottom left scatters light equally in the azimuth direction. However, the top of fig 9A, shows the asymmetric scattering nature of nonspherical particles, showing pristine particles (top right) and highly irregular particles (top left). The asphericity factor, shown in Fig. 1.9(B) plotted as a function of particle diameter, is a measure of the degree of asymmetry in the scattered light, measured by the SID-2 detectors. Azimuthally, spheres scatter light equally in all directions (Fig. 1.9(A)); however, nonspherical particles do not. Therefore, the asphericity factor allows discrimination between water and ice particles. Fig. 1.9(B) shows that the asphericity factor clearly discriminates between water ($Af < 4$) and ice particles ($Af > 4$). Interestingly, the SID-2 measurements, shown in Fig. 1.9(B), were obtained at temperatures of about $-30°$C. Other instruments, unable to distinguish small particle phase, would have assumed that all measurements, at such temperatures, were of ice particles. Instruments unable to resolve small particles, would classify, for the case shown in Fig. 1.9, water spheres as ice particles, and later assume these to be solid ice, when estimating the total amount of ice mass. The SID series of instruments have also been used to discriminate between sea salt and mineral dust particles (Cotton et al., 2010).

Instruments such as the SID are an important addition, as they are able to measure, at high resolution, the 2-D scattering pattern, from which fundamental information on the surface roughness, concavity, shape, size and phase of particles can be extracted as functions of atmospheric state. These fundamental properties determine the light scattering properties of particles, and this aspect will be discussed in the next section.

In Section 1.4.1, the radiative properties of aerosol could be defined in terms of a characteristic size, called the effective radius. There have been attempts to define the radiative properties of cirrus, using a similar definition. However, Figs. 1.5, 1.6 and 1.7 show that ice crystals are highly variable in both shape and size; therefore the shape of the PSD will also be very different. Therefore, any definition of effective radius applied to cirrus, would need to be independent of particle shape and PSD. In the case of cirrus, the characteristic size is expressed through the effective dimension, D_e, or diameter, of the PSD. This concept was first proposed by Foot (1988), and is given by the following expression:

$$D_e = \frac{3}{2} \frac{IWC}{\rho \langle S(\vec{q}) \rangle} \tag{1.13}$$

where ρ is the density of solid ice and is assumed to have a value of $0.92\,\mathrm{g\,cm^{-3}}$. The other terms have been previously defined. There are, however, many definitions of D_e, as can be found in the following set of papers (Francis, 1995; McFarquhar and Heymsfield, 1998; Fu et al., 1999; Mitchell, 2002; Mitchell et al., 2011). The expression for D_e, can be rewritten in terms of integral quantities, the column integrated IWC, called the ice water path (IWP) and the optical depth τ (see

Eq. (1.10)). For very large ice crystals, the dimensions of the particle are much greater than the incident wavelength, for this case the volume extinction coefficient is twice the orientation-averaged cross-section (van de Hulst, 1957), since $Q_{\mathrm{ext}} = 2.0$. Applying these integral quantities, to Eq. (1.13), the following expression is derived, for D_{e}.

$$D_{\mathrm{e}} = \frac{3\,IWP}{\rho\tau} \tag{1.14}$$

This expression means that D_{e} is itself, a fundamental optical quantity that describes light extinction in clouds (Mitchell et al., 1996; Wyser and Yang, 1998; Kokhanovsky, 2004; Mitchell, 2002; Mitchell et al., 2011). The concept of D_{e} does not, however, apply to the asymmetry parameter, since this quantity depends on the assumed geometry of the ice crystal, as shown in (Kokhanovsky and Macke, 1997; Wyser and Yang, 1998). Equation (1.14) is in part an expression of geometric optics, that is D_{e} only has any physical meaning when the size of the ice crystal is much larger than the incident wavelength. In the limit of geometric optics, the inverse relationship between D_{e} and the mass extinction coefficient (defined as the ratio of K_{ext} to IWC) holds well, as demonstrated by Wyser and Yang (1998), but begins to break down at wavelengths in the infrared (Mitchell, 2002; Baran 2005; Mitchell et al., 2011). The study of Baran (2005) showed that the D_{e} concept begins to break down at the wavelength of about 4 μm, and completely breaks down by a wavelength of 30 μm. Moreover, for some PSDs common to cirrus, D_{e} and IWC do not uniquely define the radiative properties of cirrus, as cirrus PSDs are not uniquely the same (Mitchell 2002; Deleon and Haigh, 2007; Mitchell et al., 2011).

A further difficulty with the concept of D_{e} is that, in the case of aggregating ice crystals, mass $\propto D^2$, where D is the maximum dimension of ice crystals (Westbrook et al., 2004). Moreover, the orientation-averaged cross-section, in the denominator of Eq. (1.13), is approximately D^2. Therefore, in regions of ice crystal aggregation, Eq. (1.13) predicts that D_{e} becomes a constant value (Baran et al., 2011b). Clearly, in these regions, D_{e} is not applicable. Moreover, the work of (Mitchell 2002; Deleon and Haigh, 2007; Baran, 2009; Mitchell et al., 2011; Baran et al., 2011b) suggest that D_{e} should not generally be applied to represent the radiative effect of cirrus in GCMs or in the remote sensing of cirrus. An alternative formulation to representing the radiative properties of cirrus as functions of other variables is described in the next section.

1.5 Idealized geometrical models of mineral dust aerosol and ice crystals and their single-scattering properties

1.5.1 Aerosol models and their light scattering properties

As Fig. 1.4 demonstrates, aerosol such as dust or for that matter volcanic ash (Johnson et al., 2012), that appears in the Earth's atmosphere is not spherical, but irregular, and may also possess sharp edges, with roughened surfaces. The composition of atmospheric aerosol may also vary, from homogeneous to inhomogeneous and porous. For the smallest sizes of aerosol i.e., in the sub-micrometre range, at visible wavelengths, calculating their light scattering properties is achievable using

Lorenz–Mie theory or T-matrix applied to spheres or rotationally symmetric particles, respectively. However, for the larger sizes of aerosol, greater than about 1 μm, the aerosol properties previously described, make calculating their light scattering properties difficult. For such larger sizes, the use of Lorenz–Mie theory should be avoided, even for irradiance calculations (Nousiainen, 2009). The typical idealized geometrical shapes, generally assumed for calculating the single scattering properties of atmospheric aerosols, are shown in Fig. 1.10. The aerosol models shown in Fig. 1.10, are the (a) spheroid, (b) tri-axial ellipsoid (Meng et al., 2010), (c) Gaussian random sphere (Muinonen et al., 1996), (d) shifted hexahedra (Bi et al., 2010), (e) hexagonal columns combined with polyhedral particles (Osborne et al., 2011), and large polyhedral particles (Kokhanovsky, 2003).

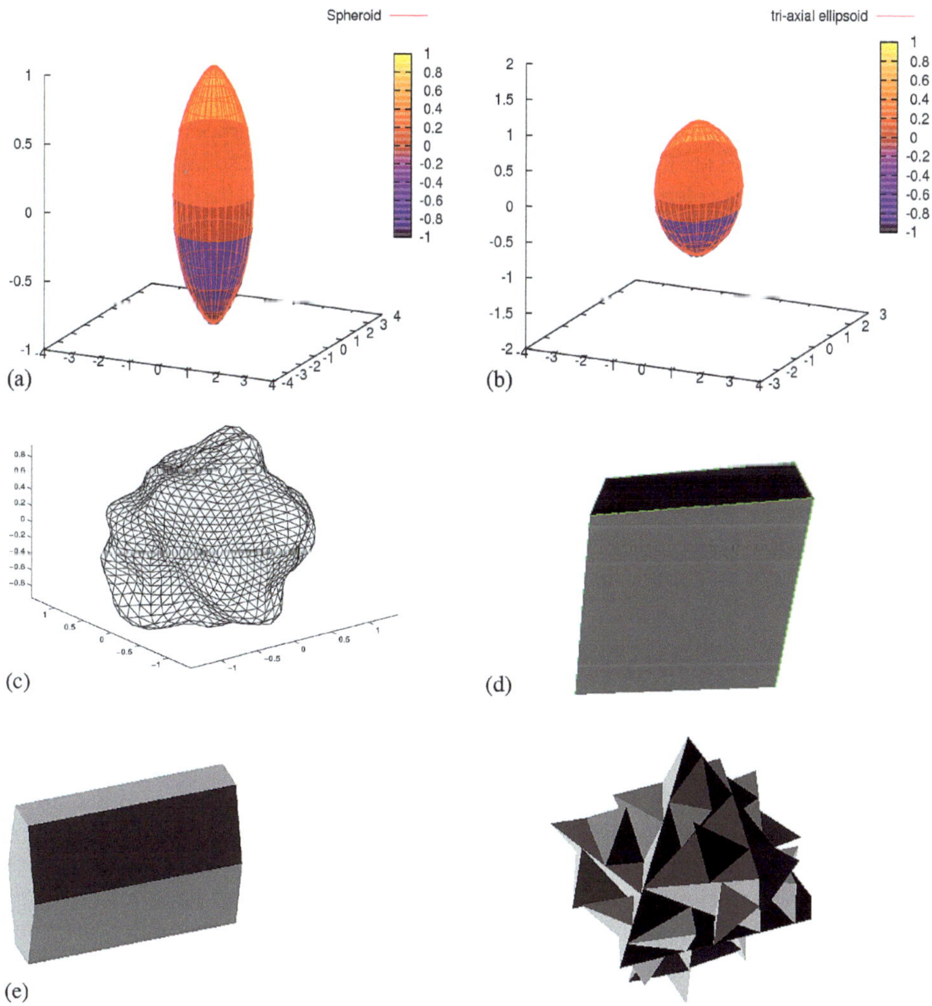

Fig. 1.10. Idealized geometrical models proposed to represent mineral and volcanic dust aerosols, showing the (a) spheroid, (b) tri-axial ellipsoid, (c) Gaussian random sphere, (d) shifted hexahedra, and (e) shape mixture of hexagons and polyhedral particles.

The simplest nonspherical particle shape shown in Fig. 1.10 (a) is the spheroid, prolate or oblate. The spheroid is a rotationally symmetric particle that is fully described by two morphological parameters. These are, the axis of spheroid rotation, called a, and the axis that is perpendicular to the axis of spheroid rotation, called b (Mishchenko et al., 2002). The ratio of a to b is called the aspect ratio,ε. Thus, for $\varepsilon < 0$ the spheroids are prolate, and for $\varepsilon > 1$, the spheroids are oblate. The T-matrix method can be readily applied to polydispersions of these rotationally symmetric particles, as demonstrated by Mishchenko and Travis (1994). In that paper, they show that differences between equivalent spheres and spheroids, in terms of the phase matrix elements, can be significant, and that the variation of the angle dependent quantities, depends on effective radius and shape of the scattering particle. Differences between equivalent spheres and spheroids in terms of the scalar quantities, Q_{ext}, ω_0, and g were not as dramatic, as differences found for the phase matrix elements. An important conclusion of the paper is that even for moderately aspherical particles of aspect ratio 1.5, differences between spheres and spheroids can still be as large as a factor of 2.5. Clearly, if spheres are used to represent nonspherical aerosols, in remote sensing their properties, then large errors may result. The T-matrix method has also been applied to layered-spheroids; see the review and references therein, by Mishchenko et al. (1996). More recently, the modified T-matrix approach, called the Sh-matrix method, has also been applied to layered-spheroids by Petrov et al. (2007), as well as the extended boundary condition method by Farafonov and Voshchinnikov (2012).

For aerosol scattering calculations, the concept of the 'equivalent sphere' may be invoked as a justification for using Lorenz–Mie theory. The 'equivalent sphere', means that the equal surface area, volume, or surface area to volume ratio, as the nonspherical particle, has been assumed. As Mishchenko and Travis (1994) demonstrate, this concept, in terms of the phase matrix elements, does not apply and so, for radiance calculations, should be avoided. Moreover, the same is true for irradiance or flux calculations as well (Nousiainen, 2009). There is no longer any justification for using Lorenz–Mie theory to interpret radiance or irradiance measurements of atmospheric aerosol; spheres can be considered to be purely a figment of the imagination.

Although spheroids are nonspherical they do not possess irregularity and sharp edges (see Fig. 1.4). In order to overcome this problem, Dubovic et al. (2006) introduced a shape mixture of spheroids of varying aspect ratio, to try and replicate the irregularity of actual nonspherical aerosols. The spheroid shape mixture model of Dubovic et al. (2006) consists of randomly oriented oblate and prolate spheroids of aspect ratio 0.3 to 3.0, respectively. The size parameter range covered is from about 0.012 to approximately 625, which covers realistic measurements of aerosol PSDs. To cover this large size-parameter range the methods of T-matrix, IGO and geometrical optics were applied, to compute the scattering phase matrix and scalar optical properties, which were then integrated to obtain the bulk scattering properties. The bulk scattering properties were calculated assuming that the real part of the complex refractive index varies between 1.33 and 1.6, and the imaginary part between 0.0005 and 0.5. The spheroid shape mixture model has been compared against laboratory measurements of the scattering phase matrix elements of mineral dust aerosol (Volten et al., 2001), and ground-based measurements (Dubovic et al.,

2006). The model was shown to reproduce well the scattering matrices of Volten et al. (2001), and fitted well to ground-based angular measurements of the total and linearly polarized intensity (Dubovic et al., 2006).

The advantage of the spheroid model is that it has just two free morphological parameters. However, the spheroid retains symmetry, not apparent in actual images of large aerosol see Fig. 1.4. A further way to reduce the symmetry of spheroids is to consider tri-axial ellipsoids, Fig. 1.10(b); this introduces an additional free morphological parameter, but reduces the symmetrical properties of spheroids (Bi et al., 2009). The work of Bi et al. (2009) showed that a weighted shape mixture of mineral dust ellipsoids, with optimized three-dimensional aspect ratios, could also replicate the scattering phase matrix elements measured by Volten et al. (2001). There is now available a single-scattering database of dust-like tri-axial ellipsoids that can be applied between the ultraviolet and far-infrared regions of the electromagnetic spectrum (Meng et al., 2010). Of course, by definition, the database of Meng et al. (2010) also includes the case of spheroids.

The single-scattering database of Meng et al. (2010) was produced by applying the methods of Lorenz–Mie theory, the T-matrix method, the discrete dipole approximation, and IGOM, to compute the scattering phase matrix elements and scalar optical properties for 42 particle shapes, 69 complex refractive indices, and 471 size parameters. The database also comes supplied with software to interpolate the single-scattering properties for shapes, refractive indices, and size parameters specified by the user.

The Gaussian random sphere, shown in Fig. 1.10(c), was introduced by Muinonen et al. (1996) and Muinonen (2000), to capture the irregularity of atmospheric aerosol. The Gaussian random sphere (GRS) is a statistical shape, represented by its radius r, specified as a function of θ and ϕ. Since the Gaussian random sphere is a statistical shape, it also depends on the mean radius, the relative standard deviation and the log radius, summed over spherical harmonic functions that are weighted with complex coefficients. Although elongated and flattened spheroids have been successful in generally replicating the scattering phase matrix measurements of Volten et al. (2001). However, features in the angle-dependent quantities specific to the irregularity of the aerosol, such as the flat phase function at backscattering angles and the minimum in the linearly depolarized phase function, are best modelled with the Gaussian random sphere (Veihelmann et al., 2006).

All the aerosol models discussed so far consider the aerosol surfaces to be smooth, without possessing sharp edges. To replicate sharp edges and the irregularity of mineral dust aerosol Bi et al. (2010) have introduced non-symmetric hexahedra, an example is given in Fig. 1.10(d). Hexahedra are three-dimensional symmetric objects with six faces. The symmetry of the hexahedra is reduced by randomly tilting its faces, while keeping its centre fixed. The methods of discrete dipole approximation and IGOM are used to calculate the scattering phase matrix and scalar optical properties of the randomized hexahedra, for size parameters ranging from 0.5 to 3000.0. The randomized hexahedra removes its symmetric properties, resulting in scattering phase functions with high side scattering and diminished backscattering, relative to regular hexahedral particles. The single-scattering properties of the randomized hexahedra also replicated the mineral dust scattering phase matrix measurements of Volten et al. (2001), whilst the regular hexahedra do not (Bi et al., 2010). Therefore, some form of randomization is required in order to replicate scattering phase matrix measurements of mineral dust aerosol.

The polyhedral particle shown in Fig. 1.10(e) was initially introduced by Macke et al. (1996a) to represent the irregularity of atmospheric ice crystals. This polyhedral particle is commonly known as the polycrystal, and it is generated by randomizing a second generation Koch fractal. The aspect ratio of the polycrystal is invariant with respect to size. The polycrystal has been applied to study the scattering phase matrix elements of large mineral dust particles by Kokhanovsky (2003). It was shown that by introducing the irregularity of the polycrystal, the scattering phase matrix measurements of Volten et al. (2001) could be replicated to high accuracy for a number of complex refractive indices.

The polycrystal model was adopted by Osborne et al. (2011), combined with the hexagonal column of aspect ratio unity (Fig. 1.10(e)), called the 'irregular' model, to replicate aircraft measurements of the transmitted short-wave scattered intensity of mineral dust aerosol, between the scattering angles of about 5° to 95°. The irregular model of Osborne et al. (2011) comprises hexagonal columns and polycrystals in the size range between 0.12 and 1.0 μm, and 1.2 and 20.0 μm, respectively. Hexagonal columns were used to replicate the sharp edges found on small mineral dust aerosol, and the size parameters of these particles are sufficiently small, that the halo features at 22° and 46° are not apparent on the scattering phase function. As the aerosol becomes large in size, the polyhedral particles supposedly replicate the irregularity of the larger-sized mineral dust or volcanic particles. To compute the single-scattering properties of the hexagonal columns and polyhedral particles, the methods of T-matrix and RTDF were used, respectively. The phase function and scalar optical properties were integrated over measured *in situ* PSDs to derive the bulk scattering properties of mineral dust aerosol (Johnson and Osborne 2011). In the paper by Osborne et al. (2011), they demonstrate that the irregular model best fitted their aircraft measurements, relative to spheres and the Dubovic et al. (2006) model. The irregular model has also been applied to simulate the scattering properties of volcanic silicate aerosol by (Johnson et al., 2012; Marenco et al., 2011; Newman et al., 2012; Turnbull et al., 2012). For large mineral dust particles, randomized particles, replicating the irregularity observed in actual mineral dust particles, should generally be preferred to model their single-scattering properties. However, Fig. 1.4 also shows that aggregation of aerosol particles can also take place, and therefore an idealized model of aerosol aggregates should also be considered.

Example bulk phase functions, computed at a wavelength of 0.55 μm, for the irregular model of Osborne et al. (2011), spheres and the Dubovic et al. (2006) model are shown in Fig. 1.11, plotted as a function of scattering angle. The figure shows that the irregular model has higher side scattering than for spheres and the Dubovic et al. (2006) model, at scattering angles between about 20° and about 100°; however, it is decreased for the Dubovic et al. (2006) model, at backscattering angles, relative to the sphere and irregular models. Fig. 1.12 highlights this behaviour in the scattering phase function more clearly. In Fig. 1.12, the spheroid and irregular phase functions are normalized by the sphere phase function, plotted against scattering angle. Also shown in this figure, is the phase function for spheroids, assumed to have an invariant aspect ratio of 1.7, with respect to size.

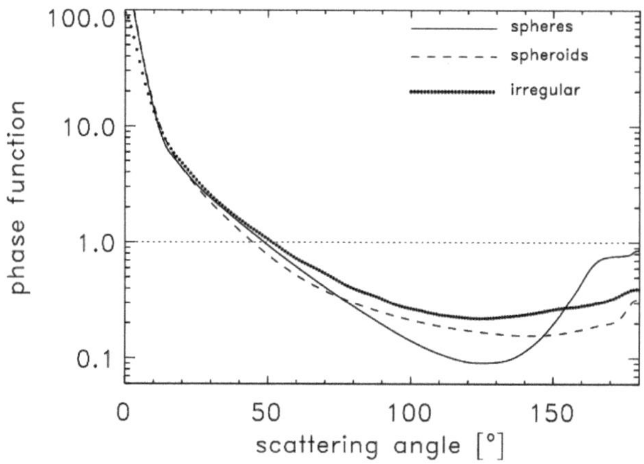

Fig. 1.11. The scattering phase function as a function of scattering angle, assuming three mineral dust aerosol models, which are spheres shown as the grey line, the Dubovic et al. (2006) model is shown as the dashed line, and the irregular model is shown as the full line (after Osborne et al., 2011).

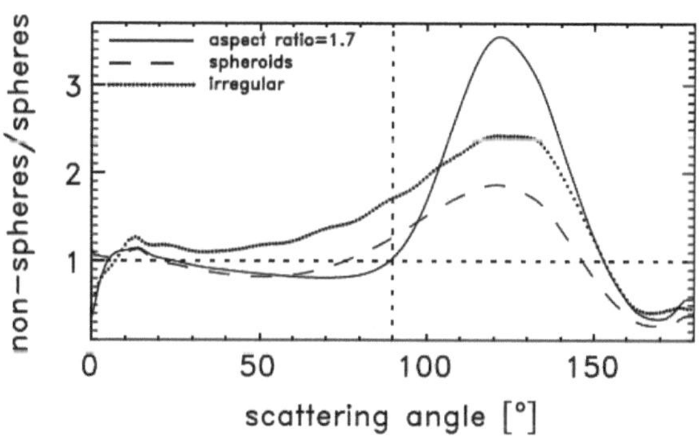

Fig. 1.12. The normalized scattering phase functions as a function of scattering angle, the non-spherical phase functions have been normalized by the sphere phase function, for the prolate spheroid of aspect ratio 1.7 (full line), the Dubovic et al. (2006) model (dashed line), and the irregular model (dotted line) (after Osborne et al., 2011).

The figure shows that, especially at backscattering angles, the differences, assuming one fixed ratio and a distribution of aspect ratios are large. Note also, the decrease in the forward peak of the phase function for the irregular model, at scattering angles less than 15°. Therefore, the irregular model shown in Fig. 1.12 will have a smaller g value relative to the sphere or spheroid models.

The g values for the sphere, Dubovic et al. (2006) model and irregular model, shown in Fig. 1.11, were determined in the short-wave region to be 0.73, 0.73 and 0.62, respectively (Johnson and Osborne 2011). The decrease in g value for the irregular model, relative to the other two models is about 15%. The impact of these modelled g values on the short-wave fluxes at the TOA and surface are shown in Fig. 1.13, as a function of time. The short-wave fluxes have been modelled using the Edwards–Slingo flexible, spherical harmonic, radiation code; for a full description

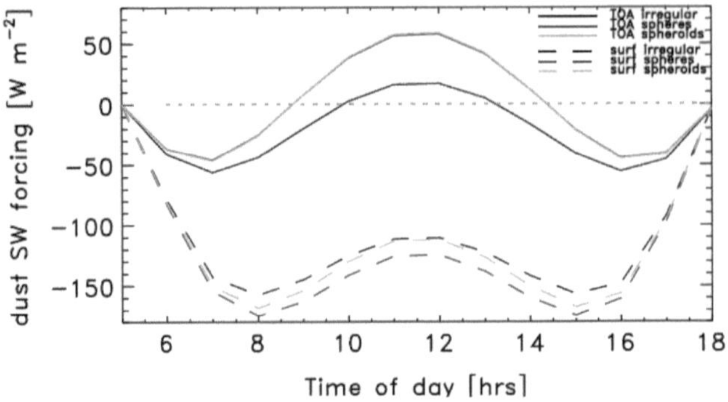

Fig. 1.13. The modelled top-of-atmosphere (TOA) and surface (surf) short-wave radiative forcing (0.3–3.0 μm) as a function of time for the irregular model, spheres and the Dubovic et al. (2006) model. The TOA and surf short-wave forcing for each assumed model is shown by the key in the top-right of the figure (after Osborne et al., 2011).

of this radiative transfer model see Edwards and Slingo (1996). The two-stream version of the Edwards–Slingo code has been used, assuming a fixed aerosol optical depth of unity. The short-wave forcing is defined as the difference between the clear-sky radiative forcing and the short-wave flux with atmosphere and aerosol. The short-wave forcing at the surface is defined as the change in the downwelling forcing, due to the presence of aerosol.

The figure shows, that the surface forcing, is negative for all models throughout the day. The daily average for irregular particles is about $-60\,\mathrm{W\,m^{-2}}$, whilst for spheres and the Dubovic et al. (2006) model it is about the same, $\sim -110\,\mathrm{W\,m^{-2}}$, almost a factor of 2 greater than the irregular model. The short-wave TOA forcing is greatest (largest negative values) at 0700 to 0800 and 1600 to 1700 local time, when the solar zenith angle was between 57° and 70°. The sign of the TOA forcing changes sign, at about local noon for spheres and the Dubovic et al. (2006) model, but is positive, for a significant period, for the irregular model. The daily mean TOA forcing efficiencies, for the irregular, sphere and Dubovic et al. (2006) models were computed to be $-13\,\mathrm{W\,m^{-2}}$, $-38\,\mathrm{W\,m^{-2}}$ and $-35\,\mathrm{W\,m^{-2}}$, respectively. Clearly, the g values of aerosols are an important quantity to constrain, as the short-wave forcing between irregulars and symmetric particles, with smooth surfaces, can be very significant.

As previously mentioned, volcanic ash is also an important type of aerosol, which needs to be considered. The polyhedral particle was adopted by Johnson et al. (2012), to estimate the mass concentration of a volcanic plume, using the CAS instrument, that had drifted over the UK during spring 2010, from the Eyjafjallajökull eruption. As discussed in Johnson et al. (2012), the response of the CAS instrument to the scattering properties of the volcanic aerosol, depends on the integral of the phase function, between the scattering angles of 4° and 12°. To obtain the mass concentration, therefore depends on the assumed scattering phase function. Figure 1.14(a) to (c), shows a number of phase functions, plotted against scattering angle, for three different randomizations applied to each particle, assuming six different volcanic aerosols. The Monte Carlo ray-tracing method

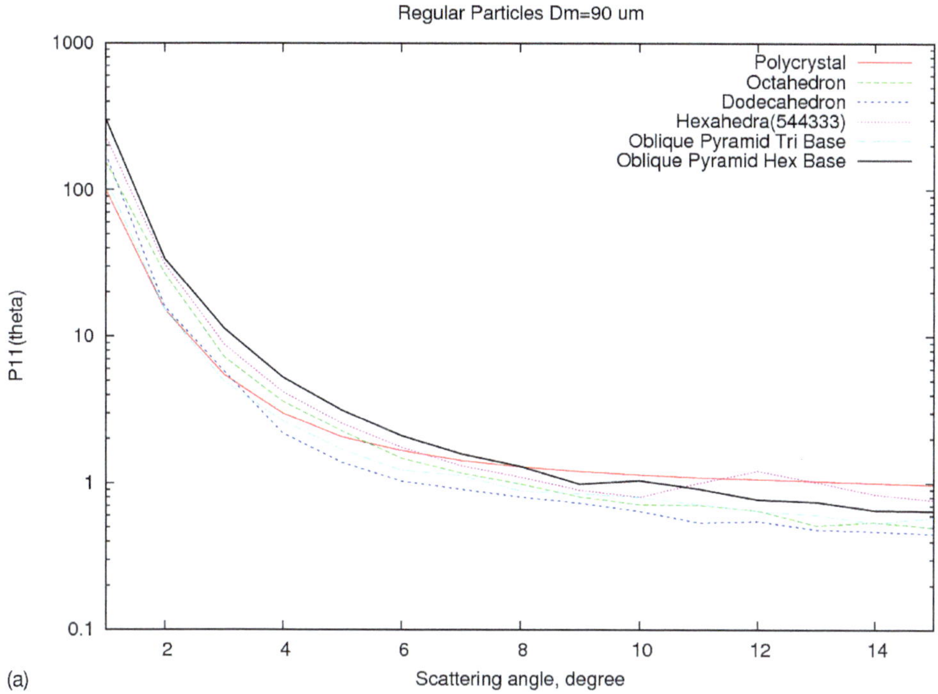
Regular Particles Dm=90 um
Polycrystal
Octahedron
Dodecahedron
Hexahedra(544333)
Oblique Pyramid Tri Base
Oblique Pyramid Hex Base
P11(theta)
1000
100
10
1
0.1
2
4
6
8
10
12
14
Scattering angle, degree

(a)

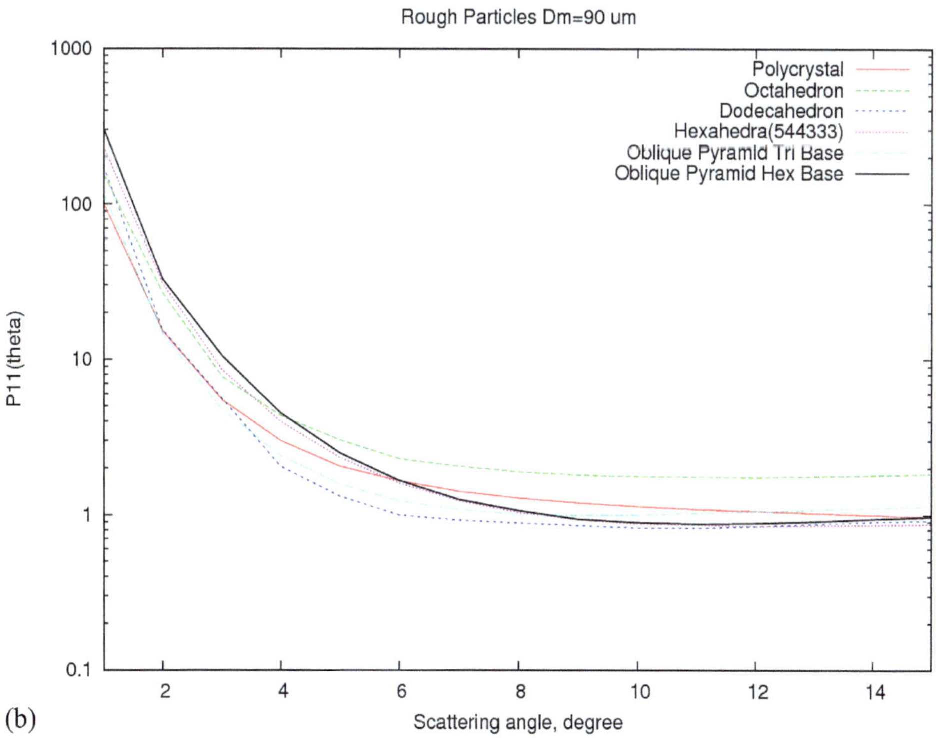
Rough Particles Dm=90 um
Polycrystal
Octahedron
Dodecahedron
Hexahedra(544333)
Oblique Pyramid Tri Base
Oblique Pyramid Hex Base
P11(theta)
1000
100
10
1
0.1
2
4
6
8
10
12
14
Scattering angle, degree

(b)

Fig. 1.14.

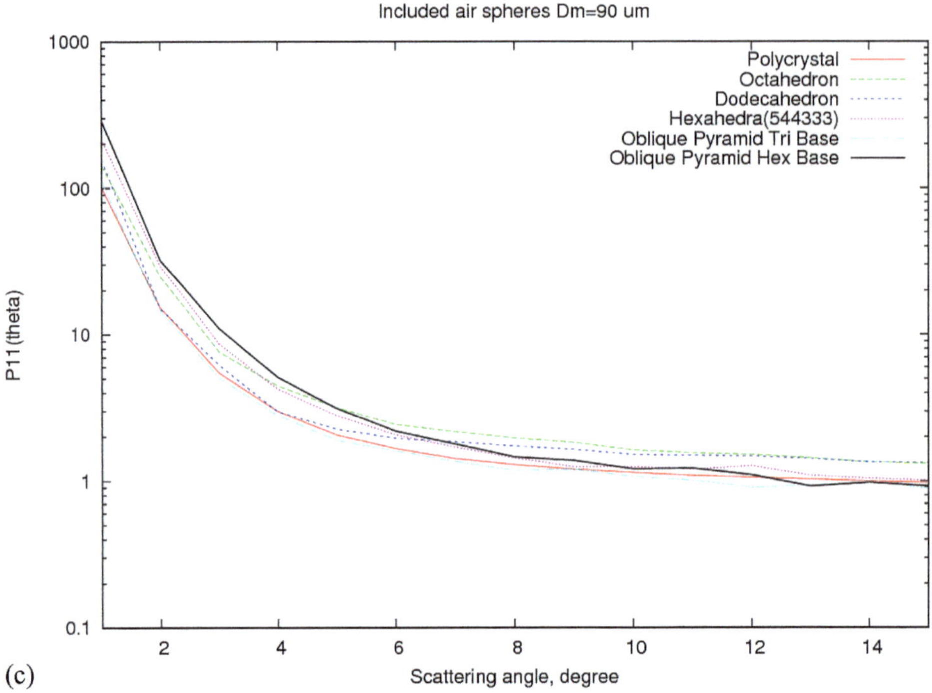

(c)

Fig. 1.14. The scattering phase function plotted against the scattering angle, assuming three different randomizations, applied separately to each model. The randomizations applied are regular models, i.e., no randomizations, (b) distorted particles and (c) spherical air inclusions. Each model is assumed to have a maximum dimension, D_{m}, of 90 μm; the calculation assumed an incident wavelength and complex refractive index of 0.55 μm and $1.55 + \mathrm{i}0.0011$, respectively.

described by Macke et al. (1996a) has been used to calculate the phase functions shown in Fig. 1.14(a) to (c). In Fig. 1.14(c) the aerosol particle has been 'distorted'; this means that at each refraction and reflection event the ray-paths are randomly tilted with respect to their original direction. This ray distortion has the effect of reducing and, at high levels of distortion, of completely removing any optical features that may be present on the phase function, such as haloes and bows. For high distortion values, featureless phase functions can be generated, and the effect of distortion on the phase function is assumed to be similar to surface roughness.

The volcanic aerosols assumed, in Fig. 1.14(a) to (c), are the polycrystal, octahedron, dodecahedron, hexahedra 5, 4, 4, 3, 3, oblique pyramid with triangular base, and an oblique pyramid but with a hexagonal base. The phase functions were calculated at the wavelength of 0.55 μm, assuming a complex refractive index of $1.55 + \mathrm{i}0.001$ (Osborne et al., 2011). The figure shows that at the scattering angles between 4° and 12° the phase functions can diverge significantly from the polycrystal, depending on randomization. Therefore, estimating volcanic ash mass concentrations using *in situ* sampling instruments such as CAS, or any other single-scattering instrument, depends on assumed particle shape and randomization. To quantify the uncertainty in the estimated mass, using single-scattering instruments, the shape and randomization must be taken into account by using a variety of idealized volcanic shapes.

The next section discusses idealized geometric models that have been proposed to represent the single-scattering properties of atmospheric ice crystals.

1.5.2 Ice crystal models and their light scattering properties

Figures 1.5, 1.6 and 1.7 show the wide range of possible ice crystal shapes that might occur in the Earth's atmosphere. As previously discussed, for ice crystal size, much less than about $100\,\mu$m, the shapes were 'indeterminate', due to the limiting resolving power of the CPI instrument. However, for larger ice crystal sizes, single pristine, spatial and aggregates of ice crystals appeared frequently in the CPI images. Due to the uncertainties surrounding the actual shapes of small ice crystals that might exist in cirrus, there are now a number of theoretical models that have been proposed to represent small ice crystals. To represent the scattering properties of large aggregated ice crystals, a number of habit mixture models have been proposed. To begin this survey of ice crystal models, we start with idealized single ice crystal models.

1.5.2.1 Single ice crystal models

Idealized single ice crystal models, that have been proposed to represent the scattering properties of atmospheric ice particles, are shown in Fig. 1.15. The uncommon, hexagonal ice column and hexagonal ice plate are shown in Fig. 1.15(a) and (b), respectively. The more common, six-branched bullet-rosette is shown in Fig. 1.15(c). As noted previously, ice crystals may also contain air cavities, and an example of a bullet-rosette model, with air cavities in its component branches, is shown in Fig. 1.15(d) (from P. Yang, personal communication). These simple single shapes have well defined three-dimensional structures, which help to simplify light scattering calculations. However, as shown in Figs 1.5, 1.6 and 1.7, ice crystals are usually indeterminate spatial and/or aggregated, and single models that have been proposed to represent these crystals, are shown in Fig. 1.15(e) to (j). To represent ice crystal randomization, shown in Figs. 1.5, 1.6, and 1.7, the 'polycrystal' was proposed by Macke et al. (1996a), previously discussed in Section 1.5.1, and this model is shown in Fig. 1.15(e). To represent the compact ice aggregates, the hexagonal ice aggregate shown in Fig. 1.15(f), was proposed by Yang and Liou (1998), and this model consists of eight arbitrarily attached hexagonal columns, the overall aspect ratio of this model also remains invariant with respect to size. Ice aggregates may also appear spatial rather than compact, and to represent these ice crystals, Baran and Labonnote (2006) proposed the eight-chain aggregate, Fig. 1.15(g), which was a re-transformation of the Yang and Liou (1998) compact hexagonal ice aggregate model. A 'radiatively equivalent' ice crystal model, Fig. 1.15(h), called the Inhomogeneous Hexagonal Monocrystal (IHM), was proposed by Labonnote et al. (2001) to represent the observed radiative properties of cirrus. The IHM was included with spherical air bubbles and aerosol to replicate the observed radiative properties of cirrus, as well as their polarization properties. Um and McFarquhar (2007, 2009)

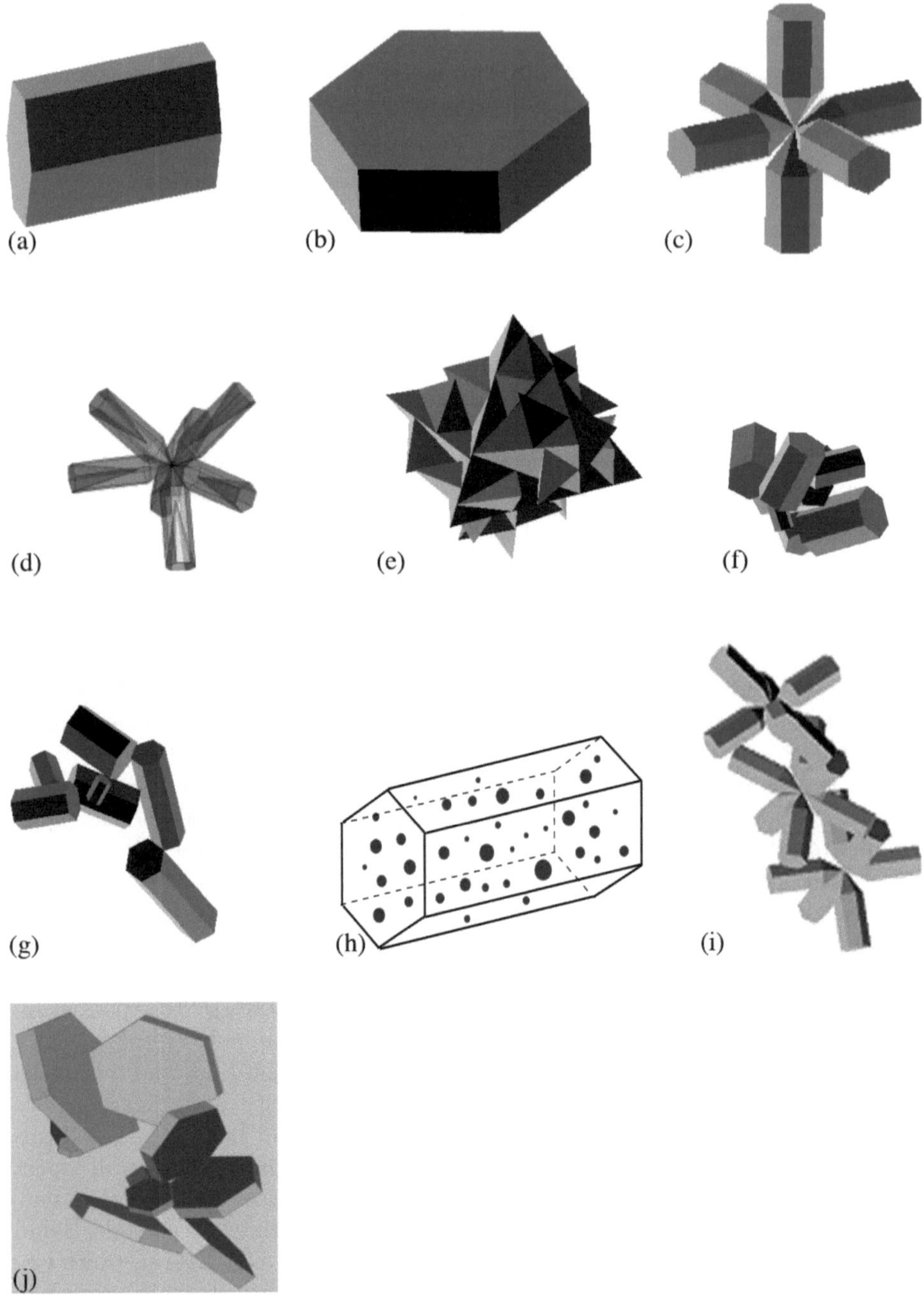

Fig. 1.15. Examples of idealized ice crystal models that have been proposed to represent single ice crystals in light scattering calculations. The models shown are (a) the hexagonal ice column, (b) the hexagonal ice plate, (c) the six-branched bullet-rosette, (d) the bullet-rosette with air cavities in each branch (Ping Yang, personal communication), (e) the polycrystal, (f) the hexagonal ice aggregate, (g) the chain hexagonal ice aggregate, (h) the IHM model, (i) the rosette-chain (G. McFarquhar and J. Um, personal communication), and (j) the chain of hexagonal plates (P. Yang, personal communication).

and Xie et al. (2011), have proposed models of rosette and plate aggregates, to represent these ice crystal shapes that occur in Figs. 1.5, 1.6, and 1.7. Figure 1.15(i) and Fig. 1.15(j) show idealized models of the rosette aggregate (Um and McFarquhar, 2007), and hexagonal plate aggregate (P. Yang, personal communication), respectively.

These more idealized aggregate ice crystal models shown in Fig. 1.15(e) to (j) or their 'radiative equivalents' represent the larger ice crystals. However, as shown in Figs. 1.6 and 1.7, there also exist small ice crystals of maximum dimensions much less than about 100 μm, and the shapes of these much smaller ice crystals are not at present known. To represent the shapes of small ice crystals various idealized geometrical models have been proposed, and some of these are shown in Fig. 1.16. Due to the 'rounded' nature of the small ice crystals, imaged by the CPI, shown in Figs. 1.6 and 1.7, simple spheroids have been proposed by Asano and Sato (1980), the so-called 'quasi-spherical' models. However, indentations can sometimes been seen on the CPI images of small ice crystals, and a model based on the Chebyshev polynomial has been proposed by McFarquhar et al. (2002), to account for these images, shown in Fig. 1.16(a) (from G. McFarquhar, personal communication). The

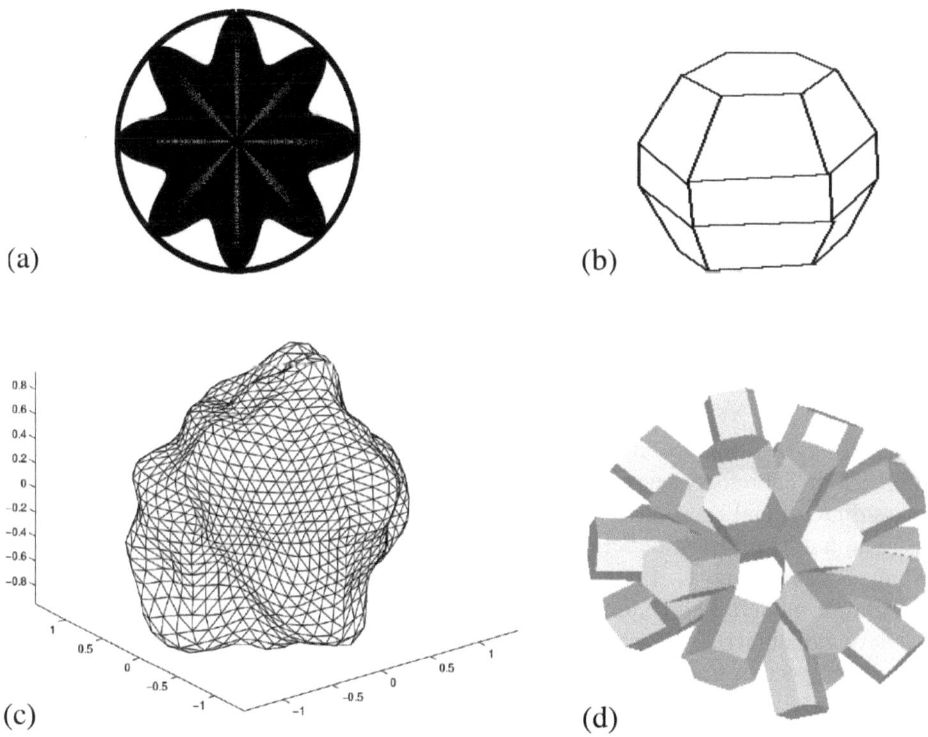

Fig. 1.16. Examples of idealized geometrical ice crystal models that have been proposed to represent small ice crystals in light scattering calculations. The models shown are (a) the Chebyshev model (G. McFarquhar, personal communication), (b) the droxtal, (c) the Gaussian random sphere and (d) the bucky ball model (G. McFarquhar and J. Um, personal communication).

droxtal ice crystal shown in Fig. 1.16(b) has been proposed by Yang et al. (2003), and this shape may exist in freezing fog (Ohtake, 1970) and wave cloud (Zhang et al., 2004) (from P. Yang, personal communication). The Gaussian random sphere shown in Fig. 1.16(c) has been proposed by Nousiainen and McFarquhar (2004) (from T. Nousiainen and G. McFarquhar, personal communication). This model has also been applied to simulate the properties of Saharan dust particles as previously discussed in Section 1.5.1. The more recent, budding bucky ball model, shown in Fig. 1.16(d), has been proposed by Um and McFarquhar (2011), and this model was based on a germinating ice crystal analogue model, described in Ulanowski et al. (2006) (from G. McFarquhar and J. Um, personal communication). However, the recent cloud chamber results of López and Ávila (2012), suggest that pure quasi-spherical models, may not actually exist, due to the protrusions of sharp edges, from their surfaces. Even small ice crystals are indeed, nonspherical to a high degree (López and Ávila, 2012). However, the more recent nonspherical models proposed to represent small ice crystals, cannot as yet be discounted. Clearly, further laboratory scattering studies and measurements made by SID-3 are required, to constrain the scattering properties of small ice crystals. The phase function of the small ice crystals imaged by López and Ávila (2012) should be measured and compared against idealized small ice crystal model predictions of the phase function.

As can be seen from Figs. 1.15 and 1.16, there are a large number of single ice crystal models that have been proposed to represent the variability of ice crystal shape found in cirrus. In the next subsection ice crystal habit mixture models are reviewed.

1.5.2.2 Habit mixture models of cirrus

Figs. 1.5, 1.6 and 1.7 indicate that cirrus is not wholly composed of single ice crystal shapes, but of shape mixtures or an ensemble collection of ice crystals. Habit mixture models have been proposed by (Volkivitsky et al., 1980; McFarquhar et al., 1999; Liou et al., 2000; Rolland et al., 2000; Baran et al., 2001b; McFarquhar et al., 2002; Baum et al., 2005; Baran and Labonnote 2007; Bozzo et al, 2008; Mitchell et al., 2008; Baum et al., 2011). However, as noted in Section 1.4, the mass of aggregating ice crystals is proportional to the square of their maximum dimension. Therefore, ice crystal aggregating models should be shown to follow this mass-dimensional relationship.

The habit mixture model of Baum et al. (2005) is shown in Fig. 1.17 (from P. Yang, B. Baum and G. Hong, personal communication), hereinafter referred to as the Baum model. The Baum model is currently used as the operational model to retrieve global cirrus properties using the space-based passive Moderate Resolution Imaging Spectroradiometer (MODIS) instrument (see, for example, Hong et al. (2007) and Lee et al. (2009)). The Baum model comprises of a size-dependent weighted habit mixture, as shown in Fig. 1.17, The smallest ice crystals in the PSD, are represented by droxtals; then for larger sizes, the habit mixture consists of six-branched bullet-rosettes, solid hexagonal columns and solid hexagonal plates, and hollow columns; and the largest ice crystals are represented by eight-branched compact hexagonal ice aggregates. However, due to the compact nature of the hexagonal ice aggregate, this model predicts that ice mass is proportional to the

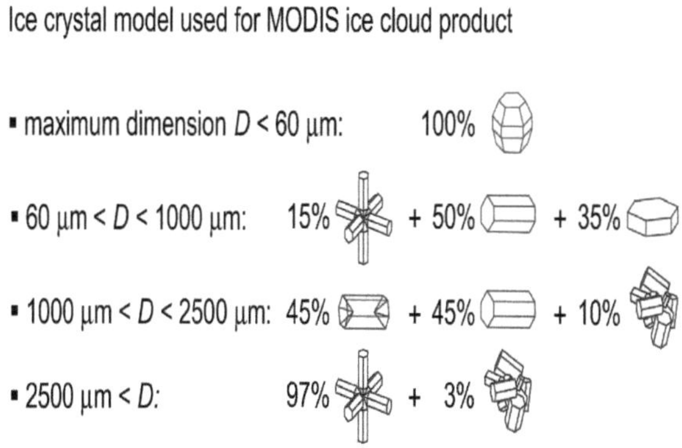

Fig. 1.17. The habit mixture model of Baum et al. (2005) (from P. Yang, B. Baum and G. Hong, personal communication).

cube of its maximum dimension. Therefore, in regions of ice crystal aggregation, it will over-predict IWC. For this reason, and amongst others discussed in Baum et al. (2011), a new habit mixture model has been proposed by Baum et al. (2011), hereinafter referred to as Baum2.

The Baum2 habit mixture model is shown in Fig. 1.18 (P. Yang, personal communication), and this new model, now consists of three additional habits, which are hollow bullet rosettes, with air cavities in each component branch, small spatial hexagonal plate aggregates and large spatial hexagonal plate aggregates. Without any dependence on temperature, this more generalized habit mixture model is able to replicate, generally within a factor of 2, many global *in situ* estimates of IWC. However, temperature-dependent, habit mixture models have also been developed by Baum et al. (2011), such as convective and non-convective models, as well as polar models. All these models appear to conserve ice crystal mass reasonably well, which means they must follow the generally observed ice aggregation power law relationship.

An alternative to the Baum and Baum2 habit mixture model, called the 'ensemble model', has been proposed by Baran and Labonnote (2007). The CPI images shown in Figs. 1.5, 1.6 and 1.7 indicate that ice crystals, as a function of depth from cloud-top, become progressively complex and spatial. The ensemble model, attempts to replicate this generally observed aggregation process, and the model is shown in Fig. 1.19. The ensemble model is composed of six elements, which become progressively more complex as a function of maximum dimension, D. The first ensemble member consists of a solid hexagonal ice column of aspect ratio unity, which is shown in Fig. 1.19(a). With increasing D, the ensemble model becomes progressively more complex and spatial, by arbitrarily attaching other hexagonal ice column elements to each other, until the last ensemble member, a spatial ten element chain, shown in Fig. 1.19(f), is formed. It should be noted here, that unlike the single ice crystal models previously described in subsection 1.5.2.1, such as the polycrystal and eight-branched compact hexagonal ice aggregate, the overall aspect ratio of each ensemble member does not remain invariant with respect to D. Each of the ensemble members, shown in Fig. 1.19, is distributed throughout a

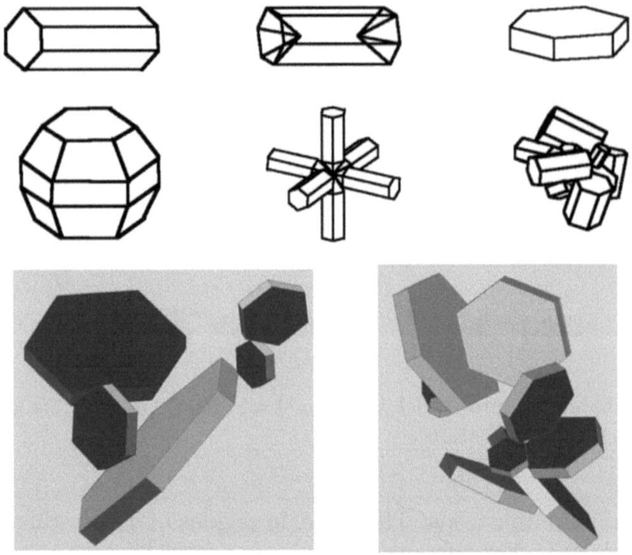

Fig. 1.18. The habit mixture model of Baum et al. (2011) (from Ping Yang, personal communication).

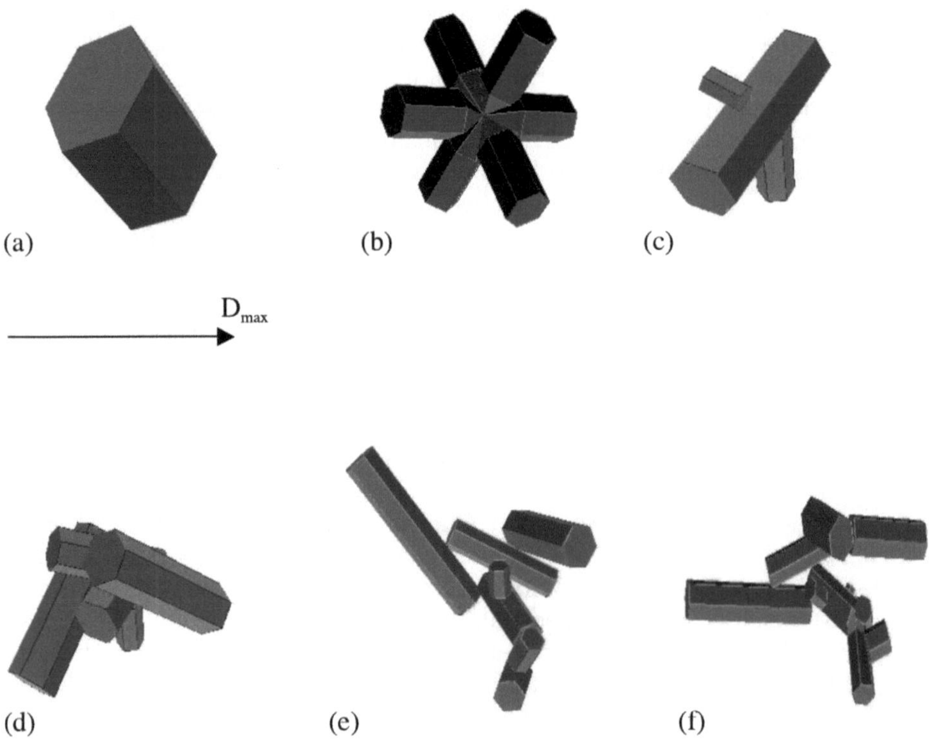

Fig. 1.19. The ensemble model of Baran and Labonnote (2007) (after Baran and Labonnote, 2007).

PSD, which is divided into six equal intervals. The first member of the ensemble is distributed into the first interval, and the last ensemble member is distributed into the sixth interval.

The single-scattering properties predicted by the Baum and Baum2 models are integrated using *in situ* derived PSDs, whereas the ensemble model is integrated using a moment estimation parameterization of the PSD. This parameterized PSD has been proposed by Field et al. (2007). The parameterized PSD is based on many thousands of *in situ* measured PSDs, at temperatures between 0°C and −60°C. The moments of the *in situ* PSDs are parameterized, by relating the second moment (i.e., IWC) to higher moments, through power law relationships, depending on the in-cloud temperature. Thus, from the IWC and in-cloud temperature, the original PSD is estimated. Relating the PSD to IWC, is a desirable link, as the IWC is a prognostic variable of a GCM. This parameterized PSD, is based merely on the measured size, and is thus independent of assumed ice particle shape, a further desirable property if it is to be consistently applied to the single-scattering properties of ice crystals.

Other parameterized PSDs that are available in the literature, tend to rely on specific mass-dimensional relationships, which depend on the particular shapes and specific pre-factors and exponents, estimated during field campaigns. Therefore, such parameterized PSDs cannot be consistently applied to ice crystal models, as these models, may predict different mass-dimensional relationships to the ones used to construct the parameterized PSD. A further advantage of the Field et al (2007) parameterization is that it is based on linear algebra; this means that the resolution of the PSD can be as fine as the user wishes, making integration of the single-scattering properties accurate. A further desirable feature of this parameterization is that it circumvents the problem of ice crystal shattering. It does this by ignoring ice crystal sizes less than 100 μm, measured by the *in situ* probes. The parameterization, for particle sizes less than 100 μm, approximates the PSD by an exponential, for particle sizes greater than 100 μm, the parameterization assumes a gamma function.

The ensemble model of Baran and Labonnote (2007), when combined with the Field et al. (2007) PSD, has been demonstrated to have predictive value in replicating, to within measurement uncertainty, mid-latitude and tropical cirrus IWC, and total solar optical depth (Baran et al., 2009; Baran et al., 2011a). This predictive property, of the ensemble model, means that it too, can be readily applied in GCM radiation schemes without regard to location; this is an important consideration in GCM climate change experiments (Baran et al., 2010; Baran 2012a).

A further habit mixture model has been proposed by Mitchell et al. (2008), based on many mid-latitude and tropical *in situ* measurements of area-dimensional and mass-dimensional relationships, for a number of ice particle shapes, such as hexagonal ice columns, hexagonal ice plates, bullet-rosettes and ice aggregates. This scheme is also linked to a parameterized PSD, generated from the IWC and cloud temperature, and has been applied to study the radiative effect of small ice crystals in GCMs (Mitchell et al., 2008).

Ice crystal aggregating models, if they follow the generally observed mass-dimensional relationship can be used to forward model the radar reflectivity or microwave radiances. Indeed, the Baum and Baran and Labonnote (2007) models have been shown to have predictive value in forward modelling the CloudSat radar reflectivity at 94 Ghz (Hong et al., 2008; Baran et al., 2011b). Moreover, Baran et al. (2011b) have proposed a radar reflectivity forward model at 94 GHz for application to GCMs, so that CloudSat, or any other space-based radar instrument operating at 94 GHz, can be simulated directly using only GCM prognostic variables.

The ice aggregating models discussed in this section should be physically consistent across the electromagnetic spectrum. That is, the same ice crystal model should be used to forward model solar, infrared, and microwave radiances or radar reflectivity, without the need to apply different ice crystal models to different regions of the spectrum. With the advent of the space-based A-train that near-simultaneously samples cirrus across the electromagnetic spectrum, this requirement of physical consistency across the spectrum, is a prerequisite of any model; see also Baran and Francis (2004). The next subsection discusses the single-scattering properties of ice crystals.

1.5.2.3 The single-scattering properties of ice crystals

Figures 1.20 and 1.21 show the P_{11} and P_{12} elements of the scattering phase matrix, respectively, predicted by some of the ice crystal models discussed in subsection 1.5.2.2. The ice crystal models, shown in Fig. 1.20, are the randomly oriented hexagonal ice plate of aspect ratio 0.5, 50% hollow hexagonal ice plate of aspect ratio 0.5, six-branched bullet-rosette, very distorted six-branched bullet-rosette, very distorted hexagonal ice aggregate, the IHM model, and the ensemble model. Also shown in Fig. 1.20 is the calculated g value, for each of the models. The calculations shown in Fig. 1.20 and 21 assume an incident wavelength of 0.55 μm, and a complex refractive index of $1.33 + i1.0 \times 10^{-12}$, where i is the imaginary part of the refractive index. The calculations were performed using the method of Monte Carlo ray-tracing described by Macke et al. (1996a).

Fig. 1.20 shows that, smooth ice crystal models, such as the hexagonal ice plate, and six-branched bullet rosette, predict phase functions, which have the familiar 22° and 46° haloes. As well as the ice bow feature, at scattering angles between about 135° amd 160°, and reflection peak, at the exact backscattering angle of 180°. The 50% hollow hexagonal ice plate has more forward scattering, relative to the smooth particles. This is why the g parameter predicted by the 50% hollow plate has increased relative to the smooth particles; see also Schmitt et al. (2006) and Yang et al. (2008)). Moreover, the 22° halo and ice bow features are either reduced or almost removed on the phase function of the 50% hollow hexagonal plate, respectively (Schmitt et al., 2006; Yang et al., 2008).

In contrast to the smooth ice crystal models, the distorted or rough ice crystals, such as the very distorted six-branched bullet-rosette and very distorted hexagonal ice aggregate models, produce phase functions which are featureless (Yang and Liou 1998; Ulanowski et al., 2006). Moreover, the distorted ice crystal models predict g values, which can be considerably smaller than their smooth or hollow counterparts. Therefore, depending on the process dominating the randomization of the ice crystal, the asymmetry parameter, may increase or decrease. The IHM model, predicts much diminished haloes and a featureless phase function, at scattering angles

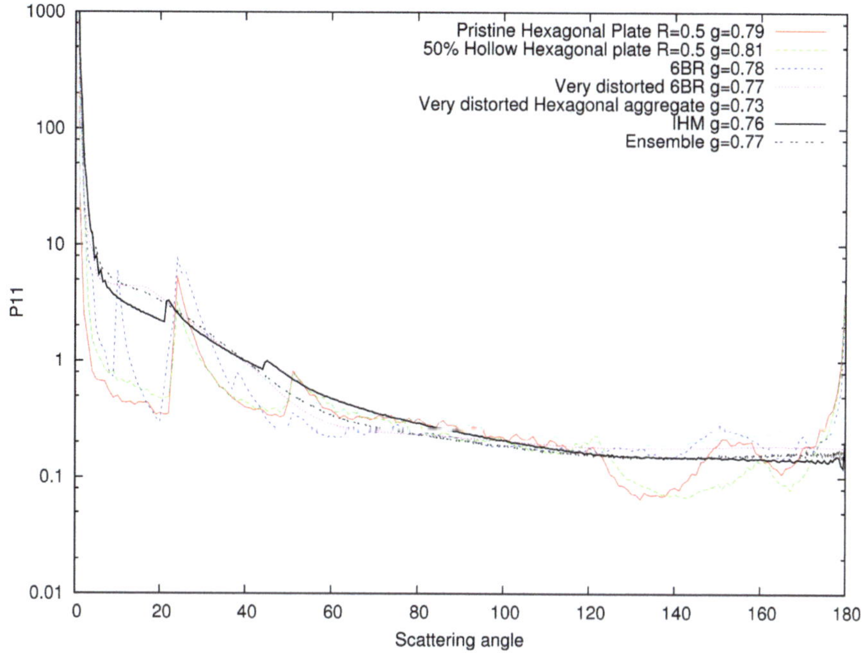

Fig. 1.20. The scattering phase function plotted against scattering angle for a variety of ice crystal models. The models are shown by the key in the top right-hand side of the figure, together with their predicted values of g.

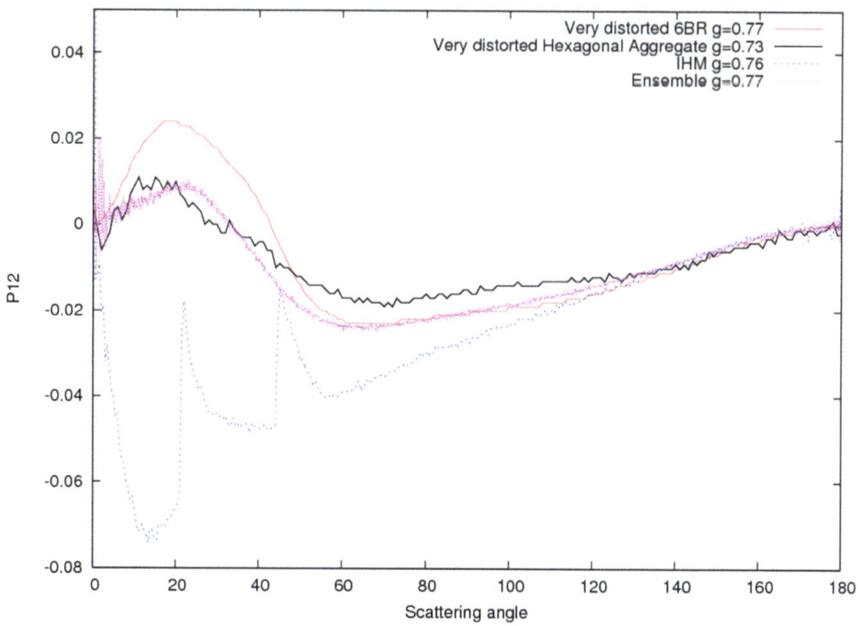

Fig. 1.21. The P_{12} element of the scattering phase matrix plotted against the scattering angle for a number of ice crystal models. The models are shown by the key in the top right-hand side of the figure, together with their predicted values of g.

greater than about 46°. This is because the IHM is composed of spherical air and aerosol bubbles which, due to multiple scattering between the spherical bubbles, diminish haloes and produce a generally featureless phase function (Labonnote et al., 2001). The ensemble model of Baran and Labonnote (2007), predicts a phase function, which is completely featureless and almost flat at backscattering angles, due to a randomization of ray paths and spherical air bubble inclusions within the crystal volume (Macke et al., 1996a; Macke et al., 1996b; Shcherbakov et al., 2006).

However, a recent paper by Gayet et al. (2012) reports on Polar Nephelometer (PN) measurements; the PN measures the scattering phase function between the polar angles of about 15° to 162°. The phase function measurements were obtained in a mid-latitude anvil cloud, toward the cloud-top at temperatures of about −58°C. The ice crystals that produced the measured phase functions were chains of aggregates and chains of quasi-spherical ice crystals. The PN measurements revealed that, although there were no halo features present on the phase functions, there was, an ice bow-like feature at around scattering angles of 140° to 160°. Moreover, the presence of this backscattering feature lowers the side-scattering of the phase function, relative to completely randomized particles, with no features present at all. The averaged asymmetry parameter of the *in situ* measured phase function was estimated by Gayet et al. (2012) to be about 0.78 ± 0.04. The measurements obtained by Gayet et al. (2012) show that it is not sufficient to measure phase functions over narrow ranges of scattering angle, but rather measurements must be obtained that cover both forward scattering and backscattering angles, only with such measurements can inferences be made about the asymmetry parameters of ice crystals.

The unusual phase functions reported by Gayet et al. (2012) have implications not only for the energy balance of anvils but also for remote sensing. Clearly, the occurrence of such phase functions needs to be further quantified.

The variation in g, predicted by the models, shown in Fig. 1.20, is considerable. The difference in g, between the smallest and largest g values, is about 12%. This 12% difference is radiatively very significant, as the following example illustrates. According to asymptotic theory, for the case of a semi-infinite atmosphere, the reflection, r_∞, depends on ω_0 and g, which are related to r_∞, through the similarity principle S, given by (van de Hulst, 1980):

$$S = \sqrt{\frac{(1-\varpi_0)}{(1-\varpi_0 g)}} \tag{1.15}$$

$$r_\infty = \frac{1-S}{1+S} \tag{1.16}$$

If we assume that $\varpi_0 = 0.9$, and $g = 0.73$ and 0.81, the highest and lowest g values in Fig. 1.20. Then, the difference in r_∞ is about 6%, between the highest and lowest g values. In terms of flux units, this difference is approximately $55\,\mathrm{W\,m^{-2}}$. These differences are significant, and this is why in climate models, the assumed value of g is so important. Therefore, it is important to constrain values of g, predicted by theory, by using *in situ* measurements, such as those provided by SID-3 and the PN instruments.

In the literature, there does appear to be a cluster of g values around about 0.75 ± 0.03. However, theoretically, it is possible to predict g values which are much lower than 0.75 (Mishchenko and Macke, 1997; Ulanowski et al., 2006). Moreover, examination of spatial light scattering patterns of ice crystals, measured by SID-3, in a variety of mid-latitude cirrus, suggest very rough ice crystals, which imply significantly lower g values relative to their smooth counterparts (Z. Ulanowski, personal communication). In a more recent paper, Mauno et al. (2011) found that differences between simulated and measured ground-based short-wave fluxes could not be accounted for by uncertainties in the assumed ice crystal model shapes or shapes of the PSD. However, they found that by reducing their ice crystal model values of g, the average value of which at $0.55\,\mu$m was about 0.78, by about 10%, improved their agreement, for some periods, between model and measurements. They speculated that this 10% reduction in g could be due to ice crystal surface roughness or inclusions or irregularities on the ice crystal not accounted for in their model.

On the other hand, Yang et al. (2011), find that ignoring the vertical profile of ice crystal shapes, and the shape of the PSD, can lead to significant errors in the downwelling and upwelling short-wave fluxes, if only simple shapes, such as columns, are considered in GCMs. Moreover, they find that the error in ignoring the vertical structure is similar to the impact of scaling the asymmetry parameter from high to moderate values, i.e., reducing their g values by about 6%. If their g values are reduced by extreme amounts, such as by 13%, then differences between simple columns and rough columns could be as large as $25\,\mathrm{W\,m^{-2}}$. However, further randomizations need to be considered in simulations, such as the ones performed by Yang et al. (2011), apart from surface roughness. These include concavities and porosity, as well as irregularity.

To illustrate the degree to which each of these processes either increases or decreases g, for a particular shape, we consider the hexagonal plate, from Fig. 1.20, assuming the same wavelength and complex refractive index as used in that figure. For this shape, the ray-tracing code of Macke et al. (1996a) is applied, and the modification to the original Macke et al. (1996b) code by Shcherbakov et al. (2006), which included air bubbles or aerosol bubbles within the volume of the ice crystal, is used to fully randomize the crystal, apart from distortion.

Table 1.1 illustrates the effect on the value of g, by applying the processes of hollowing the crystal by 50%, distorting the ice crystal and then including the crystal with spherical air bubbles. As in Fig. 1.20, the pristine crystal has a g value of 0.79, and hollowing the crystal then increases the value of g to 0.81. Applying severe distortion to the crystal, decreases g, for both the solid and hollow crystals, by about 5% and 7%, respectively. If the crystals are then included, with spherical air bubbles, which increases multiple scattering and therefore decreases g, then the value for g in both cases decreases by about 12%. The inclusions of spherical air bubbles decreases g further, by about 5% or 6%. The difference between the two extreme g values of 0.81 and 0.71, in terms of reflected flux, using Eq. (1.16), is about $66\,\mathrm{W\,m^{-2}}$.

Clearly, Table 1.1 demonstrates that it is possible to generate extreme differences in the value of g, using the same ice crystal model. However, whether such extreme g values really do occur in the natural atmosphere, requires further

Table 1.1. The asymmetry parameter, g, tabulated as a function of randomization (process), assuming the hexagonal ice plate of aspect ratio 0.5. The processes applied are pristine, 50% hollow, very distorted (VD), 50% hollow plus very distorted, and finally 50% hollow plus very distorted plus spherical air inclusions.

Process	g
Pristine	0.79
50% Hollow	0.81
Very distorted (VD)	0.74
50% Hollow + VD	0.75
50% Hollow + VD + Air Inclusions	0.71

measurement, using many more *in situ* measurements of the ice crystal scattering patterns, covering a considerable degree of scattering angle space. It is also important to determine, which process or processes acting on the ice crystal, is or are occurring, as a function of atmospheric state (i.e., temperature and humidity) and possibly vertical velocities, as these processes will ultimately determine the real value of g (Ulanowski et al., 2010, 2011; Gayet et al., 2011; Gayet et al., 2012). The unusual in situ-measured ice crystal phase function reported by Gayet et al. (2012) has been theoretically interpreted by Baran et al. (2012) as being chiefly due to the quasi-spherical ice crystals dominating the PN measurements, rather than to the underlying aggregate shape. This is evidence that retrieving ice crystal shape using remotely sensed measurements may not be accurate.

The phase functions predicted by the very distorted ice crystal models and ice crystal models with inclusions are very similar at backscattering angles, as shown in Fig. 1.20. Intensity alone measurements, could not distinguish between these different models, and yet their geometries are very different.

Figure 1.21 shows the linearly polarized P_{12} element of the scattering phase matrix, predicted by the very distorted six-branched bullet-rosette, very distorted hexagonal ice aggregate, IHM and ensemble model. The figure shows that, for scattering angles greater than about 60°, the gradient in the linear polarization is quite different, for the various models, and between about 10° and 60°, the predictions are significantly different. This figure illustrates the importance of linear polarization measurements, which can be used to further constrain ice crystal models (Labonnote et al., 2001; Baran and Labonnotte 2006; Sun et al., 2006; Mishchenko et al., 2007). However, given the range of ice crystal complexity, as shown in Figs 1.5, 1.6 and 1.7, there might be no unique intensity or polarized signature, for a given ice crystal shape or ensemble of shapes. Since, what is measured by radiometers is intensity, which is a convolution of the ice crystal size, shape and complexity. As illustrated by Fig. 1.20, the very distorted models predict scattered intensities that are very similar to each other, and so for these, the shape information is lost. The measured intensity is not, therefore, a unique signature of shape, unless the crystals are perfect geometrically, and smooth. This point is illustrated by Fig. 1.22.

Figure 1.22 shows the phase functions predicted by the ensemble model, after progressively randomizing the model, from pristine to distorted, eventually becoming highly distorted with spherical air bubble inclusions. As previously discussed, as the ice crystals become progressively more randomized, the optical features such as the 22° and 46° halo are removed, and the backscattering is decreased, with an increase in side scattering. The phase function of the most randomized ensemble model becomes featureless and at backscattering angles, is flat. Figure 1.22 illustrates that retrieving ice crystal 'shape' is an entirely meaningless proposition, and what should be retrieved instead, is a randomization. It is the randomization, which has the greatest impact on g, unless the ice crystal is perfectly smooth (Baran et al., 2012). This ensemble model phase function, shown in Fig. 1.22, can be readily applied to any inversion scheme to retrieve randomization, rather than shape. Of course, any other model could be randomized, and those phase functions applied to retrieval schemes (Doutriaux-Boucher et al., 2000; Baran et al., 2001b; Ulanowski et al., 2006; Baran and Labonnote 2007; Baum et al., 2011).

In the literature the scalar optical properties are often plotted as functions of D_e, however, in this chapter, they are plotted as a function of IWC and cloud temperature. Figure 1.23(a) and and Fig. 1.23(b) show the scalar optical properties, ω_0 and g, predicted by the ensemble model at the wavelength of 1.6 μm, plotted as a function of IWC and cloud temperature. The scalar optical properties were calculated using the ray-tracing code of Macke et al. (1996a); the bulk scattering properties were then derived, by integrating the scalar optical properties over 20662 PSDs using the Field et al. (2007) parameterization. The IWC and cloud temper-

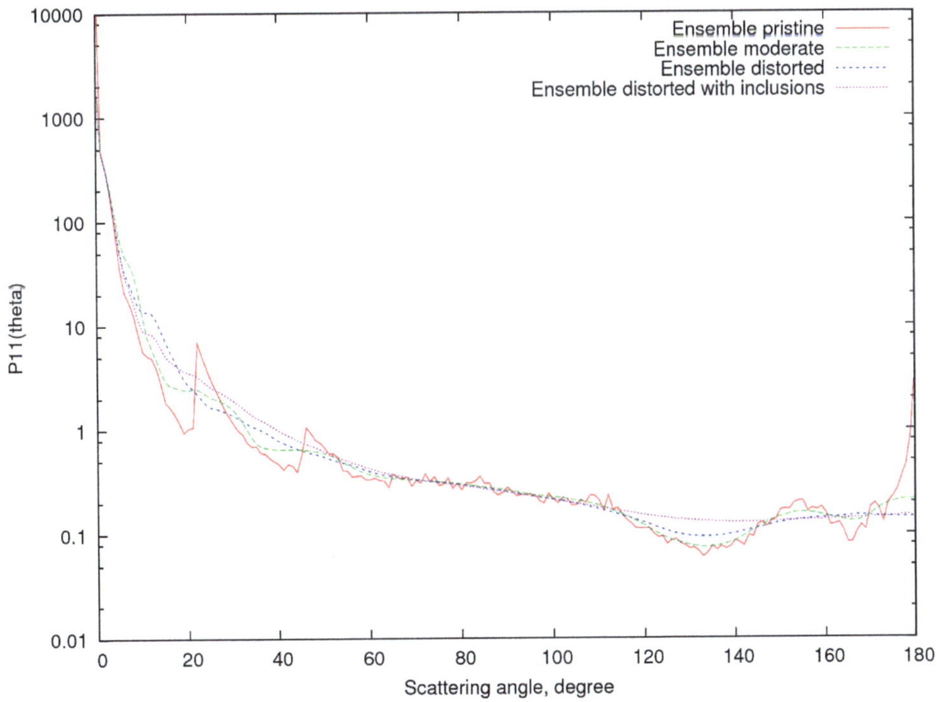

Fig. 1.22. The scattering phase function plotted against the scattering angle for the ensemble model, assuming various randomizations. The randomizations are shown by the key in the top right-hand side of the figure. The incident wavelength and complex refractive index are the same as used in Fig. 1.20.

ature values, which were used to generate the PSDs, were obtained in a variety of cirrus, and these measurements are described in Baran et al. (2011b).

Figure 1.23(a) and Fig. 1.23(b) show that the physical behaviour of ω_0 and g is sensible. At low IWC and cold temperatures, the ω_0 and g values are high and low, respectively. This is because at low IWC and cold temperatures, the shape of the PSD is narrow, which means that it is dominated by small ice crystals. Therefore,

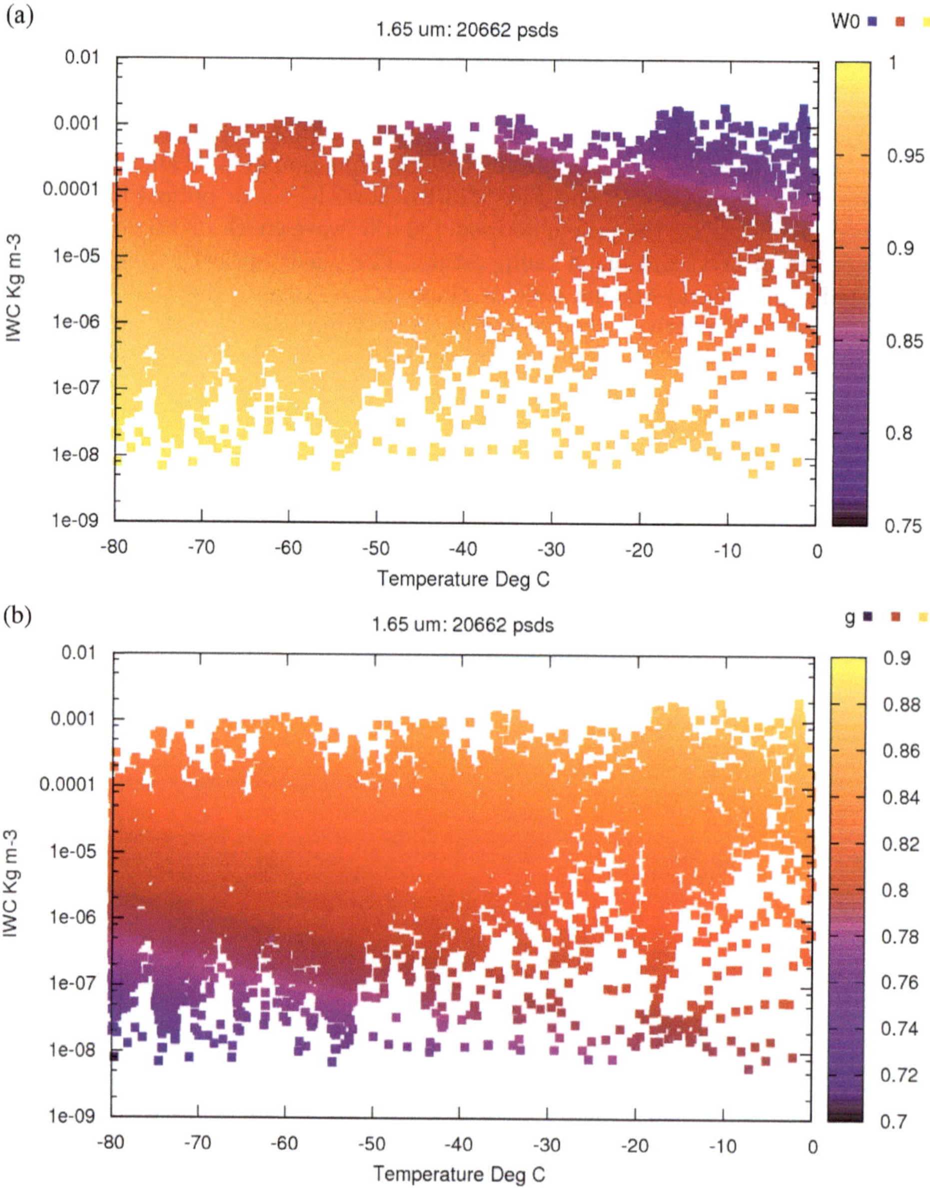

Fig. 1.23. The ensemble model predicted scalar optical properties at the wavelength of 1.6 μm, plotted as a function of IWC and cloud temperature, using tropical PSDs. The scalar optical properties shown are (a) ω_0, and (b) g.

these small ice crystals have low g values and scatter incident radiation efficiently. At warm temperatures and high IWC, the reverse is true. In this case, the ω_0 and g values become low and high, respectively. Again, this is due to the shape of the PSD, which becomes broad, for values of high IWC and warm temperatures, and so the PSD has larger ice crystals occurring more frequently than previously. Thus, larger ice crystals will absorb more incident radiation, and consequently ω_0 will become lower, whilst the g values become large, due to the greater absorption, which increases the diffracted component. As Fig. 1.23(a) and Fig. 1.23(b) demonstrate, the bulk scalar optical properties vary in both the horizontal and vertical directions, though they depend mostly on IWC, and only weakly on cloud temperature.

The scalar optical properties shown in Fig. 1.23(a) and Fig. 1.23(b) can be readily parameterized into climate models, thus linking directly, GCM prognostic variables to the bulk scattering properties of the cloud, without the need for D_e. Indeed, this has been achieved by Mitchell et al. (2008), Baran et al. (2010) and Baran (2012a). These new parameterizations demonstrate that there is no need to link the optical properties in GCMs via properties such as effective diameter, which is still the general approach adopted in parameterizing bulk scalar optical properties in climate models (Edwards et al., 2007; Fu, 2007; Hong et al., 2009; Gu et al., 2011).

As previously discussed the ensemble model predicts that the ice crystal mass is proportional to the square of its maximum dimension. This means that the ensemble model can predict the radar reflectivity of aggregating ice crystals. An example of this is shown in Fig. 1.24. The radar reflectivity is calculated, using the Rayleigh–Gans approximation, and is given by (Baran et al., 2011b);

$$Z_e = 10^{18} C \int_{D_{\min}}^{D_{\max}} \frac{36\pi^3}{\lambda^4 \rho_i^2} \left| \frac{\varepsilon - 1}{\varepsilon + 1} \right|^2 m^2(D) f(D) n(D)\, \mathrm{d}D \qquad (1.17)$$

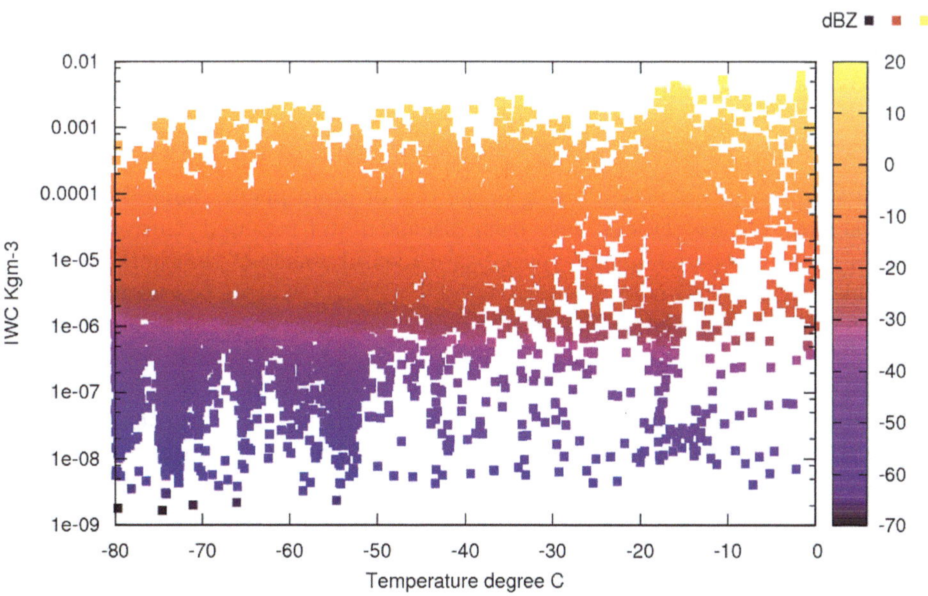

Fig. 1.24. Same as Fig. 1.23 but for the radar reflectivity (dBZ$_e$) at 94 GHz (after Baran et al., 2011b).

where the units of Z_e are in $\mathrm{mm}^6\,\mathrm{m}^{-3}$, the constant $C = \lambda^4/(|K|^2\pi^5)$, and λ is the incident wavelength in m, $|K|^2$ is the dielectric factor, assumed to have a value of 0.75 at 94 GHz (this value of $|K|^2$ has been assumed, since it is the value used to calibrate the CloudSat radar), and the choice of the dielectric factor is dictated by convention, to ensure that for water droplets $Z = \int N(D)D^6\,\mathrm{d}D$, where $N(D)$ is the droplet size distribution function. The constant C has a value of $4.520 \times 10^{-13}\,\mathrm{m}^4$ and $n(D)$ is the PSD, in units of m^{-4}. The units of the integrand are in SI ($\mathrm{m}^6\,\mathrm{m}^{-3}$) units, to convert these to $\mathrm{mm}^6\,\mathrm{m}^{-3}$, the integrand must be multiplied by a factor of 10^{18}. In Eq. (1.17) ρ_i is the density of solid ice, assumed to be $920\,\mathrm{kg\,m}^{-3}$, and ε is the dielectric constant of solid ice, and $f(D)$ is the form factor. The form factor represents the deviation from the Rayleigh approximation, as the size parameter increases beyond unity. The form factor has been previously computed by Westbrook et al. (2006) and Westbrook et al. (2008) for aggregating ice crystals. Since the form factor presented in Westbrook et al. (2008) has been computed for aggregating ice crystals, the same form factor is used in this chapter, and applied to Eq. (1.17). The mass of ice crystals, in Eq. (1.17), is represented by m(D). The predicted ensemble model, mass-dimensional relationship, is $0.04D^2$, in SI units (Baran et al., 2011b).

The radar reflectivity Z_e, is generally expressed in decibels, given by 10 log $10Z_e$. The ensemble model tropical radar reflectivity is shown in Fig. 1.24, as a function of IWC and cloud temperature, derived in the same manner as Fig. 1.23(a) and Fig. 1.23(b). Similar to Fig. 1.23, the radar reflectivity behaves in the same way as the scalar optical properties. Again, radar reflectivity can be parameterized without the need for an effective diameter, and the parameterization can be incorporated into a GCM.

The point is, however, that it is possible to construct a high- and low-frequency scattering model of cirrus that is applicable across the electromagnetic spectrum, without the need for a hierarchy of cirrus scattering models applied to particular regions of the electromagnetic spectrum.

1.6 Conclusion

In this chapter, the light scattering properties of atmospheric mineral dust, volcanic aerosol and ice crystals have been discussed and reviewed. Current climate models, when compared against space-based measurements of the short-wave flux at TOA, can be in error by as much as $50\,\mathrm{W\,m}^{-2}$. This error is still in part due to an incomplete understanding of how incident light interacts with atmospheric particulates. Therefore, to reduce the uncertainty in climate model predictions of climate change, under the scenario of increased industrial emissions, it is vital to understand the basic interaction between light and atmospheric particulates.

It is still common practice to parametrize mineral dust particles in climate models using scalar optical properties derived from Lorenz–Mie theory. However, as Fig. 1.4 demonstrates, mineral dust aerosols are nonspherical, and so are volcanic ash particles (Johnson et al., 2012). Moreover, these particles are irregular, with deformations, and exhibit rough surfaces. There have been attempts to model these particles as systems of spheroids of varying aspect ratios. However, the work of Osborne et al. (2011) has shown that, for the case of heavily laden Saharan mineral aerosol, the often-applied habit mixture model of spheroids fails to reproduce the measured transmitted radiance as a function of scattering angle at various wavelengths in the solar region. Although, such models may be appropriate for small aerosol or not so heavily laden aerosol, for larger particles the model is not sufficiently general. Therefore, models that exhibit more irregularity on their surface, rather than smooth surfaces, need to be investigated further, as well as aggregated aerosol, which to date has been completely ignored.

The impact, on the short-wave upwelling and downwelling fluxes, of assuming habit mixture smooth spheroid models or irregular models is very significant. This is because both models predict very different asymmetry parameters; these differences in the asymmetry parameter can cause differences of about $-22\,\mathrm{W\,m^{-2}}$ in the averaged daily TOA radiative forcing. This difference in TOA radiative forcing is not dissimilar to the differences shown in Fig. 1.1(b). Therefore, constraining, the general irregularity and asymmetry parameter of mineral dust aerosol is important, if Fig. 1.1(b) is to be further improved. In this respect, new instrumentation such as SID-3 is important, which will help to characterize the irregularity of small particles. Moreover, although the SID series of instruments is useful for characterizing the forward scattering properties of aerosol, there are still insufficient measurements of the backscattering properties of dust and ash aerosol, at sufficient angular resolution in scattering angle space, to be able to fully constrain light scattering models of dust and ash aerosols. Instruments that combine features of SID-3 with the scattering angle range of the Polar Nephelometer are required to fully understand the light scattering patterns of atmospheric particulates.

However, polarization measurements, either *in situ* or space-based, will also help to constrain aerosol models, as the scattering phase matrix elements, other than P_{11}, are particularly sensitive to assumptions about particle irregularity.

In recent times, the volcanic eruption of Eyjafjallajökull closed a number of European airports as the volcanic plume advected over them. They closed because aircraft, at that time, could not fly into volcanic material. In order to relax this constraint on aircraft, it is necessary to know the mass concentration contained in the plume. To estimate mass concentration, single scattering microphysical probes, have and are being developed, to estimate the mass concentration to some uncertainty. However, here again, to estimate the mass concentration requires knowledge of the light scattering properties of volcanic ash. Figures 1.14(a) to (c) demonstrate, how different the scattering phase functions can be between different models of volcanic ash. Therefore, light scattering models of volcanic ash are required, that span the resonance (i.e., size parameters around unity) and geometric optics regions, so that estimates of mass concentration can be better estimated using single-scattering microphysical probes. Here too, polarization measurements must be further utilized, as the backscattering properties of ice and ash may be different, thereby enabling potential discrimination between ash and ice. If ice is not successfully discriminated, then this could, potentially, lead to significant errors in the estimates of volcanic mass concentration.

A further uncertainty is the complex refractive index of volcanic ash, as highlighted by Newman et al. (2012) and Baran (2012b). Although, there are measurements of their complex refractive index in existence, these are now some 40 years old. New measurements are required, and should be encouraged, so that uncertainties in light scattering calculations can be further reduced. Currently, as regards aerosol, there exist no determinations of the complex refractive indices of these materials at submillimetre frequencies (i.e., $> 300\,\text{GHz}$) as highlighted by Baran (2012b). If new instrumentation in the submillimetre region of the electromagnetic spectrum is to be exploited, then determination of the dielectric properties of mineral dust and volcanic aerosol is a necessity (Baran, 2012b).

It is now commonplace to find measurements of cirrus that span the electromagnetic spectrum of importance to the energetics of the Earth–atmosphere system. Therefore, there is no further point in constructing theoretical light scattering models of cirrus, that apply only, to particular regions of the electromagnetic spectrum (Baran and Francis, 2004). Moreover, there is also little point in constructing theoretical light scattering models of cirrus based on single idealized ice crystals, since cirrus is composed of habit mixtures. Furthermore, single geometrical models do not, generally, exhibit the correct mass-dimensional relationships, in the presence of ice aggregation. Therefore, what is required are theoretical light scattering models of cirrus that are physically consistent across the electromagnetic spectrum, and that satisfy observed mass-dimensional relationships for aggregating ice crystals (i.e., mass $\propto D^2$). This mass $\propto D^2$ condition means that ice crystal models should be spatial rather than compact, and cloud physics and radiation research, cannot be considered as two separate disciplines; rather, they are intrinsically coupled (Baran and Labonnote, 2007; Mitchell et al., 2008; Baran et al., 2009; Baran et al., 2010; Mitchell et al., 2011; Baran et al., 2011a; Baran et al., 2011b; Baran, 2012a). The short-wave TOA flux calculations, shown in Fig. 1.1(a), have been accomplished using decoupled cloud physics and radiation schemes. However, coupled cloud-physics and radiation schemes are to be preferred, as these directly link model prognostic variables with radiation measurements (Baran and Labonnote, 2007; Mitchell et al., 2008; Baran et al., 2010; Baran, 2012a).

The short-wave flux differences between a climate model and measurements, shown in Fig. 1.1(a), can be as large as $-40\,\text{W}\,\text{m}^{-2}$. Assuming different ice crystal models, to calculate the asymmetry parameter, can lead to short-wave flux differences as large as approximately $66\,\text{W}\,\text{m}^{-2}$, as demonstrated in this chapter. Clearly, such a difference, due merely to changing the asymmetry parameter, can be comparable to other differences in climate models, due to other model parameters, apart from the scalar optical properties. It is therefore important to further constrain possible values for the asymmetry parameter. In this regard, further measurements using the Polar Nephelometer (Gayet et al., 2011; Gayet et al., 2012) and SID-3 (Ulanowski et al., 2010) appear particularly useful.

However, to take advantage of the Polar Nephelometer and SID-3 measurements, especially at visible wavelengths, requires improved treatments of electromagnetic scattering. Currently, there are now electromagnetic and physical optics methods that bridge the gap between the resonance and geometric optics regimes. The method outlined in Bi et al. (2011) is particularly promising, as this incorporates concepts of electromagnetic theory and physical optics, inclusive of edge effects. Moreover, it is essentially independent of size parameter through the innovative use of beam tracing, but becomes limited by the shape of the particle. But a method that encompasses both the resonance and geometric optics regimes still eludes researchers. The traditional approach to electromagnetic scattering is to solve the Maxwell equations until the size or frequency becomes so large that the equation systems cannot be solved. However, other areas of research such as numerical–asymptotic methods, applied to high-frequency scalar wave scattering, appear particularly interesting, and are currently receiving considerable attention; for further information, see the review papier of Chandler-Wilde et al. (2012).

This chapter, it is hoped, has demonstrated the need to understand the light scattering properties of atmospheric particulates, and how important this need is, if climate model predictions are to be further improved and constrained. In the areas of remote sensing, this need is ever greater, as more regions of the electromagnetic spectrum are beginning to be explored, such as the wavelength or frequency resolved solar, infrared and far-infrared regions (see, for example, Baran and Francis (2004); Cox et al. (2010)) and submillimetre regions (Evans et al., 2005; Buehler et al., 2007; Baran 2012b).

Acknowledgements

The author would like to thank J. Mulcahy of the Met Office for providing Fig. 1.1(b), R. A, Burgess, Atmospheric Sciences, School of Earth, Atmospheric and Environmental Sciences, University of Manchester, for providing Figs. 1.4(a) and (b) and A. J. Heymsfield for providing Figs. 1.5, and 1.6. Z. Ulanowski of the University of Hertfordshire is thanked for Fig. 1.8. P. Yang is thanked for Figs. 1.15 (d) and (j), and Fig. 1.16(b) and Fig. 1.18. G. McFarquhar and J. Um are thanked for Figs. 1.15(i), and 1.16(d). T. Nousiainen and G. McFarquhar are thanked for Fig. 1.16(c). P. Yang, B. Baum and G. Hong are thanked for Fig. 1.17. R. Cotton, Met Office, is thanked for providing Fig. 1.9.

References

Asano, S., and M. Sato, 1980, Light scattering by randomly oriented spheroidal particles, *Appl. Opt.*, **19**, 962–974.

Asano, S., 1983, Light scattering by horizontally oriented spheroidal particles, *Appl. Opt.*, **22**, 1390–1396.

Baker, B. B., and R. P., Lawson., 2006, In-situ observations of the microphysical properties of wave, cirrus, and anvil clouds. Part I: Wave clouds, *J. Atmos. Sci.*, **63**, 3160–3185.

Balkanski, Y., M. Schulz, T. Claquin, and S. Guibert, 2007, Reevaluation of mineral aerosol radiative forcings suggests a better agreement with satellite and AERONET data, *Atmospheric Chemistry and Physics*, **7**, 81–95.

Baran, A. J., J.-F., Gayet, and V. Shcherbakov, 2012, On the interpretation of an unusual *in situ* measured ice crystal scattering phase function, *Atmos. Chem. Phys. Discuss.*, **12**, 12485–12502, doi:10.5194/acpd-12-12485-2012.

Baran, A. J., 2012a, From the single-scattering properties of ice crystals to climate prediction: A way forward. Atmos. Res., **112**, 45–69.

Baran, A. J., 2012b, A new application of a multi-frequency submillimetre radiometer in determining the microphysical and macrophysical properties of volcanic plumes: A sensitivity study, *J. Geophys. Res.*, **117**, D00U18, doi:10.1029/2011JD016781

Baran, A. J., P. J. Connolly, A. J. Heymsfield, A. Bansemer, 2011a, Using *in situ* estimates of ice water content, volume extinction coefficient, and the total solar optical depth obtained during the tropical ACTIVE campaign to test an ensemble model of cirrus ice crystals, *Q. J. R. Meteorol. Soc.*, **137**, 199–218.

Baran, A. J, A. Bodas-Salcedo, R. Cotton, and C. Lee, 2011b, Simulating the equivalent radar reflectivity of cirrus at 94 GHz using an ensemble model of cirrus ice crystals: a test of the Met office global numerical weather prediction model, *Q. J. R. Meteor. Soc.*, **137**, 1547–1560.

Baran, A. J., J. Manners, and P. R. Field, 2010, 13th Conference on Atmospheric Radiation, Portland, Oregon, 28th June–2nd July, 2010.

Baran, A. J., 2009, A review of the light scattering properties of cirrus, *Journal of Quantitative Spectroscopy and Radiative Transfer*, **110**, 1239–1260.

Baran, A. J., P. J. Connolly, and C. Lee, 2009, Testing an ensemble model of cirrus ice crystals using mid-latitude *in situ* estimates of ice water content, volume extinction coefficient, and the total solar optical depth, *Journal of Quantitative Spectroscopy and Radiative Transfer*, 110, 1579–1598.

Baran, A. J., and L.-C. Labonnote, 2007, A self-consistent scattering model for cirrus. 1: The solar region, *Q. J. R. Meteor. Soc.*, **133**, 1899–18912.

Baran, A. J., and L.-C. Labonnote, 2006, On the reflection and polarization properties of ice cloud, *Journal of Quantitative Spectroscopy and Radiative Transfer*, **100**, 41–54.

Baran, A. J., 2005, The dependence of cirrus infrared radiative properties on ice crystal geometry and shape of the size-distribution function, *Q. J. R. Meteor. Soc.*, **131**, 1129–1142.

Baran, A., and P. N. Francis, 2004, On the radiative properties of cirrus cloud at solar and thermal wavelengths: A test of model consistency using high-resolution airborne radiance measurements, *Q. J. R. Meteor. Soc.*, **130**, 763–778.

Baran, A. J., 2003, Simulation of infrared scattering from ice aggregates by use of a size-shape distribution of circular ice cylinders, *Applied Optics*, **42**, 2811–2818.

Baran, A. J., P. Yang, and S. Havemann, 2001a, Calculation of the single-scattering properties of randomly oriented hexagonal ice columns: A comparison of the T-matrix and the finite-difference time-domain methods, *Applied Optics*, **40**, 4376–4386.

Baran, A. J., P. N. Francis, L.-C. Labonnote, and M. Doutriaux-Boucher, 2001b, A scattering phase function for ice cloud: Tests of applicability using aircraft and satellite multi-angle multiwavelength radiance measurements of cirrus, *Q. J. R. Meteor. Soc.*, **127**, 2395–2416.

Baran, A. J., and J. Foot, 1994, New application of the operational sounder HIRS in determining a climatology of sulphuric acid aerosol from the Pinatubo eruption, J. Geophys. Res. 99, doi: 10.1029/94JD02044.

Baker, B. A., and R. P. Lawson, 2006, In-situ observations of the microphysical properties of wave, cirrus and anvil clouds. Part I. Wave clouds, *J. Atmos. Sci.*, **63**, 3160–3185.

Baum, B. A., P. Yang, A. J. Heymsfield, C. G. Schmitt, Y. Xie, A. Bansemer, Y.-X Hu, and Z. Zhang, 2011, Improvements in shortwave bulk scattering and absorption models for the remote sensing of ice clouds, *J. Appl. Meteor. Climatol.*, **50**, 1037–1056.

Baum, B. A., A. J. Heymsfield, P. Yang, and S. T. Bedka, 2005, Bulk scattering properties for the remote sensing of ice clouds. Part I. Microphysical data and models. *J. Appl. Meteor.*, **44**, 1885–1895.

Bi, L., P. Yang, G. W. Kattawar, Y. X. Hu, and B. A. Baum, 2011, Scattering and absorption of light by ice particles: Solution by a new physical-geometric optics hybrid method, *Journal of Quantitative Spectroscopy and Radiative Transfer*, **112**, 1492–1508.

Bi, L., P. Yang, G. W. Kattawar, and R. Kahn, R, 2010, Modeling optical properties of mineral aerosol particles by using nonsymmetric hexahedra, *Applied Optics*, **49**, 334–342.

Bi, L., P. Yang, G. W. Kattawar, and R. Kahn, 2009, Single-scattering properties of triaxial ellipsoidal particles for a size parameter range from the Rayleigh to geometric-optics regimes, *Applied Optics*, **48**, 114–126.

Bluth, G. S., S. D. Doiton, C. C. Schnetzler, A. Krueger, and L. S. Walter, 1992, Global tracking of the SO_2 clouds from the June 1991 Mount Pinatubo eruptions, *Geophys. Res. Lett.*, **19**, 151–154.

Bohren, C. F., and D. R. Huffman, 1983, *Absorption and Scattering of Light by Small Particles*, New York: Wiley.

Borovoi, A. G., and I. A. Grishin, 2003, Scattering matrices for large ice crystal particles, *Journal of the Optical Society of America A.*, **20**, 2071–2080.

Bozzo, A., T. Maestri, R. Rizzi, and E. Tosi, 2008, Parameterization of single-scattering properties of mid-latitude cirrus clouds for fast radiative transfer models using particle mixtures, *Geophys. Res. Lett.*, **35**, L16809.

Bredow, J. W., R. Porco, M. S. Dawson, C. L. Betty, S. Self, and T. Thordarson, 1995, A multifrequency laboratory investigation of attenuation and scattering from volcanic ash clouds, *IEEE Transactions on Geoscience and Remote Sensing*, **33**, 1071–1082.

Bréon, F.-M., and B. Dubrulle, 2004, Horizontally oriented plates in clouds, *J. Atmos. Sci.*, **61**, 2888–2898.

Buehler, S. A., C. Jiménez, K. F. Evans, P. Eriksson,B. Rydberg, A. J. Heymsfield, C. J. Stubenrauch, U. Lohmann, C. Emde,V. O. John, T. R. Sreerekhai, and C. P. Davis, 2007, A concept for a satellite mission to measure cloud ice water path, ice particle size, and cloud altitude, *Q. J. R. Meteor. Soc.*, **133**, 109–128.

Chandler-Wilde, S. N., S. Langdon, and D. Hewitt, 2012, Review of numerical-asymptotic methods applied to high frequency scattering, *Acta Numerica*, Cambridge University Press, UK, doi:10.1017/S09624929.

Chen G., P. Yang, and G. W. Kattawar, 2008, Application of the pseudospectral time-domain method to the scattering of light by nonspherical particles, *Journal of the Optical Society of America A*, **25**, 785–790.

Chepfer, H., G. Brogniez, P. Goloub, F. M. Bréon, and P. H. Flamant, 1999, Observations of horizontally oriented ice crystals in cirrus clouds with POLDER-1/ADEOS-1, *Journal of Quantitative Spectroscopy and Radiative Transfer*, **63**, 521–543.

Clarke, A. J. M., E. Hesse, Z. Ulanowski, and P. H. Kaye, 2006, A 3D implementation of ray tracing combined with diffraction on facets: Verification and a potential application, *Journal of Quantitative Spectroscopy and Radiative Transfer*, **100**, 103–114.

Connolly, P. J., C. P. R. Saunders, M. W. Gallagher, K. N. Bower, M. J. Flynn, T. W. Choularton, J. Whiteway, and R. P. Lawson, 2005, Aircraft observations of the influence of electric fields on the aggregation of ice crystals, *Q. J. R. Meteorol. Soc.*, **128**, 1–19.

Cooper, W. A., 1978, Cloud physics investigation by the University of Wyoming in HIPLEX 1977, Bureau of Reclamation Rep., AS, 119, 321.

Cotton, R., S. Osborne, Z. Ulanowsk, P. Hirst, P. H. Kaye, and R. S. Greenway, 2010, The ability of the Small Ice Detector (SID-2) to characterise cloud particle and aerosol morphologies obtained during flights of the FAAM BAe-146 research aircraft, *J. Oceanic. Atmos. Tech.*, **27**, 290–303.

Cox, C. V., J. E. Harries, J. P. Taylor, P. D. Green, A. J. Baran, J. C. Pickering, A. E. Last, and J.E. Murray, 2010, Measurement and simulation of mid- and far-infrared spectra in the presence of cirrus, *Q. J. R. Meteorol. Soc.*, **136**, 718–739.

De Leon, R. R., and J. D. Haigh, 2007, Infrared properties of cirrus clouds in climate models, *Q. J. R. Meteor. Soc.*, **133**, 273–282.

Doutriaux-Boucher, M., J. C. Buriez, G. Brogniez, L. C.-Labonnote, and A. J. Baran, 2000, Sensitivity of retrieved POLDER directional cloud optical thickness to various ice particle models, *Geophys. Res. Lett.*, **27**, 109–112.

Draine, B.T., and P. J. Flatau, 1994, Discrete dipole approximation for scattering calculations, *J. Opt. Soc. Am. A*, **11**, 1491–1499.

Dubovic, O., A. Sinyuk, T. Lapyonok, B.N. Holben, M. Mishchenko, P. Yang, T.F. Eck, H. Volten, O. Muñoz, B. Veihelmann, W.J. van der Zande, J.-F. Leon, M. Sorokin, and I. Slutsker, 2006, Application of spheroid models to account for aerosol particle nonsphericity in remote sensing of desert dust, *J. Geophys. Res.*, **111**, D11208, doi:10.1029/2005JD006619.

Edwards, J. M., S. Havemann, J.-C. Thelen, and A. J. Baran, 2007, A new parameterization for the radiative properties of ice crystals: Comparison with existing schemes and impact in a GCM, *Atmos. Res.*, **83**, 19–35.

Edwards, J. M., and T. Slingo, 1996, Studies with a flexible new radiation code. I: Choosing a configuration for a large-scale model, *Q. J. R. Meteorol. Soc.*, **122**, 689–719.

Evans, K. F., J. R. Wang, P. E. Racette, G. Heymsfield, and L. Li., 2005, Ice cloud retrievals and analysis with the compact scanning sunmillimeter imaging radiometer and the cloud radar system during CRYSTAL FACE, *J. Appl. Meteor.*, **44**, 839–859.

Farafonov, V. G., and N. V. Voshchinnikov, 2012, Light scattering by multilayered spheroidal particle, *Applied Optics*, **51**, 1586–1597.

Feng, Q., P. Yang, G. W. Kattawar, C. N. Hsu, Si-Chee Tsay, I. Laszlo, 2009, Effects of particle nonsphericity and radiation polarization on retrieving dust properties from MODIS observations, *Aerosol Science*, **40**, 776–789.

Field, P. R., A. J. Heymsfield, and A. Bansemer, 2008, Determination of the combined ventilation factor and capacitance for ice crystal aggregates from airborne observations in a tropical anvil cloud, *J. Atmos. Sci.*, **65**, 376–391.

Field, P. R., A. J. Heymsfield, and A. Bansemer, 2007, Snow size distribution parameterization for midlatitude and tropical ice cloud, *J. Atmos. Sci.*, **64**, 4346–4365.

Field, P. R., A. J. Heymsfield, and A. Bansemer, 2006, Shattering and particle interarrival times measured by optical array probes in ice clouds, *J. Atmos. Ocean. Tech.*, **23**, 1357–1371.

Field, P. R., R. J. Hogan, P. R. A. Brown, A. J. Illingworth, T. W. Choularton, and R. J. Cotton, 2005, Parametrization of ice-particle size distribution functions for mid-latitude stratiform cloud, *Q. J. R. Meteorol. Soc.*, **131**, 1997–2017.

Field, P. R., R. Wood, P. R. A. Brown, P. H. Kaye, E. Hirst, R. Greenaway, and J. A. Smith, 2003, Ice particle interarrival times measured with a fast FSSP, *J. Atmos. Ocean. Tech.*, **20**, 249–261.

Foot, J. S., Some observations of the optical properties of clouds. II: Cirrus, 1988, *Q. J. R. Meteor. Soc.*, **114**, 141–164.

Francis, P. N., 1995, Some aircraft observations of the scattering properties of ice crystals, *J. Atmos. Sci.*, **52**, 1142–1154.

Fu, Q., 2007, A new parameterization of an asymmetry factor of cirrus clouds for climate models, *J. Atmos. Sci.*, **64**, 4140–4150.

Fu Q., W. B. Sun, and P. Yang, 1999, Modeling of scattering and absorption by nonspherical cirrus ice particles at thermal infrared wavelengths, *J. Atmos. Sci.*, **56**, 2937–2947.

Ganesh, M., and S. C. Hawkins, 2010, Three dimensional electromagnetic scattering T-matrix computations, *Journal of Computational and Applied Mathematics*, **234**, 1702–1709.

Gayet, J.-F., G. Mioche, L. Bugliaro, A. Protat, A. Minikin, M.Wirth, A. Dörnbrack, V.Shcherbakov, B. Mayer, A. Garnier, and C. Gourbeyre, 2012, On the observation of unusual high concentration of small chain-like aggregate ice crystals and large ice water contents near the top of a deep convective cloud during the CIRCLE-2 experiment: *Atmos. Chem. Phys.*, **12**, 727–744.

Gayet, J.-F., G. Mioche, V. Shcherbakov, C. Gourbeyre, R. Busen, and A. Minikin, 2011, Optical properties of pristine ice crystals in mid-latitude cirrus clouds: a case study during CIRCLE-2 experiment, *Atmos. Chem. Phys.*, **11**, 2537–3544.

Ghobrial, S. I., and S. M. Sharief, 1987, Microwave attenuation and cross polarization in dust storms, *IEEE Trans. Antennas. Propagat.*, **AP-35**, 418–425.

Grenfell, T. C., S. P. Neshyba, and S. G. Warren, 2005, Representation of a nonspherical ice particle by a collection of independent spheres for scattering and absorption of radiation: 3. Hollow columns and plates, *J. Geophys. Res.*, **110**, Art. D17203.

Grenfell, T. C., and S. G. Warren, 1999, Representation of a nonspherical ice particle by a collection of independent spheres for scattering and absorption of radiation, *J. Geophys. Res.*, **104**, 31697–31709.

Guignard, A., C. J. Stubenrauch, A. J. Baran, and R. Armante, 2012, Bulk microphysical properties of semi-transparent cirrus from AIRS: a six years global climatology and statistical analysis in synergy with CALIPSO and CloudSat, *Atmos. Chem. Phys.*, **12**, 503–525.

Gu, Y., K. N. Liou, S. C. Ou, and R. Fovell, 2011, Cirrus cloud simulations using WRF with improved radiation parameterization and increased vertical resolution, *J. Geophys. Res.*, **116**, D06119, doi:10.1029/2010JD014574.

Hansen, J.E., and L.D. Travis, 1974, Light scattering in planetary atmospheres. Space Sci. Rev.,16, 527-610.

Harrison, R. G., K. A. Nicoll, Z. Ulanowski, and T. A. Mather, 2010, Self charging of the Eyjafjallajökull volcanic ash plume, *Environ. Res. Lett.*, **5**, doi:10.1088/1748-9326/5/2/024004.

Havemann, S., A. J. Baran, and J. M. Edwards, 2003, Implementation of the T-matrix method on a massively parallel machine: a comparison of hexagonal ice cylinder single-scattering properties using the T-matrix and improved geometric optics methods, *Journal of Quantitative Spectroscopy and Radiative Transfer*, **79–80**, 707–720.

Havemann, S., and A. J. Baran, 2001, Extension of T-matrix to scattering of electromagnetic plane waves by non-axisymmetric dielectric particles: Application to hexagonal cylinders, *Journal of Quantitative Spectroscopy and Radiative Transfer*, **70**, 139–158.

Haywood, J. M., B. T. Johnson, S. R. Osborne, A. J. Baran, M. Brooks, S. F. Milton, J. Mulcahy, D. Walters, R. P. Allan, A. Klaver, P. Formenti, H. E. Brindley, S. Christopher, and P. Gupta, 2011, Motivation, rationale and key results from the GERBILS Saharan dust measurement campaign, *Q. J. R. Meteorol. Soc.*, **137**, 1106–1116.

Haywood, J. M., R. P. Allan, I. Culverwell, T. Slingo, S. Milton, J. Edwards, and N. Clerbaux, 2005, Can desert dust explain the outgoing longwave radiation anomaly over the Sahara during July 2003, *J. Geophys. Res.*, **110**, D05105, doi:10.1029/2004JD005232.

Haywood, J., P. Francis, S. Osborne, M. Glew, N. Loeb, E. Highwood, D. Tanré, G. Myhre, P. Formenti, and E. Hirst, 2003, Radiative properties and direct radiative effect of Saharan dust measured by the C-130 aircraft during SHADE: 1. Solar spectrum, *J. Geophys. Res.*, **108** (D18), 8577, doi:10.1029/2002JD002687.

Hesse, E., 2008, Modelling diffraction during ray tracing using the concept of energy flow lines, *Journal of Quantitative Spectroscopy and Radiative Transfer*, **109**, 1374–1383.

Heymsfield, A. J., 2007, On measurements of small ice particles in clouds, *Geophys. Res. Lett.*, Art. No. L23812.

Heymsfield, A. J., 2003, Properties of tropical and midlatitude ice cloud particle ensembles. Part II: Applications for mesoscale and climate models, *J. Atmos. Sci.*, **60**, 2592–2611.

Heymsfield, A. J., and L. M. Miloshevich, 2003, Parametrizations for the cross-sectional area and extinction of cirrus and stratiform ice cloud particles, *J. Atmos. Sci.*, **60**, 936–956.

Heymsfield, A. J., A. Bansemer, P. R. Field, S. L. Durden, J. L. Stith, J. E. Dye, W. Hall, and C. A. Grainger, 2002, Observations and parameterizations of particle size distributions in deep tropical cirrus and stratiform precipitating clouds: Results from *in situ* observations in TRMM field campaigns, *J. Atmos. Sci.*, **59**, 3457–3491.

Heymsfield, A. J., and L. M. Miloshevich, 1995, Relative humidity and temperature influences on cirrus formation and evolution: Observations from wave clouds and FIRE II, *J. Atmos. Sci.*, **52**, 4302–4326.

Heymsfield, A. J., and C. R. Platt, 1984, A parameterization of the particle size spectrum of ice clouds in terms of ambient temperature and their ice water content, *J. Atmos. Sci.*, **41**, 846–855.

Highwood, E. J., J. M. Haywood, M. D. Silverstone, S. M. Newman, and J. P. Taylor, 2003, Radiative properties and direct effect of Saharan dust measured by the C-130 aircraft during SHADE. 2. Terrestrial spectrum, *J. Geophys. Res.*, **108**, doi:10.1029/2002JD002552.

Hong, G., P. Yang, B. A. Baum, A. J. Heymsfield, and K. M. Xu, 2009, Parameterization of shortwave and longwave properties of ice clouds for use in climate models, *Journal of Climate*, **22**, 6287–6312.

Hong G, P. Yang, B. A. Baum, and A. J. Heymsfield, 2008, Relationship between ice water content and equivalent radar reflectivity for clouds consisting of nonspherical ice particles, *J. Geophys. Res.*, **113**, D20205, doi: 10.1029/2008JD009890.

Hong, G., P. Yang, B. C. Gao, B. A. Baum, Y. X. Hu, M. D. King, and S. Platnick, 2007, High cloud properties from three years of MODIS Terra and Aqua collection-4 data over the tropics, *J. Appl. Met and Climat.*, **46**, 1840–1856.

Hu, Y., M. Vaughan, Z. Liu, B. Lin, P. Yang, D. Flittner, B. Hunt, R. Kuehn, J. Huang, D. Wu, S. Rodier, K. Powell, C. Trepte, and D. Winker, 2007, The depolarization–attenuated backscatter relation: CALIPSO lidar measurements vs. theory, *Optics Express*, **15**, 5327–5332.

IPCC Intergovernmental Panel on Climate Change, (2007). *Climate Change 2007 – The Physical Science Basis: Contribution of Working Group I to the Fourth Assessment Report of the IPCC*, Cambridge: Cambridge University Press.

Ivanova, D., D. L. Mitchell, W. P. Arnott, and M. Poellot, 2001, A GCM parameterization for bimodal size spectra and ice mass removal rates in mid-latitude cirrus clouds, *J. Atmos. Res.*, **59**, 89–113.

Johnson, B., K. Turnbull, P. Brown, R. Burgess, J. Dorsey, A. J. Baran, H. Webster, J. Haywood, R. Cotton, Z. Ulanowski, E. Hesse, A. Woolley, and P. Rosenberg, 2012, Insitu observations of volcanic ash clouds from the FAAM aircraft during the eruption of Eyjafjallajökull in 2010, *J. Geophys. Res.*, **117**, D00U24, doi:10.1029/2011JD016760.

Johnson, B, and S. R. Osborne, 2011, Physical and optical properties of mineral dust aerosol measured by aircraft during the GERBILS campaign, *Q. J. R. Meteorol. Soc.*, **137**, 1117–1130.

Kahnert, F. M., 2003, Numerical methods in electromagnetic scattering theory, *Journal of Quantitative Spectroscopy and Radiative Transfer*, **79–80**, 775–824.

Kahnert, F. M., J. J. Stamnes, and K. Stamnes, 2002, Using simple particle shapes to model the stokes scattering matrix of ensembles of wavelength-sized particles with particles with complex shapes: possibilities and limitations, *Journal of Quantitative Spectroscopy and Radiative Transfer*, **74**, 167–182.

Kaye, P. H., E. Hirst, R. S. Greenaway, Z. Ulanowski, E. Hesse, P. J. DeMott, C. Saunders, and P. Connolly, 2008, Classifying atmospheric ice crystals by spatial light scattering, *Optics Letters*, **33**, 1545–1547.

Kokhanovsky, A., 2004, Optical properties of terrestrial clouds, *Earth Sci. Revs.*, **64**, 189–241.

Kokhanovsky, A. A., 2003, Optical properties of irregularly shaped particles. *J. Phys. D – Appl. Phys.*, **36**, 915–923.

Kokhanovsky, A. A., 1998, On light scattering in random media with large densely packed particles, *J. Geophys. Res.*, **103**, 6089–6096.

Kokhanovsky, A., and A. Macke, 1997, The dependence of the radiative characteristics of optically thick media on the shape of particles, *Journal of Quantitative Spectroscopy and Radiative Transfer*, **63**, 393–407.

Korolev, A. V., E. F. Emery, J. W. Strapp, S. G. Cober, G. A. Isaac, M. Wasey, and D. Marcotte, 2011, Small ice particles in tropospheric clouds: fact or artefact? Airborne icing instrumentation evaluation experiment, *Bull. Amer. Meteor. Soc.*, **92**, 967–973.

Korolev, A. V., and G. A. Isaac, 2003, Roundness and aspect ratios of particles in ice clouds, *J. Atmos. Sci.*, **60**, 1795–1808.

Korolev, A., G. A. Isaac, and J. Hallett, 2000, Ice particle habits in stratiform clouds, *Q. J. R. Meteor. Soc.*, **126**, 2873–2902.

Korolev, A. V., G. A. Isaac, and J. Hallett, 1999, Ice particle habits in Arctic clouds, *Geophys. Res. Lett.*, **26**, 1299–1302.

Labonnote, L.-C., G. Brogniez, J. C. Buriez, M. Doutriaux-Boucher, J. F. Gayet, and Macke A., 2001, Polarized light scattering by inhomogeneous hexagonal monocrystals: Validation with ADEOSPOLDER measurements, *J. Geophys. Res.*, **106**, 12139–12153.

Lambert, A., R. G. Grainger, J. J. Remedios, C. D. Rodgers, M. Corney, and F. W. Taylor, 1993, Measurements of the evolution of the Mt. Pinatubo aerosol cloud by ISAMS, *Geophys. Res. Lett.*, **20**, 1287–1290.

Lawson, R. P., 2011, Effects of ice particle shattering on the 2D-S probe, *Atmos. Meas. Tech.*, **4**, 1361–1381.

Lawson, R. P., D. O'Connor, P. Zmarzly, K. Weaver, B. Baker, Q. X. Mo, and H. Jonsson, 2006, The 2D-S (Stereo) probe: Designs and preliminary tests of a new airborne, high-speed, high-resolution imaging probe, *J. Atmos. Ocean. Tech.*, **23**, 1462–1477.

Lawson, R. P., B. A. Baker, and B. L. Pilson, 2003, In-situ measurements of microphysical properties of mid-latitude and anvil cirrus and validation of satellite retrievals. In Proc. 30th Symposium on Remote Sensing of Environment. 2003. Honolulu, Hawaii.

Lee, J., P. Yang, A. E. Dessler, B.-C. Gao, and S. Platnick, 2009, Distribution and radiative forcing of tropical thin cirrus clouds, *J. Atmos. Sci.*, **66**, 3721–3731.

Lee, Y. K., P. Yang, M. I. Mishchenko, B. A. Baum, Y. Hu, H.-L. Huang, W. J. Wiscombe, and A. J. Baran, 2003, On the use of circular cylinders as surrogates for hexagonal pristine ice crystals in scattering calculations at infrared wavelengths, *Applied Optics*, **42**, 2653–2664.

Levy, R. C., L. A. Remer, and Y. J. Kaufman, 2004, Effects of neglecting polarization on the MODIS aerosol retrieval over land, *IEEE Transactions on Geoscience and Remote Sensing*, **42**, 2576–2583.

Liou, K. N., 2002, *An Introduction to Atmospheric Radiation*, 2nd edition, Academic Press, New York.

Liou, K. N., Y. Takano, and P. Yang, 2000, Light scattering and radiative transfer in ice crystal clouds: Applications to climate research. In: *Light Scattering by Non-spherical Particles: Theory, Measurements and Geophysical Applications*, Eds. M. I. Mishchenko, J. W. Hovenier, and L. D. Travis, Academic Press, New York, Chapter 15.

Liou, K. N., and Y. Takano, 1994, Light scattering by nonspherical particles: Remote sensing and climatic implications, *J. Atmos. Res.*, **31**, 271–298.

Long, C. S., and L. L. Stowe, 1994, Using the NOAA/AVHRR to study stratospheric aerosol optical thicknesses following the Mt. Pinatubo Eruption, *Geophys. Res. Lett.*, **21**, 2215–2218.

López, M. L., and E. E. Ávila, 2012, Deformations of frozen droplets formed at −40°C, *Geophys. Res. Lett.*, **39**, L01805, doi:10.1029/2011GL050185.

Macke, A., J. Mueller, and E. Raschke, 1996a, Single scattering properties of atmospheric ice crystal, *J. Atmos. Sci.*, **53**, 2813–2825.

Macke, A., M. I. Mishchenko, and B. Cairns, 1996b, The influence of inclusions on light scattering by large particles, *J. Geophys. Res.*, **101**, 23311–23316.

Mackowski, D. W., 2002, Discrete dipole moment method for calculation of the T matrix for nonspherical particles, *J. Opt. Soc. Am. A*, **19**, 881–893.

Magono, C., and C. W. Lee C, 1966, Meteorological classification of natural snow crystals, *J. Faculty of Sci., Hokkaidô University*, **2**, 321–335.

Mano, Y., 2000, Exact solution of electromagnetic scattering by a three-dimensional hexagonal ice column obtained with the boundary-element method, *Applied Optics*, **39**, 5541–5546.

Marenco, F., B. Johnson, K. Turnbull, S. Newman, J. Haywood, H. Webster, and H. Ricketts, 2011, Airborne lidar observations of the 2010 Eyjafjallajökull volcanic ash plume, *J. Geophys. Res.*, **116**, doi:10.1029/2011JD016396.

Mauno, P., G. M. McFarquhar, P. Räisänen, M. Kahnert, M. S. Timlin, and T. Nousiainen, 2011, The influence of observed cirrus microphysical properties on shortwave radiation: A case study over Oklahoma, *J. Geophys. Res.*, **116**, D22208, doi:10.1029/2011JD016058.

McCormick, M.P., and E. Veiga, 1992, SAGE II measurements of early Pinatubo aerosols, *Geophys. Res. Lett.*, **19**, 155–158.

McFarquhar G. M., J. Um, M. Freer, D. Baumgardner, G. L. Kok, and G. Mace, 2007, Importance of small ice crystals to cirrus properties: Observations from the Tropical Warm Pool Cloud Experiment (TWP-ICE), *Geophys. Res. Lett., Art. No.* L13803.

McFarquhar, G. M., P. Yang, A. Macke, and A. J. Baran, 2002, A new parameterization of single scattering solar radiative properties for tropical anvils using observed ice crystal size and shape distributions, *J. Atmos. Sci.*, **59**, 2458–2478.

McFarquhar, G. M., A. J. Heymsfield, A. Macke, J. Iaquinta, and S. M. Aulenbach, 1999, Use of observed ice crystal sizes and shapes to calculate the mean-scattering properties and multispectral radiance, CEPEX April 4 1993 case study, *J. Geophy. Res.*, **104**, 31763–31779.

McFarquhar, G. M., and A. J. Heymsfield, 1998, The definition and significance of an effective radius for ice clouds, *J. Atmos. Sci.*, **55**, 2039–2052.

McFarquhar, G. M., and A. J. Heymsfield, 1996, Microphysical characteristics of three cirrus anvils sampled during the Central Equatorial Pacific Experiment (CEPEX), *J. Atmos. Sci.*, **53**, 2401–2423.

Meng, Z., P. Yang, G. Kattawar, L. Bi, K. Liou, and I. Laszlo, 2010, Single-scattering properties of tri-axial ellipsoidal mineral dust aerosols: A database for application to radiative transfer calculations, *Journal of Aerosol Science*, **41**, 501–512.

Mishchenko, M. I., B. Cairns, G. Kopp, C. F. Schueler, B. A. Fafaul, J. E. Hansen, R. J. Hooker, T. Itchkawich, H. B. Maring, and L. D. Travis, 2007, Accurate monitoring of terrestrial aerosols and total solar irradiance – Introducing the glory mission, *Bulletin of the American Meteorological Society*, **88**, 677–691.

Mishchenko, M. I., L. D. Travis, and A. A. Lacis, 2002, *Scattering, Absorption, and Emission of Light by Small Particles*, Cambridge University Press, UK.

Mishchenko, M. I., 2000, Calculation of the amplitude matrix for a nonspherical particle in fixed orientation, *Applied Optics*, **39**, 1026–1031.

Mishchenko, M. I., and K. Sassen, 1998, Depolarization of lidar returns by small ice crystals, An application to contrails, *Geophys. Res. Lett.*, **25**, 309–312.

Mishchenko, M. I., and L. D. Travis, 1998, Capabilities and limitations of a current FORTRAN implementation of the T-matrix method for randomly oriented, rotationally symmetry scatterers, *Journal of Quantitative Spectroscopy and Radiative Transfer*, **60**, 309–324.

Mishchenko, M. I., and A. Macke A., 1997, Asymmetry parameters of the phase function for isolated and densely packed spherical particles with multiple internal inclusions in the geometric optics limit, *Journal of Quantitative Spectroscopy and Radiative Transfer*, **57**, 767–794.

Mishchenko, M. I., L. D. Travis, and D. W. Mackowski, 1996, T-matrix computations of light scattering by nonspherical particles: A review, *Journal of Quantitative Spectroscopy and Radiative Transfer*, **55**, 535–575.

Mishchenko, M. I., and J. W. Hovenier, 1995, Depolarization of light backscattered by randomly oriented nonspherical particles, *Opt. Lett.*, **20**, 1356–1358.

Mishchenko, M. I., A. A. Lacis, and L. D. Travis, 1994, Errors induced by the neglect of polarization in radiance calculations for Rayleigh-scattering atmospheres, *Journal of Quantitative Spectroscopy and Radiative Transfer*, **51**, 491–510.

Mishchenko, M. I., and L. D. Travis, 1994, Light scattering by polydispersion of randomly oriented spheroids with sizes comparable to wavelengths of observation, *Applied Optics*, **33**, 7206–7225.

Mishchenko, M. I., 1991, Light scattering by randomly oriented axially symmetric particles, *J. Opt. Soc. Am. A*, **8**, 871–882.

Mitchell, D. L., R. P. Lawson, and B. Baker, 2011, Understanding effective diameter and its application to terrestrial radiation in ice clouds, *Atmos. Chem. Phys.*, **11**, 3417–3429.

Mitchell, D. L., P. Rasch, D. Ivanova, G. M. McFarquar, and T. Nousiainen, 2008, Impact of small ice crystal assumptions on ice sedimentation rates in cirrus clouds and GCM simulations, *Geophys. Res. Lett.*, **35**, L09806.

Mitchell, D. L., A. J. Baran, W. P. Arnott, and C. Schmitt, 2006, Testing and comparing the modified anomalous diffraction approximation, *J. Atmos. Sci.*, **63**, 2948–2962.

Mitchell, D. L., 2002, Effective diameter in radiation transfer, *J. Atmos. Sci.*, **59**, 2330–2346.

Mitchell, D. L., W. P. Arnott, C. Schmitt, A. J. Baran, S. Havemann, and Q. Fu, 2001, Photon tunneling contributions to extinction for laboratory grown hexagonal columns, *Journal of Quantitative Spectroscopy and Radiative Transfer*, **70**, 761–776.

Mitchell, D. L., Y. Liu, and A. Macke, 1996, Modeling cirrus clouds. Part II: Treatment of radiative properties, *J. Atmos. Sci.*, **53**, 2967–2988.

Muinonen, K. *Light scattering by stochastically shaped particles*, 2000, In: Mishchenko MI, Hovenier JW, Travis LD, editors, *Light Scattering by Monspherical Particles*, San Diego, Academic Press, p. 323–352.

Muinonen, K., T. Nousiainen, P. Fast, K. Lumme, and J. I. Peltoniemi, 1996, Light scattering by Gaussian random particles: ray optics approximation. *Journal of Quantitative Spectroscopy and Radiative Transfer*, **55**, 577–601.

Muinonen, K., 1989, Scattering of light by crystals: a modified Kirchhoff approximation, *Applied Optics*, **28**, 3044–3050.

Nazaryan, H., M. P. McCormick, and W. P. Menzel, 2008, Global characterization of cirrus clouds using CALIPSO data, *J. Geophys. Res.*, **113**, D16211.

Neshyba, S. P., T. C. Grenfell, and S. G. Warren, 2003, Representation of a nonspherical ice particle by a collection of independent spheres for scattering and absorption of radiation: 2. Hexagonal columns and plates, *J. Geophys. Res.*, **108**, Art. 4448.

Newman, S., L. Clarisse, D. Hurtmans, F. Marenco, B. Johnson, K. Turnbull, S. Havemann, A. J. Baran, D. O'Sullivan, and J. Haywood, 2012, A case study of observations of volcanic ash from the Eyjafjallajökull eruption. Part 2: airborne and satellite radiative measurements, *J. Geophys. Res.*, **117**, D00U13, doi:10.1029/2011JD016780.

Nicoll, K. A., R. G. Harrison, and Z. Ulanowski, 2011, Observations of Saharan dust layer electrification, *Environ. Res. Lett.*, **6**, doi:10.1088/1748-9326/6/1/014001.

Noel, V., and H. Chepfer, 2010, A global view of horizontally oriented crystals in ice clouds from Cloud-Aerosol Lidar and Infrared Pathfinder Satellite Observation (CALIPSO), *J. Geophys. Res.*, 115, art. D00H23.

Noel, V., and K. Sassen, 2005, Study of planar ice crystal orientations in ice clouds from scanning polarization lidar observations, *J. Appl. Met.*, **44**, 653–664.

Noel, V., and H. Chepfer H, 2004, Study of ice crystal orientation in cirrus clouds based on satellite polarized radiance measurements, *J. Atmos. Sci.*, **61**, 2073–2081.

Nousiainen, T., 2009, Optical modeling of mineral dust particles: A review, *Journal of Quantitative Spectroscopy and Radiative Transfer*, **110**, 1261–1279.

Nousiainen, T., and G. M. McFarquhar, 2004, Light scattering by small quasi-spherical ice crystals, *J. Atmos. Sci.*, **61**, 2229–2248.

Nousiainen, T., and K. Vermeulen, 2003, Comparison of measured single-scattering matrix of feldspar particles with T-matrix simulations using spheroids, *Journal of Quantitative Spectroscopy and Radiative Transfer*, **79–80**, 1031–1042.

Ohtake, T., 1970, Unusual crystal in ice fog, *J. Atmos. Sci.*, **27**, 509–511.

Osborne, S. R., A. J. Baran, B. T. Johnson, J. M. Haywood, E. Hesse, and S. Newman, 2011, Short-wave and long-wave radiative properties of Saharan dust aerosol, *Q. J. R. Meteorol. Soc.*, **137**, 1149–1167, doi: 10.1002/qj.771.

Otto, S., T. Trautmann, and M. Wendisch, 2011, On realistic size equivalence and shape of spheroidal Saharan mineral dust particles applied in solar and thermal radiative transfer calculations, *Atmos. Chem. Phys.*, **11**, doi:10.5194/acp-11-4469-2011.

Penner, J. E., M. Andreae, H. Annegarn, L. Barrie, J. Feichter, D. Hegg, A. Jayaraman, R. Leaitch, D. Murphy, J. Nganga, and G. Pitari, 2001, Aerosols, their direct and indirect effects, in: *Climate Change 2001: The Scientific Basis*, edited by: J. T. Houghton, Y. Ding, D. J. Griggs, M. Noguer, P. J. Van der Linden, X. Dai, K. Maskell, and C. A. Johnson, Report to Intergovernmental Panel on Climate Change from the Scientific Assessment Working Group (WGI), Cambridge University Press, UK, 289–416.

Petrov, D., Y. Shkuratov, and G. Videen, 2008, Sh-matrices method applied to light scattering by finite circular cylinders, *Journal of Quantitative Spectroscopy and Radiative Transfer*, **109**, 1474–1495.

Petrov, D, Y. Shkuratov, E. Zubko, and G. Videen, 2007, Sh-matrices method as applied to scattering by particles with layered structure, *Journal of Quantitative Spectroscopy and Radiative Transfer*, **106**, 437–54.

Pollack, J., O. Toon, and B. Khare, 1973, Optical properties of some terrestrial rocks and glasses, *Icarus*, **19**, 372–389.

Prospero, J. M., W. M. Landing, and M. Schulz, 2010, African dust deposition to Florida: Temporal and spatial variability and comparisons to models, *J. Geophys. Res.*, **115**, D13304, doi:10.1029/2009JD012773.

Purcell, E. M., and C. R. Pennypacker, 1973, Scattering and absorption of light by nonspherical dielectric grains, *Astrophys. J.*, **186**, 705–714.

Ramachandran, S., V. Ramaswamy, G. L. Stenchikov, and A. Robock, 2000, Radiative impacts of the Mt. Pinatubo volcanic eruption: Lower stratospheric response, *J. Geophys. Res.*, **105**, 24,409–24,429.

Rolland, P., K. N. Liou, M. D. King, S. C. Tsay, G. M. McFarquhar, 2000, Remote sensing of optical and microphysical properties of cirrus clouds using MODIS channels: methodology and sensitivity to assumptions, *J. Geophys. Res.*, **105**, 11721-11738.

Sassen, K., Z. Wang, and D. Liu, 2008, Global distribution of cirrus clouds from CloudSat/Cloud-Aerosol lidar and infrared pathfinder satellite observations (CALIPSO) measurements, *J. Geophys. Res.*, **113**, D00A12.

Saunders, C. P. R., and N. M. A. Wahab, 1975, The influence of electric fields on the aggregation of ice crystals, *J. Meteorol. Soc. Japan*, **53**, 121–126.

Schmitt, C. G., and A. J. Heymsfield, 2007, On the occurrence of hollow bullet rosette- and column-shaped ice crystals in midlatitude cirrus, *Journal of the Atmospheric Sciences*, **64**, 4514-4519.

Schmitt, C. G., J. Iaquinta, A. J. Heymsfield, 2006, The asymmetry parameter of cirrus clouds composed of hollow bullet rosette-shaped ice crystals from ray-tracing calculations, *J. Appl. Meteor. Climatol.*, **45**, 973–981.

Shcherbakov, V., G.-F. Gayet, B. Baker, and P. Lawson, 2006, Light scattering by single natural ice crystals, *J. Atmos. Sci.*, **63**, 1513–1525.

Schumann, U., B. Weinzerl, O. Reitebuch, H. Schlager, A. Minikin, C.Forster, R. Baumann, T. Sailer, K. Graf, H. Mannstein, C. Voigt, S. Rahm, R. Simmet, M. Scheibe, M. Lichtenstern, R. Stock, H. Rüba, D. Schäuble, A. Tafferner, M. Rautenhaus, T. Gerz, H. Ziereis, M. Krautstrunk, C. Mallaun, J.-F. Gayet, K. Lieke, K. Kandler, M. Ebert, S. Weinbruch, A. Stohl, J. Gasteiger, S. Gross, V. Freudenthaler, M. Wiegner, A. Ansmann, M. Tesche, H. Olafsson, and K. Sturm, 2011, Airborne observations of the Eyjafjalla volcano ash cloud over Europe during air space closure in April and May 2010, *Atmos. Chem. Phys.*, **11**, 2245-2279, doi:10.5194/acp-11-2245-2011.

Soden, B. J., R. T. Wetherald, G. L. Stenchikov, and A. Robock, 2002, Global cooling following the eruption of Mt. Pinatubo: A test of climate feedback by water vapor, *Scicncc*, **206**, 727 730.

Stoelinga, M. T., J. D. Locatelli, and C. P. Woods, 2007, The occurrence of 'irregular' ice particles in stratiform clouds, *J. Atmos. Sci.*, **64**, 2740–2750.

Stowe, L., R. M. Carey, and P. P. Pelligrino, 1992, Monitoring the Mt. Pinatubo aerosol layer with NOAA 11 AVHRR data, *Geophys. Res. Lett.*, **19**, 159–162.

Strong, A. E., and L. L. Stowe, 1993, Comparing stratospheric aerosols from el Chichon and Mount Pinatubo using AVHRR data, *Geophys. Res. Lett.*, **20**, 1183–1186.

Stubenrauch, C. J, A. Chédin, G. Rädel, N. A. Scott, and S. Serrar. 2006, Cloud properties and their seasonal and diurnal variability from TOVS Path-B, *J. Climate*, **19**, 5531–5553.

Sun, W., N. G. Loeb, and P. Yang, 2006, On the retrieval of ice cloud particle shape from POLDER measurements, *Journal of Quantitative Spectoscopy and Radiative Transfer*, **101**, 435–447.

Sun, W. B., Q. Fu, and Z. Chen, 1999, Finite-difference time-domain solution of light scattering by dielectric particles with perfectly matched layer absorbing boundary conditions, *Applied Optics*, **38**, 3141–3151.

Tanré, D., M. Herman, and Y. J. Kaufman, 1996, Information on aerosol size distribution contained in solar reflected spectral radiances, *J. Geophys. Res.*, **101**, 19043–19060.

Thelen, J.-C., and J. M. Edwards, 2012, Short-wave radiances: comparison between SEVIRI and the unified model, *Q. J. R. Meteorol. Soc.*, in press.

Turnbull, K., B. Johnson, F. Marenco, J. Haywood, A. Woolley, et al., 2012, A case study of observations of volcanic ash from the Eyjafjallajökull eruption; *in situ* airborne observations, *J. Geophys. Res.*, **117**, D00U12, doi:10.1029/2011JD016688.

Ulanowski, Z., P.H. Kaye, E. Hirst, and R. Greenaway, 2011, Retrieving the size of particles with rough surfaces from 2D scattering patterns. 13th Int. Conf. on Electromagnetic & Light Scatt., Taormina. In: *Atti Accad. Pelorit. Pericol. Cl. Sci. Fis. Mat. Nat. 89*, Suppl. 1, C1V89S1P087. doi: 10.1478/C1V89S1P087.

Ulanowski, Z., P. H. Kaye, E. Hirst, and R. S. Greenway, 2010, Light scattering by ice particles in the Earth's atmosphere and related laboratory measurements, In Electromagnetic and Light Scattering by Nonspherical Particles XII, Helsinki 2010.

Ulanowski, Z., J. Bailey, P. W. Lucas, J. H. Hough, and E. Hirst, 2007, Alignment of atmospheric mineral dust due to electric field, *Atmos. Chem. Phys.*, **7**, 6161–6173.

Ulanowski, Z., E. Hesse, P. H. Kaye, and A. J. Baran, 2006, Light scattering by complex ice-analogue crystals, *Journal of Quantitative Spectroscopy and Radiative Transfer*, **100**, 382–392.

Um, J., and G. M. McFarquhar, 2011, Dependence of the single-scattering properties of small ice crystals on idealized shape models, *Atmospheric Chemistry and Physics*, **11**, doi:10.5194/acp-11-1-2011.

Um, J., and G. M. McFarquhar, G. M., 2009, Single-scattering properties of aggregates of plates, *Q. J. Roy. Meteor. Soc.*, **135**, 291–304.

Um, J., and G. M. McFarquhar, 2007, Single-scattering properties of aggregates of bullet rosettes in cirrus, *J. Appl. Meteorol. Climatol.*, **46**, 757–775.

van de Hulst, H. C., 1980, *Multiple Light Scattering: Tables, Formulas, and Applications*, Academic Press, New York.

van de Hulst, H. C., 1957, *Light Scattering by Small Particles*, Wiley, New York.

Veihelmann, B., T. Nousiainen, M. Kahnert, W. J. van der Zande, 2006, Light scattering by small feldspar particles simulated using the Gaussian random sphere geometry, *Journal of Quantitative Spectroscopy and Radiative Transfer*, **100**, 393–405.

Volkovitskiy, O. A., L. N. Pavlova, and A. G. Petrushin, 1980, Scattering of light by ice crystals, *Atmos. Ocean. Phys.*, **16**, 90–102.

Volten, H., O. Munoz, E. Rol, J. F. de Haan, W. Vassen, J. W. Hovenier, K. Muinonen, and T. Nousiainen, 2001, Scattering matrices of mineral aerosol particles at 441.6 nm and 632.8 nm, *J. Geophys. Res.*, **106**, 17,375–17,401.

Wahab, N. M. A., 1974, Ice crystal interactions in electric fields. PhD;UMIST.

Waterman, P. C., 1971, Symmetry, unitarity, and geometry in electromagnetic scattering, *Physical Review, D*, **3**, 825–839.

Weickmann, H., 1947, Die Eisphase in der Atmosphare, Royal Aircraft Establishment, Farnborough.

Weinman, J. A., and M. J. Kim, 2007, A simple model of the millimeter-wave scattering parameters of randomly oriented aggregates of finite cylindrical ice hydrometeors, *J. Atmos. Sci.*, **64**, 634–644.

Wendling, P., R. Wendling, and H. K. Weickmann, 1979, Scattering of solar radiation by hexagonal ice crystals, *Applied Optics*, **18**, 2663–2671.

Westbrook, C. D., A. J. Illingworth, E. J. Connor, and R. J. Hogan, 2009, Doppler lidar measurements of oriented planar ice crystals falling from supercooled and glaciated layer clouds, *Q. J. R. Meteor. Soc.*, **136**, 260–276.

Westbrook C. D., R. C. Ball, and P. R. Field, 2008, Notes and correspondence corrigendum: Radar scattering by aggregate snowflakes, *Q. J. R. Meteorol. Soc.*, **134**, 547–548.

Westbrook, C. D., R. C. Ball, and P. R. Field, 2006, Radar scattering by aggregate snowflakes, *Q. J. R. Meteorol. Soc.*, **132**, 897–914.

Westbrook, C. D., R. C. Ball, P. R. Field, and A. J. Heymsfield, 2004, Theory of growth by differential sedimentation, with application to snowflake formation, *Phys. Rev., E,* **70**, Art. No. 021403.

Whitby, K. T., 1978, The physical characteristics of sulfur aerosols, *Atmos. Environ.*, **12**, 135–159.

Whitby, K. T., The physical characteristics of sulfur aerosols, *Atmos. Environ.*, **12**, 135–159.

Wriedt, T., 2009, Light scattering theories and computer codes, *Journal of Quantitative Spectroscopy and Radiative Transfer*, **110**, 833–843.

Wriedt, T., and Doicu, A., 1998, Formulation of the extended boundary condition method for three-dimensional scattering using the method of discrete sources, *Journal of Modern Optics*, **45**, 199–213.

Wylie, D. P, and W. P. Menzel, 1999, Eight years of cloud statistics using HIRS, *J. Climate*, **12**, 170–184.

Wyser, K., and P. Yang, 1998, Average ice crystal size and bulk short-wave single-scattering properties of cirrus clouds, *Atmos. Res.*, **49**, 315–335.

Xie, Y. P., Yang, G. W. Kattawar, B. A. Baum, and Y. Hu, 2011, Simulation of the optical properties of plate aggregates for application to the remote sensing of cirrus clouds, *Applied Optics*, **50**, 1065–1081.

Yang, H., S. Dobbie, R. Herbert, P. Connolly, M. Gallagher, S. Ghosh, S. M. R. K. Al-Jumur, and J. Clayton, 2011, The effect of observed vertical structure, habits, and size distributions on the solar radiative properties and cloud evolution of cirrus clouds, *Q. J. R. Meteorol. Soc.*, doi:10.1002/qj.973.

Yang., P., Z. Zhang, G. W. Kattawar, S. G. Warren, B. A. Baum, H.-L. Huang, Y.-X. Hu, D. Winker, and J. Iaquinta, 2008, Effect of cavities on the optical properties of bullet rosettes: Implications for active and passive remote sensing of ice cloud properties, *J. Appl. Meteor. Climatol.*, **47**, 2311–2330.

Yang, P., B. A. Baum, A. J. Heymsfield, Y. X. Hu, H. L. Huang, S. C. Tsay, and S. Ackerman, 2003, Single-scattering properties of droxtals, *Journal of Quantitative Spectroscopy and radiative Transfer*, **79–80**, 1159–1180.

Yang, P., K. N. Liou, M. I. Mishchenko, and B. C. Gao, 2000, An efficient finite-difference time domain scheme for light scattering by dielectric particles: application to aerosols, *Applied Optics*, **39**, 3727–3737.

Yang, P., and K. N. Liou, 1998, Single-scattering properties of complex ice crystals in terrestrial atmosphere, *Contr. Atmos. Phys.*, **71**, 223–248.

Yang P., and K. N. Liou, 1997, Light scattering by hexagonal ice crystals, solution by a ray-by-ray integration algorithm, *J. Opt. Soc. Am. A.*, **14**, 2278–2289.

Yang, P., and K. N. Liou, 1996, Geometric-optics-integral-equation method for light scattering by nonspherical ice crystals, *Applied Optics*, **35**, 6568–6584.

Yee, K. S., 1966, Numerical solution of initial value boundary problems involving Maxwell's equations in isotropic media, *IEEE Trans. Antennas and Propagat.*, **14**, 302–307.

Yoshida, R., H. Okamoto, Y. Hagihara, and H. Ishimoto, 2010, Global analysis of cloud phase and ice crystal orientation from Cloud-Aerosol Lidar and Infrared Pathfinder Satellite Observation (CALIPSO) data using attenuated backscattering and depolarization ratio, *J. Geophys. Res.*, **115**, art. D00H32.

Yuan, J., R. A. Houze Jr, and A. J. Heymsfield, 2011, Vertical structures of anvil clouds of tropical mesoscale convective systems observed by CloudSat, J. Atmos. Sci., 68, 1653–1674.

Zhang, Z. B., P. Yang, G. W. Kattawar, S.-C. Tsay, B. A. Baum, Y. X. Hu, A. J. Heymsfield, and J. Reichardt, 2004, Geometrical optics solution to light scattering by droxtal ice crystals, Applied Optics, 43, 2490–2499.

Zhang, X. Y., R. Arimoto, G. H. Zhu, T. Chen, and G. Y. Zhang, 1998, Concentration, size-distribution and deposition of mineral aerosol over Chinese desert regions, *Tellus*, **50B**, 317–330.

Zhao, Y., G. Mace, and J. M. Comstock, 2011, The occurrence of particle size distribution bimodality in midlatitude cirrus as inferred from ground-based remote sensing data, *J. Atmos. Sci.*, **68**, 1162–1175.

Zubko, E., Y. Shkuratov, M. Mishchenko, and G. Videen, 2008, Light scattering in a finite multi-particle system, *Journal of Quantitative Spectroscopy and Radiative Transfer*, **109**, 2195–2206.

2 Physical-geometric optics hybrid methods for computing the scattering and absorption properties of ice crystals and dust aerosols

Lei Bi and Ping Yang

2.1 Introduction

Exact solutions and reasonable approximations of the optical properties of nonspherical particles in the atmosphere (particularly, coarse mode mineral dust particles, ice crystals within cirrus clouds, and aviation-induced contrails) are fundamental to numerous climate studies and remote sensing applications (Chýlek and Coakley, 1974; Haywood and Boucher, 2000; Ramanathan et al., 2001; Sokolik et al., 2001; Kaufman et al., 2002; Liou et al., 2000; Liou, 2002; Baum et al., 2005; Baran, 2009; Yang et al., 2010). The morphologies of realistic aerosols (Reid et al., 2003) and ice crystal habits (Heymsfield and Iaquinta, 2000) are extremely diverse. For simplicity, light scattering simulations reported in the literature are limited to a small set of well-defined nonspherical geometries such as hexagonal columns or plates, aggregates of columns or plates, bullet rosettes, circular cylinders, and ellipsoids (Asano and Yamamoto, 1975; Mishchenko and Travis, 1998; Yang et al., 2005; Bi et al., 2008; Meng et al., 2010; Xie et al., 2011). Rigorous solutions of elastic light scattering for the defined nonspherical particles are obtained by solving either Maxwell's equations or their mathematical equivalents (Kahnert, 2003). The most commonly used light scattering computational methods are the T-matrix (Mishchenko et al., 2000 and references therein; Mishchenko et al., 2002), the finite-difference time-domain (FDTD) (Yee, 1966; Yang and Liou, 1996a; Sun et al., 1999), the pseudo-spectral time-domain (PSTD) (Liu, 1999; Tian and Liu, 2000; Chen et al., 2008), and the discrete dipole approximation (DDA) (Purcell and Pennypacker, 1973; Kahnert, 2003) methods. The use of these methods has significantly advanced the knowledge of the optical properties of nonspherical particles. However, unlike the Lorenz–Mie theory for spherical particles, the aforementioned methods are applicable to a limited range of particle size parameters $\chi \in (0, \chi_{\max}]$, and the maximum value $\chi_{\max}$ varies with the selected method, the defined particle shape, the refractive index, the number of particle orientations, and the computational resources. In practice, an exact solution to a light scattering process by a generally irregular particle can be efficiently obtained only when the size parameter is in either the Rayleigh or the resonance (i.e., the particle size is on the order of the incident wavelength) regime. This is particularly true when a large number of simulations (e.g., random orientations and a series of sizes and refractive indices) are involved. Thus, the challenge is to develop approximate methods that can handle size parameter ranges beyond the resonance regime.

Numerous studies have attempted to develop methods based on the principles of geometric optics or ray optics to approximately calculate the single-scattering properties of large size nonspherical particles (Cai and Liou, 1982; Muinonen, 1989; Takano and Liou, 1989; Macke, 1993; Macke et al., 1996a,b; Macke and Mishchenko, 1996; Yang and Liou, 1996b; Yang and Liou, 1997; Borovoi et al., 2002; Borovoi and Grishin, 2003; Bi et al., 2011a,b). Yang and Liou (2006) briefly reviewed the development of the geometric-optics method. Rigorously speaking, the term 'geometric-optics' is inaccurate because the geometric-optics principle fails to deal with diffraction, a wave nature of light, which is commonly accounted for in physical optics or semi-classical analysis. Here, the meaning of diffraction is in a broad sense rather than considering only Fraunhofer diffraction by a projected area (see Nussenzveig, 1992). As noted in most texts, in addition to diffraction, the geometric-optics principle fails to consider 'interference', which is classified as another physical-optics effect. However, the present convention assumes interference is considered in generalized geometric optics (Born and Wolf, 1959), because the resultant field values in the ray-tracing calculation are computed from the superposition of the fields in conjunction with individual rays and include the phase information. Hereafter, we use the term 'physical-geometric optics hybrid (PGOH)' to emphasize the hybrid nature of the method, and the nomenclature has previously been used in publications (Bi et al., 2010a, 2011b). The literature also has references to the PGOH method as the physical-optics approximation (Ravey and Mazeron, 1982; Mazeron and Muller, 1996) or the ray-wave approximation (Priezzhev et al., 2009).

Physical optics deals with the Fraunhofer diffraction effect arising from the incomplete incident wave front due to blocking by a particle as well as the diffraction of rays exiting the particle surface. The diffraction effect of outgoing rays is not considered in the early and some later developments of the geometric-optics method (e.g., Wendling et al., 1979; Cai and Liou, 1982; Macke, 1993). The origin of the PGOH that considers the diffraction of rays from the particle to the radiation zone may be traced to the work of Ravey and Mazeron (1982), who utilized the Kirchhoff approximation in electrodynamics to compute the single-scattering properties of spheroids. Inside the framework of the PGOH, geometric-optics principles are employed to obtain either the internal field within a scattering particle or the field on the external surface of the particle via the superposition of the electromagnetic field vectors associated with all the rays. The field is represented in a form that includes all information about the amplitude, the phase, and the polarization state. For practical applications, the PGOH has an advantage over exact methods in two aspects: (1) the PGOH is applicable to the large size parameter region where rigorous methods attempting to solve Maxwell's equations are inefficient or inapplicable; and, (2) the PGOH has an intuitive physical insight about light scattering processes that is useful in identifying the relationship between the optical properties and microphysical properties of a scattering system. However, the PGOH is an approximate method useful for moderate or large size parameters and, consequently, is less accurate for smaller size parameters than those methods solving Maxwell's equations. Whenever the size parameter is smaller than $\sim$20, rigorous methods are necessary to solve Maxwell's equations.

This chapter is designed to elaborate the fundamentals of the geometric-optics method and to discuss its accuracy and application regime based on the current status of knowledge in modeling the optical properties of ice crystals and mineral dust aerosols. Specifically, we intend to:

- review the conceptual basis and various modifications of the geometric-optics method reported in the literature;
- describe advanced numerical techniques such as the beam-splitting approach and the line-integration method;
- establish rigorous mathematical formulation of inhomogeneous waves in the case of absorptive particles, the amplitude variation over wave front of propagating beams, and the analytical integration of the geometric-optics near-field; and
- illustrate the accuracy and efficiency of the PGOH in the computation of the single-scattering properties of ice crystals and mineral dust aerosols.

2.2 Conceptual Basis

In the geometric optics framework for light scattering by a particle, the presence of the particle blocks a partial wave front of a propagating plane wave according to its projected area. As shown in Fig. 2.1, the blocked wave front, via interacting with a particle, undergoes a series of reflection and transmission (i.e., refraction) processes on the particle surface generating various induced or secondary waves exiting the particle, and the remaining incomplete wave front (i.e., the original wave front subtracted by the blocked wave front) induces Fraunhofer diffraction in the radiation zone. As stated in Babinet's principle (van de Hulst, 1981), the contribution of the surrounding incomplete wave front to scattering is equivalent to the diffraction of a localized incident plane wave within the projected area or shadow. Based on this insight, early developments of geometric optics assumed that the scattered far-field was contributed by geometric rays and Fraunhofer diffraction. Furthermore, without the consideration of the interference between diffraction and scattered beams, the extinction cross-section is assumed to be twice the projected area. This concept was easily implemented for spheres and randomly oriented cylinders and ellipsoids (Liou and Hansen, 1971; Macke and Mishchenko, 1996). A comparison between the geometric-optics-based phase matrix and its counterparts computed from the Lorenz–Mie theory and the T-matrix method shows that geometric-optics provides reasonably accurate results when the particle size parameter is larger than approximately 100 (Liou and Hansen, 1971; Macke and Mishchenko, 1996). An extension of the method, usually known as the conventional geometric optics method (CGOM), to faceted particles such as hexagonal ice crystals reveals a unique feature known as the delta-transmission that is not present in the cases of spheres and ellipsoids but may be quite pronounced for particles with parallel faces (Takano and Liou, 1989; Mishchenko and Macke, 1998). In addition, isolated points exist in the phase matrix for oriented particles and an artificial halo phenomenon is observed for particles of moderate sizes (Mishchenko and Macke, 1998, 1999). The fundamental physical reason for the phenomena is the wave front of outgoing beams from faceted particles has no curvature causing caustics in the radiation region, and, thus, the diffraction effects of beams exiting the particle surface must be considered.

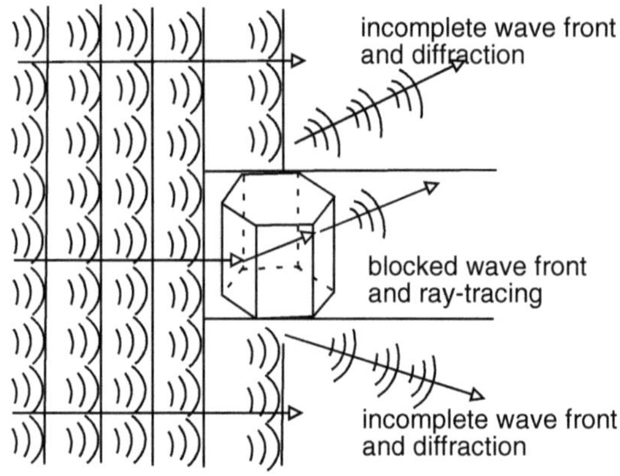

Fig. 2.1. Conceptual figure of incomplete wave front and diffraction, and blocked wave front and ray-tracing.

Various approaches have been developed to take into account the diffraction associated with an exiting beam. In a simplified formulation, the diffraction effect of outgoing beams is assumed to behave similarly to Fraunhofer diffraction applied to the blocking wave front. In the more rigorous theory of electrodynamics, the scattered far-field is related to the near-field, either through the Fredholm volume integral equation or the Kirchhoff surface integration equation. The establishment between the far-field and the near-field provides a straightforward approach to include the diffraction effect. In this case, the approximate nature of the method is attributed to the application of the geometric-optics principles to the near-field calculation. The surface- and volume-integration-based PGOH methods do not lead to the same optical properties because the geometric-optics near-field is not accurate. In the near-to-far-field transformation based on the Kirchhoff surface integral, the diffraction arises from the mapping of either the incident field on the illuminated side or the negative of the incident field on the non-illuminated side of the particle to the radiation region and is similar to Babinet's principle. However, if the far electric field is obtained from the Fredholm volume-integral equation, Fraunhofer diffraction is inherently combined with the external reflection and cannot be explicitly separated (Bi et al., 2010a). The physical-optics approximation for a conducting particle is usually based on the surface integral approach because the internal field is zero. By assuming that a scattering particle is an extremely absorptive dielectric particle, a simple formula can be derived from the volume-integral equation to account for the combined effect of diffraction and external reflection.

The two alternative methods to obtain the geometric-optics-based near-field are: (1) to trace narrow geometric rays (hereafter, the ray-tracing method); and, (2) to trace broad beams or ray tubes (hereafter, the beam-tracing method). In the ray-tracing method, the incident wave front is imagined as consisting of a bundle of separate rays with very small cross-sections (Fig. 2.2(a)). In the beam-tracing method, the wave front is divided into several parts with each part incident on a single facet of the particle, as shown in Fig. 2.2(b). The single-scattering properties (i.e., the extinction efficiency, the single-scattering albedo, and the phase matrix) have been formulated from the different near-to-far field transformations using the ray-tracing or beam-tracing methods. Based on the ray-tracing method, Yang and Liou (1996b, 1997) derived the optical properties of hexagonal ice crystals from the Kirchhoff surface integral equation and developed the formalism of the optical properties of hexagonal ice crystals from the Fredholm volume integral equation. Borovoi and Grishin (2003) developed an algorithm based on the beam-tracing technique for non-absorbing hexagonal ice crystals from the theory of vector Fraunhofer diffraction (Jackson, 1999). Bi et al. (2011b) reported the integration of the geometric-optics-based near-field in terms of the beam-tracing method based on the Fredholm volume integral equation with no simplification. Specifically, the geometric-optics-based near-field is obtained by superimposing the electric field associated with various ray tubes. Analytical integrations are carried out for each ray tube by transforming all the integrations into summations associated with vertices of ray cross-sections. Furthermore, in the case of an absorptive particle, the inhomogeneity of waves associated with various orders of reflection and refraction events is fully considered via an algorithm (Yang and Liou, 2009a,b) applicable to arbitrary refractive indices.

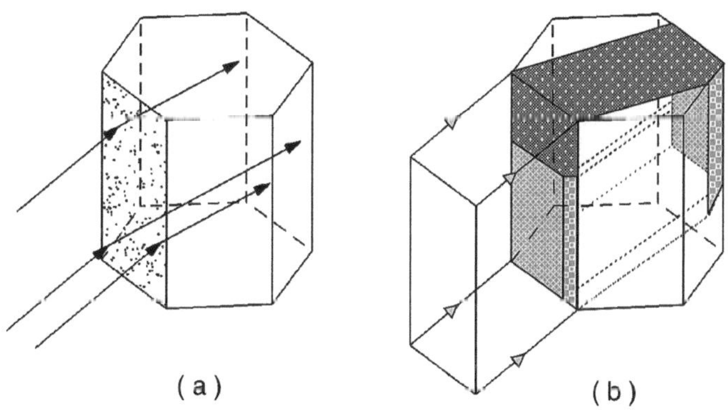

Fig. 2.2. (a) A partial wave front impinging on a facet. (b) A partial wave front refracted into the particle and split into three parts impinging subsequently on three facets; the initial cross-section has the same pattern as the associated cross-section.

To obtain the angular distribution of scattered light for randomly oriented particles, a rigorous PGOH method (i.e., an approach of exactly integrating the geometric-optics near-field to obtain the far-field) demands tremendous computational effort, particularly when the PGOH is implemented with the ray-by-ray integration method (Yang and Liou, 1997). In practice, a simplified PGOH algorithm, the intensity mapping algorithm (Yang and Liou, 1996), has been developed for randomly oriented particles. In this algorithm, the phase matrix elements are calculated by incorporating the diffraction effect into the phase matrix obtained from the CGOM. Interference among rays is neglected based on the assumption that the interference becomes less important if random orientations are considered. A combination of the simplified algorithm for the phase matrix calculation and a method based on the Fredholm volume integral equation (Yang and Liou, 1997) for the calculation of the extinction and absorption efficiency, the improved geometric optics method (IGOM), is used in some applications to derive the single-scattering properties of ice crystals (Yang et al., 2005) and mineral dust aerosols (Yang et al., 2007).

Rigorously speaking, the diffraction associated with beams propagating inside of the particle must be incorporated similar to the diffraction effect associated with outgoing beams; however, at this time, no known studies are underway to investigate the issue due to the inherent complexity. Based on the Fresnel–Huygens principle, van de Hulst (1981) states that for the existence of a ray (i.e., a pencil of light, or a localized wave), with a length l and at a wavelength λ, the width of its base area needs to be significant in comparison with $\sqrt{\lambda l}$. Qualitatively, this speculation can be regarded as a constraint to the lower size parameter limit for the applicability of the geometric-optic principles in the near-field calculation. As a numerical example, we consider a linearly polarized plane wave incident normally on a basal face of a circular cylinder (Fig. 2.3(a)). Shown in Fig. 2.3(b) and 2.3(c) are the intensities of the total electric field on the two basal faces of a circular cylinder simulated from the Amsterdam DDA (ADDA) computational program (Yurkin and Hoekstra, 2011). According to geometric optics principles, the intensity should be the same at any arbitrary location on the cross-section. From Fig. 2.3(a), it is evident that the variation of the intensity on the cylinder cross-section facing the incoming radiation is small, indicating the geometric-optics to be approximately valid. However, a pronounced diffraction-like pattern is observed at the end cross-section, which implies the geometric-optics to be essentially invalid. This example illustrates that the accuracy of the geometric-optics principle in the near-field calculation depends on the particle geometry, orientation, and size parameter. If the particle is absorptive, the contribution of higher-order beams to the optical properties is insignificant, and thus, the contribution from the first-order reflection and refraction dominates. The beam cross-section associated with the first-order reflection or refraction is relatively large in comparison with its higher-order beam counterparts making the geometric-optics method more accurate for absorptive particles than for non-absorptive particles. In practical calculations, the ray con-

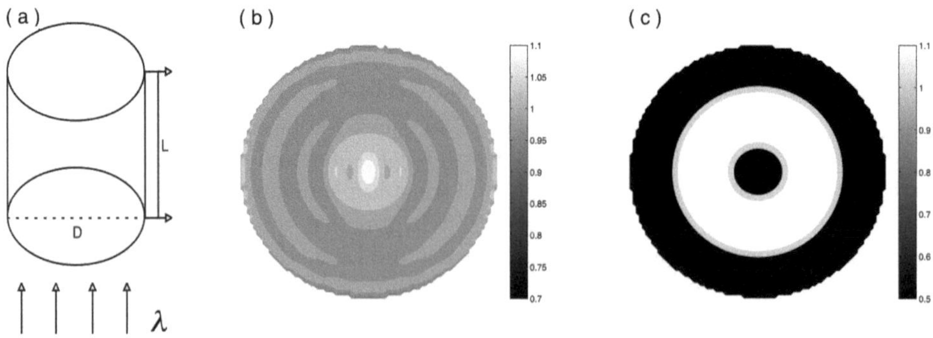

Fig. 2.3. Intensity of the total electric field on the initial (middle panel) and end (right panel) side of a circular cylinder simulated from the DDA method. The size parameter, defined in terms of diameter, is $kD = 50$, where k is the wave number. The aspect ratio (length L/diameter D) is 5. The refractive index is 1.05.

cept within the particle is reasonable when the size parameter is approximately 20 (the value will increase if the particle surface has some fine variations or the facets composing the particle are small). However, in special cases such as that shown in Fig. 2.3, the size parameters need to be of a diameter much larger than 20 in order to distinguish rays within particles. Comparisons between the PGOH-simulated optical properties and the ADDA simulations also support the aforementioned van de Hulst speculation.

Based on the van de Hulst speculation, rays with very narrow cross-sections may not exist. In fact, the technique of tracing narrow rays can be used numerically. For example, in Fig. 2.3a, there is physically only one ray. However, a number of rays may be employed to do the ray tracing to calculate the near field. The use of multiple narrow rays instead of a large cross-section ray is valid only when the large cross-section ray exists according to the van de Hulst speculation.

The tunneling or edge effect (Nussenzveig, 1992) associated with tunneling rays is not included in the PGOH. Tunneling rays are those passing the particle and interacting with the particle through a tunneling process found within the framework of wave optics. In the theory of light scattering of spheres, the semi-classical scattering analysis justifies the existence of both surface waves and tunneling rays and formulates their contribution to the scattering of light (Nussenzveig, 1992). Fu et al. (1999) investigated the Poynting vector both near to and inside a hexagonal particle by using the FDTD method and illustrated the extra contribution of tunneling rays to the extinction and absorption of the particle. The ray-tracing process fails to consider tunneling; consequently, without the contribution of the associated semi-classical scattering effects, discontinuities appear between the transitions from the exact solutions of the extinction and absorption efficiencies to those computed from the PGOH. The justification of the contribution of tunneling rays to the extinction of light by the DDA method will be detailed in Section 2.5. The semi-empirical formulas considering the edge-effect contribution to the extinction and the absorption efficiencies are also discussed.

2.3 Geometric-optics-based near-field

In this section, we focus on a combination of the beam-tracing and field-tracing processes within an absorbing particle to obtain the geometric-optics-based near-field (note that a non-absorbing particle is a special case not needing additional treatment). A detailed ray-tracing discussion can be found in Yang and Liou (2009a,b).

2.3.1 Effective refractive index and Snell's law

When a wave is incident on a local planar surface, within the framework of the geometric-optics approximation, both reflection and refraction occur. Snell's law determines the beam direction change, and the Fresnel formulas give the amplitude and polarization of the electromagnetic field associated with the reflected and refracted beams. In this section, we focus on the ray/beam-tracing process, which requires only Snell's law, given by

$$\sin\theta_t = \sin\theta_i/m\,, \tag{2.1}$$

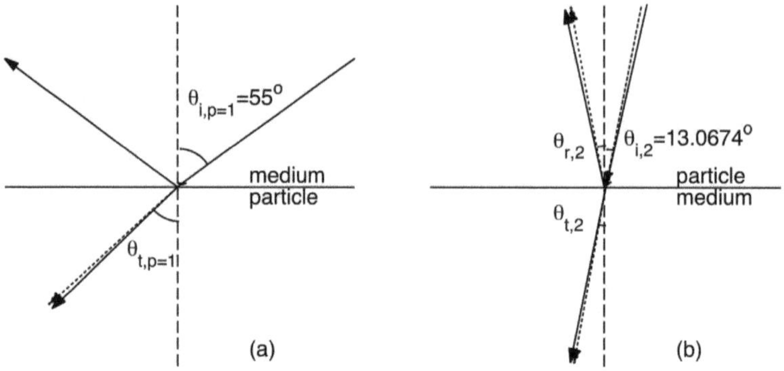

Fig. 2.4. Diagram to illustrate the difference between the trajectory of rays using the real part of the refractive index (solid) and the real part of effective refractive index (dotted). The angle (not shown in the diagram) between the interface for the first-order reflection/refraction (a) and that for the second-order reflection/refraction (b) is 60°. The refractive index of ice is selected to be 1.0925 + i0.248.

where θ_t is the refraction angle, θ_i is the incident angle, and m is the refractive index. When m is complex (i.e., the particle is absorptive), θ_t is complex and the refracted plane wave is inhomogeneous, i.e., in general, the surface of constant amplitude does not coincide with the surface of constant phase. The inhomogeneous wave in the absorbing medium takes the following form,

$$\vec{E}(\vec{r}, t) = \vec{E}_0 \exp\left\{i\left[\vec{k}\cdot\vec{r} - \omega t\right]\right\}, \tag{2.2}$$

where $\vec{k}$ is a complex vector, ω is the circular frequency, and $\vec{E}_0$ is the amplitude. From Maxwell's equations, we have

$$\vec{k}\cdot\vec{E}_0 = 0, \quad \vec{k}\cdot\vec{H}_0 = 0, \tag{2.3}$$

where $\vec{H}_0$ is the amplitude of the magnetic field. The complex transversality shown in Eq. (2.3) implies that both $\vec{E}_0$ and $\vec{H}_0$ have nonzero components along the propagation direction. Based on the phase-match condition, the effective refractive index can be defined in formulating the inhomogeneous refracted waves. The real part of the effective refractive index N_r determines the propagation direction of the constant-phase surface of the refracted wave, and the imaginary part N_i accounts for the attenuation of the associated amplitude. Specifically, the inhomogeneous refractive wave can be written as

$$\vec{E}(\vec{r})\exp(-kN_i l)\exp(ikN_r l), \tag{2.4}$$

where l is the propagating distance from the position on the interface. We have assumed an $e^{-i\omega t}$ time dependence of the harmonic electromagnetic field, which implies a positive imaginary part of refractive index. In general, the effective refractive index is not the same at different orders of interaction. For clarity, let the subscript index p $(= 1, 2, 3, \ldots)$ indicate the pth order reflection/refraction event; $p = 1$ corresponds to the external reflection and refraction (from medium to particle); and, $p > 2$ indicates the internal reflection and refraction (from particle to medium). The effective refractive index for the pth order of reflection/refraction event is denoted as N_p whose real and imaginary parts are $N_{r,p}$ and $N_{i,p}$. The generalized Snell's law for the first-order reflection/refraction is given by (Yang and Liou, 2009a)

$$\sin\theta_{t,p=1} = \sin\theta_{i,p=1}/N_{r,p=1}\,, \tag{2.5}$$

$$N_{r,p=1} = \sqrt{\left\{m_r^2 - m_i^2 + \sin^2\theta_{i,p=1} + \sqrt{(m_r^2 - m_i^2 - \sin^2\theta_{i,p=1})^2 + 4m_r^2 m_i^2}\right\}/2}\,, \tag{2.6}$$

$$N_{i,p=1} = \cos\theta_{t,p=1} \times \sqrt{\left\{-(m_r^2 - m_i^2 - \sin^2\theta_{i,p=1}) + \sqrt{(m_r^2 - m_i^2 - \sin^2\theta_{i,p=1})^2 + 4m_r^2 m_i^2}\right\}/2}\,, \tag{2.7}$$

where $\theta_{i,p=1}$ is the incident angle (the reflection angle $\theta_{r,p=1} = \theta_{i,p=1}$) and $\theta_{t,p=1}$ is the refraction angle. The effective refractive index is equal to the original refractive index when $m_i = 0$ or $\theta_{i,p} = 0$, i.e., the incident wave is perpendicular to the interface. In all other cases, the effective refractive index depends on the incident angle; therefore, the use of the real part of the refractive index to determine the refracted angle introduces some uncertainty. Similarly, for successive internal reflections, the effective refractive indices can be defined (Yang and Liou, 2009a) as,

$$N_{r,p+1} = \times \sqrt{\left\{m_r^2 - m_i^2 + N_{r,p}^2 \sin^2\theta_{i,p+1} + \sqrt{(m_r^2 - m_i^2 - N_{r,p}^2 \sin^2\theta_{i,p+1})^2 + 4m_r^2 m_i^2}\right\}/2}\,, \tag{2.8}$$

$$N_{i,p+1} = \cos\theta_{r,p+1} \times \sqrt{\left\{-(m_r^2 - m_i^2 - N_{r,p}^2 \sin^2\theta_{i,p+1}) + \sqrt{(m_r^2 - m_i^2 - N_{r,p}^2 \sin^2\theta_{i,p+1})^2 + 4m_r^2 m_i^2}\right\}/2}\,, \tag{2.9}$$

where the reflection angle $\theta_{r,p+1}$ is related to the incident angle and the refraction angle as,

$$N_{r,p}\sin\theta_{i,p+1} = N_{r,p+1}\sin\theta_{r,p+1} = \sin\theta_{t,p+1}. \tag{2.10}$$

Note that the reflection angle is not equal to the incident angle, and Eqs. (2.8) and (2.9) are iterative formulas. Based on the effective refractive index, the incident, reflection, and refraction angles are real, and the ray-tracing process can be performed for an arbitrary complex refractive index. For detailed physical insight regarding the ray-tracing procedure in the case of an absorptive particle, the reader is recommended to refer to Dupertuis et al (1994), Chang et al (2005), and Yang and Liou (2009a) for in-depth discussions. As a numerical example, Fig. 2.4 is a comparison of ray-tracing of the first- and second-order reflection/refraction based on the real part of the refractive index and the real part of the effective refractive index, and the differences in the ray paths are evident.

2.3.2 Beam-tracing technique

The ray-tracing process, which represents rays as rectilinear lines (see Fig. 2.1), can be applied to arbitrarily shaped geometries; however, for faceted particles, the beam-tracing method implemented with broad cross-sections is more efficient than ray-tracing. The incident partial wavefront intercepted by the particle is split into several parts according to the facets facing the incoming wave. As a portion of the wave front (or localized wave) impinges on one of the facets, the subsequent electromagnetic interaction leads to outgoing reflected and inwardly propagating refracted beams. The first-order refracted beams from medium to particle and the higher-order internally reflected beams may split during their subsequent propagation processes. An appropriate beam-splitting algorithm must describe the mechanism of the split internal beams and specify the geometries of internal ray paths (or ray tubes). The geometry of the scattering particle is assumed to be convex and any externally reflected beams and higher-order refracted beams exiting the particle surface cannot be blocked by the particle itself and are not involved in the subsequent beam-tracing calculation. Therefore, the beam-splitting algorithm is unaffected by beams propagating outside the particle. Similar studies of splitting beams according to the particle geometry have been reported by Popov (1996) and Borovoi and Grishin (2003). As the beam-tracing calculation for concave particles is much more complicated than for convex particles, this chapter only reviews the beam-splitting process for convex faceted particles reported in Bi et al. (2011b).

A beam of light is defined by its propagation direction and its initial beam cross-section. In the case of a faceted particle, the cross-sections of all involved beams are in the shape of a polygon. To identify various beams generated in the beam-tracing process, the direction of one internal beam, the first-order refracted beam or higher-order reflected beam, leaving some interface of the pth-order reflection/refraction is specified by $\hat{e}_p$, and the vertices of the beam cross-section on the interface of electromagnetic interaction are denoted as $\vec{r}_{p,i}$ $(i = 1,\ N_V)$, where N_V is the number of vertices. The sequence of the vertices is arranged in a counterclockwise direction with respect to the outward normal direction of the local facet. When $p = 1$, the coordinates of the initial cross-section of the first-order refracted beam $\vec{r}_{1,i}$ $(i = 1,\ N_V)$ are the coordinates of the vertices of the facet where the beam is refracted into the particle.

To consider the split of an internal beam specified by $\hat{e}_p$ and $\vec{r}_{p,i}$, the first step is to determine those particle facets intercepting the beam. We generate N_V number of rectilinear rays, starting from the positions of N_V vertices and propagating in the direction of $\hat{e}_p$. Assume that a number of N_V rays strike a number of M_{v} different facets. The symbols $\hat{\tau}_i$ $(i = 1,\ M_{\mathrm{v}})$ are assigned to denote the normal directions of the facets. The beam will not split and impinge on a single facet when $M_v = 1$; whereas, the beam splits when $M_v \geq 2$. If $M_v \geq 2$, the beam can be split into two parts based on the information of two facets with normal directions $\hat{\tau}_1$ and $\hat{\tau}_2$. Figure 2.4 shows an initial beam cross-section with $\mathrm{N}_V = 4$. An arbitrary position within the initial beam cross-section can be written as,

$$\vec{r} = c_u \vec{u}_p + c_v \vec{v}_p \,, \tag{2.11}$$

where c_u and c_v are two arbitrary coefficients with respect to two basic vectors, $\vec{u}_p$ and $\vec{v}_p$. Vectors $\vec{u}_p$ and $\vec{v}_p$ can be defined as

$$\vec{u}_p = \vec{r}_{p,2} - \vec{r}_{p,1}\,, \quad \vec{v}_p = \vec{r}_{p,N} - \vec{r}_{p,1}\,. \tag{2.12}$$

Note that $\vec{u}_p$ and $\vec{v}_p$ are neither normalized nor necessarily orthogonal. The coordinates (c_u, c_v) of the points on the initial beam cross-section that will strike the intersection line of two planes associated with the selected two facets, whose outward normal directions are $\hat{\tau}_1$ and $\hat{\tau}_2$, must satisfy the following condition,

$$c_u w_u + c_v w_v = d_1 - d_2\,, \tag{2.13}$$

where d_1 and d_2 represent the propagation distances from $\vec{r}_{p,1}$ in the direction of $\vec{e}_p$ to the planes of the two selected facets. w_u and w_v are given by

$$w_u = \left(\frac{\vec{u}_p \cdot \hat{\tau}_1}{\vec{e}_p \cdot \hat{\tau}_1} - \frac{\vec{u}_p \cdot \hat{\tau}_2}{\vec{e}_p \cdot \hat{\tau}_2}\right), \quad w_v = \left(\frac{\vec{v}_p \cdot \hat{\tau}_1}{\vec{e}_p \cdot \hat{\tau}_1} - \frac{\vec{v}_p \cdot \hat{\tau}_2}{\vec{e}_p \cdot \hat{\tau}_2}\right). \tag{2.14}$$

All the coordinate points (c_u, c_v) that satisfy Eq. (2.13) define a straight line to split the original beam cross-section into two sub-beams. The intersection points between the straight line given by Eq. (2.13) and the polygon-shaped boundary can be written in the form of

$$\vec{r} = \vec{r}_{p,j} + (\vec{r}_{p,j+1} - \vec{r}_{p,j}) l_j\,, \quad \text{if } l_j \in [0,1]\,, \quad (j = 1,\ N_\mathrm{v}) \tag{2.15}$$

where l_j are defined as:

$$l_1 = (d_1 - d_2)/w_u\,, \quad l_N = (w_v - d_1 + d_2)/w_v\,, \tag{2.16}$$

$$l_j = \begin{cases} \dfrac{\hat{n}_p \cdot \left[(\vec{r}_{p,1} + l_1 \vec{u} - \vec{r}_{p,j}) \times \vec{Q}\right]}{\hat{n}_p \cdot \left[(\vec{r}_{p,j+1} - \vec{r}_{p,j}) \times \vec{Q})\right]}, & |w_v| \le |w_u| \\[2ex] \dfrac{\hat{n}_p \cdot \left[(\vec{r}_{p,1} + (1 - l_N)\vec{v} - \vec{r}_{p,j}) \times \vec{Q}\right]}{\hat{n}_p \cdot \left[(\vec{r}_{p,j+1} - \vec{r}_{p,j}) \times \vec{Q})\right]}, & |w_v| > |w_u| \end{cases}, \quad j = 2,\ N_\mathrm{v} - 1\,, \tag{2.17}$$

and where

$$\vec{Q} = \begin{cases} \vec{v}_p - \dfrac{w_v}{w_u}\vec{u}_p, & |w_v| \le |w_u| \\[2ex] \dfrac{w_u}{w_v}\vec{v}_p - \vec{u}_p, & |w_v| > |w_u| \end{cases}. \tag{2.18}$$

In Eq. (2.17), $\hat{n}_p$ is the outward or inward normal direction of the particle facet where the initial beam cross-section locates (for simplicity, it is defined to be inward). To avoid the occurrence of singularity in the evaluation of l_j and $\vec{Q}$, different formulas have been used in the computation based on comparing the absolute values of w_u and w_v. Because the beam cross-section is always convex in cases of convex faceted particles, only two l_j in the 0 to 1 range can give two solutions based on Eq. (2.15), and for an example, see the case shown in Fig. 2.5. At this point, it is straightforward to split the original beam into two sub-beams by regrouping the vertices of the original beam cross-section and the two intersection points given by Eq. (2.15). Generally speaking, each sub-beam may impinge on multiple facets, and the process must be repeated for each sub-beam until each next-order sub-beam impinges on a single facet. Once the initial beam cross-section is divided, the vertex coordinates of the end cross-section of each sub-beam can be obtained. The initial and end beam cross-sections define an internal ray tube. An example of splitting a refracted beam from one facet of the particle into three sub-beams with each sub-beam incident on a single facet is illustrated in Fig. 2.2(b). All sub-beams belonging to different ray tubes undergo internal reflections at different facets, corresponding to the emergence of the next-order reflected beams. The beam-tracing process is flexible in order to consider arbitrarily shaped convex faceted particles. The computational program is designed to read the geometry specified by the coordinates of vertices.

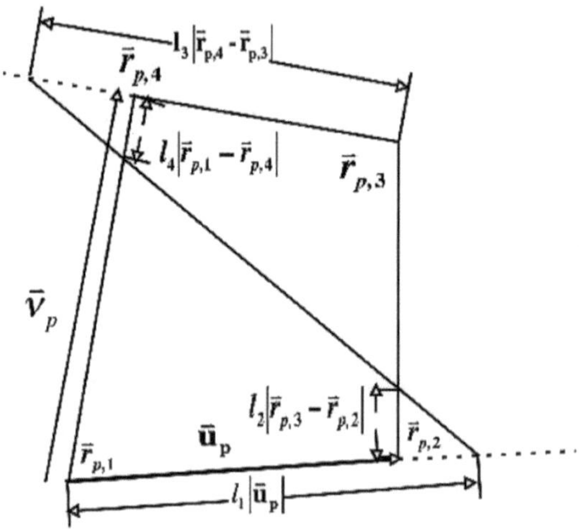

Fig. 2.5. Diagram of the splitting algorithm applied to an initial beam cross-section.

A recursive subroutine is developed to implement the beam-tracing process. Included in the recursive subroutine are an algorithm for splitting an input beam and a follow-up loop that calls the recursive subroutine itself with each sub-beam as the input. One condition to terminate the beam-tracing process is that the area of a beam cross-section must be smaller than a prescribed value. We note that all the sub-beams must be slightly scaled with a scaling factor of 0.9999 to allow the computer program to be stable especially for the case of l_j very close to 0 or 1. The efficiency of the algorithm and the required computer memory depend on the number of facets and the particle orientation but are not very sensitive to the particle size. The programming feature based on recursive subroutines is an additional complexity in the beam-tracing process because only a single beam is traced at each step. The recursive subroutine is unnecessary in the traditional ray-tracing algorithm for simple geometries (e.g., cubes or hexagonal columns), because for a single incident ray, only one internal ray emerges in conjunction with each subsequent reflection and refraction event. For complex particle geometries (e.g., hollow columns or aggregates), the recursive subroutine is required because the outgoing rays may be blocked by the particle itself. To avoid the recursive structure of the code, a Monte Carlo ray-tracing algorithm (Takano and Liou, 1995; Yang and Liou, 1998) may be used, but the accuracy of the final optical properties may be decreased.

In practical calculations, if a series of sizes of a well-defined faceted particle of fixed overall shape and aspect ratio are involved in the computation, the computational efficiency may be increased because the geometries of all the ray-tubes are the same for different sizes. In this case, the beam-tracing process can be performed only once for a chosen size, and the beams for different sizes can be obtained by scaling the geometries of the beams already obtained. If we consider a randomly oriented particle, the algorithm is readily parallelized such that each individual processor deals with a single orientation and the wall-clock time is reduced.

2.3.3 Field-tracing

The field-tracing requires the determination of the amplitude, phase, and polarization of the electromagnetic field in the beam-tracing process. The final internal field is the superposition of the fields associated with all the ray-tubes, and the surface field is the superposition of the fields associated with the beams that exit the particle. In a more specific form, we intend to establish the relationship between the electromagnetic field of each beam with that of the incident beam. Such a relationship contains information about the process how the particle scatters and absorbs the incident light, and, thus accounts for the optical properties of the particle. In the conventional ray-tracing technique, the electric fields are traced without calculating the magnetic field. Because of the convenience of dealing with the inhomogeneous waves for absorptive particles, we consider both the electric and magnetic fields while tracing the fields. The magnetic near-field must be considered if the Kirchhoff surface integral equation is applied to obtain the far-field in the radiation zone. In the description of the field-tracing process, we focus only on the first-order reflection and refraction (from air to particle) and the second-order reflection and refraction (from particle to medium). The calculation of the higher-order interactions is similar to that of the second-order interaction and will not be addressed in detail, but we will explain the additional complexity involved in the process of beam splitting.

Figure 2.6(a) defines the propagation direction ($\hat{e}^{\text{inc}}$) of an incident wave in the laboratory coordinate system and two orthogonal directions ($\hat{\theta}^{\text{inc}}$ and $\hat{\varphi}^{\text{inc}}$) used to specify the polarization state of incident field $\vec{E}^{\text{inc}}(\vec{r})$. The electric field of a linearly polarized incident plane wave can be written in the form

$$\vec{E}^{\text{inc}}(\vec{r}) = \begin{bmatrix} E_{\varphi}^{\text{inc}} \\ E_{\theta}^{\text{inc}} \end{bmatrix} \exp\!\left(ik\hat{e}^{\text{inc}} \cdot \vec{r}\right), \tag{2.19}$$

where E_{θ}^{inc} and $E_{\varphi}^{\text{inc}}$are the amplitudes of the electric field components decomposed with respect to the $\hat{\theta}^{\text{inc}}$ and $\hat{\varphi}^{\text{inc}}$ directions and $\vec{r}$ is the position vector. The

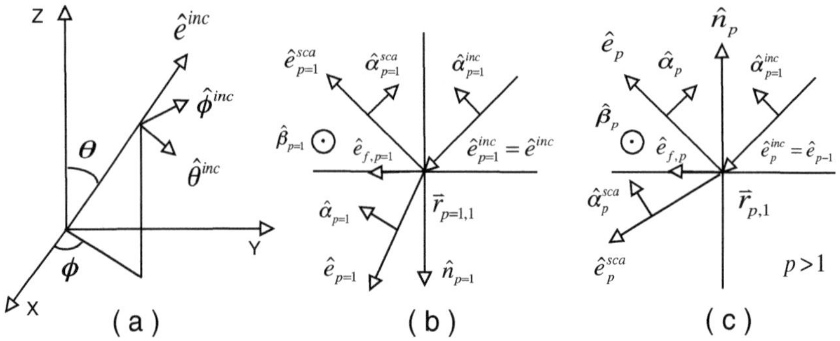

Fig. 2.6. (a) The direction of the incident ray in the laboratory coordinate system. (b) The vectors defined to describe the first-order interaction. (c) The vectors defined to describe higher-order interactions.

corresponding magnetic field is obtained through

$$\vec{H}^{\rm inc}(\vec{r}) = \frac{1}{ik}\vec{\nabla} \times \vec{E}^{\rm inc}(\vec{r}) = \begin{bmatrix} E_\theta^{\rm inc} \\ -E_\varphi^{\rm inc} \end{bmatrix} \exp\left(ik\hat{e}^{\rm inc} \cdot \vec{r}\right) . \tag{2.20}$$

Figure 2.6(b) defines a set of unit vectors to specify the propagation direction and the polarization configuration of the incident wave, the reflected wave, and the refracted wave at the first-order interaction (p=1, from the medium to the particle). The plane of incidence is the plane spanned by the incident direction $\hat{e}^{\rm inc}$ and the local inward normal direction $\hat{n}_{p=1}$. In the particular case where $\hat{n}_{p=1} \times \hat{e}^{\rm inc} = 0$, the plane of incidence is defined by $\hat{\varphi}^{\rm inc}$ and $\hat{n}_{p=1}$. The unit vectors $\hat{e}^{\rm sca}_{p=1}$ and $\hat{e}_{p=1}$ represent the propagation directions of the reflected (i.e., scattered wave exits the particle) and refracted waves (internal waves). The unit vector perpendicular to the incident plane can be determined by

$$\hat{\beta}_{p=1} = \begin{cases} -(\hat{n}_{p=1} \times \hat{e}^{\rm inc})/\left|\hat{n}_{p=1} \times \hat{e}^{\rm inc}\right| , & \hat{n}_{p=1} \times \hat{e}^{\rm inc} \neq 0 \\ \hat{\theta}^{\rm inc} , & \hat{n}_{p=1} \times \hat{e}^{\rm inc} = 0 \end{cases} , \tag{2.21}$$

and the unit vectors $\hat{\alpha}^{\rm inc}_{p=1}$, $\hat{\alpha}^{\rm sca}_{p=1}$, and $\hat{\alpha}_{p=1}$ parallel to the incident plane are

$$\hat{\alpha}^{\rm inc}_{p=1} = \hat{e}^{\rm inc} \times \hat{\beta}_{p=1} . \tag{2.22}$$

$$\hat{\alpha}^{\rm sca}_{p=1} = \hat{e}^{\rm sca}_{p=1} \times \hat{\beta}_{p=1} , \tag{2.23}$$

$$\hat{\alpha}_{p=1} = \hat{e}_{p=1} \times \hat{\beta}_{p=1} . \tag{2.24}$$

$\hat{e}_{f,1}$ is a unit vector on the interface within the incident plane, defined by

$$\hat{e}_{f,1} = \hat{n}_{p=1} \times \vec{\beta}_1 . \tag{2.25}$$

Similar to Fig. 2.6(b), Fig. 2.6(c) shows relevant defined vectors at successive higher-order interactions ($p > 1$, from the particle to the medium).

Because the medium is non-absorbing (the transverse wave condition is implied), the electric field associated with beams that exit the particle can be expressed by

$$\vec{E}_p^{\rm sca}(\vec{r}_{p,1}) = \hat{\alpha}_p^{\rm sca} E_{p,\alpha}^{\rm sca}(\vec{r}_{p,1}) + \hat{\beta}_p^{\rm sca} E_{p,\beta}^{\rm sca}(\vec{r}_{p,1})\,. \tag{2.26}$$

Due to the occurrence of inhomogeneous waves within the particle for the complex refractive index, the electric and magnetic field have nonzero components along the propagation direction. Therefore, the polarized electric field associated with beams inside of the particle has three components and can be written as

$$\vec{E}_p(\vec{r}_{p,1}) = \hat{\alpha}_p E_{p,\alpha}(\vec{r}_{p,1}) + \hat{\beta}_p E_{p,\beta}(\vec{r}_{p,1}) + \hat{e}_p E_{p,\gamma}(\vec{r}_{p,1}). \tag{2.27}$$

The task of field tracing is to establish the relationship between the incident field given by Eq. (2.19) and the field associated with beams at each order of reflection and refraction event given by Eqs. (2.26) and (2.27). Symbolically, we have

$$\begin{bmatrix} E_{p,\alpha}^{\rm sca}(\vec{r}_{p,1}) \\ E_{p,\beta}^{\rm sca}(\vec{r}_{p,1}) \end{bmatrix} = \mathbf{U}_p^{\rm sca} \begin{bmatrix} E_\varphi^{\rm inc} \\ E_\theta^{\rm inc} \end{bmatrix}, \tag{2.28}$$

$$\begin{bmatrix} E_{p,\alpha}(\vec{r}_{p,1}) \\ E_{p,\beta}(\vec{r}_{p,1}) \\ E_{p,\gamma}(\vec{r}_{p,1}) \end{bmatrix} = \mathbf{U}_p \begin{bmatrix} E_\varphi^{\rm inc} \\ E_\theta^{\rm inc} \end{bmatrix}, \tag{2.29}$$

where $\mathbf{U}_p^{\rm sca}$ and $\mathbf{U}_p$ are 2×2 and 3×2m atrices, respectively.

Referring to the plane of incidence, the two components of the electric field at the position of $\vec{r}_{p,1}$ can be obtained

$$\begin{bmatrix} \mathrm{E}_{p=1,\alpha}^{\rm inc}(\vec{r}_{p=1,1}) \\ \mathrm{E}_{p=1,\beta}^{\rm inc}(\vec{r}_{p=1,1}) \end{bmatrix} = \Lambda \begin{bmatrix} \mathrm{E}_\varphi^{\rm inc} \\ \mathrm{E}_\theta^{\rm inc} \end{bmatrix} \exp\left(ik\hat{e}^{\rm inc} \cdot \vec{r}_{p=1,1}\right), \tag{2.30}$$

where Λ is a rotation matrix, given by

$$\Lambda = \begin{bmatrix} \hat{\alpha}_{p=1}^{\rm inc} \cdot \hat{\varphi}^{\rm inc} & \hat{\alpha}_{p=1}^{\rm inc} \cdot \hat{\theta}^{\rm inc} \\ -\hat{\alpha}_{p=1}^{\rm inc} \cdot \hat{\theta}^{\rm inc} & \hat{\alpha}_{p=1}^{\rm inc} \cdot \hat{\varphi}^{\rm inc} \end{bmatrix}. \tag{2.31}$$

The refracted field is assumed to be the superposition of fields corresponding to the transverse electric (TE) and transverse magnetic (TM) modes. The TE mode is defined in the case of

$$E_{p=1,\alpha}^{\rm inc} = H_{p=1,\beta}^{\rm inc} = 0, E_{p=1,\beta}^{\rm inc} = H_{p=1,\alpha}^{\rm inc} \neq 0\,, \tag{2.32}$$

whereas, the TM mode corresponds to

$$E_{p=1,\beta}^{\rm inc} = H_{p=1,\alpha}^{\rm inc} = 0, E_{p=1,\alpha}^{\rm inc} = -H_{p=1,\beta}^{\rm inc} \neq 0\,. \tag{2.33}$$

Therefore, based on the electromagnetic boundary conditions, we obtain the wave reflected to the medium (i.e., scattered light) and the wave transmitted to the particle in the form of

$$\begin{bmatrix} H^{\text{sca}}_{p=1,\beta}\ (\vec{r}_{p=1,1}) \\ E^{\text{sca}}_{p=1,\beta}\ (\vec{r}_{p=1,1}) \end{bmatrix} = \begin{bmatrix} R_{p=1,M} & 0 \\ 0 & R_{p=1,E} \end{bmatrix} \begin{bmatrix} H^{\text{inc}}_{p=1,\beta}\ (\vec{r}_{p=1,1}) \\ E^{\text{inc}}_{p=1,\beta}\ (\vec{r}_{p=1,1}) \end{bmatrix}, \tag{2.34}$$

$$\begin{bmatrix} H_{p=1,\beta}\ (\vec{r}_{p=1,1}) \\ E_{p=1,\beta}\ (\vec{r}_{p=1,1}) \end{bmatrix} = \begin{bmatrix} T_{p=1,M} & 0 \\ 0 & T_{p=1,E} \end{bmatrix} \begin{bmatrix} H^{\text{inc}}_{p=1,\beta}\ (\vec{r}_{p=1,1}) \\ E^{\text{inc}}_{p=1,\beta}\ (\vec{r}_{p=1,1}) \end{bmatrix}. \tag{2.35}$$

The reflection coefficients, $R_{p=1,E}$ and $R_{p=1,M}$, and transmission coefficients, $T_{p=1,E}$ and $T_{p=1,M}$, are given by (Yang and Liou, 2009b)

$$R_{p=1,E} = \frac{\cos\theta_{i,1} - (N_{r,1}\cos\theta_{t,1} + iN_{n,1})}{\cos\theta_{i,1} + N_{r,1}\cos\theta_{t,1} + iN_{n,1}}, \tag{2.36}$$

$$R_{p=1,M} = \frac{m^2\cos\theta_{i,1} - (N_{r,1}\cos\theta_{t,1} + iN_{n,1})}{m^2\cos\theta_{i,1} + N_{r,1}\cos\theta_{t,1} + iN_{n,1}}, \tag{2.37}$$

$$T_{p=1,E} = \frac{2\cos\theta_{i,1}}{\cos\theta_{i,1} + N_{r,1}\cos\theta_{t,1} + iN_{n,1}}, \tag{2.38}$$

$$T_{p=1,M} = \frac{2m^2\cos\theta_{i,1}}{m^2\cos\theta_{i,1} + N_{r,1}\cos\theta_{t,1} + iN_{n,1}}, \tag{2.39}$$

where $N_{n,1} = N_{i,1}/\cos\theta_{t,1}$. A combination of the TE and TM modes yields,

$$\vec{E}^{\text{sca}}_{p=1}(\vec{r}_{p=1,1}) = E^{\text{sca}}_{p=1,\beta}(\vec{r}_{p=1,1})\hat{\beta}_{p=1} - H^{\text{sca}}_{p=1,\beta}(\vec{r}_{p=1,1})\hat{\alpha}^{\text{sca}}_{p=1}, \tag{2.40}$$

$$\vec{H}^{\text{sca}}_{p=1}(\vec{r}_{p=1,1}) = H^{\text{sca}}_{p=1,\beta}(\vec{r}_{p=1,1})\hat{\beta}_{p=1} + E^{\text{sca}}_{p=1,\beta},(\vec{r}_{p=1,1})\hat{\alpha}^{\text{sca}}_{p=1}, \tag{2.41}$$

$$\vec{E}_{p=1}(\vec{r}_{p=1,1}) = E_{p=1,\beta}(\vec{r}_{p=1,1})\hat{\beta}_{p=1} - \frac{1}{ikm^2}\vec{\nabla}\times\left[H_{p=1,\beta}(\vec{r}_{p=1,1})\hat{\beta}_{p=1}\right], \tag{2.42}$$

$$\vec{H}_{p=1}(\vec{r}_{p=1,1}) = H_{p=1,\beta}(\vec{r}_{p=1,1})\hat{\beta}_{p=1} + \frac{1}{ik}\vec{\nabla}\times\left[E_{p=1,\beta}(\vec{r}_{p=1,1})\hat{\beta}_{p=1}\right]. \tag{2.43}$$

Using Eqs. (2.28)–(2.42), we obtain

$$U^{\text{sca}}_{p=1} = \begin{bmatrix} R_{p=1,M} & 0 \\ 0 & R_{p=1,E} \end{bmatrix} \Lambda \exp\left(ik\hat{e}^{\text{inc}}\cdot\vec{r}_{p=1,1}\right), \tag{2.44}$$

$$U_{p=1} = \begin{bmatrix} T_\alpha & 0 \\ 0 & T_\beta \\ T_\gamma & 0 \end{bmatrix} \Lambda \exp\left(ik\hat{e}^{\text{inc}}\cdot\vec{r}_{p=1,1}\right), \tag{2.45}$$

where

$$T_\alpha = \frac{2(N_{r,1} + iN_{n,1}\cos\theta_{t,1})}{m^2\cos\theta_{i,1} + [N_{r,1}\cos\theta_{t,1} + iN_{n,1}]}, \tag{2.46}$$

$$T_\beta = \frac{2\cos\theta_{i,1}}{\cos\theta_{i,1} + [N_{r,1}\cos\theta_{t,1} + iN_{n,1}]}, \tag{2.47}$$

$$T_\gamma = \frac{i2N_{n,1}\cos\theta_{i,1}\sin\theta_{t,1}}{m^2\cos\theta_{i,1} + [N_{r,1}\cos\theta_{t,1} + iN_{n,1}]}. \tag{2.48}$$

From Eqs. (2.42) and (2.43), it is evident that the TE mode generates a nonzero magnetic field along the propagation direction; whereas, the TM mode generates a nonzero electric field along the propagation direction. Once the electromagnetic field at one vertex position of the beam cross-section is known, the electric field at an arbitrary position can be obtained by considering the variances of the phase and amplitude of the electromagnetic field. Therefore, we always calculate the electromagnetic field at the position of the first vertex of the beam cross-section. The electric field in an arbitrary position within the beam cross-section (represented by $\vec{r} = \vec{r}_{p=1,1} + \vec{w}_1$) is written as

$$\begin{bmatrix} E^{\text{sca}}_{p=1,\alpha}\left(\vec{r}_{p=1,1} + \vec{w}_1\right) \\ E^{\text{sca}}_{p=1,\beta}\left(\vec{r}_{p=1,1} + \vec{w}_1\right) \end{bmatrix} = \begin{bmatrix} E^{\text{sca}}_{p=1,\alpha}\left(\vec{r}_{p=1,1}\right) \\ E^{\text{sca}}_{p=1,\beta}\left(\vec{r}_{p=1,1}\right) \end{bmatrix} \exp[ikN_{r,1}\hat{e}_{p=1}\cdot\vec{w}_1]\exp\left[-kN_{i,1}\vec{A}_{p=1}\cdot\vec{w}_1\right], \quad (2.49)$$

$$\begin{bmatrix} E_{p=1,\alpha}(\vec{r}_{p=1,1} + \vec{w}_1) \\ E_{p=1,\beta}(\vec{r}_{p=1,1} + \vec{w}_1) \\ E_{p=1,\gamma}(\vec{r}_{p=1,1} + \vec{w}_1) \end{bmatrix} = \begin{bmatrix} E_{p=1,\alpha}(\vec{r}_{p=1,1}) \\ E_{p=1,\beta}(\vec{r}_{p=1,1}) \\ E_{p=1,\gamma}(\vec{r}_{p=1,1}) \end{bmatrix} \exp[ikN_{r,1}\hat{e}_{p=1}\cdot\vec{w}_1]\exp\left[-kN_{i,1}\vec{A}_{p=1}\cdot\vec{w}_1\right], \quad (2.50)$$

where $\vec{A}$ is a vector, defined to count for the amplitude variation, which is zero for the first-order refracted beam. If beam splitting occurs, the field associated with the first vertex of each sub-beam can be obtained accordingly. In this case, $\vec{w}_1$ is the difference between the first vertex of the sub-beam and the original beam.

To consider the second-order reflection and refraction when one of the first-order refracted beams is incident on a single facet, the field of the first-order refracted beam can be represented with respect to the plane of incidence containing the incident direction $\hat{e}_{p=1}$ and the inward normal direction of $\hat{n}_{p=2}$ (see Fig. 2.6(c)). A combination of the TE and TM modes gives the incident field of

$$\begin{bmatrix} H^{\text{inc}}_{p=2,\beta}(\vec{r}_{p=2,1}) \\ E^{\text{inc}}_{p=2,\beta}(\vec{r}_{p=2,1}) \end{bmatrix} = \begin{bmatrix} \hat{\beta}_2\cdot\hat{\beta}_1 & N_{r,1}(\hat{\alpha}_{t,1}\cdot\hat{\beta}_2)+iN_{n,1}(\hat{e}_{f,1}\cdot\hat{\beta}_2) \\ -\dfrac{N_{r,1}(\hat{\alpha}_{t,1}\cdot\hat{\beta}_2)+iN_{n,1}(\hat{e}_{f,1}\cdot\hat{\beta}_2)}{m^2} & \hat{\beta}_2\cdot\hat{\beta}_1 \end{bmatrix} \times \begin{bmatrix} H_{p=1,\beta}\left(\vec{r}_{p=1,1}\right) \\ E_{p=1,,\beta}\left(\vec{r}_{p=1,1}\right) \end{bmatrix} \exp(ik\delta_{2,1})\exp(-k\rho_{2,1}), \quad (2.51)$$

where $k\delta_{2,1}$ is the phase associated with the first vertex of the beam on the interface of the second-order interaction and $\delta_{2,1} = \hat{e}^{\text{inc}}\cdot\vec{r}_{1,1} + |\vec{r}_{2,1} - \vec{r}_{1,1}|$, and the second exponential determines the amplitude decrease where $\rho_{2,1} = N_{i,1}\left|\vec{r}_{2,1} - \vec{r}_{1,1}\right|$. The second-order reflected and refracted fields are given by

$$\begin{bmatrix} H_{p=2,\beta}\left(\vec{r}_{p=2,1}\right) \\ E_{p=2,,\beta}\left(\vec{r}_{p=2,1}\right) \end{bmatrix} = \begin{bmatrix} R_{p=2,M} & 0 \\ 0 & R_{p=2,E} \end{bmatrix} \begin{bmatrix} H^{\text{inc}}_{p=2,\beta}\left(\vec{r}_{p=2,1}\right) \\ E^{\text{inc}}_{p=2,\beta}\left(\vec{r}_{p=2,1}\right) \end{bmatrix}, \tag{2.52}$$

$$\begin{bmatrix} H^{\text{sca}}_{p=2,\beta}\left(\vec{r}_{p=2,1}\right) \\ E^{\text{sca}}_{p=2,\beta}\left(\vec{r}_{p=2,1}\right) \end{bmatrix} = \begin{bmatrix} T_{p=2,M} & 0 \\ 0 & T_{p=2,E} \end{bmatrix} \begin{bmatrix} H^{\text{inc}}_{p=2,,\beta}\left(\vec{r}_{p=2,1}\right) \\ E^{\text{inc}}_{p=2,\beta}\left(\vec{r}_{p=2,1}\right) \end{bmatrix}, \tag{2.53}$$

where

$$R_{p=2,M} = \frac{N_{r,1}\cos\theta_{i,2} - iN_{n,1}\hat{n}_1\cdot\hat{n}_2 - m^2\cos\theta_{t,2}}{N_{r,2}\cos\theta_{r,2} + iN_{n,2} + m^2\cos\theta_{t,2}}, \tag{2.54}$$

$$R_{p=2,E} = \frac{N_{r,1}\cos\theta_{i,2} - iN_{n,1}\hat{n}_1\cdot\hat{n}_2 - \cos\theta_{t,2}}{N_{r,2}\cos\theta_{r,2} + iN_{n,2} + \cos\theta_{t,2}}, \tag{2.55}$$

$$T_{p=2,M} = \frac{N_{r,2}\cos\theta_{r,2} + iN_{n,2} + N_{r,1}\cos\theta_{i,2} - iN_{n,1}\hat{n}_1\cdot\hat{n}_2}{N_{r,2}\cos\theta_{r,2} + iN_{n,2} + m^2\cos\theta_{t,2}}, \tag{2.56}$$

$$T_{p=2,E} = \frac{N_{r,2}\cos\theta_{r,2} + iN_{n,2} + N_{r,1}\cos\theta_{i,2} - iN_{n,1}\hat{n}_1\cdot\hat{n}_2}{N_{r,2}\cos\theta_{r,2} + iN_{n,2} + \cos\theta_{t,2}}, \tag{2.57}$$

where $N_{n,2} = N_{i,2}/\cos\theta_{r,2}$. The electric fields after a combination of TM and TE modes associated with the reflected and refracted waves are given by

$$\vec{E}_{p=2}\left(\vec{r}_{p=2,1}\right) = E_{p=2,\beta}\left(\vec{r}_{p=2,1}\right)\hat{\beta}_2 - \frac{1}{ikm^2}\vec{\nabla}\times\left[H_{p=2,\beta}\hat{\beta}_2\right]\left(\vec{r}_{p=2,1}\right), \tag{2.58}$$

$$\vec{E}^{\text{sca}}_{p=2}(\vec{r}_{p=2,1}) = E^{\text{sca}}_{p=2,\beta}(\vec{r}_{p=2,1})\hat{\beta}_2 - H^{\text{sca}}_{p=2,\beta}(\vec{r}_{p=2,1})\hat{\alpha}^{\text{sca}}_2\,. \tag{2.59}$$

Based on Eq. (2.58) and (2.59), we obtain

$$\begin{aligned} U_{p=2} = {} & \begin{bmatrix} -\dfrac{N_{r,2} + iN_{n,1}\cos\theta_{r,2}}{m^2} & 0 \\ 0 & 1 \\ \dfrac{iN_{n,2}\sin\theta_{r,2}}{m^2} & 0 \end{bmatrix} \begin{bmatrix} R_{M,2} & 0 \\ 0 & R_{E,2} \end{bmatrix} \\ & \times \begin{bmatrix} \hat{\beta}_2\cdot\hat{\beta}_1 & N_{r,1}(\hat{\alpha}_{t,1}\cdot\hat{\beta}_2) + iN_{n,1}(\hat{e}_{f,1}\cdot\hat{\beta}_2) \\ -\dfrac{N_{r,1}(\hat{\alpha}_{t,1}\cdot\hat{\beta}_2) + iN_{n,1}(\hat{e}_{f,1}\cdot\hat{\beta}_2)}{m^2} & \hat{\beta}_2\cdot\hat{\beta}_1 \end{bmatrix} \\ & \times \begin{bmatrix} -T_{p=1,M} & 0 \\ 0 & T_{p=1,E} \end{bmatrix} \Lambda\exp(ik\delta_{2,1})\exp(-k\rho_{2,1}) \end{aligned} \tag{2.60}$$

and

$$U^{\mathrm{sca}}_{p=2} = \begin{bmatrix} -1 & 0 \\ 0 & 1 \end{bmatrix} \begin{bmatrix} T_{p=2,M} & 0 \\ 0 & T_{p=2,E} \end{bmatrix}$$
$$\times \begin{bmatrix} \hat{\beta}_2 \cdot \hat{\beta}_1 & N_{r,1}(\hat{\alpha}_{t,1} \cdot \hat{\beta}_2) + iN_{n,1}(\hat{e}_{f,1} \cdot \hat{\beta}_2) \\ -\dfrac{N_{r,1}(\hat{\alpha}_{t,1} \cdot \hat{\beta}_2) + iN_{n,1}(\hat{e}_{f,1} \cdot \hat{\beta}_2)}{m^2} & \hat{\beta}_2 \cdot \hat{\beta}_1 \end{bmatrix}$$
$$\times \begin{bmatrix} -T_{p=1,M} & 0 \\ 0 & T_{p=1,E} \end{bmatrix} \Lambda \exp(ik\delta_{2,1}) \exp(-k\rho_{2,1})$$

At an arbitrary position (represented by $\vec{r} = \vec{r}_{p=2,1} + \vec{w}_2$) within the beam cross-section, the reflected and refracted electric fields are

$$\begin{bmatrix} E_{p=2,\alpha}(\vec{r}_{p=2,1} + \vec{w}_2) \\ E_{p=2,\beta}(\vec{r}_{p=2,1} + \vec{w}_2) \end{bmatrix} = \begin{bmatrix} E_{p=2,\alpha}(\vec{r}_{p=2,1}) \\ E_{p=2,\beta}(\vec{r}_{p=2,1}) \end{bmatrix} \exp(ikN_{r,2}\hat{e}_{p=2} \cdot \vec{w}_{p=2}) \exp(-\vec{A}_{p=2} \cdot \vec{w}_{p=2}) , \tag{2.61}$$

$$\begin{bmatrix} E^{\mathrm{sca}}_{p=2,\alpha}(\vec{r}_{p=2,1} + \vec{w}_2) \\ E^{\mathrm{sca}}_{p=2,\beta}(\vec{r}_{p=2,1} + \vec{w}_2) \end{bmatrix} = \begin{bmatrix} E^{\mathrm{sca}}_{p=2,\alpha}(\vec{r}_{p=2,1}) \\ E^{\mathrm{sca}}_{p=2,\beta}(\vec{r}_{p=2,1}) \end{bmatrix} \exp(ikN_{r,2}\hat{e}_{p=2} \cdot \vec{w}_{p=2}) \exp(-\vec{A}_{p=2} \cdot \vec{w}_{p=2}) , \tag{2.62}$$

with $\vec{A}_{p=2}$ nonzero except for special cases. The phase variation governed by Snell's law is independent of the history of ray tracing and can be considered in terms of the real part of the refractive index and the propagating direction. The amplitude variation is dependent on the ray-tracing history and $\vec{A}_p$ can be obtained from $\vec{A}_{p-1}$ as follows

$$\vec{A}_p = \frac{(\vec{A}_p \cdot \hat{v}_p)(\hat{u}_p \cdot \hat{v}_p) - \vec{A}_p \cdot \hat{u}_p}{(\hat{u}_p \cdot \hat{v}_p)^2 - 1} \hat{u}_p + \frac{(\vec{A}_p \cdot \hat{u}_p)(\hat{u}_p \cdot \hat{v}_p) - \vec{A}_p \cdot \hat{v}_p}{(\hat{u}_p \cdot \hat{v}_p)^2 - 1} \hat{v}_p , \tag{2.63}$$

$$N_{i,p-1}(\vec{A}_p \cdot \hat{u}_p)\hat{u}_p = N_{i,p-2}(\vec{A}_{p-1} \cdot \hat{u}_{p-1})\hat{u}_{p-1} + \left[N_{i,p-1} - N_{i,p-2}(\hat{e}_{p-1} \cdot \hat{u}_{p-1})(\vec{A}_{p-1} \cdot \hat{u}_{p-1})\right] \frac{\hat{n}^u_{p-1}}{\hat{e}_{p-1} \cdot \hat{n}^u_{p-1}} , \tag{2.64}$$

$$N_{i,p-1}(\vec{A}_p \cdot \hat{v}_p)\hat{v}_p = N_{i,p-2}(\vec{A}_{p-1} \cdot \hat{v}_{p-1})\hat{v}_{p-1} + \left[N_{i,p-1} - N_{i,p-2}(\hat{e}_{p-1} \cdot \hat{v}_{p-1})(\vec{A}_{p-1} \cdot \hat{v}_{p-1})\right] \frac{\hat{n}^v_{p-1}}{\hat{e}_{p-1} \cdot \hat{n}^v_{p-1}} , \tag{2.65}$$

$$\vec{n}^{u}_{p-1} = \hat{e}_{p-1} - (\hat{e}_{p-1} \cdot \hat{u}_{p-1})\hat{u}_{p-1}\,, \tag{2.66}$$

$$\vec{n}^{v}_{p-1} = \hat{e}_{p-1} - (\hat{e}_{p-1} \cdot \hat{v}_{p-1})\hat{v}_{p-1}\,. \tag{2.67}$$

A similar procedure applies to higher-order internal reflection. Note that Eq. (2.12) in Bi et al. (2011b) should be written in the form of Eqs. (2.63)–(2.67).

2.4 Physical optics and scattered far-field

After the near-field is determined via the beam–tracing calculation, the scattered field in the radiation zone (far-field region) can be found through electromagnetic relationships. This approach allows the geometric optics of obtaining the far-field to be more physical than in the CGOM, and the effect incorporated into the final results is referred to as the physical-optics correction. In this section, we discuss several alterative methods, not mathematically equivalent, for establishing the geometric-optics-based near-field and the physical-optics-based far-field.

2.4.1 Fredholm volume integral equation

Yang and Liou (1997) appear to have been the first to use the Fredholm volume-integral equation to obtain the far-field from the geometric-optics-based near-field. In Yang and Liou (1996b), the ray tracing is based on rays with small circular cross-sections with radii on the order of $\lambda/2\pi$. The variations of the phase and the amplitude within the ray cross-sections are negligible as the ray cross-sections are small, and a ray tube within the particle can be assumed to be a circular cylinder, although the initial and end cross-sections may not be perpendicular to the propagation direction (see Fig. 2.1 in Yang and Liou, 1997). With the simplifications, the far-field corresponding to the near-field in each ray tube is obtained in a straightforward manner, and the method is referred to as ray-by-ray integration (RBRI). The disadvantage of the RBRI algorithm is that the number of rays increases with an increase in the size parameter which, consequently, leads to a significant demand on computational resources. However, if only the first-order refracted beam, whose contribution to the scattering and the absorption dominates in the case of strongly absorptive particles is considered, the resultant amplitude scattering matrix may be expressed in terms of a surface integral on the illuminated particle facets and the corresponding numerical computation can be quite efficient (Yang et al., 2001; Bi et al., 2011a). The algorithm based on the volume integration is further developed by analytically integrating the near-field in exactly defined ray tubes instead of small circular cylinders (Bi et al., 2011b). In this case, the number of ray tubes depends only on the particle geometry and orientation. The simplified algorithm successfully improves both the computational efficiency and numerical accuracy.

The volume-integral equation that relates the total electric field within the particle to the induced scattered field in the radiation zone (i.e., the far-field region) is given by (Saxon, 1973; Yang and Liou, 1997),

$$\vec{E}^s(\vec{r})|_{kr\to\infty} = \frac{k^2\exp(ikr)}{4\pi r}\iiint\limits_v (m^2-1)\left\{\vec{E}(\vec{r}\,') - \hat{r}[\hat{r}\cdot\vec{E}(\vec{r}\,')]\right\}\exp(-ik\hat{r}\cdot\vec{r}\,')\,d^3\vec{r}\,', \tag{2.68}$$

where v is the particle volume (the domain of non-unity refractive index), $\hat{r}$ is the direction of the scattered light to the observation position , and m is the complex refractive index with non-negative imaginary part. The scattered electric field $\vec{E}^s(\vec{r})$ in the radiation region is a spherical wave, which is evident from the factor before the volume integral in Eq. (2.68). Furthermore, $\vec{E}^s(\vec{r})$ is locally transverse with respect to the scattering direction $\hat{r}$ and can be decomposed into two components parallel and perpendicular to scattering planes in the form

$$\vec{E}^s(\vec{r}) = E^s_\alpha(\vec{r})\hat{\alpha}^s + E^s_\beta(\vec{r})\hat{\beta}^s , \tag{2.69}$$

where $\hat{\alpha}^s$ and $\hat{\beta}^s$ are two unit vectors parallel and perpendicular to the scattering plane, as shown in Fig. 2.7. Applying the dot products $\hat{\alpha}^s\cdot$ and $\hat{\beta}^s\cdot$ on both sides of Eq. (2.68) yields the following vector equation:

$$\begin{bmatrix} E^s_\alpha \\ E^s_\beta \end{bmatrix}_{kr\to\infty} = \frac{k^2\exp(ikr)}{4\pi r}\iiint\limits_v (m^2-1)\begin{bmatrix} \hat{\alpha}^s\cdot\vec{E}(\vec{r}\,') \\ \hat{\beta}^s\cdot\vec{E}(\vec{r}\,') \end{bmatrix} \exp(-ik\hat{r}\cdot\vec{r}\,')\,d^3\vec{r}\,' . \tag{2.70}$$

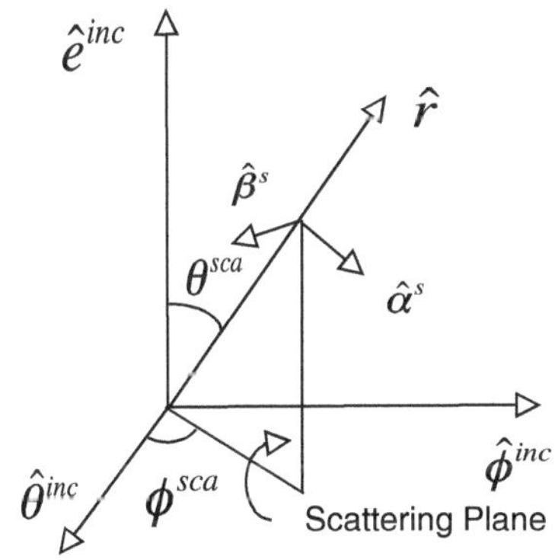

Fig. 2.7. Definition of scattering plane, scattering angle, and scattering azimuthal angles.

The internal electric field in Eq. (2.70) can be formally written as a double summation with each term arising from beams associated with different orders of reflection/refraction events,

$$\vec{E}(\vec{r}\,') = \sum_{p=1}^{\infty}\sum_{q} E^q_{p,\alpha}(\vec{r}\,')\hat{\alpha}^q_p + E^q_{p,\beta}(\vec{r}\,')\hat{\beta}^q_p + E^q_{p,\gamma}(\vec{r}\,')\hat{e}^q_p . \tag{2.71}$$

The summation in terms of the index p corresponds to different orders of reflection and refraction events; whereas, the index q refers to a number of beams for the pth-order interaction. The maximum value of q depends on the particle geometry and orientation and is not given explicitly. Hereafter, variables with index q have the same physical meaning as their counterparts without q defined in previous sections. For non-absorptive particles, based on the transverse-wave condition, the third term in Eq. (2.70) is zero. Substituting Eq. (2.71) into Eq. (2.70) yields

$$\begin{bmatrix} E_\alpha^s \\ E_\beta^s \end{bmatrix}_{kr\to\infty} = \frac{k^2 \exp(ikr)}{4\pi r} \sum_{p=1}^{\infty} \sum_q \iiint_{V_p^q} (m^2-1) K_p^q \begin{bmatrix} E_{p,\alpha}^q \\ E_{p,\beta}^q \\ E_{p,\gamma}^q \end{bmatrix} \exp(-ik\hat{r}\cdot\vec{r}\,')\, d^3\vec{r}\,', \tag{2.72}$$

where v_p^q is the volume associated with one pth order internal ray tube and K_p^q is a 2-by-3 matrix given by

$$\mathrm{K}_p^q = \begin{bmatrix} \hat{\alpha}^s\cdot\hat{\alpha}_p^q & \hat{\alpha}^s\cdot\hat{\beta}_p^q & \hat{\alpha}^s\cdot\hat{e}_p^q \\ \hat{\beta}^s\cdot\hat{\alpha}_p^q & \hat{\beta}^s\cdot\hat{\beta}_p^q & \hat{\beta}^s\cdot\hat{e}_p^q \end{bmatrix}. \tag{2.73}$$

The internal field in Eq. (2.72) was calculated in the field-tracing section. To obtain the amplitude scattering matrix, we express the electric field in Eq. (2.72) in terms of the incident electric field in the form

$$\begin{bmatrix} E_{p,\alpha}^q(\vec{r}\,') \\ E_{p,\beta}^q(\vec{r}\,') \\ E_{p,\gamma}^q(\vec{r}\,') \end{bmatrix} = U_p^q \exp\left(ik\delta_{p,1}^q - k\rho_{p,1}\right) \exp\left[ik(N_{r,p-1}^q\hat{e}_{p-1} + iN_{i,p-1}^q\vec{A}_p^q)\cdot\vec{w}_p^q\right] \times \exp(ikN_p^q l) \begin{bmatrix} E_\alpha^{\mathrm{inc}} \\ E_\beta^{\mathrm{inc}} \end{bmatrix}. \tag{2.74}$$

Here U_p^q is 3-by-2 matrix and obtained in the field-tracing process. Eq. (2.72) is transformed to

$$\begin{bmatrix} E_\alpha^s \\ E_\beta^s \end{bmatrix}_{kr\to\infty} = \frac{k^2 \exp(ikr)}{4\pi r} \sum_{p=1}^{\infty} \sum_q (m^2-1)\, K_p^q U_p^q \mathit{\Gamma} \exp\left(ik\delta_{p,1}^q - k\rho_{p,1}^q\right) \times \iiint_{V_p^q} \exp\left[ik(N_{r,p-1}^q\hat{e}_{p-1}^q + iN_{i,p-1}^q\vec{A}_p^q)\cdot\vec{w}_p^q\right] \exp(ikN_p^q l) \exp(-ik\hat{r}\cdot\vec{r}\,') d^3\vec{r}\,' \begin{bmatrix} E_\phi^{\mathrm{inc}} \\ E_\theta^{\mathrm{inc}} \end{bmatrix}. \tag{2.75}$$

The amplitude scattering matrix is readily given by

$$\begin{bmatrix} S_2 & S_3 \\ S_4 & S_1 \end{bmatrix} = \frac{-ik^3}{4\pi} \sum_{p=1}^{\infty} \sum_q (m^2-1)\, K_p^q U_p^q \mathit{\Gamma} \exp\left(ik\delta_{p,1}^q - k\rho_{p,1}^q\right) I_p^q\,, \tag{2.76}$$

where

$$I_p^q = \iiint_{V_p^q} \exp\left[ik(N_{r,p-1}^q\hat{e}_{p-1}^q + iN_{i,p-1}^q\vec{A}_p^q)\cdot\vec{w}_p^q\right] \exp\left[ik\left(N_{r,p}^q + iN_{i,p}^q\right) l\right] \times \exp(-ik\hat{r}\cdot\vec{r}\,') d^3\vec{r}\,', \tag{2.77}$$

and

$$\mathit{\Gamma} = \begin{bmatrix} \hat{\theta}^{\mathrm{inc}}\cdot\hat{\beta}^s & \hat{\varphi}^{\mathrm{inc}}\cdot\hat{\beta}^s \\ -\hat{\varphi}^{\mathrm{inc}}\cdot\hat{\beta}^s & \hat{\theta}^{\mathrm{inc}}\cdot\hat{\beta}^s \end{bmatrix}. \tag{2.78}$$

Γ is a rotational matrix that transforms the components of the incident electric field with respect to $\hat{\theta}^{\text{inc}}$ and $\hat{\varphi}^{\text{inc}}$ into those referring to the scattering plane (see Fig. 2.7). To efficiently calculate the amplitude scattering matrix, the integral I_p^q must be analytically evaluated and Eq. (2.77) transformed into the form

$$I_p^q = \iint\limits_{S_p^q} \exp\Big[ik(N_{r,p-1}^q \hat{e}_{p-1}^q + iN_{i,p-1}^q \vec{A}_p^q)\cdot\vec{w}_p^q\Big] \exp\big[-ik\hat{r}\cdot(\vec{r}_{p,1}^{\,q} + \vec{w}_p^q)\big] \left|\hat{e}_p^q \cdot \hat{n}_p^q\right| d^2\vec{w}_p^q$$
$$\times \int_0^{l_m} \exp\big[ik(N_{r,p}^q + iN_{i,p}^q - \hat{r}\cdot\hat{e}_p^q)l\big]\, dl\,, \tag{2.79}$$

where

$$l_m = \left|\vec{r}_{p+1,1}^{\,q} - \vec{r}_{p,1}^{\,q}\right| + \frac{\vec{w}_{p+1}^q \cdot \hat{n}'}{\hat{e}_p^q \cdot \hat{n}'}. \tag{2.80}$$

In Eq. (80), $\hat{n}'$ is a vector in the plane composed of $\vec{w}_p^q$ and $\vec{w}_{p+1}^q$ and perpendicular to $\vec{w}_p^q$. Solving the integration in terms of l and employing the following identities

$$\vec{w}_{p+1}^q = \vec{w}_p^q + \frac{\vec{w}_{p+1}^q \cdot \hat{n}'}{\hat{e}_p^q \cdot \hat{n}'}\hat{e}_p^q\,, \tag{2.81}$$

$$\vec{A}_{p+1}^q \cdot \vec{w}_{p+1}^q = \frac{N_{i,p-1}}{N_{i,p}}\vec{A}_p^q \cdot \vec{w}_p^q + \frac{\vec{w}_{p+1}^q \cdot \hat{n}'}{\hat{e}_p^q \cdot \hat{n}'}\,, \tag{2.82}$$

$$\hat{e}_p^q \cdot \vec{w}_{p+1}^q - \hat{e}_p^q \cdot \vec{w}_p^q + \frac{\vec{w}_{p+1}^q \cdot \hat{n}'}{\hat{e}_p^q \cdot \hat{n}'}\,, \tag{2.83}$$

results in an explicit expression for Eq. (2.77)

$$I_p^q = \frac{4\pi}{k^2}\frac{1}{ik(N_p^q - \hat{r}\cdot\hat{e}_p^q)}\Big[\left|\hat{e}_p^q \cdot \hat{n}_{p+1}^q\right| \bar{D}_{p+1}^q \exp\!\big(ikN_p^q\left|\vec{r}^{\,q}_{\,p+1} - \vec{r}^{\,q}_{\,p}\right|\big) - \left|\hat{e}_p^q \cdot \hat{n}_p^q\right| D_p^q\Big]\,, \tag{2.84}$$

where the term D_p^q is given by Bi et al. (2011)

$$D_p^q = \frac{k^2}{4\pi}\exp\left(-ik\hat{r}\cdot\vec{r}_{p,1}^{\,q}\right)\int \exp\left\{ik\left(N_{r,p}^q\hat{e}_p^q - \hat{r} + iN_{i,p}^q\vec{A}_p^q\right)\cdot\vec{w}_p^q\right\} d^2\vec{w}_p^q$$
$$= \frac{ik}{4\pi}\sum_{j=1}^{N}\frac{(\vec{r}_{p,j+1} - \vec{r}_{p,j})\cdot\Big[(N_{r,p}^q\hat{e}_p^q - \hat{r} + iN_{i,p}^q\vec{A}_p^q)\times(-\hat{n}_p^q)\Big]}{(N_{r,p}^q\hat{e}_p^q - \hat{r} + iN_{i,p}^q\vec{A}_p^q)\cdot(N_{r,p}^q\hat{e}_p^q - \hat{r} + iN_{i,p}^q\vec{A}_p^q) - [(N_{r,p}^q\hat{e}_p^q - \hat{r} + iN_{i,p}^q\vec{A}_p^q)\cdot\hat{n}_p^q]^2}$$
$$\times \frac{\sin\Big[k(N_{r,p}^q\hat{e}_p^q - \hat{r} + iN_{i,p}^q\vec{A}_p^q)\cdot\left(\vec{r}_{p,j+1}^{\,q} - \vec{r}_{p,j}^{\,q}\right)\Big]}{k(N_{r,p}^q\hat{e}_p^q - \hat{r} + iN_{i,p}^q\vec{A}_p^q)\cdot\left(\vec{r}_{p,j+1}^{\,q} - \vec{r}_{p,j}^{\,q}\right)/2}\exp\big[-ik\hat{r}\cdot\left(\vec{r}_{p,j+1}^{\,q} + \vec{r}_{p,j}^{\,q}\right)/2\big]$$
$$\times \exp\Big[ik(N_{r,p}^q\hat{e}_p^q + iN_{i,p}^q\vec{A}_p^q)\cdot\left(\vec{r}_{p,j+1}^{\,q} + \vec{r}_{p,j}^{\,q} - 2\vec{r}_{p,1}^{\,q}\right)/2\Big]\,, \tag{2.85}$$

and the term $\bar{D}_{p+1}^q$ is defined by

$$\bar{D}_{p+1}^q = \frac{k^2}{4\pi}\exp(-ik\hat{r}\cdot\vec{r}_{p+1,1}^{\,q})\int \exp\Big\{ik\left(N_{r,p}^q\hat{e}_p^q - \hat{r} + iN_{i,p}^q\vec{A}_{p+1}^q\right)\cdot\vec{w}_{p+1}^q\Big\}\, d^2\vec{w}_{p+1}^q\,, \tag{2.86}$$

which can be calculated similar to Eq. (2.85). In Bi et al. (2011b), Eq. (2.86) is assumed to be Eq. (2.85) by replacing p with $p+1$ because the effective refractive index of higher-order is not rigorously taken into account.

2.4.2 Kirchhoff surface integral equation

The use of the Kirchhoff surface-integral equation to calculate the scattered field in the radiation zone from the geometric-optics near-field can be traced to studies by Ravey and Mazeron (1982), Muinonen (1989), and Yang and Liou (1996b). Instead of the conventional straight-line rays (with small beam cross-section), we employ the beam-tracing technique to the faceted particle case described in Section 2.2 in order to calculate the geometric-optics-based near-field and formulate the optical properties based on the Kirchhoff surface-integral equation. In the radiation region, the Kirchhoff surface-integral in an asymptotic form can be written as

$$\vec{E}^{\text{sca}}(\vec{r})_{kr\to\infty} = \frac{\exp(ikr)}{-ikr}\frac{k^2}{4\pi}\iint \vec{Z}\,\exp(ik\hat{r}\cdot\vec{r}\,')\,d^2\vec{r}\,' \tag{2.87}$$

where

$$\vec{Z} = \hat{r}\times\left[\hat{n}_s\times\vec{E}(\vec{r}\,')\right] - \hat{r}\times\hat{r}\times\left[\hat{n}_s\times\vec{H}(\vec{r}\,')\right], \tag{2.88}$$

where $\hat{n}_s$ is outward normal direction. Based on the transverse-wave condition, Eq. (2.87) can be written in a vector form,

$$\begin{bmatrix} E_\alpha^{\text{sca}} \\ E_\beta^{\text{sca}} \end{bmatrix}_{kr\to\infty} = \frac{\exp(ikr)}{-ikr}\frac{k^2}{4\pi}\iint \begin{bmatrix} \hat{\alpha}^s\cdot\vec{Z} \\ \hat{\beta}^s\cdot\vec{Z} \end{bmatrix} \exp(-ik\hat{r}\cdot\vec{r}\,')d^2\vec{r}\,'?., \tag{2.89}$$

where

$$\begin{bmatrix} \hat{\alpha}^s\cdot\vec{Z} \\ \hat{\beta}^s\cdot\vec{Z} \end{bmatrix} = \begin{bmatrix} \vec{E}\times\hat{\beta}^s + \vec{H}\times\vec{\alpha}^s \\ -\vec{E}\times\hat{\alpha}^s + \vec{H}\times\hat{\beta}^s \end{bmatrix}\cdot\hat{n}_s \tag{2.90}$$

If remaining consistent throughout, $\vec{E}$ and $\vec{H}$ may be either the total field or the scattered field. In the present context, they represent the total electric field and total magnetic field or the superposition of the incident field on the illuminated side of the particle and the fields from various outgoing beams. Symbolically,

$$\vec{E} = \vec{E}^{\text{inc}}(\text{illuminated faces}) + \sum_{p=1}^{\infty}\sum_{q}\vec{E}_p^{q,\text{sca}}(\text{outgoing beams}). \tag{2.91}$$

The far-scattered field corresponding to the first term in Eq. (2.91), essentially the diffraction contribution, is given by

$$\begin{aligned}\begin{bmatrix} E_\alpha^{dif} \\ E_\beta^{dif} \end{bmatrix} &= \frac{\exp(ikr)}{-ikr}\sum_{q=1}(-\hat{n}_{p=1}^q) \\ &\cdot\begin{bmatrix} \hat{\alpha}^{\text{inc}}\times\hat{\beta}^s - \hat{\beta}^{\text{inc}}\times\hat{\alpha}^s & \hat{\beta}^{\text{inc}}\times\hat{\beta}^s + \hat{\alpha}^{\text{inc}}\times\hat{\alpha}^s \\ -(\hat{\beta}^{\text{inc}}\times\hat{\beta}^s + \hat{\alpha}^{\text{inc}}\times\hat{\alpha}^s) & \hat{\alpha}^{\text{inc}}\times\hat{\beta}^s - \hat{\beta}^{\text{inc}}\times\hat{\alpha}^s \end{bmatrix}\begin{bmatrix} E_\alpha^{\text{inc}} \\ E_\beta^{\text{inc}} \end{bmatrix} \\ &\times \bar{D}_{p=1}^q\exp(ik\hat{e}^{\text{inc}}\cdot\vec{r}_{p=1,1}^{\,q}),\end{aligned} \tag{2.92}$$

where $\bar{D}^q_{p=1}$ is a defined integration on the area of the qth illuminated facet S_q given by

$$\bar{D}^q_{p=1} = \frac{k^2}{4\pi}\exp(-ik\hat{r}\cdot\vec{r}^{\,q}_{p=1,1})\iint_{S_q}\exp\{ik(\hat{e}^{\rm inc}-\hat{r})\cdot\vec{w}^{\,q}_{p=1}\}\,d^2\vec{w}^{\,q}_{p=1}. \tag{2.93}$$

The fields associated with outgoing scattered beams can be represented as

$$\begin{aligned}\vec{E}^{q,sca}_p(\vec{r}^{\,q}_{p,1}+\vec{w}^{\,q}_p) &= \left[E^{q,sca}_{p,\alpha}(\vec{r}^{\,q}_{p,1})\hat{\alpha}^{q,sca}_p + E^{q,sca}_{p,\beta}(\vec{r}^{\,q}_{p,1})\hat{\beta}^{q,sca}_p\right]\exp(ik\delta^q_p - k\rho^q_p)\\ &\quad\times\exp\left[ik\left(\hat{e}^{q,sca}_p + i\vec{A}^q_p\right)\cdot\vec{w}^{\,q}_p\right]),\end{aligned} \tag{2.94}$$

$$\begin{aligned}\vec{H}^{q,sca}_p(\vec{r}^{\,q}_{p,1}+\vec{w}^{\,q}_p) &= \left[E^{q,sca}_{p,\beta}(\vec{r}^{\,q}_{p,1})\hat{\alpha}^{q,sca}_p - E^{q,sca}_{p,\alpha}(\vec{r}^{\,q}_{p,1})\hat{\beta}^{q,sca}_p\right]\exp(ik\delta^q_p - k\rho^q_p)\\ &\quad\times\exp\left[ik\left(\hat{e}^{q,sca}_p + iN^q_{p-1}\vec{A}^q_p\right)\cdot\vec{w}^{\,q}_p\right]).\end{aligned} \tag{2.95}$$

Their counterparts in the radiation zone are

$$\begin{aligned}\begin{bmatrix}E^{\rm sca}_\alpha\\ E^{\rm sca}_\beta\end{bmatrix}^{ray} &= \frac{\exp(ikr)}{-ikr}(-\hat{n}^q_p)\\ &\quad\cdot\begin{bmatrix}\hat{\alpha}^{q,sca}_p\times\hat{\beta}^s - \hat{\beta}^{q,sca}_p\times\hat{\alpha}^s & \hat{\beta}^{q,sca}_p\times\hat{\beta}^s + \hat{\alpha}^{q,sca}_p\times\hat{\alpha}^s\\ -(\hat{\beta}^{q,sca}_p\times\hat{\beta}^s + \hat{\alpha}^{q,sca}_p\times\hat{\alpha}^s) & \hat{\alpha}^{q,sca}_p\times\hat{\beta}^s - \hat{\beta}^{q,sca}_p\times\hat{\alpha}^s\end{bmatrix}\\ &\quad\times\begin{bmatrix}E^{q,sca}_{p,\alpha}\\ E^{q,sca}_{p,\beta}\end{bmatrix}\exp(ik\delta^q_p - k\rho^q_p)\bar{D}^q_p,\end{aligned} \tag{2.96}$$

where $\bar{D}^q_p$ is given by Eq. (2.72). At this step, the final amplitude scattering matrix can be written as

$$\begin{aligned}\begin{bmatrix}S_{11} & S_{12}\\ S_{21} & S_{22}\end{bmatrix} &= \sum_q(-\hat{n}^q_1)\\ &\quad\cdot\begin{bmatrix}\hat{\phi}^i\times\hat{\beta}^s - \hat{\theta}^i\times\hat{\alpha}^s & \hat{\theta}^i\times\hat{\beta}^s + \hat{\phi}^i\times\hat{\alpha}^s\\ -(\hat{\theta}^i\times\hat{\beta}^s + \hat{\phi}^i\times\hat{\alpha}^s) & \hat{\phi}^i\times\hat{\beta}^s - \hat{\theta}^i\times\hat{\alpha}^s\end{bmatrix}\Gamma\bar{D}^q_{p=1}\exp(-ik\hat{e}^{\rm inc}\cdot\vec{r}^{\,q}_{p=1,1})\\ &\quad+\sum_{p=1}^{\infty}\sum_q(-\hat{n}^q_p)\\ &\quad\cdot\begin{bmatrix}\hat{\alpha}^{q,sca}_p\times\hat{\beta}^s - \hat{\beta}^{q,sca}_p\times\hat{\alpha}^s & \hat{\beta}^{q,sca}_p\times\hat{\beta}^s + \hat{\alpha}^{q,sca}_p\times\hat{\alpha}^s\\ -(\hat{\beta}^{q,sca}_p\times\hat{\beta}^s + \hat{\alpha}^{q,sca}_p\times\hat{\alpha}^s) & \hat{\alpha}^{q,sca}_p\times\hat{\beta}^s - \hat{\beta}^{q,sca}_p\times\hat{\alpha}^s\end{bmatrix}U^{q,sca}_p\Gamma\\ &\quad\times\exp(ik\delta^q_p - k\rho^q_p)\,\bar{D}^q_p.\end{aligned} \tag{2.97}$$

The explicit expression of the amplitude scattering matrix associated with the diffraction and external reflection can be found in Bi et al. (2011a).

Similarly, Borovoi and Grishin (2003) employed the beam-splitting technique to the calculation of the near-field for a hexagonal particle. To obtain the scattered field in the radiation zone, the spirit of vector Fraunhofer diffraction described in Jackson (1999) is adopted, i.e., the diffraction of an outgoing beam behaves like the diffraction of a plane wave by an aperture with the same shape as that of the beam cross-section. Mathematically, instead of Eq. (2.87), the following equation is used to transform the near-field to the far-field:

$$\vec{E}^{\rm sca}(\vec{r})_{kr\to\infty} = \frac{ike^{ikr}}{2\pi r}\hat{r} \times \int_s \hat{n}_s \times \vec{E}(\vec{r}') \exp(-ik\hat{r}\cdot\vec{r}')\, d\bar{s}\,. \tag{2.98}$$

In Borovoi and Grishin's study, the particle is assumed to be non-absorptive or weakly absorptive, thus, the amplitude variance is reasonably negligible. A similar approach was reported by Popov (1996) from studying light scattering by hexagonal ice crystals; unfortunately, no detailed treatments of inhomogeneous waves for absorptive particles and commonly defined optical properties were included. Instead of using vector Fraunhofer diffraction, Priezzhev et al. (2009) obtained the scattered field in the radiation zone from scalar Fraunhofer diffraction. Priezzhev's algorithm allows for the calculation of the phase function and not the phase matrix. In Section 2.2, the amplitude variance over the beam cross-section has been taken into account through an iterative formula with respect to $\vec{A}_p$. Therefore, the optical properties for absorptive particles can be established in a similar framework of vector Fraunhofer diffraction.

2.4.3 Intensity mapping algorithm

The intensity-mapping algorithm is aimed at incorporating the ray-spreading effect into the phase matrix obtained in the CGOM. In a simplified form (Yang and Liou, 1996b), the amplitude scattering matrix associated with a ray is given by

$$\begin{bmatrix} S_{11} & S_{12} \\ S_{21} & S_{22} \end{bmatrix} = -\frac{k^2}{4\pi}\exp(ik\delta)\begin{bmatrix} \Xi_2 & \Xi_3 \\ \Xi_4 & \Xi_1 \end{bmatrix}\begin{bmatrix} \tilde{S}_{11} & \tilde{S}_{12} \\ \tilde{S}_{21} & \tilde{S}_{22} \end{bmatrix}\begin{bmatrix} \cos\phi_t & \sin\phi_t \\ -\sin\phi_t & \cos\phi_t \end{bmatrix}, \tag{2.99}$$

where δ is the phase of the ray, $\tilde{S}_{ij}$are the elements of the amplitude scattering matrix computed from the CGOM that transform the incident electric field to the scattered field associated with a outgoing ray (with respect to plane A in Fig. 2.8), φ_t is the rotation angle from the plane A to the scattering plane B composed of the observation direction and the direction of the incident light, and the matrix Ξ accounts for the spreading of light from the direction of the outgoing ray to the observation direction and is given by

$$\Xi_1 = h\cos\varphi_t\,, \tag{2.100}$$

$$\Xi_2 = h\cos\theta\cos\theta_t\cos\varphi_t + h\sin\theta\sin\theta_t\,, \tag{2.101}$$

$$\Xi_3 = -h\cos\theta\sin\varphi_t\,, \tag{2.102}$$

$$\Xi_4 = h\cos\theta_t\sin\varphi_t\,, \tag{2.103}$$

where

$$h = \pi a^2(1+\cos\Theta)\frac{J_1(ka\sin\Theta)}{ka\sin\Theta}\,. \tag{2.104}$$

In Eq. (2.104), a is the radius of the ray cross-section and Θ is the angle of ray-spreading. Eq. (2.104) is similar to the Fraunhofer diffraction of a beam passing an aperture with the radius of a. To proceed, we neglect the phase information δ to incorporate the diffraction effect into the phase matrix computed from the CGOM. The matrices on the two sides of the matrix $\tilde{S}_{ij}$ are Jones matrices, whose corresponding Mueller matrices are in the shapes of,

$$H = \begin{bmatrix} H_{11} & H_{12} & H_{13} & 0 \\ H_{21} & H_{22} & H_{23} & 0 \\ H_{31} & H_{32} & H_{33} & 0 \\ 0 & 0 & 0 & H_{44} \end{bmatrix}. \tag{2.105}$$

and

$$L = \begin{bmatrix} 1 & 0 & 0 & 0 \\ 0 & \cos(2\phi_t) & \sin(2\phi_t) & 2 \\ 0 & -\sin(2\phi_t) & \cos(2\phi_t) & 0 \\ 0 & 0 & 0 & 1 \end{bmatrix}. \tag{2.106}$$

The relationship between the elements of the H matrix and the four elements of the Ξ matrix is similar to the relationship between the amplitude scattering matrix and the phase matrix given by Boren and Huffman (1983). Some elements in Eq. (2.105) are zero because Ξ_i are real rather than complex numbers.

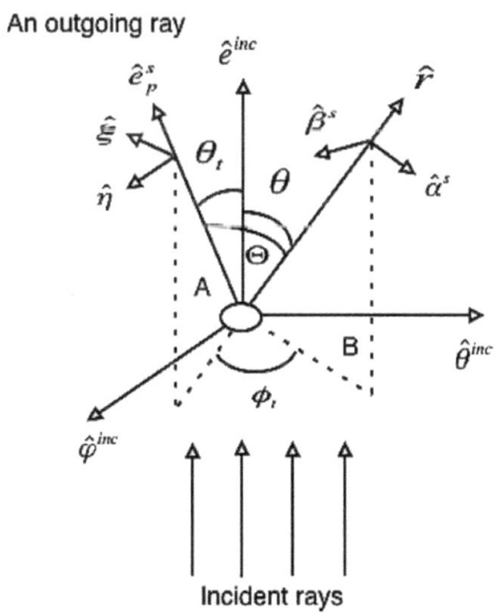

Fig. 2.8. Diagram to show the spreading of light from the direction of an outgoing ray to the observation position according to the diffraction.

The resultant phase matrix is given by

$$P(\theta, \phi) = \iint H(\theta, \theta_t, \varphi, \varphi_0)\tilde{P}(\theta_t, \varphi_0)L(\varphi_t = \varphi - \varphi_0) \sin\theta_t \, d\theta_t \, d\varphi_0, \tag{2.107}$$

where L is associated with the rotation of the scattering plane and H accounts for the effect of ray-spreading. To distinguish various algorithms in the calculation of the phase matrix, the literature refers to the method using the intensity-mapping algorithm as the IGOM.

2.5 Extinction and absorption

In addition to the phase matrix, the extinction and single-scattering albedo are optical parameters critical to radiative transfer simulation. We describe the algorithm used to compute the extinction and absorption cross-sections, and discuss both the edge effect beyond the computational capability of the PGOH method and a semi-empirical approach to incorporate the edge effect into the extinction and absorption efficiencies (Bi et al., 2010b, 2011b).

2.5.1 PGOH cross-sections

The extinction cross-section for an oriented particle can be found from the volume integral equation,

$$C_{\text{ext}} = \text{Im}\left[\frac{k}{\left|\vec{E}^{\text{inc}}\right|^2}(m^2-1)\iiint\limits_{v} \vec{E}\left(\vec{r}'\right)\cdot\vec{E}^{\text{inc}*}\left(\vec{r}'\right)d^3\vec{r}'\right]. \tag{2.108}$$

After substituting the geometric-optics internal field into the above equation, the extinction cross-section derived is the same as that derived from the optical theorem (Bohren and Huffman, 1983):

$$C_{\text{ext}} = \frac{2\pi}{k^2}\text{Re}\left[S_{11}(\hat{e}^{\text{inc}}) + S_{22}(\hat{e}_{\text{inc}})\right]. \tag{2.109}$$

Here, the two diagonal elements of the amplitude scattering matrix are obtained from the volume-integral equation. Physically, the previously derived result takes into account the interference between the diffraction and the fields associated with forward scattered beams but neglects the edge effect associated with tunneling rays. The process has been justified by separating the contribution of tunneling rays to the extinction from the total extinction cross-section in the circular cylinder case (Bi et al., 2010b).

Given the internal field, the absorption cross-section is expressed as (Hage et al., 1991)

$$C_{\text{abs}} = \frac{k}{\left|\vec{E}^{\text{inc}}\right|^2}\varepsilon_i\iiint\limits_{v} \vec{E}\left(\vec{r}'\right)\cdot\vec{E}^{*}\left(\vec{r}'\right)d^3\vec{r}', \tag{2.110}$$

where ε_i is the imaginary part of permittivity. If the interference between the internal beam fields is neglected, the absorption cross-section can be obtained by integrating the electric fields with each ray tube generated in the beam-tracing process. The final expression for the absorption cross-section is (Bi et al., 2011b)

$$\begin{aligned}C_{\text{abs}} &= \frac{1}{2}\sum_{p=1}^{\infty}\sum_{q} N_{r,p}^{q}\exp(-2k\rho_{p,1}^{q}) \\ &\times \left(\left|U_{p,11}^{q}\right|^2 + \left|U_{p,12}^{q}\right|^2 + \left|U_{p,21}^{q}\right|^2 + \left|U_{p,22}^{q}\right|^2 + \left|U_{p,31}^{q}\right|^2 + \left|U_{p,32}^{q}\right|^2\right) \\ &\times \left(\left|\hat{e}_p^1 \cdot \hat{n}_p^q\right| \tilde{D}_p^q - \exp\left(-2N_{i,p}^q k \left|\vec{r}_{p+1}^q - \vec{r}_p^q\right|\right) \left|\hat{e}_{p+1}^q \cdot \hat{n}_{p+1}^q\right| \tilde{D}_{p+1}^q\right), \end{aligned} \tag{2.111}$$

where

$$\begin{aligned}\tilde{D}_p^q &= \iint_s \exp\left(-2kN_{i,p-1}^q \vec{w}_p^q \cdot \vec{A}_p^q\right) d^2\vec{w}_p^q \\ &= \frac{1}{2kN_{i,p}^q}\sum_{j=1}^{N} \frac{\left(\vec{r}_{p,j+1}^q - \vec{r}_{p,j}^q\right)\cdot\left(\vec{A}_p^q \times (-\hat{n}_p^q)\right)}{\left|\vec{A}_p^q\right|^2 - (\vec{A}_p^q \cdot \hat{n}_p^q)^2} \frac{\sin\left(2kN_{i,p-1}^q \vec{A}_p^q \cdot \left(\vec{r}_{p,j+1}^q - \vec{r}_{p,j}^q\right)\right)}{2kN_{i,p-1}^q \vec{A}_p^q \cdot \left(\vec{r}_{p,j+1}^q - \vec{r}_{p,j}^q\right)} \\ &\times \exp\left\{-kN_{i,p-1}^q \vec{A}_p^q \cdot (\vec{r}_{p,j+1} + \vec{r}_{p,j} - 2\vec{r}_{p,1})\right\}. \end{aligned} \tag{2.112}$$

To interpret the physical meaning implied in Eq. (2.111), let us express the energy passing across the initial beam cross-section of a pth order internal beam in the form

$$\begin{aligned}F &= \frac{1}{2}N_{r,p}^q \exp(-2k\rho_{p,1}^q) \left|\hat{e}_p^q \cdot \hat{n}_p^q\right| \tilde{D}_p^q \\ &\times \left(\left|U_{p,11}^{q}\right|^2 + \left|U_{p,12}^{q}\right|^2 + \left|U_{p,21}^{q}\right|^2 + \left|U_{p,22}^{q}\right|^2 + \left|U_{p,31}^{q}\right|^2 + \left|U_{p,32}^{q}\right|^2\right). \end{aligned} \tag{2.113}$$

A comparison between Eqs. (2.111) and (2.113) provides straightforward physical proof that the absorption cross-section is associated with the energy lost in all the internal ray tubes. The tunneling contribution to the absorption cross-section is not considered in Eq. (2.111).

2.5.2 Tunneling/edge effect

To incorporate the tunneling effect into both the PGOH extinction and absorption cross-sections, we must numerically justify the contribution of tunneling rays. The process is only understood for spheres, in which case an analytical solution exists. van de Hulst (1981) has attempted to obtain geometric-optics results from a Lorenz–Mie formula based on the localization principle, which states that 'a term of the order n in Mie coefficients corresponds to a ray passing the origin at a distance of $(n + 1/2)\lambda/2\pi$'. Based on the localization principle, the edge effect for a circular cylinder/disk (Fig. 2.3a) can be separated from the DDA results (Bi et al., 2010b).

In the DDA method, the geometry of the particle is discretized to a number of small volumes termed 'dipoles". The basic equation for the DDA method is given by

$$\vec{E}_l^{\text{inc}} = \alpha_l^{-1}\vec{P}_l - \sum_{m \neq l} G_{lm}\vec{P}_m\,, \tag{2.114}$$

where $\vec{E}_l^{\text{inc}}$is the electric vector at the dipole of the index l, α_l^{-1} is the inverse of polarizability, $\vec{P}_l^{\text{inc}}$indicates the electric dipole moment, and G_{ml} is the dyadic Green's function. The solution to Eq. (2.114) is semi-rigorous in the sense that it solves Maxwell's equations, and the numerical errors can be reduced so that the true solution is approached by increasing the number of dipoles that represent the particle geometry. Once the vector of polarizability at each dipole is determined, the extinction efficiency and absorption efficiency can be obtained. The incident electric field can be expanded in terms of multipole fields similar to those in the Lorenz–Mie theory. We truncate the summation to an edge term of the order of $n = [ka - 1/2]$, where a is the radius of the circular cylinder and the new incident field is noted as $\vec{E}_{\text{no\_edge}}^{\text{inc}}$. The DDA equation can be written as

$$\vec{E}_{l,\text{no\_edge}}^{\text{inc}} = \alpha_l^{-1}\vec{P}_l - \sum_{m \neq l} G_{lm}\vec{P}_m \,. \tag{2.115}$$

Physically, the solution to Eq. (2.115) corresponds to the solution of the PGOH. Based on the principle of the PGOH, the solution to the extinction efficiency is (Bi et al., 2010b),

$$Q_{\text{ext}} = 2\,\text{Re}\left\{1 - \frac{4m\,\exp\{i(m-1)kL\}}{(m+1)^2 - (m-1)^2\,\exp(i2mkL)}\right\}. \tag{2.116}$$

Figure 2.9 shows a comparison of the extinction efficiency factors computed based on Eqs. (2.114)–(2.116). The general agreement between the results simulated from the DDA method with the edge effect separate and the PGOH results supports the explanation of the edge effect based on the localization principle.

The edge effect contribution to the extinction and the absorption efficiency for a sphere can be written in the form of (Nussenzveig, 1992)

$$\Delta Q_{\text{ext}} = f_{\text{ext}} x^{-2/3} \,, \tag{2.117}$$

$$\Delta Q_{\text{abs}} = f_{\text{abs}} x^{-2/3} \,, \tag{2.118}$$

where x is the size parameter, $f_{\text{ext}} = 1.99239$, and f_{abs} can be expressed in integral terms. To incorporate the edge effect into the PGOH, we first simulate the extinction and absorption efficiencies for a moderate size particle by a rigorous method. By comparing of the rigorous result with that from the PGOH, we can determine the coefficients of the edge effect contribution. We assume the coefficients are independent of the size parameter but dependent on the particle shape. In principle, the method based on Eqs. (2.117) and (2.118) for nonspherical particles is semi-empirical, but in practice, the size parameter must be sufficiently large in order for the ray to represent waves. The determination of f_{ext} in absorptive particles is critical to avoid a highly oscillated Q_{ext} curve. f_{ext} is weakly refractive-index dependent and f_{abs} is strongly refractive-index dependent. The semi-empirical methods including edge effects for spheroids and ellipsoids can be found in studies by Jones (1957), Fournier and Evans (1991), Yang et al. (2007), and Bi et al. (2008). Liou et al. (2011) have investigated the edge effect contribution to the extinction efficiency, the absorption efficiency, and the asymmetry factor, and the coefficients are determined from the ratio of two defined volumes. Note that the extinction efficiencies derived from different PGOH algorithms may have subtle differences, which could lead to some uncertainties of the semi-empirical factors.

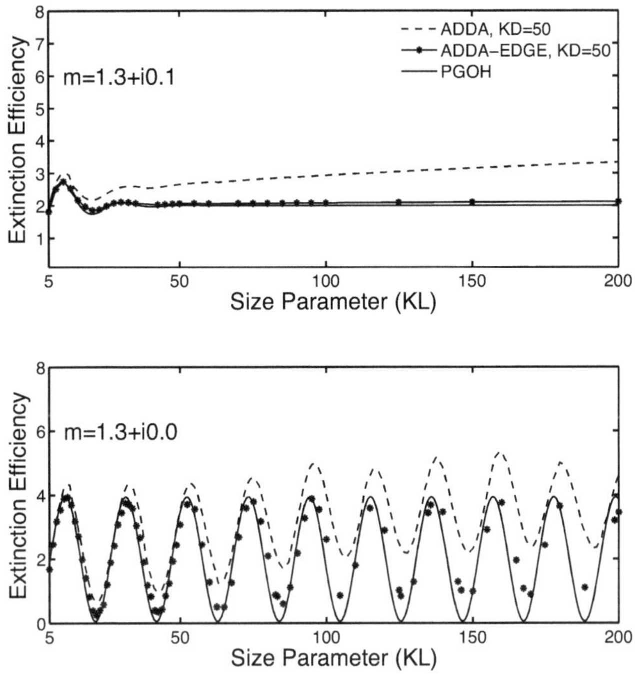

Fig. 2.9. The comparison of the extinction efficiency factor from two selected refractive indices computed from the DDA method, the DDA method without the edge effect, and the PGOH method. The lower panel data is from Bi et al. (2010b).

2.6 Numerical examples for ice crystals and mineral dusts

The PGOH method has been applied in investigations of the optical properties (i.e., the extinction efficiency, the single-scattering albedo, and the phase matrix) of large ice crystals. Conventionally, the PGOH is applied to study randomly oriented ice crystals and seldom to study the optical properties of an ice particle with a fixed orientation due to the existence of isolated points. Developments have improved the method by taking into account the diffraction effects of light beams; however, the algorithm is very CPU-time-consuming when applied to large size parameters. An algorithm developed by Borovoi and Grishin (2003) adopted the beam-splitting technique to the near-field calculation and is suitable to study preferably oriented ice crystals. Borovoi and Grishin's algorithm is used to study the scattering of light within the visible range where the particle is either non-absorptive or weakly absorptive. In principle, the algorithm is applicable to any convex faceted particles with arbitrary refractive indices at fixed orientations or to orientations described by a distribution function. A comparison of the present PGOH method with the one developed based on vector Fraunhofer diffraction with the consideration of absorption is interesting; however, the comparison will not be discussed here. We will show some representative numerical results obtained from the Kirchhoff surface integral equation, the Fredholm volume integral equation, and the IGOM. Note, unlike the other methods, the IGOM is only applicable to randomly oriented particles.

Figure 2.10 shows the refractive indices of ice ranging from 0.2 to 15 μm and compiled by Warren and Brandt (2008). We select several refractive indices at representative wavelengths to illustrate the phase matrices computed from the PGOH methods. For integrated single-scattering properties (i.e., the extinction efficiency, the single-scattering albedo, and the asymmetry factor), we present results over a complete range of the spectrum.

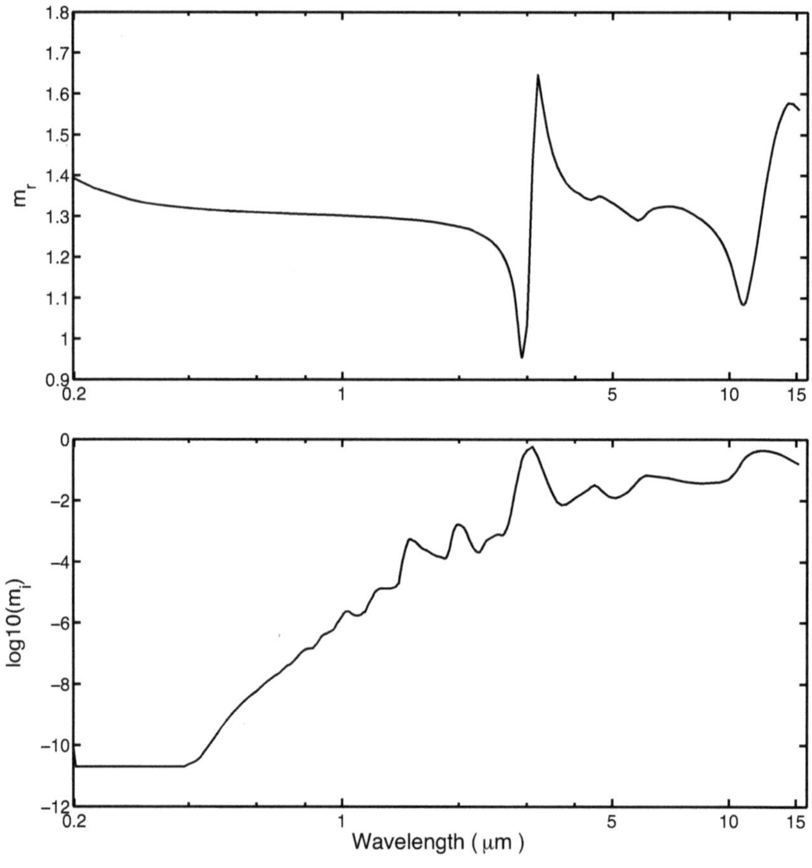

Fig. 2.10. Refractive indices of ice crystals compiled by Warren and Brandt (2008).

Figure 2.11 illustrates a comparison of the phase matrix elements simulated from the Fredholm volume integral equation ($PGOH_v$), the Kirchhoff surface integration equation ($PGOH_s$), and the ADDA method for an oriented hexagonal ice crystal at the wavelength of 0.66 μm and a refractive index of $1.3078 + i1.66 \times 10^{-8}$. A small imaginary refractive index part indicates negligible absorption. The size parameter defined in terms of the semi-width is 25 and the aspect ratio is unity (i.e., the height is equal to the diameter). In the ADDA simulation, the number of dipoles per wavelength is 15 and the lattice dispersion relation is used to describe the polarization relation. The phase matrix elements are averaged with respect to the scattering azimuthal angle. For simplicity, six phase matrix elements (P_{ij})

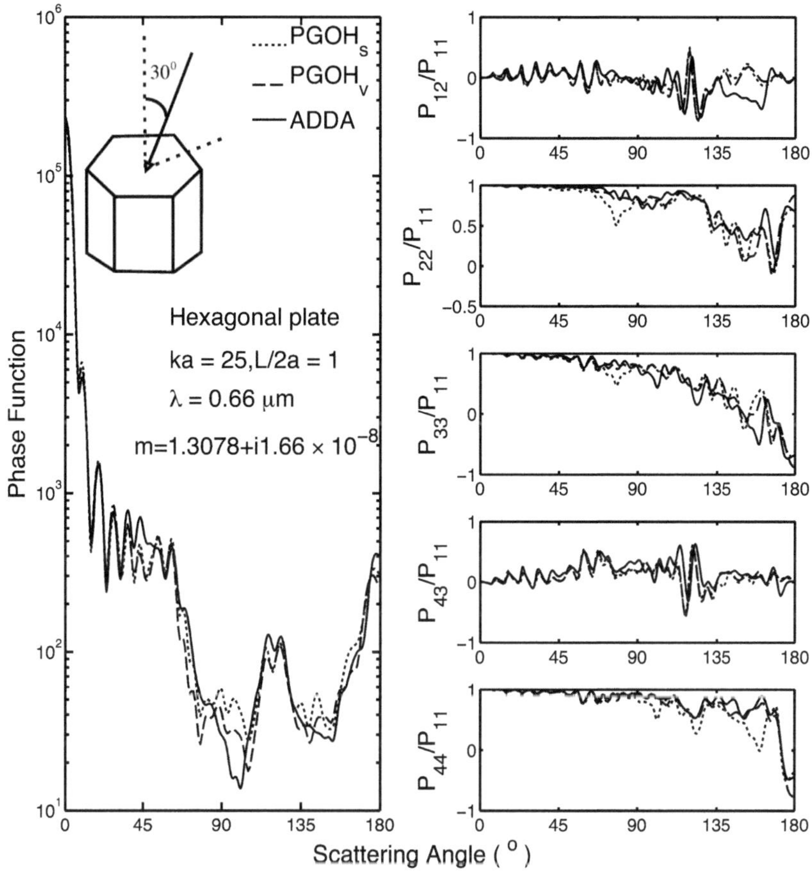

Fig. 2.11. Comparison of the phase matrix elements computed from the $PGOH_s$, $PGOH_v$, and ADDA.

are demonstrated, although the phase matrix pattern is not in the Lorenz–Mie structure for an oriented ice crystal. From Fig. 2.11, it is evident that both the $PGOH_s$ and $PGOH_v$ yield results with reasonable accuracy. Figure 2.12 is similar to Fig. 2.11 except that the wavelength is at 3.2 μm where the ice is quite absorptive and the diffraction and external reflection dominate the scattering pattern. The phase function peak at 120 degrees is evidently from the external reflection. From the comparison, $PGOH_v$ appears to be more accurate than $PGOH_s$. Hereafter, for simplicity, we will only demonstrate the results computed from the $PGOH_v$.

Figure 2.13 illustrates the comparison of phase matrix elements computed from the IGOM, the PGOH, and the DDA methods for a randomly oriented hexagonal ice crystal. General agreement between the three results is identified, although there are differences (e.g., the phase function near the backscattering directions). As evident from the comparison, the PGOH mimics the oscillation feature of the optical properties of ice crystals whereas the IGOM does not. In the IGOM, the interference among rays is neglected, which explains the anomaly. Note that ice haloes are not observed because the size parameter is not sufficiently large. If the

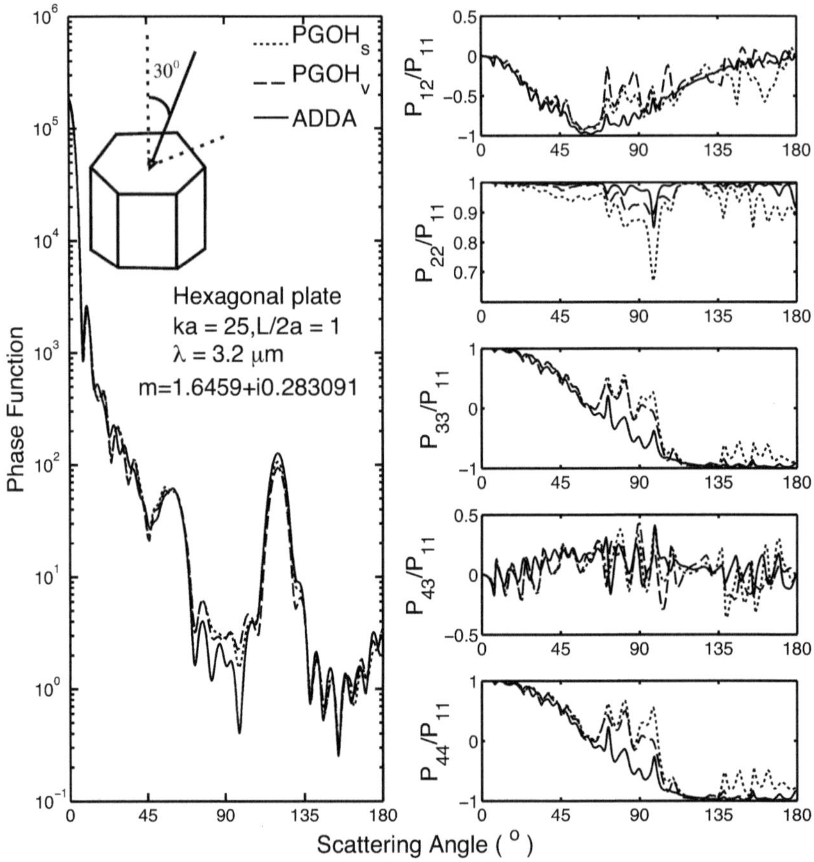

Fig. 2.12. Similar to Fig. 2.11, but at a wavelength of 3.2 μm.

CGOM is employed for the computation, ice haloes will exist no matter how large the size parameter, but the explanation is nonphysical.

Figure 2.14 shows the comparison of phase matrix elements simulated from the IGOM and the PGOH for large randomly oriented ice crystals. The size parameter is 500, which is beyond the modeling capability of the DDA method. The RBRI algorithm developed in Yang and Liou (1997) is time consuming for large size parameters; therefore, the comparison between the RBRI and FDTD, at that time, was only carried out for 2-D randomly oriented hexagonal particles. The PGOH is much more efficient than the RBRI, and, thus, makes the simulation of 3-D randomly oriented ice crystals possible. The phase matrix is averaged in terms of 360 scattering planes. 2560 incident angles corresponding to two Euler angles are specified for the orientation-average computation. The comparison between the IGOM results and the PGOH results supports the feasibility of physical simplifications in the IGOM algorithm described in Section 3 by Yang and Liou (1996b). Due the difficulty of defining the solid angle in the backscattering direction, the IGOM is less accurate than the PGOH in the computation of backscattering optical properties, and the differences may be seen in Fig. 2.14. The curves from the IGOM

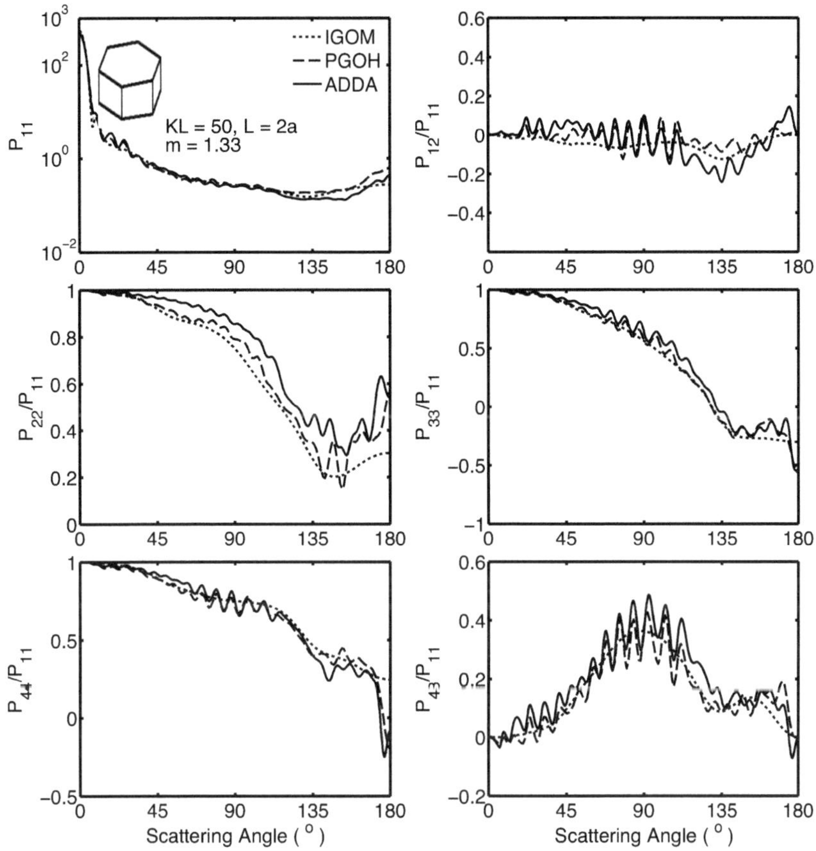

Fig. 2.13. Phase matrix elements for randomly oriented ice crystals simulated from the IGOM, PGOH, and ADDA.

are much smoother than those from the PGOH. The optical properties computed from the PGOH and the IGOM are closer in accuracy when a size distribution is applied to obtain the back-scattering properties.

The present PGOH algorithm can be applied to arbitrarily shaped faceted particles. Figure 2.15 shows the PGOH phase matrix elements for a droxtal ice crystal at two wavelengths. The definition of droxtal geometry can be found in Yang et al (2003), and Zhang et al. (2004). The diameter of the droxtal is 8 μm, and the size parameters at the wavelengths of 0.2 μm and 1.0 μm are approximately 251 and 50. At the small size parameter, oscillations of the phase functions, P12 and P43, are evident (similar FDTD simulations can be found in Yang and Liou, 2006). At the large size parameter, the geometric optics features are identified (e.g., the peaks of the phase function), and can be compared with the results obtained from the IGOM in Zhang et al. (2004).

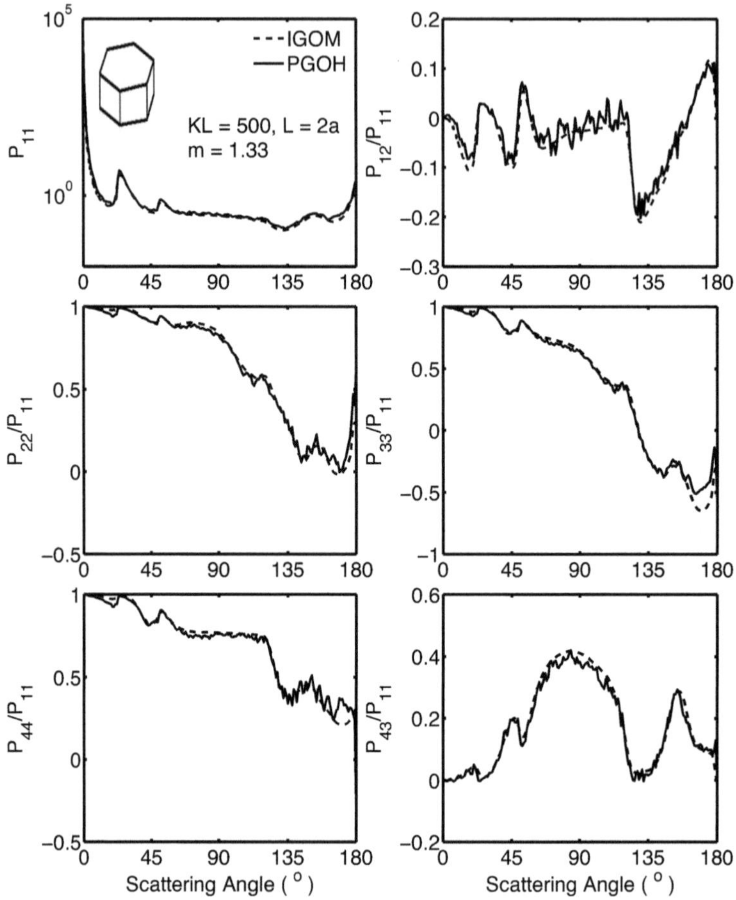

Fig. 2.14. Phase matrix elements for large randomly oriented hexagonal ice crystals simulated from the IGOM and PGOH.

Figure 2.16 shows the extinction efficiency, the single-scattering albedo, and the asymmetry factor for hexagonal ice plates and droxtals in a spectral range of wavelengths from 0.2 to 15 μm. The diameters of the ice plates and droxtals are 5 μm and the particles are assumed to be randomly oriented in space. The PGOH is employed for the calculation when the wavelength is smaller than 1.03 μm (the size parameter is on the order of 30.5), and the ADDA method is used for the remaining spectral regime. This figure shows a combination of the PGOH method and the ADDA method for the computation of the optical properties of ice crystals in a complete range of wavelengths, and the differences in the optical properties between the two shapes are evident. The edge effect is included in the PGOH results.

Unlike ice crystals, mineral dust aerosols rarely have particular shapes. Optical modeling of mineral dust aerosols is usually based on a few simple randomly oriented nonspherical geometries, such as spheroids/ellipsoids (Dubovik et al., 2002; Yang et al., 2007; Bi et al., 2008; Nousiainen, 2009; Meng et al., 2010; Merikallio et al., 2011) and non-symmetric hexahedra (Bi et al., 2010). The PGOH with

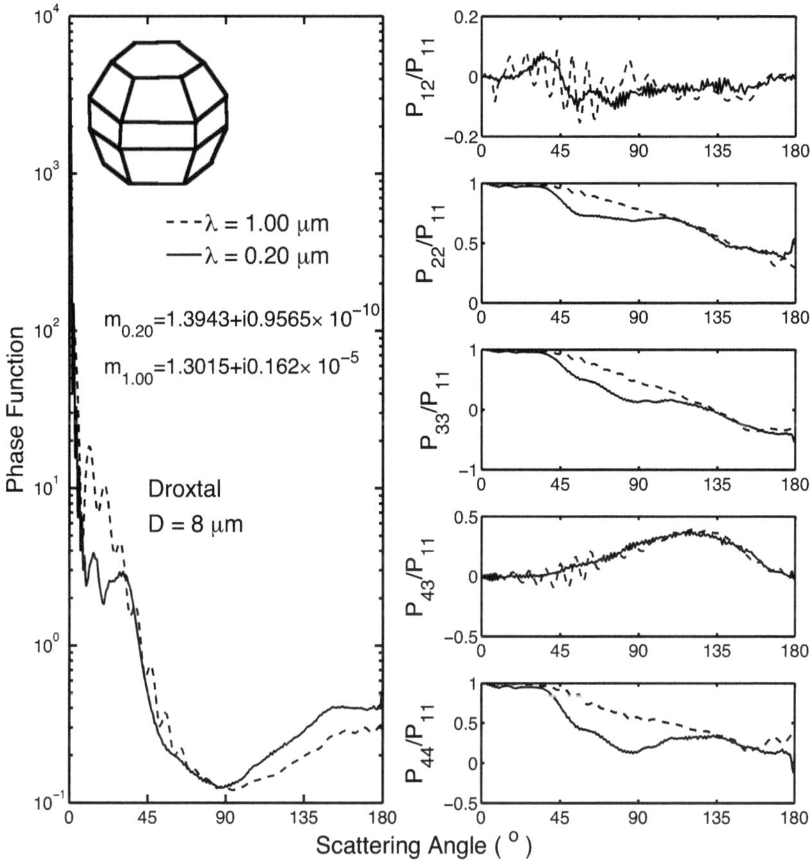

Fig. 2.15. Phase matrix elements for a randomly oriented droxtal particle at two wavelengths and simulated by the PGOH.

the beam-splitting technique is applicable to non-symmetric hexahedra, which are faceted, but not to ellipsoids with curved surfaces. For ellipsoids, the IGOM must be used.

Figure 2.17 shows the phase matrix elements of randomly oriented non-symmetric hexahedra. The size parameter defined in terms of the radius of a surface-area-equivalent sphere is 10. For such a small size parameter, the PGOH can produce quite similar results with those simulated from the DDA method. The element P_{12} in the range of 30°–90° has relatively large differences.

To illustrate the applicability of a combination of the ADDA and the IGOM to modeling dust particles, Fig. 2.18 shows the comparison of the phase matrix elements for Pinatubo aerosol samples simulated based on randomly oriented non-symmetric hexahedra and actual measurements (Volten, 2006). In the comparison, three non-symmetric hexahedra are used to match the theoretical results and the

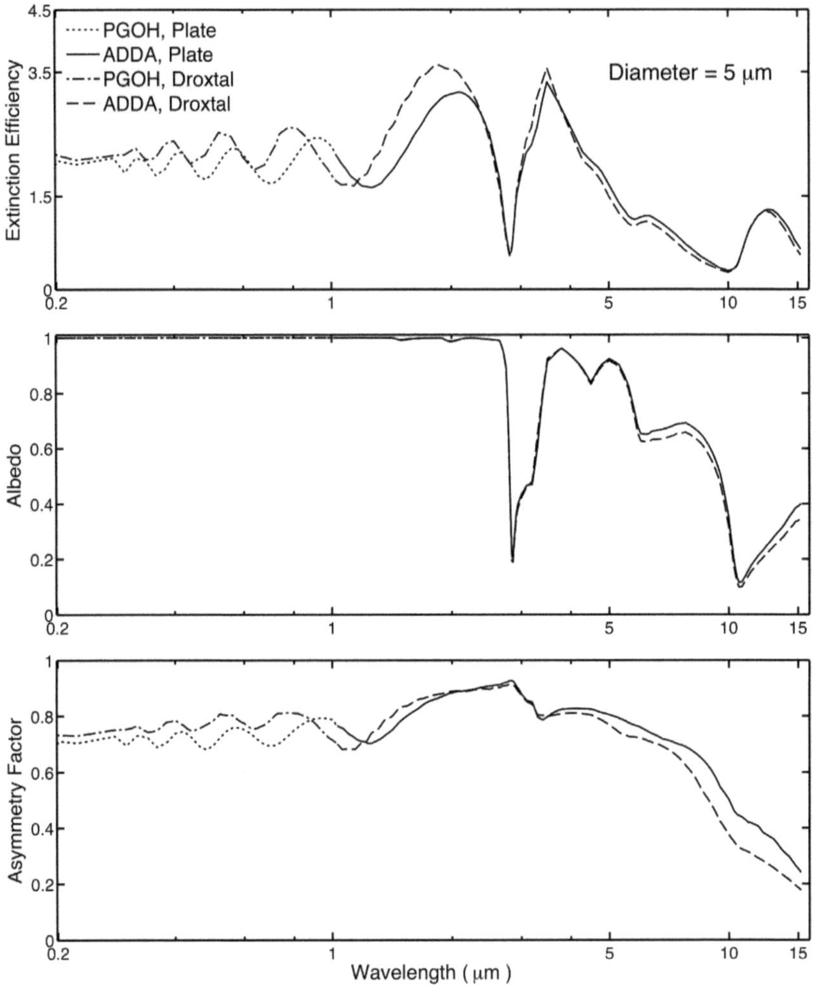

Fig. 2.16. Extinction efficiency, single-scattering albedo, and asymmetry factors computed from the PGOH and ADDA for hexagonal ice plates and droxtals with a diameter of 5 μm in a spectral range from 0.2 to 15 μm. The aspect ratio of a plate is unity (i.e., the height is equal to the diameter).

measurements. The effective radius and effective variance indicated in Fig. 2.18 are defined as (Hansen and Travis, 1974)

$$r_{\text{eff}} = \frac{\int_{r_1}^{r_2} r^3 n(r)\, dr}{\int_{r_1}^{r_2} r^2 n(r)\, dr}, \tag{2.119}$$

$$\sigma_{\text{eff}} = \sqrt{\frac{\int_{r_1}^{r_2} (r - r_{\text{eff}})^2 r^2 n(r)\, dr}{(r_{\text{eff}})^2 \int_{r_1}^{r_2} r^2 n(r)\, dr}}\,. \tag{2.120}$$

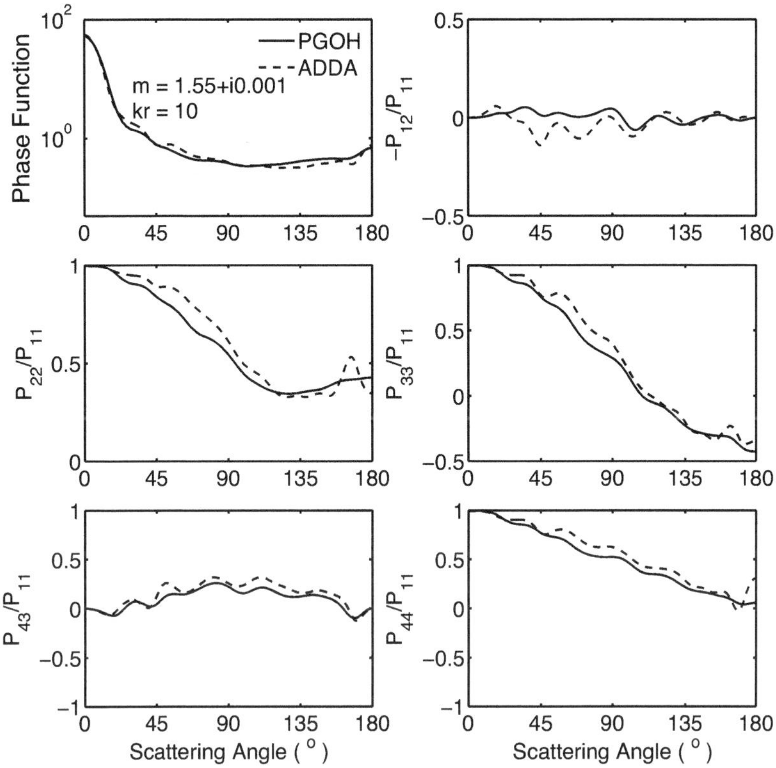

Fig. 2.17. Phase matrix for a randomly oriented non-symmetric hexahedra.

It was found that the shape-averaged phase matrix elements are not very sensitive to relative weights of the shape. The study demonstrated both the theoretical possibility of using irregularly faceted particles to model realistic aerosols with complicated geometries without facets and the nonspherical model to be much better than the spherical model in reproducing laboratory measurements (also see Kokhanovsky, 2002).

Currently, the optical modeling of nonspherical dust aerosols is generally based on spheroids (Mishchenko and Travis, 1998; Dubovik, et al., 2002). Yi et al. (2011) have investigated the uncertainties of particle shapes and refractive indices in atmospheric flux calculations based on the single-scattering database of tri-axial ellipsoids developed by Meng et al. (2010). Other efforts have attempted to simulate the optical properties of nonspherical aerosols using quite complicated geometries (e.g., Kalashnikova and Sokolik, 2004).

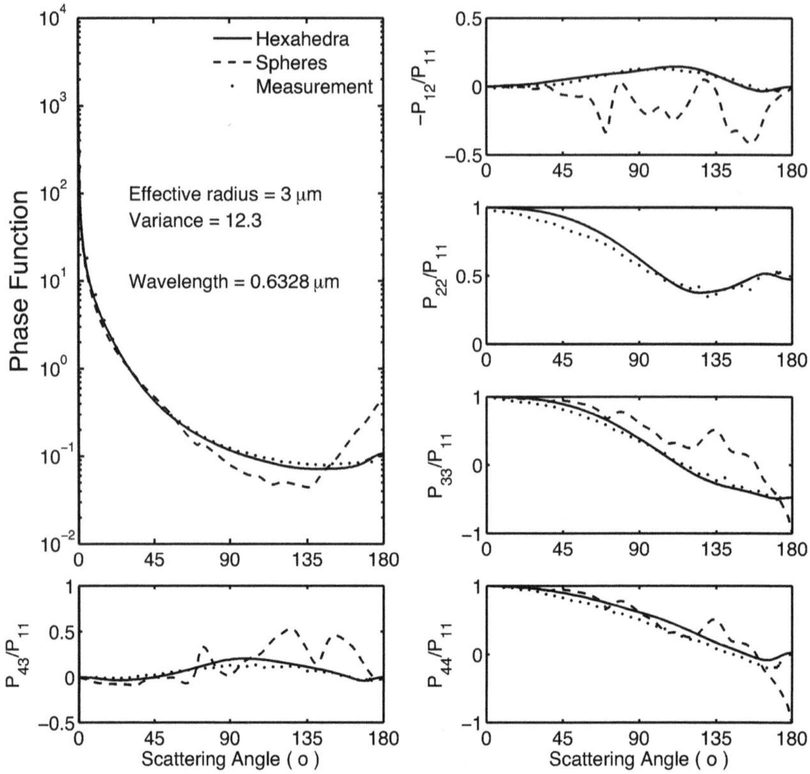

Fig. 2.18. Comparison of simulated results from hexahedra and sphere models and laboratory measured data (data from Bi et al., 2010).

2.7 Summary

In this chapter, we have reviewed the conceptual basis and theoretical development of the PGOH methods for the computation of the single-scattering properties of atmospheric nonspherical particles. The PGOH has been demonstrated to be a computationally efficient method to solve light scattering by dielectric particles whose characteristic dimensions are much larger than the incident wavelength. The derived solution is expected to be more accurate when the particle size parameter increases, and is unlikely to be reliable at small size parameters as the ray concept fails. An estimation of the lower limit of the size parameter is near 20.

Several physical-geometric optics hybrid algorithms exist: CGOM, IGOM, $PGOH_s$, and $PGOH_v$. The CGOM is based on a straightforward combination of Fraunhofer diffraction and angular scattering pattern from geometric-optics and an assumption of the extinction efficiency of two. The IGOM incorporates the ray-spreading effect into the CGOM phase matrix. The $PGOH_s$ and $PGOH_v$ are more rigorous PGOH algorithms because they adopt no additional simplifications beyond the geometric-optics approximation. In regard to the application regimes, the CGOM is only applicable to large randomly oriented particles (the size parameter should be larger than ~100); the IGOM is only applicable to randomly oriented particles but can be used for lower size parameters and transitions to the CGOM when the size parameter increases; the $PGOH_s$ and $PGOH_v$ are applicable to oriented ice crystals and randomly oriented particles, but, although less efficient than the CGOM and IGOM, have better accuracy. In principle, the CGOM and the IGOM are more semi-empirical than the $PGOH_s$ and $PGOH_v$. In an attempt to forward a better understanding of geometric-optics methods, we have listed some key references in Table 2.1. For practical electromagnetic scattering calculations involving a dielectric particle, we summarize the major strengths of the present PGOH formalism ($PGOH_s$ and $PGOH_v$) as:

- no limitations exist on the maximum particle size parameter, but a conservative estimate of the lower size parameter is $x > 20$;
- the algorithm is applicable to particles of arbitrary orientation and very efficient for studying oriented ice crystals;
- the algorithm is reasonably accurate in the description of backscattering properties for lidar applications (Zhou et al., 2012); and
- the algorithm allows light scattering computation in the case of an arbitrary refractive index in the optical regime.

The applicability of the PGOH algorithm to the study of the single-scattering properties of ice crystals within cirrus clouds can be justified by the fact that ice crystals are large in size in comparison with the incident wavelengths in the optical spectrum and have faceted geometry. The applicability of faceted model particles to mineral dust aerosols is explored based on the objective of using 'simple geometries to represent irregular realistic particles without any particular geometry'

Table 2.1. Some references pertinent to the CGOM, IGOM, and PGOH methods.

CGOM	IGOM	PGOH
Liou and Hansen, 1971	Yang and Liou, 1996	Ravey and Mazeron, 1982
Jacobowitz, 1971	Yang and Liou, 1998	Mazeron and Muller, 1996
Wednling et al., 1979	Yang et al., 2007	Popov, 1996
Coleman and Liou, 1981	Zhang et al., 2004	Muinonen, 1989
Cai and Liou, 1982	Bi et al., 2009	Yang and Liou, 1996
Takano and Jayaweera, 1985	Bi et al., 2010	Yang and Liou, 1997
Takano and Liou, 1989	Meng et al., 2010	Yang et al., 2003
Macke, 1993 (at http://www.ifm-geomar.de /index.php?id=981)		Borovoi, 2003
		Priezzhev, 2009
		Liou et al., 2011
Macke et al., 1996a,b		Bi et al., 2011a
Hess and Wiegner, 1994		Bi et al., 2011b
Muinonen et al., 1997 (http://www.atm.helsinki.fi /~tpnousia/siris.html)		
Borovoi, 2002		
Yang and Liou, 2009b		

(Kahnert et al., 2002). The comparison of measurements and simulations suggests the feasibility of the approach suggested by Kahnert et al. (2002).

Acknowledgments

The authors thank M. A. Yurkin and A. G. Hoekstra for the use of their ADDA code. Ping Yang acknowledges support from the U.S. National Science Foundation (ATM-0803779) and NASA (NNX11AK37G).

References

Asano, S., and G. Yamamoto, 1975: Light scattering by randomly oriented spheroidal particles, *Appl. Opt.*, **14**, 29–49.

Baran, A., J., 2009: A review of the light scattering properties of cirrus, *J. Quant. Spectrosc. Radiative Transfer*, **110**, 1239–1260.

Baum, B. A., P. Yang, A. J. Heymsfield, S. Platnick, M. D. King, Y. X. Hu, and S. M. Bedka, 2005: Bulk scattering properties for the remote sensing of ice clouds. II: Narrowband models, *J. Appl. Meteor.*, **44**, 1896–1911.

Bi, L., P. Yang, G. W. Kattawar, and R. Kahn, 2008: Single-scattering properties of triaxial ellipsoidal particles for a size parameter range from Rayleigh to geometric-optics regimes, *Appl. Opt.*, **48**, 114–126.

Bi, L., P. Yang, G. W. Kattawar, B. A. Baum, Y.-X. Hu, D. M. Winker, R. S. Brock, and J. Q. Lu, 2009: Simulation of the color ratio associated with the backscattering of radiation by ice crystals at the wavelengths of 0.532 and 1.064 μm, *J. Geophys. Res.*, **114**, D00H08, doi:10.1029/2009JD011759

Bi, L., P. Yang, G. W. Kattawar, and R. Kahn, 2010a: Modeling optical properties of mineral aerosol particles by using non-symmetric hexahedra, *Appl. Opt.*, **49**, 334–342.

Bi, L., P. Yang, and G. W. Kattawar, 2010b: Edge-effect contribution to the extinction of light by dielectric disk and cylindrical particles, *Appl, Opt.*, **49**, 4641–4646.

Bi, L., P. Yang, G. W. Kattawar, Y. Hu, and B. A. Baum, 2011a: Diffraction and external reflection by dielectric faceted particles, *J. Quant. Spectrosc. Radiat. Transfer*, **112**, 163–173.

Bi, L., P. Yang, G. W. Kattawar, Y. Hu and B. A. Baum, 2011b: Scattering and absorption of light by ice particles: solution by a new physical-geometric optics hybrid method, *J. Quant. Spectrosc. Radiat. Transfer*, **112**, 1492–1508.

Bohren, C. F., and D. R. Huffman, 1983: *Absorptin and Scattering of Light by Small Particles*, New York: John Wiley & Sons.

Born, M., and E. Wolf, 1959: *Principles of Optics*, Oxford: Pergamon Press.

Borovoi, A., I. Grishin, E. Naats, and U. Oppel, 2002: Light backscattering by hexagonal ice crystals, *J. Quant. Spectrosc. Radiat. Transfer*, **72**, 403–417.

Borovoi, A. G., and I. A. Grishin, 2003: Scattering matrices for large ice crystal particles, *J. Opt. Soc. Am. A.*, **20**, 2071–2080.

Cai, Q., and K. N. Liou, 1982: Polarized light scattering by hexagonal ice crystals: theory, *Appl. Opt.*, **21**, 3569–3580.

Chang, P. C., J. G., Walker, and K. I. Hopcraft, 2005: Ray tracing in absorbing media, *J. Quant. Spectrosc. Radiative Transfer*, **96**, 327–341.

Chen, G., P. Yang, and G. W. Kattawar, 2008: Application of the pseudospectral time-domain method to the scattering of light by nonspherical particles, *J. Opt. Soc. Am. A.*, **25**, 785–790.

Chýlek, P., and J. Coakley, 1974: Aerosols and climate, *Science*, **183**, 75–77.
Coleman, R., and K. N. Liou, 1981: Light scattering by hexagonal ice crystals, *J. Atmos. Sci.*, **38**, 1260–1271.
Dubovik, O., B. N. Hilben, T. Lapyonok, A. Sinyuk, M. I. Mishchenko, P. Yang, and I. Slutsker, 2002: Non-spherical aerosols retrieval method employing light scattering by spheroids, *Geophys. Res. Letter*, **29**, 541–544.
Dupertuis M. A., M. Proctor, and B. Acklin, 1994: Generalization of complex Snell–Descartes and Fresnel laws, *J. Opt. Soc. Am. A.*, 11,1159–1166.
Fournier, G. R., and Evans, B. T., 1991: Approximations to extinction efficiency for randomly oriented spheroids, *Appl. Opt.*, **30**, 2042–2048.
Fu, Q., W. B. Sun, and P. Yang, 1999: Modeling of scattering and absorption by non-spherical cirrus ice partiles at thermal infrared wavelengths, *J. Climat.*, **25**, 223–2237.
Hage, J. I., J. M. Greenberg, and R. T. Wang, 1991: Scattering from arbitrary shaped particles: theory and experiment, *Appl. Opt.*, **30**, 1141–1152.
Hansen, J. E., and L. D. Travis, 1974: Light scattering in planetary atmospheres, *Space Sci. Rev.*, **16**, 527–610.
Haywood, J., and O. Boucher, 2000: Estimates of the direct and indirect radiative forcing due to troposphere aerosols: a review, *Rev. Geophys.*, **38**, 513–544.
Hess, M., and Wiegner, M., 1994: COP: a data library of optical properties of hexagonal ice crystals, *Appl. Opt.*, **33**, 7740–7746.
Heymsfield, A. J., and J. Iaquinta, 2000: Cirrus crystal terminal velocities, *J. Atmos. Sci.*, **57**, 916–938.
Jackson, J. D., 1999: *Classical Electrodynamics*, 3rd ed. New York: John Wiley & Sons.
Jacobowitz, H., 1971: A method for computing the transfer of solar radiation through clouds of hexagonal ice crystals, *J. Quant. Spectrosc. Radiat. Transfer*, **11**, 691–695
Jones, D. S. 1957: High-frequency scattering of electromagnetic waves, *Proceedings of the Royal Society of London Series A*, **240**, 206–213.
Kahnert, F. M., 2003: Numerical methods in electromagnetic scattering theory, *J. Quant. Spectrosc. Radiat. Transfer*, **79–80**, 775–824.
Kahnert, F. M., J. J. Stamnes, and K. Stamnes, 2002: Using simple particle shapes to model the Stokes scattering matrix of ensembles of wavelength-sized particles with complex shapes: possibilities and limitations, *J. Quant. Spectrosc. Radiat. Transfer*, **74**, 167–182.
Kalashnikova, O. V., and I. N. Sokolik, 2004: Modeling the radiative properties of non-spherical soil-derived mineral aerosols, *J. Quant. Spectrosc. Radiat. Transfer*, **87**, 137–166.
Kaufman, Y. J., D. Tanre, and O. Boucher, 2002: A satellite view of aerosols in the climate system, *Nature*, **419**, 215–222.
Kokhanovsky, A., 2003: Optical properties of irregularly shaped particles, *J. Phys. D*, 36, 915–923
Liou, K. N. 2002: *An Introduction to Atmospheric Radiation*, San Diego: Academic Press.
Liou, K. N., and J. E. Hansen, 1971: Intensity and polarization for single scattering by polydisperse spheres: a comparison of ray optics and Mie theory, *J. Atmos. Sci.*, **28**, 995–1004.
Liou, K. N., Y. Takano, and P. Yang, 2000: Light scattering and radiative transfer by ice crystal clouds: applications to climate research, in *Light Scattering by Nonspherical Particles: Theory, Measurements, and Geophysical Applications*, M. I. Mishchenko, J.W. Hovenier, and L. D. Travis (eds), pp. 417–449. San Diego: Academic Press.
Liou, K. N., Y. Takano, P. Yang, 2011: Light absorption and scattering by aggregates: application to black carbon and snow grains, *J. Quant. Spectrosc. Radiat. Transfer*, **112**, 1581–1594.

Liu, Q. H., 1999: PML and PSTD algorithm for arbitrary lossy anisotropic media, *IEEE Microw. Guid. Wave Lett.*, **9**, 48–50.

Macke, A., 1993: Scattering of light by polyhedral ice crystals, *Appl. Opt.*, **32**, 2780–2788.

Macke, A., J. Mueller, and E. Raschke, 1996a: Single scattering properties of atmospheric ice crystal. *J. Atmos. Sci.*, **53**, 2813–2825.

Macke, A., M. I. Mishchenko, and B. Cains, 1996b: The influence of inclusions on light scattering by large ice particles, *J. Geophys. Res.*, **101**, 23,311–23,316.

Macke, A., and M. I. Mishchenko, 1996: Applicability of regular particle shapes in light scattering calculations for atmospheric ice particles, *Appl. Opt.*, **35**, 4291–4296.

Mazeron, P., and S. Muller, 1996: Light scattering by ellipsoids in a physical optics approximation, *Appl. Opt.*, **35**, 3726–3735.

Meng, Z. K., P. Yang, G. W. Kattawar, L. Bi, K. Liou and I. Laszlo, 2010, Single-scattering properties of tri-axial ellipsoidal mineral dust aerosols: A database for application to radiative transfer calculations, *Aerosol Science*, doi:10.1016/j.jaerosci.2010.02.008

Merikallio S, H. Lindqvist, T. Nousiainen, M. Kahnert, 2011: Modelling light scattering by mineral dust using spheroids: assessment of applicability, *Atmos. Chem. Phys.*, **11**, 5347–63,

Mishchenko, M. I., and A. Macke, 1998: Incorporation of physical optics effects and δ-function transmission, *J. Geophys. Res.*, **103**, 1799–1805.

Mishchenko, M. I., and A. Macke, 1999: How big should hexagonal ice crystals be to produce halos? *Appl. Opt.*, **38**, 1626–1629.

Mishchenko, M. I., and L. D. Travis, 1998: Capabilities and limitations of a current FORTRAN implementation of the T-matrix method for randomly oriented, rota- tionally symmetry scatterers, *J. Quant. Spectrosc. Radiat. Transfer*, **60**, 309–324.

Mishchenko, M. I., W. J. Wiscombe, J. W. Hovenier, and L. D. Travis, 2000: Overview of scattering by nonspherical particles, in *Light Scattering by Nonspherical Particles: Theory, Measurements, and Geophysical Applications*, M. I. Mishchenko, J. W. Hovenier, and L. D. Travis (eds), pp. 29–60, San Diego: Academic Press.

Mishchenko, M. I., L. D. Travis, and A. A. Lacis, 2002: *Scattering, Absorption, and Emission of Light by Small Particles.* Cambridge, UK: Cambridge University Press.

Muinonen, K., 1989: Scattering of light by crystals: a modified Kirchhoff approximation. *Appl. Opt.*, **28**, 3044–3050.

Muinonen, K., L. Lamberg, P. Fast, and K. Lumme, 1997: Ray optics regime for Gaussian random spheres, *J. Quant. Spectrosc. Radiat. Transfer*, **57**, 197–205.

Nousiainen, T., 2009: Optical modeling of mineral dust particles: A review. *J. Quant. Spectrosc. Radiat. Transfer*, **110**, 1261–1279, doi:10.1016/j.jqsrt.2009.03.002.

Nussenzveig, H. M., 1992: *Diffraction Effects in Semicalssical Scattering*, Cambridge, UK: Cambridge University Press.

Popov A. A., 1996: New method for calculating the characteristics of light scattering by spatially oriented atmospheric crystals, *Proc SPIE*, **2822**, 186–194.

Priezzhev, A. V., S. Yu. Nikitin, and A. E. Lugovtsov, 2009: Ray-wave approximation for the calculation of laser light scattering by transparent dielectric particles, mimicking red blood cells or their aggregates, *J. Quant. Spectrosc. Radiat. Transfer*, **110**, 1535–1544.

Purcell, E. M., and C. R. Pennypacker, 1973: Scattering and absorption of light by nonspherical dielectric grains, *Astrophys. J.*, **186**, 705–714.

Ramanathan, V., P. J. Crutzen, J. T. Kiehl, and D. Rosenfeld, 2001: Aerosols, climate, and the hydrological cycle, *Science*, **294**, 2119–2124.

Ravey, J. C., and P. Mazeron, 1982: Light scattering in the physical optics approximation: application to large spheroids, *J. Opt. (Paris)*, **13**, 273–282.

Reid, E. A., J. S. Reid, M. M. Meier, M. R. Dunlap, S. S. Cliff, A. Broumas, K. Perry, and H. Maring, 2003: Characterization of African dust transported to Puerto Rico by individual particle and size segregated bulk analysis. *Journal of Geophysical Research – Atmospheres*, **108**, 85–91.

Saxon, D. S., Lectures on the scattering of light, in Proceedings of the UCLA Inter- national Conference on Radiation and Remote Sensing of the Atmosphere, J. G. Kuriyan (ed.), pp. 227–308. Western Periodicals, North Hollywood, CA.

Sokolik, I.N., D. Winker, G. Bergametti, D. Gillette, G. Carmichael, Y. J. Kaufman, L. Gomes, L. Schuetz, and J. Penner, 2001: Introduction to special section on mineral dust: outstanding problems in quantifying the radiative impact of mineral dust, *J. Geophys. Res.*, **106**, 18015–18027.

Sun, W., Q. Fu, and Z. Chen, 1999: Finite-difference time-domain solution of light scattering by dielectric particles with perfectly matched layer absorbing boundary conditions, *Appl. Opt.*, **38**, 3141–3151.

Takano, Y., and K. Jayaweera, 1985: Scattering phase matrix for hexagonal ice crystals computed from ray optics, *Appl. Opt.*, **24**, 3254–3263.

Takano, Y., and K. N., Liou, 1989: Solar radiative transfer in cirrus clouds. Part I. Single-scattering and optical properties of hexagonal ice crystals, *J. Atmos. Sci.*, **46**, 3–19.

Takano, Y., and K. N. Liou, 1995: Radiative transfer in cirrus clouds. Part III: Light scattering by irregular ice crystals, *J. Atmos. Sci.*, **52**, 818–837.

Tian, B., and Q. H. Liu, 2000: Nonuniform fast cosine transform and Chebyshev PSTD algorithm, *Prog. Electromagn. Res.*, **28**, 259–279.

van de Hulst, H. C., 1981: *Light Scattering by Small Particles*, New York: Dover.

Volten, H., O. Muñoz, J. W. Hovenier, and L. B. F. M. Waters, 2006: An update of the Amsterdam light scattering database, *J. Quant. Spectrosc. Radiat. Transfer*, **100**, 437–443.

Warren, S. G., and R. E. Brandt, 2008: Optical constants of ice from the ultraviolet to the microwave: A revised compilation, *J. Geophys. Res.*, **113**, D14220, doi:10.1029/2007JD009744.

Wendling, P., R. Wendling, and H. K. Weickmann, 1979: Scattering of solar radiation by hexagonal ice crystals, *Appl. Opt.*, **18**, 2663–2671.

Xie, Y., P. Yang, G. W. Kattawar, B. A. Baum, and Y. Hu, 2011: Simulation of the optical properties of plate aggregates for application to the remote sensing of cirrus clouds, *Appl. Opt.*, **50**, 1065–1081.

Yang, P., and K. N. Liou, 1996a: Finite-difference time domain method for light scattering by small ice crystals in three-dimensional space, *J. Opt. Soc. Am. A*, **13**, 2072–2085

Yang, P., and K. N. Liou, 1996b: Geometric-optics-integral-equation method for light scattering by nonspherical ice crystals, *Appl. Opt.*, **35(33)**, 6568–6584.

Yang, P., and K. N. Liou, 1997: Light scattering by hexagonal ice crystals: solution by a ray-by-ray integration algorithm, *J. Opt. Soc. Amer. A.*, **14**, 2278–2288.

Yang, P., and K. N. Liou, 1998: Single-scattering properties of complex ice crystals in terrestrial atmosphere, *Contr. Atmos. Phys.*, **71**, 223–248.

Yang, P., B.-C. Gao, B. A. Baum, W. Wiscombe, M. I. Mischenko, D. M. Winker, and S. L. Nasiri, 2001: Asymptotic solutions of optical properties of large particles with strong absorption, *Appl. Opt.*, **40**, 1532–1547.

Yang, P., B. A. Baum, A. J. Heymsfield, Y. X. Hu, H.-L. Huang, S.-Chee Tsay, and S. Ackerman, 2003: Single-scattering properties of droxtals, *J. Quant. Spectrosc. Radiat. Transfer*, **79–80**, 1159-1180.

Yang, P., H. L. Wei, H. L. Huang, B. A. Baum, Y. X. Hu, G. W. Kattawar, M. I. Mishchenko, and Q. Fu, 2005: Scattering and absorption property database for nonspherical ice particles in the near- through far-infrared spectral region, *Appl. Opt.*, **44(26)**, 5512–5523.

Yang, P. and K. N. Liou, 2006: Light scattering and absorption by nonspherical ice crystals, in *Light Scattering Reviews*: Single and Multiple Light Scattering,, edited by A. A. Kokhanovsky, Springer-Praxis, pp. 31–71.

Yang, P., Q. Feng, G. Hong, G. W. Kattawar, W. J. Wiscombe, M. I. Mishchenko, O. Dubovik, I. Laszlo, and I. N. Sokolik, 2007, Modeling of the scattering and radiative properties of nonspherical dust particles, *J. Aerosol. Sci.*, **38**, 995–1014.

Yang, P., and K. N. Liou, 2009a: Effective refractive index for determining ray propagation in an absorbing dielectric particle, *J. Quant. Spectrosc. Radiat. Transfer*, **110**, 300–306.

Yang, P., and K. N. Liou, 2009b: An 'exact' geometric-optics approach for computing the optical properties of large absorbing particles, *J. Quant. Spectrosc. Radiat. Transfer*, **110**, 1162–1177.

Yang, P., G. Hong, A. Dessler, S. S. C. Ou, K.-N. Liou, P. Minnis, and Harshvardhan, 2010: Contrails and induced cirrus – Optics and radiation, *Bull. Amer. Meteorol. Soc.*, **90**, 473–478.

Yee, S.K., 1966: Numerical solution of initial boundary value problems involving Maxwell's equations in isotropic media, *IEEE Trans. Antennas Propag.*, **14**, 302–307.

Yi, B., C. N. Hsu, P. Yang, and S.-H. Tsay, 2011: Radiative transfer simulation of dust-like aerosols: Uncertainties from particle shape and refractive index, *J. Aerosol. Sci.*, **42**, 631–644.

Yurkin, M.A. and A. G. Hoekstra, 2011: The discrete-dipole-approximation code ADDA: capabilities and known limitations, *J. Quant. Spectrosc. Radiat. Transfer*, **112**, 2234–2247.

Zhang, Z., P. Yang, G. W. Kattawar, S.-C. Tsay. B. A. Baum, H.-L. Huang, Y. X. Hu, A. J. Heymsfield, and J. Reichardt, 2004: Geometrical-optics solution to light scattering by droxtal ice crystals, *Appl. Opt.*, **43**, 2490–2499.

Zhou, C., P. Yang, A. E. Dessler, Y.-X. Hu, and B. A. Baum, 2012: Study of horizontally oriented ice crystals with CALIPSO observations and comparison with Monte Carlo radiative transfer simulations, *J. Appl. Meteor. Climatol.*, **51**, 1426–1439.

3 Light scattering by large particles: physical optics and the shadow-forming field

Anatoli G. Borovoi

3.1 Introduction

There are a lot of excellent books and papers considering the theory and various approximations to the problem of light scattering by spherical and nonspherical particles (see, for example, van de Hulst, 1981; Bohren and Huffman, 1983; Kokhanovsky, 1999; Mishchenko et al., 2002; and numerous references therein). These works start from the fundamental Maxwell equations and then the desired solutions are derived from the Maxwell equations as some series. Finally, these series are summarized by a computer code. However, such procedures are effective for relatively small nonspherical particles and the maximum particle size occurs to be strongly dependent on computer power. At present, this particle size limit is reached at, say, under the condition: (particle size)/(incident wavelength) < 20. Otherwise such calculations become too computationally expensive.

Fortunately, the problem of light scattering by large particles can be effectively attacked from the opposite side. Namely, there is a classical asymptotics of the Maxwell equations called geometric optics. Here propagation and scattering of the electromagnetic fields are associated with propagation of photons or, equivalently, geometric-optics rays similarly to evident propagation of classical-mechanics particles. Ray-tracing simulation of such propagation and scattering is a common code for this case. In particular, the problem of light scattering by atmospheric ice crystals and coarse aerosol particles has been widely studying by ray-tracing codes already for many years. A survey of these works can be found, for example, in (Liou, 2002; Yang and Liou, 2006; Bi and Yang, 2013; Baran,2013).

In parallel to the numerical calculations of the scattering matrices by means of ray-tracing algorithms, a number of attempts were made in such calculations to take into account the wave properties of light, i.e. interference and diffraction. These works concerned mainly the problem of light scattering by atmospheric particles. Here different algorithms of taking into account the wave properties of light had obtained different names in the literature depending on the volume of the light wave properties included. Thus, if the scattering matrix calculated by a ray-tracing code was supplemented with the forward scattering peak described by the Fraunhofer diffraction from an effective circle (Cai and Liou, 1982), it was called GOM-1 method (geometric-optics method). This procedure extended to any scattering directions (Muinonen, 1989) was called the modified Kirchhoff approximation. In the next GOM-2 or IGOM method (Yang and Liou, 1995; Yang and Liou, 1996), the authors replaced the geometric-optics rays inside a crystal by thin ray tubes called the wavelets or localized waves. Then the electromagnetic fields on ice crystal surfaces were found numerically by tracing these localized waves. Later the localized waves were used also for the integral over a crystal volume (Yang and Liou, 1997) in the method called RBRI (ray-by-ray integration). Though the wave properties of light were included in the GOM-2 and RBRI methods with a reasonable accuracy, these methods proved to be computationally costly because of great number of the ray tubes needed to cover ice crystal facets. Only recently, this drawback of the GOM-2 and RBRI methods was eliminated in the method (Bi et al., 2011; Bi and Ping, 2013) called PGOH (physical-geometric optics hybrid). Here the localized waves were replaced by the real plane-parallel beams of light propagating inside the ice crystals. It is worth noting that the same method for the case of nonabsorbing particles was developed by the author too (Borovoi and Grishin, 2003). In our algorithm, the plane-parallel beams inside the crystals originated from the illuminated crystal facets are numerically calculated.

This variety of terminologies and names of methods is caused by the fact that while geometric optics is a well-defined field of physics, physical optics has not been commonly defined. The first purpose of this chapter is to systemize the approaches achieved in the problem of light scattering by large particles and to define strictly a concept of physical-optics approximations.

The second and more important purpose of this chapter is to draw the attention of the light scattering community to the fact that the main feature peculiar to light scattering by large particles is the appearance of the so-called shadow-forming field. We are going to convince a reader that the shadow-forming field exists in reality. The shadow-forming field is strictly determined at any distance from scattering particles and the shadow-forming field is the same full member of any superposition of scattered waves as, say, the reflected/refracted fields. If one assumes the concept of the shadow-forming field, a lot of various phenomena like the δ-function transmission (Takano and Liou, 1989; Mishchenko and Macke, 1998), the extinction paradox (van de Hulst, 1981), Babinet's principle, etc. becomes physically obvious. Moreover, we demonstrate that a number of important results can be obtained without any analytical calculations if we consider the shadow-forming field in the near zone of the large particles instead of the common wave zone.

3.2 Physical-optics approximations in the problem of light scattering

3.2.1 Light scattering by use of the Maxwell equations

The Maxwell equations in the problem of light scattering by an arbitrary particle can be reduced to the following differential equation for the electric field $\mathbf{E}(\mathbf{r})$

$$(L - V)\mathbf{E} = 0 \tag{3.1}$$

where $L = -rotrot + k^2$ is the propagation operator for a free space, $k = 2\pi/\lambda$, λ is the wavelength in the free space, $V(\mathbf{r}) = k^2[1 - m^2(\mathbf{r})]$, and $m(\mathbf{r})$ is the complex-valued refractive index of the particle in the point $\mathbf{r} = (x, y, z)$. The function $V(\mathbf{r})$ becomes zero outside the volume occupied by a particle. Therefore this function $V(\mathbf{r})$ is sometimes more convenient to determine a size, shape and structure of a particle as compared with the refractive index $m(\mathbf{r})$.

The differential equation (3.1) is equivalent to the volume integral equation

$$\mathbf{E}(\mathbf{r}) = \mathbf{E}_0(\mathbf{r}) + \int G(\mathbf{r}, \mathbf{r}')V(\mathbf{r}')\mathbf{E}(\mathbf{r}')\, d\mathbf{r}' \tag{3.2}$$

where

$$G(\mathbf{r}, \mathbf{r}') = L^{-1} = \left(\frac{\nabla_r \nabla_{r'}}{k^2} - \hat{\mathbf{1}}\right) \frac{e^{ik|\mathbf{r}-\mathbf{r}'|}}{4\pi|\mathbf{r} - \mathbf{r}'|}$$

is the volume Green function and

$$\mathbf{E}_0(\mathbf{r}) = \mathbf{E}^0\, e^{ik\mathbf{n}_0\mathbf{r}} \tag{3.3}$$

is the plane wave incident on the particle in the direction $\mathbf{n}_0$ where $|\mathbf{n}_0| = 1$.

Equation (3.2) corresponds to the general superposition of the total field $\mathbf{E}(\mathbf{r})$ into the incident and scattered fields

$$\mathbf{E}(\mathbf{r}) = \mathbf{E}_0(\mathbf{r}) + \mathbf{E}_s(\mathbf{r}) \tag{3.4}$$

According to Eq. (3.2), the scattered wave is the integral over the volume occupied by the particle

$$\mathbf{E}_s(\mathbf{r}) = \int G(\mathbf{r}, \mathbf{r}')V(\mathbf{r}')\mathbf{E}(\mathbf{r}')\, d\mathbf{r}' \tag{3.5}$$

Two general statements are now noticeable. First, if we know the scattered field inside the particle, it is simply found outside the particle by means of the integral (3.5). Second, if we know the scattered field at any surface S surrounding the particle, it is found outside the surface by means of the surface integral with the surface Green function $G_S(\mathbf{r}, \mathbf{r}')$, see, e.g., (Jackson, 1999)

$$\mathbf{E}_s(\mathbf{r}) = \int_S G_S(\mathbf{r}, \mathbf{r}')\mathbf{E}(\mathbf{r}')\, d\mathbf{r}' \tag{3.6}$$

At far distance from a particle $R = |\mathbf{r} - \mathbf{r}_0| \to \infty$, where $\mathbf{r}_0$ is a point chosen as a center of the particle, the scattered field is transformed into the divergent spherical wave

$$\mathbf{E}_s(R, \mathbf{n}) = \frac{1}{R}\, e^{ikR + ik\mathbf{n}_0\mathbf{r}_0}\mathbf{J}(\mathbf{n}, \mathbf{n}_0)\mathbf{E}^0 \tag{3.7}$$

where $\mathbf{n} = (\mathbf{r} - \mathbf{r}_0)/|\mathbf{r} - \mathbf{r}_0|$ is the scattering direction, and the matrix $\mathbf{J}$ of 2×2 dimensions is responsible for polarization of the transverse electromagnetic wave. We shall call the matrix as the Jones matrix for brevity though there are a number of other names.

For the quadratic values of the field, i.e. for the Stokes vectors **I**, we have the similar equation, where the Jones matrix is replaced by the (4×4) Mueller matrix **M**

$$\mathbf{I}_s(R,\mathbf{n}) = \frac{1}{R^2}\mathbf{M}(\mathbf{n},\mathbf{n}_0)\mathbf{I}_0 \tag{3.8}$$

3.2.2 Geometric optics versus the Maxwell equations

Geometric optics is a well-known asymptotics to the Maxwell equations that is indispensable for the case of small wavelengths $\lambda \to 0$. Here the electric field is represented as the series over powers of the wavelength $\lambda = 2\pi/k$ (Born and Wolf, 1959)

$$\mathbf{E}(\mathbf{r}) = e^{ikL(\mathbf{r})}\sum_{j=0}^{\infty}\frac{\mathbf{e}_j(\mathbf{r})}{(ik)^j} \tag{3.9}$$

The scalar function $L(\mathbf{r})$ is called the eikonal and the vector functions $\mathbf{e}_j(\mathbf{r})$ are the jth order amplitudes. Substitution of the series (3.9) into the Maxwell equations results in the following equation for the eikonal

$$(\nabla L(\mathbf{r}))^2 = n^2(\mathbf{r}) \tag{3.10}$$

where $n(\mathbf{r}) = \operatorname{Re} m(\mathbf{r})$ is the real part of the refractive index. Solutions of Eq. (3.10) are a set of ray trajectories. The surfaces perpendicular to the ray trajectories are the wave fronts where phases of the electromagnetic waves are constant. The vector amplitudes $\mathbf{e}_j(\mathbf{r})$ are tangent to the wave fronts obeying their own equations. Here only the zeroth amplitude $\mathbf{e}_0(\mathbf{r})$ is of physical importance since it provides a conservation of the energy flux along any ray tube.

There is a remarkable analogy that classical mechanics is obtained from quantum mechanics by means of the same mathematical procedure. Such an analogy is a powerful instrument to explain various physical regularities appearing in both the electrodynamics and quantum mechanics.

In particular, within the geometric optics, light propagation can be treated as a motion of photons along the ray trajectories. These trajectories are curved inside an inhomogeneous medium. If a photon meets an interface, i.e. a jump of the refractive index, its trajectory becomes broken in accordance to the well-known reflection/refraction laws. In this case, the incident trajectory sets are transformed into other sets and so on. As a result, a geometric-optics field $\mathbf{E}_g(\mathbf{r})$ becomes the superposition that summarizes different trajectory sets

$$\mathbf{E}_g(\mathbf{r}) = \sum_l \mathbf{e}_0^{(l)}(\mathbf{r})\, e^{ikL^{(l)}(\mathbf{r})} \tag{3.11}$$

These photons trajectories are easy simulated by computer ray-tracing codes that can include the phases or eikonals as well. Sometimes the phases of light are of no interest and then we arrive at the geometric optics without interference that coincides completely with propagation of classical-mechanics particles.

3.2.3 Light scattering by use of geometric optics

Thus, the problem of light scattering by a large particle within the framework of geometric optics consists of obvious findings of ray trajectories or ray tubes according to the eikonal equation (3.10). Fig. 3.1 shows such kinds of the findings for a homogeneous nonspherical particle. Here, for an optically soft particle in Fig. 3.1(a), we can neglect light reflection and the scattering directions are concentrated near the incident direction unlike the case of the optically hard particle depicted in Fig. 3.1(b).

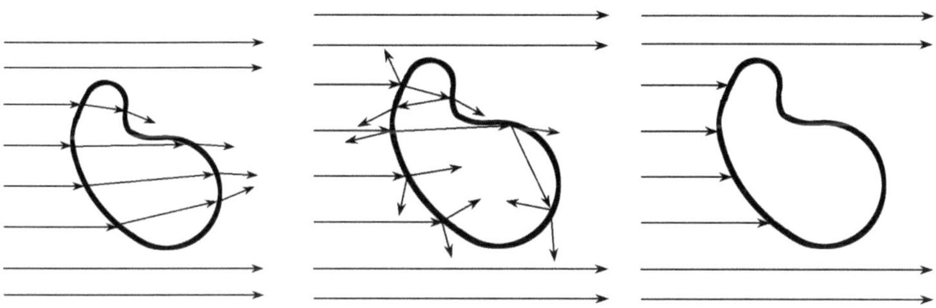

Fig. 3.1. Light scattering by a large homogeneous nonspherical particle for: (a) an optically soft particle $|n-1| \ll 1$; (b) an optically hard particle, (c) a perfectly absorbing particle (no reflection/refraction).

The geometric-optics field $\mathbf{E}_g(\mathbf{r})$ of Eq. (3.11) is often needed at far distances from an object

$$R \gg a \tag{3.12}$$

where a is a characteristic particle size. At this distance, a particle is seen as a point source of light with variable radiance relative to the scattering direction $\mathbf{n} = (\mathbf{r} - \mathbf{r}_0)/|\mathbf{r} - \mathbf{r}_0|$ as illustrated in Fig. 3.2. Here all photons leaving a particle with the same propagating or scattering direction $\mathbf{n}$ are collected in one point $\mathbf{n}$ situated on the scattering direction sphere. This sphere is constructed by means of a replacement of the 3-D variable $\mathbf{r}$ by two variables: the distance $R = |\mathbf{r} - \mathbf{r}_0|$ and scattering direction $\mathbf{n} = (\mathbf{r} - \mathbf{r}_0)/|\mathbf{r} - \mathbf{r}_0|$ that are used in the general Eqs.(3.7) and (3.8), too. In literature, there is no name for the distance defined by the simple inequality of Eq. (3.12) where the variables R and $\mathbf{n}$ are effective. In this chapter we call it as the remote zone, for brevity.

It is worthwhile to note that, for such a summation in the remote zone, we use the fields presented by the superposition (3.11), i.e. both the amplitude $\mathbf{e}_0^{(l)}$ and the phase $kL^{(l)}$ of a summand are taken into account. The scattered fields obtained have the same structure as the exact Eqs. (3.7) and (3.8), i.e. we have

$$\mathbf{E}_g(R, \mathbf{n}) = \frac{1}{R} e^{ikR + ik\mathbf{n}_0\mathbf{r}_0} \mathbf{J}_g(\mathbf{n}, \mathbf{n}_0)\mathbf{E}^0 \tag{3.13}$$

$$\mathbf{I}_g(R, \mathbf{n}) = \frac{1}{R^2} \mathbf{M}_g(\mathbf{n}, \mathbf{n}_0)\mathbf{I}_0 \tag{3.14}$$

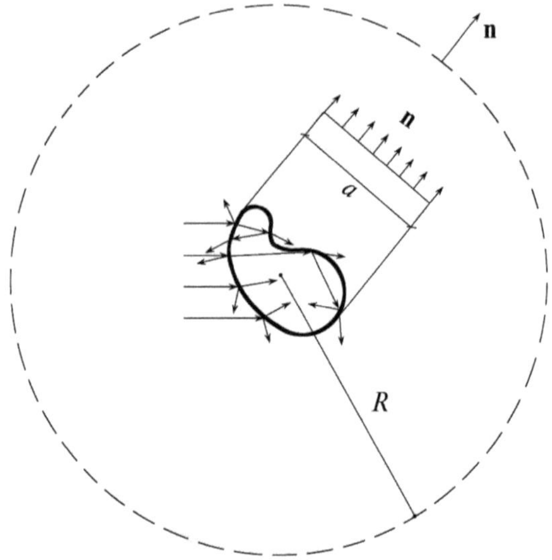

Fig. 3.2. Scattering of light by a large particle to the remote zone.

Thus, the Jones and Mueller matrices appeared in the remote zone within the framework of geometric optics are some approximations to the exact matrices **J** and **M**. The geometric-optics Mueller matrices $\mathbf{M}_g(\mathbf{n}, \mathbf{n}_0)$ are usually calculated numerically by some ray-tracing techniques. In such calculations, the phase factors $\exp(ikL^{(l)})$ of the terms can be often omitted. It means that interference has been ignored and light scattering is considered as totally equivalent to scattering of classical-mechanics particles. Such kinds of the Mueller matrices obtained for light scattering by ice crystals are surveyed in (Liou, 2002; Yang and Liou, 2006).

3.2.4 What is physical optics? Diffraction and interference

Historically, physical optics appeared as an understanding that light is not an ensemble of corpuscles but it is a wave. Two phenomena were reasons for this understanding: interference and diffraction.

One may state that there is no common definition of what the physical optics is. In all textbooks, interference and diffraction are discussed after an evident concept of ray trajectories. Therefore it may be inferred that physical optics is an extension of geometric optics by inclusion of both interference and diffraction. However, such a definition has the following drawback. If we assume that geometric optics is strictly defined by Eqs. (3.9)–(3.11), then we shall see that interference has been already included in geometric optics. Indeed, interference means taking into account phases of waves, but this is the eikonal that just determines a phase of an electromagnetic wave. Moreover, in a lot of experimental schemes in textbooks, for example, reflection and transmission of light through thin plane-parallel plates are successfully treated by use of ray trajectories where a phase along rays explains the interference phenomena in the plates discussed in these textbooks.

So, it is only diffraction that is not included in the geometric optics equations. Diffraction means a violation of the geometric-optics law that light propagates only along ray tubes. Therefore diffraction can be generally defined as a penetration of light energy from any ray tube to the neighbor tubes. Thus, we can state that physical optics is an extension of geometric optics by inclusion of only diffraction. This is the second step to the strict Maxwell equations as compared with the first step of geometric optics. However, at present, there is no common mathematical procedure to take diffraction into account. Therefore there is no a mathematical definition of physical optics. *Anybody who extends geometric optics by means of diffraction can state that he uses a physical-optics approximation.*

In optics, it is common to associate diffraction with transmission of light through an aperture in a black thin screen where a size of the aperture is much larger than the wavelength $a \gg \lambda$ (see Fig. 3.3). In radiophysics, the term of diffraction is treated more widely. Here the terms of diffraction and scattering are equivalent. Indeed, from the standpoint of the Maxwell equations, if we have either a particle of finite size or a finite aperture in an infinite screen, in both cases we should use the same mathematical methods and the results obtained would be similar either quantitatively or, at least, qualitatively.

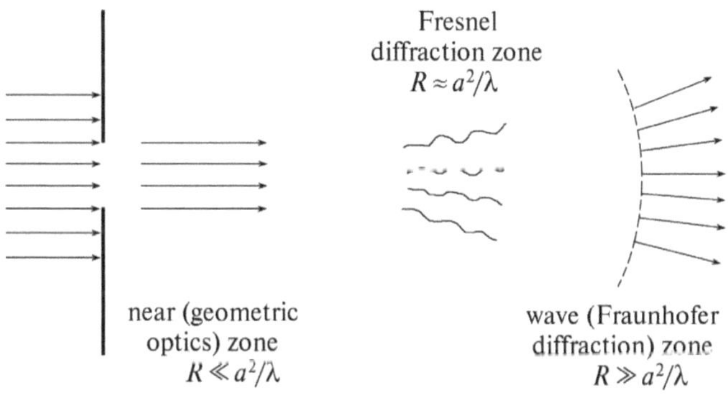

Fig. 3.3. Diffraction by a large aperture at different distances.

Main qualitative features of diffraction are well seen in the classical experimental scheme where a plane electromagnetic wave is incident on a black screen with a large aperture $a \gg \lambda$ (Fig. 3.3). Mathematically, this problem is described by just Eq. (3.6) that was written above for a scattering problem. A classical approach to a solution of this problem is the Kirchhoff approximation. In this approximation, the surface integral of Eq. (3.6) is taken only over the aperture and the unknown electromagnetic field inside the aperture is replaced by the incident field $\mathbf{E}_0(\mathbf{r})$ or, that is the same, by its geometrical optics value $\mathbf{E}_{0g}(\mathbf{r})$

$$\mathbf{E}_s(\mathbf{r}) \approx \int_S G_S(\mathbf{r}, \mathbf{r}')\mathbf{E}_0(\mathbf{r}')\, d\mathbf{r}' = \int_S G_S(\mathbf{r}, \mathbf{r}')\mathbf{E}_{0g}(\mathbf{r}')\, d\mathbf{r}' \tag{3.15}$$

It is well known that the electromagnetic wave just after the screen becomes a plane-parallel beam propagating in the same direction as the incident wave. By the way, let us notice that this plane-parallel beam can be created in space by other ways, too. For example, it can be generated by a laser. Also it can be created by reflection/refraction of a plane wave from ice crystal facets, etc. So, in general, we arrive at the problem of propagation of any plane-parallel beams.

As known, diffraction does not practically distorts the beam and the light propagates inside its plane-parallel geometric-optics ray tube of the transversal size a until the distance $R \ll a^2/\lambda$. Therefore this distance $R \ll a^2/\lambda$ is called the near zone or the geometric-optics zone.

At distance $R \approx a^2/\lambda$, geometric optics is violated by the Fresnel diffraction that means a smoothing of the sharp geometric-optics edges of the beams. In addition, some transversal inhomogeneities of light amplitudes appear across the beam. This distance $R \approx a^2/\lambda$ is called the Fresnel diffraction zone.

And at far distance $R \gg a^2/\lambda$, the Fraunhofer diffraction transforms the beam into a diverging spherical wave corresponding to the general Eq. (3.7). This distance is called the wave zone or the Fraunhofer diffraction zone.

3.2.5 Physical-optics approximations

Thus, for a mathematical definition of a physical-optics approximation, we have to take into account diffraction in the geometric-optics scattered fields shown in Fig. 3.1(a–c). Assume that a scattering particle is sufficiently large $a \gg \lambda$ to justify the ray structure of the scattered fields drawn in Fig. 3.1. Then a substitution of the total geometric-optics field $\mathbf{E}_g(\mathbf{r})$ consisting of the incident $\mathbf{E}_0(\mathbf{r}) = \mathbf{E}_{0g}(\mathbf{r})$ and the scattered geometric-optics field $\mathbf{E}_{sg}(\mathbf{r})$

$$\mathbf{E}_g(\mathbf{r}) = \mathbf{E}_{0g}(\mathbf{r}) + \mathbf{E}_{sg}(\mathbf{r}) \tag{3.16}$$

into the general integrals of Eqs. (3.5) and (3.6) that are taken either over the particle volume

$$\mathbf{E}_s(\mathbf{r}) \approx \int G(\mathbf{r},\mathbf{r}')V(\mathbf{r}')\mathbf{E}_g(\mathbf{r}')\,d\mathbf{r}' \tag{3.17a}$$

or over a surface surrounding this particle in the near zone

$$\mathbf{E}_s(\mathbf{r}) \approx \int_S G_S(\mathbf{r},\mathbf{r}')\mathbf{E}_g(\mathbf{r}')\,d\mathbf{r}' \tag{3.17b}$$

results in a desired approximation for the scattered field.

Consider the properties of this scattered field. For the first, this approximation should lead to the same geometric-optics scattered fields if an observing point is situated in the near zone. This fact is provided by the asymptotic transformation of the Maxwell equations into the geometric optics equations within the near zone.

Then, moving an observation point $\mathbf{r}$ away from the particle, an observer should pass the distances where the geometric-optics spherical wave of Eqs. (3.13) and (3.14) is already formed according to Fig. 3.2, but the Fresnel diffraction is not essential yet. In this region, light propagates along the conical ray tubes shown in Fig. 3.4. For example, if a particle is a large ice crystal of a fixed orientation, its geometric-optics Mueller matrix is a superposition of the Dirac δ-functions $\delta(\mathbf{n}-\mathbf{n}_j)$ on the scattering direction sphere (Borovoi et al., 2005). Here the points $\mathbf{n}_j$ indicate the propagating directions of plane-parallel beams leaving the crystal. Localization of these points $\mathbf{n}_j$ on the scattering direction sphere does not depend of the distance R.

Coming to the Fresnel diffraction zone $R \approx a^2/\lambda$, propagation of light along conical tubes is violated as shown in Fig. 3.4 that is similar to Fig. 3.3. In the Fresnel diffraction zone, light energy is exchanged between the neighbor ray tubes like the transverse diffusion. In radiophysics, such a treatment of the Fresnel diffraction is well known, see, e.g., Nussenzveig (1992). And, finally, this redistribution of light energy among the conical ray tubes is completed in the wave zone $R \gg a^2/\lambda$. Here light propagates again along the conical ray tubes as shown in Fig. 3.4 that is described by the exact Mueller matrix of Eq. (3.8). Now it is obvious that the exact Mueller matrix is a result of smoothing geometric-optics Mueller matrix of Eq. (3.14) because of the Fraunhofer diffraction. In particular, for the aforementioned case of an ice crystal of a fixed orientation, the Dirac δ-functions are smoothed into the Fraunhofer diffraction patterns of the outgoing plane-parallel beams of given transversal shape.

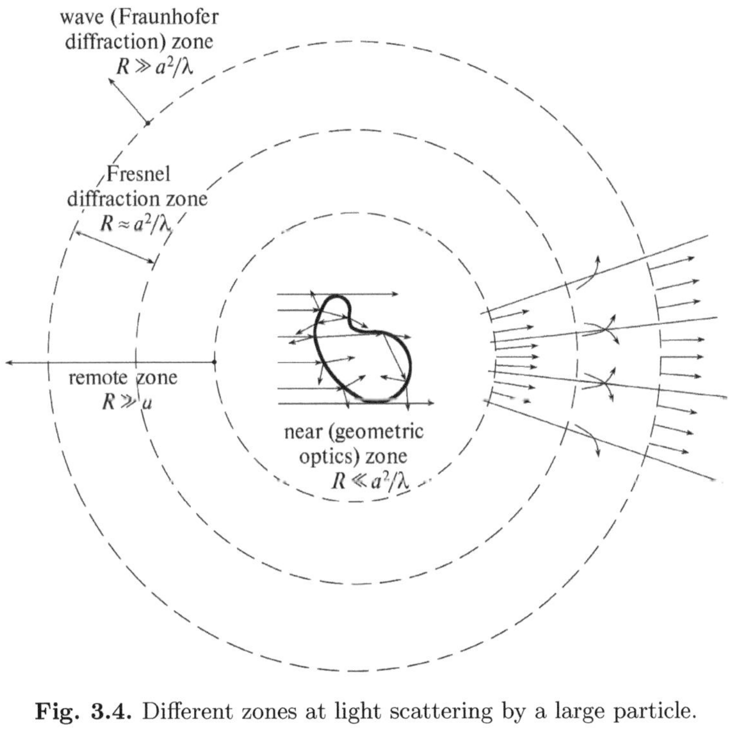

Fig. 3.4. Different zones at light scattering by a large particle.

Thus, the substitution of the geometric optics field found in the near zone of the particle into general equations (3.5) and (3.6) has taken into account diffraction in both the Fresnel and wave zones of the particle. We have obtained the following conclusion:

Conclusion 1. *Equations (3.17) are the desired mathematical definition of the simplest physical-optics approximation. Such an approximation is the direct extension of the classical Kirchhoff approximation (Eq. (3.15)) from a large aperture to any large 3-D scatterers.*

Factually, this approximation was used by majority of authors considering light scattering by large particles. For example, this approximation was used in both GOM-2 (Yang and Liou, 1995) and GPOH (Bi et al., 2011) methods. Also the author (Borovoi and Grishin, 2003) called this approximation GALP (general approximation for large particles), etc. Now we propose to refuse the complicated terminologies and assume that this is only the simplest approximation of physical optics directly generalizing the classical Kirchhoff approximation.

Of course, other more complicated approximations of physical optics can be defined, too. Let us indicate two such possibilities. For the first, assume that a large particles of typical sizes $L \gg \lambda$ includes more small either volume or surface inhomogeneities of sizes $a \gg \lambda$. Here, if $a^2/\lambda \leq L$, either the Fresnel or Fraunhofer diffraction from these inhomogeneities appears inside the particle. Fig. 3.5 illustrates existence of the Fresnel diffraction inside a long cylinder. Here a thin front layer of the cylinder marked by a thin curve forms the Fresnel diffracted field both inside and outside of the similar rear layer. In such cases, our approximation of Eqs. (3.17a) and (3.17b) should be replaced by other, more complicated, equations.

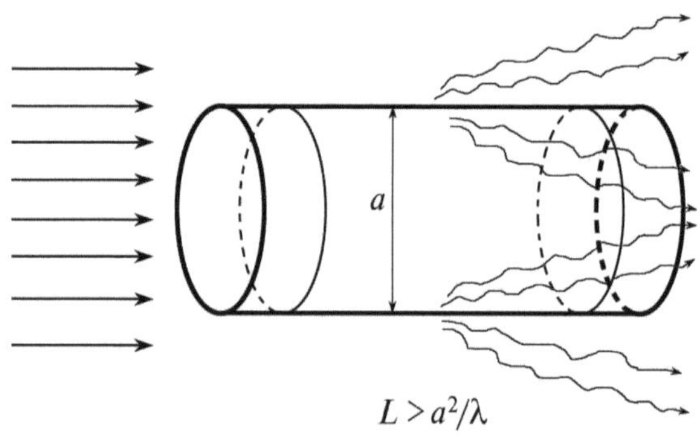

Fig. 3.5. Fresnel diffraction inside a particle.

For the second, from the standpoints of the Maxwell equations, any geometric-optics field considered on a surface of a large particle is only the first term of certain asymptotic series. It is reasonable to add the next term of this series as against Eq. (3.16)

$$\mathbf{E}'_g(\mathbf{r}) = \mathbf{E}_{0g}(\mathbf{r}) + \mathbf{E}_{sg}(\mathbf{r}) + \mathbf{E}_{edge}(\mathbf{r}) \tag{3.18}$$

Substitution of this field into Eq. (3.17b) leads to another physical-optics approximation. In such an approach, the case of a homogeneous sphere was mainly studied because of its relative simplicity. In particular, the spherical Earth surface was considered in radiophysics to estimate the radiowave beyond-the-horizon communications (Fock, 1965). Analogously, in quantum mechanics there is a problem to estimate the small diffraction term at large scattering angles for high-energy incident particles (Landau and Lifshitz, 1991). In both cases, the scattered field is considered in the region where a contribution from the geometric-optics field $\mathbf{E}_{sg}$ is small and the term $\mathbf{E}_{edge}$ becomes essential. In optics, this term was recently used to study a transition from geometric optics to the exact Mie solution for spherical particles (Liou et al., 2010). The term $\mathbf{E}_{edge}(\mathbf{r})$ was associated with the edge effect, tunneling effect, surface waves, etc (Nussenzveig, 1992). It is worthwhile to note, that the similar terms were also studied in radiophysics for estimation of radar signals reflected from scattering objects of more complicated shapes (Ufimtsev, 2007). Recently (Bi et al., 2010; Bi and Yang, 2013), light scattering by the long cylinder depicted in Fig. 3.5 was calculated by use of the discrete dipole approximation (DDA). Here a deviation of the scattered field calculated by DDA from the field corresponding to the physical-optics approximation of Eq. (3.17) was also associated with the edge effect. Since terminology in this field of optics is not generally accepted, we would prefer to interpret this deviation more directly as an influence of the Fresnel diffraction inside a large particle on the scattered field.

3.3 The shadow-forming field

3.3.1 Does the shadow-forming field exist in reality?

The main feature of the problem of light scattering by large $a \gg \lambda$ particles is appearance of the so-called shadow-forming field. The shadow-forming field is a necessary component of any field scattered by a large particle. To make sure of that it is enough to look at Fig. 3.1 representing the total field $\mathbf{E}(\mathbf{r})$ in the near zone. In all Fig. 3.1(a) to 3.1(c) we watch the obvious shadow on the background of the incident plane-parallel rays. This shadow pattern is pure for a perfectly absorbing particle in Fig. 3.1(c), otherwise it is superposed with refracted/reflected rays in Figs. 3.1(a) and 3.1(b). Mathematically, the shadow-forming field appears when we use the general superposition of the total field into the incident and scattered components

$$\mathbf{E}(\mathbf{r}) = \mathbf{E}_0(\mathbf{r}) + \mathbf{E}_s(\mathbf{r}) \tag{3.19}$$

The scattered field will be further considered in the physical-optics approximation of Eq. (3.17). Consequently, the total geometric-optics fields depicted in Fig. 3.1 describe the reality. In particularly, in the simplest case of perfectly absorbing particles shown in Fig. 3.1(c) it is obvious that the total field $\mathbf{E}(\mathbf{r})$ is equal to zero inside the shadow and it is equal to the incident wave outside the shadow. In this case the scattered field of Eq. (3.19) is reduced to the field

$$E_{sh}(\mathbf{r}) = \begin{cases} -\mathbf{E}_0(\mathbf{r}) & \textit{inside the particle shadow in the near zone} \\ 0 & \textit{outside the particle shadow in the near zone} \end{cases} \tag{3.20}$$

that is called the shadow-forming field since its superposition with the incident field transforms the total field into zero.

As seen in Figs. 3.1(a) and 3.1(b), the shadow-forming field appears for arbitrary particles as well, but here the scattered field becomes a superposition of the shadow-forming field and the refracted field $\mathbf{E}_r(\mathbf{r})$ created by a set of the refracted/reflected rays in the near zone

$$\mathbf{E}_s(\mathbf{r}) = \mathbf{E}_{sh}(\mathbf{r}) + \mathbf{E}_r(\mathbf{r}) \tag{3.21}$$

According to Fig. 3.3, the shadow-forming field propagates in the near zone as a plane-parallel beam with a transverse shape corresponding to a shadow or projection of a particle. So, its energy flux is equal to

$$\Phi_{sh} = s \tag{3.22}$$

where s is area of the shadow. At large distances from the particle where the Fresnel diffraction does not appear $a \ll R \ll a^2/\lambda$, i.e in the so-called remote zone, this plane-parallel beam can be represented, if needed, in the variables of the distance $R = |\mathbf{r} - \mathbf{r}_0|$ and of the scattering direction $\mathbf{n} = (\mathbf{r} - \mathbf{r}_0)/|\mathbf{r} - \mathbf{r}_0|$. In these variables, the intensity of the plane-parallel beam is equal to

$$I(R, \mathbf{n}) = \frac{s}{R^2} \delta(\mathbf{n} - \mathbf{n}_0) \tag{3.23}$$

where δ is the Dirac delta-function. Then, according to Figs. 3.3 and 3.4, Eq. (3.23) is violated at the distances $R \approx a^2/\lambda$ since the Fresnel diffraction forces to move light energy from the central ray tube to the neighbor tubes. At last, in the wave zone $R \gg a^2/\lambda$, this redistribution of light energy among conical ray tubes is completed and the final angular distribution of intensity is described by the Fraunhofer diffraction pattern for a given shadow size and shape.

One may ask a question: does the shadow-forming field exist in reality? In the other words, does the shadow-forming field exist for small particles, say $a \approx \lambda$? The full-length answer is as following. For the first, Eq. (3.20) gives a strict mathematical definition of the shadow-forming field. Though Eq. (3.20) defines the field in the near zone, it is determined at any distance by means of the surface Green function in Eq. (3.6). For the second, any field can be decomposed in a superposition of any components that we like. For example, if we define any component $\mathbf{E}_1$ in an exact scattered field $\mathbf{E}_s(\mathbf{r}) = \mathbf{E}_1(\mathbf{r}) + \mathbf{E}_2(\mathbf{r})$, the rest component $\mathbf{E}_2$ will compensate an approximation used for the first component.

Thus, the shadow-forming component defined by Eq. (3.20) can be used in any superposition of fields including the case of small particles, too. However, such a procedure is not expedient for small particles because it would result in an unjustified complexity both physically and mathematically. As for the large particles $a \gg \lambda$, the shadow-forming component is a realistic component of the scattered fields and it is a powerful instrument to study the scattering problem as it will be shown below.

3.3.2 Conservation of the partial energy fluxes

In the theory of light scattering, there is a simple and useful law of conservation for partial energy fluxes of scattered waves that is not widely explored in the literature. To emphasize its importance, we describe this law in a separate section. Let us surround a scattering particle by an arbitrary closed surface. The total field $\mathbf{E}(\mathbf{r})$ can be represented as a finite sum of arbitrary components

$$\mathbf{E}(\mathbf{r}) = \sum \mathbf{E}_j(\mathbf{r}) \tag{3.24}$$

Denote an energy flux of one of these components through this surface as Φ_j. It is well known that if a field $\mathbf{E}_j(\mathbf{r})$ is generated by some source, the flux of this field through any surrounding closed surface is conserved for any distance from the source and for any surface. Therefore the flux Φ_j is a constant through the near zone, Fresnel diffraction zone and wave zone.

For a superposition of two waves $\mathbf{E}_1 + \mathbf{E}_2$ the flux consists of three terms

$$\Phi = \Phi_1 + \Phi_2 + \Phi_{12} \tag{3.25}$$

where Φ_1 and Φ_2 are fluxes of the components and the flux Φ_{12} is formed by interference of these fields. Since three fluxes Φ, Φ_1 and Φ_2 are constants independently of the shape of the surrounding surface, the interference flux Φ_{12} should be also a constant though the interference pattern between two fields is often a rapidly oscillating function on a surface. These results are directly generalized for any number of components in Eq. (3.24)

$$\Phi = \sum_j \Phi_j + \sum_{j \neq l} \Phi_{jl} \tag{3.26}$$

where all terms are constants independently of a distance and shape of a surrounding surface.

3.3.3 Scattering and extinction cross-sections

Now let us come back to the general superposition of the total field $\mathbf{E}(\mathbf{r}) = \mathbf{E}_0(\mathbf{r}) + \mathbf{E}_s(\mathbf{r})$ where $\mathbf{E}_0(\mathbf{r})$ and $\mathbf{E}_s(\mathbf{r})$ are the incident and exact scattered fields, respectively. Intensity of the incident field is common to assume as unity ($|\mathbf{E}_0(\mathbf{r})|^2 = 1$). By definition, the scattering cross-section σ_s is the energy flux of the scattered field $\mathbf{E}_s(\mathbf{r})$ over any surface surrounding a particle

$$\sigma_s = \Phi_s \tag{3.27}$$

The scattered wave is produced by the incident wave, therefore its energy appears because of extraction of the same energy from the incident wave. A part of the energy can be also absorbed inside a particle if absorption takes place. This absorption is characterized by the absorption cross-section σ_a. The absorption is a sink of energy inside a particle. Therefore the value σ_a is determined mathematically as the flux of the total field $\mathbf{E}(\mathbf{r})$ over a surrounding surface with the negative sign

$$\sigma_a = -\Phi \tag{3.28}$$

Total extraction of energy from the incident wave is characterized by the sum

$$\sigma_e = \sigma_a + \sigma_s \tag{3.29}$$

that is called the extinction cross-section σ_e.

There is a fact that is important for future discussion. Namely, extinction can be also treated as a result of interference between the incident and scattered waves. Indeed, let us substitute Eq. (3.25) in Eq. (3.28)

$$\sigma_a = -\Phi = -(\Phi_0 + \Phi_s + \Phi_{0s}) \tag{3.30}$$

Here the flux of the incident field is equal to zero $\Phi_0 = 0$ because a source of the incident wave is situated outside of the surface surrounding a particle. The flux of the scattered wave is the scattering cross-section $\Phi_s = \sigma_s$ by definition. As a result, we get an alternative definition of the extinction cross-section. This value σ_e proves to be just the interference flux Φ_{0s}

$$\sigma_e = -\Phi_{0s} \tag{3.31}$$

Two interesting conclusions appear from Eq. (3.31):

Conclusion 2. *The total extraction of energy from the incident wave is equal to the interference flux Φ_{0s} between the incident and scattered waves.*

Conclusion 3. *If we know the scattered wave $\mathbf{E}_s(\mathbf{r})$ on any surface surrounding a particle, this wave $\mathbf{E}_s(\mathbf{r})$ already contains information about absorption inside the particle.*

Conclusion 2 is a direct consequence of the energy conservation law presented factually by Eq. (3.30). Indeed, if there is no absorption $\sigma_a = -\Phi = 0$ we get

$$\sigma_s = \Phi_s = -\Phi_{0s} \tag{3.32}$$

i.e. a carry-over of energy by the scattered wave is compensated by its extraction from the incident wave because of the interference. Then Conclusion 3 includes absorption as $\sigma_a + \Phi_s = -\Phi_{0s}$.

According to the previous section, the interference flux Φ_{0s} is conserved for any surrounding surface. As an example, let us calculate analytically the flux Φ_{0s} if a surrounding surface is a sphere situated in the wave zone $R \gg a^2/\lambda$. Here the scattered wave becomes a diverging transverse electromagnetic wave due to Eq. (3.7)

$$\mathbf{E}_s(R, \mathbf{n}) = \frac{1}{R} e^{ikR + ik\mathbf{n}_0\mathbf{r}_0} \mathbf{J}(\mathbf{n}, \mathbf{n}_0)\mathbf{E}^0 = \frac{1}{R} e^{ikR + ik\mathbf{n}_0\mathbf{r}_0} \mathbf{E}^{(s)}(\mathbf{n}, \mathbf{n}_0) \tag{3.33}$$

To calculate the flux Φ_{0s}, we arrive at a well-known problem of interference between plane and spherical waves. Intensity of the total field on the plane $z = const$ perpendicular to the incident direction $\mathbf{n}_0$ is equal to

$$I(z, \boldsymbol{\rho}) = |\mathbf{E}^0 \, e^{ikz} + \mathbf{E}^{(s)}(\mathbf{n}(\boldsymbol{\rho}), \mathbf{n}_0) \, e^{ikR(\boldsymbol{\rho})}/R(\boldsymbol{\rho})|^2 \tag{3.34}$$

where $\boldsymbol{\rho} = (x, y)$ are 2-D coordinates on the plane. Here the quadratic values determine the fluxes of the incident and scattered waves, respectively, and the cross terms determine the interference flux Φ_{0s} over this plane as the following integral

$$\Phi_{0s} = 2 \, \mathrm{Re} \int \left[\mathbf{E}^{*0} \mathbf{E}^{(s)}(\mathbf{n}, \mathbf{n}_0) \, e^{ik[R(\boldsymbol{\rho}) - z]} \right] R^{-1}(\boldsymbol{\rho}) \, d\boldsymbol{\rho} \tag{3.35}$$

The exponential in the integrand describes the well-known oscillating and alternating in sign Fresnel rings (Born and Wolf, 1959). An integral of them is reduced to contribution mainly from the central Fresnel ring that is formally provided by the following integral.

$$\int e^{ikR(\boldsymbol{\rho})} R^{-1}(\boldsymbol{\rho}) \, d\boldsymbol{\rho} = \int_0^{2\pi} d\varphi \int_z^{\infty} dR \, e^{ikR} = i\lambda \, e^{ikz} \tag{3.36}$$

Within the central Fresnel ring, the scattering amplitude varies negligibly and it can be replaced by its value $\mathbf{E}^{(s)}(\mathbf{n}_0, \mathbf{n}_0)$ at the forward scattering direction. Then application of Eq. (3.36) to Eq. (3.35) results in the famous optical theorem

$$\sigma_e = -\Phi_{0s} = 2\lambda \operatorname{Im}[\mathbf{E}^0 \mathbf{E}^{(s)*}(\mathbf{n}_0\,, \mathbf{n}_0)] \tag{3.37}$$

We would comment the optical theorem by the next conclusion:

Conclusion 4. *In the wave zone $R \gg a^2/\lambda$, extraction of energy from the incident wave takes place only in the vicinity of the forward scattering direction. Therefore the extinction cross-section is strictly expressed through the amplitude of the scattered wave taken in the forward scattering direction.*

3.3.4 Cross-sections for large optically hard particles

All equations of the previous section are quite general; they are applicable for both small and large particles. In this section, we are going to show that the cross-sections σ_s, σ_a, and σ_e for large particles are found without tedious calculations if we consider the partial energy fluxes not in the wave zone but in the near zone.

Let us begin from the simplest case of a perfectly absorbing particle depicted in Fig. 3.1(c). Here the scattered field is reduced to the shadow-forming field $\mathbf{E}_s(\mathbf{r}) = \mathbf{E}_{sh}(\mathbf{r})$. The total field $\mathbf{E}(\mathbf{r})$ in the near zone behind the particle is as following

$$\mathbf{E}(z, \boldsymbol{\rho}) = \mathbf{E}^0\, e^{ikz}(1 - \eta(\boldsymbol{\rho})) \tag{3.38}$$

where z and $\boldsymbol{\rho}$ are longitudinal and transverse coordinates, respectively, and the function $\eta(\boldsymbol{\rho})$ outlines the shadow by means of the equation

$$\eta(\boldsymbol{\rho}) = \begin{cases} 1 & \textit{inside the particle shadow} \\ 0 & \textit{outside the particle shadow} \end{cases} \tag{3.39}$$

The energy flux of the total field over a plane $z = const$behind the particles is equal to

$$\Phi = \int |\mathbf{E}(z, \boldsymbol{\rho})|^2\, d\boldsymbol{\rho} = \int \left(1 - 2\eta(\boldsymbol{\rho}) + \eta^2(\boldsymbol{\rho})\right)\, d\boldsymbol{\rho} \tag{3.40}$$

Here the first addend is the flux of the incident wave through the plane $z = const$ behind a particle. If we supplement the given plane $z = const$ with another plane $z' = const$ situated before the particle, the flux of the incident wave through the both planes surrounding the particle becomes zero $\Phi_0 = 0$ as it is true for any surrounding surface. The shadow-forming field by definition is equal to zero on any plane $z' = const$ situated before the particle. Therefore the third term of Eq. (3.40) gives the scattering cross-section without any calculations

$$\sigma_s = \Phi_s = \Phi_{sh} = s \tag{3.41}$$

where s is the shadow area. We note that the simple result of Eq. (3.41) is often obtained in the literature by a tedious integration of the Fraunhofer diffraction patterns of the wave zone for some given particle shapes.

The second term in Eq. (3.40) corresponds to interference between the fields. Therefore, due to Eq. (3.31), it gives the extinction cross-section

$$\sigma_e = -\Phi_{0s} = -\Phi_{0sh} = 2s \tag{3.42}$$

Finally, Eq. (3.29) determines the absorption cross-section

$$\sigma_a = \sigma_e - \sigma_s = s \tag{3.43}$$

Eqs. (3.41)–(3.43) obtained within the framework of the physical-optics approximation have obvious physical interpretation. Namely, Eq. (3.43) proves the evident fact that all photons incident on a perfectly absorbing particle should be absorbed. Their energy flux is equal to the particle projection $\sigma_a = s$ as seen in Fig. 3.1(c). Then, Eq. (3.41) gives the energy flux of the shadow-forming field and its value of s is also evident from the definition of the shadow-forming field by Eqs. (3.20) and (3.38).

It is interesting to comment Eq. (3.42) for the interference flux giving the double projection area $2s$. From the standpoint of the definition of the extinction cross-section by the equation $\sigma_e = \sigma_a + \sigma_s$, the result of $\sigma_e = 2s$ is trivial. But anyone may ask a question: why does the same scattered field $\mathbf{E}_s(\mathbf{r})$ produce two fluxes: s and $2s$? The answer is worth mentioning. Indeed, the factor of (-1) in Eq. (3.20) defining the shadow-forming field can be interpreted also as a phase shift of π relative to the incident one. While this shift is not important for the flux Φ_s, it is quite important for the interference flux Φ_{0s}. In Conclusion 3 of the previous section, we stated that any scattered field always contains information about absorption inside a particle. In this case, it is just the phase shift of π in the scattered field that managed to take into account absorption inside the particle.

Let us come back to other cases depicted in Fig. 3.1. Here the scattered field consists already of two components that are the shadow-forming and refracted fields according to Eq. (3.21). Therefore the new energy fluxes Φ_r, Φ_{0r}, and $\Phi_{sh,r}$ appear for the scattering and extinction cross-sections

$$\sigma_s = \Phi_s = \Phi_{sh} + \Phi_r + \Phi_{sh,r} \tag{3.44}$$

$$\sigma_e = -\Phi_{0s} = -(\Phi_{0sh} + \Phi_{0r}) \tag{3.45}$$

The fluxes $\Phi_{sh} = s$ and $\Phi_{0sh} = -2s$ are already found due to Eqs. (3.41) and (3.42). Consider the interference flux Φ_{0r} produced by the incident and refracted fields. As seen from the structure of the refracted field in the near zone shown in Figs. 3.1(a) and 3.1(b), the phases of the field $\mathbf{E}_r(\mathbf{r})$ are rapidly varying values on any plane $z = const$ where the phase of the incident field is constant. Therefore the interference pattern is a rapidly oscillating with alternative signs and rather chaotic value along the plane $z = const$. Usually an integral of such a pattern, i.e. the flux Φ_{0r}, is close to zero

$$\Phi_{0r} \approx 0 \tag{3.46}$$

The flux $\Phi_{sh,r}$ is an integral of the same interference pattern but it is taken only over shadow region on the plane $z = const$ behind a particle. Here the scale of the chaotic interference pattern is also much smaller than the particle size a. Therefore we get the same

$$\Phi_{sh,r} \approx 0 \tag{3.47}$$

Substitution of Eq. (3.46) in Eq. (3.45) gives an important result

$$\sigma_e = \sigma_s + \sigma_a = 2s \tag{3.48}$$

i.e. the extinction cross-section for large optically hard particles is equal to the double area $2s$ independently of the presence of absorption. This energy flux $\sigma_e = 2s$ is extracted from the incident wave by means of interference between the shadow-forming and incident fields that also takes place for the case of perfectly absorbing particles.

The equation for the scattering cross-section σ_s obtained from Eq. (3.48)

$$\sigma_s = 2s - \sigma_a \tag{3.49}$$

has the following interpretation. A half of the flux $2s$ extracted from the incident wave is carried over by the flux of the shadow-forming field $\Phi_{sh} = \sigma_{sh} = s$ as before. Another half is carried over by the refracted field whose flux is equal to $\Phi_r = s$ if there is no absorption. Indeed, it is easily seen in Figs. 3.1(a) and 3.1(b) that the energy fluxes of the incident field over particle projection are completely transformed into the flux of the refracted field. Therefore, absorption along the geometric-optics rays inside a particle subtracts the value of σ_a from the refracted field, resulting in the equation $\Phi_r = \sigma_r = s - \sigma_a$. In the case of absolute absorption depicted in Fig. 3.1(c), the cross-section of the refracted field becomes zero.

Summarizing the results obtained in this section, we can formulate the following conclusion.

Conclusion 5. *Extraction of energy from the incident wave in the case of optically hard particles is made by only the shadow-forming field resulting in the extinction cross-section* $\sigma_e = 2s$*. Though the refracted field carries over the energy flux of* $\Phi_r = s - \sigma_a$*, this field does not take part in the energy extraction because of smallness of the interference flux* Φ_{0r}*.*

It is worth noting that Eq. (3.48) in the literature is called the extinction paradox (van de Hulst, 1981) because the cross-section is doubled in comparison with the shadow area. This paradox is commonly explained by penetration of light from an illuminated region to shadow. In addition, proving that the cross-section of this process is equal to s, Babinet's principle should be involved. Both steps of the explanation look rather artificial. Our concept of the shadow-forming field looks more consistent for such explanations. Indeed, our shadow-forming field does not appear from Babinet's principle but it is a direct consequence of the quite general principle of field superpositions. Its energy flux is equal to s that is obvious in the near zone. Then, in the process of propagation from the near zone to the Fresnel diffraction zone, the shadow-forming field is spread in the transverse direction like any plane-parallel beam. A single peculiarity is that this beam differs from the incident field by the factor of (-1) that is equivalent to the phase shift of π. Therefore a well-known transversal spreading of the beam because of either Fresnel or Fraunhofer diffraction multiplied by the negative sign can be more artificially treated as penetration of light from an illuminated area to shadow as was used in the common explanation of the extinction paradox.

It is important to note that the results of this section are obtained under conditions of Eqs. (3.46) and (3.47). The opposite case is discussed in the next sections.

3.3.5 Cross-sections for large optically soft particles

In the case of optically soft particles, i.e. $|n-1| \ll 1$, one can ignore any curving of rays inside a particle as depicted in Fig. 3.6(a). Here light propagation in the near zone $z \ll a^2/\lambda$ is reduced to accumulation of phase shifts along any straight ray $\boldsymbol{\rho} = const$, where z and $\boldsymbol{\rho}$ are the longitudinal and transverse coordinates, respectively. The total field behind the particle is described by the intuitively evident equation

$$\mathbf{E}(z,\boldsymbol{\rho}) = \mathbf{E}^0 \exp\left\{ik \int_0^z m(z',\boldsymbol{\rho})\,dz'\right\} = \mathbf{E}_0(z,\boldsymbol{\rho})u(\boldsymbol{\rho}) \tag{3.50}$$

where $m(\mathbf{r})$ is the complex-valued refractive index; $\mathbf{E}_0(z,\boldsymbol{\rho}) = \mathbf{E}^0 \exp(ikz)$ is the incident wave; $z' = 0$ and $z' = z$ are planes situated before and behind a particle, respectively; and $u(\boldsymbol{\rho})$ is the complex-valued scalar amplitude of the total field

$$u(\boldsymbol{\rho}) = \exp\left\{ik \int_0^z \left[m(z',\boldsymbol{\rho}) - 1\right] dz'\right\} \equiv e^{i\phi(\boldsymbol{\rho})} \tag{3.51}$$

Now we can use the scalar field $u(\boldsymbol{\rho})$ instead of the vector field $\mathbf{E}(z,\boldsymbol{\rho})$ since they differ by the factor of $\mathbf{E}^0 \exp(ikz)$. The phase $\phi(\boldsymbol{\rho})$ accumulates additional phase shifts caused by the particle. If there is absorption, the phase is complex-valued

$$\phi(\boldsymbol{\rho}) = \varphi(\boldsymbol{\rho}) + i\chi(\boldsymbol{\rho}) \tag{3.52}$$

The total field of Eq. (3.51) is equal to the incident field $u_0(\rho) = 1$ outside the particle shadow and it differs from unity inside the shadow.

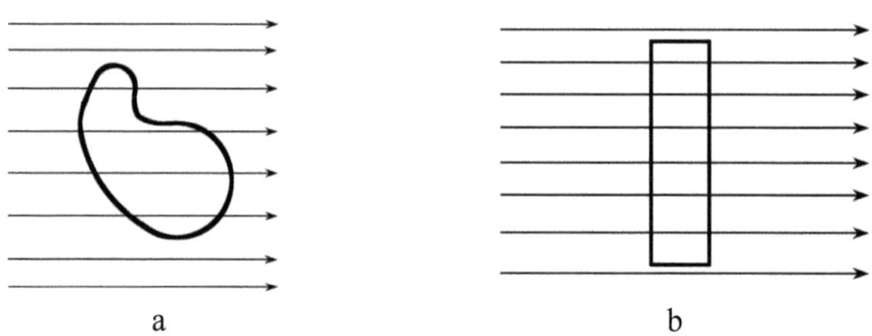

Fig. 3.6. Light scattering by large optically soft particles in the near zone.

It is worth noting that, strictly speaking, Eq. (3.50) does not correspond to geometric optics. Here we summarize the phase shifts along the ray trajectories similarly to the eikonal equation (3.10) but the curved geometric-optics trajectories are replaced by straight rays. Thus, the geometric-optics orthogonality between trajectories and wave fronts is violated. To justify such an approximation, we recall that it is a common approach in physics. In particular, in quantum mechanics it is called the straight path approximation. In radiophysics, it is called the phase screen approximation. In optics, van de Hulst proposed the term of the anomalous diffraction approximation. The last term is worst since it is associated already with scattered fields in the wave zone while an essence of this approximation is just an assumption of the straight rays inside a particle. In our papers (Borovoi, 2006) we prefer to use the term of the straight-ray approximation.

Now the physical-optics approximation is reduced to substitution of the field defined by Eq. (3.50) in Eqs. (3.17). The superpositions of the fields widely used in the previous sections are applicable to the scalar field $u(\boldsymbol{\rho})$ as well. Thus, by definition, the scattered wave is found by subtraction of the incident wave $u_0(\boldsymbol{\rho}) = 1$ from Eq. (3.51)

$$u_s(\boldsymbol{\rho}) = \left[e^{i\phi(\boldsymbol{\rho})} - 1\right] \eta(\boldsymbol{\rho}) \tag{3.53}$$

Though the term in the square brackets is already equal to zero outside the shadow, the factor $\eta(\boldsymbol{\rho})$ defined by Eq. (3.39) is included for convenience. Now Eq. (3.53) can be treated as the superposition of the scattered wave consisting of the refracted and shadow-forming fields. These fields occur to be equal to

$$u_{sh}(\boldsymbol{\rho}) = -\eta(\boldsymbol{\rho}) \tag{3.54}$$

$$u_r(\boldsymbol{\rho}) = e^{i\phi(\boldsymbol{\rho})}\eta(\boldsymbol{\rho}) \tag{3.55}$$

It is clear that the shadow-forming fields for either optically soft particles obtained by Eq. (3.54) or optically hard particles defined by Eq. (3.20) are just the same fields. Therefore all results obtained in the previous section concerning the shadow-forming field and its interference flux Φ_{0sh} remain the same

$$\Phi_{0sh} = -2s\,; \quad \Phi_{sh} = s \tag{3.56}$$

It means that the shadow-forming field carries over an energy flux of s; and its interference with the incident wave extracts the double flux $2s$. Moreover, if the real part of the phase $\phi(\boldsymbol{\rho})$ in the refracted wave of Eq. (3.55) is a function quickly changing across the shadow, the interference flux between the refracted and incident waves $\Phi_{0r}$is small like Eq. (3.46). In this case, the cross-sections for both optically hard and optically soft particles are the same.

The difference between these two kinds of particle appears only if the real phase $\varphi(\boldsymbol{\rho})$ in Eq. (3.55) changes slowly across the shadow. By the way, note that since the fields of Eqs. (3.54) and (3.55) are restricted by shadow region, the fluxes Φ_{0r} and $\Phi_{sh,r}$ differ by only the sign

$$\Phi_{0r} = -\Phi_{sh,r} \tag{3.57}$$

The limit case where the phase $\varphi(\boldsymbol{\rho})$ is real and it does not vary across the shadow is of most importance. It corresponds to a perpendicular plate without absorption giving the constant phase shift $\varphi(\boldsymbol{\rho}) = \varphi_0$ as shown in Fig. 3.6(b). The scattering cross-section of this plate is found from Eq. (3.53) as following

$$\sigma_s = \Phi_s = \Phi_r + \Phi_{sh,r} + \Phi_{sh} = |e^{i\varphi_0} - 1|^2 s = (1 - 2\,\cos\varphi_0 + 1)s \tag{3.58}$$

In spite of triviality of Eq. (3.58) describing a simplest interference phenomenon, this equation leads to important physical conclusions. In the last expression of Eq. (3.58), the first and third terms are energy fluxes of the refracted Φ_r and shadow-forming Φ_{sh} fields, respectively, resulting in the flux of $2s$. The second term is the interference flux between the refracted and shadow-forming fields. This interference term causes oscillations of the scattering cross-section between zero and $4s$, i.e.

$$0 \leq \sigma_s \leq 4s \tag{3.59}$$

The boundary values of the scattering cross-section σ_s together with the phase shifts and values of all fields inside the shadow are presented below

$$\varphi_0 = N2\pi \qquad \sigma_s = 0 \qquad u_s = 0 \qquad u_r = 1 \qquad u_{sh} = -1 \qquad u = 1 \tag{3.60}$$

$$\varphi_0 = (2N+1)\pi \qquad \sigma_s = 4s \qquad u_s = -2 \qquad u_r = -1 \qquad u_{sh} = -1 \qquad u = -1 \tag{3.61}$$

The case $\varphi_0 = N2\pi$, where $N = 0, 1, 2, \ldots$, corresponds to an invisible plate. The invisibility is provided by the fact that the refracted field occurs to be in antiphase with the shadow-forming field. As a result, their sum gives the zeroth scattered field. Consequently, the total field behind the particle coincides with the incident field.

In the case $\varphi_0 = (2N+1)\pi$, on the contrary, the refracted field is in phase with the shadow-forming field. The amplitude of the scattered field is doubled and, consequently, the scattering cross-section is quadrupled. It is interesting to note that the total field behind the particle occurs to be in antiphase with the incident field.

If absorption takes place, the absorption and extinction cross-sections are equal to

$$\begin{aligned} \sigma_a &= s - \Phi_s = (1 - |u_s|^2)s = \left[1 - |e^{i\varphi_0 - \chi_0} - 1|^2\right] s \\ &= (1 + e^{-2\chi_0} - 2\,e^{-\chi_0}\,\cos\varphi_0)s \end{aligned} \tag{3.62}$$

$$\begin{aligned} \sigma_e &= -\Phi_{0s} = \Phi_{sh,s} = -2s\,\mathrm{Re}\,u_s = -2s\,\mathrm{Re}(e^{i\varphi_0 - \chi_0} - 1) \\ &= (1 - e^{-\chi_0}\cos\varphi_0)2s \end{aligned} \tag{3.63}$$

Owing to Eq. (3.63), we see that absorption only decreases the amplitude of oscillations for the extinction cross-section between the values of 0 and $4s$ inherent to the transparent particles. Similarly to Eq. (3.59), we get the general inequality

$$0 \leq \sigma_e \leq 4s \tag{3.64}$$

Though Eqs. (3.58)–(3.64) are obtained for a plate, the results are easy generalized for a particle of any shape shown in Fig. 3.6(a). Indeed, such a particle can be mentally divided into a lot of narrow tubes with cross areas of $d\boldsymbol{\rho}$ and phase shifts of $\phi(\boldsymbol{\rho})$. Therefore all cross-sections of a particle of arbitrary shape are found as integrals of the corresponding functions of Eqs. (3.58), (3.62), and (3.63), for example

$$\sigma_e = -2\,\mathrm{Re}\int u_s(\boldsymbol{\rho})\,d\boldsymbol{\rho} = -2\,\mathrm{Re}\int (e^{i\phi(\boldsymbol{\rho})} - 1)\,d\boldsymbol{\rho} \tag{3.65}$$

Consequently, the inequality (3.64) remains valid for particles of arbitrary shapes, too.

We arrive at the following conclusion.

Conclusion 6. *This is the interference between the shadow-forming and refracted fields that forces the extinction cross-section to oscillate between zero and quadrupled shadow area. The quadrupled extinction cross-section $4s$ appears if the total field behind a particle is in antiphase with the incident field and has the same amplitude.*

3.3.6 Can the extinction efficiency exceed number 4?

Though Conclusion 6 is obtained by use of the case of optically soft particles, the principle formulated seems to be quite general. Indeed, let us consider a coherent electromagnetic wave emitted by an arbitrary source. One may ask a question: what should the disturbance of the incident wave in the near zone of a large scatterer (with arbitrary refractive index and the projection area s) be to get the maximum extinction cross-section? The answer is as follows. We need to provide the total field to be in antiphase with the incident wave and to have just the same amplitude over all projection area. In this case, the interference flux between the incident and scattered waves is maximal. If these two conditions are not satisfied, the interference flux having the physical meaning of the extinction cross-section should be decreased.

As an illustration, consider light scattering by a large optically hard plate depicted in Fig. 3.7. Here the refracted field behind the particle can be also in phase with the shadow-forming field as in Fig. 3.6(b). But the amplitude of the refracted field is decreased because of reflection. Moreover, the coherence between the refracted and shadow-forming fields takes place over a certain area that is less than s. Consequently, the extinction efficiency of this plate should be less than 4.

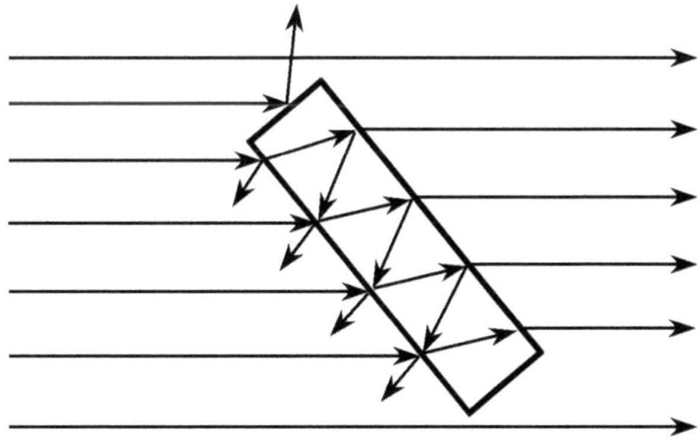

Fig. 3.7. Light scattering by a large optically hard plate.

This example of the optically hard plate shows that the geometric-optics field created around any large particle with arbitrary refractive index is not capable of providing the full coherence between the refracted and shadow-forming fields. Consequently, the extinction efficiency should be less than 4.

As has been demonstrated in this chapter, the refracted and shadow-forming fields are often not coherent, i.e. the interference flux $\Phi_{r,sh}$ is negligible. In this case, this is only the shadow-forming field that provides the extinction efficiency $\sigma_e/s = 2$ corresponding to the extinction paradox. If the extinction efficiency deviates noticeably from the number 2, this case was called the anomalous diffraction (van de Hulst, 1981). We see that the case of the anomalous diffraction is characterized by noticeable interference between the refracted and shadow-forming fields resulting in oscillations of the extinction efficiency about the number 2 between zero and number 4.

The same results are also evident in the wave zone of a particle where the optical theorem of Eq. (3.37) takes place. Indeed, if the shadow-forming and refracted components are partly coherent in the near zone, in the wave zone they are comparable in the forward scattering direction as it follows from the Fraunhofer diffraction equations. The scattered field in the forward scattering direction and, consequently, the extinction cross-section happen to be sensitive to any change of the refractive field. It causes the oscillations of the extinction cross-section corresponding to the anomalous diffraction. Otherwise, if the phase of the refracted field is a quickly oscillating function in the near zone, as compared to the shadow-forming counterpart, in the wave zone the refracted field spreads over large solid angles. As a result, the contribution of the refracted field to the total scattered field in the forward scattering direction is negligible. Here the shadow-forming component becomes predominant that extracts from the incident field the standard energy flux $2s$. It is worth mentioning that these results for a sphere can be obtained from the Mie series as well (Lock and Yang, 1991).

Let us recall that we dealt with the physical-optics approximation defined by Eq. (3.17) where it is assumed that the scattered field in the near zone of a large particle is the geometric-optics field. It is worth noting that, recently, calculations of the extinction efficiency for a large long cylinder depicted in Fig. 3.5 were performed by use of the discrete-dipole approximation (Bi et al., 2010; Bi and Yang, 2013). Here the extinction efficiency reaches the maximum value of about 5.5. The authors associate this result with the edge effect (Nussenzveig, 1992). Following the terminology used in this chapter (see Section 3.2.5), we would better refer these results to the effect of the Fresnel diffraction inside a particle.

3.4 Conclusions

In this chapter we have considered the cross-sections for particles of arbitrary shapes and internal structure in the limit of short wavelengths $\lambda \to 0$. Two concepts have come to be useful in this study: the shadow-forming field and the partial energy fluxes. We show that these concepts applied not to the common wave zone of the particles but to the near zone allow us to get several general results without tedious calculations. The conclusions reached in the text are repeated below.

1. Equations (3.17) are the desired mathematical definition of the simplest physical-optics approximation. Such an approximation is the direct extension of the classical Kirchhoff approximation (Eq. (3.15)) from a large aperture to any large 3-D scatterers.
2. The total extraction of energy from the incident wave is equal to the interference flux Φ_{0s}between the incident and scattered waves.
3. If we know the scattered wave $\mathbf{E}_s(\mathbf{r})$ on any surface surrounding a particle, this wave $\mathbf{E}_s(\mathbf{r})$ already contains information about absorption inside the particle.
4. In the wave zone $R \gg a^2/\lambda$, extraction of energy from the incident wave takes place only in the vicinity of the forward scattering direction. Therefore the extinction cross-section is strictly expressed through the amplitude of the scattered wave taken in the forward scattering direction.
5. Extraction of energy from the incident wave in the case of optically hard particles is made by only the shadow-forming field resulting in the extinction cross-section $\sigma_e = 2s$. Though the refracted field carries over the energy flux of $\Phi_r = s - \sigma_a$, this field does not take part in the energy extraction because of smallness of the interference flux Φ_{0r}.
6. This is the interference between the shadow-forming and refracted fields that forces the extinction cross-section to oscillate between zero and quadrupled shadow area. The quadrupled extinction cross-section $4s$ appears if the total field behind a particle is in antiphase with the incident field and has the same amplitude.

Acknowledgments

I am grateful to Alexander Kokhanovsky who suggested the writing of this chapter and encouraged the author in the course of the work. This work is partly supported by the Russian Foundation for Basic Research under the grant No. 12-05-00675a.

References

Baran, A. J, 2013: Light scattering by irregular particles in the Earth's atmosphere, this volume.

Bi, L., P. Yang, and G. W. Kattawar, 2010: Edge-effect contribution to the extinction of light by dielectric disk and cylindrical particles, *Appl, Opt.*, **49**, 4641–4646.

Bi, L., P. Yang, G. W. Kattawar, Y. Hu, and B. A. Baum, 2011: Scattering and absorption of light by ice particles: solution by a new physical-geometric optics hybrid method, *J. Quant. Spectrosc. Radiat. Transfer*, **112**, 1492–1508

Bi L., and P. Yang, 2013: Physical-geometric optics hybrid methods for computing the scattering and absorption properties of ice crystals and dust aerosols, this volume.

Bohren, C. F., and D. R. Huffman, 1983: *Absorptin and Scattering of Light by Small Particles*, New York: John Wiley & Sons.

Born, M., and E. Wolf, 1959: *Principles of Optics,* Oxford: Pergamon Press.

Borovoi, A.G., and I. A. Grishin, 2003: Scattering matrices for large ice crystal particles, *JOSA A*, **20**, 2071–2080

Borovoi, A. G., N. V. Kustova, and U. G. Oppel, 2005: Light backscattering by hexagonal ice crystal particles in the geometrical optics approximation, *Opt. Engineering*, **44**, 071208(10).

Borovoi, A.G., 2006: Multiple scattering of short waves by uncorrelated and correlated scatterers. In A. A. Kokhanovsky, *Light Scattering Reviews*, vol. 1, Chichester: Springer-Praxis, 181–252.

Cai, Q., and K. N. Liou. 1982: Polarized light scattering by hexagonal ice crystals: theory, *Appl. Opt.*, **21**, 3569–3580.

Fock, V. A., 1965: *Electromagnetic Diffraction and Propagation Problems*, Oxford: Pergamon Press.

Jackson, J. D., 1999: *Classical Electrodynamics*, 3rd edn, New York: John Wiley & Sons.

Kokhanovsky, A. A., 1999: *Light Scattering Media Optics: Problems and Solutions*, Chichester: Wiley-Praxis (2nd edn: 2001, 3rd edn: 2004).

Landau, L. D., and E. M. Lifshitz, 1991: *Quantum Mechanics: Non-relativistic Theory*, 3rd edn, Oxford: Pergamon Press.

Lock, J. A., and L. Yang, 1991: Interference between diffraction and transmission in the Mie extinction efficiency, *JOSA A*, **8**, 1132–1134.

Liou, K. N. 2002: *An Introduction to Atmospheric Radiation*, San Diego: Academic Press.

Liou K. N., Y. Takano, and P. Yang, 2010: On geometric optics and surface waves for light scattering by spheres, *J. Quant. Spectrosc. Radiat. Transfer*, **111**, 1980–1989.

Mishchenko, M. I., and A. Macke, 1998: Incorporation of physical optics effects and δ-function transmission. *J. Geophys. Res.*, **103**, 1799–1805.

Mishchenko, M. I., L. D. Travis, and A. A. Lacis, 2002: *Scattering, Absorption, and Emission of Light by Small Particles.* Cambridge, UK: Cambridge University Press.

Muinonen, K., 1989: Scattering of light by crystals: a modified Kirchhoff approximation. *Appl. Opt.*, **28**, 3044–3050.

Nussenzveig, H. M., 1992: *Diffraction Effects in Semiclassical Scattering*, Cambridge, UK: Cambridge University Press.

Takano Y., and K. N. Liou, 1989: Solar radiative transfer in cirrus clouds. Part 1. Single-scattering and optical properties of hexagonal ice crystals, *J. Atmos. Sci.*, **46**, 3–19.

Ufimtsev, P. Ya., 2007: *Fundamentals of the Physical Theory of Diffraction*, New York, John Wiley & Sons.

van de Hulst, H. C., 1981: *Light Scattering by Small Particles*, New York: Dover.

Yang, P., and K. N. Liou, 1995: Light scattering by hexagonal ice crystals: comparison of finite-difference time domain and geometric optics models, *J. Opt. Soc. Am.*, **A12**, 162–176.

Yang, P., and K. N. Liou, 1996: Geometric-optics-integral-equation method for light scattering by nonspherical ice crystals, *Appl. Opt.*, **35**(33), 6568–6584.

Yang, P., and K. N. Liou, 1997: Light scattering by hexagonal ice crystals: solutions by a ray-by-ray integration algorithm, *J. Opt. Soc. Am.*, **A14**, 2278–2289.

Yang, P., and K.N. Liou, 2006: Light scattering and absorption by nonspherical ice crystals. In A. A. Kokhanovsky, *Light Scattering Reviews*, vol. 1, Chichester: Springer-Praxis, pp. 31–72.

4 A pseudo-spectral time domain method for light scattering computation

R. Lee Panetta, Chao Liu, and Ping Yang

4.1 Introduction

Atmospheric particles, for example ice crystals, dust, soot, or various chemical crystals, play a significant role in the atmosphere by scattering and absorbing radiation, principally in two bands: incident solar, with peak at about 0.5 μm, and terrestrial thermal emission, with peak at about 10 μm. Knowledge of aerosol scattering properties is a fundamental but challenging aspect of radiative transfer studies and remote sensing applications. In this chapter we consider only scattering by single homogeneous particles, but in the atmosphere particles occur both individually and as constituents of such aerosols as homogeneous or heterogeneous aggregates with other particles and sometimes coated with liquids. The pseudo-spectral time domain method (PSTD) for calculating scattering properties that we discuss, like a number of other methods currently in use, can be used to investigate scattering properties of a wide variety of aerosols, homogeneous or heterogeneous, singly or in aggregate.

Even with a narrow focus on single scattering by homogeneous particles, there are significant obstacles remaining to a comprehensive understanding of scattering properties, given the complexity introduced by considerations of particle shape, size, and refractive index. Much of what we know of this complexity comes from numerical work, and the estimation of errors can become quite challenging in the absence of either a known exact solution or observations. The 'gold standard' in single scattering is provided by the Lorenz–Mie theory (Mie, 1908). It provides an exact solution of the scattering problem for a single spherically homogeneous particle of arbitrary size, thereby giving a way of assessing in one special case the faithfulness of numerical methods developed to treat particles of different shapes and compositions, as well as methods designed to work in special particle size regimes.

There is of course no guarantee that a method working well in homogenous and spherically symmetric cases will necessarily work well in general cases, but if the method has no built-in preference for spherically symmetric problems (as might be the case, for example, with a spectral method based on spherical harmonics), and the tests applied also have no such prejudice, we have done the best we can to justify proceeding to use the method in more general cases. Similar remarks may be made concerning such comparisons as we make below between different methods.

We emphasize that precisely because the number of exact solutions is very limited, there is considerable value in having more than one numerical method that performs well. Our purpose here is to argue that the PSTD is a method that performs well over wide and atmospherically relevant ranges of particle sizes and indices of refraction, but not to argue that it is the only method that should be used. Confidence in any given numerical result is gained when more than one method produces the same result. Inevitably it will emerge that one method has advantages in one regime and another method has advantages in another: for the PSTD, the regime in which it appears to show competitive advantage is when indices of refraction exceed approximately 1.2, especially as the particle size gets large.

In scattering calculations, what is crucially important is the relation between the size of the particle and the wavelength of the incident light. For a spherical particle of radius a and incident wavelength λ, or incident wavenumber $k = 2\pi/\lambda$, the *size parameter* x is defined by

$$x = k\,a = \frac{2\pi\,a}{\lambda}\,;$$

the second form is the ratio of the circumference to the wavelength. In the regime of very large particles, $x \gg 1$, ray theories and geometric optics are useful and computationally relatively inexpensive, while in the regime of Rayleigh scattering, $x \ll 1$, computations are also relatively inexpensive. In the intermediate case, recourse must be made to numerical solution of some form of Maxwell's equations. In this case, cpu demands typically grow rapidly as x increases, especially for indices of refraction m that become significantly larger than 1. Given the current computational resources available to most researchers, the effective bound for all but truly heroic efforts begins to be felt for particles with $x \sim 100$. Computations can get challenging with smaller x if the index of refraction increases much beyond 1, as we will show in Section 4.6. Our interest is in pushing this technology-imposed boundary and we will present results indicating that the PSTD method shows promise of helping us to do so.

Numerical simulation of single-particle scattering has a history of over a century of work, and a proper survey is well beyond our scope here. We will only briefly mention here a few methods that are relevant to the results we will present using the pseudo-spectral method. The discrete dipole approximation (DDA) (Purcell and Pennypacker, 1973; Draine and Flatau, 1994; Yurkin and Hoekstra, 2007, 2011) and the finite-difference time domain method (FDTD) (Yee, 1966; Yang and Liou, 1996a), are two methods which can be used for scatterers with arbitrary shapes, and have been widely applied to simulate single-scattering properties of atmospheric particles, e.g. hexagonal columns (Yang and Liou, 1996a), droxtals (Yang et al., 2003), tri-axial ellipsoids (Bi et al., 2009), and other shapes. Both DDA and FDTD discretize the three-dimensional spatial domain, with dipoles or grid cells, and solve Maxwell's equations. However, even with parallelized implementations (Yurkin et al., 2007b; Brock et al., 2005) on multi-processors, they are applicable for only particles with small-to-moderate size parameters, say x a few multiples of 10, and become computationally expensive and impractical for large ones. To the best of our knowledge, the maximum size parameter of spheres with refractive index significantly larger than 1.0 that has been simulated using DDA is 130 (Yurkin et al., 2007b, using a refractive index of 1.2). Furthermore, because of the high requirement for the spatial resolution (10 to 20 grid cells per wavelength in the particle) and numerical dispersion, the FDTD technique is difficult to apply for particles with size parameter over 100. (We will illustrate the dispersion in Section 4.4 below in the case of one dimension.) If results involving averaging over random orientation are required for nonspherical particles, both methods become prohibitively time-consuming (given current hardware) for averaging over tens to hundreds of particle orientations.

Another powerful approach is the T-matrix method (Waterman, 1965; Waterman, 1971; Mischenko and Travis, 1998; Mishchenko et al., 1996). The central idea in the approach is to represent the incident and scattered fields as expansions in vector spherical harmonic series, with the T-matrix being a transform matrix mapping the sequences of expansion coefficients for incident waves to sequences of expansion coefficients of scattered waves. Once the matrix is given, all the far-field scattering properties are derived from analytical formulas. Getting the matrix itself involves calculation of various integral properties that depend on the particle doing the scattering. In the special case that the particle is a homogeneous sphere, the T-matrix approach reduces to the Lorenz–Mie solution. Using extended precision arithmetic, (Mischenko and Travis, 1998) showed T-matrix results for spheroids or circular cylinders with size parameters over 100. The calculation of the T-matrix, in principle possible for particles of any size or shape, can run into numerical difficulties in dealing with particles that have large aspect ratios or surfaces with concave regions. Aside from such situations, the approach is widely regarded as a good source of 'reference solutions', and we will make use of it as appropriate in discussion of PSTD results.

The conventional geometric-optics method (CGOM) (Macke et al., 1996) and the improved geometric-optics method (IGOM) (Yang and Liou, 1996b, 1997) have been developed to simulate light scattering by moderate-to-large-sized particles. Although significant improvements have been included in IGOM, including consideration of edge effects (Jones, 1957; Bi et al., 2009, 2010), in these approaches the near fields are approximated with the ray-tracing method, making this an inappropriate method for small- to moderate-size particles. The recently developed physical-geometric optical hybrid method (PGOH) (Bi et al., 2011) is suitable for calculating the optical properties of particles with complex refractive indices and a wide range of size parameters. By employing a beam-splitting technique instead of the ray-tracing algorithm, virtually no limitation exists on the maximum particle size parameter for the PGOH method. However, its accuracy becomes greatly compromised for particles with size parameters smaller than 30–40. The context in which we will find geometric-optics useful is in discussing PSTD simulations of scattering by particles with large size parameter and concave surfaces.

Before mentioning previous work with PSTD in light scattering problems, it is useful to mention that pseudo-spectral methods have a long history starting in fluid dynamical studies the early 1970s (Kreiss and Oliger, 1972; Orszag, 1972). They now have achieved considerable sophistication and have extensive use in numerical studies of many partial differential equations of mathematical physics: their principle advantage is their ability to approximate derivatives much more accurately and efficiently than is possible with finite difference methods, as we illustrate in special cases below. The terminology *pseudo*-spectral was introduced by Orszag (1972) to distinguish the method from true, or fully spectral, methods in which all calculations are carried out in wavenumber space. Fully spectral methods are made prohibitively expensive by the presence of quadratic nonlinearities in the equations of fluid motion. It was the breakthrough observation of Orszag (1972) that a combination of approaches, computation of derivatives by Fourier transform methods and computation of nonlinear terms by grid point multiplication, could yield a significant improvement in numerical performance over finite difference methods, provided that such an efficient Fourier transform algorithm as the fast Fourier transform (FFT) is available. The term pseudo-spectral has since come to mean any of a class of methods that generalize the Fourier interpolation method that we outline in Section 4.3.

The use of PSTD in electromagnetic scattering problems was pioneered by (Liu, 1994, 1997, 1998, 1999), (Yang et al., 1997), and (Yang and Hesthaven, 1999), and has been applied in a number of forms to model the transient electromagnetic field by solving Maxwell's equations. The pseudo-spectral method based on trigonometric polynomials (Liu, 1998) and Chebyshev polynomials (Yang et al., 1997) has been used to give a better approximation for the spatial derivatives, and the multi-domain PSTD method in general curvilinear coordinates has been developed to solve problems with complex structures in a manner avoiding the Gibbs phenomenon (Yang and Hesthaven, 1999, 2000; Tian and Liu, 2000). Chen et al., (2008) have successfully used the PSTD to calculate the single-scattering properties of atmospheric particles, treating spheres with a maximum size parameter of 80 (refractive index of 1.31), and have also shown the PSTD to be a robust method for light scattering problems of nonspherical particles such as hollow hexagonal columns and hexagonal aggregates. Based on the work of Chen (2007) and Chen et al., (2008), Liu et al., (2012a) improved and parallelized the PSTD implementation, using an exponential filter in wavenumber space to eliminate the Gibbs phenomenon and stabilize the simulation in a manner that we explain below. At the stage of this writing, the applicability of PSTD has been demonstrated for spheres with size parameter up to 200 (Liu et al., 2012a), as well randomly oriented nonspherical particles with the same size parameter.

The central difference between the PSTD and the FDTD methods, which are otherwise closely related, is in the treatment of spatial differentiation. Each method can be formulated in terms of a spatial grid. For purposes of finite difference calculations of derivatives, the FDTD is often formulated in terms of 'cells' centered on grid points of the grid, and different field components (electric or magnetic) are considered to be evaluated at centers of either edges or walls of these cells (Yee, 1966; Taflove and Hagness, 2005). In the FDTD these derivatives are most commonly approximated using centered second-order finite difference methods, resulting in a second-order accurate scheme for computing spatial derivatives. The complexity of cell wall edge versus center field representation is swept away in the PSTD method, in which all field variables are evaluated at the grid points that are the centers of cells in the FDTD formulation, and the notions of cells and walls are not used. In place of a finite difference approximation to spatial derivatives, the PSTD method uses pairs of Fourier transforms at each time step and results in what is known as a 'spectrally accurate' approximation to the spatial derivatives. The principle purpose of this chapter is to explain in some detail the notion of spectral differentiation, and the meaning of spectral accuracy, in a way that makes clear both the connection between the spectral and finite difference methods, and the reason for the increased accuracy provided by the PSTD. This is the content of Section 4.3. A second aim of the chapter is to explain how the Gibbs phenomenon, mentioned in Chen et al., (2008) as representing a difficulty with pseudo-spectral methods, can be handled. This is the focus of Section 4.4.

Before turning to the technical discussion of spectral differentiation and the treatment of the Gibbs phenomenon, we outline in Section 4.2 the background of numerical simulation in which the pseudo-spectral method is embedded, sketching the main steps in a time-domain numerical simulation of single particle scattering optical properties. After the discussion in Sections 4.3 and 4.4, we present in Section 4.5 results validating the PSTD method by comparison with Lorenz–Mie and T-matrix solutions, and finish with a comparison in Section 4.6 of the PSTD and DDA methods that indicates some of the potential of PSTD methods to push into regimes of size parameter and index of refraction that are beyond the current reach of DDA.

4.2 Conceptual background

The general scattering problem simply stated is: given the properties of a wave field incident on a dielectric particle, determine the properties of the scattered wave field that results from the interaction of the incident field with the particle. Because of the application to remote sensing we have in mind, we are interested in the properties of this scattered wave field at great distance from the scattering body, the *far field* properties. By definition, the far field refers to distances r from the scatterer for which the scatterer appears to be essentially a point and kr is large enough that the scattered wave field is well approximated as an outward propagating spherical wave.

It would be far too expensive of cpu time to use numerical methods like the FDTD, PSTD, or DDA to compute the solution in a domain extending out into the far field. Fortunately, it is unnecessary to do so because the far field may be expressed, in a Green's function approach, as the distant response to a distribution of near-field sources of charges and currents. Our scattering calculation thus proceeds in two stages: the 'near-field' response to an incident wave field is calculated with a high degree of accuracy (and the bulk of the cpu time). With data gathered from this calculation a far less cpu intensive 'near-to-far-field' transformation is carried out, the result being the far-field approximation used to calculate the scattering data.

In the subsections below, we first give brief descriptions of the scattering properties of interest and computational boundary conditions, followed by some equally brief discussions of issues important in time-domain calculations, namely two commonly used near-to-far-field transformation methods. The discussion of the distinction between finite difference and spectral methods in their treatment of differentiation will be given in the following section.

4.2.1 Scattering properties of interest

The central quantity from which all others of interest may be derived is a matrix relating components of the Stokes vectors of the incident and far-field scattered waves.

The stokes vector and phase matrix

The electric field associated with a monochromatic plane wave may be written

$$\vec{E}(\vec{x}) = \vec{\mathcal{E}}_0 \exp^{i(\vec{k}\cdot\vec{x}-\omega t)} . \tag{4.1}$$

The direction of propagation is given by the unit vector $\hat{k}$, where

$$\vec{k} = k\,\hat{k}\,;$$

the amplitude k is variously called the wavenumber or propagation constant.

The constant vector $\vec{\mathcal{E}}_0$ has nonzero components only in directions orthogonal to the direction of propagation $\hat{k}$: if orthogonal unit vectors $\vec{n}_1$ and $\vec{n}_2$ in the plane orthogonal to $\hat{k}$ are chosen, then

$$\vec{\mathcal{E}}_0 = A_1\,\mathrm{e}^{\mathrm{i}\theta_1}\,\vec{n}_1 + A_2\,\mathrm{e}^{\mathrm{i}\theta_2}\,\vec{n}_2.$$

The vector $\vec{\mathcal{E}}_0$ is thus specified by the four real numbers A_1, θ_1, A_2, θ_2. (Note that the phase angles θ_1 and θ_2 neither are individually measurable nor have intrinsic physical significance. It is only the difference $\theta_1 - \theta_2$ that has intrinsic physical significance, so there are in fact only three significant quantities: A_1, A_2, and $\theta_1 - \theta_2$. The wave can also be described in terms of a related set of four real numbers called Stokes parameters that are measurable:

$$I = |\vec{\mathcal{E}}_0|^2 = A_1^2 + A_2^2, \tag{4.2}$$
$$Q = A_1^2 - A_2^2, \tag{4.3}$$
$$U = 2\,A_1\,A_2\,\cos(\theta_1 - \theta_2), \tag{4.4}$$
$$V = 2\,A_1\,A_2\,\sin(\theta_1 - \theta_2). \tag{4.5}$$

These numbers are the four components of the Stokes vector $\vec{S}$. The component I evidently gives the intensity of the wave; the pair (Q, U) together determine the linear polarization, and V determines the circular polarization (see, e.g., Jackson, 1999). Again, there are only three independent quantities, since $I^2 = Q^2 + U^2 + V^2$. Only the linear polarization parameters Q and U are dependent on the particular choice of the orthonormal pair $(\vec{n}_1; \vec{n}_2)$ in the plane orthogonal to the propagation direction.

In the immediate vicinity of the scatterer, the electromagnetic field can have quite complex structure, but for an observer at a large distance r from the scatterer the field is well approximated by a simple outgoing wave. The interaction with the particle being linear, the Stokes vector for the outgoing wave can be related to that of the incoming wave by a matrix multiplication that can be written in more than one form. For instance, using a spherical coordinate system centered on the particle and considering an observation point at scattering direction given by zenith and azimuth angles (θ, ϕ), the linear relation can be written

$$\vec{S}_s(\theta,\phi) = \frac{1}{k^2r^2}\,\mathbf{F}(\theta,\phi)\,\vec{S}_i \qquad (k\,r \gg 1)\,. \tag{4.6}$$

This form results in a matrix $\mathbf{F}$ whose component F_{11} has the property (see, e.g., Liou, 2002) that integration over all scattering angles produces

$$\sigma_{\text{sca}} = \frac{1}{k^2}\int_0^{2\pi}\int_0^{\pi} F_{11}(\theta,\phi)\,\sin\theta\,\mathrm{d}\theta\,\mathrm{d}\phi\,,$$

where σ_{sca} is the *scattering cross-section* of the scatterer: this is the area that if oriented perpendicular to the incident wave would intercept an amount of energy equal to that scattered in all directions by the scatterer.

We will use the scattering cross-section as part of a normalization of the matrix in (4.6), rewriting that equation in terms of the *phase matrix* $\mathbf{P}$:

$$\vec{S}_s = \frac{\sigma_{\text{sca}}}{4\pi r^2}\,\mathbf{P}\,\vec{S}_i, \qquad (k\,r \gg 1)\,. \tag{4.7}$$

(The terminology 'phase matrix' is traditional: the matrix is defined by the relation (4.7) and has nothing to do with the phase of a plane wave.)

For a general scatterer with no geometric symmetries, there are sixteen nonzero elements P_{ij} in the matrix $\mathbf{P}$ (only seven of which are independent), but for a spherical scatterer the scattering matrix is independent of the azimuthal angle ϕ and has a particularly simple block-diagonal form with only four independent nonzero entries:

$$\mathbf{P}_{\text{sphere}} = \begin{bmatrix} P_{11} & P_{12} & 0 & 0 \\ P_{12} & P_{11} & 0 & 0 \\ 0 & 0 & P_{33} & P_{34} \\ 0 & 0 & -P_{34} & P_{33} \end{bmatrix} \qquad \text{(all quantities functions of } \theta)\,.$$

Explicit expressions for these elements are given by Lorenz–Mie theory in the case of a homogeneous sphere. In the general case of a scatterer with no special symmetries another variable enters the problem, the orientation of the scatterer with respect to the incident wave field. But in many applications in remote sensing, where scattering is done by an ensemble of aerosols at random orientations, with the aerosols spatially separated by distances considerably greater than a wavelength so that multiple scattering effects may be neglected, it becomes useful to consider the scattering matrix that results from averaging over 'random' orientations (i.e., assuming a uniform probability distribution over orientation angles). In this case it can be shown by taking advantage of symmetry arguments that what results is a scattering matrix $\mathbf{P}_{\text{avg}}$ having a similar block-diagonal form but now six independent nonzero entries:

$$\mathbf{P}_{\text{avg}} = \begin{bmatrix} P_{11} & P_{12} & 0 & 0 \\ P_{12} & P_{22} & 0 & 0 \\ 0 & 0 & P_{33} & P_{34} \\ 0 & 0 & -P_{34} & P_{44} \end{bmatrix} \qquad \text{(all quantities functions of } \theta)\,. \tag{4.8}$$

(see, e.g., van de Hulst, 1957).

A caution. The intensity I_s of the scattered wave is determined in the following way (omitting the prefactor $\sigma_{\text{sca}}/(4\pi r^2)$):

$$\begin{aligned} I_s &\approx P_{11}I_i + P_{12}Q_i + P_{13}U_i + P_{14}V_i \\ &= \left[P_{11} + P_{12}\frac{Q_i}{I_i} + P_{13}\frac{U_i}{I_i} + P_{14}\frac{V_i}{I_i}\right] I_i \,. \end{aligned} \tag{4.9}$$

In our discussion we refer to $P_{11} = P_{11}(\theta)$ that appears in (4.8) as the 'phase function', but sometimes this term is used for the entire term in the square brackets in (4.9).

Although all elements of $\mathbf{P}(\theta)$ are important in applications, we will be most interested in $P_{11}(\theta)$, and its dependence on size parameter and index of refraction, in our comparisons of results using the various numerical methods. Figure 4.1 gives two versions of a planar representation of P_{11} illustrating size effects. The figure shows scattering by two spheres, each having an index of refraction of $m = 1.311$ (ice) and incident visible wavelength $0.532\,\mu$m. The smaller sphere has $x = 1$ and the larger has $x = 10$, and the data for the figure were calculated using Lorenz–Mie theory. The quantity $P_{11}(\theta)$ is always positive, and being an azimuthal average is only defined for $0 \leq \theta \leq \pi$. For graphic display it can be extended to a function $r(\theta)$ of full range in θ using symmetry to produce a curve in the (r, θ) plane:

$$r_1(\theta) = \begin{cases} P_{11}(\theta) & 0 \leq \theta \leq \pi \\ P_{11}(2\pi - \theta) & \pi \leq \theta \leq 2\pi \,. \end{cases}$$

The upper panel in the figure shows that the increase in size introduces pronounced asymmetry in the forward direction. (In the Rayleigh scattering regime, with $x \ll 1$, the curve would have lobes symmetrical about the line $\theta = \pi/2$.) This representation makes it hard to see much beyond the strong shift to forward scattering. To show more detail, the data are replotted in the lower panel by taking logarithms of

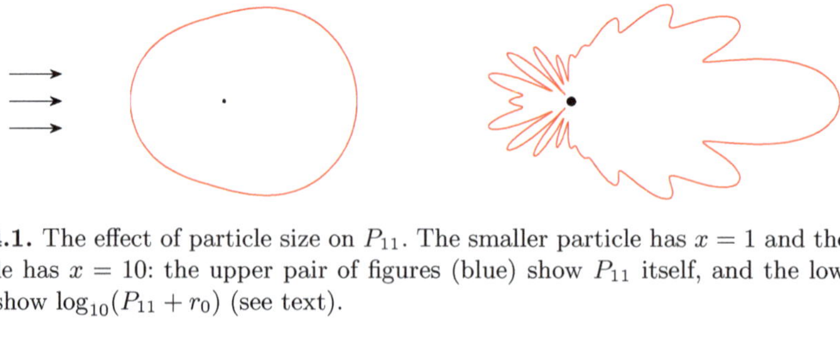

Fig. 4.1. The effect of particle size on P_{11}. The smaller particle has $x = 1$ and the larger particle has $x = 10$: the upper pair of figures (blue) show P_{11} itself, and the lower pair (red) show $\log_{10}(P_{11} + r_0)$ (see text).

the data in the upper panel, with an offset r_0 in the logarithm to keep the resultant numbers positive and make the representation intelligible:

$$r_1(\theta) = \log_{10}[P_{11}(\theta) + r_0] .$$

In our comparisons in later sections, a different graphical representation of the phase matrix elements is used, this time using semilog axes. The semilog representation for this case is shown in Figure 4.2.

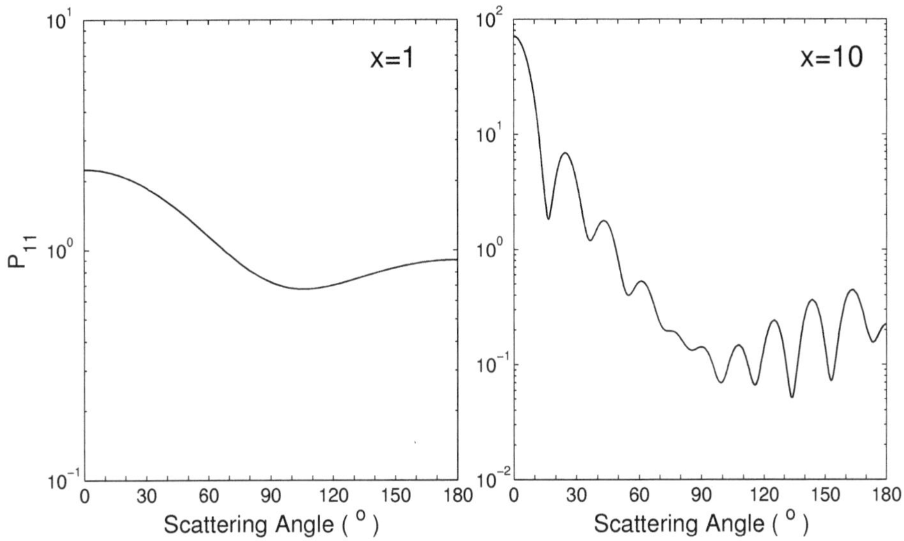

Fig. 4.2. The representation of the same values of P_{11} shown in Fig. 4.1. This is the representation that will be used in subsequent figures.

Other key quantities: cross-sections, efficiencies, and the asymmetry factor

As mentioned above, the scattering cross-section $\sigma_{\rm sca}$ is the area oriented perpendicular to the incident wave that would intercept an amount of energy equal to that scattered in all directions by the scatterer. The *scattering efficiency* $Q_{\rm sca}$ is the non-dimensional number that expresses the ratio of this area to the projected area of the scatterer on a plane normal to the incident wave:

$$Q_{\rm sca} = \frac{\sigma_{\rm sca}}{\text{projected area}} .$$

Similar definitions give the absorption $Q_{\rm abs}$ and extinction $Q_{\rm ext}$ efficiencies using their respective cross-sections. Energy conservation requires that

$$Q_{\rm ext} = Q_{\rm sca} + Q_{\rm abs} ,$$

so any two of these efficiencies determine the third. A related variable is the fraction of extinction due to scattering, called the *single-scattering albedo* (SSA):

$$SSA = \frac{Q_{\rm sca}}{Q_{\rm sca} + Q_{\rm abs}} = \frac{1}{1+\eta} \qquad \left(\eta = \frac{Q_{\rm abs}}{Q_{\rm sca}}\right).$$

As the particle gets less and less absorptive its SSA approaches 1.

The pronounced asymmetry in scattering amplitudes between the forward and backward directions that is evident in Fig. 4.1 is quantified by the *asymmetry factor* g, defined by

$$g = \frac{1}{2}\int_0^\pi P_{11}(\theta) cos\theta sin\theta \, d\theta \qquad \begin{cases} > 0 & \text{for preferentially forward scattering} \\ = 0 & \text{for isotropic scattering} \\ < 0 & \text{for preferentially backward scattering}. \end{cases}$$

Figure 4.3 shows the values of $Q_{\rm ext}$ and g over a range of size parameters that includes the cases shown in Figs. 4.1 and 4.2. The index of refraction and incident wavelength are the same as in those figures. The fact that the extinction efficiency exceeds 1, i.e. that the scattering cross-section exceeds the projected cross-section, for a sphere with size parameter above 2 is due to diffraction of the incident wave around the sphere.

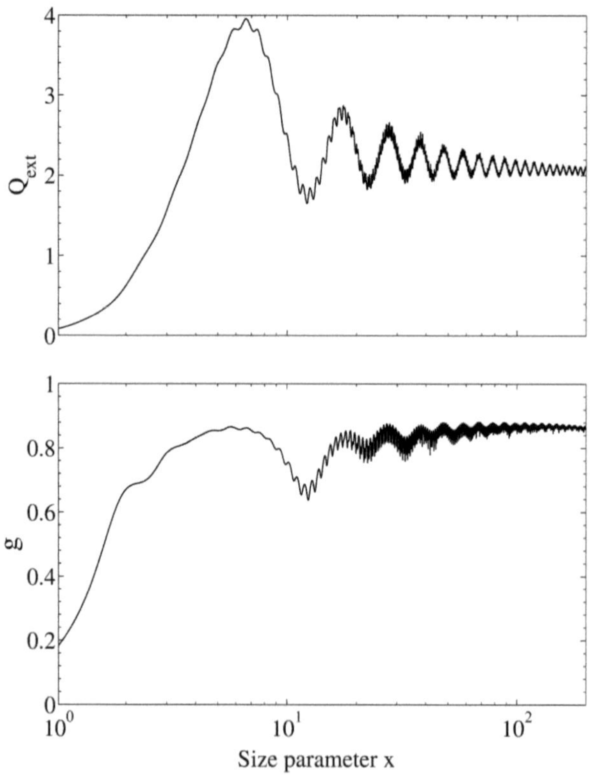

Fig. 4.3. The extinction efficiency $Q_{\rm ext}$ (upper panel) and asymmetry factor g (lower panel) for spherical particles over the range of size parameters $1 \leq x \leq 200$.

4.2.2 Near-field calculations

The near field numerical simulations using Maxwell's equations are carried out in a computational domain that necessarily has a boundary at finite distance, but the boundary is only a feature of the computational approach and does not represent any physical feature of the scattering problem. Care must be taken at the boundary of the computational domain so that there is negligibly small reflection of what should be a purely outgoing scattered signal. For this purpose it is common now in numerical simulations to introduce at computational boundaries what is called a 'perfectly matched' boundary layer. Such a layer is used in both finite difference and pseudo-spectral methods: by proper adjustment of the optical characteristics of the layer, waves incident on it from any direction are absorbed without reflection. The earliest version of the method (Berenger, 1994) was developed for use in FDTD simulations and is now known as the PML ('perfectly matched layer') method. The layer was constructed in a mathematical manner that made physical interpretation difficult, and applicability to more general unstructured grid simulations unclear. These deficiencies were removed in the reformulated 'uniaxial' PML, or UPML by Gedney (1996). The layer matching methods are now seen more generally in the context of stretched coordinate methods (see Johnson (2010) and its references). We use an implementation of the UPML in our PSTD simulations. (The DDA method, which is based on a distribution of dipoles and is not formulated in terms of Maxwell's equations in the time domain, does not require a special boundary layer treatment.)

In Fig. 4.4, which shows a two-dimensional cross-section of the computational domain, the UPML boundary layer is indicated by the dark gray border. The scatterer (here a light gray sphere) is at the center of the computational domain, and the white region between the scatterer and the UPML is meant to represent a region with $\epsilon = 1.0$ ('free space'). The relative sizes of areas in the sketch do not correspond to the relative sizes in our simulations: these relative sizes will be indicated in the discussion of the near-to-far-field transformation below.

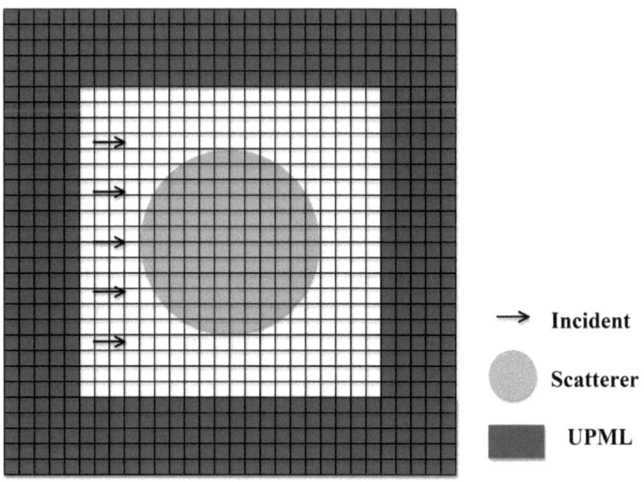

Fig. 4.4. The three regions of the computational domain: scatterer, free space, and 'perfectly matched layer' (relative areas not to scale).

In the presence of both current densities $\vec{J}$ and charge densities ρ, Maxwell's equations written in Gaussian units are

$$\frac{\epsilon}{c}\frac{\partial \vec{E}}{\partial t} = \nabla \times \vec{H} - \frac{4\pi}{c}\vec{J}, \tag{4.10}$$

$$\frac{\mu}{c}\frac{\partial \vec{H}}{\partial t} = -\nabla \times \vec{E}, \tag{4.11}$$

$$\nabla \cdot \vec{E} = \frac{4\pi}{\epsilon}\rho, \qquad \nabla \cdot \vec{H} = 0. \tag{4.12}$$

Here ϵ is the permittivity of the dielectric medium, μ is the permeability (from here on assumed to have the vacuum value of 1 everywhere), c is the speed of light in vacuum, $\vec{E}$ and $\vec{H}$ are the electric and magnetic fields, and $\vec{J}$ is the current density. The permittivity ϵ in absorptive (sometimes called 'lossy') media is a complex parameter that is related to the complex refractive index m by

$$\epsilon = \epsilon_R + i\,\epsilon_I = m^2\,. \tag{4.13}$$

We will not consider the presence of current densities or free charges in any of the calculations we present in this chapter, and assume $\rho = 0$. But for this section alone we include current densities in the statement of the equations and their 'frequency domain' formulations immediately to follow, in order to discuss an approximation that saves computer memory that is presented later in this section.

In a 'frequency domain' approach, the time-evolution equations are Fourier transformed in time to get expressions in terms of temporal frequency ω. That is, for each ω, complex-valued solutions are sought of the form

$$\vec{E} = \vec{\mathcal{E}}(\vec{x})\,\mathrm{e}^{-i\,\omega\,t}\,, \qquad \vec{H} = \vec{\mathcal{H}}(\vec{x})\,\mathrm{e}^{-i\,\omega\,t}\,, \qquad \vec{J} = \vec{\mathcal{J}}(\vec{x})\,\mathrm{e}^{-i\,\omega\,t}\,,$$

where $\vec{\mathcal{E}}$, $\vec{\mathcal{H}}$ and $\vec{\mathcal{J}}$ are complex-valued functions of space. (As usual, physical solutions are found by taking real parts.) Then Maxwell's equations transform to

$$-i\,\omega\,\frac{\epsilon}{c}\vec{\mathcal{E}} = \nabla \times \vec{\mathcal{H}} - \frac{4\pi}{c}\vec{\mathcal{J}}\,, \tag{4.14}$$

$$-i\,\omega\,\frac{1}{c}\vec{\mathcal{H}} = -\nabla \times \vec{\mathcal{E}}\,, \tag{4.15}$$

$$\nabla \cdot \vec{\mathcal{E}} = 0, \qquad \nabla \cdot \vec{\mathcal{H}} = 0\,. \tag{4.16}$$

In the absence of free charges or current densities, this system can be easily seen to lead to an elliptic system of partial differential equations (Helmholtz equations for plane waves), and can be solved using any of a number of elliptic solvers. Pseudo-spectral methods may be used here as well, but discussion of this approach is beyond our scope.

An approximation to save memory

While the scattering problem we consider does not directly involve current densities, it does involve dielectric particles with complex indices of refraction. This fact introduces complex numbers into calculations, and effectively doubles the demands on computer memory, since all field variables must then have both real and imaginary parts. As discussed in Yang and Liou (1996a), it is possible to get around this difficulty in the case of monochromatic incident waves by making an approximation to Maxwell's equations that is exact at precisely the frequency of the incident wave. A way to view the approximation is to note the formal similarity between a term introduced by a nonzero imaginary part of the index of refraction and a current density term in the frequency-domain formulation. We outline the argument here, focusing just on the formal nature of the approximation itself and refer the reader to (Yang and Liou, 1996a) for more discussion.

With the complex permitivity decomposed into real and imaginary parts as in (4.13) above, the frequency-domain equation (4.14) becomes

$$-i\,\omega\,\frac{\epsilon_R}{c}\vec{\mathcal{E}} = \nabla\times\vec{\mathcal{H}} - \left(\frac{\omega\,\epsilon_I}{c}\vec{\mathcal{E}} + \frac{4\pi}{c}\vec{\mathcal{J}}\right) \qquad (\omega \text{ arbitrary})\,.$$

Thus the presence of a nonzero imaginary part of the permitivity at a point *formally* behaves (at one frequency) as would an 'effective current density' there. In the absence of any true current or charge densities ($\vec{\mathcal{J}} = 0$), this frequency domain equation has the simpler form

$$-i\,\omega\,\frac{\epsilon_R}{c}\vec{\mathcal{E}} = \nabla\times\vec{\mathcal{H}} - \frac{\omega\,\epsilon_I}{c}\vec{\mathcal{E}}\,. \tag{4.17}$$

Now suppose we want to do a scattering calculation with an incident monochromatic wave having frequency ω_0, and consider instead the modified equation that differs only in the last term:

$$-i\,\omega\,\frac{\epsilon_R}{c}\vec{\mathcal{E}} = \nabla\times\vec{\mathcal{H}} - \frac{\omega_0\,\epsilon_I}{c}\vec{\mathcal{E}}\,. \tag{4.18}$$

(The choice of ω_0 will be given below.) Solutions to the frequency domain equations (4.17) and (4.18) will in general be different, but will agree at $\omega = \omega_0$. The equation (4.18) is the Fourier transform of the evolution equation

$$\frac{\epsilon_R}{c}\frac{\partial\vec{E}}{\partial t} = \nabla\times\vec{H} - \frac{\omega_0\epsilon_I}{c}\vec{E}\,, \tag{4.19}$$

an evolution equation that has only real coefficients. This approximate equation, equivalent to the one derived in Yang and Liou (1996a), is used in place of equation (4.10), with $\vec{J} = 0$, in the PSTD calculations discussed in this chapter. The natural choice $\omega_0 = k\,c$ is made, where k is the wavenumber of the incident wave.

The new equation has purely real coefficients, so if we use it there will be no need to introduce complex numbers into the numerical simulations and we effectively halve the memory requirement of computations. It is true that modification of this one equation gives a model which is exact at *only the one frequency* $\omega = \omega_0$, and is approximate at other frequencies. Since the light scattering problem is linear, and we are only interested in the one frequency ω_0, we may ignore errors at other frequencies. The comparisons we give below with the Mie solution in the case of spheres validate our expectation that the approximation is good at the frequency of our interest.

Scattered and incident fields

In the time-domain simulations, the total fields that appear in equations (4.19, 4.11) with $\mu = 1$ are decomposed in terms of scattered and incident fields,

$$\vec{E} = \vec{E}_{\text{inc}} + \vec{E}_{\text{sca}}, \qquad \vec{H} = \vec{H}_{\text{inc}} + \vec{H}_{\text{sca}},$$

where the incident fields satisfy

$$\frac{1}{c}\frac{\partial \vec{E}_{\text{inc}}}{\partial t} = \nabla \times \vec{H}_{\text{inc}}, \tag{4.20}$$

$$\frac{1}{c}\frac{\partial \vec{H}_{\text{inc}}}{\partial t} = -\nabla \times \vec{E}_{\text{inc}}, \tag{4.21}$$

$$\nabla \cdot \vec{E}_{\text{inc}} = 0, \qquad \nabla \cdot \vec{H}_{\text{inc}} = 0. \tag{4.22}$$

The equations satisfied by the scattered field are then

$$\frac{\partial \vec{E}_{\text{sca}}}{\partial t} = \frac{c}{\epsilon_R}\nabla \times \vec{H}_{\text{sca}} - \omega_0 \frac{\epsilon_I}{\epsilon_R}\vec{E}_{\text{sca}} + \left[\left(\frac{1-\epsilon_R}{\epsilon_R}\right)\frac{\partial}{\partial t} - \omega_0 \frac{\epsilon_I}{\epsilon_R}\right]\vec{E}_{\text{inc}}, \tag{4.23}$$

$$\frac{\partial \vec{H}_{\text{sca}}}{\partial t} = -c\nabla \times \vec{E}_{\text{sca}}, \tag{4.24}$$

and at each time step in the numerical integration the exact values for the expressions involving $\vec{E}_{\text{inc}}$ are used, so that once again the right-hand sides of the equations involve only spatial derivatives. The distinguishing feature of the PSTD method is how it evaluates these spatial derivatives, and will be discussed in the following section: the choice of time-stepping methods is a separate consideration beyond the scope of this chapter. All results using the PSTD that we discuss here were obtained using the standard second-order accurate centered difference time-stepping method, sometimes called the leapfrog method.

Using the PSTD, the equations (4.23, 4.24) are solved in the region of the computational domain interior to the UPML region (see Fig. 4.4), and in the UPML region the equations are augmented by the UPML expressions that match impedances across the layer boundary in such a way as to prevent any reflection as the outgoing waves enter the layer, and furthermore damp the entering waves sufficiently rapidly that they never re-emerge upon reflection at the outer boundary of the computational domain.

The particular form of the incident wave we use will be described when we discuss near-to-far-field transformations below.

4.2.3 Near-to-far-field transformation

There are two methods commonly in use to compute the scattered field far from the scattering object: the 'volume integral method' (VIM) and the 'surface integral method' (SIM). The two methods are mathematically equivalent, but impose substantially different computational burdens. Each method assumes first that

$$\vec{\mathcal{E}}(\vec{x}) = \vec{\mathcal{E}}_{\text{inc}}(\vec{x}) + \vec{\mathcal{E}}_{\text{outgoing}}(\vec{x}),$$

where the origin of the coordinate system is at the center of the scatterer, and that $|\vec{x}|$ is great enough that the response to a monochromatic plane wave input is a monochromatic outward-propagating spherical wave. The natural formulation of this is in the frequency domain, and the assumption (the 'Sommerfield radiation', or outward energy propagation, condition) is that

$$\vec{\mathcal{E}}_{\text{outgoing}}(\vec{x}) \sim \frac{\mathrm{e}^{\mathrm{i}\,k\,|\vec{x}|}}{|\vec{x}|}\,.$$

Adoption of this assumption provides the necessary outer boundary condition in using identities of Green (essentially integration-by-parts arguments) to write solutions of the vector Helmholtz equation at positions outside some surface $\mathcal{S}$ enclosing the scatterer in one of two integral forms:

- an integral over the surface $\mathcal{S}$:

$$\vec{\mathcal{E}}(\vec{x}) = \frac{\mathrm{i}k\,\mathrm{e}^{\mathrm{i}\,k\,|\vec{x}|}}{4\,\pi\,|\vec{x}|}\,\hat{x}\times\iint_{\mathcal{S}}\{\vec{n}_{\mathcal{S}}\times\vec{\mathcal{E}}(\vec{x}\,')-\hat{x}\times[\hat{n}_{\mathcal{S}}\times\vec{\mathcal{H}}(\vec{x}\,')]\}\mathrm{e}^{-\mathrm{i}\,k\,\hat{x}\cdot\vec{x}\,'}\,\mathrm{d}\,\vec{x}\,'\,, \quad (4.25)$$

  where

$$\hat{x} = \frac{\vec{x}}{|\vec{x}|}\,.$$

  This is the 'surface integral method'.
- an integral over the volume $\mathcal{V}$ enclosed by the surface $\mathcal{S}$

$$\vec{\mathcal{E}}(\vec{x}) = \frac{k^2\,\mathrm{e}^{\mathrm{i}\,k\,|\vec{x}|}}{4\pi\,|\vec{x}|}\iiint_{\mathcal{V}}[\epsilon(\vec{x}\,')-1]\{\vec{\mathcal{E}}(\vec{x}\,')-[\vec{\mathcal{E}}(\vec{x}\,')\cdot\hat{x}]\,\hat{x}\}\,\mathrm{e}^{-\mathrm{i}\,k\,\hat{x}\cdot\vec{x}\,'}\,\mathrm{d}\vec{x}\,'\,. \quad (4.26)$$

  where $\epsilon = m^2$ is the complex permittivity of the particle. This is the 'volume integral method'.

It would take us too far afield to reproduce the mathematical arguments that lead to these particular expressions: see Umashankar and Taflove (1982) for the SIM and Goedecke and O'Brien (1988) for the VIM.

The way in which either of these methods is used is to extract data $\vec{\mathcal{E}}(\vec{x}\,')$, $\vec{\mathcal{H}}(\vec{x}\,')$ needed for the integrals from the near-field calculations and perform the indicated integrations to get the far fields: thus they are each called 'near-to-far-field' transformations. Neither integral method is as expensive of cpu time as is the numerical simulation of Maxwell's equations. Details differ with choices of parameters, but for a size parameter $x = 200$, with 512^3 gridpoints in the computational domain ('equivalent resolution', in a sense to be explained below, for a PSTD with 256 Fourier coefficients in each direction), typically a calculation using the SIM uses 2%–5% of the total cpu time. Thus the natural computational strategy is to take $\mathcal{S}$ very close to the scatterer, and the UPML just outside it, minimizing the domain for the cpu-intensive near-field calculations.

The comparison in computational times just sketched is slightly different when considering the VIM, which does demand a more cpu time (on the order of 5%–10% of the total cpu time), the reason for which can be appreciated easily from dimensional considerations. However, it is also true that in the SIM the data must be very carefully gathered on the lower dimensional object $\mathcal{S}$. Our experience has been that with the required care taken, the SIM is the better method to use, and all of the computations we report below use that method. Again on the basis of experimentation, we have found that setting the distance from the scatterer to the UPML to be on the order of $6\Delta x$ or $8\,\Delta x$, with the distance from the scatterer to $\mathcal{S}$ to be $2\Delta x$, works well in the cases we have considered. (Here and below Δx is the grid-spacing in each dimension of a uniform mesh in the computational domain.)

What then determines the total amount of time a simulation must be run is the amount of time that it takes to get good *frequency* data on the surface $\mathcal{S}$ using our time-domain integration method. For this purpose we use as incident signal a 'Gaussian pulse',

$$\vec{E}_{\text{inc}}(\vec{x},t) = \vec{E}_0\,P(\hat{k}\cdot\vec{x} - c\,t), \qquad \text{with} \qquad P(s) = \mathrm{e}^{-\frac{4\,s^2}{\lambda^2}}\cos(|\vec{k}|\,s)\,. \tag{4.27}$$

Here $\hat{k} = \frac{\vec{k}}{|\vec{k}|}$: this Gaussian has e-folding width λ, and at a given point $\vec{x}_0$ in space the electromagnetic field has time behavior like

$$\vec{E}(\vec{x}_0,t) \sim \mathrm{e}^{-\frac{4\,(t-t_0)^2}{T}}\cos(\omega\,t), \qquad T = \frac{\lambda}{c}\,.$$

The time t_0 is chosen (e.g. $t_0 = 5T$) so that the pulse has exponentially small amplitude at the start of the numerical integration.

Doing the Fourier transformation

In order to use the near-to-far-field transformation, the single-frequency response in the near-field time-domain calculations (PSTD or FDTD) must be extracted. As opposed to using some kind of FFT method, which would require storing all the temporal data over a long time integration before doing the FFT, we choose a method much more sparing of memory. The method can be appreciated by considering a simple example. Suppose $G(x,t)$ is a time-domain signal whose frequency transform $\mathcal{G}(x,\omega)$ at some frequency ω is desired. For any finite time interval of length T_* we can make the estimate

$$\mathcal{G}(x,\omega) \approx \frac{1}{T_*}\int_0^{T_*} G(x,t)\,\mathrm{e}^{-\mathrm{i}\omega\,t}\,\mathrm{d}t\,, \tag{4.28}$$

with the estimate improving in accuracy with increasing integration length T_*. The time-discrete version of this is

$$\mathcal{G}_N(x,\omega) = \frac{1}{N}\sum_{n=1}^{N} G_n(x)\,\mathrm{e}^{-\mathrm{i}\omega n\Delta t}\,, \tag{4.29}$$

where $G_n(x) = G(x, n\Delta t)$ and $N\,\Delta t = T_*$. Now we make the simple observation that when T_* increases to $(N+1)\Delta t$,

$$\mathcal{G}_{N+1}(x,\omega) = \left(\frac{N}{N+1}\right)\mathcal{G}_N(x,\omega) + \frac{1}{N+1}G_{N+1}(x)\,\mathrm{e}^{-\mathrm{i}\omega(N+1)\Delta t}\,, \tag{4.30}$$

which allows us to update estimates of $\mathcal{G}(x,\omega)$ as we integrate in time. Thus we need save only the data required by our time-stepping method, and we need only run that method long enough for our transforms $\mathcal{G}_N$ to become constant, that is for the successive iterates to satisfy (uniformly in space) a criterion set for being constant. That this should happen eventually can be understood by considering equation (4.30), and remembering that the incident wave packet has a narrow width in time as it travels in free space. So as N increases, there comes point beyond which the incremental update becomes exponentially small. However, when exactly this decay sets in is not easy to estimate *a priori*, since the full interaction time of the packet with the particle is not easily approximated, and our best guide has been experimentation. We have found that the total time of integration needed for the Fourier transform to converge is between four and five times the amount of time that the packet would take to cross a distance in free space equal to one diameter of the particle.

4.3 Derivatives: finite difference versus spectral

We present here a unifying view of the connections between finite difference approximations of various orders and spectral approximations, and discuss the behavior of errors as resolution increases in each case: it is shown that in the case of smooth functions the decrease of errors in spectral approximations to derivatives is dramatically more rapid as resolution increases than is the case with finite difference approximations. We confine attention to a case in which complications are minimal, a one-dimensional case and assume functions to be *periodic* on some finite interval $x \in [0, L]$: these restrictions in no way affect the main points we make. In this discussion we assume functions are as smooth as needed to make the appeals to Taylor series arguments valid. We necessarily relax this assumption later in the discussion of the Gibbs phenomenon.

As a focus for discussion, consider a linearly polarized wave propagating in a homogeneous medium: if we let $u = E_z$ and $v = H_y$, and all other components of the fields be zero, we have

$$\frac{\epsilon}{c}\frac{\partial u}{\partial t} = \frac{\partial v}{\partial x}, \tag{4.31}$$

$$\frac{\mu}{c}\frac{\partial v}{\partial t} = \frac{\partial u}{\partial x}. \tag{4.32}$$

Taking the time derivative of the first equation and substituting from the second gives either of the following two equivalent forms

$$\frac{\partial^2 u}{\partial t^2} - c_*^2\frac{\partial^2 u}{\partial x^2} = 0, \tag{4.33}$$

$$\left(\frac{\partial}{\partial t} + c_*\frac{\partial}{\partial x}\right)\left(\frac{\partial}{\partial t} - c_*\frac{\partial}{\partial x}\right)u = 0, \tag{4.34}$$

where $c_* = c/\sqrt{\epsilon\,\mu}$. It is clear from the second form that waves uniformly translating toward either positive or negative values of x at speed c_* are solutions. We consider a simple waveform moving to the right, which satisfies

$$\frac{\partial u}{\partial t} = c_* \frac{\partial u}{\partial x}.$$

As stated above, our comparison of finite difference and pseudo-spectral methods will be carried out using the same (leapfrog) time-stepping method in each case, and focus on the two different numerical treatments for the derivative on the right-hand side of the equation.

In finite difference methods functions are represented in terms of their values at grid points, with better representation of the functions being made possible by increasing the number of grid points. With N grid points on an equally spaced grid, the points may be denoted

$$x_j = (j-1)\frac{L}{N} = (j-1)\Delta x, \qquad j = 1, 2, \ldots, N.$$

(Notice that it is not necessary to include what would be an N-plus-first grid point $x_{N+1} = N\,\Delta x = L$ because of the periodicity assumption.) In this section we will always assume N is an even integer.

The familiar centered difference approximation

$$\frac{\partial u}{\partial x}(x_j) \approx \frac{u(x_{j+1}) - u(x_{j-1})}{2\Delta x} \tag{4.35}$$

is a second-order approximation: as Δx approaches zero (i.e. as N increases without bound) the error in the approximation goes to zero quadratically (halving Δx results in a reduction in the error by a factor of $(1/2)^2 = 1/4$. A simple argument based on Taylor series establishes this.

In the context of higher-order and spectral approximations there is an illuminating way to view the expression on the right-hand side of (4.35). If we want to get a second order accurate approximation to the derivative of u at x_j, we construct a second-degree polynomial that should (based on our information at grid points) be 'close to' u near x_j, and calculate its derivative at x_j. Specifically, we consider the (unique) *second-order polynomial interpolant* $U(x)$ through the points $(x_m, u(x_m))$ for $m = j-1, j, j+1$, and then define the approximation to be the value of the *derivative* of this polynomial interpolant , evaluated at the point x_j. As a little algebra easily verifies, the interpolant can be written as a sum of a set of three basic quadratic polynomials weighted by the values of the function at grid points

$$U(x) = u(x_{j-1})l_{j-1}(x) + u(x_j)\,l_j(x) + u(x_{j+1})l_{j+1}(x), \tag{4.36}$$

where the $l_i(x)$ are the second-order Lagrange polynomials

$$\begin{aligned} l_{j-1}(x) &= \frac{(x-x_j)(x-x_{j+1})}{(x_{j-1}-x_j)(x_{j-1}-x_{j+1})}, \\ l_j(x) &= \frac{(x-x_{j-1})(x-x_{j+1})}{(x_j-x_{j-1})(x_j-x_{j+1})}, \\ l_{j+1}(x) &= \frac{(x-x_{j-1})(x-x_j)}{(x_{j+1}-x_{j-1})(x_{j+1}-x_j)}. \end{aligned}$$

By inspection we see that each Lagrange polynomial is quadratic, vanishes at two of the three grid points, and has unit value at the third:

$$l_i(x_m) = \delta_{i,m} \,.$$

So $U(x)$ is clearly the interpolant. Simply differentiating the polynomial and evaluating it at the grid point x_j shows

$$\frac{\mathrm{d}}{\mathrm{d}x}U(x_j) = \frac{u(x_{j+1}) - u(x_{j-1})}{2\Delta x} \,,$$

and we see that the second order accurate centered differencing is equivalent to the approximation using second order interpolating polynomials

$$\frac{\partial u}{\partial x}(x_j) = \frac{\mathrm{d}U}{\mathrm{d}x}(x_j) + O(\Delta x^2) \,.$$

(This construction of an interpolant is done at each grid point separately, so we could have written $U_j(x)$ and $\mathrm{d}U_j/\mathrm{d}x$ instead of $U(x)$ and $\mathrm{d}U/\mathrm{d}x$, but with little gain in understanding.)

The value in taking this view is that it is easy to anticipate how to get, for example, a fourth-order accurate finite difference approximation to the derivative at the grid-point x_j: find the unique fourth-order polynomial through the five points $(x_m, u(x_m))$, $j-2 \le m \le j+2$, and then evaluate the derivative at the gridpoint x_j. The interpolating polynomial will then be a sum of *quartic* Lagrange polynomials weighted by values of the function at the grid points: the second-order interpolant formula (4.36) is replaced by

$$U(x) = \sum_{m=j-2}^{j+2} u(x_m) l_m(x), \tag{4.37}$$

where for $j-2 \le m, p \le j+2$,

$$l_m(x) = \prod_{p \ne m} \frac{(x - x_p)}{(x_m - x_p)} \,. \tag{4.38}$$

In this case again the derivative approximation is a linear combination of the derivatives of the Lagrange polynomials evaluated at the gridpoint:

$$\frac{\partial u}{\partial x}(x_j) = \sum_{m=j-2}^{j+2} u(x_m) l'_m(x_j) + O(\Delta x^4) \,. \tag{4.39}$$

(It can be easily checked that the method based on the sum term does in fact give fourth-order accuracy.) Figure 4.5 shows the Lagrange polynomials for the case of the second-order and fourth-order interpolations. (The x-axis in each case is labeled in multiples of Δx, and so covers only a part of the $[0, L]$ interval.) Each individual polynomial is zero at all grid points except one, where it has unit value, and so is 'concentrated' on a single grid point.

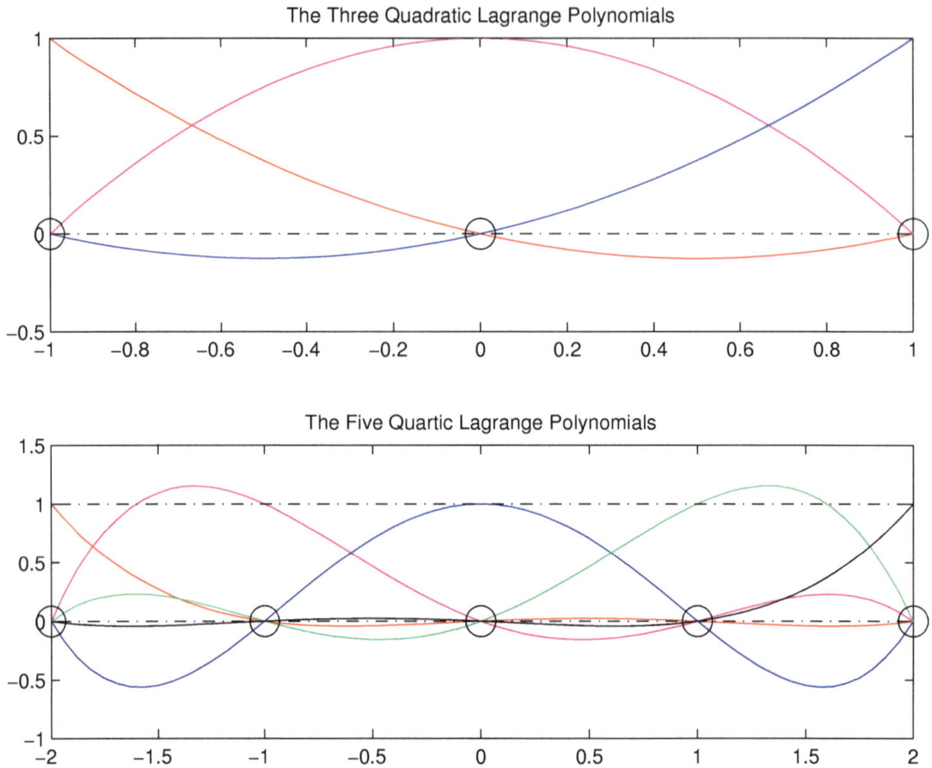

Fig. 4.5. The sets of $l_m(x)$ for second- and fourth-order polynomial interpolation. Horizontal axes centered on gridpoint x_j, scaled in multiples of Δx: circles indicate nearby gridpoints x_m.

In practice, each of these finite difference methods can be written as a matrix multiplication: if $\mathbf{u}$ and $\mathbf{u}'$ denote column vectors of gridpoint values, the finite difference calculation of order p is accomplished at all gridpoints simultaneously by the matrix multiplication

$$\mathbf{u}' = \mathbf{D}_p \mathbf{u}\,, \tag{4.40}$$

where $\mathbf{D}_p$ is a banded diagonal matrix with entries determined, at the start of calculations, by the weights for grid point values implied by the derivatives of the Lagrange polynomials. The width of the band is related to the order p of the derivative approximation, and advantage of this structure can be taken to optimize performance as the number of grid points increases. In the limiting case, for a fixed number of grid points $N + 1$, we could choose to use a Lagrange polynomial of order N to get a numerical differentiation scheme that is Nth-order accurate: the expression for the interpolant would be similar to (4.37), but use information from all gridpoint values: the interpolant could be written in the form

$$U(x) = \sum_{m=j-N/2}^{j+N/2-1} u(x_m) l_m(x)\,, \tag{4.41}$$

and the $\{l_m(x)\}$ would be the collection of Nth-degree polynomial analogues of (4.38). This approximation, by involving all gridpoint values at once, is highly *non-local*, as compared with the local centered difference approximation. A gain in order, hence accuracy, comes about through this non-locality, but the 'band' about the diagonal fills the matrix. A consequence of going to a dense matrix is that the matrix multiplication in (4.40) now involves $O(N^2)$ operations, rather than the $O(N)$ operations in the multiplication corresponding to the second-order accurate centered difference method: high accuracy comes with a computational cost.

Going to higher order is one way to increase accuracy, and another way is to simply increase the number of gridpoints and keep the same order of accuracy. Inevitably, higher order involves greater demands on cpu time and memory, and so pursuit of greater accuracy in calculations involves a trade-off between increasing the order of the scheme and increasing the number of gridpoints keeping the order of the scheme fixed.

Figure 4.6 illustrates derivative calculations using second- and fourth-order difference schemes as the number of grid points N is increased. The function being differentiated is a simple Gaussian $G(x) = \exp^{-(x-1)^2/\sigma^2}$: the upper panels show both the Gaussian itself (blue curve) as well as the numerical derivatives and the exact values. The lower panels give one way of seeing how the error in the calculation is reduced as the number of gridpoints increases for the second- and fourth-order finite difference methods. We give more insight into the behavior of error with increasing numbers of grid points below, after introducing the spectral method.

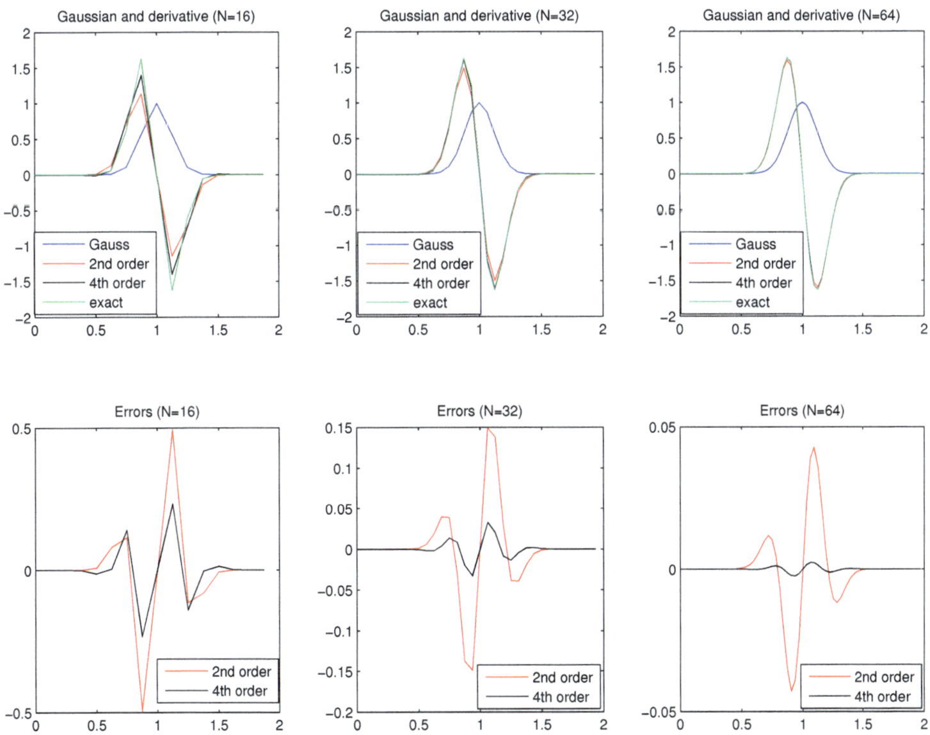

Fig. 4.6. Error in derivative calculations as the number of gridpoints increases.

In the spectral method that we discuss here, the analog to the collections of basic Lagrange interpolating polynomials concentrated at grid points, one collection for each grid point, is essentially a single set of trigonometric polynomials that is used for all grid points. That is we consider as interpolation basis a single set of highly non-local (concentrated at no single grid point) complex exponentials, and write the interpolant:

$$U(x) = \sum_{k=-N/2}^{N/2-1} \tilde{U}_k \, e^{i \hat{k} x}, \qquad \hat{k} = \frac{2\pi}{L} k. \tag{4.42}$$

The N coefficients $\tilde{U}_k$ are determined by the requirement that $U(x)$ be an interpolant of the values of u at grid points. That is, once again we require $U(x_j) = u(x_j)$, and hence that

$$u(x_j) = \sum_{k=-N/2}^{N/2-1} \tilde{U}_k \, e^{i \hat{k} x_j}, \qquad j = 1, 2, \ldots N. \tag{4.43}$$

Given the $u(x_j)$, this is a system of N equations in the N unknowns $\tilde{U}_k$. It can be shown using simple algebra and properties of the complex exponential that the solution is

$$\tilde{U}_k = \frac{1}{N} \sum_{j=1}^{N} u(x_j) \, e^{-i \hat{k} x_j}, \qquad -\frac{N}{2} \le k \le \frac{N}{2} - 1. \tag{4.44}$$

The association between the sequence of grid point values $\{u(x_j)\}$ and the sequence of Fourier amplitudes $\{\tilde{U}_k\}$ is the one established by the discrete Fourier transform and its inverse, and transforming between the two sequences can be done efficiently using a fast Fourier transform (FFT) algorithm.

Notice that this approach associates a natural maximum wavenumber $K = N/2$ with a number of grid points, natural on the assumption of equal spacing of grid points. In terms of wavelengths, the smallest wavelength included in the interpolant is the '$2\,\Delta x$' wave. Conversely, the equally spaced N-point grid is called the 'equivalent spatial grid' for the $N/2$-wave spectral representation.

Approximating the derivative is done in the same manner as in the case of polynomial interpolations. In this case the derivative of the interpolant is especially easily calculated:

$$U'(x) = \sum_{k=-N/2}^{N/2-1} i \hat{k} \, \tilde{U}_k \, e^{i \hat{k} x}. \tag{4.45}$$

We see that, unlike the situation with the Lagrange polynomials, the derivative is easily expressible in terms of the same basis functions that are used in the interpolant itself. Thus, approximations to derivatives at grid points may be calculated by (i) finding the $\tilde{U}_k$ using an FFT, (ii) constructing a new sequence $\tilde{D}_k = i\,\hat{k}\tilde{U}_k$, and (iii) using an inverse fast Fourier transform (IFFT) construct the sum indicated in (4.45) to get the derivative values at gridpoints.

One important feature to note is that this approximation, as well as its calculated derivative, is *exact* in the case that the function $u(x)$ only involves oscillations on scales larger than $2\Delta x$, or discrete Fourier coefficients up to index $m = N/2$. This means that in the general case that u involves variation at smaller scales, as long as u is a smooth function the error in spectral differentiation is entirely due to omission of high wavenumbers in the interpolative representation. If the amplitude in these higher wavenumbers is small, so will be the error. We return to this point below.

One of the features of the spectral method that can make it so attractive emerges when we look into the order of the approximation. It turns out that, if the function being approximated is very smooth (infinitely differentiable) the approximation is better than order Δx^p *for any* p: this better-than-any-polynomial-order accuracy is called *spectral accuracy*. The reason underlying this potential for high accuracy is that there is a relation between the amount of smoothness (differentiability) of a function and the rate of decay of spectral coefficients: very smooth functions have spectral coefficients whose amplitude decays as a function of wavenumber faster than any integral power. Thus, as a function of the truncation wavenumber $K \sim \Delta x^{-1}$, the error decays faster than any negative power of K, hence any positive power of Δx. A detailed exposition can be found in (Tadmor, 1986); a less technical discussion can be found in the excellent (Trefethen, 2000), on which we have based some of the presentation in this section, including the particular test function we consider next.

The distinction between polynomial and spectral accuracy can be illustrated by considering numerical calculations of the derivative of the function $y = \mathrm{e}^{-\sin(x)}$ whose graph is shown in Fig. 4.7.

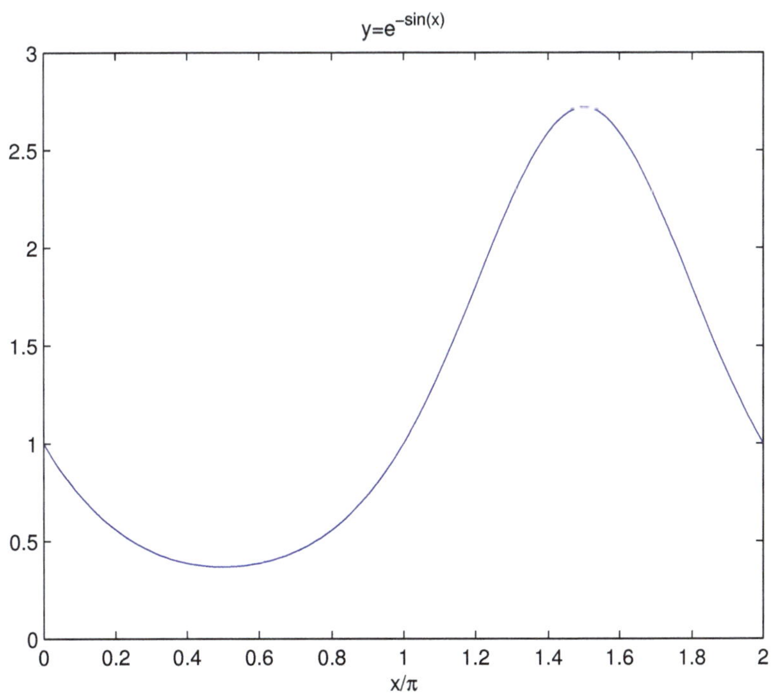

Fig. 4.7. $y = \mathrm{e}^{-\sin(x)}$

This function is sufficiently complicated that no polynomial or finite term spectral approximation can be truly exact: there is always some error and the question becomes of how rapidly this error shrinks with increasing N using each of the numerical approximations. There is a numerical limit, of course, to how small the error can be that is determined by the limits of numerical representation on any given computer, the 'machine epsilon' ε (not to be confused here with ϵ, the standard symbol for permittivity of a dielectric medium): once the approximation error is reduced to this level, no further improvement is possible.

Figure 4.8 shows how the error (calculated as the sum of squares of errors at grid points) shrinks as the number of grid points N increases, i.e. as $\Delta x \sim N^{-1}$ decreases. The use of a log-log plot makes power-law decay show as linear: clearly evident are power-law decay rates in the case of the second- and fourth-order schemes. The figure also makes clear the dramatically more rapid rate of decay of errors shown by the spectral method: the level of machine accuracy is reached quite quickly on a computer with a machine epsilon of $2^{-52} \approx 2.22 \times 10^{-16}$.

The rapid attainment of a high order of accuracy means that, for a fixed level of error to be tolerated in a numerical calculation, the Δx required may be (considerably) larger for a spectral method than a finite difference method. This not only means that less demand is made on memory resources, but in a time-dependent

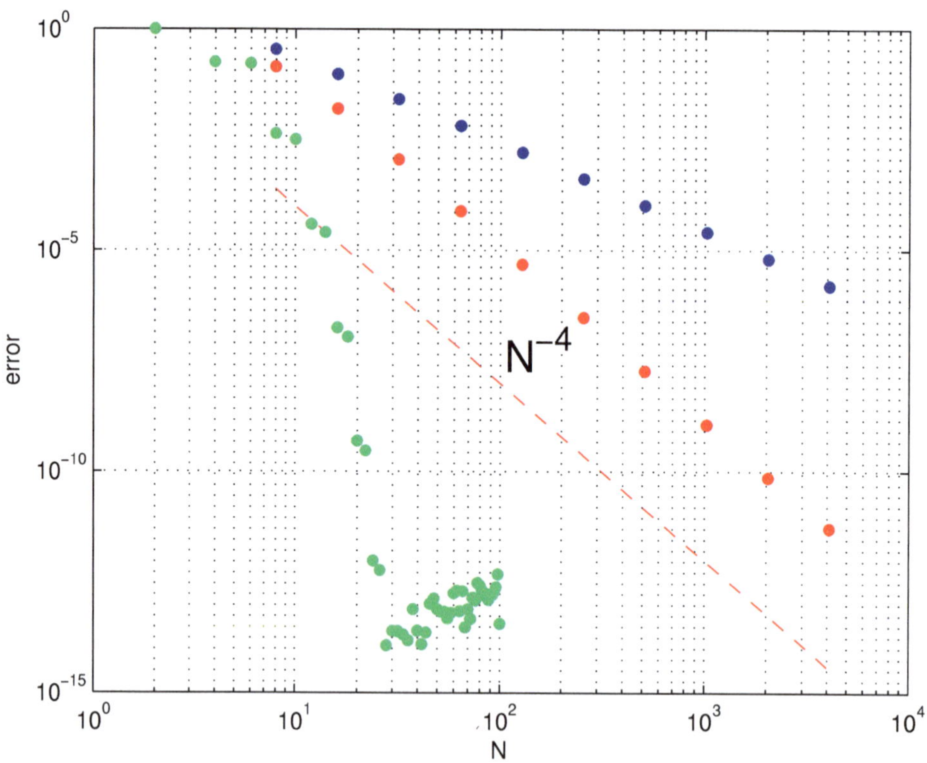

Fig. 4.8. Errors as N increases for second-order (blue), four-order (red), and spectral (green) derivative calculations.

calculation in which the time step must satisfy a CFL condition (see, e.g., Morton and Mayers, 2005) set by the choice of Δx, a larger time step may be used with a spectral method than with a finite difference method that satisfies the same error tolerance. This is a second important way in which a spectral method can save cpu resources.

A third source of savings in cpu time comes from special features of complex exponentials as they occur in Fourier transform calculations. It will be recalled that in the case of finite difference methods, the cost of going to highly non-local polynomial interpolation to get higher-order accuracy comes at the $O(N^2)$ cost of having to do matrix multiplication with dense matrices. On the face of it, the expression (4.45) would suggest a similar $O(N^2)$ cost, since tracing back through the definition of the $\tilde{U}_k$ means that the expression (4.45), when applied at grid points, could be rewritten in the form

$$U'(x_j) = \sum_{l=1}^{N} E_{j\,l}\, U(x_l)\,, \tag{4.46}$$

where the matrix with entries $E_{j\,l}$ that involve products of wavenumbers and complex exponentials is dense. However, fast Fourier transforms take advantage of properties of complex exponentials to calculate the $\tilde{U}_m$ as well as the inverse transform, and the entire derivative calculation can be done, not by the matrix multiplication, but instead by a pair of calls to fast Fourier transform routines in what is, for N a product of a small number of primes, in fact $O(N \log N)$ operations, a significantly smaller number than $O(N^2)$ when N gets large. A very common choice is $N = 2^n$ for some integer N. Versions of FFTs, in dimensions 1–3, are included in most computational mathematics software packages, and vendors of large computer systems typically provide versions that are tuned optimally to their systems.

Another benefit that spectral methods can bring to wave propagation problems is the much reduced 'numerical dispersion', when compared to the situation with finite difference methods. This can be illustrated by considering numerical solutions to the simple wave equation

$$\begin{aligned} \frac{\partial u}{\partial t} &= -2\pi\,\frac{\partial u}{\partial x}\,, \\ u(x,0) &= \mathrm{e}^{-\sin(x)}\,, \\ u(0,t) &= u(2\pi,t) \qquad \text{(periodicity in } x\text{)}\,. \end{aligned} \tag{4.47}$$

The exact solution is $u(x,t) = \mathrm{e}^{-\sin(x-2\pi\,t)}$. Because of the periodicity of the sine function, integrating the solution for an interval of time equal to an integer should reproduce the initial condition, and using this fact makes checking the numerical solution at integral multiples of time a simple way to illustrate error properties. Figure 4.9 shows results of integrations over different periods and using different numbers of grid points, comparing spectral results with second-order finite difference results. The time-stepping method was the same in each case (a simple second-order accurate leapfrog method), and the only difference was in the calculation of the spatial derivative.

With $N = 16$, the finite difference scheme is clearly suffering numerical dispersion after just one period of integration, while the spectral scheme shows only a little error of any sort (note that the peak is in the correct position even at such coarse resolution). To get error even comparable with the spectral results requires doubling the number of grid points and the dispersion is still evident in the failure to correctly position the maximum. Integration at the same doubled spatial resolution, but for 10 periods, shows the clear difficulty the finite difference method is having. Quadrupling the number of grid points improves the 10-period solution using the finite difference method, but at this resolution the 100-period solution is again clearly inferior to the spectral approximation. The essential reason why the spectral method propagates this wave so well is that the individual Fourier components propagate independently in this linear problem, and the spatial derivative calculation is exact for all the represented components.

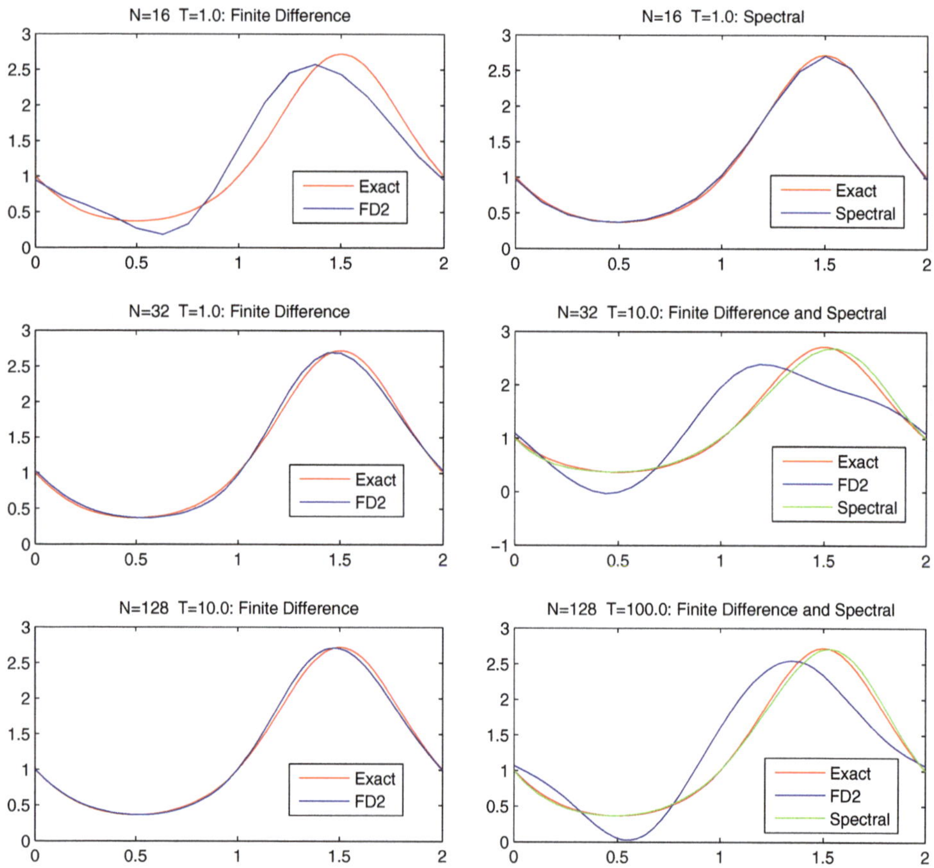

Fig. 4.9. Solutions to equation (4.47) using second-order finite difference and spectral methods, for varying resolutions and time durations.

4.4 The Gibbs phenomenon

There is of course one difficulty with spectral methods that is well known: they work best with smoothly varying functions and do not easily handle functions that have discontinuities, where they show the 'Gibbs phenomenon'. Spectral methods require inclusion of high wavenumbers (small-scale oscillations) to represent rapidly varying features of functions, and if a function has variations, even at only one location, on very small scales, high wavenumbers are required in the Fourier representation and their omission through truncation will have deleterious effects everywhere.

An extreme example is what happens at a simple jump discontinuity. While the spectral representation using an infinite Fourier series (which includes components of arbitrarily small wavelength) is excellent away from a simple discontinuity, if the series is truncated the result has error-containing oscillations, greatest near the discontinuity but present everywhere in sufficient amplitude to drastically reduce the overall error to $O(1/N)$. And even though with increase in N significant amplitude in these oscillations can be confined closer and closer to the discontinuity as the number of waves included increases, there is always an overshoot and undershoot of the representation in the immediate vicinity of the jump, and the amount of the overshoot is never reduced in any finite truncation. The phenomenon is shown in Fig. 4.10, for the case of a simple 'sawtooth' function $S(x)$. The figure has $S(x)$ along with approximations using wavenumbers up to M for $M = 2^p, p = 2, 4, 6, 7$. (The reader will recall that the maximum wavenumber that can be represented using N equally spaced grid points is $M = N/2$.) The evident agreement of the partial sums with each other right at the jump reflects the fact that the Fourier series of a function with an isolated jump discontinuity converges at the location of the discontinuity to the average value of the left- and right-hand limits. Notice that while the error for any choice of M is worst near the jump, even at large M there is error evident far from the jump in the form of a small-wavelength signal. In a time-dependent calculation, the possibility exists for the largest errors, originally located near the jump, to propagate away from it.

Since scattering calculations involve changes that are effectively jump discontinuities in indices of refraction at particle boundaries, it is *a priori* important to have a way of minimizing errors introduced by the Gibbs phenomenon. What leads to these errors is the presence of significant amplitudes in the high wave number Fourier components: with 'infinite' resolution these high-wavenumber components destructively interfere, but with any sum involving only finitely many of them the destructive interference is incomplete and the result is the oscillatory error behavior away from the jump in Fig. 4.10. A number of 'filtering' treatments have been applied to the high wavenumber modal amplitudes, essentially replacing $\tilde{U}_k$ with $g(k)\,\tilde{U}_k$, where the function $g(k)$, defined for non-negative k, has the properties

$$g(k) \qquad \begin{cases} \approx 1 & \text{for small } k \\ \to 0 & \text{`rapidly' for } k \text{ approaching } K\,. \end{cases} \tag{4.48}$$

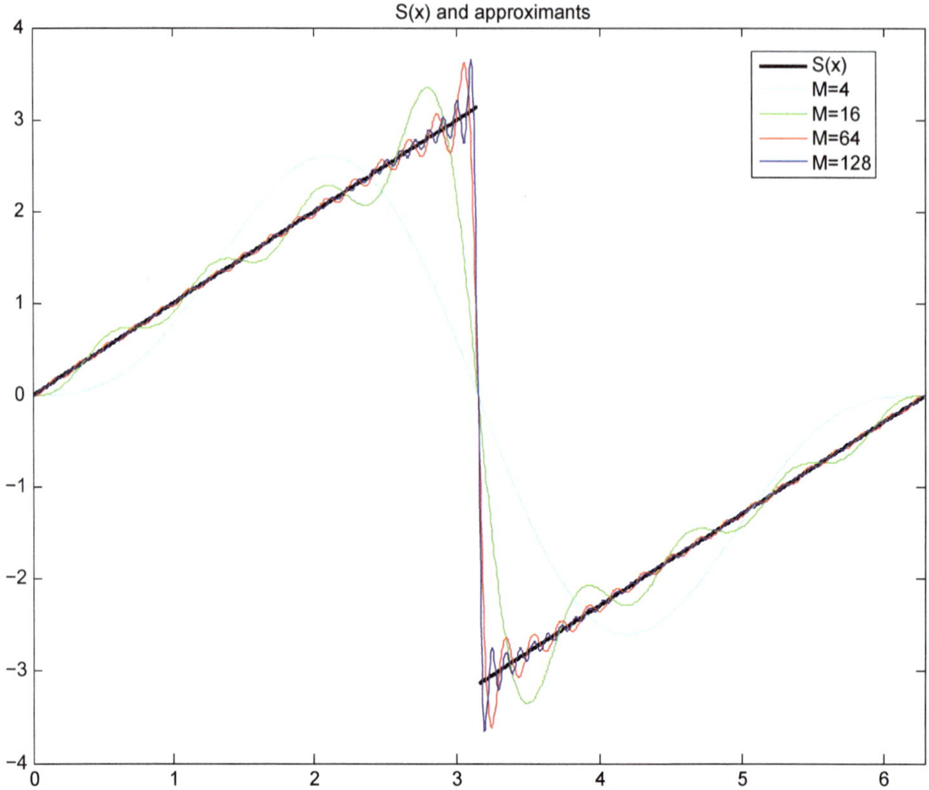

Fig. 4.10. The Gibbs phenomenon at a simple jump discontinuity. M is the truncation wavenumber in the Fourier series partial sum.

A choice of g that has a number of desirable properties is the 'exponential filter'. If

$$g_0(k) = \exp\left\{-\gamma\frac{k}{K}\right\}, \qquad \text{where} \quad \gamma = -\ln(\varepsilon)\,,$$

(ε is again the machine epsilon) then $g_0(0) = 1$ and $g_0(K) = \varepsilon$, but the drop in g with increasing wavenumber k occurs too quickly. Taking a power p of the exponential's argument,

$$g_p(k) = \exp\left\{-\gamma\left(\frac{k}{K}\right)^p\right\}$$

postpones the approach to ε. That is, successively higher powers p give filters that stay near 1 for successively greater wavenumbers before dropping quickly to ϵ, as shown in Fig. 4.11.

Then if the function $u(x)$ has Fourier amplitudes $\tilde{u}_k$, the filtered version $u_p(x)$ of $u(x)$, using exponent p is defined by

$$u_p(x) = \sum_{k=-N/2}^{N/2-1} g_p(|\,k\,|)\,\tilde{u}_k\,\mathrm{e}^{\mathrm{i}\,\hat{k}\,x} = \mathit{IFFT}\left(\{g_p(|\,k\,|)\,\tilde{u}_k\}\right)\,.$$

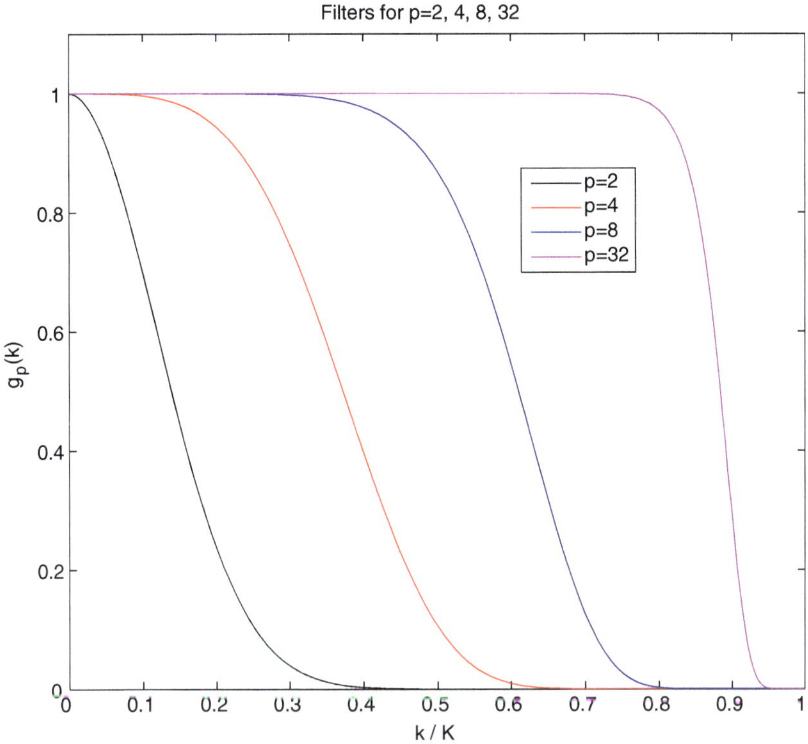

Fig. 4.11. Filter behavior as p varies.

Figure 4.12 shows the results of filtering for different choices of truncation wavenumber M and exponent p. A low value of p gives a filtered version which has very few oscillations away from the discontinuity, especially as the M increases. This is shown in the upper panel of the figure, which has results with $p = 2$. However, with such a small value of p the jump is spread over a comparatively wide band around the true jump. The middle panel shows what happens when p is increased to 8: the jump is less spread but the high wavenumber oscillations are starting to be evident again, which is understandable because the filter has come to resemble more a sharp cut-off at a particular wavenumber (see Fig. 4.11 again). The bottom panel compares the error in the $p = 0$ (unfiltered) approximation and in the $p = 8$ approximation, in each case using $M = 128$ waves. This panel uses a stretched vertical scale to show the reduction in the oscillatory error away from the jump that the exponential filter provides. (Note that because of the magnification chosen the amplitude of the error close to the jump is off the scale.)

It can be shown (see, e.g., Gottlieb and Shu, 1997) that the use of an exponential filter of order p will produce $O(N^{1-p})$ accuracy everywhere away from the jump: the Gibbs phenomenon pollution away from the jump can be essentially removed.

This has been a general discussion of the Gibbs phenomenon and its treatment using simple exponential filters: the matter of how to choose a filter, exponential with a certain order or another kind of filter, for a given scattering problem is a subject of current research. It was found by Liu et al., (2012b) that simply

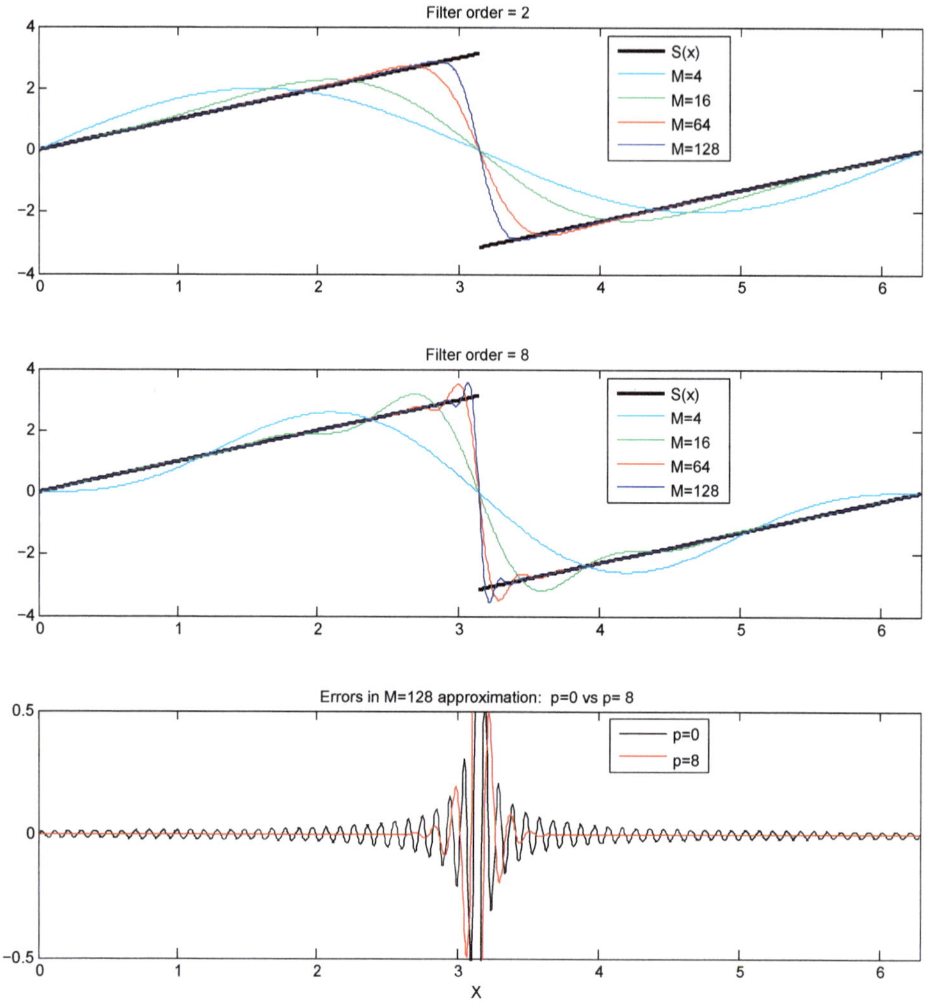

Fig. 4.12. Filtering using the exponential filter and various values of M. In comparing the panels, note that the the vertical scale in the bottom panel has been stretched to show the difference in error behavior away from the jump between the unfiltered and filtered $M = 128$ approximations.

truncating the Fourier expansion at $M = 0.9\,K$ or $0.95\,K$, which is comparable to taking a very high order of p, produced results suitable for their validation tests that were not noticeably different from those obtained using an exponential filter. In fact, the PSTD results shown in Sections 4.5 and 4.6 below were obtained using such a simple truncation.

4.5 Some PSTD results

In this section we show two kinds of results. The first kind establishes the validity of the PSTD method by considering scattering problems for which either an exact solution or another highly reliable method is available. In the case of spherical particles the exact solution is the Lorenz–Mie solution, and in the case of spheroids the reliable method is the T-matrix method. The second kind of results involve cases in which neither of these approaches to validation is available. In the section that follows this one we discuss the relative performance characteristics of the PSTD and DDA methods. All the calculations in this section and the following one were performed on a single 8-cpu node of an IBM iDataplex cluster with 2.8-GHz processor at the Texas A&M Supercomputing Facility.

4.5.1 Comparison with Lorenz–Mie calculations

As a first demonstration we use PSTD to calculate light scattering by spheres, considering size parameters ranging from 10 to 200, and three realistic refractive indices of ice: at wavelengths of visible ($0.670\,\mu$m), near infrared ($4.05\,\mu$m) and infrared ($11.45\,\mu$m) using values of refractive index $1.308 + 1.93 \times 10^{-8}\,$i, $1.358 + 1.336 \times 10^{-2}\,$i and $1.162 + 3.537 \times 10^{-1}\,$i, respectively (Warren and Brandt, 2008). The three refractive indices represent the non-absorptive, weakly absorptive and strongly absorptive cases.

Table 4.1 shows the computational times and spatial resolutions used for each size and refractive index using the PSTD. The spatial resolution is defined to be $\lambda/\Delta x$, the number of grid points per wavelength. Even with the parallelized implementations, the computational burden increases significantly with increase in particle size. For small size parameter, high spatial resolution is necessary (but affordable) to give an accurate representation of the particle shape, whereas, for

Table 4.1. Computational time and spatial resolution for numerical simulation of light scattering by spheres.

	Visible ($m=1.308+1.930\times10^{-8}\,$i)		Near-IR ($m=1.358+1.336\times10^{-2}\,$i)		IR ($m=1.162+3.537\times10^{-1}\,$i)	
x	time(s)	$\lambda/\Delta x$	time(s)	$\lambda/\Delta x$	time(s)	$\lambda/\Delta x$
10	2.9×10^2	26.7	1.8×10^2	22.9	1.5×10^2	22.9
20	2.5×10^3	24.6	2.2×10^3	26.5	6.5×10^2	21.5
30	5.5×10^3	18.5	2.6×10^3	18.5	1.5×10^3	16.4
40	2.0×10^4	23.3	1.1×10^4	20.8	4.6×10^3	17.0
60	1.5×10^4	14.5	1.3×10^4	14.5	6.8×10^3	12.2
80	2.0×10^4	10.4	4.2×10^4	14.1	2.1×10^4	11.7
100	1.1×10^5	14.3	5.8×10^4	11.3	4.3×10^4	10.6
120	1.5×10^5	12.0	1.5×10^5	12.0	6.5×10^4	9.45
140	1.8×10^5	11.9	1.9×10^5	10.3	1.2×10^5	8.46
160	2.2×10^5	8.97	1.8×10^5	8.03	1.5×10^5	8.03
180	2.9×10^5	8.53	3.0×10^5	8.53	2.6×10^5	8.53
200	3.4×10^5	7.68	3.5×10^5	7.68	3.0×10^5	7.68

large size parameters, the particle surface radii of curvature become much larger compared to the wavelength, i.e. effectively smoother, and relatively coarse spatial resolution may be taken. The ratio of wavelength to grid-point spacing is no more than 30 for small particles, and no more than 10 for spheres with size parameters larger than 140. There are between five and seven grid points (depending on index of refraction) per intra-particle wavelength for simulations with size parameter of 200, a number of grid points that is much smaller number than that needed for FDTD. Table 4.1 shows that the computational time is not monotonic as function of the size parameter: this is because, with different spatial resolutions used, the number of grid points N is not monotonic for these simulations. In a calculation using single 8-cpu node, the most time-consuming simulation took 3.5×10^5 seconds, i.e. approximately 4 days for a sphere with size parameter of 200 and refractive index of $1.358 + 1.336 \times 10^{-2}\,\mathrm{i}$, while the computational time was no more than 4.3×10^4 seconds (about 12 hours) for $x < 100$.

With the exact solutions given by the Lorenz–Mie method, we can evaluate the overall performance of PSTD quantitatively. For this purpose six different diagnostic quantities related to scattering properties are calculated:

- relative error (RE) in extinction efficiency, single-scattering albedo (for the two absorptive cases), asymmetry factor, and phase function at 180°,
- root-mean-square relative error (RMSRE) of $P_{11}(\theta)$, and
- root-mean-square absolute error (RMSAE) of the ratio $P_{12}(\theta)/P_{11}(\theta)$.

Here the 'relative error' RE for a value calculated by a numerical method is defined in terms of the 'true' value as follows:

$$RE = \left| \frac{\mathit{approximate\ value} - \mathit{true\ value}}{\mathit{true\ value}} \right| .$$

The quantity $P_{11}(180°)$ is an important parameter for lidar applications, while both the RMSRE and RMSAE can give a good overall measure of the accuracy of the phase matrix elements simulated by a numerical method.

Figure 4.13 shows how the diagnostic quantities for the PSTD vary as functions of size parameter for the three refractive indices at visible, near-infrared (near-IR) and infrared (IR) wavelengths. For most cases, the relative errors for Q_{ext}, SSA, and g are no more than 2%. Most numerical methods have difficulty giving an accurate approximation for backward scattering, especially scattering at exactly 180°, and PSTD is no exception (weak backward scattering for large particles may be several orders of magnitude smaller than the forward scattering). The relative errors of $P_{11}(180°)$ are extremely large for same cases: e.g. over 100% for a sphere with size parameter of 160 at the visible wavelength, whereas most relative errors for $P_{11}(180°)$ are smaller than 50%. The RMSREs of P_{11} are smaller than 50% and the RMSAEs of P_{12}/P_{11} are all less than 30% (except for a sphere with size parameter of 180 at the visible wavelength). The errors of P_{11} and P_{12}/P_{11} for the absorptive cases are smaller than those of the non-absorptive ones, essentially because the backward scattering is smoothed out by the absorption. The relative errors for the integral scattering properties are highly irregular, and no clear pattern is found as x and m are varied. However, the RMSREs of P_{11} and the RMSAEs of P_{12}/P_{11} generally increase with the increase of the size parameter, because the phase matrix elements of spheres become more oscillatory for large x.

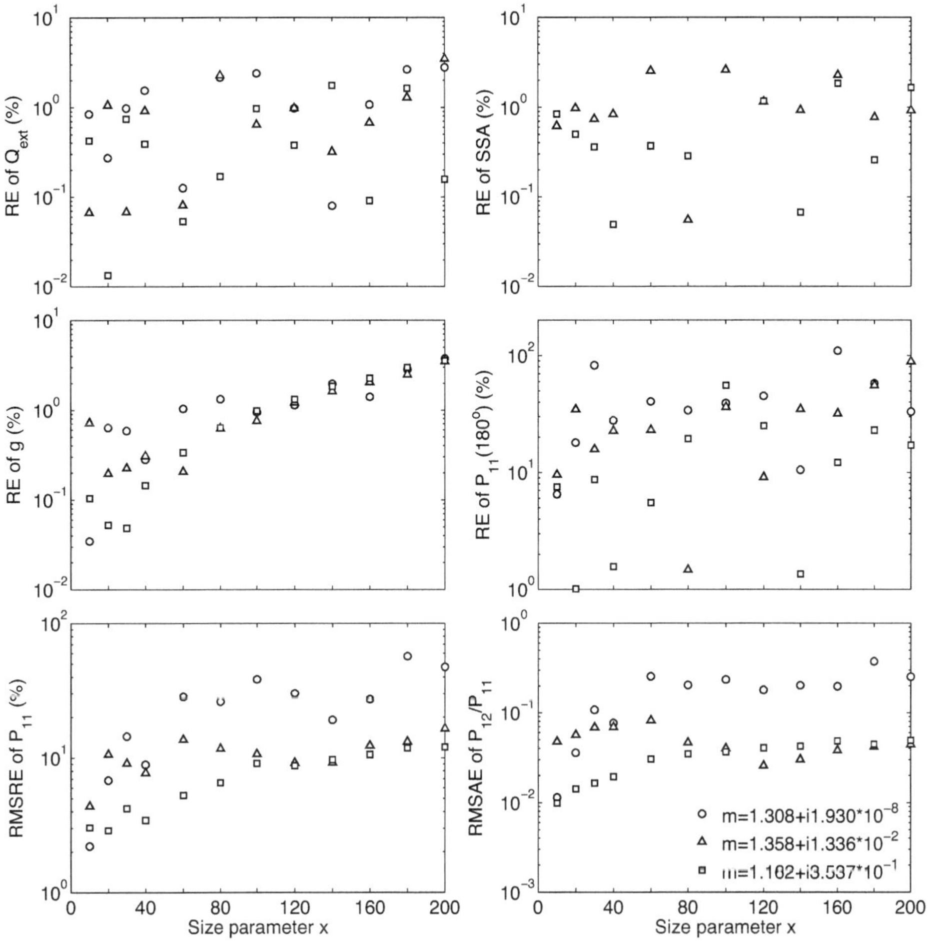

Fig. 4.13. The REs, RMSREs and RMSAEs for spheres with different x at three wavelengths: visible ($0.67\,\mu$m, $m = 1.308 + 1.93 \times 10^{-8}$ i); near-IR ($4.05\,\mu$m, $m = 1.358 + 1.336 \times 10^{-2}$ i) and IR ($11.45\,\mu$m, $m = 1.162 + 3.537 \times 10^{-1}$ i).

Overall, PSTD appears to give accurate and reliable results for light scattering by a sphere in a case for which the product of the size parameter and the real part of the refractive index $x\,\mathrm{Re}(m)$ reaches approximately 270. Particles of this size can be treated with other methods (Mie or T-matrix) if the particle has appropriate symmetry, but not if the particle is significantly irregular. The PSTD has no such symmetry requirements. DDA methods may also be used for large particles without symmetry requirements, but as will also be seen below, the index of refraction must be close to 1.

Large particles

Figure 4.14 illustrates the normalized phase functions given by PSTD and Lorenz–Mie theories for spheres with size parameter of 200 and the three refractive indices.

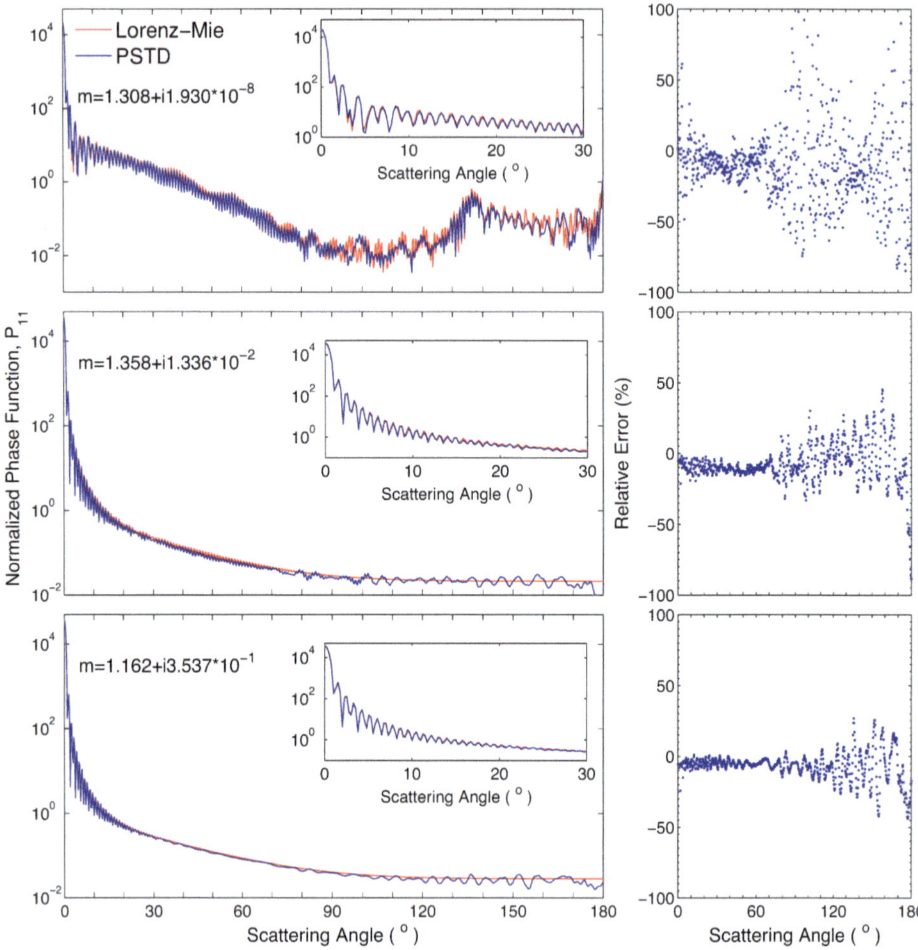

Fig. 4.14. The normalized phase function and relative errors for spheres with $x = 200$ at three wavelengths: visible ($0.67\,\mu$m, $m = 1.308 + 1.93 \times 10^{-8}\,$i); near-IR ($4.05\,\mu$m, $m = 1.358 + 1.336 \times 10^{-2}\,$i) and IR ($11.45\,\mu$m, $m = 1.162 + 3.537 \times 10^{-1}\,$i).

The relative errors of the normalized phase functions are given in the right panel. Even with such a large size parameter, for which the phase function oscillates significantly, the PSTD results agree quite well with the exact solutions given by the Lorenz–Mie theory in the forward scattering directions: the inserts are provided to emphasize this point. The relative errors for the forward scattering are mostly less than 30%, but much more significant errors arise for backward scattering than forward. While the performance is not as good in the non-forward directions, it should be borne in mind that the FDTD method has great difficulty reaching such large particle size, and the DDA runs into trouble with refractive indices that exceed 1.2 (see results in the next section). For the absorptive cases, the backward scattering of the large spheres becomes very smooth, whereas PSTD results are more oscillatory.

Larger m_R

The real parts of the refractive indices for ice crystals and aerosols at different wavelengths are usually under 2, but they become very large at microwave wavelengths. For example, the refractive index of water at the wavelength 3.2 cm is $8.2252 + 1.680\,i$ (Yang et al., 2004). It is quite challenging to calculate the optical properties of particles with large refractive index accurately (Sun and Fu, 2000; Yang et al., 2004; Zhai et al., 2007).

Figure 4.15 shows results from a PSTD calculation for a sphere with $x = 40$ and refractive index $8.2252 + 1.680\,i$. The relative errors of the phase function and the absolute errors of the ratios of other phase matrix elements to it are shown in the right column. With very smooth backward scattering that PSTD approximates accurately, the errors are basically in the forward directions (scattering angles θ in the range $0° - 40°$), although the relative errors are no more than 30%. The absolute errors for the ratios are also very small, less than 0.4 for P_{12}/P_{11} and P_{33}/P_{11}, although the absolute errors do reach nearly 0.8 for P_{34}/P_{11}.

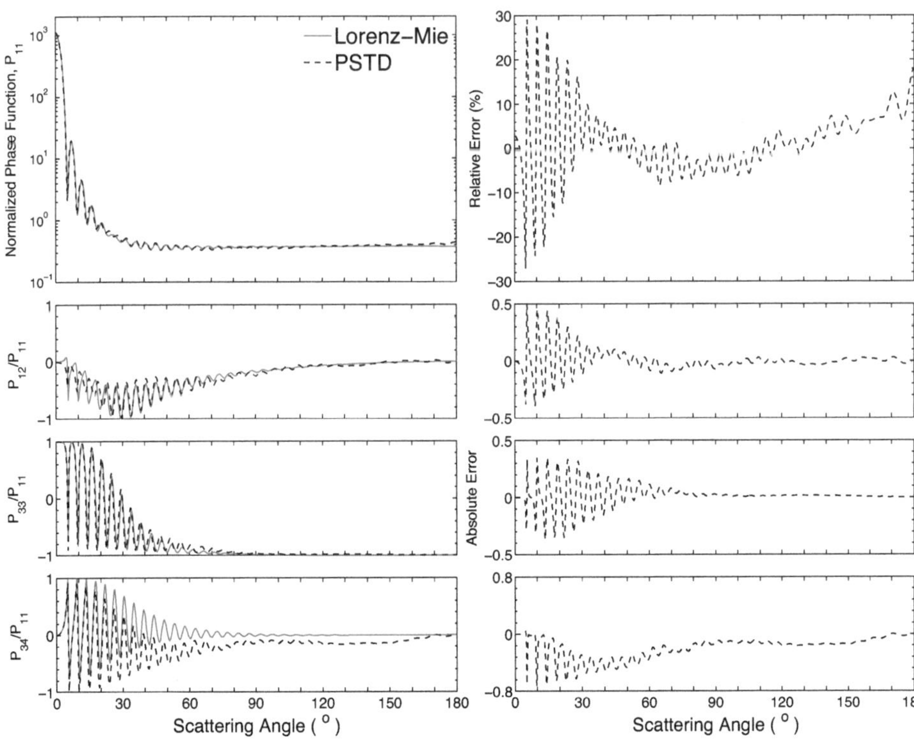

Fig. 4.15. The normalized phase function and the ratios of other nonzero phase matrix elements to it for spheres with size parameter of 40 and refractive index of $m = 8.2252 + 1.680\,i$ computed by PSTD. The relative errors of the phase function and the absolute errors of the ratios are shown in the right panel.

4.5.2 Comparison with T-matrix calculations

The central purpose in developing the PSTD technique is to be able to use it in the study of scattering by randomly oriented nonspherical particles. We give now some examples of calculations for randomly oriented spheroids using PSTD, and compare the results with those from solutions using the T-matrix method (Mishchenko et al., 1996).

Figure 4.16 shows Q_{ext} and g for oblate spheroids as functions of the size parameter defined in the form of $2\pi a/\lambda$, where a is the equatorial radius. The aspect ratio a/b is equal to 2.0, where b is the semi-length of the shorter symmetry axis (see the figure insert). The refractive index of ice at wavelength of $0.67\,\mu$m ($m = 1.308{+}1.93{\times}10^{-8}\,$i) is used for the simulation, and scattering of the spheroids with 16 orientations are simulated and averaged for the randomly oriented properties (using 32 orientations produced no significantly different results). The solid lines in the figure are given by the T-matrix theory, and the dots are the PSTD results with their relative errors shown in the right column. The ratios of wavelength to Δx used are over 100 for the small particles, while they can be reduced to only about 10 for spheroid with size parameter of 100. The figure shows excellent agreement between the results given by the PSTD and T-matrix methods for size parameter from 1 to 100. The relative errors of Q_{ext} are no more than 1.2%, and those of g are less than 0.8%, with the errors for the asymmetry factor generally increasing (but not monotonically) with increasing particle size.

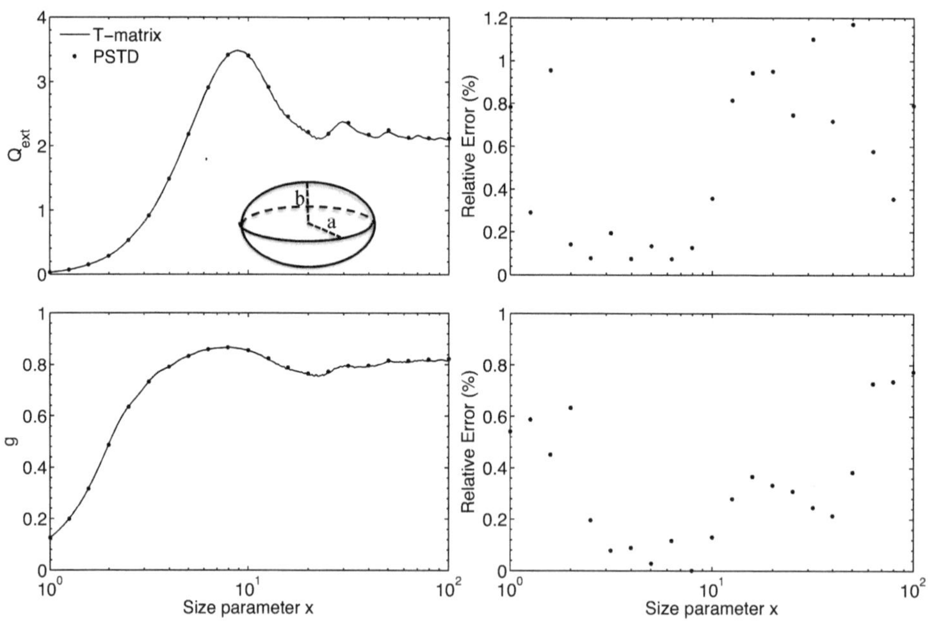

Fig. 4.16. Q_{ext} and g for spheroids as functions of size parameter, and the refractive index of the spheroids is $m = 1.308 + 1.93 \times 10^{-8}\, i$. Relative errors are shown in the right column.

Figure 4.17 shows the phase function and the ratios of other nonzero phase matrix elements to it for a randomly oriented spheroid having size parameter 100. The PSTD results show good agreement with those given by the T-matrix theory, except some errors (with relative errors less than 25%) for P_{11} at scattering angles from 160° to 180°. PSTD approximates the scattering properties of the randomly oriented spheroid much more accurately than those of the spherical cases, because the oscillations of the phase matrix elements for individual orientations cancel each other out in the averaging, with the result being relatively smooth curves. (For the same reason randomly oriented asymmetrical nonspherical particles also present relatively smooth phase matrix element curves, as we will show in Figs. 4.18 and 4.19 below.)

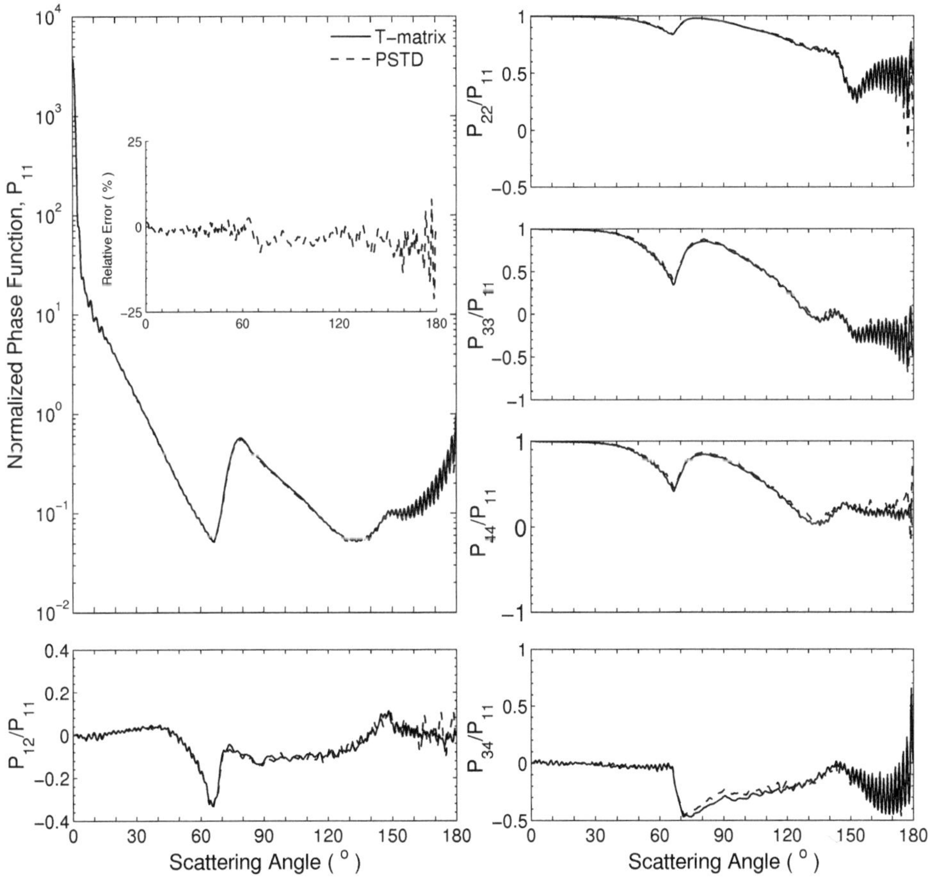

Fig. 4.17. The normalized phase function and the ratios of other nonzero phase matrix elements to it for randomly oriented spheroid with size parameter of 100 and refractive index of $1.312 + 1.489\,i \times 10^{-9}$ given by the T-matrix and PSTD methods.

4.5.3 Two less-symmetric examples

The PSTD can of course be applied to calculate scattering from particles of a much wider range of shapes than the highly symmetrical cases just considered. In conclusion to this section, we show results from calculations of scattering by more asymmetrical particles. (For these cases there are no exact solutions available, so we cannot provide error estimates.) Figure 4.18 shows results for a hexagonal column, a model for an ice crystal that may occur in a cirrus cloud, calculated by the PSTD method and the IGOM method. Figure 4.19 shows results using a fractal model for a dust particle (see the inset in the figure). The incident wavelength for the ice crystal is $0.532\,\mu$m and the index of refraction is $m = 1.3117 + 1.489 \times 10^{-9}\,\mathrm{i}$; for the dust particle the incident wavelength is $0.6328\,\mu$m and the refractive index is $m = 1.55 + 1 \times 10^{-3}\,\mathrm{i}$. The fractal particle was constructed using an algorithm, due to Macke et al. (1996), that starts with a regular tetrahedron and goes through a specified number of iterations, at each iteration adding a scaled-down tetrahedron to all the faces of the particle at that stage. For details, see Macke et al. (1996) or Liu et al. (2013). In the case of each figure, what is shown is the result of averaging over a number of orientations. In the case of the hexagonal column, advantage is taken of the symmetries of the hexagon and the number of orientations is just 48, but in the fractal there are no symmetries and the number of orientations used is 256.

The hexagonal columns in Figs. 4.18(a) and (b) have the size parameters $kL = 100$ and $kL = 200$, respectively, where L is the length of the columns. The width-to-length ratio $2a/L$ is chosen to be 1, and a is the semi-width of the hexagonal cross-section. With a size parameter of 100 in Fig. 4.18(a), both the phase functions from PSTD and IGOM show weak scattering peaks at scattering angles 22° and 46°, and the peaks become very strong when the size parameter increases to 200, as evident from the P_{11} curves shown in Fig. 4.18(b). The agreement of the IGOM approximations to the PSTD results becomes better as the size parameter increases. The ratios of other phase matrix elements to the phase functions given by the PSTD and IGOM also show similar overall patterns. The PSTD solutions show small oscillations with scattering angle, oscillations that are not obtained by IGOM. This is because the PSTD is able to take into account phase interference in the electromagnetic field. The size parameter of the fractal particle shown in Fig. 4.19 is based on the equivalent-projected-area sphere, and the value of 30 is used. Again, we can see that the PSTD and IGOM results agree quite well.

As this brief survey indicates, the PSTD appears to perform well in reproducing integral scattering properties and phase matrices given by the analytical Lorenz–Mie and T-matrix theories, over a range of size parameters and refractive indices. In the next section, we discuss some recent work comparing the PSTD with the DDA method in the outer region, in terms of size parameter and index of refraction, of what is currently numerically feasible with the DDA.

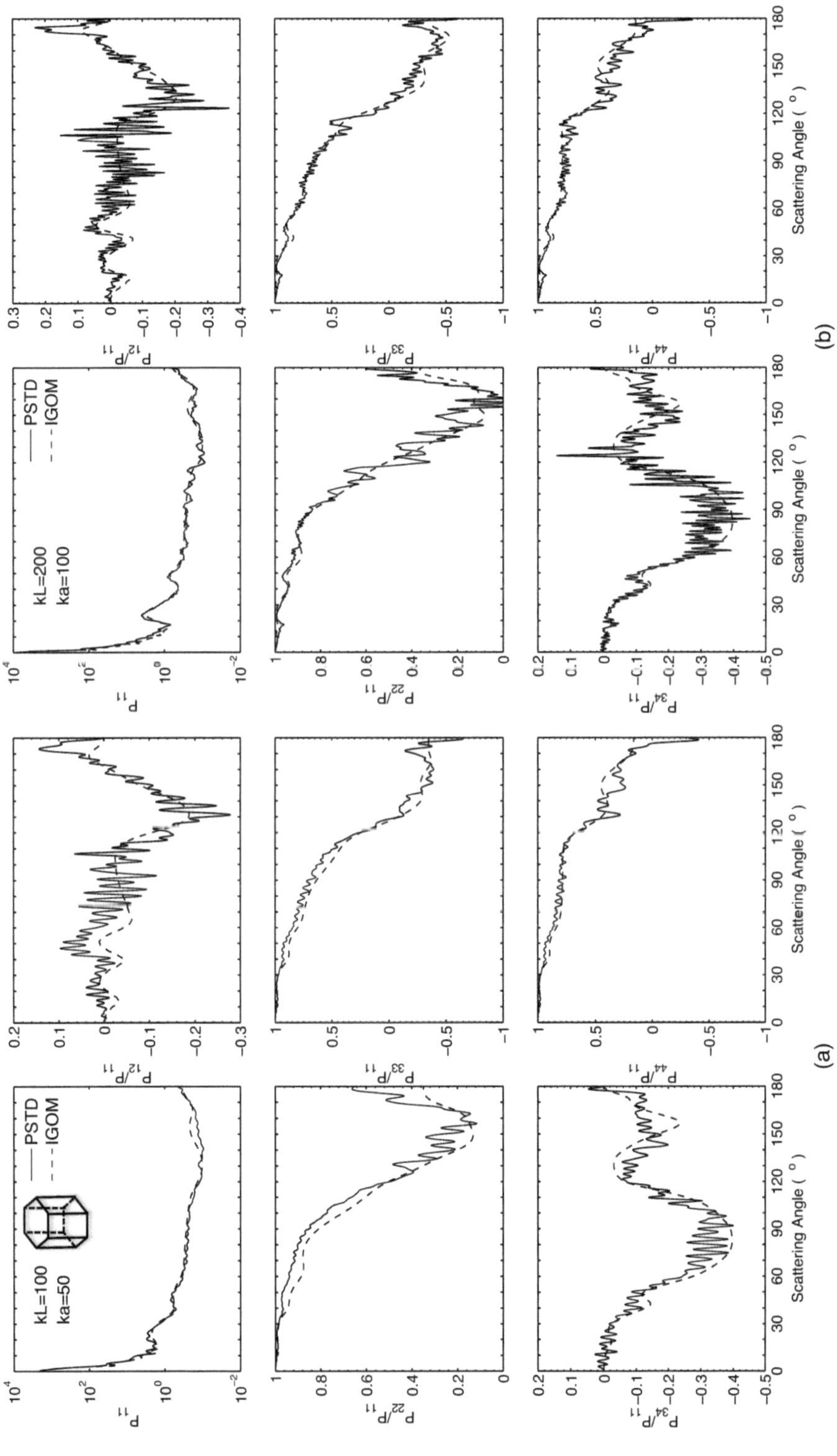

Fig. 4.18. Hexagonal column: incident wavelength 0.532 μm, refractive index $m = 1.3117 + 1.489 \times 10^{-9}$ i, and (a) $kL = 100$, and $ka = 50$; (b) $kL = 200$, and $ka = 100$. Calculations done using PSTD and IGOM.

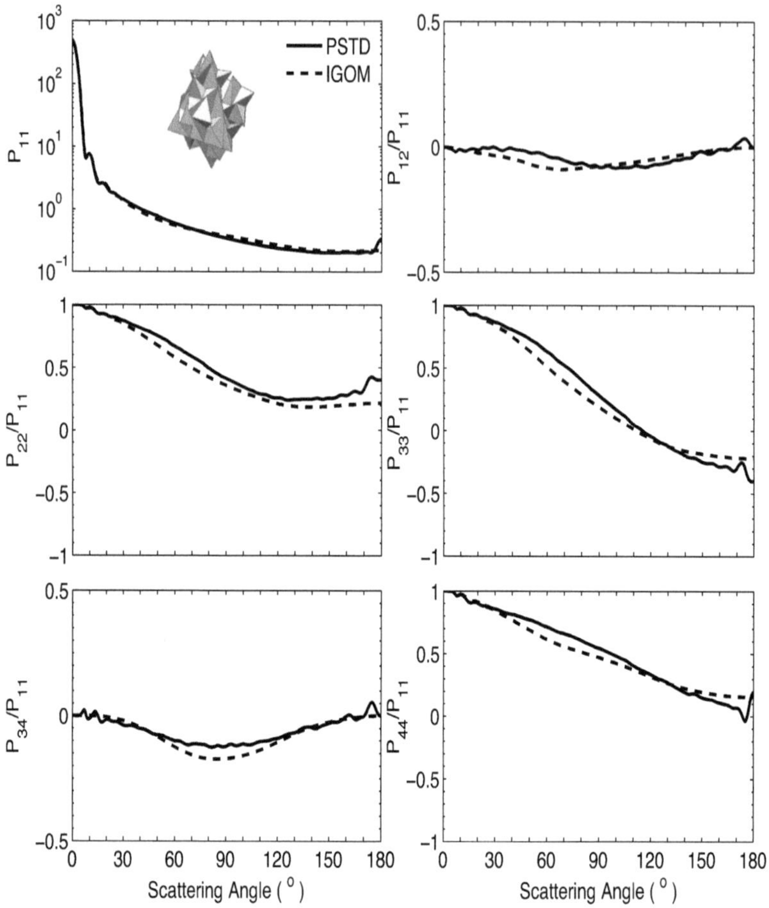

Fig. 4.19. Fractal particle: incident wavelength 0.6328 μm, refractive index $m = 1.55 + 1.0 \times 10^{-3}\,i$, $ka = 30$ (see text). Calculations done using PSTD and IGOM.

4.6 Comparison with DDA

The discrete dipole approximation (DDA) has been extensively applied for atmospheric particles (e.g. Bi et al., 2009, 2010; Yang et al., 2005; Meng et al., 2010), and a number of DDA implementations have been developed in the past decades (Draine and Flatau, 1994; Zubko et al., 1999; Yurkin and Hoekstra, 2007). DDA was compared with FDTD by Yurkin et al., 2007a, who showed that the two methods perform comparably around the refractive indices of 1.4. DDA is faster for smaller refractive indices, whereas FDTD is the more efficient method for larger ones.

A comparison between PSTD and DDA was carried out by Liu et al. (2012b), using a version of a DDA method known as the Amsterdam DDA (ADDA) v.0.79 code that was parallelized with MPI to run on a cluster of processors. The PSTD implementation was parallelized using OpenMP, which supports shared-memory parallel programming (Liu et al., 2012b), but can only be used with a collection of processors on a single node. MPI codes have the advantage that they can run across multiple nodes, but intrinsic to their operation is a cpu overhead incurred in inter-processor communication. This overhead cost is greater when the processors in communication are on different nodes, but is still not negligible when they are on the same node. So comparing performance an MPI code with an OpenMP code must involve some consideration of this issue. Liu et al. (2012b) estimated that the overhead cost was not more than 20%, and hence did not affect their conclusion that the PSTD not only outperforms DDA for larger refractive indices, but also for large spheres with smaller refractive indices. We summarize some of the results of Liu et al. (2012b) here. As mentioned above, all calculations were carried out on a single node of an iDataplex cluster; the node had 8 Nehalem-based 2.8 GHz processors.

Simulations were performed for spheres with size parameters from 10 to 100 and real part of refractive indices from 1.2 to 2.0. The excellent performance of DDA for light scattering by particles with refractive index close to 1 (the vacuum value) is well known (Yurkin and Hoekstra, 2007, 2011): consistently, Liu et al. (2012b) found that for particles with indices of refraction very close to 1, the PSTD is comparatively more expensive of cpu time. The balance was found to shift, as we show below, as indices of refraction significantly larger than one are considered. The results we discuss are for refractive indices that are 1.2 or larger, and all refractive indices are purely real.

The scatterers considered were spherical, so the true values are known. In our comparison of the two methods, we choose accuracy criteria for Q_{ext} and $P_{11}(\theta)$ as follows: the relative error of Q_{ext} should be less than 1%, and the root mean square of the relative errors of the phase function $P_{11}(\theta)$ should be less than 25%. The phase matrix in one scattering plane is calculated with the scattering angle θ varying from 0° to 180° in steps of 0.25°. For each method, and for a given choice of size parameter and index of refraction, the resolution was increased until the calculation using that method met the accuracy criteria. (Increasing the resolution in a DDA simulation means increasing the number of dipoles and decreasing the space between them.) The DDA code has default settings for the dipole polarizability and iterative method, and those were used in the comparison runs. The convergence criterion of the iterative solver was set to be 10^{-3} (larger than the default value 10^{-5}); this weaker convergence criterion proved sufficient to achieve the accuracy required for the comparison of methods. With the same accuracy achieved by the two methods, the computational time becomes the most direct way to describe the overall performance of the methods.

Table 4.2 lists the most important computational parameters and the resultant accuracy of the simulated results in reference to those from Lorenz–Mie theory. It includes the spatial resolutions, the computational times, the relative errors of Q_{ext}, and the RMSREs of the normalized phase function. (For the DDA the spatial resolution is the number of dipoles per wavelength.) In the interest of keeping the table legible, we have not included a pair of columns to show the amounts of memory required by each simulation and method. We simply report here two examples, with fixed index of refraction $m = 1.2$, of memory usage at two different size parameters. For $x = 10$, a case in which the DDA ran slightly more than twenty times faster than the PSTD, the PSTD required about 70 MB of memory and the DDA only

Table 4.2. Parameters and performance results for the comparison of PSTD and DDA for spheres with different x and m. 'NR' means no result (see text). Note (*): the DDA for a sphere with $x = 100$ and $m = 1.2$ did not converge with the default iteration method (quasi-minimal residual), and the bi-conjugate stabilized method was used instead.

		Time (s)		$\lambda/\Delta x$		$RE(Q_{\text{ext}})$ (%)		$RMSRE(P_{11})$ (%)	
m	x	PSTD	DDA	PSTD	DDA	PSTD	DDA	PSTD	DDA
1.2	10	2.1×10^1	1.0×10^0	13	10	0.34	0.071	5.6	0.74
	20	4.4×10^1	2.0×10^0	7.7	7.5	0.0083	0.54	8.5	13
	30	3.0×10^3	1.2×10^1	20	6.7	0.83	0.25	4.2	16
	40	3.9×10^4	1.2×10^2	30	7.5	1.0	0.43	25	19
	60	2.5×10^4	2.3×10^3	18	8.4	0.91	0.20	15	13
	80	1.0×10^4	7.3×10^4	9.2	9.4	0.26	0.62	19	19
	100(*)	2.3×10^4	2.7×10^4	9.3	10	0.050	0.25	18	13
1.4	10	2.3×10^2	2.0×10^0	22	15	0.30	0.69	6.1	12
	20	3.3×10^3	1.1×10^3	22	25	0.78	0.98	10	22
	30	3.8×10^2	9.8×10^3	11	17	0.87	0.74	19	25
	40	6.7×10^3	1.8×10^4	18	18	0.99	0.68	18	15
	60	2.9×10^3	NR	18	NR	1.0	NR	21	NR
	80	(1.2×10^4)	NR	(9.2)	NR	(0.32)	NR	(38)	NR
	100	8.9×10^4	NR	13	NR	0.47	NR	23	NR
1.6	10	4.9×10^1	5.4×10^1	12	25	0.85	0.76	14	7.1
	20	(1.1×10^3)	(3.2×10^4)	(20)	(40)	(5.4)	(5.7)	(44)	(45)
	30	8.3×10^2	4.4×10^4	13	30	0.78	0.75	25	15
	40	2.7×10^3	(2.4×10^5)	14	(20)	0.23	(1.5)	24	(33)
	60	(3.2×10^4)	NR	(18)	NR	(0.035)	NR	(29)	NR
1.8	10	2.7×10^2	6.4×10^2	26	35	0.92	0.88	10	8.8
	20	1.5×10^3	(3.0×10^3)	23	(40)	0.85	(2.7)	10	(19)
	30	3.0×10^3	(9.5×10^4)	19	(25)	0.70	(5.4)	15	(52)
	40	1.5×10^4	NR	21	NR	0.63	NR	19	NR
	60	1.7×10^4	NR	15	NR	0.28	NR	22	NR
2.0	10	5.1×10^1	2.0×10^3	13	40	0.90	0.45	16	16
	20	5.6×10^2	(5.0×10^4)	16	(35)	0.58	(8.9)	13	(35)
	30	1.3×10^3	(5.1×10^5)	14	(25)	0.21	(2.0)	21	(55)
	40	(3.4×10^3)	NR	(14)	NR	(2.3)	NR	(26)	NR

20 MB. When the size parameter was increased to 100 the PSTD took only 85% of the time required for the DDA and used 5 GB, while the DDA used 16 GB of memory. This was generally the trend: as the DDA struggled to deal with increasing index of refraction and particle size, its demands on memory became increasingly larger than those of PSTD.

For some cases, a method would fail to reach the prescribed accuracy even with a very fine spatial resolution (40 dipoles per wavelength for DDA or 30 grid points per wavelength for PSTD), and results of these simulations are indicated with parentheses. The computations that are too time-consuming (the time limit for any given calculation was set at 7 days, i.e. 6.048×10^5 s) to reach even lower accuracy for DDA are marked as 'NR' (no result) in the tables, whereas PSTD was able to handle all x and m chosen. To reach the prescribed accuracy, the spatial resolutions used for PSTD range from 10 to 30 grid points per wavelength, with most simulations using spatial resolution less than 20. However for DDA, the spatial resolution, under 20 dipoles per wavelength for refractive indices of 1.2 and 1.4, rose from 25 to 40 for larger refractive indices. Both PSTD and DDA require much more computational time for large x, because of the increase in the total grid point or dipole number in the computational domain. PSTD reached the prescribed accuracy for most cases within 24 hours (i.e. 86 400 s), except for spheres with $x = 100$ and $m = 1.4$, whereas for DDA the computational time increased significantly with both size parameter and refractive index. Convergence for most cases with large refractive indices could not be achieved by DDA. A pattern becomes clear in the calculations: the two methods show significant differences in capability for different x and m. For small refractive indices ($m = 1.2$ or 1.4), there is a critical size parameter above which PSTD is more efficient, and this critical value decreases from 80 to 30 as the refractive index increases from 1.2 to 1.4. For refractive indices larger than 1.4, PSTD is almost more than an order magnitude faster, and DDA encounters a challenge even for size parameters around 30.

In the next two figures we give some details of the comparison calculations by showing how the accuracy in P_{11} and P_{12}/P_{11} changes, for a fixed particle size $x = 30$, as the index of refraction varies from 1.2 to 2.0. Included in the panels in the left column is the ratio of cpu times

$$\rho = \frac{PSTD\ time}{DDA\ time}$$

required for the calculations.

Figure 4.20 compares the phase functions of spheres with the same size parameter of 30 and different refractive indices in the left panels, and the relative errors of P_{11} are illustrated in the right panels. From the top to the lower panel, the refractive index increases from 1.2 to 2.0 in steps of 0.2. The PSTD and DDA results, as well as the exact solutions given by the Mie theory, are shown. At size parameter 30, the DDA is more efficient than PSTD only for the sphere with refractive index of 1.2 with the ratio ρ greater than 1; the PSTD is more efficient for m larger than 1.2. We can see that the PSTD and DDA results themselves are comparable for spheres with refractive indices of 1.2, 1.4 and 1.6, while the relative errors given by DDA for $m = 1.8$ and 2.0 become significantly larger than those given by the PSTD. Similar results are shown in Fig. 4.21 for P_{12}/P_{11}. More detailed results and discussion can be found in (Liu et al., 2012b).

We summarize the data in Table 4.2 with a 'regime diagram' in Fig. 4.22. It is a representation of the (x, m) plane, in which green symbols indicate parameter choices (x, m) for which the DDA seems to be the preferable method, based on cpu time needed to meet accuracy criteria, and red symbols indicate choices for which the PSTD was preferable. The value entered at a location in the diagram is the time ratio ρ. Cases in which the PSTD produced results meeting the accuracy criteria but the DDA did not are indicated by open rather than solid circles.

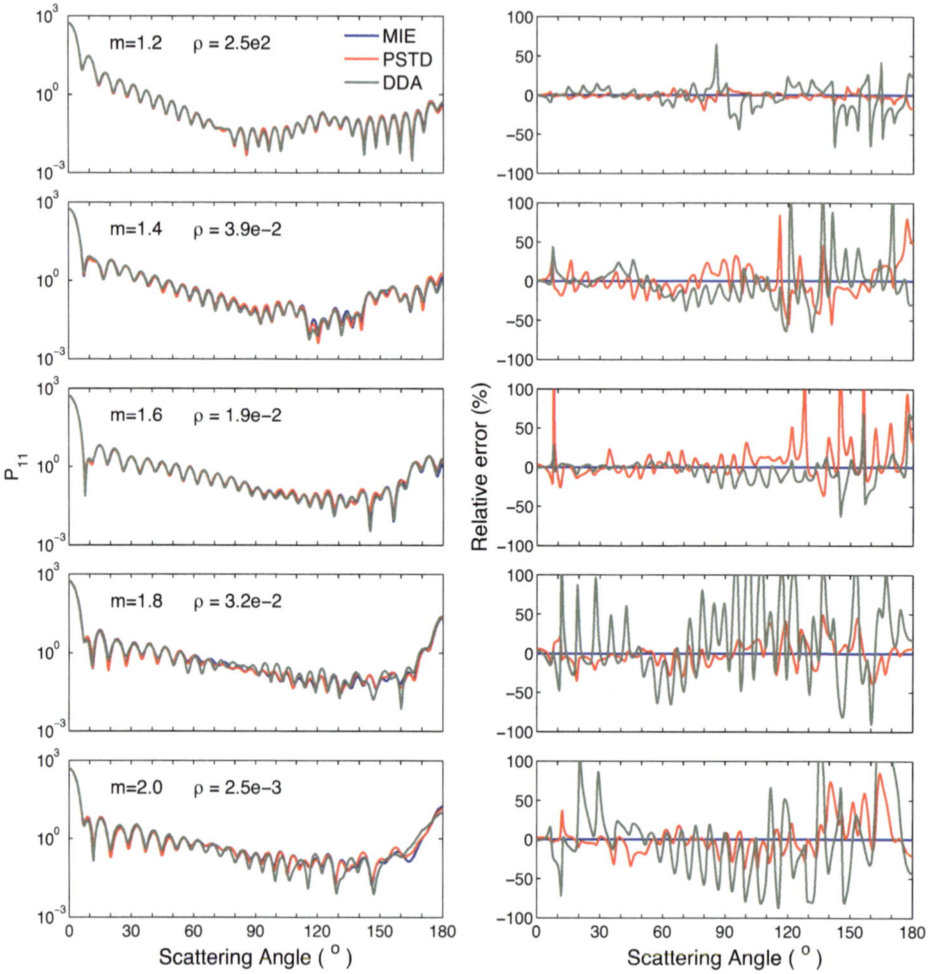

Fig. 4.20. Comparison of calculations of P_{11} using PSTD and DDA for spheres with $x = 30$ and indices of refraction ranging from 1.2 to 2.0. In addition to the index of refraction m, the ratio ρ of PSTD to DDA cpu times is displayed in each panel of the left column.

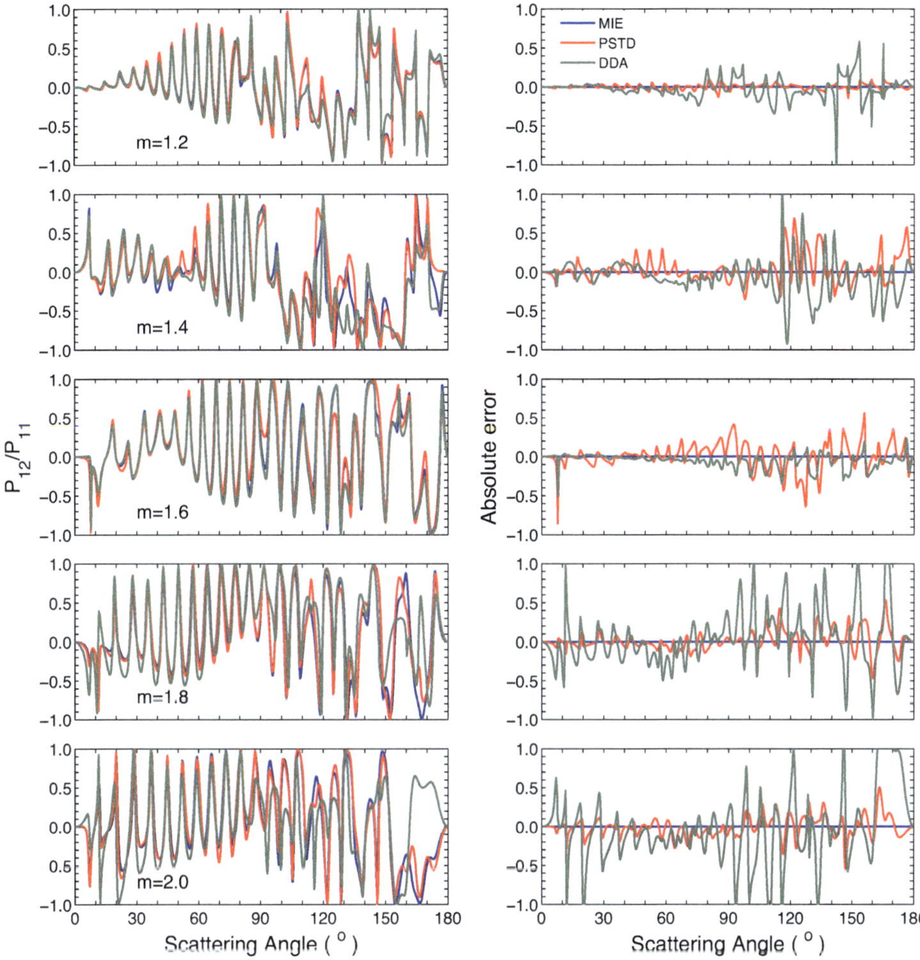

Fig. 4.21. As in Fig. 4.20, but for P_{12}/P_{11}: comparison calculations for PSTD and DDA for spheres with $x = 30$ and indices of refraction ranging from 1.2 to 2.0.

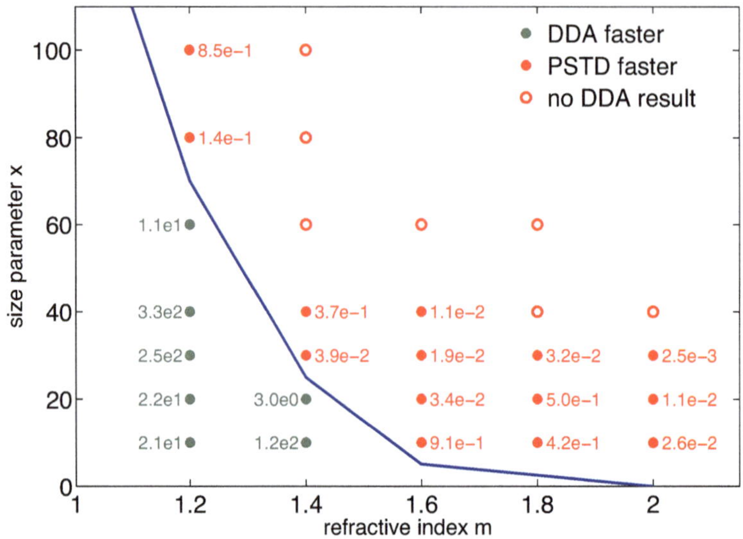

Fig. 4.22. A diagram illustrating the relative performance of the PSTD and the ADDA for spherical particles with different x and m. Numbers in the figure are the ratios ρ of PSTD to DDA cpu time required for the scattering calculation at indicated (x, m). Open circles indicate that a PSTD result was calculated, but the DDA calculation failed to converge. See text for more details.

4.7 Summary

This chapter reviewed the theoretical development and numerical performance of the pseudo-spectral time-domain approach to calculating the single-scattering properties of atmospheric aerosols. The key features of the pseudo-spectral method that were discussed were significant computational economy due to the high order of accuracy in computation of spatial derivatives, with associated benefits in choice of time-step size, and the relatively small numerical dispersion. A discussion was given of the manner in which the Gibbs phenomenon can be handled in the case of solution discontinuities. Selected numerical results were presented to validate the PSTD in highly symmetric cases for which exact solutions are known, and some examples of results with asymmetrical particles were shown. Comparisons with the highly successful DDA, focusing on the case of spherical particles, indicate that the PSTD cannot really compete with the DDA for spherical particles with modest size parameters and indices of refraction with real part close to 1. But the PSTD appears to have an advantage as size parameters increase, especially when indices of refraction have real parts that are above 1.4. Some comparisons of PSTD and DDA calculations for spheroids were given by Liu et al., (2012b).

As emphasized in the introduction, it is not to be expected, or even desired, that a single method of calculation should be regarded as superior in all aspects to all others, and we make no such claims for the PSTD here. Furthermore, none of the cpu times that we quote have absolute meaning in the context of a computer technology that is always rapidly advancing. What seems out of reach computationally today may be an unremarkable undertaking in two years. But we see no reason why the relative performance strengths that we have found will not remain representative for a longer time. We particularly emphasize that while many of the comparison calculations we have discussed here have been with simple particle shapes, and those simple shapes happen because of their symmetry to be within the reach of other methods, there are no special requirements of particle shape or symmetry that are made by the PSTD method itself. On the basis of results shown here, as well as results recently obtained for scatterers with inhomogeneous composition, we believe that the PSTD method shows real promise for pushing the boundary of what is feasible in the regime of large size, large index of refraction, and complex geometry.

Acknowledgments

The authors thank M. I. Mishchenko for his T-matrix code, and M. A. Yurkin and A. G. Hoekstra for their ADDA code. We also thank Lei Bi for helpful conversations on the T-matrix and DDA methods, and an anonymous reviewer who carefully read the manuscript and provided a number of suggestions that led to improvements. This research was supported by a National Science Foundation (NSF) Grant (ATMO-0803779). All computations were carried out at the Texas A&M University Supercomputing Facility and we gratefully acknowledge the assistance of Facility staff in porting the codes.

References

Berenger, J.-P., 1994: A perfectly matched layer for the absorption of electromagnetic waves, *J. Comput. Phys.*, **114**, 185–200.

Bi, L., P. Yang, G. W. Kattawar, and R. Kahn, 2009: Single-scattering properties of triaxial ellipsoidal particles for a size parameter range from the Rayleigh to geometric-optics regimes, *Appl. Opt.*, **48**, 114–126.

Bi, L., P. Yang, G. W. Kattawar, and R. Kahn, 2010: Modeling optical properties of mineral aerosol particles by using nonsymmetric hexahedra, *Appl. Opt.*, **49**, 334–342.

Bi, L., P. Yang, G. W. Kattawar, Y. Hu, and B. A. Baum, 2011: Scattering and absorption of light by ice particles: Solution by a new physical-geometric optics hybrid method, *J. Quant. Spectrosc. Radiat. Transfer*, **112**, 1492–1508.

Brock, R. S., X.-H. Hu, P. Yang, and J. Lu., 2005: Evaluation of a parallel FDTD code and application to modeling of light scattering by deformed red blood cells, *Opt. Express*, **13**, 5279–5292.

Chen, G., 2007: Modeling of the optical properties of nonspherical particles in the atmosphere, Ph.D. dissertation, Texas A&M University.

Chen, G., P. Yang, and G. W. Kattawar, 2008: Application of the pseudospectral time-domain method to the scattering of light by nonspherical particles, *J. Opt. Soc. Am. A*, **25**, 785–790.

Draine, B. T., and P. J. Flatau, 1994: Discrete-dipole approximation for scattering calculations, *J. Opt. Soc. Am. A*, **11**, 1491–1499.

Gedney, S. D., 1996: An anisotropic perfectly matched layer – absorbing medium for the truncation of FDTD lattices, *IEEE Trans. Antennas Propag.*, **44**, 1630–1639.

Goedecke, G. H., and S. G. O'Brien, 1988: Scattering by irregular inhomogeneous particles via the digitized Green's function algorithm, *Appl. Opt.*, **27**, 2431–2438.

Gottlieb, D., and C.-W. Shu, 1997: On the Gibbs phenomenon and its resolution, *SIAM Rev.*, **39**, 644–668.

Jackson, J. D., 1999: *Classical Electrodynamics* (3rd edn), John Wiley & Sons.

Johnson, S. G., 2010: Notes on perfectly matched layers (PML), http://math.mit.edu/∼stevenj/18.369/pml.pdf, online MIT course notes.

Jones, D. S., 1957: High-frequency scattering of electromagnetic waves, *Proc. R. Soc. Lond. A*, **240**, 206–213.

Kreiss, H.-O., and J. Oliger, 1972: Comparison of accurate methods for the integration of hyperbolic equations, *Tellus*, **24**, 199–215.

Liou, K. N., 2002: *An Introduction to Atmospheric Radiation*, Academic Press.

Liu, C., R. L. Panetta, and P. Yang, 2012a: Application of the pseudo-spectral time domain method to compute particle single-scattering properties for size parameters up to 200, *J. Quant. Spectrosc. Radiat. Transfer*, **113**, 1728–1740.

Liu, C., L. Bi, R. L. Panetta, P. Yang, and M. A. Yurkin, 2012b: Comparison between the pseudo-spectral time domain method and the discrete dipole approximation for light scattering simulations, *Opt. Express*, **20**, 16763–16776.

Liu, C., R. L. Panetta, P. Yang, A. Macke, and A. J. Baran, 2013: Modeling the scattering properties of mineral aerosols using concave fractal polyhedra, *Appl. Opt.*, **52**, 640–652.

Liu, Q. H., 1994: Transient electromagnetic modeling with the generalized k-space (GKS) method, *Microwave Opt. Tech. Lett.*, **7**, 842–848.

Liu, Q. H., 1997: The PSTD algorithm: a time-domain method requiring only two cells per wavelength, *Microwave Opt. Tech. Lett.*, **15**, 158–165

Liu, Q. H., 1998: The pseudospectral time-domain (PSTD) algorithm for acoustic waves in absorptive media, *IEEE T. Ultrason Ferr.*, **45**, 1044–1055.

Liu, Q. H., 1999: PML and PSTD algorithm for arbitrary lossy anisotropic media, *IEEE Microw. Guided. W.*, **9**, 48–50.

Macke, A., J. Mueller, and E. Raschke, 1996: Single scattering properties of atmospheric ice crystals, *J. Atmos. Sci.*, **53**, 2813–2825.

Meng, Z., P. Yang, G. W. Kattawar, L. Bi, K. N. Liou, and I. Laszlo, 2010: Single-scattering properties of tri-axial ellipsoidal mineral dust aerosols: A database for application to radiative transfer calculations, *J. Aerosol. Sci.*, **41**, 501–512.

Mie, G., 1908: Beitrge zur optik trber medien, speziell kolloidaler metallsungen, *Ann. Phys.*, **25**, 377–445.

Mishchenko, M. I., L. D. Travis, and D. W. Mackowski, 1996: T-matrix computations of light scattering by nonspherical particles: a review, *J. Quant. Spectrosc. Radiat. Transfer*, **55**, 535–575.

Mishchenko, M. I., and L. D. Travis, 1998: Capabilities and limitations of a current FORTRAN implementation of the T-matrix method for randomly oriented, rotationally symmetric scatterers, *J. Quant. Spectrosc. Radiat. Transfer*, **60**, 309–324.

Morton, K. W., and D. F. Mayers, 2005: *Numerical Solution of Partial Differential Equations*, Cambridge University Press.

Orszag, S. A., 1972: Comparison of pseudospectral and spectral approximation, *Studies in Applied Mathematics*, **51**, 253–259.

Purcell, E. M., and C. R. Pennypacker, 1973: Scattering and absorption of light by nonspherical dielectric grains, *Astrophys J.*, **186**, 705–714.

Sun, W., and Q. Fu, 2000: Finite-difference time-domain solution of light scattering by dielectric particles with large complex refractive indices, *Appl. Opt.*, **39**, 5569–5578.

Tadmor, E., 1986: The exponential accuracy of Fourier and Chebyshev differencing methods, *SIAM J. Numer. Anal.*, **23**, 1–10.

Taflove, A., and S. C. Hagness, 2005: *Computational Electrodynamics: the Finite-difference Time-domain Method*, Artech House.

Tian, B., and Q. H. Liu, 2000: Nonuniform fast consine transform and Chebyshev PSTD algorithms, *Prog. in Electromagn. Res.*, **28**, 253–273.

Trefethen, L. N., 2000: *Spectral Methods in Matlab*, Society for Industrial and Applied Mathematics.

Umashankar, K., and A. Taflove, 1982: A novel method to analyze electromagnetic scattering of complex objects, *IEEE Trans. Electromagn. Compat.*, **24**, 397–405.

van de Hulst, H. C., 1957: *Light Scattering by Small Particles*, John Wiley.

Warren, S. G., and R. E. Brandt, 2008: Optical constants of ice from the ultraviolet to the microwave: A revised compilation, *J. Geophys. Res.*, **113**, doi:10.1029/2007JD009744.

Waterman, P. C., 1965: Matrix formulation of electromagnetic scattering, *Proc. IEEE*, **53**, 805–812.

Waterman, P. C., 1971: Symmetry, unitarity, and geometry in electromagnetic scattering, *Phys. Rev. D*, **3**, 825–839.

Yang, B., D. Gottlieb, and J. S. Hesthaven, 1997: Spectral simulation of electromagnetic wave scattering, *J. Comput. Phys.*, **134**, 216–230.

Yang, B., and J. S. Hesthaven, 1999: A pseudospectral method for time-domain computation of electromagnetic scattering by bodies of revolution, *IEEE Trans. Antennas Propag.*, **47**, 132–141.

Yang, B., and J. S. Hesthaven, 2000: Multidomain pseudospectral computation of Maxwell's equations in 3-D general curvilinear coordinates, *Applied Numerical Mathematics*, **33**, 281–289.

Yang, P., and K. N. Liou, 1996a: Finite-difference time domain method for light scattering by small ice crystals in three-dimensional space, *J. Opt. Soc. Am. A*, **13**, 2072–2085.

Yang, P., and K. N. Liou, 1996b: Geometric-optics-integral-equation method for light scattering by nonspherical ice crystals, *Appl. Opt.*, **35**, 6568–6584.

Yang, P., and K. N. Liou, 1997: Light scattering by hexagonal ice crystals: solutions by a ray-by-ray integration algorithm, *J. Opt. Soc. Am. A*, **14**, 2278–2289.

Yang, P., B. A. Baum, A. J. Heymsfield, Y. X. Hu, H.-L. Huang, S.-C. Tsay, and S. Ackerman, 2003: Single-scattering properties of droxtals, *J. Quant. Spectrosc. Radiat. Transfer*, **79–80**, 1159–1169.

Yang, P., G. W. Kattawar, K. N. Liou, and J. Q. Lu, 2004: Comparison of cartesian grid configurations for application of the finite-difference time-domain method to electromagnetic scattering by dielectric particles, *Appl. Opt.*, **43**, 4611–4624.

Yang, P., H. Wei, H.-L. Huang, B. A. Baum, Y. X. Hu, G. W. Kattawar, M. I. Mishchenko, and Q. Fu, 2005: Scattering and absorption property database for nonspherical ice particles in the near- through far- infrared spectral region, *Appl. Opt.*, **44**, 5512–5523.

Yee, K. S., 1966: Numerical solution of initial boundary value problems involving Maxwell'ïs equations in isotropic media, *IEEE Trans. Antennas Propag.*, **14**, 302–307.

Yurkin, M. A., A. G. Hoekstra, R. S. Brock, and J. Q. Lu, 2007a: Systematic comparison of the discrete dipole approximation and the finite difference time domain method for large dielectric scatterers, *Opt. Express*, **15**, 17902–17911.

Yurkin, M. A., V. P. Maltsev, and A. G. Hoekstra, 2007b: The discrete dipole approximation for simulation of light scattering by particles much larger than the wavelength, *J. Quant. Spectrosc. Radiat. Transfer*, **106**, 546–557.

Yurkin, M. A., and A. G. Hoekstra, 2007: The discrete dipole approximation: An overview and recent developments, *J. Quant. Spectrosc. Radiat. Transfer*, **106**, 558–589.

Yurkin, M. A., and A. G. Hoekstra, 2011: The discrete-dipole-approximation code ADDA: Capabilities and known limitations, *J. Quant. Spectrosc. Radiat. Transfer*, **112**, 2234–2247.

Zhai, P.-W., C. Li, G. W. Kattawar, and P. Yang, 2007: FDTD far-field scattering amplitudes: Comparison of surface and volume integration methods, *J. Quant. Spectrosc. Radiat. Transfer*, **106**, 590–594.

Zubko, E. S., M. A. Kreslavskii, and Y. G. Shkuratov, 1999: The role of scatterers comparable to the wavelength in forming negative polarization of light, *Solar Sys. Res.*, **33**, 296–301.

5 Application of non-orthogonal bases in the theory of light scattering by spheroidal particles

Victor Farafonov

5.1 Introduction

The theory of light scattering by single particles and their ensembles has important applications in various areas of science and technology, e.g. in optics of the atmosphere, radio physics, astrophysics and biophysics as well as in environmental monitoring, analysis of the Earth's climate changes and so on. So far in such applications one used to employ the Mie theory that provides the solution to the light scattering problem for a sphere (van de Hulst, 1957; Bohren and Huffman, 1983). In this solution the fields are represented by their expansions in terms of vector spherical wave functions that form an orthogonal basis for the problem including the boundary conditions on the spherical particle surface. As a result the Mie solution is rather simple and allows one to extensively apply numerical modelling due to the ease of calculations of the spherical wave functions in a very wide range of parameter values. However, real particles tend to differ substantially in shape from spheres, and one needs to consider the light scattering by nonspherical particles (Mishchenko *et al.*, 2000, 2002).

After excluding spheres, the simplest finite size particles appear to be prolate and oblate spheroids. But even in this case the corresponding vector spheroidal wave functions are not orthogonal on spheroidal coordinate surfaces (Morse and Feshbach, 1953). Hence in solving even these rather simple problems by the separation of variables method, there arise infinite systems of linear algebraic equations (ISLAEs) relative to the unknown field expansion coefficients (Asano and Yamamoto, 1975; Sinha and McPhie, 1977; Farafonov and Slavyanov, 1980; Farafonov, 1983; Schultz *et al.*, 1998). The same result occurs when applying the extended boundary condition method with a spherical basis, i.e. with the field expansions in terms of spherical wave functions (Barber and Yeh, 1975; Barber and Hill, 1990; Farafonov *et al.*, 2010). This method with a spheroidal basis has only recently been developed (Farafonov, 2001; Kahnert, 2003a), and numerical results were obtained just in a few works (Il'in *et al.*, 2007; Farafonov *et al.*, 2007; Farafonov and Voshchinnikov, 2012). In the general case, in solving the light scattering problem for a nonspherical particle, the requirement to satisfy the boundary conditions practically always leads to an ISLAE relative to the unknown expansion coefficients (Kahnert, 2003b; Farafonov and Il'in, 2006). Even if, as in the Mie theory, the basis chosen is orthogonal on any spherical coordinate surface, it is not orthogonal on the boundary of the nonspherical particle where the boundary conditions are imposed. Therefore, to solve the light scattering problem correctly one has to prove both solvability of the arising ISLAEs and convergence of the field expansions everywhere up to the particle surface. These questions have been considered in the case of a spheroidal basis in the recent work of Farafonov (2011).

This review deals with application of non-orthogonal bases to the problem of the light scattering by dielectric and perfectly conducting spheroids. In Section 5.2 we discuss the differential and integral formulations of the problem and describe an original solution for the both kinds of spheroids. In contrast to the standard approaches our solution assumes representation of the field by a sum of two components where one component is independent of the azimuthal angle while averaging of the other component over this angle gives zero. Commutativity of the operator $L_z = \partial/\partial\varphi$ and the integral operator T corresponding to the light scattering problem allows one to solve the problem for each of the components independently. This commutativity also provides separation of variables (here for the azimuthal angle only) in the light scattering problem for a spheroid, i.e. each term of the Fourier series can be found separately. Another feature of our approach is the use of scalar potentials properly chosen for each of the components. When solving the problem for the axisymmetric component of the field, one can apply the Abraham potentials that reduce the vector problem to a scalar one. In the problem for the non-axisymmetric component one can utilize a combination of the scalar potentials U and V usually introduced when solving the light scattering problem for a circular cylinder and a sphere, respectively. The former potential is the z-component of the electric or magnetic Hertz vector (Stratton, 1941; Farafonov and Il'in, 2006), the latter one is the Debye potential being the product of the radius-vector magnitude and the r-component of the corresponding Hertz vector. Note that application of the potentials U and V is equivalent to the use of $\boldsymbol{M}^{\mathrm{z}}_{\nu}$, $\boldsymbol{M}^{\mathrm{r}}_{\nu}$ and $\boldsymbol{N}^{\mathrm{z}}_{\nu}$, $\boldsymbol{N}^{\mathrm{r}}_{\nu}$ as the vector function basis for the transverse magnetic (TM) and electric (TE) modes, respectively (Farafonov and Il'in, 2006; Farafonov, 2011). The axisymmetric and non-axisymmetric problems are solved by the separation of variables method where the potentials are represented by their expansions in terms of the spheroidal wave functions. Substitution of the expansions into the boundary conditions leads to ISLAEs relative to the unknown coefficients of the scattered field expansion. All characteristics of the scattered radiation (cross-sections, phase function, etc.) are expressed through these coefficients. Thus, to solve the light scattering problem one needs to solve the ISLAEs arisen and to calculate the required characteristics using the expansion coefficients obtained.

Section 5.3 is devoted to analysis of the ISLAEs typical of the light scattering problems for spheroids. The analysis is based on the obtained estimates of integrals of products of the spheroidal angular functions (SAFs) and their derivatives and the derived asymptotics of the spheroidal radial functions (SRFs) for large index values. It is found that excluding the case of a segment and a disk, the ISLAEs are completely quasi-regular, and the properties of such systems are discussed. As a result, we prove that excluding the cases mentioned above the ISLAEs have the only solution that can be found by the reduction method. This has the practical importance as in numerical calculations we always truncate the infinite systems. We also show that for a selected spheroidal basis the field expansions converge everywhere up to the scatterer surface.

The light scattering by extremely prolate and oblate spheroids is considered in Section 5.4. In these cases, the ISLAEs can be solved explicitly in the first approximation with respect to the small parameter b/a (i.e. for the large semiaxis ratio $a/b \gg 1$). The principal term of the scattered field asymptotics is found to coincide with the so-called quasi-statical approximation where the field inside a scatterer is approximated by the incident wave taking into account polarizability when the scatterer is essentially smaller than the wavelength. Thus, the quasi-static approximation is a generalization of the Rayleigh–Gans and Rayleigh approximations. We also consider asymptotics of the radiation scattered by extremely prolate perfectly conducting spheroids. The principal term in the case the incident TM mode wave when the electric field is parallel to the spheroidal symmetry axis is of order $O(1)$ under the condition of the linear antenna excitation $d = n\lambda/2$ (the oscillator length is equal to an integer number of half-wavelengths). Otherwise, the strength of the scattered field is inversely proportional to the logarithm of the aspect ratio $1/\ln a/b$. The principal term of the scattered field asymptotics for the TE mode and the second term for the TM mode are proportional to the square of the small parameter $(b/a)^2$ and are derived explicitly. Numerical calculations have completely confirmed these analytical results.

In Section 5.5 we consider the light scattering by extremely oblate perfectly conducting spheroids. The main difficulty is here related with the fact that in the particular case of the disk there appear some additional conditions at its edge. These are the Meixner conditions whose sense is that the edge of the perfectly conducting disk should not radiate. Initially, our solution does not satisfy these conditions, and hence it should be improved. We suggest a new solution that involves the solution for the disk and hence automatically satisfies the Meixner conditions. Numerical calculations performed have shown that the improved solution allows one to treat spheroids of a large aspect ratio a/b. This solution is also applicable to perfectly conducting disks.

In Section 5.6 'Conclusions', we formulate the main results obtained in the Chapter. In Appendix A, various integrals of the SAFs and their derivatives are represented by sums containing the coefficients of the SAF expansions in terms of the associated Legendre functions of the first kind. Some relations between the integrals are presented as well. This representation is very efficient for numerical calculations and is useful for analysis of the integrals. A Fortran code used for illustrative calculations of light scattering by homogeneous spheroids is available at the DOP site http://www.astro.spbu.ru/DOP/6-SOFT/SPHEROID/.

5.2 Light scattering problem for a spheroidal particle

The behaviour of the electromagnetic field in any medium is described by the macroscopic Maxwell equations which are in the CGS system as follows (Jackson, 1975):

$$\begin{aligned} \nabla \times \boldsymbol{E} &= -\frac{1}{c}\frac{\partial \boldsymbol{B}}{\partial t}\,, & \nabla \cdot \boldsymbol{D} &= 4\pi \varrho\,, \\ \nabla \times \boldsymbol{H} &= \frac{4\pi}{c}\boldsymbol{j} + \frac{1}{c}\frac{\partial \boldsymbol{D}}{\partial t}\,, & \nabla \cdot \boldsymbol{B} &= 0\,, \end{aligned} \tag{5.1}$$

where $\boldsymbol{E}$ and $\boldsymbol{D}$ are the electric field and displacement, $\boldsymbol{H}$ and $\boldsymbol{B}$ the magnetic field and induction, ϱ and $\boldsymbol{j}$ the free charge and current densities, c is the speed of light in vacuum.

The Maxwell equations are supplemented by the constitutive equations that describe the properties of the medium where the electromagnetic field is considered. Below we deal with the media characterized by the following equations:

$$\boldsymbol{D} = \tilde{\varepsilon}\,\boldsymbol{E}\,, \qquad \boldsymbol{B} = \mu\,\boldsymbol{H}\,, \qquad \boldsymbol{E} = \sigma\,\boldsymbol{j}\,, \tag{5.2}$$

where $\tilde{\varepsilon}$ and μ are the dielectric permittivity and the magnetic permeability of a medium, and σ is its specific conductivity.

Due to linearity of Eqs. (5.1)–(5.2) no generality is lost if we consider further only the harmonic fields, i.e. the fields with the time-dependence given by $\mathrm{e}^{-\mathrm{i}\omega t}$ (Bohren and Huffman, 1983). We also assume that there are no free charges ($\varrho = 0$).

5.2.1 Differential and integral formulations of the light scattering problem

To find the field of radiation scattered by a particle, one must supplement the equations presented above with the boundary conditions at the scatterer surface (continuity of the tangential components of the field) and at infinity (the Sommerfeld condition about existence of divergent waves only).

Let us denote the known field of incident radiation by $\boldsymbol{E}^{(0)}$, $\boldsymbol{H}^{(0)}$, the unknown fields of scattered radiation by $\boldsymbol{E}^{(1)}$, $\boldsymbol{H}^{(1)}$ and of radiation inside the scatterer by $\boldsymbol{E}^{(2)}$, $\boldsymbol{H}^{(2)}$.

Then the light scattering problem can be written as follows:

$$\Delta \boldsymbol{E}^{(1)} + k_1^2\,\boldsymbol{E}^{(1)} = 0\,, \qquad \boldsymbol{r} \in R^3 \setminus \bar{D}\,, \tag{5.3}$$

$$\Delta \boldsymbol{E}^{(2)} + k^2\,\boldsymbol{E}^{(2)} = 0\,, \qquad \boldsymbol{r} \in D\,, \tag{5.4}$$

$$\nabla \cdot \boldsymbol{E}^{(1)} = 0\,, \qquad \nabla \cdot \boldsymbol{E}^{(2)} = 0\,, \tag{5.5}$$

$$\left(\boldsymbol{E}^{(0)} + \boldsymbol{E}^{(1)}\right) \times \boldsymbol{n} = \boldsymbol{E}^{(2)} \times \boldsymbol{n}, \qquad \boldsymbol{r} \in S\,, \tag{5.6}$$

$$\left(\boldsymbol{H}^{(0)} + \boldsymbol{H}^{(1)}\right) \times \boldsymbol{n} = \boldsymbol{H}^{(2)} \times \boldsymbol{n}\,, \qquad \boldsymbol{r} \in S, \tag{5.7}$$

$$\lim_{r\to\infty} r \left(\frac{\partial \boldsymbol{E}^{(1)}}{\partial r} - \mathrm{i}k_1 \boldsymbol{E}^{(1)}\right) = 0\,, \tag{5.8}$$

where $k = k_1\sqrt{\varepsilon\mu}$ is the wavenumber in the medium, $\varepsilon = \tilde{\varepsilon} + \mathrm{i}4\pi\sigma/\omega$ the complex dielectric permittivity, $k_1 = \omega/c$ the wavenumber in vacuum, ω the radiation frequency, $\boldsymbol{n}$ the outer normal to the surface S of the particle having the volume D, $\boldsymbol{r}$ the radius-vector, $r = |\boldsymbol{r}|$. The magnetic fields $\boldsymbol{H}^{(1)}$, $\boldsymbol{H}^{(2)}$ are determined from the known electric fields $\boldsymbol{E}^{(1)}$, $\boldsymbol{E}^{(2)}$ by using the Maxwell equations

$$\boldsymbol{H} = \frac{1}{\mathrm{i}\mu k_1}\boldsymbol{\nabla} \times \boldsymbol{E}\,. \tag{5.9}$$

Sometimes it is more convenient to present the problem in the integral form by using the Stratton–Chu formula. All solutions to the Maxwell equations inside the domain D (here inside a particle) are known to satisfy (Colton and Kress, 1984)

$$\begin{aligned}\boldsymbol{\nabla} \times \int_S \boldsymbol{n} \times \boldsymbol{E}(\boldsymbol{r}')\,G(\boldsymbol{r},\boldsymbol{r}')\,\mathrm{d}s' - \frac{1}{\mathrm{i}k_1\varepsilon}\boldsymbol{\nabla} \times \boldsymbol{\nabla}\\ \times \int_S \boldsymbol{n} \times \boldsymbol{H}(\boldsymbol{r}')\,G(\boldsymbol{r},\boldsymbol{r}')\,\mathrm{d}s' = \begin{cases} -\boldsymbol{E}(\boldsymbol{r}), & \boldsymbol{r} \in D\,,\\ 0, & \boldsymbol{r} \in R^3 \setminus \bar{D}\,,\end{cases}\end{aligned} \tag{5.10}$$

where $G(\boldsymbol{r},\boldsymbol{r}')$ is the Green function of the scalar Helmholtz equation for free space

$$G(\boldsymbol{r},\boldsymbol{r}') = \frac{\mathrm{e}^{\mathrm{i}k_1|\boldsymbol{r}-\boldsymbol{r}'|}}{4\pi|\boldsymbol{r}-\boldsymbol{r}'|}\,. \tag{5.11}$$

For the solutions to the Maxwell equations outside $\bar{D}$ that also satisfy the radiation condition at infinity (5.8), one has integral equations similar to Eqs. (5.10)

$$\begin{aligned}\boldsymbol{\nabla} \times \int_S \boldsymbol{n} \times \boldsymbol{E}(\boldsymbol{r}')\,G(\boldsymbol{r},\boldsymbol{r}')\,\mathrm{d}s' - \frac{1}{\mathrm{i}k_1\varepsilon}\boldsymbol{\nabla} \times \boldsymbol{\nabla}\\ \times \int_S \boldsymbol{n} \times \boldsymbol{H}(\boldsymbol{r}')\,G(\boldsymbol{r},\boldsymbol{r}')\,\mathrm{d}s' = \begin{cases} 0, & \boldsymbol{r} \in D,\\ \boldsymbol{E}(\boldsymbol{r}), & \boldsymbol{r} \in R^3 \setminus \bar{D}\,.\end{cases}\end{aligned} \tag{5.12}$$

If one applies these integral equations to the incident $\boldsymbol{E}^{(0)}$ and scattered $\boldsymbol{E}^{(1)}$ fields, adds the equations and takes into account the boundary conditions (5.6)–(5.7), the surface integral equation formulation of the light scattering problem can be obtained

$$\begin{aligned}\boldsymbol{\nabla} \times \int_S \boldsymbol{n} \times \boldsymbol{E}^{(2)}(\boldsymbol{r}')\,G(\boldsymbol{r},\boldsymbol{r}')\,\mathrm{d}s' - \frac{1}{\mathrm{i}k_1\varepsilon}\boldsymbol{\nabla} \times \boldsymbol{\nabla}\\ \times \int_S \boldsymbol{n} \times \boldsymbol{H}^{(2)}(\boldsymbol{r}')\,G(\boldsymbol{r},\boldsymbol{r}')\,\mathrm{d}s' = \begin{cases} -\boldsymbol{E}^{(0)}(\boldsymbol{r}), & \boldsymbol{r} \in D,\\ \boldsymbol{E}^{(1)}(\boldsymbol{r}), & \boldsymbol{r} \in R^3 \setminus \bar{D}\,.\end{cases}\end{aligned} \tag{5.13}$$

Usually, the first step is to solve the integral equation for the domain D and to determine the internal field $\boldsymbol{E}^{(2)}$. After that the scattered field $\boldsymbol{E}^{(1)}$ can be easily found from the equation for the domain $R^3 \setminus \bar{D}$.

5.2.2 Original solution to the problem for a dielectric spheroid

We solve this light scattering problem by the separation of variables method in the spheroidal coordinates (ξ, η, φ) connected with the Cartesian coordinates (x, y, z) as follows (Flammer, 1957; Komarov *et al.*, 1976):

$$x = \frac{d}{2}\,(\xi^2 \mp 1)^{\frac{1}{2}}\,(1-\eta^2)^{\frac{1}{2}}\,\cos\varphi\,,$$
$$y = \frac{d}{2}\,(\xi \mp 1)^{\frac{1}{2}}\,(1-\eta^2)^{\frac{1}{2}}\,\sin\varphi\,, \tag{5.14}$$
$$z = \frac{d}{2}\,\xi\,\eta\,,$$

where d is the focal distance of the spheroid under consideration. For the prolate spheroidal coordinates, $\xi \in [1,\infty)$, $\eta \in [-1,1]$, $\varphi \in [0,2\pi)$ and the upper sign is selected, for the oblate spheroidal coordinates, $\xi \in [0,\infty)$, $\eta \in [-1,1]$, $\varphi \in [0,2\pi)$ and the lower sign is used. The metric coefficients are

$$h_\xi = \frac{d}{2}\left(\frac{\xi^2 \mp \eta^2}{\xi^2 \mp 1}\right)^{\frac{1}{2}}, \quad h_\eta = \frac{d}{2}\left(\frac{\xi^2 \mp \eta^2}{1-\eta^2}\right)^{\frac{1}{2}}, \quad h_\varphi = \frac{d}{2}\left[(\xi^2 \mp 1)(1-\eta^2)\right]^{\frac{1}{2}}. \tag{5.15}$$

Note that the transition from the prolate spheroidal coordinates to the oblate ones is done by the replacements $d \to -id$ and $\xi \to i\xi$.

A plane wave of arbitrary polarization incident at the angle α to the symmetry axis of the spheroid (the z axis) can be represented by a superposition of two mode waves:

(1) TE mode

$$\begin{aligned} \boldsymbol{E}^{(0)} &= \boldsymbol{i}_{\rm y}\ \exp[ik_1(x\sin\alpha + z\cos\alpha)]\,, \\ \boldsymbol{H}^{(0)} &= -\sqrt{\frac{\varepsilon_1}{\mu_1}}\ [(\boldsymbol{i}_{\rm x}\cos\alpha - \boldsymbol{i}_{\rm z}\sin\alpha)\exp[ik_1(x\sin\alpha + z\cos\alpha)]\,; \end{aligned} \tag{5.16}$$

(2) TM mode

$$\begin{aligned} \boldsymbol{E}^{(0)} &= (\boldsymbol{i}_{\rm x}\cos\alpha - \boldsymbol{i}_{\rm z}\sin\alpha)\ \exp[ik_1(x\sin\alpha + z\cos\alpha)]\,, \\ \boldsymbol{H}^{(0)} &= \sqrt{\frac{\varepsilon_1}{\mu_1}}\ \boldsymbol{i}_{\rm y}\ \exp[ik_1(x\sin\alpha + z\cos\alpha)]\,, \end{aligned} \tag{5.17}$$

where $\boldsymbol{i}_{\rm x}, \boldsymbol{i}_{\rm y}, \boldsymbol{i}_{\rm z}$ are the unit vectors of the Cartesian system.

The differential formulation of the problem for a spheroid looks like that for an arbitrary shape particle (5.3)–(5.8) with the exception of the boundary conditions which become as follows:

$$\left.\begin{aligned} E_\eta^{(0)} + E_\eta^{(1)} &= E_\eta^{(2)}\,, \qquad & H_\eta^{(0)} + H_\eta^{(1)} &= H_\eta^{(2)}\,, \\ E_\varphi^{(0)} + E_\varphi^{(1)} &= E_\varphi^{(2)}\,, \qquad & H_\varphi^{(0)} + H_\varphi^{(1)} &= H_\varphi^{(2)}\,, \end{aligned}\right\}_{\xi=\xi_0} \tag{5.18}$$

where ξ_0 is the value of the radial coordinate corresponding to the particle surface.

The approach under consideration has two features. First, the fields are represented by the sums

$$\boldsymbol{E}^{(i)} = \boldsymbol{E}_1^{(i)} + \boldsymbol{E}_2^{(i)}\,, \qquad \boldsymbol{H}^{(i)} = \boldsymbol{H}_1^{(i)} + \boldsymbol{H}_2^{(i)}\,, \tag{5.19}$$

where $\boldsymbol{E}_1^{(i)}$ and $\boldsymbol{H}_1^{(i)}$ do not depend on the azimuthal angle φ, while averaging of $\boldsymbol{E}_2^{(i)}$ and $\boldsymbol{H}_2^{(i)}$ over this angle gives zero.

Second, we apply original basis functions $\boldsymbol{M}_\nu$, $\boldsymbol{N}_\nu$ (or corresponding scalar potentials – Farafonov and Il'in (2006)) to represent the fields $\boldsymbol{E}_2^{(i)}$ and $\boldsymbol{H}_2^{(i)}$. Consideration of the axisymmetric components $\boldsymbol{E}_1^{(i)}$ and $\boldsymbol{H}_1^{(i)}$ is often useful and can be simplified due to application of the Abraham potentials.

The possibility to separately consider the problems for the field components introduced by Eqs. (5.19) is provided by commutativity of the operator T corresponding to the light scattering problem and the operator $L_z = \partial/\partial\varphi$. To prove that, we consider T in the integral form

$$T\boldsymbol{E} = \mathrm{rot} \int_S \boldsymbol{n} \times \boldsymbol{E}(\boldsymbol{r}')\, G(\boldsymbol{r}, \boldsymbol{r}')\, \mathrm{d}s' - \frac{1}{ik_1\varepsilon}\, \mathrm{rot}\, \mathrm{rot} \int_S \boldsymbol{n} \times \boldsymbol{H}(\boldsymbol{r}')\, G(\boldsymbol{r}, \boldsymbol{r}')\, \mathrm{d}s'\,, \tag{5.20}$$

where we use the problem formulation (5.13).

One can write T in the spheroidal coordinates and express $\boldsymbol{H}$ through the electric field (see Eq. (5.9))

$$\begin{aligned} T\boldsymbol{E} &= \mathrm{rot} \int_{-1}^{1} \int_0^{2\pi} \boldsymbol{i}_{\xi'} \times \boldsymbol{E}(\boldsymbol{r}')\, G(\boldsymbol{r}, \boldsymbol{r}')\, h_{\eta'} h_{\varphi'}\, \mathrm{d}\eta'\, \mathrm{d}\varphi' \\ &+ \frac{1}{k_1^2 \mu\varepsilon}\, \mathrm{rot}\, \mathrm{rot} \int_{-1}^{1} \int_0^{2\pi} \boldsymbol{i}_{\xi'} \times \mathrm{rot}' \boldsymbol{E}(\boldsymbol{r}')\, G(\boldsymbol{r}, \boldsymbol{r}')\, h_{\eta'} h_{\varphi'}\, \mathrm{d}\eta'\, \mathrm{d}\varphi'\,. \end{aligned} \tag{5.21}$$

When considering $L_z T\boldsymbol{E}$, we take into account that the Lame coefficients h_ξ, h_η, h_φ for the spheroidal coordinates are independent of the angle φ (see Eqs. (5.15)), i.e. $\partial/\partial\varphi\, \mathrm{rot} = \mathrm{rot}\, \partial/\partial\varphi$ and get

$$\begin{aligned} L_z T\boldsymbol{E} &= \mathrm{rot} \int_{-1}^{1} \int_0^{2\pi} \boldsymbol{i}_{\xi'} \times \boldsymbol{E}(\boldsymbol{r}')\, \frac{\partial G(\boldsymbol{r}, \boldsymbol{r}')}{\partial\varphi}\, h_{\eta'} h_{\varphi'}\, \mathrm{d}\xi'\, \mathrm{d}\varphi' \\ &+ \frac{1}{k_1^2 \mu\varepsilon}\, \mathrm{rot}\, \mathrm{rot} \int_{-1}^{1} \int_0^{2\pi} \boldsymbol{i}_{\xi'} \times \mathrm{rot}'\, \boldsymbol{E}(\boldsymbol{r}')\, \frac{\partial G(\boldsymbol{r}, \boldsymbol{r}')}{\partial\varphi}\, h_{\eta'} h_{\varphi'}\, \mathrm{d}\xi'\, \mathrm{d}\varphi'\,. \end{aligned} \tag{5.22}$$

We have $\frac{\partial}{\partial\varphi} G(\boldsymbol{r}, \boldsymbol{r}') = -\frac{\partial}{\partial\varphi'} G(\boldsymbol{r}, \boldsymbol{r}')$, and integration by parts over φ', gives

$$\begin{aligned} L_z T\boldsymbol{E} &= \mathrm{rot} \int_{-1}^{1} \int_0^{2\pi} \boldsymbol{i}_{\xi'} \times \frac{\partial \boldsymbol{E}(\boldsymbol{r}')}{\partial\varphi'}\, G(\boldsymbol{r}, \boldsymbol{r}')\, h_{\eta'} h_{\varphi'}\, \mathrm{d}\xi'\, \mathrm{d}\varphi' \\ &\quad + \frac{1}{k_1^2 \mu\varepsilon}\, \mathrm{rot}\, \mathrm{rot} \int_{-1}^{1} \int_0^{2\pi} \boldsymbol{i}_{\xi'} \times \mathrm{rot}'\, \frac{\partial \boldsymbol{E}(\boldsymbol{r}')}{\partial\varphi'}\, G(\boldsymbol{r}, \boldsymbol{r}') \\ &\quad \times h_{\eta'} h_{\varphi'}\, \mathrm{d}\xi'\, \mathrm{d}\varphi' = T L_z \boldsymbol{E}\,. \end{aligned} \tag{5.23}$$

Here we kept in mind that $\boldsymbol{E}(\boldsymbol{r}')$ and $G(\boldsymbol{r}, \boldsymbol{r}')$ were 2π-periodic functions of φ' and hence all terms appearing after the integration by parts outside the integrals are equal to zero.

Thus, the light scattering problem for a spheroid allows the separation of variables for the azimuthal angle φ, and hence each term of the Fourier series (i.e. expansion in the trigonometric functions of this angle) of the fields including the components $E_1^{(i)}$ and $H_1^{(i)}$ can be found separately.

Solution to the axisymmetric problem

When the electromagnetic field does not depend on the azimuthal angle, one can introduce the Abraham potentials

$$P = h_\varphi E_\varphi , \qquad Q = h_\varphi H_\varphi . \tag{5.24}$$

Other components of $\boldsymbol{E}$ and $\boldsymbol{H}$ are expressed through P and Q as follows:

$$\begin{aligned} E_\xi &= -\frac{i}{k_1\varepsilon}\frac{1}{h_\eta h_\varphi}\frac{\partial Q}{\partial \eta}, & H_\xi &= \frac{i}{k_1\mu}\frac{1}{h_\eta h_\varphi}\frac{\partial P}{\partial \eta}, \\ E_\eta &= \frac{i}{k_1\varepsilon}\frac{1}{h_\xi h_\varphi}\frac{\partial Q}{\partial \xi}, & H_\eta &= -\frac{i}{k_1\mu}\frac{1}{h_\xi h_\varphi}\frac{\partial P}{\partial \xi}. \end{aligned} \tag{5.25}$$

It is known that the azimuthal components of the vectors $\boldsymbol{E}$ and $\boldsymbol{H}$ independent of the angle φ can be expressed through the spheroidal functions with the index $m = 1$ (Komarov *et al.*, 1976). When a plane wave is scattered by a prolate spheroid, such components of the incident, scattered and internal fields are expanded as follows:

$$\begin{aligned} E_{1\varphi}^{(0)} &= \sum_{l=1}^{\infty} a_l^{(0)}\, S_{1l}(c_1,\eta)\, R_{1l}^{(1)}(c_1,\xi)\,, \\ H_{1\varphi}^{(0)} &= \sum_{l=1}^{\infty} b_l^{(0)}\, S_{1l}(c_1,\eta)\, R_{1l}^{(1)}(c_1,\xi)\,, \end{aligned} \tag{5.26}$$

$$\begin{aligned} E_{1\varphi}^{(1)} &= \sum_{l=1}^{\infty} a_l^{(1)}\, S_{1l}(c_1,\eta)\, R_{1l}^{(3)}(c_1,\xi)\,, \\ H_{1\varphi}^{(1)} &= \sum_{l=1}^{\infty} b_l^{(1)}\, S_{1l}(c_1,\eta)\, R_{1l}^{(3)}(c_1,\xi)\,, \end{aligned} \tag{5.27}$$

$$\begin{aligned} E_{1\varphi}^{(2)} &= \sum_{l=1}^{\infty} a_l^{(2)}\, S_{1l}(c_2,\eta)\, R_{1l}^{(1)}(c_2,\xi)\,, \\ H_{1\varphi}^{(2)} &= \sum_{l=1}^{\infty} b_l^{(2)}\, S_{1l}(c_2,\eta)\, R_{1l}^{(1)}(c_2,\xi)\,, \end{aligned} \tag{5.28}$$

where $S_{ml}(c,\eta)$ are the prolate SAFs with the normalizing coefficients $N_{ml}(c)$, $R_{ml}^{(j)}(c,\xi)$ the prolate SRFs of the jth kind, and $c_i = k_i d/2$ is a dimensionless parameter ($i = 1,\, 2$).

The fields represented by the expansions (5.26)–(5.28) satisfy the Maxwell equations, and the boundary conditions (5.6)–(5.7) allow one to find the unknown expansion coefficients $a_l^{(1)}$, $b_l^{(1)}$, and $a_l^{(2)}$, $b_l^{(2)}$.

Let us determine the coefficients of the incident field (5.26). For the TE mode, we have

$$E_{1\varphi}^{(0)} = \frac{1}{2\pi}\int_0^{2\pi} \boldsymbol{E}^{(0)}\cdot \boldsymbol{i}_\varphi\, \mathrm{d}\varphi = \frac{1}{2\pi}\int_0^{2\pi} \mathrm{e}^{ik_1(x\sin\alpha + z\cos\alpha)}\cos\varphi\, \mathrm{d}\varphi\,. \tag{5.29}$$

The expansion of the scalar plane wave in terms of spheroidal wave functions is as follows (Komarov *et al.*, 1976):

$$\begin{aligned} \mathrm{e}^{\mathrm{i}k_1(x\sin\alpha+z\cos\alpha)} &= 2\sum_{m=0}^{\infty}\sum_{l=m}^{\infty}\mathrm{i}^l(2-\delta_m^0)\,N_{ml}^{-2}(c_1) \\ &\quad\times S_{ml}(c_1,\cos\alpha)\,S_{ml}(c_1,\eta)\,R_{ml}^{(1)}(c_1,\xi)\,\cos m\eta\,. \end{aligned} \tag{5.30}$$

From orthogonality of the trigonometric functions, we get

$$E_{1\varphi}^{(0)} = 2\sum_{l=1}^{\infty}\mathrm{i}^l\,N_{1l}^{-2}(c_1)\,S_{1l}(c_1,\cos\alpha)\,S_{1l}(c_1,\eta)\,R_{1l}^{(1)}(c_1,\xi)\,. \tag{5.31}$$

The azimuthal component of the magnetic field is obtained analogously

$$H_{1\varphi}^{(0)} = \sqrt{\frac{\varepsilon_1}{\mu_1}}\frac{\cos\alpha}{2\pi}\int_0^{2\pi}\mathrm{e}^{\mathrm{i}k_1(x\sin\alpha+z\cos\alpha)}\sin\varphi\,\mathrm{d}\varphi = 0\,. \tag{5.32}$$

Thus, the coefficients $a_l^{(0)}$ and $b_l^{(0)}$ for the TE mode are

$$\begin{aligned} a_l^{(0)} &= 2\mathrm{i}^l\,N_{1l}^{-2}(c_1)\,S_{1l}(c_1,\cos\alpha)\,, \\ b_l^{(0)} &= 0\,. \end{aligned} \tag{5.33}$$

In the case of the TM mode (see Eqs. (5.16)–(5.17)), we have

$$\begin{aligned} a_l^{(0)} &= 0\,, \\ b_l^{(0)} &= 2\sqrt{\frac{\varepsilon_1}{\mu_1}}\,\mathrm{i}^l\,N_{1l}^{-2}(c_1)\,S_{1l}(c_1,\cos\alpha)\,. \end{aligned} \tag{5.34}$$

From Eqs. (5.24)–(5.25) it follows that the Abraham potentials are determined independently from each other and for the TE mode only the potential P is not equal to zero (i.e. $b_l^{(1)} = b_l^{(2)} = 0$), while for the TM mode only the potential Q is not zero (i.e. $a_l^{(1)} = a_l^{(2)} = 0$).

Let us substitute the field expansions in the boundary conditions. For the TE mode, we have

$$\left\{ \begin{aligned} &\sum_{l=1}^{\infty}\left(a_l^{(0)}\,R_{1l}^{(1)}(c_1,\xi_0) + a_l^{(1)}\,R_{1l}^{(3)}(c_1,\xi_0)\right)\,S_{1l}(c_1,\eta) \\ &\qquad = \sum_{l=1}^{\infty}a_l^{(2)}\,R_{1l}^{(1)}(c_2,\xi_0)\,S_{1l}(c_2,\eta)\,, \\ &\frac{1}{\mu_1}\sum_{l=1}^{\infty}\left\{a_l^{(0)}\left[(\xi_0^2-1)^{\frac{1}{2}}R_{1l}^{(1)}(c_1,\xi_0)\right]' \right. \\ &\qquad \left. +a_l^{(1)}\left[(\xi_0^2-1)^{\frac{1}{2}}R_{1l}^{(3)}(c_1,\xi_0)\right]'\right\}S_{1l}(c_1,\eta) \\ &\qquad = \frac{1}{\mu_2}\sum_{l=1}^{\infty}a_l^{(2)}\left[(\xi_0^2-1)^{\frac{1}{2}}R_{1l}^{(1)}(c_2,\xi_0)\right]'\,S_{1l}(c_2,\eta)\,, \end{aligned} \right. \tag{5.35}$$

where the prime means differentiation by ξ at $\xi = \xi_0$. Due to orthogonality of the prolate SAFs, multiplication of these equations by $N_{1n}^{-1}(c_2)\, S_{1n}(c_2, \eta)$ and integration over η from -1 to 1 give

$$\left\{ \begin{aligned} &\sum_{l=1}^{\infty} \left(a_l^{(0)}\, R_{1l}^{(1)}(c_1, \xi_0) + a_l^{(1)}\, R_{1l}^{(3)}(c_1, \xi_0) \right) N_{1l}(c_1)\, \delta_{nl}^{(1)}(c_2, c_1) \\ &\qquad = \sum_{l=1}^{\infty} a_l^{(2)}\, R_{1n}^{(1)}(c_2, \xi_0)\, N_{1n}(c_2, \eta)\,, \\ &\frac{1}{\mu_1} \sum_{l=1}^{\infty} \Big\{ a_l^{(0)} \left[(\xi_0^2 - 1)^{\frac{1}{2}}\, R_{1l}^{(1)}(c_1, \xi_0) \right]' \\ &\qquad + a_l^{(1)} \left[(\xi_0^2 - 1)^{\frac{1}{2}}\, R_{1l}^{(3)}(c_1, \xi_0) \right]' \Big\}\, N_{1l}(c_1)\, \delta_{nl}^{(1)}(c_2, c_1) \\ &\qquad = \frac{1}{\mu_2} \sum_{l=1}^{\infty} a_l^{(2)} \left[(\xi_0^2 - 1)^{\frac{1}{2}}\, R_{1l}^{(1)}(c_2, \xi_0) \right]' N_{1n}(c_2)\,. \end{aligned} \right. \tag{5.36}$$

The integrals of the products of the prolate SAFs denoted by $\delta_{nl}^{m}(c_2, c_1)$ are presented in Appendix A.

Now we exclude the unknown coefficients $a_n^{(2)}$ and introduce some notation for the vectors $\boldsymbol{Z}_0 = \{z_{0l}\}_{l=1}^{\infty}$, $\boldsymbol{F}_0 = \{f_{0l}\}_{l=1}^{\infty}$, the unit matrix $I = \{\delta_l^n\}_{n,l=m}^{\infty}$, and the matrices $\Delta = \{\delta_{nl}^{(m)}(c_2, c_1)\}_{n,l=m}^{\infty}$, $R_{0,1} = \{r_{ml}^{(1),(3)}(c_1)\ \delta_l^n\}_{n,l=m}^{\infty}$, $R_2 = \{r_{ml}^{(1)}(c_2)\ \delta_l^n\}_{n,l=m}^{\infty}$, where

$$\begin{aligned} z_{0l} &= a_l^{(1)}\, R_{1l}^{(3)}(c_1, \xi_0)\, N_{1l}(c_1)\,, \\ f_{0l} &= a_l^{(0)}\, R_{1l}^{(1)}(c_1, \xi_0)\, N_{1l}(c1) \\ &= 2\mathrm{i}^l\, N_{1l}^{-1}(c_1)\, S_{1l}(c_1, \cos\alpha)\, R_{1l}^{(1)}(c_1, \xi_0)\,, \\ r_{ml}^{(j)}(c_i) &= \frac{R_{ml}^{(j)'}(c_i, \xi_0)}{R_{ml}^{(j)}(c_i, \xi_0)}\,. \end{aligned} \tag{5.37}$$

Then the system (5.36) can be written in the matrix form

$$\begin{aligned} &\left[\xi_0(\mu_2 - \mu_1)\Delta + (\xi_0^2 - 1)(\mu_2 \Delta R_1 - \mu_1 R_2 \Delta) \right] \boldsymbol{Z}_0 \\ &\qquad + \left[\xi_0(\mu_2 - \mu_1)\Delta + (\xi_0^2 - 1)(\mu_2 \Delta R_0 - \mu_1 R_2 \Delta) \right] \boldsymbol{F}_0 = 0\,. \end{aligned} \tag{5.38}$$

For the TM mode, we get

$$\begin{aligned} &\left[\xi_0(\varepsilon_2 - \varepsilon_1)\Delta + (\xi_0^2 - 1)(\varepsilon_2 \Delta R_1 - \varepsilon_1 R_2 \Delta) \right] \boldsymbol{Z}_0 \\ &\qquad + \left[\xi_0(\varepsilon_2 - \varepsilon_1)\Delta + (\xi_0^2 - 1)(\varepsilon_2 \Delta R_0 - \varepsilon_1 R_2 \Delta) \right] \boldsymbol{F}_0 = 0\,, \end{aligned} \tag{5.39}$$

where in the expression for z_{0l} one should replace $a_l^{(1)}$ with $b_l^{(1)}$ while f_{0l} should remain the same. Note that these ISLAEs are similar and can be transformed into each other by the replacements $\mu_i \to \varepsilon_i$ and $\varepsilon_i \to \mu_i$.

To obtain similar results for an oblate spheroid, one should make the standard replacements $c \to -\mathrm{i}c$ ($d \to -\mathrm{i}d$) and $\xi \to \mathrm{i}\xi$ as well as replace the prolate spheroidal wave functions with the oblate ones. It is convenient to introduce the parameter f equal to 1 for prolate spheroids and to -1 for oblate ones. If the magnetic permeability is the same everywhere, the ISLAEs (5.38) and (5.39) can be rewritten as follows:

$$(\tilde{R}_2 - R_1)\boldsymbol{Z}_0 + (\tilde{R}_2 - R_0)\boldsymbol{F}_0 = 0, \tag{5.40}$$

$$\begin{aligned}&\left[\xi_0(\varepsilon - 1)I + (\xi_0^2 - f)(\varepsilon R_1 - \tilde{R}_2)\right] \boldsymbol{Z}_0 \\ &\quad + \left[\xi_0(\varepsilon - 1)I + (\xi_0^2 - f)(\varepsilon R_0 - \tilde{R}_2)\right] \boldsymbol{F}_0 = 0,\end{aligned} \tag{5.41}$$

where $\varepsilon{=}\varepsilon_2/\varepsilon_1$ is the relative dielectric permittivity, $\tilde{R}_2 = \Delta^{-1} R_2 \Delta$ (see properties of the integrals of SAFs products in Appendix A).

Solution to the non-axisymmetric problem

The second components in the sums (5.19) are represented as follows:

(1) for the TE mode

$$\begin{aligned}\boldsymbol{E}_2^{(i)} &= \mathrm{rot}\left(U^{(i)}\,\boldsymbol{i}_z + V^{(i)}\boldsymbol{r}\right), \\ \boldsymbol{H}_2^{(i)} &= \frac{1}{\mathrm{i}\,\mu_i k_0}\,\mathrm{rot}\,\mathrm{rot}\left(U^{(i)}\,\boldsymbol{i}_z + V^{(i)}\,\boldsymbol{r}\right);\end{aligned} \tag{5.42}$$

(2) for the TM mode

$$\begin{aligned}\boldsymbol{E}_2^{(i)} &= -\frac{1}{\mathrm{i}\,\mu_i k_0}\,\mathrm{rot}\,\mathrm{rot}\left(U^{(i)}\,\boldsymbol{i}_z + V^{(i)}\,\boldsymbol{r}\right), \\ \boldsymbol{H}_2^{(i)} &= \mathrm{rot}\left(U^{(i)}\,\boldsymbol{i}_z + V^{(i)}\,\boldsymbol{r}\right),\end{aligned} \tag{5.43}$$

where the scalar potentials $U^{(i)}$ and $V^{(i)}$ are expanded in terms of spheroidal wave functions

$$\begin{aligned}U^{(0)} &= \sum_{m=1}^{\infty}\sum_{l=m}^{\infty} a_{ml}^{(0)}\, S_{ml}(c_1,\eta)\, R_{ml}^{(1)}(c_1,\xi)\, \cos m\varphi\,, \\ V^{(0)} &= \sum_{m=1}^{\infty}\sum_{l=m}^{\infty} b_{ml}^{(0)}\, S_{ml}(c_1,\eta)\, R_{ml}^{(1)}(c_1,\xi)\, \cos m\varphi\,,\end{aligned} \tag{5.44}$$

$$\begin{aligned}U^{(1)} &= \sum_{m=1}^{\infty}\sum_{l=m}^{\infty} a_{ml}^{(1)}\, S_{ml}(c_1,\eta)\, R_{ml}^{(3)}(c_1,\xi)\, \cos m\varphi\,, \\ V^{(1)} &= \sum_{m=1}^{\infty}\sum_{l=m}^{\infty} b_{ml}^{(1)}\, S_{ml}(c_1,\eta)\, R_{ml}^{(3)}(c_1,\xi)\, \cos m\varphi\,,\end{aligned} \tag{5.45}$$

$$\begin{aligned}U^{(2)} &= \sum_{m=1}^{\infty}\sum_{l=m}^{\infty} a_{ml}^{(2)}\, S_{ml}(c_2,\eta)\, R_{ml}^{(1)}(c_2,\xi)\, \cos m\varphi\,, \\ V^{(2)} &= \sum_{m=1}^{\infty}\sum_{l=m}^{\infty} b_{ml}^{(2)}\, S_{ml}(c_2,\eta)\, R_{ml}^{(1)}(c_2,\xi)\, \cos m\varphi\,,\end{aligned} \tag{5.46}$$

where summation over the azimuthal index begins with $m = 1$ since averaging over this angle has to give zero. Note that these expansions are equivalent to the expansions of the fields in terms of the vector spheroidal wave functions $\boldsymbol{M}^{\mathrm{z}}_{ml}$, $\boldsymbol{M}^{\mathrm{r}}_{ml}$ and $\boldsymbol{N}^{\mathrm{z}}_{ml}$, $\boldsymbol{N}^{\mathrm{r}}_{ml}$ (see Farafonov and Il'in, 2006).

By using the equation

$$\begin{gathered}\boldsymbol{i}_{\mathrm{y}}\,\mathrm{e}^{ik_1(x\sin\alpha+z\cos\alpha)} = \mathrm{rot}\Bigg(\frac{2}{k_1\sin\alpha}\sum_{m=0}^{\infty}\sum_{l=m}^{\infty}\mathrm{i}^{(l-1)}(2-\delta_m^0)\\ N_{ml}^{-2}(c_1)\,S_{ml}(c_1,\cos\alpha)\,S_{ml}(c_1,\eta)\,R_{ml}^{(1)}(c_1,\xi)\,\cos m\varphi\,\boldsymbol{i}_z\Bigg)\,,\end{gathered} \tag{5.47}$$

we get for the incident field

$$\begin{aligned} a_{ml}^{(0)} &= \frac{4\mathrm{i}^{l-1}}{k_1\sin\alpha}N_{ml}^{-2}(c_1)\,S_{ml}(c_1,\cos\alpha)\,,\\ b_{ml}^{(0)} &= 0 \end{aligned} \tag{5.48}$$

for both polarizations of the plane wave (see Eqs. (5.16)–(5.17)).

The field expansions (5.44)–(5.46) satisfy the Maxwell equations, and the boundary conditions (5.6)–(5.7) allow one to find the unknown coefficients $a_{ml}^{(1)}$, $b_{ml}^{(1)}$ and $a_{ml}^{(2)}$, $b_{ml}^{(2)}$. After substitution of Eq. (5.42) into the boundary conditions and laborious transformations described in detail by Voshchinnikov and Farafonov (1993), we get

$$\left.\begin{aligned} &\eta U + \xi\frac{d}{2}V = \eta\,U^{(2)} + \xi\frac{d}{2}V^{(2)},\\ &\frac{\partial}{\partial\xi}\left(\xi U + f\eta\frac{d}{2}V\right) = \frac{\partial}{\partial\xi}\left(\xi\,U^{(2)} + f\eta\frac{d}{2}V^{(2)}\right),\\ &\varepsilon_1\left(\xi U + f\eta\frac{d}{2}V\right) = \varepsilon_2\left(\xi\,U^{(2)} + f\eta\frac{d}{2}V^{(2)}\right),\\ &\frac{1}{\mu_1}\left[\frac{\partial}{\partial\xi}\left(\eta U + \xi\frac{d}{2}V\right) + \left(1-\frac{c_1^2}{c_2^2}\right)\frac{1-\eta^2}{\xi^2-f}\right.\\ &\qquad\left.\frac{\partial}{\partial\eta}\left(\xi U + f\eta\frac{d}{2}V\right)\right] = \frac{1}{\mu_2}\frac{\partial}{\partial\xi}\left(\eta\,U^{(2)} + \xi\frac{d}{2}V^{(2)}\right)\,. \end{aligned}\right\}_{\xi=\xi_0} \tag{5.49}$$

Similar expressions for the TM mode coincide with Eqs. (5.49) after the replacements $\mu_i \to \varepsilon_i$ and $\varepsilon_i \to \mu_i$. When the magnetic permeability is the same everywhere, the first and third equations can be rewritten as follows:

$$\left.\begin{aligned} &U = U^{(2)},\\ &\frac{\partial}{\partial\xi}\left(\xi U + f\eta\frac{d}{2}V\right) = \frac{\partial}{\partial\xi}\left(\xi\,U^{(2)} + f\eta\frac{d}{2}V^{(2)}\right),\\ &V = V^{(2)},\\ &\frac{1}{\varepsilon_1}\left[\frac{\partial}{\partial\xi}\left(\eta U + \xi\frac{d}{2}V\right) + \left(1-\frac{c_1^2}{c_2^2}\right)\frac{1-\eta^2}{\xi^2-f}\right.\\ &\qquad\left.\frac{\partial}{\partial\eta}\left(\xi U + f\eta\frac{d}{2}V\right)\right] = \frac{1}{\varepsilon_2}\frac{\partial}{\partial\xi}\left(\eta\,U^{(2)} + \xi\frac{d}{2}V^{(2)}\right)\,. \end{aligned}\right\}_{\xi=\xi_0} \tag{5.50}$$

In the systems (5.49) and (5.50) we use the parameter f introduced above.

In addition to the notation (5.37) let us define

$$\begin{aligned}
&\boldsymbol{Z}_1 = \{z_{1l}^{(m)}\}_{l=m}^{\infty}, && \boldsymbol{Z}_2 = \{z_{2l}^{(m)}\}_{l=m}^{\infty}, \\
&\boldsymbol{X}_1 = \{x_{1l}^{(m)}\}_{l=m}^{\infty}, && \boldsymbol{X}_2 = \{x_{2l}^{(m)}\}_{l=m}^{\infty}, \\
&\boldsymbol{F}_m = \{f_l^{(m)}\}_{l=m}^{\infty}, && \Gamma(c_i, c_j) = \{\gamma_{nl}^{(m)}(c_i, c_j)\}_{n,l=m}^{\infty}, \\
&K(c_i, c_j) = \{\kappa_{nl}^{(m)}(c_i, c_j)\}_{n,l=m}^{\infty}, && \Sigma(c_i, c_j) = \{\sigma_{nl}^{(m)}(c_i, c_j)\}_{n,l=m}^{\infty},
\end{aligned} \tag{5.51}$$

where

$$\begin{aligned}
z_{1l}^{(m)} &= k_1 a_{ml}^{(1)} N_{ml}(c_1) R_{ml}^{(3)}(c_1, \xi_0), \qquad & z_{2l}^{(m)} &= c_1 b_{ml}^{(1)} N_{ml}(c_1) R_{ml}^{(3)}(c_1, \xi_0), \\
x_{1l}^{(m)} &= k_1 a_{ml}^{(2)} N_{ml}(c_2) R_{ml}^{(1)}(c_2, \xi_0), & x_{2l}^{(m)} &= c_1 b_{ml}^{(2)} N_{ml}(c_2) R_{ml}^{(1)}(c_2, \xi_0), \\
f_l^{(m)} &= k_1 a_{ml}^{(0)} N_{ml}(c_1) R_{ml}^{(1)}(c_1, \xi_0) = \frac{4\mathrm{i}^{l-1}}{\sin\alpha} \frac{S_{ml}(c_1 \cos\alpha) R_{ml}^{(1)}(c_1, \xi_0)}{N_{ml}(c_1)}.
\end{aligned} \tag{5.52}$$

The integrals of products of the SAFs and their derivatives $\gamma_{nl}^{(m)}$, $\kappa_{nl}^{(m)}$, and $\sigma_{nl}^{(m)}$ are given in Appendix A.

After substitution of the expansions (5.44)–(5.46) into the boundary conditions (5.49), multiplication by $N_{mn}^{-1}(c_2) S_{mn}(c_2, \eta) \cos m\varphi$, and integration over η from -1 to 1 and over φ from 0 to 2π, orthogonality of the functions $\cos m\varphi$ provides the following ISLAFs written by us in the matrix form ($m = 1, 2$):

$$\left\{\begin{aligned}
&\Gamma(c_2, c_1)(\boldsymbol{Z}_1 + \boldsymbol{F}_m) + \xi_0 \Delta(c_2, c_1) \boldsymbol{Z}_2 \\
&\qquad = \Gamma(c_2, c_2) \boldsymbol{X}_1 + \xi_0 \boldsymbol{X}_2, \\
&\Delta(c_2, c_1)[(I + \xi_0 R_1) \boldsymbol{Z}_1 + (I + \xi_0 R_0) \boldsymbol{F}_m] \\
&\qquad + f \Gamma(c_2, c_1) R_1 \boldsymbol{Z}_2 = (I + \xi_0 R_2) \boldsymbol{X}_1 + f \Gamma(c_2, c_2) R_2 \boldsymbol{X}_2, \\
&\varepsilon_1 [\xi_0 \Delta(c_2, c_1)(\boldsymbol{Z}_1 + \boldsymbol{F}_m)] + f \Gamma(c_2, c_1) \boldsymbol{Z}_2 \\
&\qquad = \varepsilon_2 (\xi_0 \boldsymbol{X}_1 + f \Gamma(c_2, c_2) \boldsymbol{X}_2), \\
&\Gamma(c_2, c_1)[R_1 \boldsymbol{Z}_1 + R_0 \boldsymbol{F}_m] + \Delta(c_2, c_1)(I + \xi_0 R_1) \boldsymbol{Z}_2 \\
&\qquad + \left(1 - \frac{c_1^2}{c_2^2}\right) \frac{1}{\xi_0^2 - f} [\xi_0 K(c_2, c_1)(\boldsymbol{Z}_1 + \boldsymbol{F}_m) + f \Sigma(c_2, c_1) \boldsymbol{Z}_2] \\
&\qquad = \Gamma(c_2, c_1) R_2 \boldsymbol{X}_1 + (I + \xi_0 R_2) \boldsymbol{X}_2.
\end{aligned}\right. \tag{5.53}$$

The first and third equations give the unknown vectors

$$\begin{aligned}
\boldsymbol{X}_1 &= Q(c_2, c_1) \left[\left(\frac{\xi_0^2}{\varepsilon} I - f \Gamma^2(c_1, c_1) \right) (\boldsymbol{Z}_1 + \boldsymbol{F}_m) + \left(\frac{1}{\varepsilon} - 1 \right) f \xi_0 \Gamma(c_1, c_1) \boldsymbol{Z}_2 \right], \\
\boldsymbol{X}_2 &= Q(c_2, c_1) \left[\left(\xi_0 I - \frac{f}{\varepsilon} \Gamma^2(c_1, c_1) \right) \boldsymbol{Z}_2 + \left(1 - \frac{\xi_0}{\varepsilon} \Gamma(c_1, c_1) \right) (\boldsymbol{Z}_1 + \boldsymbol{F}_m) \right],
\end{aligned} \tag{5.54}$$

where $Q(c_i, c_j) = \left[\xi_0^2 \Delta(c_i, c_j) - f \Gamma^2(c_i, c_j)\right]^{-1}$. We used the properties of the infinite matrices which elements are the integrals of products of the SAFs and their derivatives (see Appendix A).

Substitution of the vectors $\boldsymbol{X}_1$ and $\boldsymbol{X}_2$ into the second and fourth equations gives an ISLAE relative to the vectors $\boldsymbol{Z}_1$ and $\boldsymbol{Z}_2$

$$\left\{\begin{aligned} &\left[\xi_0(\tilde{R}_2 - R_1) - \xi_0\left(1 - \frac{1}{\varepsilon}\right)A\right]\boldsymbol{Z}_1 + f\left[\Gamma(\tilde{R}_2 - R_1) - \left(1 - \frac{1}{\varepsilon}\right)\Gamma\right]\boldsymbol{Z}_2 \\ &\qquad + \left[\xi_0(\tilde{R}_2 - R_0) - \xi_0\left(1 - \frac{1}{\varepsilon}\right)A\right]\boldsymbol{F}_m = 0\,, \\ &\left[\Gamma(\tilde{R}_2 - R_1) - \xi_0\left(1 - \frac{1}{\varepsilon}\right)B\right]\boldsymbol{Z}_1 + \left[\xi_0(\tilde{R}_2 - R_1) - f\left(1 - \frac{1}{\varepsilon}\right)B\Gamma\right]\boldsymbol{Z}_2 \\ &\qquad + \left[\Gamma(\tilde{R}_2 - R_1) - \xi_0\left(1 - \frac{1}{\varepsilon}\right)B\right]\boldsymbol{F}_m = 0\,, \end{aligned}\right. \tag{5.55}$$

where

$$\begin{aligned} A &= \xi_0\,(I + \xi_0\,\tilde{R}_2)\,Q - f\,\Gamma\tilde{R}_2\Gamma\,Q = \Omega_2\,, \\ B &= \xi_0\,\Gamma\tilde{R}_2\,Q - (I + \xi_0\,\tilde{R}_2)\,\Gamma\,Q + \frac{1}{\xi_0^2 - f}\,K = \Omega_1\,, \end{aligned} \tag{5.56}$$

and the matrices $\Delta, \Gamma, K, \Sigma, Q$ depend on the parameter c_1 only.

For the TM mode, the boundary conditions (5.50) become more simple, and using the notation (5.37) and (5.51), one can obtain an ISLAE relative to the vectors $\boldsymbol{Z}_1$ and $\boldsymbol{Z}_2$

$$\left\{\begin{aligned} &\xi_0\,(\tilde{R}_2 - R_1)\,\boldsymbol{Z}_1 + f\,\Gamma\,(\tilde{R}_2 - R_1)\,\boldsymbol{Z}_2 + \xi_0\,(\tilde{R}_2 - R_0)\,\boldsymbol{F}_m = 0\,, \\ &\left[\Gamma\left(\frac{1}{\varepsilon}\tilde{R}_2 - R_1\right) - \left(1 - \frac{1}{\varepsilon}\right)\frac{\xi_0}{\xi_0^2 - f}\,K\right]\boldsymbol{Z}_1 \\ &\qquad + \left[\frac{1}{\varepsilon}(I + \xi_0\,\tilde{R}_2) - (I + \xi_0\,R_1) - f\left(1 - \frac{1}{\varepsilon}\right)\frac{\xi_0}{\xi_0^2 - f}\,K\Gamma\right]\boldsymbol{Z}_2 \\ &\qquad + \left[\Gamma\left(\frac{1}{\varepsilon}\tilde{R}_2 - R_0\right) - \left(1 - \frac{1}{\varepsilon}\right)\frac{\xi_0}{\xi_0^2 - f}\,K\right]\boldsymbol{F}_m = 0\,. \end{aligned}\right. \tag{5.57}$$

An analytical study of the ISLAE arisen in the solution to the axisymmetric (5.40)–(5.41) and non-axisymmetric (5.55)–(5.57) problems of light scattering by spheroids is presented in Section 5.3.

5.2.3 Perfectly conducting spheroids

The model of a perfectly conducting spheroid is used in radio physics to study the effects of electromagnetic radiation scattering by metallic bodies. In this case the boundary conditions for a spheroidal body are

$$\left.\begin{aligned} E_\eta^{(0)} + E_\eta^{(1)} &= 0\,, \\ E_\varphi^{(0)} + E_\varphi^{(1)} &= 0\,. \end{aligned}\right\}_{\xi=\xi_0} \tag{5.58}$$

The approach to solution of the diffraction problem remains the same. Note that the parameter c_2 is absent in this problem.

The axisymmetric component of radiation scattered by a perfectly conducting spheroid is obtained directly:

(1) for the TE mode

$$\begin{aligned} a_l^{(1)} &= -2\mathrm{i}^l \, \frac{R_{1l}^{(1)}(c_1,\xi_0)}{R_{1l}^{(3)}(c_1,\xi_0)} \, N_{1l}^{-2}(c_1)\, S_{1l}(c_1,\cos\alpha)\,, \\ b_l^{(1)} &= 0\,; \end{aligned} \tag{5.59}$$

(2) for the TM mode

$$\begin{aligned} a_l^{(1)} &= 0\,, \\ b_l^{(1)} &= -2\mathrm{i}^l \, \frac{[(\xi_0^2-1)^{\frac{1}{2}} R_{1l}^{(1)}(c_1,\xi_0)]'}{[(\xi_0^2-1)^{\frac{1}{2}} R_{1l}^{(3)}(c_1,\xi_0)]'} \, N_{1l}^{-2}(c_1)\, S_{1l}(c_1,\cos\alpha)\,. \end{aligned} \tag{5.60}$$

For the non-axisymmetric components, the coefficients of the expansions (5.45) are derived from the boundary conditions (5.58) that can be rewritten using the potentials $U^{(i)}$ and $V^{(i)}$ as follows ($U = U^{(0)} + U^{(1)}$, $V = V^{(0)} + V^{(1)}$):

$$\left.\begin{aligned} \eta\, U + \xi\, \frac{d}{2}\, V = 0, \\ \frac{\partial}{\partial \xi}(\xi\, U + f\,\eta\, \frac{d}{2}\, V) = 0 \end{aligned}\right\}_{\xi=\xi_0} \tag{5.61}$$

for the TE mode, and

$$\left.\begin{aligned} \xi\, U + \eta\, \frac{d}{2}\, V = 0, \\ \frac{\partial}{\partial \xi}(\eta\, U + f\,\xi\, \frac{d}{2}\, V) = 0 \end{aligned}\right\}_{\xi=\xi_0} \tag{5.62}$$

for the TM mode, respectively.

Substitution of Eqs. (5.44) and (5.45) into Eqs. (5.61) and (5.62), respectively, and exclusion of one of the unknown vectors give for the TE mode

$$\begin{cases} [\xi_0 I - f\Gamma (I+\xi_0 R_1)^{-1}\, \Gamma R_1]\, \boldsymbol{Z}_2 + \xi_0 \Gamma (I+\xi_0 R_1)^{-1}\, (R_1 - R_0)\, \boldsymbol{F}_m = 0\,, \\ \boldsymbol{Z}_1 = -(I+\xi_0 R_1)^{-1}\, [f\Gamma R_1\, \boldsymbol{Z}_2 + (I+\xi_0 R_0)\, \boldsymbol{F}_m] \end{cases} \tag{5.63}$$

and for the TM mode

$$\begin{cases} [\xi_0 I - f\Gamma (I+\xi_0 R_1)^{-1}\, \Gamma R_1]\, \boldsymbol{Z}_1 + [\xi_0 I - f\Gamma (I+\xi_0 R_1)^{-1}\, \Gamma R_0)]\, \boldsymbol{F}_m = 0\,, \\ \boldsymbol{Z}_2 = -(I+\xi_0 R_1)^{-1}\, (f\Gamma R_1\, \boldsymbol{Z}_1 + \Gamma R_0\, \boldsymbol{F}_m)\,. \end{cases} \tag{5.64}$$

Note that these ISLAEs differ just in their right-hand parts and hence the problems for both polarizations can be solved simultaneously.

The infinite systems (5.63) and (5.64) are generally similar to those for dielectric spheroids. However, light scattering by a perfectly conducting disk, being a particular case of the oblate spheroid with $\xi_0 = 0$, has some important features. There arise additional boundary conditions at the disk edge that are called the

Meixner conditions (Meixner, 1950). The direct transition $\xi_0 \to 0$ in Eqs. (5.63) and (5.64) leads to wrong solutions

$$\boldsymbol{Z}_2 = 0\,, \qquad \boldsymbol{Z}_1 = -\boldsymbol{F}_m \tag{5.65}$$

and

$$\boldsymbol{Z}_1 = -R_1^{-1} R_0\, \boldsymbol{F}_m\,, \qquad \boldsymbol{Z}_2 = 0\,, \tag{5.66}$$

as they do not satisfy the Meixner conditions. Diffraction of electromagnetic radiation by extremely oblate perfectly conducting spheroids and necessary improvements of the solution presented above are considered in detail in Section 5.5.

5.2.4 Spherical particles

Such a particle is a particular case of the spheroid when one makes the transitions $\xi_0 \to \infty$, $c \to 0$, and $c\xi_0 \to kr_0$. Then the spheroidal functions can be replaced by the spherical ones (Komarov *et al.*, 1976)

$$\begin{aligned} &R_{ml}^{(1)}(c_1, \xi_0) \to j_l(k_1 r_0), \quad R_{ml}^{(3)}(c_1, \xi_0) \to h_l^{(1)}(k_1 r_0)\,, \\ &R_{ml}^{(1)}(c_2, \xi_0) \to j_l(k_2 r_0), \quad S_{ml}(c, \eta) \to P_l^m(\cos\theta)\,, \end{aligned} \tag{5.67}$$

and the matrices of the ISLAEs simplify

$$\begin{aligned} R_0 = c_1 J_0, \quad R_1 = c_1 H, \quad \tilde{R}_2 = c_2 J_2, \quad Q = \frac{1}{\xi_0^2} I\,, \\ A = \tilde{R}_2 = c_2 J_2\,, \quad B = \frac{c_2}{\xi_0}(\Gamma J_2 - J_2 \Gamma) + \frac{1}{\xi_0^2}(K - \Gamma) = O\left(\frac{1}{\xi_0}\right)\,, \end{aligned} \tag{5.68}$$

where $J_{0,2} = \{j_l'(k_{1,2}\, r_0) \,/\, j_l(k_{1,2}\, r_0)\}_{l=m}^{\infty}$, $H = \left\{h_l^{(1)'}(k_1 r_0) \,/\, h_l^{(1)}(k_1 r_0)\right\}_{l=m}^{\infty}$ are the diagonal matrices, and the matrices Γ and K have nonzero elements just above and below the main diagonal, respectively.

Keeping in mind Eqs. (5.67)–(5.68) and the behaviour of the vectors $\boldsymbol{Z}_0 = O(1)$, $\boldsymbol{F}_0 = O(1)$, $\boldsymbol{Z}_1 = O(1)$, $\boldsymbol{Z}_2 = O(c_1)$, and $\boldsymbol{F}_m = O(1)$ (see Eqs. (5.37), (5.52)), one can solve the ISLAEs analytically. For instance, for a TM mode plane wave incident at a dielectric spheroid, we get (see Eqs. (5.41) and (5.57))

$$\left\{ \begin{aligned} \boldsymbol{Z}_0 &= -\,(H - J_2)^{-1}\,(J_1 - J_2)\,\boldsymbol{F}_0\,, \\ \boldsymbol{Z}_1 &= -\,(H - \varepsilon J_2)^{-1}\,(J_1 - \varepsilon J_2)\,\boldsymbol{F}_m\,, \\ \boldsymbol{Z}_2 &= -\,c_1\Big[(I + k_1 r_0 H) - \frac{1}{\varepsilon}(I + k_2 r_0 J_2)^{-1}\Big]\Big\{\Big[\Gamma(J_1 - J_2) + \\ &\quad + \Big(1 - \frac{1}{\varepsilon} k_1 r_0 K\Big)\Big]\boldsymbol{F}_m + \Big[\Gamma(H - J_2) + \Big(1 - \frac{1}{\varepsilon}\Big) k_1 r_0 K\Big]\,\boldsymbol{Z}_1\Big\}\,, \end{aligned} \right. \tag{5.69}$$

where the inverse matrices are easily calculated as they are diagonal. In the expressions for the coefficients $a_{ml}^{(2)}$ the parameter c_1 is absent.

Thus, the solution given by the suggested approach differs from that given by the Mie theory because the coordinate system is not properly chosen (the z axis does not coincide with the direction of the plane wave propagation) and other scalar potentials are selected. Note that nevertheless the ISLAEs are solved analytically.

5.2.5 Characteristics of the radiation scattered by a spheroid

The plane of incidence is defined as the plane including both the x and z axes (the symmetry axis) and is the reference plane for the incident wave, i.e. the wave vector $\boldsymbol{k}_1$ lies in the plane xz. The plane of scattering is defined in a similar way as the plane containing the z axis and the direction of the scattered radiation propagation. Then the vectors perpendicular and parallel to the planes of incidence and scattering are

$$\begin{aligned} \boldsymbol{i}_{\parallel}^{(0)} &= \cos\alpha\, \boldsymbol{i}_{\mathrm{x}} - \sin\alpha\, \boldsymbol{i}_{\mathrm{z}}\,, & \boldsymbol{i}_{\parallel}^{(1)} &= -\boldsymbol{i}_{\eta}\,, \\ \boldsymbol{i}_{\perp}^{(0)} &= -\boldsymbol{i}_{\mathrm{y}}\,, & \boldsymbol{i}_{\perp}^{(1)} &= \boldsymbol{i}_{\varphi}\,, \end{aligned} \tag{5.70}$$

where $(\boldsymbol{i}_{\xi}, \boldsymbol{i}_{\eta}, \boldsymbol{i}_{\varphi})$ are the unit vectors of the spheroidal coordinate system so that the vectors $(\boldsymbol{i}_{\parallel}^{(0)}, \boldsymbol{i}_{\perp}^{(0)}, \boldsymbol{i}_{\mathrm{k}_1})$ and $(\boldsymbol{i}_{\parallel}^{(1)}, \boldsymbol{i}_{\perp}^{(1)}, \boldsymbol{i}_{\xi})$ form a right triad of vectors. Note that in the far-field zone $r \to \infty$ and hence $\xi \to \infty$, $\eta \to \cos\theta$, $\boldsymbol{i}_{\eta} \to -\boldsymbol{i}_{\theta}$.

The relation of the incident and scattered radiation in the far-field zone is determined by the amplitude matrix as follows:

$$\begin{pmatrix} E_{\parallel}^{(1)} \\ E_{\perp}^{(1)} \end{pmatrix} = \frac{1}{-\mathrm{i}k_1 r}\, \mathrm{e}^{\mathrm{i}(k_1 r - \boldsymbol{k}_1 \cdot \boldsymbol{r})} \begin{pmatrix} A_2 & A_3 \\ A_4 & A_1 \end{pmatrix} \begin{pmatrix} E_{\parallel}^{(0)} \\ E_{\perp}^{(0)} \end{pmatrix}. \tag{5.71}$$

The representation of the fields by their potentials and the asymptotics of the spheroidal radial functions for the large values of their argument allow one to obtain for the TM mode and $r \gg 1$

$$\begin{aligned} \boldsymbol{E}^{(1)} &= \frac{\mathrm{e}^{\mathrm{i}k_1 r}}{-\mathrm{i}k_1 r} \boldsymbol{A}^{(1)} = \frac{\mathrm{e}^{\mathrm{i}k_1 r}}{-\mathrm{i}k_1 r} \left\{ -\sum_{m=1}^{\infty} \sum_{l=m}^{\infty} \mathrm{i}^{-l}\, b_{ml}^{(1)}\, \frac{m S_{ml}(c_1, \cos\theta)}{\sin\theta}\, \sin m\varphi\, \boldsymbol{i}_{\varphi} \right. \\ &\quad + \left[-\sum_{l=1}^{\infty} \mathrm{i}^{-l}\, b_l^{(1)}\, S_{1l}(c_1, \cos\theta) + \sum_{m=1}^{\infty} \sum_{l=m}^{\infty} \mathrm{i}^{1-l} \Big(k_1\, a_{ml}^{(1)}\, S_{ml}(c_1, \cos\theta) \right. \\ &\quad \left.\left. + \mathrm{i}\, b_{ml}^{(1)}\, S'_{ml}(c_1, \cos\theta) \Big) \sin\theta\, \cos m\varphi \right] \boldsymbol{i}_{\theta} \right\} \end{aligned} \tag{5.72}$$

and similar equations for the TE mode.

In the case under consideration the elements of the amplitude matrix can be expressed through the expansion coefficients for the scalar potentials of the scattered field

$$\begin{aligned} A_1 &= -\sum_{l=1}^{\infty} \mathrm{i}^{-l}\, b_l^{(1)}\, S_{1l}(c_1, \cos\theta) + \sum_{m=1}^{\infty} \sum_{l=m}^{\infty} \mathrm{i}^{1-l} \Big(k_1\, a_{ml}^{(1)}\, S_{ml}(c_1, \cos\theta) \\ &\quad + \mathrm{i}\, b_{ml}^{(1)}\, S'_{ml}(c_1, \cos\theta) \Big) \sin\theta\, \cos m\varphi\,, \end{aligned} \tag{5.73}$$

$$\begin{aligned} A_2 &= -\sum_{l=1}^{\infty} \mathrm{i}^{-l}\, b_l^{(1)}\, S_{1l}(c_1, \cos\theta) + \sum_{m=1}^{\infty} \sum_{l=m}^{\infty} \mathrm{i}^{1-l} \Big(k_1\, a_{ml}^{(1)}\, S_{ml}(c_1, \cos\theta) \\ &\quad + \mathrm{i}\, b_{ml}^{(1)}\, S'_{ml}(c_1, \cos\theta) \Big) \sin\theta\, \cos m\varphi\,, \end{aligned} \tag{5.74}$$

$$A_3 = \sum_{m=1}^{\infty} \sum_{l=m}^{\infty} \mathrm{i}^{-l}\, b_{ml}^{(1)}\, \frac{m S_{ml}(c_1, \cos\theta)}{\sin\theta}\, \sin m\varphi\,, \tag{5.75}$$

$$A_4 = \sum_{m=1}^{\infty} \sum_{l=m}^{\infty} \mathrm{i}^{-l}\, b_{ml}^{(1)}\, \frac{m S_{ml}(c_1, \cos\theta)}{\sin\theta}\, \sin m\varphi\,. \tag{5.76}$$

Note that to get A_1 and A_3, one uses the coefficients obtained in the solution to the problem for the TE mode, while to calculate A_2 and A_4, one needs just the solution for the TM mode.

The amplitude matrix allows one to determine all characteristics of the scattered radiation (van de Hulst, 1957; Bohren and Huffman, 1983). For instance, the parameters of the dimensionless intensity of the scattered radiation i_{ij} are derived as follows:

$$i_{11} = |A_1|^2\,, \quad i_{12} = |A_4|^2\,, \quad i_{21} = |A_3|^2\,, \quad i_{22} = |A_2|^2\,, \tag{5.77}$$

where the first and second indices show the polarization of the incident and scattered radiation, respectively, so that the index 1 corresponds to the perpendicular component, while the index 2 to the component parallel to the reference plane.

The integral cross-sections of extinction and scattering for the TE mode are

$$\begin{aligned} C_{\mathrm{ext}} &= \frac{4\pi}{k_1^2}\, \mathrm{Re} \left(\boldsymbol{A}^{\mathrm{sca}} \cdot \boldsymbol{i}^{(0)} \right) \Big|_{\Theta=0} = \\ &= \frac{4\pi}{k_1^2}\, \mathrm{Re} \left[\sum_{l=1}^{\infty} \mathrm{i}^{-l}\, a_l^{(1)}\, S_{1l}\left(c_1, \cos\alpha\right) - \sum_{m=1}^{\infty} \sum_{l=m}^{\infty} \mathrm{i}^{-(l-1)} \left(k_1\, a_{ml}^{(1)}\, S_{ml}\left(c_1, \cos\alpha\right) \right.\right. \\ &\quad \left.\left. + \mathrm{i}\, b_{ml}^{(1)}\, \frac{\mathrm{d} S_{ml}\left(c_1, \cos\alpha\right)}{\mathrm{d}\cos\alpha} \right) \sin\alpha \right], \end{aligned} \tag{5.78}$$

$$\begin{aligned} C_{\mathrm{sca}} &= \frac{1}{k_1^2} \int\!\!\int_{4\pi} |\boldsymbol{A}^{\mathrm{sca}}|^2\, \mathrm{d}\Omega = \\ &= \frac{\pi}{k_1^2} \left\{ \sum_{l=1}^{\infty} 2 \left| a_l^{(1)} \right|^2 + \mathrm{Re} \sum_{m=1}^{\infty} \sum_{l=m}^{\infty} \sum_{n=m}^{\infty} \mathrm{i}^{(n-l)} \left[k_1^2\, a_{ml}^{(1)}\, a_{mn}^{(1)^*}\, \omega_{ln}^m \right.\right. \\ &\quad \left.\left. + \mathrm{i}\, k_1 \left(b_{ml}^{(1)}\, a_{mn}^{(1)^*}\, \kappa_{ln}^m - a_{ml}^{(1)}\, b_{mn}^{(1)^*}\, \kappa_{nl}^m \right) + b_{ml}^{(1)}\, b_{mn}^{(1)^*}\, \tau_{ln}^m \right] N_{mn}^{-1}(c_1)\, N_{ml}^{-1}(c_1) \right\}. \end{aligned} \tag{5.79}$$

Here $\boldsymbol{A}^{\mathrm{sca}}$ is the amplitude of the electric field of the scattered radiation, $\boldsymbol{i}^{(0)}$ the unit vector showing the polarization of the incident radiation, Ω the solid angle, $\cos\Theta = \cos\alpha\cos\theta - \sin\alpha\sin\theta\sin\varphi$ the angle between the directions of the incident and scattered radiation. The integrals of products of the SAFs and their derivatives $\omega_{ln}^m, \kappa_{ln}^m, \tau_{ln}^m$ are given in Appendix A.

The intensity of radiation scattered in the direction $\Theta = \pi$ or $\theta = \pi - \alpha,\ \varphi = \pi$ is determined by the backscattering cross-section

$$\begin{aligned} C_{\rm bk} &= \frac{4\pi}{k_1^2} \mid \boldsymbol{A}^{\rm sca} \mid^2 \Big|_{\Theta=\pi} \\ &= \frac{4\pi}{k_1^2} \Bigg| \sum_{l=1}^{\infty} \mathrm{i}^l\, a_l^{(1)}\, S_{1l}\,(c_1, \cos\alpha) - \sum_{m=1}^{\infty}\sum_{l=m}^{\infty} \mathrm{i}^{(l-1)} \Bigg(k_1\, a_{ml}^{(1)}\, S_{ml}\,(c_1, \cos\alpha) \\ &\quad - \mathrm{i}\, b_{ml}^{(1)}\, \frac{\mathrm{d}S_{1l}\,(c_1, \cos\alpha)}{\mathrm{d}\cos\alpha} \Bigg) \sin\alpha \Bigg|^2 , \end{aligned} \tag{5.80}$$

For the TM mode, one should just change $a_l^{(1)}$ for $b_l^{(1)}$.

Some results of numerical calculations based on the solution described above for homogeneous dielectric and absorbing spheroids have been presented by Voshchinnikov and Farafonov (1993), Voshchinnikov (1996), Farafonov *et al.* (1996), Voshchinnikov *et al.* (2000). For perfectly conducting spheroids, such results can be found in the papers of Farafonov (1984), and Voshchinnikov and Farafonov (1988). Below we discuss convergence of the numerical calculations. One of the tests of numerical calculations is based on the energy conservation law for non-absorbing and perfectly conducting particles. According to this law, the efficiency factors for extinction and scattering must be equal $Q_{\rm ext} = Q_{\rm sca}$. Note that these efficiency factors are normalized by using the cross-sections obtained in the geometrical optics limit $Q = C/C_{\rm GO}$. In Tables 5.1 and 5.2 we illustrate convergence of the results with an increasing number of terms N kept in the field expansions. It is well seen that the accuracy of the results depends only on the linear size of the scatterer divided by the wavelength. There is no dependence on the scatterer shape, i.e. on the aspect ratio a/b and on the particle kind (prolate or oblate).

Of special interest is to compare the solution presented with that of Asano and Yamamoto (1975) from computational point of view. The comparison can be carried out in such a way. Asano and Yamamoto (1975) summed $N = 20$ terms for $c = 5$ to get the agreement for the first five places for $Q_{\rm ext}$ and $Q_{\rm sca}$ values if $a/b = 2$ and $N = 40$ terms to get coincidence for the three places if $a/b = 10$.

Table 5.1. Efficiency factors for scattering $Q_{\rm sca}$ for a homogeneous sphere (the Mie theory) and different spheroids with the refractive index $m = 1.5 + 0.0\mathrm{i}$ and the size parameter $2\pi a/\lambda = 5$ (the parallel incidence of radiation, $\alpha = 0$)

	Sphere	Prolate spheroid		Oblate spheroid	
N		$a/b = 2$	$a/b = 10$	$a/b = 2$	$a/b = 10$
6	3.690000	7.580000	2.970000	2.380000	0.2280000
8	3.893700	7.503100	3.341100	2.351100	0.2453000
10	3.926970	7.508290	3.367370	2.35076	0.2433100
12	3.927816	7.508208	3.366813	2.350734	0.2434590
14	3.927827	7.508209	3.366825	2.350734	0.2434532
16	3.927827	7.508209	3.366824	2.350734	0.2434534
18	3.927827	7.508209	3.366824	2.350734	0.2434534

Table 5.2. Efficiency factors for extinction Q_{ext}, scattering Q_{sca}, and backscattering Q_{bk} for perfectly conducting spheroids with $a/b = 2$ and the size parameter $2\pi a/\lambda = 5$ (the parallel incidence of radiation, $\alpha = 0$)

	Prolate spheroid			Oblate spheroid		
N	Q_{ext}	Q_{sca}	Q_{bk}	Q_{ext}	Q_{sca}	Q_{bk}
8	1.892000	1.894000	0.140000	0.830000	7.780000	35.40000
10	1.895900	1.896000	0.146900	1.960000	2.400000	6.870000
12	1.896230	1.896250	0.147560	2.038000	2.049000	1.390000
14	1.896274	1.896276	0.147643	2.041100	2.041700	0.990000
16	1.896278	1.896279	0.147653	2.041295	2.041339	0.951000
18	1.896279	1.896279	0.147654	2.041310	2.041315	0.947200
20	1.896279	1.896279	0.147659	2.041311	2.041311	0.946820
24	1.896279	1.896279	0.147659	2.041311	2.041311	0.946764
28	1.896279	1.896279	0.147659	2.041311	2.041311	0.946764

Voshchinnikov and Farafonov (1993) obtained the same results taking into account $N = 10$ terms in the first case and only eight terms in the second case. Note that for the solution used computational time (for $\alpha = 0$) weakly depends on scatterer shape and is proportional to $t \propto N^2$, while for the solution of Asano and Yamamoto $t \propto N^3$ (Onaka, 1980). Thus, the solution presented is faster than that of Asano and Yamamoto by about one order of magnitude for $a/b \sim 2$ and by about two orders for $a/b \sim 10$. Obviously, it is not a result of calculation of spheroidal wave functions by using their expansions in terms of the spherical functions which was utilized by Asano and Yamamoto and is not appropriate for large values of the aspect ratio a/b (Flammer, 1957).

A comparison of the suggested solution with solutions obtained by other methods has been done, e.g., by Hovenier *et al.* (1996), Voshchinnikov *et al.* (2000), Il'in *et al.* (2002). The general conclusion was illustrated as Fig. 1 of the last paper. Note that other solutions, e.g. that of Asano and Yamamoto (1975), Barber and Yeh (1975), and Mishchenko *et al.* (1996), demand an essential increase of the term number N and hence of the computational time required to reach reliable results when the aspect ratio grows and meet problems when $a/b \leq 5$–10 (Hovenier *et al.*, 1996). For extremely prolate or oblate spheroids, other solutions require the use of extended precision calculations (Zakharova and Mishchenko, 2000). Thus, the suggested solution to the light scattering problem for spheroids has certain advantages when the scatterers essentially differ in shape from spheres. Note that a new efficient algorithm to compute the prolate radial spheroidal functions for extremely prolate spheroids has been suggested by Voshchinnikov and Farafonov (2003).

5.2.6 Diffraction of the dipole field by a spheroid

Let us assume that the moment of a dipole $\boldsymbol{D}$ is coplanar to the vector $\boldsymbol{i}_z$ directed along the symmetry axis of the spheroid, i.e. this axis and $\boldsymbol{D}$ lie in a plane. Without loss of generality we can assume that the dipole is located in the plane xz, and its moment is directed along the x or z axis. Solution to the problem of the dipole field diffraction on a spheroid is similar to the problem of the plane wave scattering.

We begin with expanding the azimuthal component of the dipole field that is independent of the angle φ. For a vertical magnetic dipole,

$$\boldsymbol{E}^{(0)} = \mathrm{i}k \,\mathrm{rot}\left(D\,\boldsymbol{i}_{\mathrm{z}} \frac{\exp(\mathrm{i}k|\boldsymbol{r}-\boldsymbol{r}_1|)}{|\boldsymbol{r}-\boldsymbol{r}_1|} \right), \tag{5.81}$$

where $\boldsymbol{r}$ and $\boldsymbol{r}_1$ are the radius-vectors to the point of observations and to the dipole position, and we have the axisymmetric component

$$\begin{aligned} E_{1\varphi}^{(0)} &= -\frac{2k^3 D}{c} \frac{(\xi_1^2-1)^{\frac{1}{2}}(1-\eta_1^2)^{\frac{1}{2}}}{\xi_1^2-\eta_1^2} \\ &\times \sum_{l=1}^{\infty} \left[\xi_1 \frac{\partial}{\partial \xi_1} - \eta_1 \frac{\partial}{\partial \eta_1} + \frac{\xi_1^2-\eta_1^2}{(\xi_1^2-1)(1-\eta_1^2)} \right] \\ &\times R_{1l}^{(1)}(c,\xi_<)\, R_{1l}^{(3)}(c,\xi_>)\, S_{1l}(c,\eta_1)\, S_{1l}(c,\eta)\, N_{1l}^{-2}(c_1)\,, \end{aligned} \tag{5.82}$$

where $\xi_< = \min(\xi,\xi_1)$, $\xi_> = \max(\xi,\xi_1)$. This result can be obtained by using the following equations:

$$\begin{aligned} E_{1\varphi}^{(0)} &= \frac{1}{2\pi}\int_0^{2\pi} (\boldsymbol{E}_0,\, \boldsymbol{i}_\varphi)\,\mathrm{d}\varphi = \frac{\mathrm{i}kD}{2\pi}\int_0^{2\pi}\left(\boldsymbol{i}_{\mathrm{x}} \frac{\partial \tilde{G}}{\partial y} - \boldsymbol{i}_{\mathrm{y}} \frac{\partial \tilde{G}}{\partial x},\, \boldsymbol{i}_\varphi \right)\mathrm{d}\varphi \\ &= -\frac{\mathrm{i}kD}{2\pi}\int_0^{2\pi}\left(\boldsymbol{i}_{\mathrm{x}} \frac{\partial \tilde{G}}{\partial y_1} - \boldsymbol{i}_{\mathrm{y}} \frac{\partial \tilde{G}}{\partial x_1},\, \boldsymbol{i}_\varphi \right)\mathrm{d}\varphi \\ &= -\frac{\mathrm{i}kD}{2\pi}\int_0^{2\pi}\left[-\sin\varphi\left(\mathrm{grad}_1\tilde{G},\, \boldsymbol{i}_{\mathrm{y}_1}\right) - \cos\varphi\left(\mathrm{grad}_1\tilde{G},\, \boldsymbol{i}_{\mathrm{x}_1}\right)\right]\mathrm{d}\varphi \\ &= -\frac{\mathrm{i}kD}{2\pi}\int_0^{2\pi}\left[-\sin\varphi \frac{\partial \tilde{G}}{\partial \varphi_1} \frac{1}{\frac{d}{2}(\xi_1^2-1)^{\frac{1}{2}}(1-\eta_1^2)^{\frac{1}{2}}} \right. \\ &\quad \left. + \cos\varphi \frac{(\xi_1^2-1)^{\frac{1}{2}}(1-\eta_1^2)^{\frac{1}{2}}}{\frac{d}{2}(\xi_1^2-\eta_1^2)} \left(\eta_1 \frac{\partial G}{\partial \eta_1} - \xi_1 \frac{\partial \tilde{G}}{\partial \xi_1} \right) \right]\mathrm{d}\varphi\,, \end{aligned} \tag{5.83}$$

where we assume that the dipole is located at the point $(\xi_1,\eta_1,0)$ and use the expansion of the Green function (Komarov *et al.*, 1976)

$$\begin{aligned} \tilde{G}(\boldsymbol{r},\boldsymbol{r}_1) &= \frac{\exp(\mathrm{i}k|\boldsymbol{r}-\boldsymbol{r}_1|)}{|\boldsymbol{r}-\boldsymbol{r}_1|} = 2\mathrm{i}k \sum_{m=0}^{\infty}\sum_{l=m}^{\infty} (2-\delta_m^0)\, N_{ml}^{-2}(c) \\ &\times S_{ml}(c_1,\eta_1)\, S_{ml}(c,\eta)\, R_{ml}^{(1)}(c,\xi_<)\, R_{ml}^{(3)}(c,\xi_>)\cos m(\varphi-\varphi_1)\,. \end{aligned} \tag{5.84}$$

For a horizontal dipole,

$$\boldsymbol{E}^{(0)} = \mathrm{i}k\ \mathrm{rot}\left[D\boldsymbol{i}_{\mathrm{x}} \frac{\exp(\mathrm{i}k|\boldsymbol{r}-\boldsymbol{r}_1|)}{|\boldsymbol{r}-\boldsymbol{r}_1|} \right], \tag{5.85}$$

and in a similar way we get

$$\begin{aligned} E_{1\varphi}^{(0)} &= \frac{2k^3 D}{c(\xi_1^2-\eta_1^2)} \sum_{l=1}^{\infty} \left[\eta_1(\xi_1^2-1)\frac{\partial}{\partial \xi_1} + \xi_1(1-\eta_1^2)\frac{\partial}{\partial \eta_1} \right] N_{1l}^{-2}(c) \\ &\times S_{ml}(c,\eta_1)\, S_{ml}(c,\eta)\, R_{ml}^{(1)}(c,\xi_<)\, R_{ml}^{(3)}(c,\xi_>)\cos m(\varphi-\varphi_1)\,. \end{aligned} \tag{5.86}$$

In both cases the azimuthal component of the magnetic field $H_{1\varphi}^{(0)} = 0$.

When one considers the vertical and horizontal electric dipoles, $E_{1\varphi}^{(0)} = 0$ and the corresponding component of the magnetic field is determined as for the magnetic dipoles.

The non-axisymmetric component of the magnetic field is (Flammer, 1957)

$$\boldsymbol{E}_2^{(0)} = \sum_{m=1}^{\infty} \sum_{l=1}^{\infty} \left(a_{ml}^{(0)} \boldsymbol{M}_{ml}^{\mathrm{z}(1)} + b_{ml}^{(0)} \boldsymbol{M}_{ml}^{\mathrm{r}(1)} \right) , \tag{5.87}$$

with the coefficients being equal to:

(a) for a vertical dipole,

$$\begin{aligned} a_{ml}^{(0)} &= -4k^2 D \, N_{ml}^{-2}(c) \, S_{ml}(c, \eta_1) \, , \\ b_{ml}^{(0)} &= 0 \, ; \end{aligned} \tag{5.88}$$

(b) for a horizontal dipole,

$$\begin{aligned} a_{ml}^{(0)} &= -4k^2 \cot \theta_1 \, N_{ml}^{-2}(c) \, S_{ml}(c, \eta_1) \, , \\ b_{ml}^{(0)} &= \frac{4k^2 D}{r_1 \sin \theta_1} \, N_{ml}^{-2}(c) \, S_{ml}(c, \eta_1) \, , \end{aligned} \tag{5.89}$$

where $(r_1, \theta_1, 0)$ are the spherical coordinates of the dipole.

For electric dipoles, the functions $\boldsymbol{M}_{ml}^{\mathrm{z}}$, $\boldsymbol{M}_{ml}^{\mathrm{r}}$ in the expansion (5.87) should be replaced by $\boldsymbol{N}_{ml}^{\mathrm{z}}$, $\boldsymbol{N}_{ml}^{\mathrm{r}}$, which is typical of the approach considered.

Eqs. (5.82)–(5.87) allow one easily to build solution to the problem of diffraction of the dipole field on a spheroid. The ISLAEs relative to the coefficients of the scattered field expansions (5.27) and (5.45) differ from the corresponding ISLAEs for the plane wave scattering by their right-hand parts. For instance, for diffraction of the field of the horizontal magnetic dipole on a perfectly conducting prolate spheroid, we have:

(a) for the axisymmetric problem (see Eqs. (5.59), (5.88))

$$\begin{aligned} a_l^{(1)} =& \frac{-2k^3 D}{c(\xi_1^2 - \eta_1^2)} \frac{R_{1l}^{(1)}(c, \xi_0)}{R_{1l}^{(3)}(c, \xi_0)} \left[\eta_1 (\xi_1^2 - 1) \frac{\partial}{\partial \xi_1} \right. \\ & \left. + \xi_1 (1 - \eta_1^2) \frac{\partial}{\partial \eta_1} \right] N_{1l}^{-2}(c) \, S_{1l}(c, \eta_1) \, R_{1l}^{(3)}(c, \xi_1) \, , \\ b_l^{(1)} =& 0 \, ; \end{aligned} \tag{5.90}$$

(b) for the non-axisymmetric problem (see Eqs. (5.60), (5.89))

$$\begin{aligned} & \left[\xi_0 \, I - \Gamma (I + \xi_0 \, R_1)^{-1} \Gamma R_1 \right] \boldsymbol{Z}_2 \\ & = -\xi_0 \, \Gamma (I + \xi_0 \, R_1)^{-1} (R_1 - R_0) \, \boldsymbol{F}_m^{(1)} \\ & \quad - \left[\xi_0 \, I - \Gamma (I + \xi_0 \, R_1)^{-1} \Gamma R_0 \right] \boldsymbol{F}_m^{(2)} \, , \end{aligned} \tag{5.91}$$

where $\boldsymbol{F}_m^{(i)} = \{ f_{ml}^{(i)} \}_{l=m}^{\infty}$ and $f_{ml}^{(1)} = k \, a_{ml}^{(0)} \, N_{ml}(c) \, R_{ml}^{(1)}(c, \xi_0) \, R_{ml}^{(3)}(c, \xi_1)$, $f_{ml}^{(2)} = c \, b_{mk}^{(0)} \, N_{ml}(c) \, R_{ml}^{(1)}(c, \xi_0) \, R_{ml}^{(3)}(c, \xi_1)$. Note that the right-hand part of the ISLAE (5.91) is similar to the corresponding part of the systems in Section 5.2.3 for both the TE and TM modes (see Eqs. (5.63)–(5.64)).

5.3 Analysis of ISLAEs arisen in the light scattering by spheroids

5.3.1 Estimates of integrals of products of the SAFs

Here we consider some integrals of products of the spheroidal angular functions and their derivatives for large values of one and more indices. To estimate the integrals, we represent the SAFs by their expansions in terms of the Legendre functions of the first kind (see Appendix A for more details). The coefficients of these expansions d_r^{ml} satisfy the recurrence relation (Komarov *et al.*, 1976)

$$A_r\, d_{r+2}^{ml} + (B_r - \lambda_{ml})\, d_r^{ml} + C_r\, d_{r-2}^{ml} = 0\,, \tag{5.92}$$

where

$$\begin{aligned} A_r =& \frac{(r+2m+2)(r+2m+1)}{(2r+2m+5)(2r+2m+3)}\, c^2, \\ B_r =& (m+r)(m+r+1) - 2c^2\, \frac{(r+m)(r+m+1)+m^2-1}{(2r+2m+3)(2r+2m-1)}, \\ C_r =& \frac{r(r-1)}{(2r+2m-1)(2r+2m-3)}\, c^2, \qquad\qquad d_{-2}^{ml},\, d_{-1}^{ml} = 0\,. \end{aligned} \tag{5.93}$$

We estimate the coefficients d_r^{mn}. For $r < n-m$, the recurrence relation (5.92) can be solved from the beginning

$$\frac{d_r^{mn}}{d_{r+2}^{mn}} = \frac{A_r}{\lambda_{mn} - B_r - C_r\, \frac{d_{r-2}^{mn}}{d_r^{mn}}}\,. \tag{5.94}$$

The eigenvalues λ_{mn} satisfy the inequality

$$n(n+1) - c^2 \le \lambda_{mn}(c) \le n(n+1) \tag{5.95}$$

that is obtained from the equations (Komarov *et al.*, 1976)

$$\frac{1}{2c}\frac{\mathrm{d}\lambda_{mn}(c)}{\mathrm{d}c} = -N_{mn}^{-2}(c) \int_{-1}^{1} S_{mn}^2(c,\eta)\,(1-\eta^2)\,\mathrm{d}\eta \tag{5.96}$$

and $\lambda_{mn}(0) = n(n+1)$. Eqs. (5.93) give the following inequalities for the coefficients:

$$A_r \le c^2, \quad C_r \le \frac{c^2}{4}, \quad (r+m)(r+m+1) - c^2 \le B_r \le (r+m)(r+m+1)\,. \tag{5.97}$$

Hereafter we assume that the parameter c is positive, which occurs for media without absorption. Otherwise, one should simply write all inequalities for moduli of the corresponding quantities.

By using the mathematical induction over the index r from the recurrence relation (5.94), one gets the inequality for $r < m-n$

$$\left|\frac{d_r^{mn}}{d_{r+2}^{mn}}\right| \le \frac{2c^2}{n(n+1)-(m+r)(m+r+1)} \tag{5.98}$$

that is valid under the condition

$$n \geq 2c^2 + \frac{1}{2}. \tag{5.99}$$

The inequality (5.98) gives for $r \leq n - m$

$$|d_r^{mn}| \leq \left(\frac{c^2}{2}\right)^{\frac{n-m-r}{2}} \frac{\Gamma(\frac{n+m+r+1}{2})\,|d_{n-m}^{mn}|}{\Gamma(\frac{n-m-r}{2}+1)\,\Gamma(n+\frac{1}{2})}. \tag{5.100}$$

If $r > n - m$, one should use the relation (5.92) and solve it backward

$$\left|\frac{d_r^{mn}}{d_{r-2}^{mn}}\right| = \frac{-C_r}{B_r - \lambda_{mn} + A_r \frac{d_{r+2}^{mn}}{d_r^{mn}}}. \tag{5.101}$$

From this relation keeping in mind the inequalities (5.95)–(5.97) and the asymptotics $d_{r+2}^{mn}/d_r^{mn} \sim c^2/(4r^2)$ for $r \to \infty$ (Komarov *et al.*, 1976), the mathematical induction over the index r gives under the condition (5.99) for $r > n - m$

$$\left|\frac{d_r^{mn}}{d_{r-2}^{mn}}\right| \leq \frac{c^2}{2[(r+m)(r+m+1) - n(n+1)]}. \tag{5.102}$$

Then for $r \geq n - m$, we get

$$|d_r^{mn}| \leq \left(\frac{c^2}{8}\right)^{\frac{r-n+m}{2}} \frac{\Gamma(n+\frac{3}{2})\,|d_{n-m}^{mn}|}{\Gamma(\frac{r-n+m}{2}+1)\,\Gamma(\frac{n+r+m+3}{2})}. \tag{5.103}$$

Without the restriction (5.99) on n, for $r \geq n-m+L$, where L is the minimum even number for which $n + L \geq 2c^2 + \frac{1}{2}$, we have

$$|d_r^{mn}| \leq \left(\frac{c^2}{8}\right)^{\frac{r-(n+L)+m}{2}} \frac{\Gamma(n+L+\frac{3}{2})\,|d_{n-m+L}^{mn}|}{\Gamma(\frac{r-(n+L)+m}{2}+1)\,\Gamma(\frac{n+L+r+m+3}{2})}. \tag{5.104}$$

In the Eq. (5.95) for $\lambda_{m,n+L}$ we should also replace $n(n+1)$ with $(n+L)(n+L+1)$.

Let us consider the integrals $\gamma_{nl}^{(m)}$. For $l \geq n \geq 2c^2 + \frac{1}{2}$, the inequalities (5.97) and (5.102) allow one to rewrite Eqs. (5.92) as follows:

$$\begin{aligned} \left|\gamma_{nl}^{(m)}\right| \leq N_{mn}^{-1}(c)\,N_{ml}^{-1}(c) \Bigg[& \sum_{r=0,1}^{n-m-2}{}' |d_r^{mn}|\,|d_{r+1}^{ml}| \frac{2(r+2m+1)(r+2m)!}{(2r+2m+1)(2r+2m+3)r!} \\ & + \sum_{r=n-m}^{l-m-1}{}' |d_r^{mn}|\,|d_{r+1}^{ml}| \frac{2(r+2m+1)(r+2m)!}{(2r+2m+1)(2r+2m+3)r!} \\ & + \sum_{r=l-m+1}^{\infty}{}' \frac{2r\,|d_r^{mn}|\,|d_{r-1}^{ml}|}{(2r+2m+1)(2r+2m-1)} \frac{(r+2m)!}{r!} \Bigg] \left(1 + \frac{2c^2}{l}\right). \end{aligned} \tag{5.105}$$

We discuss three series appeared in Eq. (5.105). The first series is estimated from the Cauchy–Schwarz inequality and a comparison with the geometric progression with the common ratio 1/2

$$
\begin{aligned}
I_1 = & N_{mn}^{-1}(c)\, N_{ml}^{-1}(c) \sum_{r=0,1}^{n-m-2}{}' \, |d_r^{mn}|\, |d_{r+1}^{ml}| \, \frac{2(r+2m+1)}{(2r+2m+1)(2r+2m+3)} \\
& \times \frac{(r+2m)!}{r!} \leq 2 \left(\frac{c^2}{2}\right)^{\frac{l-n+1}{2}} \frac{\Gamma(\frac{l+n}{2})}{\Gamma(\frac{l-n+3}{2})\, \Gamma(l+\frac{1}{2})} \, .
\end{aligned}
\tag{5.106}
$$

For the normalization factor, we used Eq. (5.295) from Appendix A.

By using Eqs. (5.101) and (5.103), one can rewrite the second series in Eq. (5.105)

$$
\begin{aligned}
I_2 \leq & \frac{|d_{n-m}^{mn}|\, |d_{l-m}^{ml}|}{N_{mn}(c)\, N_{ml}(c)} \sum_{r=n-m}^{l-m-1}{}' \left(\frac{c^2}{2}\right)^{\frac{l-n-1}{2}} \left(\frac{1}{2}\right)^{r-n+m} \frac{\Gamma(n+\frac{3}{2})}{\Gamma(\frac{r-n+m}{2}+1)} \\
& \times \frac{\Gamma(\frac{l+m+r}{2}+1)}{\Gamma(\frac{n+r+m+3}{2})\, \Gamma(\frac{l-m-r}{2}+1)\, \Gamma(l+\frac{1}{2})} \frac{2(r+2m+1)(r+2m)!}{(2r+2m+1)(2r+2m+3)r!} \, .
\end{aligned}
\tag{5.107}
$$

Let us introduce the notation

$$
\begin{aligned}
S_{l,n} = & \left(\frac{c^2}{2}\right)^{\frac{l-n-1}{2}} \sum_{r=n-m}^{l-m-1}{}' \left(\frac{1}{2}\right)^{r-n+m} \frac{\Gamma(n+\frac{3}{2})\, \Gamma(\frac{l+m+r}{2}+1)}{\Gamma(\frac{r-n+m}{2}+1)\, \Gamma(\frac{n+r+m+3}{2})} \\
& \times \frac{1}{\Gamma(\frac{l-m+r}{2}+1)\, \Gamma(l+\frac{1}{2})} \frac{r+2m+1}{2r+2m+3} \frac{2}{2r+2m+1} \frac{(r+2m)!}{r!}
\end{aligned}
\tag{5.108}
$$

and estimate $S_{l+2,n}$ using $S_{l,n}$

$$
S_{l+2,n} \leq \frac{c^2}{2} \frac{S_{l,n}}{(l+\frac{1}{2})(l+\frac{3}{2})} + \frac{1}{2} S_{l+2,n} \, . \tag{5.109}
$$

This estimate is valid under the condition

$$
n \geq 3m \tag{5.110}
$$

and due to the fact that each term of the series (5.108) is not larger than the previous term. From the inequality (5.109), we get

$$
S_{l+2,n} \leq \frac{c^2}{[(l+1)(l+3)]^{\frac{1}{2}}} S_{l,n} \, , \tag{5.111}
$$

and as a result the series (5.107) is estimated as follows:

$$
\begin{aligned}
I_2 \leq & N_{mn}(c)^{-1}\, N_{ml}(c)^{-1}\, |d_{n-m}^{mn}|\, |d_{l-m}^{ml}| \, c^{l-n-1} \left[\frac{\Gamma(n+2)}{\Gamma(l+1)}\right]^{\frac{1}{2}} S_{n+1,n} \\
\leq & \frac{n+m+1}{2n+3} c^{l-n-1} \left[\frac{\Gamma(n+2)}{\Gamma(l+1)}\right]^{\frac{1}{2}} \, .
\end{aligned}
\tag{5.112}
$$

Under the condition (5.110) the third series in Eq. (5.105) decreases faster than the geometric progression with the common ratio 1/2, and the first term of this series is smaller than the last term of the series I_2 by $c^2/2l$ times, therefore

$$I_3 \leq N_{mn}^{-1} N_{ml}^{-1} \sum_{r=l-m+1}^{\infty}{}' \, |d_r^{mn}| \, |d_{r-1}^{ml}| \, \frac{(r+2m)!}{(2r+2m+1)!} \leq \frac{c^2}{l} I_2 \, . \tag{5.113}$$

A comparison of the estimates (5.106), (5.112), and (5.113) shows that the major contribution to the integrals $\gamma_{nl}^{(m)}$ is made by the series I_2

$$\left|\gamma_{nl}^{(m)}\right| \leq \frac{n+m+1}{2n+3} \, c^{l-n-1} \left[\frac{\Gamma(n+2)}{\Gamma(l+1)}\right]^{\frac{1}{2}} \left(1+\frac{2c^2}{l}\right)^2 . \tag{5.114}$$

Similarly, one can derive the estimates of the integrals $\gamma_{nl}^{(m)}$ for $l \geq n+L \geq \max(2l^2+\frac{1}{2}, 3m)$. In this case in Eq. (5.114) n is replaced by $n+L$

$$\left|\gamma_{nl}^{(m)}\right| \leq \frac{n+L+m+1}{2(n+L)+3} \, c^{l-(n+L)-1} \left[\frac{\Gamma(n+L+2)}{\Gamma(l+1)}\right]^{\frac{1}{2}} \left(1+\frac{2c^2}{l}\right)^2 . \tag{5.115}$$

In the case of $l < n$ or $l+L < n$, the indices n and l are interchanged in Eqs. (5.114) and (5.115).

Analogously, one can estimate the other integrals ($c_2 > c_1$, $l > n$)

$$\begin{aligned}
\left|\delta_{nl}^{(m)}\right| &\leq c_2^{l-(n+L)} \left[\frac{\Gamma(n+L+1)}{\Gamma(l+1)}\right]^{\frac{1}{2}} \left(1+\frac{2c_2^2}{l}\right)^2 , \\
\left|\kappa_{nl}^{(m)}\right| &\leq \frac{(n+L+m+1)(n+L+2)}{2(n+L)+3} \, c_2^{l-(n+L)-1} \\
&\quad \times \left[\frac{\Gamma(n+L+2)}{\Gamma(l+1)}\right]^{\frac{1}{2}} \left(1+\frac{2c_2^2}{l}\right)^2 , \\
\left|\sigma_{nl}^{(m)}\right| &\leq \frac{(n+L+2)(n+L+m+1)(n+L+m+2)}{2(n+L)+3][2(n+L)+5]} \\
&\quad \times c_2^{l-(n+L)-2} \left[\frac{\Gamma(n+L+3)}{\Gamma(l+1)}\right]^{\frac{1}{2}} \left(1+\frac{2c_2^2}{l}\right)^2 ,
\end{aligned} \tag{5.116}$$

where L is the same as in Eq. (5.115). For $l < n$, in Eqs. (5.114)–(5.115) n and l are interchanged.

Estimating the integrals $\gamma_{nl}^{(m)}(-ic)$, $\delta_{nl}^{(m)}(-ic)$, $\kappa_{nl}^{(m)}(-ic)$, $\sigma_{nl}^{(m)}(-ic)$ including the oblate SAFs is similar. Under the same restrictions on l and n, the estimates (5.114)–(5.116) are obtained.

For large n, we have the following asymptotics of the integrals:

$$\begin{aligned}
\left|\gamma_{n,n-1}^{(m)}\right| &= \frac{1}{2} + O\left(\frac{1}{n}\right), & \gamma_{n,n-1}^{(m)} &= \frac{1}{2} + O\left(\frac{1}{n}\right), \\
\kappa_{n,n-1}^{(m)} &= -\frac{n}{2}\left[1 + O\left(\frac{1}{n}\right)\right], & \kappa_{n,n+1}^{(m)} &= \frac{n}{2}\left[1 + O\left(\frac{1}{n}\right)\right], \\
\sigma_{n,n-2}^{(m)} &= -\frac{n}{4}\left[1 + O\left(\frac{1}{n}\right)\right], & \sigma_{n,n}^{(m)} &= \frac{3}{4} + O\left(\frac{1}{n}\right), \\
\sigma_{n,n+2}^{(m)} &= \frac{n}{4}\left[1 + O\left(\frac{1}{n}\right)\right], & \delta_{n,n-1}^{(m)} &= 1 + O\left(\frac{1}{n}\right),
\end{aligned} \tag{5.117}$$

which follow from the relation (Komarov *et al.*, 1976)

$$S_{mn}(c,\eta) = P_n^m(\eta)\left[1 + O\left(\frac{1}{n}\right)\right]. \tag{5.118}$$

5.3.2 Asymptotics of the SRFs for large indices n

These asymptotics are built by the method of the standard equations (Komarov *et al.*, 1976). After the substitution

$$U(\xi) = (\xi^2 - 1)^{\frac{1}{2}}\, R_{mn}(c,\xi), \tag{5.119}$$

one gets the equation

$$U'' + \left[c^2 - \frac{\lambda_{mn}}{\xi^2 - 1} + \frac{1 - m^2}{(\xi^2 - 1)^2}\right] U = 0. \tag{5.120}$$

The asymptotics of the eigenvalues λ_{mn} is known (see Eq. (5.95))

$$\lambda_{mn} = n(n+1) + O(1). \tag{5.121}$$

The region of variable values can be divided in two intersecting intervals $D_1 \in [1;\xi_1)$ and $D_2 \in (\xi_2;\infty)$, where $\xi_2 < \xi_1$. In the region D_1 the standard equation is

$$W''(z) + \left[-\frac{n(n+1)}{z^2 - 1} + \frac{1 - m^2}{(z^2 - 1)^2}\right] W(z) = 0. \tag{5.122}$$

Its fundamental system of solutions is

$$W_1(z) = (z^2 - 1)^{\frac{1}{2}}\, P_n^m(z), \qquad W_2(z) = (z^2 - 1)^{\frac{1}{2}}\, Q_n^m(z), \tag{5.123}$$

where $Q_n^m(z)$ are the associated Legendre functions of the second kind.

According to the standard equation method, the solutions to Eq. (5.120) are

$$U(\xi) = \left[\frac{z(\xi)}{z'(\xi)}\right]^{\frac{1}{2}} W(z(\xi)), \tag{5.124}$$

where the function $z(\xi)$ is expanded in terms of the inverse powers of the index n

$$z(\xi) = \sum_{j=0}^{\infty} z_j(\xi)\, n^{-j}\,. \tag{5.125}$$

To determine the functions $z_j(\xi)$, we substitute Eq. (5.124) into Eq. (5.120), use Eq. (5.122) and the expansion (5.125), and consider the expressions at the same powers of n. The initial conditions are derived from the requirement of transition of the singularities of Eq. (5.120) into the singularities of Eq. (5.122).

In the first approximation, we get

$$\frac{z_0'(\xi)}{z_0^2(\xi)-1} = \frac{1}{\xi^2-1}\,, \qquad z_0(1) = 1\,, \tag{5.126}$$

and hence $z_0(\xi) = \xi$. Thus, from the behaviour of the spheroidal radial functions at $\xi = 1$ and from Eqs. (5.119) and (5.123), we have for $\xi \in D_1$

$$\begin{aligned} R_{mn}^{(1)}(c,\xi) &= C_1\, P_n^m(\xi)\left[1 + O\left(\frac{1}{n}\right)\right], \\ R_{mn}^{(2)}(c,\xi) &= C_2\, Q_n^m(\xi)\left[1 + O\left(\frac{1}{n}\right)\right] + C_3\, P_n^m(\xi)\left[1 + O\left(\frac{1}{n}\right)\right]. \end{aligned} \tag{5.127}$$

For the second interval D_2, the standard equation is

$$W''(z) + \left[c^2 - \frac{n(n+1)}{z^2}\right] W(z) = 0\,, \tag{5.128}$$

and the fundamental system of its solutions is

$$W_1(z) = (cz)^{\frac{1}{2}}\, J_{n+\frac{1}{2}}(cz)\,, \qquad W_2(z) = (cz)^{\frac{1}{2}}\, N_{n+\frac{1}{2}}(cz)\,, \tag{5.129}$$

where $N_{n+\frac{1}{2}}(cz)$ is the Bessel function of the second kind.

The approach used above gives here

$$\left[\frac{z_0'(\xi)}{z_0(\xi)}\right]^2 = \frac{1}{\xi_0^2-1}\,, \qquad \frac{z_0(\xi)}{\xi} \xrightarrow[\xi\to\infty]{} 1\,, \tag{5.130}$$

and hence

$$z_0(\xi) = \frac{1}{2}\left[\xi + (\xi^2-1)^{\frac{1}{2}}\right]. \tag{5.131}$$

Keeping in mind the asymptotics of the prolate SRFs at infinity (Komarov *et al.*, 1976), we get for $\xi \in D_2$

$$\begin{aligned} R_{mn}^{(1)}(c,\xi) &= \left[\frac{1}{2}\left(\xi(\xi^2-1)^{-\frac{1}{2}}+1\right)\right]^{\frac{1}{2}} j_n\left[\frac{c}{2}\left(\xi+(\xi^2-1)^{\frac{1}{2}}\right)\right]\left[1+O\left(\frac{1}{n}\right)\right], \\ R_{mn}^{(2)}(c,\xi) &= \left[\frac{1}{2}\left(\xi(\xi^2-1)^{-\frac{1}{2}}+1\right)\right]^{\frac{1}{2}} n_n\left[\frac{c}{2}\left(\xi+(\xi^2-1)^{\frac{1}{2}}\right)\right]\left[1+O\left(\frac{1}{n}\right)\right], \end{aligned} \tag{5.132}$$

where $j_n(z)$ and $n_n(z)$ are the spherical Bessel functions of the first and second kinds.

The asymptotic representations (5.127) and (5.132) are simultaneously valid for the intervals (ξ_2, ξ_1) and $D_1 \cap D_2$, and hence they should coincide there. The asymptotics of the functions $P_n^m(\xi)$ and $Q_n^m(\xi)$ for $n \to \infty$ and $\xi \in D_1 \cap D_2$ are as follows (Morse and Feshbach, 1953):

$$\begin{aligned} P_n^m(\xi) &= \frac{n!}{(n-m)!}\,(2\pi n)^{-\frac{1}{2}}\,(\xi^2-1)^{-\frac{1}{4}}\left(\xi+(\xi^2-1)^{\frac{1}{2}}\right)^{n+\frac{1}{2}}\left[1+O\left(\frac{1}{n}\right)\right], \\ Q_n^m(\xi) &= \frac{(-1)^m n!}{(n-m)!}\left(\frac{\pi}{2n}\right)^{\frac{1}{2}}(\xi^2-1)^{-\frac{1}{4}}\left(\xi+(\xi^2-1)^{\frac{1}{2}}\right)^{-(n+\frac{1}{2})}\left[1+O\left(\frac{1}{n}\right)\right]. \end{aligned} \tag{5.133}$$

On the other hand, from the asymptotics of the spherical Bessel functions for a large n (Morse and Feshbach, 1953), we get for $\xi \in D_1 \cap D_2$

$$\begin{aligned} R_{nm}^{(1)}(c,\xi) &= \frac{n!}{(2n+1)!}\frac{c^n}{\sqrt{2}}\,(\xi^2-1)^{-\frac{1}{4}}\left(\xi+(\xi_2-1)^{\frac{1}{2}}\right)^{n+\frac{1}{2}}\left[1+O\left(\frac{1}{n}\right)\right], \\ R_{nm}^{(2)}(c,\xi) &= -\frac{2n!}{n!}\frac{\sqrt{2}}{c^{n+1}}\,(\xi^2-1)^{-\frac{1}{4}}\left(\xi+(\xi^2-1)^{\frac{1}{2}}\right)^{-(n+\frac{1}{2})}\left[1+O\left(\frac{1}{n}\right)\right], \end{aligned} \tag{5.134}$$

A comparison of Eqs. (5.133) and (5.134) gives the unknown coefficients of the expansion (5.127)

$$C_1 = \frac{(\pi n)^{\frac{1}{2}}\,c^n\,(n-m)!}{(2n+1)!}, \quad C_2 = 2\,(-1)^{m+1}\,\frac{n^{\frac{1}{2}}\,(n-m)!\,2n!}{\pi^{\frac{1}{2}}\,c^{n+1}\,(n!)^2}, \quad C_3 = 0\,. \tag{5.135}$$

Note that, when $n \to \infty$, C_1^{-1} and C_2^{-1} determine the asymptotics of the coefficients $k_{mn}^{(1)}$ and $k_{mn}^{(2)}$ that relate the SAFs and SRFs. The asymptotic expansions of the spherical functions and eigenvalues in the cases $n \to \infty$ and $c \to 0$ coincide in the first approximation (Komarov *et al.*, 1976).

The oblate SRFs can be treated similarly. To obtain the final result, one should make the standard substitution in the equations given above

$$\begin{aligned} R_{mn}^{(1)}(-\mathrm{i}c, \mathrm{i}\xi) &= C_1\,P_n^m(\mathrm{i}\xi)\left[1+O\left(\frac{1}{n}\right)\right], \\ R_{mn}^{(2)}(-\mathrm{i}c, \mathrm{i}\xi) &= C_2\,Q_n^m(\mathrm{i}\xi)\left[1+O\left(\frac{1}{n}\right)\right], \end{aligned} \tag{5.136}$$

for $\xi \in D_1$, and

$$\begin{aligned} R_{nm}^{(1)}(-\mathrm{i}c, \mathrm{i}\xi) &= \left[\frac{1}{2}\left(\xi(\xi^2+1)^{-\frac{1}{2}}+1\right)\right]^{\frac{1}{2}} j_n\left[c_2\left(\xi+(\xi^2+1)^{\frac{1}{2}}\right)\right]\left[1+O\left(\frac{1}{n}\right)\right], \\ R_{nm}^{(2)}(-\mathrm{i}c, \mathrm{i}\xi) &= \left[\frac{1}{2}\left(\xi(\xi^2+1)^{-\frac{1}{2}}+1\right)\right]^{\frac{1}{2}} n_n\left[c_2\left(\xi+(\xi^2+1)^{\frac{1}{2}}\right)\right]\left[1+O\left(\frac{1}{n}\right)\right], \end{aligned} \tag{5.137}$$

for $\xi \in D_2$, respectively.

For the interval $D_1 \cap D_2 = (\xi_2, \xi_1)$, the asymptotic representations are

$$R^{(1)}_{mn}(-\mathrm{i}c, \mathrm{i}\xi) = \frac{n!}{(2n+1)!} \frac{c^n}{\sqrt{2}} (\xi^2+1)^{-\frac{1}{4}} \left(\xi + (\xi^2+1)^{\frac{1}{2}}\right)^{(n+\frac{1}{2})} \left[1 + O\left(\frac{1}{n}\right)\right],$$
$$R^{(2)}_{mn}(-\mathrm{i}c, \mathrm{i}\xi) = -\frac{2n!}{n!} \frac{\sqrt{2}}{c^{n+1}} (\xi^2+1)^{-\frac{1}{4}} \left(\xi + (\xi^2+1)^{\frac{1}{2}}\right)^{(n+\frac{1}{2})} \left[1 + O\left(\frac{1}{n}\right)\right]. \tag{5.138}$$

The coefficients in Eqs. (5.136) are

$$C_1 = \frac{(\pi n)^{\frac{1}{2}}\, c^n\, (n-m)!}{(2n+1)!}, \qquad C_2 = 2\,(-1)^{m+1} \frac{n^{\frac{1}{2}}\, (n-m)!\, 2n!}{\pi^{\frac{1}{2}}\, c^{n+1}\, (n!)^2}. \tag{5.139}$$

It should be emphasized that in this case the interval $[-\mathrm{i}, \mathrm{i}]$ includes the point $\xi = 0$, which requires consideration of two integrals for the radial variable intervals $D_1 = [0, \xi_1)$ and $D_2 = [\xi_2, \infty)$.

5.3.3 Properties of quasi-regular systems

The problem of the electromagnetic wave diffraction on a spheroid is here reduced to solution of an ISLAE. As analytical studies of the properties of ISLAEs are seldom in the scientific literature, we discuss both the well known and new further required properties of the ISLAEs.

The infinite system

$$x_n = \sum_{i=1}^{\infty} c_{ni}\, x_i + q_n, \qquad n = 1, 2, \dots, \tag{5.140}$$

is called regular provided

$$\sum_{i=1}^{\infty} |c_{ni}| < 1, \qquad n = 1, 2, \dots, \tag{5.141}$$

and

$$|q_n| < K \left(1 - \sum_{i=1}^{\infty} |c_{ni}|\right), \qquad n = 1, 2, \dots, \tag{5.142}$$

where K is a positive number.

When the sums (5.142) differ from the unity by a positive number

$$\sum_{i=1}^{\infty} |c_{ni}| \le p < 1, \qquad n = 1, 2, \dots, \tag{5.143}$$

the condition (5.141) transforms into a restriction for the free terms

$$|q_n| < K, \qquad n = 1, 2, \dots. \tag{5.144}$$

The system (5.140) whose coefficients satisfy the conditions (5.143) and (5.144) is called completely regular.

Theorem 1. *A regular system (5.140) has a restricted solution that can be found by the reduction method. A completely regular system has the only restricted solution that can be found by the reduction method.*

The first part of the theorem was proved by Kantorovich and Krylov (1964).

Ability to find a solution to the infinite system by the reduction method means the

$$X_n^N \xrightarrow[N\to\infty]{} X_n\,, \qquad n = 1, 2, \dots\,, \tag{5.145}$$

where X_n^N is the solution to the system of the size $N \times N$ that is obtained from the initial system by reduction. We use the term completely quasi-regular system, when the conditions (5.141) and (5.142) are satisfied starting with some row

$$\begin{aligned}
&\sum_{i=1}^{\infty} |c_{ni}| < \infty, && n = 1, 2, \dots, N\,,\\
&\sum_{i=1}^{\infty} |c_{ni}| < 1, && n = (N+1), (N+2), \dots\,,\\
&|q_n| \le K\left(1 - \sum_{i=1}^{\infty} |c_{ni}|\right).
\end{aligned} \tag{5.146}$$

Hereafter we apply the condition (5.146) in a stronger form

$$\begin{aligned}
&\sum_{i=1}^{\infty} |\,c_{ni}| < \infty, && n = 1, 2, \dots, N\,,\\
&\sum_{i=1}^{\infty} |\,c_{ni}| \le p < 1, && n = (N+1), (N+2), \dots\,,\\
&|q_n| < K\,.
\end{aligned} \tag{5.147}$$

The last condition means that the free terms are restricted. The infinite system (5.140) under the conditions (5.147) can be called completely quasi-regular one.

Theorem 2. *Either a completely quasi-regular system has the only solution that can be found by the reduction method, or the corresponding homogeneous system has non-trivial restricted solutions.*

In the book of Kantorovich and Krylov (1964) this theorem was not formulated explicitly, but it can be easily proved.

Let us rewrite a completely quasi-regular system in the form

$$\begin{cases}
x_n = \displaystyle\sum_{i=N+1}^{\infty} c_{ni}\,x_i + \left(q_n + \sum_{l=1}^{N} c_{nl}\,x_l\right), & n = N+1, \dots\,,\\[2ex]
x_n = \displaystyle\sum_{l=1}^{N} c_{nl}\,x_l + \left(q_n + \sum_{i=N+1}^{\infty} c_{ni}\,x_i\right), & n = 1, \dots, N\,.
\end{cases} \tag{5.148}$$

A solution to the first system that is completely regular relative to $x_{N+1}, x_{N+2}, \ldots$ can be represented as

$$x_n = \xi_n + \sum_{l=1}^{N} d_{nl}\, x_l, \qquad n = (N+1), \ldots , \tag{5.149}$$

where ξ_n is the solution to the system where the free terms are just q_n, and d_{nl} is the solution to the same system with the free terms c_{nl}. Here $|\xi_n| < K$, $|d_{nl}| < 1$, which follows from the general properties of the completely regular system (its solution is restricted by a constant limiting the free terms – Kantorovich and Krylov, 1964). The starting N unknowns are derived from the second system (5.148) after the substitution (5.149)

$$x_n = \sum_{l=1}^{N} c_{nl}\, x_l + \sum_{l=1}^{N} \left(\sum_{i=N+1}^{\infty} c_{ni}\, d_{il} \right) x_l + \left(q_n + \sum_{i=N+1}^{\infty} c_{ni}\, \xi_i \right), \tag{5.150}$$

where $n = 1, 2, ..., N$. The system (5.150) is finite, and hence either it has the only solution or the corresponding homogeneous system has non-trivial solutions. Therefore, either the infinite system has the only solution, or the corresponding homogeneous system has non-trivial solutions.

Let us demonstrate that the only restricted solution to a completely quasi-regular system can be found by the reduction method. For a finite system of the size $R \times R$ $(R > N)$ formed by truncation of the initial system, we have

$$\begin{cases} x_n^{(R)} = \displaystyle\sum_{i=N+1}^{R} c_{ni}\, x_i^{(R)} + \left(q_n + \sum_{l=1}^{N} c_{nl}\, x_l^{(R)} \right), & n = N+1, \ldots , R, \\ x_n^{(R)} = \displaystyle\sum_{l=1}^{N} c_{nl}\, x_l^{(R)} + \left(q_n + \sum_{i=N+1}^{\infty} c_{ni}\, x_i^{(R)} \right), & n = N+1, \ldots , N. \end{cases} \tag{5.151}$$

The first system of Eq. (5.148) can be solved by the reduction method, and hence solution to the first system of Eq. (5.151) can be represented as follows:

$$x_n^{(R)} = \xi_n^{(R)} + \sum_{l=1}^{N} d_{nl}^{(R)}\, x_l^{(R)}, \qquad n = (N+1), \ldots , R, \tag{5.152}$$

where

$$\xi_n^{(R)} \xrightarrow[R\to\infty]{} \xi_n, \qquad d_{nl}^{(R)} \xrightarrow[R\to\infty]{} d_{nl}. \tag{5.153}$$

The unknowns $x_1^{(R)}, \ldots , x_N^{(R)}$ are determined from the second system of Eqs. (5.151)

$$x_n^{(R)} = \sum_{l=1}^{N} c_{nl}\, x_l^{(R)} + \sum_{l=1}^{N} \left(\sum_{i=N+1}^{R} c_{ni}\, d_{il}^{(R)} \right) x_l^{(R)} + \left(q_n + \sum_{i=N+1}^{R} c_{ni}\, \xi_i^{(R)} \right), \tag{5.154}$$

where $n = 1, 2, \ldots , N$.

Let us compare the systems (5.150) and (5.154). They are finite, and the coefficients of the system (5.154) trend to the coefficients of the system (5.150) when

$R \to \infty$. It follows from Eqs. (5.153) and the fact that the series $\sum_{i=N+1}^{R} c_{ni}\, d_{il}^{(R)}$ and $\sum_{i=N+1}^{R} c_{ni}\, \xi_i^{(R)}$ have the majorants $\sum_{i=N+1}^{\infty} c_{ni} < 1$ and $\sum_{i=N+1}^{\infty} |c_{ni}|\, K < K$, respectively, and hence the transition to limit is possible. So, we have

$$x_n^{(R)} \xrightarrow[R\to\infty]{} x_n, \qquad n = 1, 2, \ldots, N\,. \tag{5.155}$$

From Eqs. (5.152)–(5.153) we get

$$x_n^{(R)} \xrightarrow[R\to\infty]{} x_n, \qquad n = 1, 2, \ldots\,. \tag{5.156}$$

Thus, any quasi-regular system that has the only restricted solution can be solved by the reduction method.

Let us prove an additional theorem required to perform analysis of the ISLAEs arisen in the problem of the electromagnetic wave diffraction on a spheroid.

Theorem 3. *Let us consider an infinite system*

$$z_n = \kappa_1\, z_{n+1} + \kappa_2\, z_{n-1} + \left(\sum_{i=1}^{\infty} c_{ni}\, z_i + q_n \right), \tag{5.157}$$

where $n = 1, 2, \ldots, Z_0 = 0$, and $|\kappa_1| + |\kappa_2| = p < 1$. If

$$\sum_{i=1}^{\infty} |c_{ni}| \,<\, M n^{\,(\frac{1}{2}+\varepsilon)}\,, \tag{5.158}$$

where M and ε are some positive numbers independent of n, and

$$\sum_{i=1}^{\infty} |q_n|^2 \,<\, \infty\,, \tag{5.159}$$

then either the system (5.157) has the only solutions in the space l_2, or the corresponding homogeneous system has non-trivial solutions in this space. The solution to the system (5.157) can be found by the reduction method.

Under the conditions (5.158) and (5.159), the system (5.157) is completely quasi-regular, and hence either it has the only restricted solution, or the corresponding homogeneous system has non-trivial solutions (see Theorem 2). Let us prove that the restricted solution $z = \{z_n\}$ to the system (5.157) belongs to the space l_2 (hereafter $||\ \ ||_{l_2}$ means the norm in the space l_2), i.e.

$$||z||_{l_2}^2 = \sum_{n=1}^{\infty} |z_n|^2 < \infty\,. \tag{5.160}$$

Let us introduce the notation $u = \{u_n\} = \{\sum_{i=1}^{\infty} c_{ni}\, z_i + q_n\}$. From Eqs. (5.158)–(5.159) and the triangle inequality $||a + b|| \le ||a|| + ||b||$, we have

$$||u||_{l_2} \le \left[\sum_{n=1}^{\infty} \left(\sum_{n=1}^{\infty} c_{ni} \right)^2 \right]^{\frac{1}{2}} \sup_n |z_n| + ||q||_{l_2} < \infty\,, \tag{5.161}$$

where $q = \{q_n\}$. So, $u \in l_2$, and from the system (5.157) we get the following estimate:

$$||z||_{l_2} \leq \frac{1}{1 - |\kappa_1| - |\kappa_2|} ||u||_{l_2}, \tag{5.162}$$

i.e. the restricted solution belongs to the space l_2. The inequalities (5.161) and (5.162) for $q_n = 0$ show that the restricted non-trivial solution to the homogeneous system also belongs to the space l_2.

5.3.4 Analysis of the infinite systems for perfectly conducting spheroids

Let us consider ISLAEs arisen in solution of the problem of the electromagnetic wave diffraction on a perfectly conducting spheroid. For the TE mode, the infinite systems are (the vector $\boldsymbol{Z}_2$ is not excluded)

$$\begin{cases} \Gamma\, \boldsymbol{Z}_1 + \xi_0\, \boldsymbol{Z}_2 = -\Gamma\, \boldsymbol{F}_m\,, \\ (I + \xi_0\, R_1)\, \boldsymbol{Z}_1 + f\, \Gamma\, R_1\, \boldsymbol{Z}_2 = -(I + \xi_0\, R_0)\, \boldsymbol{F}_m\,. \end{cases} \tag{5.163}$$

Eqs. (5.132)–(5.135) allow one to represent the elements of the diagonal matrices $R_{0,1}$ by

$$r_{mn}^{(1)} = \frac{R_{mn}^{(1)'}(c, \xi_0)}{R_{mn}^{(1)}(c, \xi_0)} = \frac{n}{(\xi_0^2 - 1)^{\frac{1}{2}}} \left[1 + O(n^{-1})\right], \tag{5.164}$$

$$r_{mn}^{(3)} = \frac{R_{mn}^{(3)'}(c, \xi_0)}{R_{mn}^{(3)}(c, \xi_0)} = -\frac{n}{(\xi_0^2 - 1)^{\frac{1}{2}}} \left[1 + O(n^{-1})\right]. \tag{5.165}$$

Using Eqs. (5.114), (5.117), the system (5.163) can be rewritten

$$\begin{cases} y_{2,n}^{(m)} = -\dfrac{1}{2\xi_0} \left(y_{1,n-1}^{(m)} + y_{1,n+1}^{(m)} \right) + \sum\limits_i c_{ni}^1\, y_{1,i}^{(m)} + q_{1,n}^{(m)}\,, \\ y_{1,n}^{(m)} = -\dfrac{1}{2\xi_0} \left(y_{2,n-1}^{(m)} + y_{2,n+1}^{(m)} \right) + \sum\limits_i c_{ni}^2\, y_{2,i}^{(m)} + q_{2,n}^{(m)}\,, \end{cases} \tag{5.166}$$

where we have introduced the notation

$$\begin{aligned} y_{1,2k-1}^{(m)} &= z_{1,2k}^{(m)}, & y_{1,2k}^{(m)} &= z_{2,2k-1}^{(m)}, \\ y_{2,2k-1}^{(m)} &= z_{1,2k-1}^{(m)}, & y_{2,2k}^{(m)} &= z_{2,2k}^{(m)} \end{aligned} \tag{5.167}$$

and kept in mind the fact that the infinite system (5.163) can be divided in two independent systems relative to the variables (5.167) because of the parity of the integrals of the SAFs.

The coefficients of the ISLAEs (5.166) satisfy the conditions of Theorem 3, namely

$$\sum_i |c_{ni}^1| + \sum_i |c_{ni}^2| \leq M\, n^{-1}\,, \tag{5.168}$$

$$\sum_n |q_{1,n}|^2 + \sum_n |q_{2,n}|^2 < \infty\,, \tag{5.169}$$

where the free terms are

$$q_{1,n}^{(m)} = -\sum_j \gamma_{nj}^{(m)} f_j^{(m)}, \qquad q_{2,n}^{(m)} = -\left(1 + \xi_0\, r_{mn}^{(1)}\right) f_n^{(m)}. \tag{5.170}$$

The inequality (5.168) can be easily proved. From estimates (5.114) and (5.117), we have

$$\begin{aligned}\sum_i |c_{ni}^1| \le C \left(\sum_{l=0,1}^{n-3}{}' \left|\gamma_{nl}^{(m)}\right| + \left|\gamma_{n,n-1}^{(m)} - \frac{1}{2}\right| + \left|\gamma_{n,n+1}^{(m)} - \frac{1}{2}\right| + \sum_{l=n+3}^{\infty}{}' \left|\gamma_{nl}^{(m)}\right| \frac{l}{n} \right) \\ \le C\, n^{-1}.\end{aligned} \tag{5.171}$$

The quantities $f_n^{(m)}$ involved in the free terms of the system (5.166) are estimated as follows:

$$\begin{aligned}\left|f_n^{(m)}\right| &\le \frac{4}{\sin\alpha} \left| R_{mn}^{(1)}(c, \xi_0)\, S_{mn}(c, \cos\alpha)\, N_{mn}^{-1}(c) \right| \\ &\le Const \left| j_n \left[\frac{c}{2}\left(\xi_0 + (\xi_0^2 - 1)^{\frac{1}{2}}\right)\right] \frac{1}{\sin\alpha}\, P_n^m(\cos\alpha)\, N_{mn}^{-1}(0) \right| \\ &\le Const\, \frac{(n+m)!}{(2n)!} \left[\frac{c}{2}\left(\xi_0 + (\xi_0^2 - 1)^{\frac{1}{2}}\right)\right]^n,\end{aligned} \tag{5.172}$$

where we use the relations (5.134) and the available estimate of the associated Legendre polynomials (Sinha and McPhie, 1977)

$$\left| \frac{1}{\sin\alpha}\, P_n^m(\cos\alpha) \right| \le Const\, \frac{(n+m)!}{n!}. \tag{5.173}$$

From Eqs. (5.114), (5.117) and (5.172), we get

$$\begin{aligned}\left|q_{1,n}^{(m)}\right| &\le Const\, \frac{(n+m)!}{(2n)!} \left[(n-1)!\right]^{\frac{1}{2}} \left[\frac{c}{2}\left(\xi_0 + (\xi_0^2 - 1)^{\frac{1}{2}}\right)\right]^n, \\ \left|q_{2,n}^{(m)}\right| &\le Const\, \frac{(n+m)!}{(2n)!}\, 2n \left[\frac{c}{2}\left(\xi_0 + (\xi_0^2 - 1)^{\frac{1}{2}}\right)\right]^n,\end{aligned} \tag{5.174}$$

and hence the validity of the conditions (5.169).

From Theorem 3, for a given m, the ISLAEs (5.163) as well as the systems (5.63) and (5.64) have the only solution in the space l_2 that can be found by the reduction method under the condition

$$\xi_0 > 1 \tag{5.175}$$

that is satisfied for any prolate spheroids not degenerated into a segment. Here we utilize the fact that the diffraction problem without sources and the corresponding homogeneous ISLAEs have the trivial solution only.

As squares of $z_{1,n}^{(m)}$, $z_{2,n}^{(m)}$ are summarized, from Eqs. (5.42)–(5.46) and (5.52) it follows that the parts of the solution to the diffraction problem for a plane electromagnetic wave on a perfectly conducting spheroid with a fixed $\cos m\varphi$ converge on any spheroidal surface Ω ($\xi = const$) in space $L_2(\Omega)$ everywhere up to the scatterer boundary ($\xi \geq \xi_0$). The possibility of independent solving the diffraction problem for any of the parts follows from the commutativity of the operators L_z and T (see Section 5.2.2).

For an oblate spheroid, from Eqs. (5.115), (5.134) one can derive the ISLAEs (5.166) that can be solved under the condition (5.175), i.e. under the following geometrical restriction:

$$d > 2b\,, \tag{5.176}$$

i.e. the focal distance should be larger than the minor axis of the spheroid. In this case one can draw the same conclusions as for the prolate spheroids. Note that the condition (5.175) is equivalent to the condition $a/b < \sqrt{2}$ that is required for the expansions of the internal and scattered fields to converge everywhere up to the scatterer boundary.

However, in contrast to the case of the field convergence, the condition (5.175) is not the necessary one for the ISLAEs to be uniquely solvable in the space l_2. After a more accurate investigation of the infinite systems for oblate spheroids the restriction (5.175) and hence (5.176) can be removed.

Let us introduce new variables in the systems (5.166)

$$\begin{aligned} t_{1,n}^{(m)} &= \left[q\left(\xi_0 + (\xi_0^2+1)^{\frac{1}{2}}\right)\right]^n y_{1,n}^{(m)}\,, \\ t_{2,n}^{(m)} &= \left[q\left(\xi_0 + (\xi_0^2+1)^{\frac{1}{2}}\right)\right]^n y_{2,n}^{(m)}\,, \end{aligned} \tag{5.177}$$

where $q > I$ is a number. Then the ISLAEs (5.166) can be rewritten

$$\left\{\begin{aligned} t_{1,n-1}^{(m)} =& \frac{-2\xi_0}{q\left[\xi_0 + (\xi_0^2+1)^{\frac{1}{2}}\right]}\, t_{1,n}^{(m)} - \frac{1}{q^2\left[\xi_0 + (\xi_0^2+1)^{\frac{1}{2}}\right]^2}\, t_{1,n+1}^{(m)} \\ & + \sum_i \tilde{c}_{ni}^{1}\, t_{1,i}^{(m)} + \tilde{q}_{1,n}^{(m)}\,, \\ t_{2,n-1}^{(m)} =& \frac{-2\xi_0}{q\left[\xi_0 + (\xi_0^2+1)^{\frac{1}{2}}\right]}\, t_{2,n}^{(m)} - \frac{1}{q^2\left[\xi_0 + (\xi_0^2+1)^{\frac{1}{2}}\right]^2}\, t_{2,n+1}^{(m)} \\ & + \sum_i \tilde{c}_{ni}^{2}\, t_{2,i}^{(m)} + \tilde{q}_{2,n}^{(m)}\,, \end{aligned}\right. \tag{5.178}$$

where

$$\begin{aligned} \tilde{c}_{ni}^{1,2} =& c_{ni}^{1,2}\, \frac{2\xi_0}{\left[q\left(\xi_0 + (\xi_0^2+1)^{\frac{1}{2}}\right)\right]^{n-1}}\,, \\ \tilde{q}_{kn}^{1,2} =& q_{kn}^{1,2}\, \frac{2\xi_0}{\left[q\left(\xi_0 + (\xi_0^2+1)^{\frac{1}{2}}\right)\right]^{n-1}}\,. \end{aligned} \tag{5.179}$$

Under the condition $q > 1$, we have

$$\frac{2\xi_0}{q\left[\xi_0 + (\xi_0^2+1)^{\frac{1}{2}}\right]} + \frac{1}{q^2\left[\xi_0 + (\xi_0^2+1)^{\frac{1}{2}}\right]^2} \leq \frac{1}{q} < 1\,, \qquad (5.180)$$

and the infinite systems (5.178) are completely quasi-regular. For the coefficients of these systems, the inequalities (5.168)–(5.169) are valid, and Theorem 3 can be applied as it follows from Eqs. (5.115), (5.117) and (5.172). Thus, for any perfectly conducting oblate spheroids, not degenerated into a disk ($\xi_0 = 0$), the same conclusions are valid as for the prolate spheroids. As the transition to a perfectly conducting disk is not possible within the solution suggested (see Section 5.2.3), in numerical calculations one should expect worse convergence for more flattened spheroids, which is really observed.

5.3.5 Analysis of ISLAEs arisen for dielectric spheroids

The axisymmetric component of the scattered field for perfectly conducting spheroids can be found analytically, while for dielectric spheroids one should solve ISLAEs relative to the coefficients of the corresponding expansions. For the TE wave, we have

$$(\Delta R_1 - R_2\Delta)\,\boldsymbol{Z}_0 + (\Delta R_0 - R_2\Delta)\,\boldsymbol{F}_0 = 0\,. \qquad (5.181)$$

From Eqs. (5.115), (5.117), (5.165) and (5.172), we get

$$z_{0,n} = \sum_i c_{ni}\, z_{0,i} + q_n\,, \qquad (5.182)$$

where the coefficients satisfy the conditions

$$\sum_i |c_{ni}| \leq M n^{-1}\,, \qquad \sum_n |q_n|^2 < \infty\,. \qquad (5.183)$$

For the TM wave, the ISLAEs are

$$\begin{aligned}&\left[\xi_0(\varepsilon_2-\varepsilon_1)\Delta + (\xi_0^2-1)(\varepsilon_2\Delta R_1 - \varepsilon_1 R_2\Delta)\right]\boldsymbol{Z}_0\\ &\quad + \left[\xi_0(\xi_2-\xi_1)\Delta + (\xi_0^2-1)(\varepsilon_2\Delta R_0 - \varepsilon_1 R_2\Delta)\right]\boldsymbol{F}_0 = 0\,,\end{aligned} \qquad (5.184)$$

and keeping in mind the behaviour of the coefficient they can be written in the form (5.182) with the condition (5.183).

For the oblate spheroids, we obtain similar results. So, the conclusions made above about the axisymmetric component of the scattered field in the case of perfectly conducting spheroids are valid for dielectric spheroids as well.

For the non-axisymmetric component of the scattered field, the estimates derived above allow one to write the ISLAEs for the TE mode as follows (see Eq. (5.55)):

$$\begin{cases} z_{2,n}^{(m)} + \frac{1}{2\xi_0}\left(z_{1,n-1}^{(m)} + z_{1,n+1}^{(m)}\right) = \\ \qquad x_{2,n}^{(m)} + \frac{1}{2\xi_0}\left(x_{1,n-1}^{(m)} + x_{1,n+1}^{(m)}\right) + \sum_i d_{ni}^1\, z_i + q_n^1\,, \\ z_{1,n}^{(m)} + \frac{1}{2\xi_0}\left(z_{2,n-1}^{(m)} + z_{2,n+1}^{(m)}\right) = \\ \qquad -x_{1,n}^{(m)} - \frac{1}{2\xi_0}\left(x_{2,n-1}^{(m)} + x_{2,n+1}^{(m)}\right) + \sum_i d_{ni}^2\, z_i + q_n^2\,, \\ \varepsilon_1\left[z_{1,n}^{(m)} + \frac{1}{2\xi_0}\left(z_{2,n-1}^{(m)} + z_{2,n+1}^{(m)}\right)\right] = \\ \qquad \varepsilon_2\left[x_{1,n}^{(m)} + \frac{1}{2\xi_0}\left(x_{2,n-1}^{(m)} + x_{2,n+1}^{(m)}\right)\right] + \sum_i d_{ni}^3\, z_i + q_n^3\,, \\ z_{2,n}^{(m)} + \frac{1}{2\xi_0}\left(z_{1,n-1}^{(m)} + z_{1,n+1}^{(m)}\right) = \\ \qquad -x_{2,n}^{(m)} - \frac{1}{2\xi_0}\left(x_{1,n-1}^{(m)} + x_{1,n+1}^{(m)}\right) + \frac{1}{2}\left(1 - \frac{c_1^2}{c_2^2}\right)\frac{1}{(\xi_0^2-1)^{\frac{1}{2}}} \\ \qquad \times\left[z_{1,n+1}^{(m)} - z_{1,n-1}^{(m)} + \frac{1}{2\xi_0}\left(z_{2,n+2}^{(m)} - z_{2,n-2}^{(m)}\right)\right] + \sum_i d_{ni}^4\, z_i + q_n^4\,, \end{cases} \tag{5.185}$$

where $\{z_i\}$ and $\{x_i\}$ are the sets of the unknowns related with the scattered and internal fields, respectively. The coefficients of the system (5.185) satisfy the conditions

$$\sum_i |d_{ni}^k| \le Mn^{-1}\,, \qquad \sum_n |q_n^k|^2 < \infty\,. \tag{5.186}$$

Multiplying the second equation of the system (5.185) by $-\varepsilon_1$ and ε_2 and adding of the results with the third equation give

$$\begin{cases} x_{1,n}^{(m)} + \frac{1}{2\xi_0}\left(x_{2,n-1}^{(m)} + x_{2,n+1}^{(m)}\right) = \frac{-1}{\varepsilon_1+\varepsilon_2}\left[\sum_i \left(d_{ni}^3 - \varepsilon_1 d_{ni}^2\right) z_i + q_n^3 - \varepsilon_1 q_n^2\right], \\ z_{1,n}^{(m)} + \frac{1}{2\xi_0}\left(z_{2,n-1}^{(m)} + z_{2,n+1}^{(m)}\right) = \frac{1}{\varepsilon_1+\varepsilon_2}\left[\sum_i \left(d_{ni}^3 + \varepsilon_2 d_{ni}^2\right) z_i + q_n^3 + \varepsilon_2 q_n^2\right]. \end{cases} \tag{5.187}$$

Similarly, from the first and fourth equations, we get

$$\left\{\begin{aligned} & x_{2,n}^{(m)} + \frac{1}{2\xi_0}\left(x_{1,n-1}^{(m)} + x_{1,n+1}^{(m)}\right) = \frac{1}{4}\left(1 - \frac{c_1^2}{c_2^2}\right)\frac{1}{(\xi_0^2-1)^{\frac{1}{2}}}\left[z_{1,n+1}^{(m)} - z_{1,n-1}^{(m)}\right. \\ & \qquad \left. + \frac{1}{2\xi_0}\left(z_{2,n-2}^{(m)} - z_{2,n+2}^{(m)}\right)\right] + \frac{1}{2}\left[\sum_i \left(d_{ni}^4 - d_{ni}^3\right) z_i + q_n^4 - q_n^3\right], \\ & z_{2,n}^{(m)} + \frac{1}{2\xi_0}\left(z_{1,n-1}^{(m)} + z_{1,n+1}^{(m)}\right) = \frac{1}{4}\left(1 - \frac{c_1^2}{c_2^2}\right)\frac{1}{(\xi_0^2-1)^{\frac{1}{2}}}\left[z_{1,n+1}^{(m)} - z_{1,n-1}^{(m)}\right. \\ & \qquad \left. + \frac{1}{2\xi_0}\left(z_{2,n-2}^{(m)} - z_{2,n+2}^{(m)}\right)\right] + \frac{1}{2}\left[\sum_i \left(d_{ni}^4 + d_{ni}^3\right) z_i + q_n^4 + q_n^3\right]. \end{aligned}\right. \tag{5.188}$$

From Eqs. (5.187), we have

$$\begin{aligned} z_{1,n+1}^{(m)} - z_{1,n-1}^{(m)} + \frac{1}{2\xi_0}\left(z_{2,n+2}^{(m)} - z_{2,n-2}^{(m)}\right) = \frac{1}{\varepsilon_1+\varepsilon_2}\Bigg[\sum_i \Big(d_{n+1i}^3 + \varepsilon_2 d_{n+1i}^2 \\ - d_{n-1i}^3 - \varepsilon_2\, d_{n-1i}^2\Big) z_i + q_{n+1}^3 + \varepsilon_2\, q_{n+1}^2 - q_{n-1}^3 - \varepsilon_2\, q_{n-1}^2\Bigg], \end{aligned} \tag{5.189}$$

i.e. Eqs. (5.188) actually have the same form as Eqs. (5.187). A similar result is obtained for the TM wave as well as in the case of an oblate spheroid. Thus, the infinite systems (5.187)–(5.188) for dielectric spheroids are analogous to the systems (5.166) for perfectly conducting spheroids, and the same conclusions can be drawn in both cases. Namely, the ISLAEs for any prolate or oblate spheroids not generating into a disk or a segment are completely quasi-regular and have the only solution (in the space l_2) that can be found by the reduction method. The solution to the diffraction problem for a fixed m, i.e. with the factor $\cos m\varphi$, converges on any surface $\Omega(\xi = const)$ in the space $L_2(\Omega)$ up to the surface of the scattering spheroid. Our investigation of the cases of extremely prolate and oblate dielectric spheroids and extremely prolate perfectly conducting spheroids shows that the systems arisen can be rather simply solved in the first approximation with the small parameter being the ratio of the minor to major axis, and the solution obtained gives reasonable results in a wide region of parameter values. For a perfectly conducting disk the solution should be improved (see Section 5.5). After that the solution becomes physically correct, and the Meixner conditions at the disk edge are satisfied.

5.4 Light scattering problem for extremely prolate and oblate spheroids

It is known that in the case of the axisymmetric excitation of a perfectly conducting spheroid one can find asymptotics of the scattered field for a small parameter b/a by utilizing the separation of variables method, the Abraham potentials, and some properties of the spheroidal functions. There arises the condition of the linear antenna excitation (Stratton, 1941)

$$d = n\,\frac{\lambda}{2}, \tag{5.190}$$

i.e. the length of the oscillator is equal to an integer number of half-wavelengths.

Below we demonstrate that the solution suggested is very efficient for extremely prolate dielectric and perfectly conducting spheroids as well as extremely oblate dielectric spheroids. Moreover, in these cases the solution allows one to derive the principal term of the asymptotics for the small parameter equal to the ratio of the minor to major semiaxis of the spheroid.

5.4.1 Extremely prolate spheroids

For such particles, the small parameter is related with the value of the radial spheroidal coordinate ξ_0 such that $(b/a)^2 = (\xi_0^2 - 1)/\xi_0^2$, and hence the principal term of the asymptotic expansions in $(\xi_0^2 - 1)$ and $(b/a)^2$ is the same.

The asymptotics of the prolate SRFs of the first kind for $\xi_0 \to 1$ in the region $c = O(1)$ is as follows:

$$R_{ml}^{(1)}(c, \xi_0) = C_{ml}\, P_l^m(\xi_0)\, \left[1 + O(\xi_0^2 - 1)\right] . \tag{5.191}$$

To construct the asymptotics of the prolate SRFs of the second kind, one uses the Liouville formula (Flammer, 1957)

$$R_{ml}^{(2)}(c, \xi_0) = \frac{1}{2}\, Q_{ml}(c)\, R_{ml}^{(1)}(c, \xi_0)\, \ln \frac{\xi_0 + 1}{\xi_0 - 1} + (\xi_0^2 - 1)^{-\frac{m}{2}}\, \varphi_{ml}(c, \xi_0) , \tag{5.192}$$

where $C_{ml}, Q_{ml}(c)$ are some constants, $P_l^m(\xi_0)$ the associated Legendre polynomials, and

$$\varphi_{ml}(c, \xi_0) = \begin{cases} \xi_0 \sum\limits_{r=0}^{\infty} b_r^{ml}\, (\xi_0^2 - 1)^r, & l - m = 2q - 1 , \\ \sum\limits_{r=0}^{\infty} b_r^{ml}\, (\xi_0^2 - 1)^r , & l - m = 2q . \end{cases} \tag{5.193}$$

From Eqs. (5.191)–(5.193), in the first approximation we get

$$R_0 = \frac{m}{\xi_0^2 - 1}\, I , \qquad \tilde{R}_2 = \frac{m}{\xi_0^2 - 1}\, I , \qquad R_1 = -\frac{m}{\xi_0^2 - 1}\, I . \tag{5.194}$$

Then for the matrices A and B (see Eq. (5.56)), we have

$$A = \frac{m}{\xi_0^2 - 1}\, I , \qquad B = \frac{1}{\xi_0^2 - 1}\, K . \tag{5.195}$$

Eqs. (5.194) and (5.195) allow one to solve the ISLAEs in the first approximation with respect to the parameter $(\xi_0^2 - 1)$ (see for more details, Voshchinnikov and Farafonov (1993) and Farafonov and Il'in (2006)). For the axisymmetric component, we get (see Eqs. (5.40) and (5.41)):

(1) for the TE mode

$$\boldsymbol{Z}_0 = 0 , \tag{5.196}$$

(2) for the TM mode

$$\boldsymbol{Z}_0 = (\varepsilon - 1)\, \boldsymbol{F}_0 . \tag{5.197}$$

For the non-axisymmetric component, we obtain:

(1) for the TE mode

$$\begin{aligned} \boldsymbol{Z}_1 &= -\frac{\varepsilon-1}{\varepsilon+1} Q_1 \left(-I + \frac{1}{m} \Gamma K\right) \boldsymbol{F}_m\,, \\ \boldsymbol{Z}_2 &= \frac{\varepsilon-1}{\varepsilon+1} Q_1 \left(-\Gamma + \frac{1}{m} K\right) \boldsymbol{F}_m\,, \end{aligned} \tag{5.198}$$

(2) for the TM mode

$$\begin{aligned} \boldsymbol{Z}_1 &= -\frac{\varepsilon-1}{\varepsilon+1} Q_1 \left(\Gamma^2 + \frac{1}{m} \Gamma K\right) \boldsymbol{F}_m\,, \\ \boldsymbol{Z}_2 &= \frac{\varepsilon-1}{\varepsilon+1} Q_1 \left(\Gamma + \frac{1}{m} K\right) \boldsymbol{F}_m\,, \end{aligned} \tag{5.199}$$

where $Q_1 = (I - \Gamma^2)^{-1}$. From Eqs. (5.37), (5.49) and (5.191), (5.192), in the first approximation we have

$$\begin{aligned} a_l^{(1)} &= O\left[(b/a)^4\right], & b_l^{(1)} &= O\left[(b/a)^2\right], \\ a_{ml}^{(1)} &= O\left[(b/a)^{2m}\right], & b_{ml}^{(1)} &= O\left[(b/a)^{2m}\right]. \end{aligned} \tag{5.200}$$

Thus, in the first approximation with respect to $(b/a)^2$ one should keep just the axisymmetric component for the TM mode and the non-axisymmetric component with $m = 1$, and hence the dimensionless parameters A_j are as follows (see Eqs. (5.73)–(5.76) and (5.196)–(5.199)):

$$\begin{aligned} A_1 &= \frac{\varepsilon-1}{\varepsilon+1} \left(\frac{b}{a}\right)^2 T_1\,, & A_3 &= \frac{\varepsilon-1}{\varepsilon+1} \left(\frac{b}{a}\right)^2 T_2\,, \\ A_4 &= \frac{\varepsilon-1}{\varepsilon+1} \left(\frac{b}{a}\right)^2 T_3\,, & A_2 &= \left(\frac{b}{a}\right)^2 \left(\frac{\varepsilon-1}{2} T_4 + \frac{\varepsilon-1}{\varepsilon+1} T_5\right), \end{aligned} \tag{5.201}$$

where the functions T_j do not depend on the dielectric permittivity of the particle ε, and their dependence on the azimuthal angle is known, namely: T_1, $T_5 \sim \cos\varphi$, T_2, $T_3 \sim \sin\varphi$, and T_4 does not depend on φ.

5.4.2 Extremely oblate spheroids

From the parity properties of the integrals $\delta_{nl}^{(m)}$, $\gamma_{nl}^{(m)}$, $\kappa_{nl}^{(m)}$, it follows that the ISLAEs relative to the unknown vectors $\boldsymbol{Z}_0$, $\boldsymbol{Z}_1$, and $\boldsymbol{Z}_2$ can be solved separately for their even and odd elements. Let us introduce the vectors $\boldsymbol{Z}^{\mathrm{e}} = \{z_{2(l-m)}\}_{l=m}^{\infty}$ and $\boldsymbol{Z}^{\mathrm{o}} = \{z_{2(l-m)+1}\}_{l=m}^{\infty}$ and the matrices $A^{\mathrm{e}} = \{A_{2(n-m),2(l-m)}\}_{n,l=m}^{\infty}$, $A^{\mathrm{o}} = \{A_{2(n-m)+1,2(l-m)+1}\}_{n,l=m}^{\infty}$ (in the case of R_0, $\tilde{R}_2$, R_1, Δ, Σ) and $B^{\mathrm{e}} = \{B_{2(n-m)+1,2(l-m)}\}_{n,l=m}^{\infty}$, $B^{\mathrm{o}} = \{B_{2(n-m),2(l-m)+1}\}_{n,l=m}^{\infty}$ (in the case of Γ, K). In the new notation it is obvious that the ISLAEs relative to the axisymmetric components $\boldsymbol{Z}_0^{\mathrm{e}}$ and $\boldsymbol{Z}_0^{\mathrm{o}}$ can be solved independently, and the matrices and the vector $\boldsymbol{F}_0$ should have the corresponding superscript e or o. The ISLAEe relative to the non-axisymmetric components $\boldsymbol{Z}_1^{\rm e}$, $\boldsymbol{Z}_2^{\rm o}$ and $\boldsymbol{Z}_1^{\rm o}$, $\boldsymbol{Z}_2^{\rm e}$ are solved independently as well. The matrices corresponding to the vector $\boldsymbol{Z}_1$ and the vector $\boldsymbol{F}_m$ should have the same superscript as the vector $\boldsymbol{Z}_1$, and the matrices corresponding to the vector $\boldsymbol{Z}_2$ should have the opposite superscript (i.e. the same as the vector $\boldsymbol{Z}_2$).

For extremely oblate spheroids, the aspect ratio is small ($b/a \to 0$), and the value of the radial coordinate ξ_0 is small too ($\xi_0 \to 0$). The principal term of the expansions in b/a and ξ_0 are the same as $(b/a)^2 = \xi_0^2/(\xi_0^2+1)$. Taking into account the asymptotics of the $R_{0,2}$ matrix elements for $\xi_0 \to 0$ (see Appendix), we get in the first approximation

$$\begin{aligned}
R_0^{\rm e} &= \xi_0[\Lambda^{\rm e}(-{\rm i}c_1) - (c_1^2+m^2)\,I] + O(\xi_0^3)\,,\\
R_0^{\rm o} &= \frac{1}{\xi_0}I + O(\xi_0)\,, \quad \tilde{R}_2^{\rm o} = \frac{1}{\xi_0}I + O(\xi_0)\,,\\
\tilde{R}_2^{\rm e} - R_0^{\rm e} &= -\xi_0(\varepsilon-1)\,c_1^2\,(\Gamma^2)^{\rm e} + O(\xi_0^3)\,,\\
\tilde{R}_2^{\rm o} - R_0^{\rm o} &= O(\xi_0)\,,
\end{aligned} \tag{5.202}$$

where $\Lambda^{\rm e}(-{\rm i}c_1)$ is a diagonal matrix whose elements are the eigenvalues $\lambda^{\rm e}(-{\rm i}c_1)$ for even differences $(l-m)$, including zero. From Eqs. (5.202) for the axisymmetric components, the ISLAEs (5.40) and (5.41) in the first approximation have the following solutions:

(1) for the TE mode

$$\begin{aligned}
\boldsymbol{Z}_0^{\rm e} &= -\xi_0\,(\varepsilon-1)\,(R_1^{\rm e})^{-1}\,c_1^2\,(\Gamma^2)^{\rm e}\,\boldsymbol{F}_0^{\rm e}\,,\\
\boldsymbol{Z}_0^{\rm o} &= 0\,;
\end{aligned} \tag{5.203}$$

(2) for the TM mode

$$\begin{aligned}
\boldsymbol{Z}_0^{\rm e} =& -\xi_0\frac{\varepsilon-1}{\varepsilon}\,(R_1^{\rm e})^{-1}\left[\Lambda^{\rm e}(-{\rm i}c_1) + c_1^2\,(\Gamma^2)^{\rm e}\right.\\
&\left.+(1-m^2-c_1^2)\,I\right]\,\boldsymbol{F}_0^{\rm e}\,,\\
\boldsymbol{Z}_0^{\rm o} =&(\varepsilon-1)\,\boldsymbol{F}_0^{\rm o}\,.
\end{aligned} \tag{5.204}$$

The relations (5.202) allow one to analytically solve the ISLAEs (5.55) and (5.57) relative to the non-axisymmetric components in the first approximation with respect to ξ_0 (see for more details Voshchinnikov and Farafonov, 1993; Farafonov and Il'in, 2006). The matrices A and B used in the systems (5.55) are written in the first approximation as follows:

$$A^{\rm e} = \frac{1}{\xi_0}I\,, \quad A^{\rm o} = O(\xi_0)\,, \quad B^{\rm e} = -2\,(\Gamma^{-1})^{\rm e} + K^{\rm e}\,, \quad B^{\rm o} = K^{\rm o}\,. \tag{5.205}$$

So, we get:

(1) for the TE mode

$$\begin{aligned}
\boldsymbol{Z}_1^{\rm e} =& (R_1^{\rm e})^{-1}\,(\varepsilon-1)\,\xi_0\{-c_1^2\,(\Gamma^2)^{\rm e}\\
&+(\Gamma^{\rm e})^{-1}[(\Gamma^{\rm o})^{-1} - K^{\rm e}]\}\,\boldsymbol{F}_m^{\rm e}\,,\\
\boldsymbol{Z}_2^{\rm o} =& -(\varepsilon-1)\,\xi_0\,(\Gamma^{\rm o})^{-1}\,\boldsymbol{F}_m^{\rm e}\,,\\
\boldsymbol{Z}_1^{\rm o} =& 0\,,\\
\boldsymbol{Z}_2^{\rm e} =& 0\,;
\end{aligned} \tag{5.206}$$

(2) for the TM mode

$$\begin{aligned} \boldsymbol{Z}_1^{\mathrm{e}} =& -\xi_0 \left(\frac{\varepsilon-1}{\varepsilon}\right) (R_1^{\mathrm{e}})^{-1} \left[\Lambda^{\mathrm{e}}(-\mathrm{i}c_1) - (c_1^2+m^2)\, I \right. \\ & \left. + c_1^2\, (\Gamma^2)^{\mathrm{e}} + (\Gamma^{\mathrm{e}})^{-1}\, K^{\mathrm{e}}\right] \boldsymbol{F}_m^{\mathrm{e}}\,, \\ \boldsymbol{Z}_2^{\mathrm{o}} =& \, 0\,, \\ \boldsymbol{Z}_1^{\mathrm{o}} =& \,(\varepsilon-1)\, \boldsymbol{F}_m^{\mathrm{o}}\,, \\ \boldsymbol{Z}_2^{\mathrm{e}} =& -(\varepsilon-1)(R_1^{\mathrm{e}})^{-1}(\Gamma^{\mathrm{e}})^{-1}\, \boldsymbol{F}_m^{\mathrm{o}}\,. \end{aligned} \tag{5.207}$$

From Eqs. (5.203)–(5.204), (5.206)–(5.207) and the relation $\boldsymbol{F}_m^{\mathrm{o}} \sim \xi_0$, we get the dimensionless intensity parameters of the scattered field

$$\begin{aligned} A_1 &= \frac{\varepsilon-1}{2}\frac{b}{a}\, T_1\,, & A_3 &= \frac{\varepsilon-1}{2}\frac{b}{a}\, T_2\,, \\ A_4 &= \frac{\varepsilon-1}{2}\frac{b}{a}\, T_3\,, & A_2 &= \frac{b}{a}\left[\frac{\varepsilon-1}{2\varepsilon}\, T_4 + \frac{\varepsilon-1}{2}\, T_5\right], \end{aligned} \tag{5.208}$$

where as for the extremely prolate spheroids the functions T_j being independent of the dielectric permittivity ε depend on the parameters c, α, Θ, φ. The functions are expressed through the spheroidal functions, and T_1, T_2, T_4 are even functions of $\cos\alpha$, while T_3, T_5 are odd functions. This follows from the expressions of the free terms $\boldsymbol{F}_m^{\mathrm{e}}$ and $\boldsymbol{F}_m^{\mathrm{o}}$ through the oblate SAFs which are in their turn either even or odd functions of $\cos\alpha$.

5.4.3 Justification of the quasi-static approximation

For extremely prolate and oblate spheroids, Eqs. (5.201) and (5.208) can be also considered as the Rayleigh–Gans approximation, i.e. the approximation for small optically soft particles whose dielectric permittivity is close to that of the surrounding medium

$$|\varepsilon-1| \ll 1, \qquad |\varepsilon-1|\, kl_{\max} \ll 1\,. \tag{5.209}$$

On the other hand, the Rayleigh–Gans approximation for spheroids is expressed through the elementary functions (Lopatin and Sid'ko, 1988). A comparison of these expressions for $\varepsilon \to 1$ allows one to derive the functions T_j explicitly

$$\begin{aligned} T_1 &= \frac{2c_1^3}{3}\, G(u)\, \cos\varphi\,, & T_2 &= \frac{2c_1^3}{3}\, G(u)\, \cos\theta\, \sin\varphi\,, \\ T_3 &= -\frac{2c_1^3}{3}\, G(u)\, \cos\alpha\, \sin\varphi\,, & T_4 &= \frac{2c_1^3}{3}\, G(u)\, \sin\alpha\, \sin\theta\,, \\ T_5 &= \frac{2c_1^3}{3}\, G(u)\, \cos\alpha\, \cos\theta\, \cos\varphi\,, \end{aligned} \tag{5.210}$$

where the function $G(u)$ is

$$G(u) = \frac{3}{u^3}\,(\sin u - u\cos u)\,, \tag{5.211}$$

with its argument being

$$u = c_1 | \cos\theta - \cos\alpha | \tag{5.212}$$

for extremely prolate spheroids, and

$$u = c_1 \left[\sin^2\alpha + \sin^2\theta - 2\sin\alpha\,\sin\theta\,\cos\varphi\right]^{\frac{1}{2}} \tag{5.213}$$

for extremely oblate spheroids, respectively.

Note that in the monograph of Lopatin and Sid'ko (1988) the expression for the Rayleigh–Gans approximation is given for the case when the reference plane contains both the directions of propagation of the incident and scattered radiation, while Eqs. (5.201) and (5.208) are derived in the coordinate system related with the directions of propagation of the incident wave and of the symmetry axis of the spheroid. Formulating the Rayleigh–Gans approximation in the new frame, one should keep in mind that for extremely prolate spheroids the dependence of the functions T_j on the azimuthal angle φ is known (see Eq. (5.201)) and for extremely oblate spheroids we know the parity properties of T_j as functions of $\cos\alpha$ (see Eq. (5.208)).

As Eqs. (5.210)–(5.213) coincide with similar equations of the quasi-static approximation, the internal field can be approximated as follows:

$$\boldsymbol{E}^{(2)} = K_1\,(\boldsymbol{E}^{(0)}, \boldsymbol{i}_{\rm x})\,\boldsymbol{i}_{\rm x} + K_2\,(\boldsymbol{E}^{(0)}, \boldsymbol{i}_{\rm y})\,\boldsymbol{i}_{\rm y} + K_3\,(\boldsymbol{E}^{(0)}, \boldsymbol{i}_{\rm z})\,\boldsymbol{i}_{\rm z}\,. \tag{5.214}$$

The approximation (5.214) can be applied to any tri-axial ellipsoid. For ellipsoids with at least one small semiaxis, a similar approximation was used in the paper of Seker (1986). However, as a starting point this author represented the internal field by the expansion (5.214) but in terms of the unit vectors of the spherical coordinates ($\boldsymbol{i}_r$, $\boldsymbol{i}_\theta$, $\boldsymbol{i}_\varphi$). This gave a wrong result except for the case of spheres and the correct Rayleigh–Gans approximation when all the coefficients K_j were the same.

Thus, we have proved correctness of the use of the quasi-static approximation for extremely prolate and oblate spheroids.

The characteristics of the scattered field can be found from the known fields inside and outside the particle. The dimensionless parameters of the scattered radiation intensity for extremely prolate spheroids are

$$\begin{aligned}
i_{11} &= \frac{4}{9}\left(\frac{b}{a}\right)^4 c^6\, G^2(u) \left|\frac{\varepsilon-1}{\varepsilon+1}\right|^2 \cos^2\varphi\,, \\
i_{12} &= \frac{4}{9}\left(\frac{b}{a}\right)^4 c^6\, G^2(u) \left|\frac{\varepsilon-1}{\varepsilon+1}\right|^2 \cos^2\theta\,\sin^2\varphi\,, \\
i_{21} &= \frac{4}{9}\left(\frac{b}{a}\right)^4 c^6\, G^2(u) \left|\frac{\varepsilon-1}{\varepsilon+1}\right|^2 \cos^2\alpha\,\sin^2\varphi\,, \\
i_{22} &= \frac{4}{9}\left(\frac{b}{a}\right)^4 c^6\, G^2(u) \left|\frac{\varepsilon-1}{2}\sin\alpha\,\sin\theta + \frac{\varepsilon-1}{\varepsilon+1}\cos\alpha\,\cos\theta\,\cos\varphi\right|^2 .
\end{aligned} \tag{5.215}$$

For extremely oblate spheroids, in Eq. (5.215), one should replace $(b/a)^4$ with $(b/a)^2$, $(\varepsilon-1)/(\varepsilon+1)$ with $(\varepsilon-1)/2$, and $(\varepsilon-1)/2$ with $(\varepsilon-1)/(2\varepsilon)$ and properly select the argument of the function $G(u)$ from Eq. (5.213).

The efficiency factors are calculated by integrating the scattered radiation intensity over all directions. In the general case, for extremely prolate spheroids, one should take an integral over the azimuthal angle

$$
\begin{aligned}
Q_{\text{sca}}^{\text{TE}} &= \frac{4c^4}{9\left[\left(\frac{b}{a}\right)^2 + \sin^2\alpha\right]^{\frac{1}{2}}} \left(\frac{b}{a}\right)^3 \left|\frac{\varepsilon - 1}{\varepsilon + 1}\right|^2 \int_0^{\pi} G^2(u)\,(1 + \cos^2\theta)\,\sin\theta\,\mathrm{d}\theta\,, \\
Q_{\text{sca}}^{\text{TM}} &= \frac{4c^4}{9\left[\left(\frac{b}{a}\right)^2 + \sin^2\alpha\right]^{\frac{1}{2}}} \left(\frac{b}{a}\right)^3 \left\{\left|\frac{\varepsilon - 1}{\varepsilon + 1}\right|^2 \cos^2\alpha \int\limits_0^{\pi} G^2(u)\,(1 + \cos^2\theta)\right. \\
&\left.\times \sin\theta\,\mathrm{d}\theta + 2\left|\frac{\varepsilon - 1}{2}\right|^2 \sin^2\alpha \int\limits_0^{\pi} G^2(u)\,\sin^3\theta\,\mathrm{d}\theta\right\}.
\end{aligned}
\tag{5.216}
$$

For extremely oblate spheroids, one should consider a double integral

$$
\begin{aligned}
Q_{\text{sca}}^{\text{TE}} &= \frac{4c^4}{9\pi\left[\left(\frac{b}{a}\right)^2 + \cos^2\alpha\right]^{\frac{1}{2}}} \left(\frac{b}{a}\right)^2 \\
&\times \left|\frac{\varepsilon - 1}{2}\right|^2 \int_0^{2\pi}\int_0^{\pi} G^2(u)\,(\cos^2\theta\,\sin^2\varphi + \cos^2\varphi)\,\sin\theta\,\mathrm{d}\theta\,\mathrm{d}\varphi\,, \\
Q_{\text{sca}}^{\text{TM}} &= \frac{4c^4}{9\pi\left[\left(\frac{b}{a}\right)^2 + \cos^2\alpha\right]^{\frac{1}{2}}} \left(\frac{b}{a}\right)^2 \\
&\times \left\{\left|\frac{\varepsilon - 1}{2}\right|^2 \cos^2\alpha \int_0^{2\pi}\int_0^{\pi} G^2(u)\,(\cos^2\theta\,\cos^2\varphi + \sin^2\varphi)\,\sin\theta\,\mathrm{d}\theta\,\mathrm{d}\varphi\right. \\
&\left. + \left|\frac{\varepsilon - 1}{2\,\varepsilon}\right|^2 \sin^2\alpha \int_0^{2\pi}\int_0^{\pi} G^2(u)\,(1 \quad \cos^2\theta)\,\sin\theta\,\mathrm{d}\theta\,\mathrm{d}\varphi\right\}
\end{aligned}
\tag{5.217}
$$

as the argument of the function $G(u)$ depends on the azimuthal angle.

Absorption of the electromagnetic radiation by a particle can be determined from the internal field (5.214), and in the quasi-static approximation the efficiency factors coincide with the corresponding factors derived in the Rayleigh approximation, namely for extremely prolate spheroids

$$
\begin{aligned}
Q_{\text{abs}}^{\text{TE}} &= \frac{16c}{3\left[\left(\frac{b}{a}\right)^2 + \sin^2\alpha\right]^{\frac{1}{2}}} \left(\frac{b}{a}\right)^2 \operatorname{Im}\varepsilon\,\frac{1}{|\varepsilon + 1|^2}\,, \\
Q_{abs}^{\text{TM}} &= \frac{16c}{3\left[\left(\frac{b}{a}\right)^2 + \sin^2\alpha\right]^{\frac{1}{2}}} \left(\frac{b}{a}\right)^2 \operatorname{Im}\varepsilon\left(\frac{\cos^2\alpha}{|\varepsilon + 1|^2} + \frac{1}{4}\sin^2\alpha\right)
\end{aligned}
\tag{5.218}
$$

and for extremely oblate spheroids

$$
\begin{aligned}
Q_{\text{abs}}^{\text{TE}} &= \frac{4c}{3\left[\left(\frac{b}{a}\right)^2 + \cos^2\alpha\right]^{\frac{1}{2}}} \left(\frac{b}{a}\right)^2 \operatorname{Im}\varepsilon\,, \\
Q_{\text{abs}}^{\text{TM}} &= \frac{4c}{3\left[\left(\frac{b}{a}\right)^2 + \cos^2\alpha\right]^{\frac{1}{2}}} \left(\frac{b}{a}\right)^2 \operatorname{Im}\varepsilon \left(\cos^2\alpha + \frac{1}{|\varepsilon|^2}\sin^2\alpha\right).
\end{aligned}
\tag{5.219}
$$

The extinction efficiency factors are determined as follows:

$$Q_{\text{ext}} = Q_{\text{sca}} + Q_{\text{abs}}\,. \tag{5.220}$$

The scattering matrix derived for a single particle in the quasi-static approximation differs from that in the Rayleigh approximation by the factor $G^2(u)$ only, and hence the polarization degree of the scattered radiation is the same for both the quasi-static and Rayleigh approximations.

An investigation of applicability of the quasi-static approximation for spheroids has been performed by Voshchinnikov and Farafonov (2000). It was shown that this approximation was widely applicable to extremely prolate and oblate dielectric spheroids. Besides that, these authors studied the geometry of scattering by such particles in detail. It was found that for extremely prolate spheroids the maximum intensity of the scattered radiation was formed close to the guiding lines of the cone with the opening angle 2α, where α was the incidence angle of a plane wave (Voshchinnikov and Farafonov, 1993). A similar situation occurred for infinite circular cylinders. For extremely oblate spheroids, the scattered radiation is maximal in the directions of the transmitted and reflected waves like in the case of reflection by a plane (Voshchinnikov and Farafonov, 1993). It was also shown that deep minima of the scattered radiation intensity took place in vicinity of the zeros of the function $G(u)$ that appeared in the quasi-static approximation.

5.4.4 Extremely prolate perfectly conducting spheroids

In this case the ISLAEs (5.63) and (5.64) for the non-axisymmetric components (along with the relations (5.194)) can be solved explicitly in the first approximation with respect to the parameter $(b/a)^2$:

(1) for the TE mode

$$
\begin{aligned}
\boldsymbol{Z}_1 &= Q_1\,(I + \Gamma^2)\,\boldsymbol{F}_m\,, \\
\boldsymbol{Z}_2 &= -\,2\,Q_1\,\Gamma\,\boldsymbol{F}_m\,;
\end{aligned}
\tag{5.221}
$$

(2) for the TM mode

$$
\begin{aligned}
\boldsymbol{Z}_1 &= -\,Q_1\,(I + \Gamma^2)\,\boldsymbol{F}_m\,, \\
\boldsymbol{Z}_2 &= 2\,Q_1\,\Gamma\,\boldsymbol{F}_m\,,
\end{aligned}
\tag{5.222}
$$

where $Q_1 = (I + \Gamma^2)^{-1}$.

From Eqs. (5.37), (5.52), (5.59), (5.196)–(5.199) and (5.221)–(5.222), we find that for extremely prolate spheroids the scattered field expansion coefficients $\tilde{a}_l^{(1)}$,

$\tilde{a}_{ml}^{(1)}$, $\tilde{b}_{ml}^{(1)}$ (except for the axisymmetric components of the TM mode) are related in the first approximation with respect to the parameter $(b/a)^2$ with the corresponding coefficients for extremely prolate dielectric spheroids

$$\begin{aligned} b_l^{(1),\mathrm{TM}} &= -(\varepsilon-1)\,\tilde{a}_l^{(1),\mathrm{TE}}, \\ a_{ml}^{(1),\mathrm{TE}} - a_{ml}^{(1),\mathrm{TM}} &= \frac{\varepsilon-1}{\varepsilon+1}\,\tilde{a}_{ml}^{(1),\mathrm{TE}} = -\frac{\varepsilon-1}{\varepsilon+1}\,\tilde{a}_{ml}^{(1),\mathrm{TM}}, \\ b_{ml}^{(1),\mathrm{TE}} - b_{ml}^{(1),\mathrm{TM}} &= \frac{\varepsilon-1}{\varepsilon+1}\,\tilde{b}_{ml}^{(1),\mathrm{TE}} = -\frac{\varepsilon-1}{\varepsilon+1}\,\tilde{b}_{ml}^{(1),\mathrm{TM}}. \end{aligned} \tag{5.223}$$

From Eqs. (5.72)–(5.76), (5.200), and (5.223), it follows that

$$\begin{aligned} \tilde{T}_1 &= \left(T_1 - \frac{1}{2}T_4 - T_5\right), & \tilde{T}_2 &= T_2 + T_3, \\ \tilde{T}_3 &= T_2 + T_3, & \tilde{T}_5 &= T_5 - T_1, \end{aligned} \tag{5.224}$$

where the functions T_j are determined by Eqs. (5.210).

Let us now consider the axisymmetric components of the TM mode. From the asymptotics (5.191)–(5.192), we get

$$\frac{\left[(\xi_0^2-1)^{\frac{1}{2}}\,R_{1l}^{(1)}(c,\xi_0)\right]'}{\left[(\xi_0^2-1)^{\frac{1}{2}}\,R_{1l}^{(3)}(c,\xi_0)\right]'} = \frac{1}{1+\frac{1}{2}\mathrm{i}\left[Q_{1l}(c)\ln\frac{\xi_0+1}{\xi_0-1} + O(1)\right]}. \tag{5.225}$$

This equation is valid, when $Q_{1l}(c) \neq 0$, which occurs for $c \neq l\pi/2$. When $c = n\pi/2$, the prolate SRFs are expressed through the elementary functions (Komarov *et al.*, 1976)

$$\begin{aligned} R_{1n}^{(1)}\left(\frac{n\pi}{2},\xi\right) &= \frac{\cos\left[\frac{n\pi}{2}(\xi-1)-\frac{\pi}{2}\right]}{\frac{n\pi}{2}(\xi^2-1)^{\frac{1}{2}}}, \\ R_{1n}^{(3)}\left(\frac{n\pi}{2},\xi\right) &= \frac{\exp\left\{\mathrm{i}\left[\frac{n\pi}{2}(\xi-1)-\frac{\pi}{2}\right]\right\}}{\frac{n\pi}{2}(\xi^2-1)^{\frac{1}{2}}}, \end{aligned} \tag{5.226}$$

and the corresponding ratio has the following asymptotics:

$$\frac{\left[(\xi_0^2-1)^{\frac{1}{2}}\,R_{1n}^{(1)}\left(\frac{n\xi}{2},\xi_0\right)\right]'}{\left[(\xi_0^2-1)^{\frac{1}{2}}\,R_{1n}^{(3)}\left(\frac{n\xi}{2},\xi_0\right)\right]'} = 1 + O(\xi_0^2-1). \tag{5.227}$$

In the case under consideration the principal term of the asymptotics is determined by the resonance term (5.227), and as a result we have

$$\begin{aligned} \boldsymbol{E}^{(1)} &= \frac{e^{ikr}}{-ikr}\,\frac{\cos\left[\frac{n\pi}{2}(\cos\alpha-1)-\frac{\pi}{2}\right]\,\cos\left[\frac{n\pi}{2}(\cos\theta-1)-\frac{\pi}{2}\right]}{\sin\alpha\,\sin\theta} \\ &\quad\times\left[\int_0^1 \frac{\cos^2\left[\frac{n\pi}{2}(x-1)-\frac{\pi}{2}\right]}{1-x^2}\,\mathrm{d}x\right]^{-2} \boldsymbol{i}_\theta, \end{aligned} \tag{5.228}$$

since the prolate SAFs can be represented as (Komarov *et al.*, 1976)

$$S_{1n}\left(\frac{n\pi}{2},\eta\right) = \frac{(-1)^{\frac{n-1}{2}}(n+1)!}{2^n\left(\frac{n-1}{2}\right)!\left(\frac{n+1}{2}\right)!}\,\frac{\cos\left(\frac{n\pi}{2}\eta\right)}{\sqrt{1-\eta^2}}\,, \qquad n=2q-1\,, \tag{5.229}$$

$$S_{1n}\left(\frac{n\pi}{2},\eta\right) = \frac{(-1)^{\frac{n-2}{2}}(n+2)!}{n\,\pi\,2^{n-1}\left(\frac{n-1}{2}\right)!\left(\frac{n+1}{2}\right)!}\,\frac{\sin\left(\frac{n\pi}{2}\eta\right)}{\sqrt{1-\eta^2}}\,, \qquad n=2q\,. \tag{5.230}$$

So, from Eqs. (5.72)–(5.76), (5.200), (5.210), (5.224)–(5.225) we find the principal term of the field scattered by an extremely prolate perfectly conducting spheroid

$$\begin{aligned}
\boldsymbol{E}^{(1)}_{\rm TE} &= \frac{e^{ikr}}{kr}\left(\frac{b}{a}\right)^2 \frac{2c^3}{3}\,G(u)\left\{\left[-\frac{1}{2}\sin\alpha\sin\theta + (1-\cos\alpha\cos\theta)\cos\varphi\right]\boldsymbol{i}_\varphi\right.\\
&\quad \left. + (\cos\theta-\cos\alpha)\,\sin\varphi\,\boldsymbol{i}_\theta\right\},\\
\boldsymbol{E}^{(1)}_{\rm TM} &= \frac{e^{ikr}}{kr}\left\{2i\sum_{l=1}^{\infty}\frac{\left[(\xi_0^2-1)^{\frac12}R^{(1)}_{1l}(c,\xi_0)\right]'}{\left[(\xi_0^2-1)^{\frac12}R^{(3)}_{1l}(c,\xi_0)\right]'}\,N^{-2}_{1l}(c)\,S_{1l}(c,\cos\alpha)\,S_{1l}(c,\cos\theta)\,\boldsymbol{i}_\theta\right.\\
&\quad \left. +\left(\frac{b}{a}\right)^2\frac{2c^3}{3}\,G(u)\left[(\cos\theta-\cos\alpha)\sin\varphi\,\boldsymbol{i}_\varphi + (\cos\alpha\cos\theta-1)\cos\varphi\,\boldsymbol{i}_\theta\right]\right\}.
\end{aligned} \tag{5.231}$$

For the oblique incidence of radiation including a TM mode component, the main contribution to the scattered radiation is given by the axisymmetric component of the TM mode. The asymptotics is generally inversely proportional to $\ln a/b$ (see Eq. (5.225)) and has the order $O(1)$ under the condition $c=n\pi/2$ (see Eq. (5.227)) that is equivalent to the well-known condition (5.190) of linear antenna excitation (Stratton, 1941). For the oblique incidence of a TE mode wave or for the wave propagation along the spheroid symmetry axis, the scattered field is proportional to $(b/a)^2$.

The dimensionless parameters of the scattered radiation intensity are as follows:

$$\begin{aligned}
i_{11} &= \frac{4}{9}\left(\frac{b}{a}\right)^4 c^6\,G^2(u)\left|-\frac{1}{2}\sin\alpha\,\sin\theta + (1-\cos\alpha\cos\theta)\cos\varphi\right|^2,\\
i_{12} = i_{21} &= \frac{4}{9}\left(\frac{b}{a}\right)^4 c^6\,G^2(u)\,(\cos\theta-\cos\alpha)^2\,\sin^2\varphi\,,\\
i_{22} &= \left|2i\sum_{l=1}^{\infty}\frac{\left[(\xi_0^2-1)^{\frac12}R^{(1)}_{1l}(c,\xi_0)\right]'}{\left[(\xi_0^2-1)^{\frac12}R^{(2)}_{1l}(c,\xi_0)\right]'}\,N^{-2}_{1l}(c)\,S_{1l}(c,\cos\alpha)\,S_{1l}(c,\cos\alpha)\right.\\
&\quad \left. -\left(\frac{b}{a}\right)^2\frac{2c^3}{3}\,G(u)\,(1-\cos\alpha\cos\theta)\cos\varphi\right|^2.
\end{aligned} \tag{5.232}$$

For the oblique incidence of a non-polarized wave, the scattered radiation is linearly polarized in the first approximation as the main component is related with i_{22}. For the parallel incidence, the polarization degree is equal to zero as

$$\begin{aligned} i_{11} = i_{22} &= \frac{4}{9}\left(\frac{b}{a}\right)^4 c^6\, G^2[c(1-\cos\theta)]\,(1-\cos\theta)^2\cos^2\varphi\,, \\ i_{12} = i_{12} &= \frac{4}{9}\left(\frac{b}{a}\right)^4 c^6\, G^2[c(1-\cos\theta)]\,(1-\cos\theta)^2\sin^2\varphi\,. \end{aligned} \tag{5.233}$$

The efficiency factors for scattering and backscattering are

$$\begin{aligned} Q_{\text{sca}}^{\text{TE}} &= \frac{4c^4}{9\left[\left(\frac{b}{a}\right)^2+\sin^2\alpha\right]^{\frac{1}{2}}}\left(\frac{b}{a}\right)^3\int\limits_0^{\pi} G^2(u)\Big[\frac{3}{2}(1-\cos\alpha\cos\theta)^2 \\ &\quad + \frac{1}{2}(\cos\alpha-\cos\theta)^2\Big]\sin\theta\,\mathrm{d}\theta\,, \\ Q_{\text{sca}}^{\text{TM}} &= \frac{4c^4}{9\left[\left(\frac{b}{a}\right)^2+\sin^2\alpha\right]^{\frac{1}{2}}}\sum_{l=1}^{\infty}\left|\frac{[(\xi_0^2-1)^{\frac{1}{2}}\,R_{1l}^{(1)}(c,\xi_0)]'}{[(\xi_0^2-1)^{\frac{1}{2}}\,R_{1l}^{(2)}(c,\xi_0)]'}\right|^2 \\ &\quad \times N_{1l}^{-2}(c)\,S_{1l}(c,\cos\alpha) + \frac{4c^4}{9\left[\left(\frac{b}{a}\right)^2+\sin^2\alpha\right]^{\frac{1}{2}}}\left(\frac{b}{a}\right)^3 \\ &\quad \times \int_0^{\pi} G^2(u)\left[(1-\cos\alpha\cos\theta)^2+(\cos\alpha-\cos\theta)^2\right]\sin\theta\,\mathrm{d}\theta\,, \end{aligned} \tag{5.234}$$

and

$$\begin{aligned} Q_{\text{bk}}^{\text{TE}} &= \frac{4}{9}\left[\left(\frac{b}{a}\right)^2+\cos^2\alpha\right]^2 c^4\,G^2(2c\cos\alpha)\,(3+\cos^2\alpha)^2\,, \\ Q_{\text{bk}}^{\text{TM}} &= \frac{16}{c^2}\lbrack\left(\frac{b}{a}\right)^2+\cos^2\alpha]^2\Bigg|{}_1\sum_{l=1}^{\infty}(-1)^l\ {}_1\frac{[(\xi_0^2-1)^{\frac{1}{2}}\,R_{1l}^{(1)}(c,\xi_0)]'}{[(\xi_0^2-1)^{\frac{1}{2}}\,R_{1l}^{(2)}(c,\xi_0)]'} \\ &\quad \times N_{1l}^{-2}(c)\,S_{1l}^2(c,\cos\alpha) - \left(\frac{b}{a}\right)^2\frac{c^3}{3}G(2c\cos\alpha)(1+\cos^2\alpha)\Bigg|^2\,, \end{aligned} \tag{5.235}$$

respectively.

For the parallel incidence, the expressions simplify

$$\begin{aligned} Q_{\text{sca}} &= \left(\frac{b}{a}\right)^2 c^2\,\varphi(c) = \frac{8c^4}{9}\left(\frac{b}{a}\right)^2 \\ &\quad \times \int_0^{\pi} G^2[c(1-\cos\theta)]\,(1-\cos\theta)^2\sin\theta\,\mathrm{d}\theta\,, \end{aligned} \tag{5.236}$$

$$Q_{\text{bk}} = \frac{64}{9}\,c^4\,G^2(2c)\,. \tag{5.237}$$

Note that for a small diffraction parameter $c \ll 1$ and for the parallel incidence of a plane wave, the results coincide with the Rayleigh approximation for the corresponding perfectly conducting prolate spheroid (van de Hulst, 1957).

To conclude, we consider the light scattering by randomly oriented perfectly conducting thin needles. As the scattering and backscattering cross-sections for the TE and TM modes similarly depend on the parameters, we have

$$\begin{aligned} \langle C_{\text{sca}} \rangle &= \frac{1}{2} \int_0^{\frac{\pi}{2}} \left(Q_{\text{sca}}^{\text{TE}} + Q_{\text{sca}}^{\text{TM}}\right) G_{\text{sca}} \sin\alpha \, \mathrm{d}\alpha \,, \\ \langle C_{\text{bk}} \rangle &= \frac{1}{2} \int_0^{\frac{\pi}{2}} \left(Q_{\text{bk}}^{\text{TE}} + Q_{\text{bk}}^{\text{TM}}\right) G_{\text{bk}} \sin\alpha \, \mathrm{d}\alpha \,. \end{aligned} \tag{5.238}$$

From Eqs. (5.234)–(5.235) and the orthogonality properties of the prolate SAFs, we get in the first approximation with respect to $(b/a)^2$

$$\langle C_{\text{sca}} \rangle = \frac{2\pi}{k^2} \sum_{l=1}^{\infty} \left| \frac{\left[(\xi_0^2-1)^{\frac{1}{2}} R_{1l}^{(1)}(c,\xi_0)\right]'}{\left[(\xi_0^2-1)^{\frac{1}{2}} R_{1l}^{(2)}(c,\xi_0)\right]'} \right|^2 , \tag{5.239}$$

$$\langle C_{\text{bk}} \rangle = \frac{8\pi}{k^2} \int_0^{\frac{\pi}{2}} \left| \sum_{l=1}^{\infty} \frac{\left[(\xi_0^2-1)^{\frac{1}{2}} R_{1l}^{(1)}(c,\xi_0)\right]'}{\left[(\xi_0^2-1)^{\frac{1}{2}} R_{1l}^{(2)}(c,\xi_0)\right]'} \right. \times N_{1l}^{-2}(c)\,(-1)^{l-1}\, S_{1l}^2(c,\cos\alpha) \Bigg|^2 \sin\alpha \, \mathrm{d}\alpha \,. \tag{5.240}$$

Under the condition (5.190), the principal term of the asymptotics gives the resonance term (5.225), and hence

$$\langle C_{\text{sca}} \rangle = \frac{2\pi}{k^2} \,, \tag{5.241}$$

$$\langle C_{\text{bk}} \rangle = \frac{2\pi}{k^2} \int_0^1 \frac{\cos^4\left(\frac{n\pi}{2}\eta\right)}{(1-\eta^2)^2} \, \mathrm{d}\eta \left[\int_0^1 \frac{\cos^4\left(\frac{n\pi}{2}\eta\right)}{(1-\eta^2)^2} \, \mathrm{d}\eta\right]^{-2} , \quad n = 2q-1 \,, \tag{5.242}$$

$$\langle C_{\text{bk}} \rangle = \frac{2\pi}{k^2} \int_0^1 \frac{\sin^4\left(\frac{n\pi}{2}\eta\right)}{(1-\eta^2)^2} \, \mathrm{d}\eta \left[\int_0^1 \frac{\sin^4\left(\frac{n\pi}{2}\eta\right)}{(1-\eta^2)^2} \, \mathrm{d}\eta\right]^{-2} , \quad n = 2q \,. \tag{5.243}$$

Numerical calculations of the characteristics of radiation scattered by extremely prolate perfectly conducting spheroids were performed by using the exact solution (see Section 5.2.3) and the approximation presented above.

Figures 5.1–5.2 show some results of exact calculations of the factors $Q_{\text{sca}}^{\text{TM}}$ and $Q_{\text{bk}}^{\text{TM}}$ in the case of the normal incidence. The absence of maxima under the condition $d = n\lambda$ (i.e. for even n in Eq. (5.190)) is related with the fact that the resonant term (5.228) is equal to zero as a result of oddness of the prolate SAFs. The factors $Q_{\text{sca}}^{\text{TE}}$ $Q_{\text{bk}}^{\text{TE}}$ are smaller than the corresponding factors for the TM mode by 10^3–10^4 times for $c = 1.0$ and by 5–10 times for $c = 10$, when the ratio $a/b = 10$, and by 10^6–10^8 and 10^2 times, when the ratio $a/b = 100$. For the parallel incidence, these factors are given in Tables 5.3–5.4. Note that the function $\varphi(c)$ tends to the limit $4\pi/3 \approx 4.19$ with increasing c. A comparison of the exact and approximate solutions shows that the differential characteristics of the scattered radiation are well represented by the approximate formulae, when $(c \cdot b/a) \leq 0.2$, and the integral characteristics (Q_{sca}), when $(c \cdot b/a) \leq 0.6$.

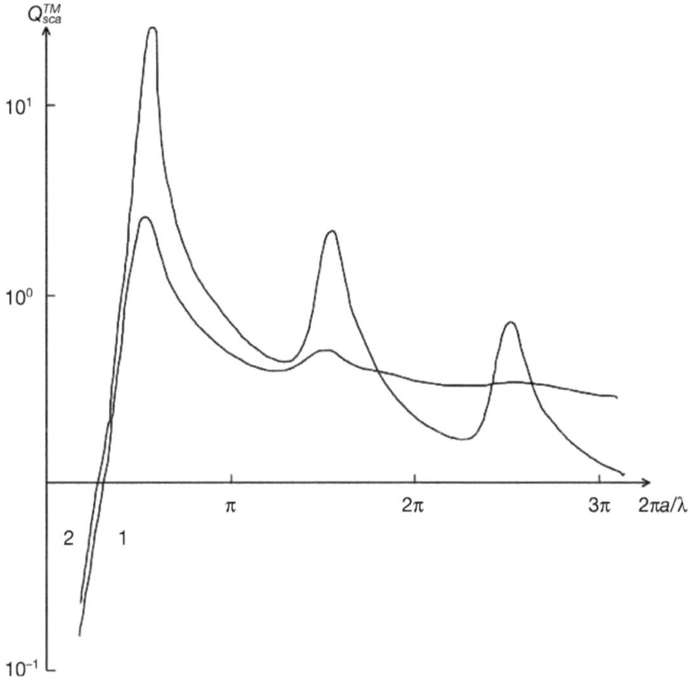

Fig. 5.1. Efficiency factors for scattering Q_{sca} for perfectly conducting prolate spheroids with the aspect ratio $a/b = 10$ (1) and 100 (2) in the dependence on the parameter $2\pi a/\lambda$. The case of the TM mode and the normal incidence ($\alpha = 90°$).

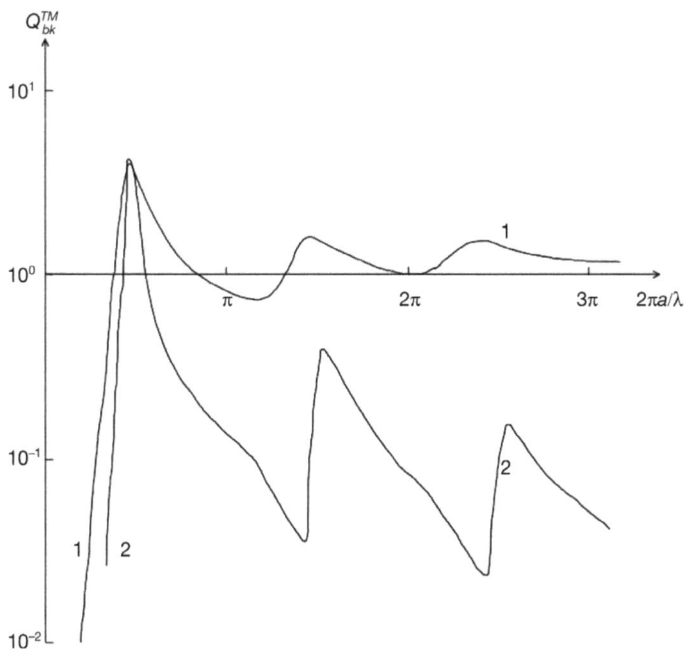

Fig. 5.2. The same as in Fig. 5.1, but for the efficiency factors for backscattering Q_{bk}.

Table 5.3. Efficiency factors for scattering Q_{sca} and backscattering Q_{bk} for perfectly conducting extremely prolate spheroids ($a/b = 100$, $\alpha = 0$)

c	$\varphi(c)$	$Q_{\text{sca}}^{\text{app}}$	Q_{sca}	$Q_{\text{bk}}^{\text{app}}$	Q_{bk}
1.0	1.473	1.47-4	1.46-4	3.03	2.99
2.0	3.290	6.58-4	6.56-4	8.63-1	8.91-1
3.0	3.493	1.05-3	1.05-3	4.05	4.02
4.0	3.673	1.47-3	1.46-3	2.90-1	2.62-1
5.0	3.807	1.90-3	1.90-3	2.46	2.52
6.0	3.843	2.31-3	2.30-3	3.16	3.09
7.0	3.905	2.73-3	2.73-3	1.74-2	2.79-2
8.0	3.943	3.15-3	3.15-3	3.53	3.58
9.0	3.960	3.56-3	3.56-3	1.97	1.88
10.0	3.992	3.99-3	3.99-3	5.25-1	5.88-1

Table 5.4. Efficiency factors for scattering Q_{sca} and backscattering Q_{bk} for perfectly conducting extremely prolate spheroids ($\alpha = 0$)

	Q_{sca}			Q_{bk}		
c	$a/b = 10$	$a/b = 50$	$a/b = 100$	$a/b = 10$	$a/b = 50$	$a/b = 100$
1.0	1.47-2	5.83-4	1.46-4	3.03	2.99	2.99
2.0	6.56-2	2.63-3	6.56-4	8.67-1	8.89-1	8.91-1
3.0	1.04-1	4.18-3	1.05-3	4.08	4.02	4.02
4.0	1.44-1	5.86-3	1.46-3	2.99-1	2.65-1	2.62-1
5.0	1.86-1	7.58-3	1.90-3	2.50	2.51	2.52
6.0	2.24-1	9.19-3	2.30-3	3.28	3.10	3.09
7.0	2.63-1	1.09-2	2.73-3	1.44-2	2.66-2	2.79-2
8.0	3.01-1	1.26-2	3.15-3	3.68	3.57	3.58
9.0	3.37-1	1.42-2	3.56-3	2.19	1.90	1.88
10.0	3.75-1	1.59-2	3.99-3	5.05-1	5.81-1	5.88-1

From Eq. (5.235) one can derive the condition of minimum of the efficiency factors for backscattering $Q_{\text{bk}}^{\text{TE}}$

$$2c \cos\alpha = u_0 \,, \tag{5.244}$$

where u_0 are the roots of the functions $G(u)$ that satisfy the condition $\tan u_0 = u_0$. With an increasing incidence angle α, the number of minima (and hence that of maxima) of the function $Q_{\text{bk}}^{\text{TE}}(c)$ decreases. For the normal incidence ($\alpha = 90°$, $G(0) = 1$), this factor increases with an increasing c in the region of applicability of the approximation.

For the parallel incidence, the dimensionless parameters of the scattered radiation intensity have minima at the points corresponding to the roots of the function $G(c(1 - \cos\theta))$ except for scattering backward (see Eq. (5.232) and Fig. 5.3). Backscattering is determined by

$$\frac{16}{9} c^6\, G^2(2c) = \frac{1}{4} (\sin 2c - 2c \cos 2c)^2 \,, \tag{5.245}$$

Table 5.5. Cross-sections for scattering $\langle C_{\rm sca}\rangle$ and backscattering $\langle C_{\rm bk}\rangle$ for perfectly conducting extremely prolate spheroids ($a/b = 100$, $k = \sqrt{2\pi}$)

	$\langle C_{\rm sca}\rangle$		$\langle C_{\rm bk}\rangle$	
c	Eq.(5.241)	Eq.(5.239)	Eq.(5.242)	Eq.(5.240)
$\pi/2$	1.0	1.0	1.27	1.27
π	1.0	1.08	1.39	1.37
$3\pi/2$	1.0	1.19	1.54	1.49
2π	1.0	1.29	1.68	1.70
$5\pi/2$	1.0	1.40	1.83	1.75
3π	1.0	1.49	1.97	2.01

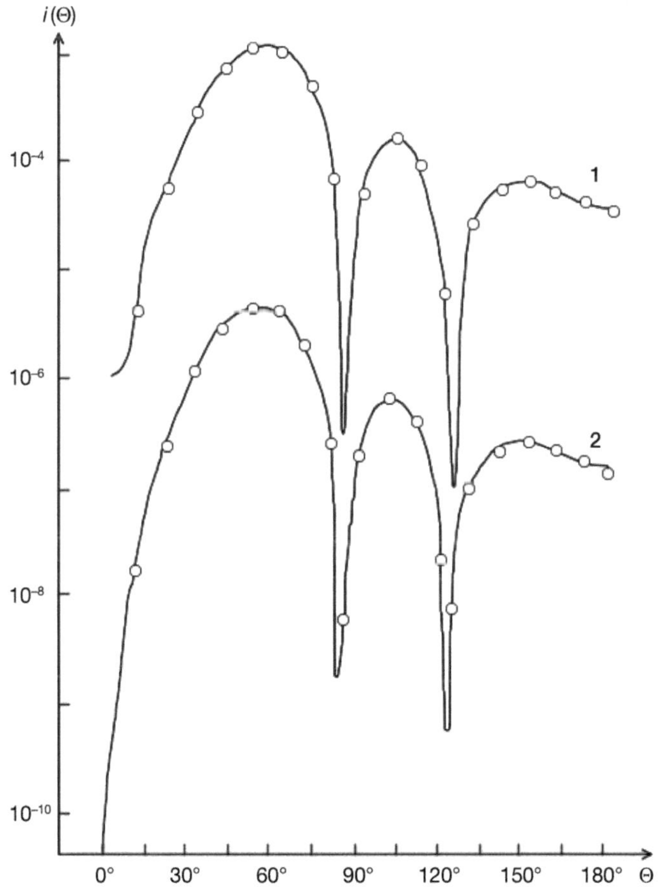

Fig. 5.3. Dimensionless phase function $i(\Theta)$ for perfectly conducting prolate spheroids with the aspect ratio $a/b = 10$ (1) and 100 (2). Circles show the results obtained with the approximate solution suggested. The case of the parallel incidence ($\alpha = 0$) of non-polarized radiation.

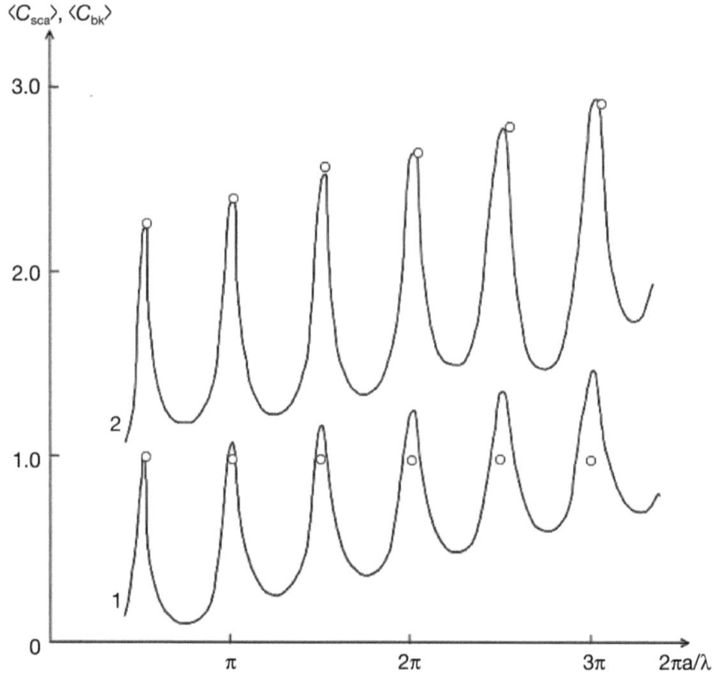

Fig. 5.4. Normalized scattering $\langle C_{\rm sca}\rangle$ (1) and backscattering $\langle C_{\rm bk}\rangle$ (2) cross-sections for randomly oriented perfectly conducting prolate spheroids with the aspect ratio $a/b = 100$. Circles show the results obtained with the approximate solution suggested. The case of the wave number $k = \sqrt{2\pi}$.

where the maximum is obtained under the condition (5.190). The same result occurs for dielectric spheroids.

For randomly oriented perfectly conducting thin needles, the maximum of the cross-sections $\langle C_{\rm sca}\rangle$ and $\langle C_{\rm bk}\rangle$ is reached under the condition (5.190) (see Fig. 5.4). Numerical calculations performed with Eqs. (5.239)–(5.243) show that the range of applicability of Eqs. (5.242)–(5.243) for the backscattering factors $\langle C_{\rm bk}\rangle$ is essentially wider than that of Eqs. (5.241) for the integral scattering factors $\langle C_{\rm sca}\rangle$ (see Table 5.5). It is explained by the fact that the contributions of the partial waves weakly decrease with an increasing semiaxis ratio ($a/b \sim 1/(\ln a/b)$), and for the cross-sections $\langle C_{\rm sca}\rangle$ the contributions are added, while for $\langle C_{\rm bk}\rangle$ they are damped.

The approximation considered above gives good results for $kb \ll 1$, $ka = O(1)$ (i.e. for $b/a \ll 1$, $c \gg O(1)$).

5.5 Scattering of a plane electromagnetic wave by extremely oblate perfectly conducting spheroids

Solution to the light scattering problem for a perfectly conducting circular disk has been known for a long time (Meixner, 1950; Meixner and Andrejewski, 1953; Andrejewski, 1953; Flammer, 1953). In these papers the solution was obtained by the separation of variables method formulated in oblate spheroidal coordinates. Difficulties arise as for the disk with $b/a = 0$ there appear additional boundary conditions at the edge (the so-called Meixner conditions) which guarantee uniqueness of the problem solution. Only in the case of the axisymmetric excitation of a perfectly conducting oblate spheroid (including the disk) when the source of the incident radiation is a dipole located at the symmetry axis of the spheroid and aligned along this axis, the problem was correctly solved by using the Abraham potentials (Meixner and Andrejewski, 1953). The solution to the problem of the plane wave scattering by perfectly conducting oblate spheroids including the disks has not been found.

5.5.1 Problem formulation

Our solution to the light scattering problem for spheroidal particles has been found to be efficient for both prolate and oblate dielectric (absorbing) spheroids. However, for extremely oblate perfectly conducting particles, the solution is not appropriate as in the limiting case of the disk it does not satisfy the Meixner conditions. Below the solutions is improved so that in the case of the perfectly conducting disk the Meixner conditions at the edge are satisfied automatically.

As before, the electric and magnetic fields are represented by sums of two components where one component is independent of the azimuthal angle φ and averaging of the other over this angle gives zero. The diffraction problem for the axisymmetric components is solved just as above by using the Abraham potentials. The non-axisymmetric components of the scattered field are represented now as follows:

$$\boldsymbol{E}^{(1)}_{2,\mathrm{TE}} = \mathrm{rot}\left(U^{(1)}\,\boldsymbol{i}_\mathrm{z} + V^{(1)}\,\boldsymbol{r} + \Pi_\mathrm{x}\,\boldsymbol{i}_\mathrm{x} + \Pi_y\,\boldsymbol{i}_\mathrm{y}\right), \tag{5.246}$$

$$\boldsymbol{E}^{(1)}_{2,\mathrm{TM}} = \frac{\mathrm{i}}{k}\,\mathrm{rot}\,\mathrm{rot}\left(U^{(1)}\,\boldsymbol{i}_\mathrm{z} + V^{(1)}\,\boldsymbol{r} + \Pi_\mathrm{x}\,\boldsymbol{i}_\mathrm{x} + \Pi_\mathrm{y}\,\boldsymbol{i}_\mathrm{y}\right). \tag{5.247}$$

The representation of the incident fields does not change (see Eqs. (5.42)–(5.44), (5.48)). The new scalar potentials Π_x and Π_y satisfy the Helmholtz equation and are expanded in terms of spheroidal wave functions

$$\Pi_\mathrm{x} = \sum_{m=0}^{\infty}\sum_{l=m}^{\infty}(c_{m+1} - c_{m-1})\,A_{ml}\,S_{ml}(-\mathrm{i}c,\eta)\,R^{(3)}_{ml}(-\mathrm{i}c,\mathrm{i}\xi)\cos m\phi\,, \tag{5.248}$$

$$\Pi_\mathrm{y} = \sum_{m=0}^{\infty}\sum_{l=m}^{\infty}(c_{m+1} - c_{m-1}),A_{ml}\,S_{ml}(-\mathrm{i}c,\eta)\,R^{(3)}_{ml}(-\mathrm{i}c,\mathrm{i}\xi)\sin m\phi\,, \tag{5.249}$$

where

$$A_{ml} = \mathrm{i}^{l-m} \frac{S'_{ml}(-\mathrm{i}c, 0)}{N^2_{ml}(-\mathrm{i}c)} \frac{R^{(1)'}_{ml}(-\mathrm{i}c, \mathrm{i}\xi_0)}{R^{(3)'}_{ml}(-\mathrm{i}c, \mathrm{i}\xi_0)} \tag{5.250}$$

for the TE mode, and

$$A_{ml} = \mathrm{i}^{l-m} \frac{S_{ml}(-\mathrm{i}c, 0)}{N^2_{ml}(-\mathrm{i}c)} \frac{R^{(1)}_{ml}(-\mathrm{i}c, \mathrm{i}\xi_0)}{R^{(3)}_{ml}(-\mathrm{i}c, \mathrm{i}\xi_0)} \tag{5.251}$$

for the TM mode, respectively. The expansions of the scattered field potentials $U^{(1)}$ and $V^{(1)}$ are as above (see Eqs. (5.45)).

Let us introduce the quantities

$$\Phi = \Pi_{\mathrm{x}} \cos\varphi + \Pi_{\mathrm{y}} \sin\varphi , \qquad \Psi = -\Pi_{\mathrm{x}} \sin\varphi + \Pi_{\mathrm{y}} \cos\varphi , \tag{5.252}$$

which are represented according to Eqs. (5.248)–(5.249) as follows ($c_{-1} = c_0 = 0$):

$$\begin{aligned} \Phi = \sum_{m=1}^{\infty} \sum_{l=m-1}^{\infty} c_m \Big[& A_{m-1,l}\, S_{m-1,l}(-\mathrm{i}c, \eta)\, R^{(3)}_{m-1,l}(-\mathrm{i}c, \mathrm{i}\xi) \\ & - A_{m+1,l}\, S_{m+1,l}(-\mathrm{i}c, \eta)\, R^{(3)}_{m+1,l}(-\mathrm{i}c, \mathrm{i}\xi) \Big] \cos m\varphi , \\ \Psi = - \sum_{m=1}^{\infty} \sum_{l=m-1}^{\infty} c_m \Big[& A_{m-1,l}\, S_{m-1,l}(-\mathrm{i}c, \eta)\, R^{(3)}_{m-1,l}(-\mathrm{i}c, \mathrm{i}\xi) \\ & - A_{m+1,l}\, S_{m+1,l}(-\mathrm{i}c, \eta)\, R^{(3)}_{m+1,l}(-\mathrm{i}c, \mathrm{i}\xi) \Big] \sin m\varphi . \end{aligned} \tag{5.253}$$

From the main integral relation for the SRFs and SAFs (Komarov *et al.*, 1976), we have

$$\begin{aligned} & S_{ml}(-\mathrm{i}c, \mathrm{i}\eta)\, R^{(1)}_{ml}(-\mathrm{i}c, \mathrm{i}\xi) = \\ & \frac{\mathrm{i}^{m-l}}{2} \int_{-1}^{1} \mathrm{e}^{\mathrm{i}c\,\xi\eta\, t} J_m \left[c(\xi^2+1)^{\frac12} (1-\eta^2)^{\frac12} (1-t^2)^{\frac12} \right] S_{ml}(-\mathrm{i}c, t)\, \mathrm{d}t , \end{aligned} \tag{5.254}$$

and by using completeness and orthogonality of the oblate SAFs, we get

$$\begin{aligned} & \mathrm{e}^{\mathrm{i}c\,\xi\eta\, t} J_m \left[c(\xi^2+1)^{\frac12} (1-\eta^2)^{\frac12} (1-t^2)^{\frac12} \right] = \\ & 2 \sum_{l=m}^{\infty} \mathrm{i}^{l-m}\, N^{-2}_{ml}(-\mathrm{i}c)\, S_{ml}(-\mathrm{i}c, \eta)\, R^{(1)}_{ml}(-\mathrm{i}c, \mathrm{i}\xi)\, S_{ml}(-\mathrm{i}c, t) . \end{aligned} \tag{5.255}$$

Now for $t = 0$ we find that

$$
\begin{aligned}
J_m(k\varrho) &= 2\sum_{l=m}^{\infty} \mathrm{i}^{l-m}\, N_{ml}^{-2}(-\mathrm{i}c)\, S_{ml}(-\mathrm{i}c,0)\, S_{ml}(-\mathrm{i}c,\eta)\, R_{ml}^{(1)}(-\mathrm{i}c,\mathrm{i}\xi)\,,\\
\mathrm{i}c\,\xi\,\eta J_m(k\varrho) &= 2\sum_{l=m}^{\infty} \mathrm{i}^{l-m}\, N_{ml}^{-2}(-\mathrm{i}c)\, S'_{ml}(-\mathrm{i}c,0)\, S_{ml}(-\mathrm{i}c,\eta)\, R_{ml}^{(1)}(-\mathrm{i}c,\mathrm{i}\xi)\,,\\
\mathrm{i}c\,\eta &\left[J_m(k\varrho) + \frac{c\xi^2(1-\eta^2)^{\frac{1}{2}}}{(\xi^2+1)^{\frac{1}{2}}} J'_m(k\varrho)\right] = \\
& 2\sum_{l=m}^{\infty} \mathrm{i}^{l-m}\, N_{ml}^{-2}(-\mathrm{i}c)\, S'_{ml}(-\mathrm{i}c,0)\, S_{ml}(-\mathrm{i}c,\eta)\, R_{ml}^{(1)'}(-\mathrm{i}c,\mathrm{i}\xi)\,,
\end{aligned}
\tag{5.256}
$$

where $k\varrho = c\,(\xi^2+1)^{\frac{1}{2}}\,(1-\eta^2)^{\frac{1}{2}}$, $J_m(z)$ is the Bessel function of the first kind.

From Eqs. (5.250)–(5.256) valid at the surface of the spheroid ($\xi = \xi_0$), we have:

(1) for the TE mode

$$
\begin{aligned}
\frac{\partial \Phi}{\partial \xi} &= \frac{\mathrm{i}c\,\eta}{2} \sum_{m=1}^{\infty} c_m \Big[J_{m-1}(k\varrho) - J_{m+1}(k\varrho)\\
&\quad + \frac{c\,\xi_0^2(1-\eta^2)^{\frac{1}{2}}}{(\xi_0^2+1)^{\frac{1}{2}}} \left(J'_{m-1}(k\varrho) - J'_{m+1}(k\varrho)\right)\Big] \cos m\varphi\,,\\
\frac{\partial \Psi}{\partial \xi} &= -\frac{\mathrm{i}c\,\eta}{2} \sum_{m=1}^{\infty} c_m \Big[J_{m-1}(k\varrho) - J_{m+1}(k\varrho)\\
&\quad + \frac{c\,\xi_0^2(1-\eta^2)^{\frac{1}{2}}}{(\xi_0^2+1)^{\frac{1}{2}}} \left(J'_{m-1}(k\varrho) + J'_{m+1}(k\varrho)\right)\Big] \sin m\varphi\,;
\end{aligned}
\tag{5.257}
$$

(2) for the TM mode

$$
\begin{aligned}
\Phi &= \frac{1}{2} \sum_{m=1}^{\infty} c_m \left[J_{m-1}(k\varrho) - J_{m+1}(k\varrho)\right] \cos m\varphi\,,\\
\Psi &= -\frac{1}{2} \sum_{m=1}^{\infty} c_m \left[J_{m-1}(k\varrho) + J_{m+1}(k\varrho)\right] \sin m\varphi\,.
\end{aligned}
\tag{5.258}
$$

The fields described by Eqs. (5.246)–(5.247), (5.45), (5.248)–(5.249) satisfy the Maxwell equations and the radiation condition at infinity. The unknown coefficients $a_{ml}^{(1)}$ and $b_{ml}^{(1)}$ can be found from the standard boundary conditions (5.58), while the coefficients c_m from the Meixner conditions at the edge of the perfectly conducting disk. The physical sense of the latter is associated with the demand of energy finiteness, i.e. the electromagnetic energy density of the scattered radiation in the vicinity of the edge must be quadratically integrable or, which is equivalent, the edge should not radiate (Meixner, 1950)

$$
j_\varrho = \left(\left[\boldsymbol{H}^{(1)} \times \boldsymbol{i}_\xi\right]\Big|_{\xi=0,\,\eta=+0} - \left[\boldsymbol{H}^{(1)} \times \boldsymbol{i}_\xi\right]\Big|_{\xi=0,\,\eta=-0},\ \boldsymbol{i}_\eta\right) = 0\,,
\tag{5.259}
$$

where j_ϱ is the radial component of the current induced on the disk. In the case under consideration, the Meixner conditions are written as follows:

$$\left.\begin{aligned} \frac{\partial \Phi}{\partial \xi} + \frac{\partial U^{(1)}}{\partial \eta} + \frac{d}{2}\frac{\partial V^{(1)}}{\partial \xi} &= 0\,, \\ \frac{\partial \Phi}{\partial \eta} - \frac{\partial U^{(1)}}{\partial \xi} + \frac{d}{2}\frac{\partial V^{(1)}}{\partial \eta} &= 0\,. \end{aligned}\right\}_{\xi=0,\,\eta=0} \qquad (5.260)$$

To obtain the conditions (5.260), one should consider the current density for the TM mode wave

$$\begin{aligned} j_\eta &= \frac{(\xi^2+1)^{\frac{1}{2}}\,(1-\eta^2)^{\frac{1}{2}}}{(\xi^2+\eta^2)} \left\{ \xi \left[\frac{(1-\eta^2)^{\frac{1}{2}}}{(\xi^2+1)^{\frac{1}{2}}} \frac{\partial \Phi}{\partial \eta} - \frac{\partial U}{\partial \xi} + \frac{d}{2}\frac{\partial V}{\partial \eta} \right] \right. \\ &\quad \left. - \eta \left[\frac{(\xi^2+1)^{\frac{1}{2}}}{(1-\eta^2)^{\frac{1}{2}}} \frac{\partial \Phi}{\partial \xi} + \frac{\partial U}{\partial \eta} - \frac{d}{2}\frac{\partial V}{\partial \xi} \right] \right\}, \\ j_\phi &= \frac{(1-\eta^2)^{\frac{1}{2}}}{(\xi^2+1)^{\frac{1}{2}}} \left\{ \frac{\partial}{\partial \xi} \left[(\xi^2+1)^{\frac{1}{2}}\,(1-\eta^2)^{\frac{1}{2}}\,\Psi \right] \right. \\ &\quad \left. - \frac{\partial}{\partial \varphi} \left[\xi \frac{(1-\eta^2)^{\frac{1}{2}}}{(\xi^2+1)^{\frac{1}{2}}} \Phi + \eta\, U + \xi \frac{d}{2} V \right] \right\}. \end{aligned} \qquad (5.261)$$

It is easy to see that the requirement $j_\varrho = -j_\eta = 0$ at the disk edge leads to the first equation in Eqs. (5.260). The second equation is an identity within the suggested solution (see Eqs. (5.246)–(5.253), (5.260)). Note that the azimuthal component of the current density j_φ changes in the edge vicinity as $R^{-\frac{1}{2}}$, where R is the distance from the point to the edge ($R^2 \sim (\xi^2 + 1 - \eta^2)$) while the radial component j_η is finite.

For a plane wave of the TE mode, we have

$$\begin{aligned} \boldsymbol{H} = &-\frac{\mathrm{i}}{k} \left[\operatorname{grad} \operatorname{div} \left(U\,\boldsymbol{i}_\mathrm{z} + V\,\boldsymbol{r} + \Pi_\mathrm{x}\,\boldsymbol{i}_\mathrm{x} + \Pi_\mathrm{y}\,\boldsymbol{i}_\mathrm{y} \right) \right. \\ &\left. -2 \operatorname{grad} V + k^2 \left(U\,\boldsymbol{i}_\mathrm{z} + V\,\boldsymbol{r} + \Pi_\mathrm{x}\,\boldsymbol{i}_\mathrm{x} + \Pi_\mathrm{y}\,\boldsymbol{i}_\mathrm{y} \right) \right]. \end{aligned} \qquad (5.262)$$

In this case, the Meixner conditions can also be written in the form (5.260), and then the first equation is an identity. The current density in the vicinity of the edge behaves like the TM mode wave.

The difference between the representations of the non-axisymmetric components of the scattered field given by Eqs. (5.246)–(5.247) and by the equations used by us above consists in appearance of $\Pi_\mathrm{x}\,\boldsymbol{i}_\mathrm{x}$ and $\Pi_\mathrm{y}\,\boldsymbol{i}_\mathrm{y}$ that allow one to find the solution for which the Meixner conditions (5.260) for a perfectly conducting disk are satisfied automatically. Note that the axisymmetric components of the scattered field satisfy the Meixner conditions at the edge of a perfectly conducting disk without any additions.

5.5.2 Derivation of the scattered field for the TE mode

In the case of the TE mode plane wave, the boundary conditions are ($U = U^{(1)} + U^{(2)}$, $V = V^{(1)}$)

$$\left.\begin{aligned} \frac{\partial}{\partial\varphi}\left(\eta U + \frac{d}{2}\xi V\right) &= \frac{\partial}{\partial\xi}\left[(\xi^2+1)^{\frac{1}{2}}(1-\eta^2)^{\frac{1}{2}}\Psi\right] - \xi\frac{(1-\eta^2)^{\frac{1}{2}}}{(\xi^2+1)^{\frac{1}{2}}}\frac{\partial\Phi}{\partial\varphi}, \\ \frac{\partial^2}{\partial\xi\partial\varphi}\left(\eta U - \frac{d}{2}\xi V\right) &= \frac{\partial^2}{\partial\xi\partial\eta}\left[(\xi^2+1)^{\frac{1}{2}}(1-\eta^2)^{\frac{1}{2}}\Psi\right] + \eta\frac{(1-\eta^2)^{\frac{1}{2}}}{(\xi^2+1)^{\frac{1}{2}}}\frac{\partial\Phi}{\partial\varphi}. \end{aligned}\right\}_{\xi=\xi_0} \tag{5.263}$$

By using Eqs. (5.253), after laborious transformations we can rewrite the systems in the form

$$\begin{aligned} \frac{\partial}{\partial\varphi}\left(\eta U + \frac{d}{2}\xi V\right) &= -\mathrm{i}\,\eta\sum_{m=1}^{\infty} m\,c_m\,J_m(k\varrho)\sin m\varphi \\ &+ \xi\frac{(1-\eta^2)^{\frac{1}{2}}}{(\xi^2+1)^{\frac{1}{2}}}\left\{\psi + \mathrm{i}c\,\xi\,\eta\sum_{m=1}^{\infty}\frac{c_m}{2}\left(J_{m-1}(k\varrho)+J_{m+1}(k\varrho)\right)\sin m\varphi\right. \\ &\left.- \frac{\partial}{\partial\varphi}\left[\Phi - \mathrm{i}c\,\eta\,\xi\sum_{m=1}^{\infty}\frac{c_m}{2}\left(J_{m-1}(k\varrho)+J_{m+1}(k\varrho)\right)\cos m\varphi\right]\right\}, \\ \frac{\partial^2}{\partial\xi\partial\varphi}\left(\xi U - \frac{d}{2}\eta V\right) &= \frac{\partial}{\partial\xi}\left[-\mathrm{i}\eta\sum_{m=1}^{\infty} m\,c_m\,J_m(k\varrho)\sin m\varphi\right] + \frac{\xi}{(\xi^2+1)^{\frac{1}{2}}} \\ &\times\left\{(1-\eta^2)^{\frac{1}{2}}\left[\psi + \mathrm{i}c\,\xi\,\eta\sum_{m=1}^{\infty}\frac{c_m}{2}\left(J_{m-1}(k\varrho)+J_{m+1}(k\varrho)\right)\sin m\varphi\right]\right. \\ &\left.+ \frac{\partial}{\partial\varphi}\frac{\eta}{(1-\eta^2)^{\frac{1}{2}}}\left[\Phi - \mathrm{i}c\,\eta\,\xi\sum_{m=1}^{\infty}\frac{c_m}{2}\left(J_{m-1}(k\varrho)-J_{m+1}(k\varrho)\right)\cos m\varphi\right]\right\}. \end{aligned} \tag{5.264}$$

We introduce the additional notation $\tilde{\boldsymbol{F}}_m = \{\tilde{f}_{ml}\}_{l=m}^{\infty}$, $\boldsymbol{G}_{1m} = \{g_{1l}^{(m)}\}_{l=m}^{\infty}$, $P^{m\mp1} = \{\varphi_{nl}^{m,m\mp1}\}_{n,l=m}^{\infty}$, $A^{m\mp1} = \{\alpha_{nl}^{m,m\mp1}\}_{n,l=m}^{\infty}$, where

$$\begin{aligned} \tilde{f}_{ml} &= f_{ml} + \mathrm{i}^{l-m+1}k\,c_m\,N_{ml}^{-1}(-\mathrm{i}c)\,S_{ml}(-\mathrm{i}c,0)\,R_{ml}^{(1)}(-\mathrm{i}c,\mathrm{i}\xi_0), \\ g_{1l}^{(m)} &= \mathrm{i}^{l-m}\,N_{ml}^{-1}(-\mathrm{i}c)\,S'_{ml}(-\mathrm{i}c,0)\,\frac{1}{R_{ml}^{(3)'}(-\mathrm{i}c,\mathrm{i}\xi_0)}. \end{aligned} \tag{5.265}$$

The integrals of products of the oblate SAFs and their derivatives, $\varphi_{nl}^{m,m\mp1}$ and $\alpha_{nl}^{m,m\mp1}$, are defined in Appendix A.

We substitute the expansions (5.246), (5.45) and (5.253)–(5.254) in the boundary conditions (5.264), multiply them by $N_{ml}^{-1}(-\mathrm{i}c)\,S_{ml}(-\mathrm{i}c,\eta)\cos m\varphi$ and integrate over φ from 0 to 2π and over η from -1 to 1.

By using the Wronskian of the oblate SRFs (Komarov *et al.*, 1976) and introducing the unknown vector $\boldsymbol{Z}_1$, we get the ISLAEs that can be written in the matrix form as follows ($m = 1, 2, \ldots$):

$$\begin{cases} [\xi_0 \, I + \Gamma(I + \xi_0 \, R_1)^{-1} \Gamma R_1] \, \boldsymbol{Z}_2 = -\xi_0 \, \Gamma \, (I + \xi_0 \, R_1)^{-1} (R_1 - R_0) \, \tilde{\boldsymbol{F}}_m \\ \quad + \dfrac{ik \, c_m \, \xi_0}{c(\xi_0^2 + 1)^{\frac{3}{2}}} \left[\dfrac{m-1}{m} \, (P^{m-1} \boldsymbol{G}_{1,m-1} - P^{m+1} \boldsymbol{G}_{1,m+1}) \right. \\ \quad \left. + \, \Gamma(I + \xi_0 R_1)^{-1} (A^{m-1} \boldsymbol{G}_{1,m-1} - A^{m+1} \boldsymbol{G}_{1,m+1}) \right] , \\ \boldsymbol{Z}_1 = -(I + \xi_0 R_1)^{-1} \left[(I + \xi_0 R_0) \, \tilde{\boldsymbol{F}}_m - \Gamma \boldsymbol{Z}_2 \right. \\ \quad \left. + \dfrac{ikc_m \xi_0}{c(\xi_0^2 + 1)^{\frac{3}{2}}} \left(A^{m-1} \boldsymbol{G}_{1,m-1} - A^{m+1} \boldsymbol{G}_{1,m+1} \right) \right] . \end{cases} \tag{5.266}$$

The unknown coefficients c_m are determined from the Meixner conditions. For a perfectly conducting disk ($\xi_0 = 0$), the systems (5.266) can be solved explicitly

$$\boldsymbol{Z}_1 = -\tilde{\boldsymbol{F}}_m \, , \qquad \boldsymbol{Z}_2 = 0 \tag{5.267}$$

and

$$\begin{aligned} a_{ml}^{(1)} = & - \frac{4\mathrm{i}^{l-1}}{k \sin \alpha} \, N_{ml}^{-2}(-\mathrm{i}c) \, S_{ml}(-\mathrm{i}c, \cos \alpha) \\ & + \mathrm{i}^{l-m+1} c_m N_{ml}^{-2}(-\mathrm{i}c) \, S_{ml}(-\mathrm{i}c, 0) \frac{R_{ml}^{(1)}(-\mathrm{i}c, \mathrm{i}\xi_0)}{R_{ml}^{(3)}(\mathrm{i}c, \mathrm{i}\xi_0)} \, , \\ b_{ml}^{(1)} = & \, 0 \, . \end{aligned} \tag{5.268}$$

After the substitution of the expansions (5.253) and (5.45) into the Meixner conditions (5.260), keeping in mind the relation (5.268), we find

$$\begin{aligned} c_m = & \sum_{l=m}^{\infty} \frac{4\mathrm{i}^{l-1}}{k \sin \alpha} \frac{S_{ml}(-\mathrm{i}c, \cos \alpha) \, S_{ml}(-\mathrm{i}c, 0)}{N_{ml}^2(-\mathrm{i}c) \, R_{ml}^{(3)}(-\mathrm{i}c, 0)} \\ & \times \left[2 \sum_{l=m}^{\infty} \frac{\mathrm{i}^{l-m+1} S_{ml}^2(-\mathrm{i}c, 0)}{N_{ml}^2(-\mathrm{i}c) \, R_{ml}^{(3)}(-\mathrm{i}c, 0)} \right. \\ & + \sum_{l=m-1}^{\infty} \mathrm{i}^{l-m+1} \left(\frac{S_{m-1,l}^2(-\mathrm{i}c, 0)}{N_{m-1,l}^2(-\mathrm{i}c) \, R_{m-1,l}^{(3)\prime}(\mathrm{i}c, 0)} \right. \\ & \left. \left. + \frac{S_{m+1,l}^2(-\mathrm{i}c, 0)}{N_{m+1,l}^2(-\mathrm{i}c) \, R_{m+1,l}^{(3)\prime}(\mathrm{i}c, 0)} \right) \right]^{-1} . \end{aligned} \tag{5.269}$$

Thus, the unknown coefficients of the scattered field expansion (5.45) are derived from the ISLAEs (5.266), where the coefficients c_m are calculated from Eq. (5.269).

5.5.3 Derivation of the scattered field for the TM mode

In the case of the TM mode plane wave, the boundary conditions are

$$\left.\begin{aligned}
&\frac{\partial}{\partial\varphi}\left(\xi U-\frac{d}{2}\eta V\right)=\eta\frac{(\xi^2+1)^{\frac{1}{2}}}{(1-\eta^2)^{\frac{1}{2}}}\frac{\partial\Phi}{\partial\varphi}\\
&\quad+\frac{\partial}{\partial\eta}\left[(\xi^2+1)^{\frac{1}{2}}(1-\eta^2)^{\frac{1}{2}}\Psi\right],\\
&(\xi^2+1)\frac{\partial^2}{\partial\xi\partial\varphi}\left(\eta U+\frac{d}{2}\xi V\right)\\
&\quad=(\xi^2+1)^{\frac{1}{2}}(1-\eta^2)^{\frac{1}{2}}\left(\eta\frac{\partial^2\Phi}{\partial\eta\partial\varphi}-\xi\frac{\partial^2\Phi}{\partial\xi\partial\varphi}\right)\\
&\quad-\frac{(\xi^2-\eta^2)}{(\xi_2+1)^{\frac{1}{2}}(1-\eta^2)^{\frac{1}{2}}}\left[\frac{\partial\Phi}{\partial\varphi}+\frac{\partial^2\Psi}{\partial\varphi^2}+c^2(\xi^2+1)(1-\eta^2)\Psi\right]\\
&\quad-(1-\eta^2)\left[\frac{\partial^2}{\partial\eta\partial\varphi}\left(\xi U-\frac{d}{2}\eta V\right)\right].
\end{aligned}\right\}_{\xi=\xi_0} \tag{5.270}$$

By using Eqs. (5.201), after laborious transformations we get

$$\left.\begin{aligned}
&\xi U-\frac{d}{2}\eta V=0,\\
&\frac{\partial}{\partial\xi}\left(\eta U+\frac{d}{2}\xi V\right)=\frac{\xi(1-\eta^2)^{\frac{1}{2}}}{(\xi^2+1)^{\frac{1}{2}}}\left[\frac{\partial}{\partial\xi}\sum_{m=1}^{\infty}\frac{c_m}{2}\right.\\
&\quad\left.\times\left(J_{m-1}(k\varrho)-J_{m+1}(k\varrho)\right)\cos m\varphi-\frac{\partial\Phi}{\partial\xi}\right].
\end{aligned}\right\}_{\xi=\xi_0} \tag{5.271}$$

In addition to notation (5.52) and (5.199), we introduce $\boldsymbol{G}_{2,m}=\{g_{2l}^{(m)}\}_{l=m}^{\infty}$, where

$$g_{2l}^{(m)}=\mathrm{i}^{l-m}N_{ml}^{-1}(-\mathrm{i}c)\,S_{ml}(-\mathrm{i}c,0)\frac{1}{R_{ml}^{(3)}(-\mathrm{i}c,0)}. \tag{5.272}$$

After the substitution of the scalar potential expansions into the boundary conditions (5.271) like in Section 5.5.2, we get the ISLAEs

$$\left\{\begin{aligned}
&\left[\xi_0 I+\Gamma(I+\xi_0)R_1)^{-1}\Gamma R_1\right]\boldsymbol{Z}_1=-\left[\xi_0 I+\Gamma(I+\xi_0 R_1)^{-1}\Gamma R_0\right]\boldsymbol{F}_m\\
&\quad-\frac{\mathrm{i}kc_m\xi_0}{c\,(\xi_0^2+1)^{\frac{3}{2}}}\Gamma(I+\xi_0 R_1)^{-1}\left(P^{m-1}\boldsymbol{G}_{2,m-1}-P^{m+1}\boldsymbol{G}_{2,m+1}\right),\\
&\boldsymbol{Z}_2=-(I+\xi_0 R_1)^{-1}\left[\Gamma\,(R_1\boldsymbol{Z}_1+R_0\boldsymbol{F}_m)\right.\\
&\quad\left.+\frac{\mathrm{i}kc_m\xi_0}{c(\xi_0^2+1)^{\frac{3}{2}}}\left(P^{m-1}\boldsymbol{G}_{2,m-1}-P_{m+1}\boldsymbol{G}_{2,m+1}\right)\right].
\end{aligned}\right. \tag{5.273}$$

To derive the coefficients c_m, we consider the case of the disk ($\xi_0 = 0$). Then the system (5.212) can be solved explicitly

$$\boldsymbol{Z}_1 = R^{-1} R_0 \boldsymbol{F}_m \,, \qquad \boldsymbol{Z}_2 = 0 \tag{5.274}$$

and hence

$$\begin{aligned} a_{ml}^{(1)} &= -\frac{4\,\mathrm{i}^{l-1}}{k \sin\alpha} N_{ml}^{-2}(-\mathrm{i}c)\, S_{ml}(-\mathrm{i}c, \cos\alpha) \\ &\quad + \mathrm{i}^{l-m+1} c_m N_{ml}^{-2}(-\mathrm{i}c)\, S_{ml}(-\mathrm{i}c, 0) \frac{R_{ml}^{(1)'}(-\mathrm{i}c, 0)}{R_{ml}^{(3)'}(\mathrm{i}c, 0)} \,, \\ b_{ml}^{(1)} &= 0. \end{aligned} \tag{5.275}$$

After the substitution of the expansions (5.253) and (5.45) into the Meixner conditions (5.260), keeping in mind the relation (5.275), we get

$$c_m = \frac{\sum\limits_{l=m}^{\infty} \frac{4\mathrm{i}^{l-1}}{k \sin\alpha} \frac{S_{ml}(-\mathrm{i}c, \cos\alpha)\, S'_{ml}(-\mathrm{i}c, 0)}{N_{ml}^2(-\mathrm{i}c)\, R_{ml}^{(3)'}(-\mathrm{i}c, 0)}}{\sum\limits_{l=m-1}^{\infty} \mathrm{i}^{l-m+1} \left(\frac{S_{m-1,l}^2(-\mathrm{i}c,0)}{N_{m-1,l}^2(-\mathrm{i}c)\, R_{m-1,l}^{(3)}(\mathrm{i}c,0)} + \frac{S_{m+1,l}^2(-\mathrm{i}c,0)}{N_{m+1,l}^2(-\mathrm{i}c)\, R_{m+1,l}^{(3)}(\mathrm{i}c,0)} \right)} \,. \tag{5.276}$$

Thus, the unknown coefficients of the scattered field expansion (5.45) are derived from the ISLAEs (5.273) where the coefficients c_m are calculated from Eq. (5.276).

The improved solution to the problem of the plane wave scattering by perfectly conducting oblate spheroids has several interesting features. Firstly, the infinite systems (5.266) and (5.273) differ from the corresponding systems of the initial solution only by the free terms. Secondly, after the limiting transition to spheres ($c \to 0$, $\xi_0 \to \infty$, $c\,\xi_0 = kr_0$, where r_0 is the sphere radius) we get the solution that coincides with the initial one as the coefficients $c_m \sim c$ and tend to zero.

The conclusions of the ISLAEs analysis performed above (see Section 5.3.4) are valid for the systems (5.266) and (5.273) as well, but the new solution allows the transition to the perfectly conducting disk.

5.5.4 Characteristics of the scattered radiation

In accordance with the solution presented above, the scattered radiation in the far-field zone is described as follows:

$$\begin{aligned}
\boldsymbol{E}_{\mathrm{TE}}^{(1)} = \frac{\mathrm{e}^{ikr}}{-\mathrm{i}kr} \Biggl\{ \Biggl[-\sum_{l=1}^{\infty} \mathrm{i}^{-l} a_l^{(1)} S_{1l}(-\mathrm{i}c, \cos\theta) \\
- \sum_{m=1}^{\infty} \Biggl(\sum_{l=m}^{\infty} \mathrm{i}^{-(l+1)} \bigl(k\, a_{ml}^{(1)} S_{ml}(-\mathrm{i}c, \cos\theta) \\
+ \mathrm{i}\, b_{ml}^{(1)} S'_{ml}(-\mathrm{i}c, \cos\theta)\bigr) \sin\theta \\
- k\, c_m \sum_{l=m-1}^{\infty} \mathrm{i}^{-(l+1)} \bigl(A_{m-1,l}\, S_{m-1,l}(-\mathrm{i}c, \cos\theta) \\
- A_{m+1,l}\, S_{m+1,l}(-\mathrm{i}c, \cos\theta)\bigr) \cos\theta \Biggr) \cos m\varphi \Biggr] \boldsymbol{i}_{\varphi} \\
+ \sum_{m=1}^{\infty} \Biggl[\sum_{l=m}^{\infty} \mathrm{i}^{-l} b_{ml}^{(1)} \frac{m\, S_{ml}(-\mathrm{i}c, \cos\theta)}{\sin\theta} \\
+ k\, c_m \sum_{l=m-1}^{\infty} \mathrm{i}^{-(l+1)} \bigl(A_{m-1,l}\, S_{m-1,l}(-\mathrm{i}c, \cos\theta) \\
+ A_{m+1,l}\, S_{m+1,l}(-\mathrm{i}c, \cos\theta)\bigr) \Biggr] \sin m\varphi\, \boldsymbol{i}_{\theta} \Biggr\}.
\end{aligned} \tag{5.277}$$

$$\begin{aligned}
\boldsymbol{E}_{\mathrm{TM}}^{(1)} = \frac{\mathrm{e}^{ikr}}{-\mathrm{i}kr} \Biggl\{ \Biggl[-\sum_{l=1}^{\infty} \mathrm{i}^{-l} b_l^{(1)} S_{1l}(-\mathrm{i}c, \cos\theta) \\
- \sum_{m=1}^{\infty} \Biggl(\sum_{l=m}^{\infty} \mathrm{i}^{-(l+1)} \bigl(k a_{ml}^{(1)} S_{ml}(-\mathrm{i}c, \cos\theta) \\
+ \mathrm{i}\, b_{ml}^{(1)} S'_{ml}(-\mathrm{i}c, \cos\theta)\bigr) \sin\theta \\
- k c_m \sum_{l=m-1}^{\infty} \mathrm{i}^{-(l+1)} \bigl(A_{m-1,l}\, S_{m-1,l}(-\mathrm{i}c, \cos\theta) \\
- A_{m+1,l}\, S_{m+1,l}(-\mathrm{i}c, \cos\theta)\bigr) \cos\theta \Biggr) \cos m\varphi \Biggr] \boldsymbol{i}_{\theta} \\
- \sum_{m=1}^{\infty} \Biggl[\sum_{l=m}^{\infty} \mathrm{i}^{-l} b_{ml}^{(1)} \frac{m S_{ml}(-\mathrm{i}c, \cos\theta)}{\sin\theta} \\
+ k\, c_m \sum_{l=m-1}^{\infty} \mathrm{i}^{-(l+1)} \bigl(A_{m-1,l}\, S_{m-1,l}(-\mathrm{i}c, \cos\theta) \\
+ A_{m+1,l}\, S_{m+1,l}(-\mathrm{i}c, \cos\theta)\bigr) \Biggr] \sin m\varphi\, \boldsymbol{i}_{\varphi} \Biggr\}.
\end{aligned} \tag{5.278}$$

From Eqs. (5.277)–(5.278), one can find the elements of the amplitude matrix, the dimensionless parameters of the scattered radiation intensity, the efficiency factors for extinction, scattering, and backscattering (see Section 5.2.5).

$$\begin{aligned} A_1 = &-\sum_{l=1}^{\infty} \mathrm{i}^{-l} a_l^{(1)} S_{1l}(-\mathrm{i}c, \cos\theta) \\ &-\sum_{m=1}^{\infty}\Bigg[\sum_{l=m}^{\infty} \mathrm{i}^{-(l+1)} \big(k a_{lm}^{(1)} S_{ml}(-\mathrm{i}c, \cos\theta) \\ &+ \mathrm{i} b_{ml}^{(1)} S'_{ml}(-\mathrm{i}c, \cos\theta)\big) \sin\theta \\ &- k c_m \sum_{l=m-1}^{\infty} \mathrm{i}^{-(l+1)} \big(A_{m-1,l} S_{m-1,l}(-\mathrm{i}c, \cos\theta) \\ &- A_{m+1,l} S_{m+1,l}(-\mathrm{i}c, \cos\theta)\big) \cos\theta\Bigg] \cos m\varphi\,, \end{aligned} \tag{5.279}$$

$$\begin{aligned} A_3 = &\sum_{m=1}^{\infty}\Bigg[\sum_{l=m}^{\infty} \mathrm{i}^{-l} b_{ml}^{(1)} \frac{m S_{ml}(-\mathrm{i}c, \cos\theta)}{\sin\theta} \\ &+ k\, c_m \sum_{l=m-1}^{\infty} \mathrm{i}^{-(l+1)} \big(A_{m-1,l} S_{m-1,l}(-\mathrm{i}c, \cos\theta) \\ &+ A_{m+1,l} S_{m+1,l}(-\mathrm{i}c, \cos\theta)\big)\Bigg] \sin m\varphi\,, \end{aligned} \tag{5.280}$$

$$\begin{aligned} A_4 = &-\sum_{m=1}^{\infty}\Bigg[\sum_{l=m}^{\infty} \mathrm{i}^{-l} b_{ml}^{(1)} \frac{m S_{ml}(-\mathrm{i}c, \cos\theta)}{\sin\theta} \\ &+ k c_m \sum_{l=m-1}^{\infty} \mathrm{i}^{-(l+1)} \big(A_{m-1,l} S_{m-1,l}(-\mathrm{i}c, \cos\theta) \\ &+ A_{m+1,l} S_{m+1,l}(-\mathrm{i}c, \cos\theta)\big)\Bigg] \sin m\varphi\,, \end{aligned} \tag{5.281}$$

$$\begin{aligned} A_2 = &-\sum_{l=1}^{\infty} \mathrm{i}^{-l} a_l^{(1)} S_{1l}(-\mathrm{i}c, \cos\theta) \\ &-\sum_{m=1}^{\infty}\Bigg[\sum_{l=m}^{\infty} \mathrm{i}^{-(l+1)} \big(k a_{ml}^{(1)} S_{ml}(-\mathrm{i}c, \cos\theta) \\ &+ \mathrm{i}\, b_{ml}^{(1)} S'_{ml}(-\mathrm{i}c, \cos\theta)\big) \sin\theta \\ &- k c_m \sum_{l=m-1}^{\infty} \mathrm{i}^{-(l+1)} \big(A_{m-1,l} S_{m-1,l}(-\mathrm{i}c, \cos\theta) \\ &- A_{m+1,l} S_{m+1,l}(-\mathrm{i}c, \cos\theta)\big) \cos\theta\Bigg] \cos m\varphi\,. \end{aligned} \tag{5.282}$$

$$\begin{aligned}
Q_{\mathrm{ext}} = {} & \frac{4}{c^2[(\xi_0^2+1)(\xi_0^2+\cos^2\alpha)]^{\frac{1}{2}}} \\
& \times R_l \Bigg\{ -\sum_{l=1}^{\infty} \mathrm{i}^{-l} a_l^{(1)}\, S_{1l}(-\mathrm{i}c, \cos\alpha) \\
& - \sum_{m=1}^{\infty} \Bigg[\sum_{l=m}^{\infty} \mathrm{i}^{-(l+1)} \big(k\, a_{ml}^{(1)}\, S_{ml}(-\mathrm{i}c, \cos\alpha) \\
& + \mathrm{i}\, b_{ml}^{(1)}\, S'_{ml}(-\mathrm{i}c, \cos\alpha)\big) \sin\alpha \\
& - k\, c_m \sum_{l=m-1}^{\infty} \mathrm{i}^{-(l+1)} \big(A_{m-1,l}\, S_{m-1,l}(-\mathrm{i}c, \cos\alpha) \\
& - A_{m+1,l}\, S_{m+1,l}(-\mathrm{i}c, \cos\alpha)\big) \cos\alpha \Bigg] \Bigg\},
\end{aligned} \tag{5.283}$$

$$\begin{aligned}
Q_{\mathrm{sca}} = {} & \frac{1}{c^2[(\xi_0^2+1)(\xi_0^2+\cos^2\alpha)]^{\frac{1}{2}}} \Bigg\{ 2 \sum_{l=1}^{\infty} \left| a_l^{(1)} \right|^2 N_{1l}^2(-\mathrm{i}c) \\
& + \sum_{m=1}^{\infty} \Bigg[\sum_{l=m}^{\infty} \sum_{n=m}^{\infty} \mathrm{i}^{(n-l)} \Big[k^2 a_{lm}^{(1)}\, a_{mn}^{(1)*}\, \omega_{ln}^{(m)} \\
& + \mathrm{i}k \left(b_{ml}^{(1)}\, a_{mn}^{(1)*}\, \kappa_{ln}^{(m)} - a_{ml}^{(1)}\, b_{mn}^{(1)*}\, \kappa_{ln}^{(m)} \right) \\
& + b_{ml}^{(1)} b_{mn}^{(1)*} \tau_{ln}^{(m)} \Big] - kc_m^* \sum_{l=m}^{\infty} \sum_{n=m-1}^{\infty} \mathrm{i}^{n-l} \\
& \times \Big[ka_{ml}^{(1)} \left(A_{m-1,n}^*\, \varrho_{ln}^{m,m-1} - A_{m+1,n}^*\, \varrho_{ln}^{m,m+1} \right) \\
& + \mathrm{i}b_{ml}^{(1)} \left(A_{m-1n}^*\, \xi_{ln}^{m,m-1} - A_{m+1,n}^*\, \xi_{ln}^{m,m+1} \right) \Big] \\
& - kc_m \sum_{l=m}^{\infty} \sum_{n=m-1}^{\infty} \mathrm{i}^{(n-l)} \\
& \times \Big[ka_{ml}^{(1)*} \left(A_{m-1,n}^*\, \varrho_{ln}^{m,m-1} - A_{m+1,n}^*\, \varrho_{ln}^{m,m+1} \right) \\
& - \mathrm{i}b_{mn}^{(1)*} \left(A_{m-1,n}\, \xi_{ln}^{m,m-1} - A_{m+1,n}\, \xi_{ln}^{m,m+1} \right) \Big] \\
& + |kc_m|^2 \sum_{l=m-1}^{\infty} \sum_{n=m-1}^{\infty} \mathrm{i}^{(n-l)} \\
& \times \Big[A_{m-1,l}\, A_{m-1,n}^* \left(2\delta_{nl} - \omega_{nl}^{(m-1)} \right) \\
& + A_{m-1,l}\, A_{m+1,n}^*\, \chi_{nl}^{m+1,m-1} \\
& + A_{m+1,l}\, A_{m-1,n}^*\, \chi_{nl}^{m+1,m-1} \\
& + A_{m-1,l}\, A_{m-1,n}^* \left(2\delta_{nl} - \omega_{nl}^{(m+1)} \right) \Big] \Bigg] \Bigg\},
\end{aligned} \tag{5.284}$$

$$
\begin{aligned}
Q_{\mathrm{bk}} = \frac{4\,(\xi_0^2+\sin^2\alpha)^2}{c^2(\xi_0^2+1)\,\xi_0^2}\,\Bigg| & \sum_{l=1}^{\infty} \mathrm{i}^l a_l^{(1)}\, S_{1l}(-\mathrm{i}c,\cos\alpha) \\
& - \sum_{m=1}^{\infty} \Bigg[\sum_{l=m}^{\infty} \mathrm{i}^{l+1} \big(k\, a_{ml}^{(1)}\, S_{ml}(-\mathrm{i}c,\cos\alpha) \\
& - \mathrm{i}\, b_{ml}^{(1)}\, S'_{ml}(-\mathrm{i}c,\cos\alpha)\big) \sin\alpha \\
& + k\, c_m \sum_{l=m-1}^{\infty} \mathrm{i}^{(l+1)} \big(A_{m-1,l}\, S_{m-1,l}(\mathrm{i}c,\cos\alpha) \\
& - A_{m+1,l}\, S_{m+1,l}(-\mathrm{i}c,\cos\alpha)\big)\cos\alpha \Bigg] \Bigg|^2 ,
\end{aligned}
\tag{5.285}
$$

where the coefficients $a_l^{(1)}$ are (see Eqs. (5.59)–(5.60))

$$
a_l^{(1)} = -2\,\mathrm{i}^l \frac{R_{ml}^{(1)}(-\mathrm{i}c,\mathrm{i}\xi_0)}{R_{ml}^{(3)}(-\mathrm{i}c,\mathrm{i}\xi_0)}\, N_{1l}^{-2}(-\mathrm{i}c)\, S_{1l}(-\mathrm{i}c,\cos\alpha)
\tag{5.286}
$$

for the TE mode and

$$
a_l^{(1)} = b_l^{(1)} = -2\,\mathrm{i}^l \frac{\left[(\xi_0^2+1)^{\frac{1}{2}}\, R_{1l}^{(1)}(-\mathrm{i}c,\mathrm{i}\xi_0)\right]'}{\left[(\xi_0^2+1)^{\frac{1}{2}}\, R_{1l}^{(3)}(-\mathrm{i}c,\mathrm{i}\xi_0)\right]'}\, N_{1l}^{-2}(-\mathrm{i}c)\, S_{1l}(-\mathrm{i}c,\cos\alpha)
\tag{5.287}
$$

for the TM mode, respectively. The integrals of products of the oblate AFSs and their derivatives can be expressed through the coefficients of the function expansions in terms of the associated Legendre functions of the first kind (see Appendix A).

The electromagnetic radiation scattered by an extremely oblate perfectly conducting spheroid and by a perfectly conducting disk coincide with the accuracy of $O(b/a)$. In the case of the disk, the expansion coefficients are calculated explicitly through the spheroidal functions from Eqs. (5.268)–(5.269) and (5.275)–(5.276).

Numerical calculations of the scattered radiation characteristics within the new solution are similar to those discussed above. For perfectly conducting oblate spheroids with a small aspect ratio a/b both solutions give nearly the same results for a given number of terms kept in the field expansions. With a growing a/b, the new solution provides better results and becomes preferable (see Table 5.6). Correctness of the results obtained with the new solution is confirmed by the fact that for the parallel incidence the TE and TM characteristics of radiation scattered by a perfectly conducting disk agree with the accuracy of 10^{-7} or better (see Table 5.7).

A consideration of data in Table 5.7 shows that in the case when the incident radiation propagates perpendicular to the disk plane, the extinction efficiency factors quickly reach the limit of geometrical optics $Q_{\mathrm{sca}} = 2$, while the backscattering cross-sections for $c \geq 3.5$ have the asymptotic dependence on the diffraction parameter $\sigma_{\mathrm{bk}}/\pi a^2 \sim c^2$. The quantities $(Q_{\mathrm{sca}} - 1)$ and $\sigma_{\mathrm{bk}}/(\pi a^2 c^2)$ as functions of

Table 5.6. Efficiency factors for extinction $Q_{\rm ext}$ and scattering $Q_{\rm sca}$ for perfectly conducting oblate spheroids with $c = 5$ ($\alpha = 0$)

N	$Q_{\rm ext}^{\rm old}$	$Q_{\rm sca}^{\rm old}$	$Q_{\rm ext}^{\rm new}$	$Q_{\rm sca}^{\rm new}$
		$a/b = 2$		
8	2.08	2.10	2.077	2.097
12	2.08328	2.08333	2.083324	2.083343
16	2.0833206	2.0833208	2.0833210	2.0833218
20	2.0833211	2.0833211	2.0833211	2.0833211
		$a/b = 5$		
20	2.114	2.093	2.1330	2.1337
24	2.1199	2.1144	2.12429	2.12445
28	2.12115	2.11987	2.12216	2.12220
32	2.12144	2.12115	2.12166	2.12168
36	2.12148	2.12134	2.121589	2.121594

Table 5.7. Efficiency factors for extinction $Q_{\rm ext}$, scattering $Q_{\rm sca}$ and backscattering $\sigma_{\rm bk}/(\pi a^2 c^2)$ for perfectly conducting disks ($\alpha = 0$)

c	$Q_{\rm ext}$	$Q_{\rm sca}$	$\sigma_{\rm bk}/(\pi a^2 c^2)$
0.5	0.0373775	0.0373775	0.235100
1.0	1.000234	1.000234	1.833080
1.5	3.374781	3.374781	3.431179
2.0	3.007385	3.007385	2.268282
2.5	2.537698	2.537698	1.614415
3.0	2.254616	2.254616	1.279257
3.5	2.071788	2.071788	1.078326
4.0	1.966437	1.966437	0.967271
4.5	1.969446	1.969446	0.971217
5.0	2.080243	2.080243	1.085286
5.5	2.138854	2.138854	1.143930
6.0	2.102719	2.102719	1.105790
6.5	2.040969	2.040969	1.042337
7.0	1.989386	1.989386	0.989802
7.5	1.971303	1.971303	0.971542
8.0	2.006665	2.006665	1.007377
8.5	2.055600	2.055600	1.056684
9.0	2.059054	2.059054	1.059945
9.5	2.031259	2.031259	1.031770
10.0	1.999405	1.999405	0.999619

the parameter c oscillate in concord around the values of 2 and 1, respectively. The comparison also gives an approximate relation

$$\frac{\sigma_{\rm bk}}{\pi a^2 c^2} \approx Q_{\rm sca} - 1. \tag{5.288}$$

The relative accuracy of this relation is better than 2% for $c \geq 3.0$, about 0.5% for $c \geq 4.0$, and 0.1% for $c \geq 6.5$.

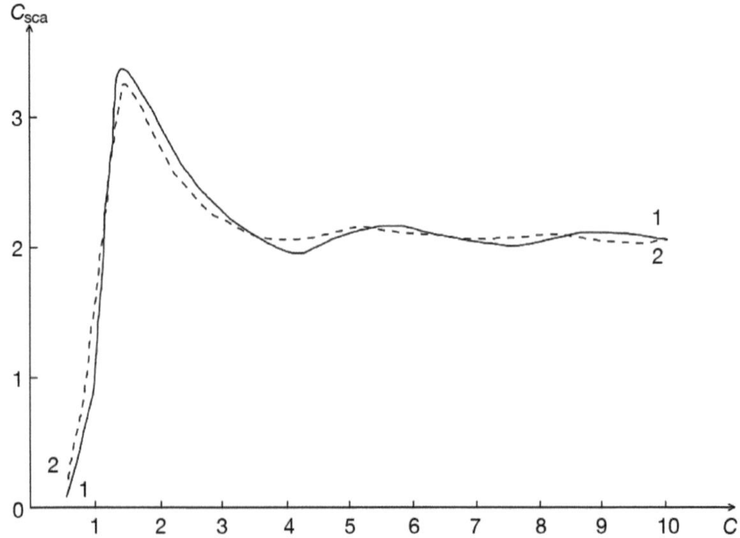

Fig. 5.5. Normalized scattering cross-sections $C_{\rm sca}$ for perfectly conducting disks (solid line) and oblate spheroids with $a/b = 5$ (dashed line) in the dependence on the parameter c for the incidence angle $\alpha = 0$.

A consideration of the extinction efficiency factors for perfectly conducting disks and oblate spheroids with the ratio $a/b = 5$ allows one to approximately determine the range of the diffraction parameter c for which the results for a disk well describe the main properties of radiation scattered by extremely oblate perfectly conducting spheroids. From Fig. 5.5 we see that the largest deviations arise for small values of c. On the other hand, starting with the c values corresponding to the first maximum of the extinction curve, the scattering characteristics of a disk well describe those of the extremely oblate perfectly conducting spheroids already with the ratio $a/b = 5$. For small values of the parameter $c < 1.5$, the agreement of the results obtained for disks and oblate spheroids becomes better with increasing particle oblateness (i.e. the ratio a/b).

5.6 Conclusions

We have considered different aspects of the light scattering problem for spheroids, including construction of the exact solution using non-orthogonal bases of wave functions and analysis of the infinite systems of linear algebraic equations (ISLAEs) arisen. The most important results are as follows.

The suggested exact solution to the problem is the most efficient one from the computational point of view provided a basis of spheroidal wave functions is utilized. This is because the solution basically transforms into an explicit solution for spheres and allows one to get the principal term of the asymptotics with respect to the small parameter b/a being the aspect ratio for extremely prolate and oblate dielectric spheroids. Note that to build the basis, we utilize both the vector functions used to solve the problem for spheres and the vector functions applied to solve the problem for infinite circular cylinders. Numerical tests show that the number of terms kept in the field expansions necessary to reach a given accuracy depends only on the maximum dimension of the spheroid and is independent of its shape characterized by the aspect ratio a/b and the spheroid kind (prolate or oblate). Thus, the solution suggested is in particular efficient for spheroids whose shape essentially differs from the spherical one.

The analysis of the ISLAEs has shown that they are quasi-regular and definitely solvable in the space l_2 for any spheroids not degenerated into a segment or a disk. They can be solved with any accuracy by the reduction method, i.e. by truncating the infinite systems, which occurs in numerical realization of the solution. Convergence of the field expansions is proved in the space $L_2(\Omega)$ for any coordinate surface $\Omega(\xi = const)$ up to the spheroid boundary. The found asymptotics of the matrix elements of the ISLAEs allow one to estimate the behaviour of the truncated systems arisen in calculations.

The quasi-static approximation, being a generalization of the Rayleigh–Gans and Rayleigh approximations, is theoretically grounded for extremely prolate and oblate dielectric spheroids, in which cases it gives the principal term of the asymptotics with respect to the small parameter b/a. For extremely prolate perfectly conducting spheroids, the principal term of the field asymptotics with respect to the parameter b/a is explicitly expressed through the spheroidal and elementary functions for the TM mode, and only through elementary functions for the TE mode. From a numerical comparison of the exact and approximate solutions, we have found the range of applicability of the latter. The optical properties of the extremely prolate and oblate perfectly conducting spheroids have been studied in some detail.

The derived solution to the problem of the plane wave diffraction by an extremely oblate perfectly conducting spheroid is improved by a special choice of the scalar potentials so that the Meixner conditions at the disk edge are fulfilled automatically. Numerical calculations have demonstrated a high efficiency of the improved solution.

Acknowledgements

The author is thankful to anonymous reviewers for many helpful remarks, to N. V. Voshchinnikov for the fruitful joint work on many tasks mentioned in this review and for numerical calculations performed and to V. B. Il'in for valuable discussions, careful reading and English translation of the manuscript. The work was partly supported by the RFBR grant 10-02-00593a.

Appendix A: Integrals of the spheroidal angular functions and other relations

The prolate spheroidal angular functions $S(c, \eta)$ are solutions to the problem (Komarov *et al.*, 1976)

$$\frac{\mathrm{d}}{\mathrm{d}\eta}(1-\eta^2)\,\frac{\mathrm{d}}{\mathrm{d}\eta}\,S(c,\eta) + \left[\lambda + c^2(1-\eta^2) - \frac{m^2}{(1-\eta^2)}\right] S(c,\eta) = 0\,, \qquad (5.289)$$

where $|S(c,\eta)| < \infty$ for $\eta = \pm 1$, $c \geq 0$, and m is an integer number. The boundary conditions can be replaced by the standard requirement that $S(c,\eta)$ is square integrable on the interval $[-1,1]$, i.e. $S(c,\eta) \in L_2[-1,\,1]$. The Sturm–Liouville problem (5.289) has an infinite set of eigenvalues λ_{ml} and eigenfunctions $S_{ml}(c,\eta)$ that for a given m form a complete orthogonal system in the space L_2[-1, 1]. The numeration of the functions $S_{ml}(c,\eta)$ is selected in such a way that they have $(l-m)$ roots in the interval $[-1,1]$, and hence $l \geq m$. The condition of normalization of the prolate SAFs is as follows:

$$S_{ml}(c,0) = P_l^m(0) = \frac{(-1)^{\frac{l-m}{2}}\,(l+m)!}{2^l(\frac{l-m}{2}))!\,(\frac{l+m}{2}))!}\,, \qquad (l-m) = 2q\,, \tag{5.290}$$

$$S'_{ml}(c,0) = P_l^{m\,\prime}(0) = \frac{(-1)^{\frac{l-m+1}{2}}\,(l+m+1)!}{2^l(\frac{l-m-1}{2}))!\,(\frac{l+m+1}{2}))!}\,, \qquad (l-m) = 2q+1\,, \tag{5.291}$$

where

$$P_l^m(\eta) = \frac{(1-\eta^2)^{\frac{m}{2}}}{2^l l!}\,\frac{\mathrm{d}^{l+m}}{\mathrm{d}\eta^{l+m}}\,(1-\eta^2)^l$$

are the associated Legendre functions of the first kind.

The normalizing factor for the prolate SAFs is introduced by

$$S_{ml}(c,\eta) = N_{ml}(c)\;\bar{S}_{ml}(c,\eta)\,, \tag{5.292}$$

and the integral of the normalized prolate SAFs is equal to 1

$$\int_{-1}^{1} \bar{S}^2_{mn}(c,\eta)\,\mathrm{d}\eta = 1\,. \tag{5.293}$$

The prolate SAFs can be expanded in terms of the Legendre polynomials

$$S_{ml}(c,\eta) = \sum_{r=0,1}^{\infty}{}' \; d_r^{ml}(c)\,P_{m+r}^m(\eta)\,, \tag{5.294}$$

where the prime means that the summation is made over the r values whose parity coincides with that of $(l-m)$.

From the normalization conditions (5.291)–(5.292), we find the normalizing factor

$$N^2_{ml}(c) = \sum_{r=0,1}^{\infty}{}' \; \left[d_r^{ml}(c)\right]^2 \frac{2}{2r+2m+1}\,\frac{(r+2m)!}{r!}\,, \tag{5.295}$$

where we used the orthogonality of the prolate SAFs.

For the oblate SAFs, one should make the substitution $c \to -ic$ in the equations given above for the prolate SAFs.

For complex values of the parameter c, the prolate and oblate SAFs are considered as analytic continuations of the functions defined above for $c \geq 0$. More information about these functions can be found in the book of Meixner and Schefke (1954) and the review of Meixner *et al.* (1980)

Below we consider some integrals of products of the normalized SAFs and their derivatives. The integrals are expressed through the coefficients of the function expansions in terms of the associated Legendre functions of the first kind. Substitution of the expansions of the spheroidal angular functions into the integrals and use of the recurrence relations for the Legendre functions allow one to obtain the following formulae:

$$\delta_{nl}^{m}(c_2,c_1) = \int_{-1}^{1} \bar{S}_{mn}(c_2,\eta)\,\bar{S}_{ml}(c_1,\eta)\,\mathrm{d}\eta = N_{mn}^{-1}(c_2)\,N_{ml}^{-1}(c_1) \times \sum_{r=0,1}^{\infty}{}' d_r^{mn}(c_2) d_r^{ml}(c_1)\,\frac{2}{2r+2m+1}\,\frac{(r+2m)!}{r!}, \tag{5.296}$$

$$\gamma_{nl}^{m}(c_2,c_1) = \int_{-1}^{1} \bar{S}_{mn}(c_2,\eta)\,\bar{S}_{ml}(c_1,\eta)\,\eta\,\mathrm{d}\eta = N_{mn}^{-1}(c_2)\,N_{ml}^{-1}(c_1) \times \sum_{r=0,1}^{\infty}{}' d_r^{mn}(c_2)\left[d_{r+1}^{ml}(c_1)\frac{r+2m+1}{2r+2m+3} + d_{r-1}^{ml}(c_1)\frac{r}{2r+2m-1}\right]\frac{2}{2r+2m+1}\,\frac{(r+2m)!}{r!}, \tag{5.297}$$

$$\kappa_{nl}^{m}(c_2,c_1) = \int_{-1}^{1} \bar{S}_{mn}(c_2,\eta)\,\frac{\mathrm{d}\bar{S}_{ml}(c_1,\eta)}{\mathrm{d}\eta}\,(1-\eta^2)\,\mathrm{d}\eta = N_{mn}^{-1}(c_2)\,N_{ml}^{-1}(c_1) \times \sum_{r=0,1}^{\infty}{}' d_r^{mn}(c_2)\left[d_{r+1}^{ml}(c_1)\frac{(r+2m+1)(r+m+2)}{2r+2m+3} - d_{r-1}^{ml}(c_1)\frac{r(r+m-1)}{2r+2m-1}\right]\frac{2}{2r+2m+1}\,\frac{(r+2m)!}{r!}, \tag{5.298}$$

$$\sigma_{nl}^{m}(c_2,c_1) = \int_{-1}^{1} \bar{S}_{mn}(c_2,\eta)\,\frac{\mathrm{d}\left(\eta\bar{S}_{ml}(c_1,\eta)\right)}{\mathrm{d}\eta}(1-\eta^2)\,\mathrm{d}\eta = N_{mn}^{-1}(c_2) \times N_{ml}^{-1}(c_1)\sum_{r=0,1}^{\infty}{}'\left[\frac{(r+2m+2)(r+2m+1)(r+m+2)}{(2r+2m+3)(2r+2m+5)} \times d_{r+2}^{ml}(c_1) + \frac{3(r+m)(r+m+1)-m^2-2}{(2r+2m-1)(2r+2m+3)} \times d_r^{ml}(c_1) - \frac{r(r-1)(r+m-1)}{(2r+2m-3)(2r+2m-1)}d_{r-2}^{ml}(c_1)\right] \times d_r^{mn}(c_2)\frac{2}{2r+2m+1}\,\frac{(r+2m)!}{r!}, \tag{5.299}$$

$$\omega_{nl}^{m}(c_2,c_1)=\int_{-1}^{1}\bar{S}_{mn}(c_2,\eta)\,\bar{S}_{ml}(c_1,\eta)\,(1-\eta^2)\,\mathrm{d}\eta=-N_{mn}^{-1}(c_2)$$
$$\times\,N_{ml}^{-1}(c_1)\sum_{r=0,1}^{\infty}{}'\left[\frac{(r+2m+2)(r+2m+1)}{(2r+2m+3)(2r+2m+5)}\right.$$
$$\times d_{r+2}^{ml}(c_1)-\frac{2[(r+m)(r+m+1)+m^2-1]}{(2r+2m-1)(2r+2m+3)} \tag{5.300}$$
$$\left.\times d_{r}^{ml}(c_1)+\frac{r(r-1)}{(2r+2m-3)(2r+2m-1)}d_{r-2}^{ml}(c_1)\right]$$
$$\times\,d_r^{mn}(c_2)\frac{2}{2r+2m+1}\,\frac{(r+2m)!}{r!}\,,$$

$$\tau_{nl}^{m}(c_2,c_1)=\int_{-1}^{1}\left[\frac{(1-\eta^2)\,\mathrm{d}\bar{S}_{mn}(c_2,\eta)}{\mathrm{d}\eta}\,\frac{\bar{S}_{ml}(c_1,\eta)}{\mathrm{d}\eta}\right.$$
$$\left.+\frac{m^2\,\bar{S}_{mn}(c_2,\eta)\,\bar{S}_{ml}(c_1,\eta)}{1-\eta^2}\right]\mathrm{d}\eta=N_{mn}^{-1}(c_2)\,N_{ml}^{-1}(c_1) \tag{5.301}$$
$$\times\sum_{r=0,1}^{\infty}{}'\,d_r^{mn}(c_2)\,d_r^{ml}(c_1)\frac{2(r+m)(r+m+1)}{2r+2m+1}\,\frac{(r+2m)!}{r!}\,,$$

$$\varrho_{nl}^{m,m\mp1}(-\mathrm{i}c_2,-\mathrm{i}c_1)=\int_{-1}^{1}\bar{S}_{mn}(-\mathrm{i}c_2,\eta)\,\bar{S}_{m\mp1,l}(-\mathrm{i}c_1,\eta)\,(1-\eta^2)^{\frac{1}{2}}\mathrm{d}\eta=$$
$$-\,N_{mn}^{-1}(-\mathrm{i}c_2)\,N_{m\mp1,l}^{-1}(-\mathrm{i}c_1)\sum_{r=0,1}^{\infty}{}'\,d_r^{mn}(-\mathrm{i}c_2)\left[\frac{d_{r+2}^{m\mp1,l}(-\mathrm{i}c_1)}{2r+2m+3}\right. \tag{5.302}$$
$$\left.-\frac{d_r^{m\mp1,l}(-\mathrm{i}c_1)}{2r+2m-1}\right]\frac{2}{2r+2m+1}\,\frac{(r+2m)!}{r!}\,,$$

$$\alpha_{nl}^{m,m\mp1}(-\mathrm{i}c_2,-\mathrm{i}c_1)=\int_{-1}^{1}\bar{S}_{mn}(-\mathrm{i}c_2,\eta)\,\left[\pm(m\mp1)\,\eta\,\bar{S}_{m\mp1,l}(-\mathrm{i}c_1,\eta)\right.$$
$$\left.+(1-\eta^2)\frac{\mathrm{d}\bar{S}_{m\mp1,l}(-\mathrm{i}c_1,\eta)}{\mathrm{d}\eta}\right]\frac{\mathrm{d}\eta}{(1-\eta^2)^{\frac{1}{2}}}=N_{mn}^{-1}(-\mathrm{i}c_2)\,N_{m\mp1,l}^{-1}(-\mathrm{i}c_1) \tag{5.303}$$
$$\times\sum_{r=0,1}^{\infty}{}'\,d_r^{mn}(-\mathrm{i}c_2)\,d_{r+1}^{m\mp1,l}(-\mathrm{i}c_1)\,\frac{2}{2r+2m+1}\,\frac{(r+2m)!}{r!}\,,$$

$$\varphi_{nl}^{m,m\mp1}(-\mathrm{i}c_2,-\mathrm{i}c_1)=\int_{-1}^{1}\bar{S}_{mn}(-\mathrm{i}c_2,\eta)\,\bar{S}_{m\mp1,l}(-ic_1,\eta)\,\eta(1-\eta^2)\,\mathrm{d}\eta=$$

$$-\,N_{mn}^{-1}(-\mathrm{i}c_2)\,N_{m\mp1,l}^{-1}(-\mathrm{i}c_1)\sum_{r=0,1}^{\infty}{}'\,d_r^{mn}(-ic_2)\left[d_{r+3}^{m\mp1,l}(-\mathrm{i}c_1)\right.$$

$$\times\frac{r+2m+1}{(2r+2m+5)(2r+2m+3)}+d_{r+1}^{m\mp1,l}(-\mathrm{i}c_1) \tag{5.304}$$

$$\times\frac{2r-4m^2-4mr+1}{(2r+2m+3)(2r+2m+1)(2r+2m-1)}-d_{r-1}^{m\mp1,l}(-\mathrm{i}c_1)$$

$$\times\left.\frac{r}{(2r+2m-1)(2r+2m-3)}\right]\frac{2}{2r+2m+1}\,\frac{(r+2m)!}{r!},$$

$$\xi_{nl}^{m,m\mp1}(-\mathrm{i}c_2,-\mathrm{i}c_1)\;=\;\int_{-1}^{1}\left[\eta\,\frac{\mathrm{d}\bar{S}_{mn}(-\mathrm{i}c_1,\eta)}{\mathrm{d}\eta}\mp m\,\frac{\bar{S}_{mn}(-\mathrm{i}c_1,\eta)}{(1-\eta^2)}\right]$$

$$\times\,\bar{S}_{m\mp1,l}(-\mathrm{i}c_1,\eta)\,(1-\eta^2)^{\frac{1}{2}}\,\mathrm{d}\eta\;=\;N_{mn}^{-1}(-\mathrm{i}c_2)\,N_{m\mp1,l}^{-1}(-\mathrm{i}c_1)$$

$$\times\sum_{r=0,1}^{\infty}{}'\,d_r^{mn}(-\mathrm{i}c_2)\left[d_{r+2}^{m\mp1,l}(-\mathrm{i}c_1)\,\frac{r+m}{(2r+2m+3)}+d_r^{m\mp1,l}(-\mathrm{i}c_1)\right. \tag{5.305}$$

$$\times\left.\frac{r+m+1}{2r+2m-1}\right]\frac{2}{2r+2m+1}\,\frac{(r+2m)!}{r!},$$

$$\chi_{nl}^{m+1,m-1}(-\mathrm{i}c_2,-\mathrm{i}c_1)=\int_{-1}^{1}\bar{S}_{m+1,n}(-\mathrm{i}c_2,\eta)\,\bar{S}_{m-1,l}(-\mathrm{i}c_1,\eta)$$

$$\times\,(1-\eta^2)\,\mathrm{d}\eta=-N_{m+1,n}^{-1}(-\mathrm{i}c_2)\,N_{m-1,l}^{-1}(-\mathrm{i}c_1)\sum_{r=0,1}^{\infty}{}'\,d_r^{m+1,n}(-\mathrm{i}c_2)$$

$$\times\left[\frac{d_{r+4}^{m-1,l}(-\mathrm{i}c_1)}{(2r+2m+7)(2r+2m+5)}+\frac{d_{r+2}^{m-1,l}(-\mathrm{i}c_1)}{(2r+2m+5)(2r+2m+1)}\right. \tag{5.306}$$

$$-\left.\frac{d_r^{m-1,l}(-\mathrm{i}c_1)}{(2r+2m+1)(2r+2m-1)}\right]\frac{2}{2r+2m+3}\,\frac{(r+2m+3)!}{r!}.$$

Further, we assume that $c_1 \geq 0$, and c_2 is a complex number. Let us consider the differential equations

$$\begin{aligned}\frac{\mathrm{d}}{\mathrm{d}\eta}(1-\eta^2)\,\frac{\mathrm{d}}{\mathrm{d}\eta}\,\bar{S}_{ml}(c_2,\eta)+\left[\lambda_{ml}(c_2)+c_2^2(1-\eta^2)-\frac{m^2}{(1-\eta^2)}\right]\bar{S}_{ml}(c_2,\eta)=0,\\ \frac{\mathrm{d}}{\mathrm{d}\eta}(1-\eta^2)\,\frac{\mathrm{d}}{\mathrm{d}\eta}\,\bar{S}_{mn}(c_2,\eta)+\left[\lambda_{mn}(c_2)+c_2^2(1-\eta^2)-\frac{m^2}{(1-\eta^2)}\right]\bar{S}_{mn}(c_2,\eta)=0\,.\end{aligned} \tag{5.307}$$

We multiply the first equation by $\bar{S}_{mn}(c_2,\eta)$ and the second one by $-\bar{S}_{ml}(c_2,\eta)$, integrate them over η from -1 to 1 and summarize the results. This gives

$$[\lambda_{ml}(c_2)-\lambda_{mn}(c_2)]\int_{-1}^{1}\bar{S}_{mn}(c_2,\eta)\,\bar{S}_{ml}(c_2,\eta)\,\mathrm{d}\eta=0\,, \tag{5.308}$$

i.e. the corresponding matrix is the unit one

$$\Delta(c_2, c_2) = \{\delta_{nl}^{(m)}(c_2, c_2)\}_{n,l=m}^{\infty} = \{\delta_{nl}^{(m)}\}_{n,l=m}^{\infty} = I\,. \tag{5.309}$$

Substitution of the expansions of the functions $\bar{S}_{mn}(c, \eta)$ and $\bar{S}_{ml}(c, \eta)$ in terms of $\{\bar{S}_{mj}(c_1, \eta)\}_{j=m}^{\infty}$ forming a complete orthogonal system into Eq. (5.308) and the symmetry relative to the indices and parameters ($\delta_{lj}^{(m)}(c_2, c_1) = \delta_{jl}^{(m)}(c_1, c_2)$) allow us to get

$$\sum_{j=m}^{\infty} \delta_{lj}^{(m)}(c_2, c_1)\, \delta_{jl}^{(m)}(c_1, c_2) = \delta_l^n\,, \tag{5.310}$$

i.e. for the matrices $\Delta(c_2, c_1) = \{\delta_{nl}^{(m)}(c_2, c_1)\}_{n,l=m}^{\infty}$, we have

$$\Delta(c_2, c_1)\, \Delta(c_1, c_2) = I\,, \tag{5.311}$$

or

$$\Delta^{(-1)}(c_2, c_1) = \Delta(c_1, c_2)\,. \tag{5.312}$$

Similarly, one can demonstrate that

$$\Pi(c_2, c_2) = \Delta(c_2, c_1)\, \Pi(c_1, c_1)\, \Delta(c_1, c_2)\,, \tag{5.313}$$

where the matrix Π is any of the matrices under consideration. As all the integrals, except for Δ, depend on the only parameter c_1, further we skip the parameters of the integrals Π.

Let us consider the integrals $\sigma_{nl}^{(m)}$. The expansion of the functions $\eta\, \bar{S}_{ml}(c_1, \eta)$ in terms of $\{\bar{S}_{mj}(c_1, \eta)\}_{j=m}^{\infty}$ is as follows:

$$\eta\, \bar{S}_{ml}(c_1, \eta) = \sum_{j=m}^{\infty} \gamma_{lj}^{(m)}\, \bar{S}_{mj}(c_1, \eta)\,. \tag{5.314}$$

Substitution of this equation into the integral $\sigma_{nl}^{(m)}$ and the relation $\gamma_{lj}^{(m)} = \gamma_{jl}^{(m)}$ give

$$\sigma_{nl}^{(m)} = \sum_{j=m}^{\infty} \kappa_{nj}^{(m)}\, \gamma_{jl}^{(m)}\,, \tag{5.315}$$

i.e.

$$\Sigma = K\, \Gamma. \tag{5.316}$$

For the integrals $\tau_{nl}^{(m)}$, integration of the first summand by parts and use of the differential equation (5.307) lead to

$$\tau_{nl}^{(m)}(c_1) = \lambda_{ml}(c_1)\, \delta_n^l + c_1^2\, \omega_{nl}^{(m)}\,. \tag{5.317}$$

Similarly, keeping in mind the properties of the SAFs, we get

$$\begin{aligned}
&\gamma_{nl}^{(m)} = \frac{1}{2}\left(\kappa_{nl}^{(m)} + \kappa_{ln}^{(m)}\right), \\
&2\,\delta_n^l - \tau_{nl}^{(m)} = \sigma_{ml} + \sigma_{ln}^{(m)}, \\
&\{\varphi_{nl}^{m,m\mp1}\} = \{\gamma_{nj}^{(m)}\}\{\rho_{jl}^{m,m\mp1}\} = \{\varphi_{ln}^{m\mp1,m}\}, \\
&\varrho_{nl}^{m,m\mp1} = \varrho_{ln}^{m\mp1,m}, \\
&\alpha_{nl}^{m,m\mp1} = \alpha_{ln}^{m\mp1,m}, \\
&\xi_{nl}^{m,m\mp1} + \xi_{ln}^{m\mp1,m} = 2\rho_{nl}^{m,m\mp1}, \\
&\{\chi_{nl}^{m-1,m+1}\} + \{\chi_{ln}^{m+1,m-1}\} = \{\rho_{nj}^{m-1,m}\}\{\rho_{jl}^{m,m+1}\}.
\end{aligned} \tag{5.318}$$

Eqs. (5.311)–(5.318) can be used for calculations of the integrals or for verification of such calculations.

To conclude, we consider some properties of the oblate spheroidal functions. The differential equation for these functions is as follows (Komarov *et al.*, 1976):

$$\frac{\mathrm{d}}{\mathrm{d}\xi}(\xi^2+1)\,\frac{\mathrm{d}}{\mathrm{d}\xi}R_{ml}(-\mathrm{i}c,\xi) + \left[-\lambda_{ml}(-\mathrm{i}c) + c^2(\xi^2+1) + \frac{m^2}{(\xi^2+1)}\right]\bar{R}_{ml}(-\mathrm{i}c,\xi) = 0\,. \tag{5.319}$$

The function substitution

$$u = (\xi^2+1)\,\frac{R'_{ml}(-\mathrm{i}c,\xi)}{R_{ml}(-\mathrm{i}c,\xi)} \tag{5.320}$$

leads to

$$u' + \frac{u^2}{\xi^2+1} + \left[-\lambda_{ml}(-\mathrm{i}c) + c^2(\xi^2+1) + \frac{m^2}{(\xi^2+1)}\right] = 0\,. \tag{5.321}$$

For the oblate SRFs, the function u in the vicinity of the point $\xi = 0$ can be represented as ($c = O(1)$)

$$u = \frac{R_{ml}^{(1)'}(-\mathrm{i}c,\xi_0)}{R_{ml}(-\mathrm{i}c,\xi_0)} = \frac{1}{\xi_0} + O(\xi_0)\,, \qquad l-m = 2q+1\,, \tag{5.322}$$

or

$$\begin{aligned}
u &= \frac{R_{ml}^{(1)'}(-\mathrm{i}c,\xi_0)}{R_{ml}(-\mathrm{i}c,\xi_0)} \\
&= \left[-\lambda_{ml}(-\mathrm{i}c) + c^2 + m^2\right]\xi_0 + O(\xi_0^3)\,, \qquad l-m = 2q\,.
\end{aligned} \tag{5.323}$$

Here we used the parity properties of the oblate SRFs of the first kind. The coefficients can be found after substitution of this equation into Eq. (5.321) and comparison of the results for the same powers of ξ_0.

Let us multiply the first and second differential equations (5.307) for the oblate SAFs depending on the parameters c_2 and c_1 by $\bar{S}_{mn}(-\mathrm{i}c_1, \eta)$ and $-\bar{S}_{ml}(-\mathrm{i}c_2, \eta)$, respectively. Then summation of the equations gives

$$\begin{aligned} &[\lambda_{ml}(-\mathrm{i}c_2) - \lambda_{mn}(-\mathrm{i}c_1)] \int_{-1}^{1} \bar{S}_{mn}(-\mathrm{i}c_2, \eta)\, \bar{S}_{ml}(-\mathrm{i}c_1, \eta)\, \mathrm{d}\eta = \\ &(c_2^2 - c_1^2) \int_{-1}^{1} \bar{S}_{mn}(-\mathrm{i}c_2, \eta)\, \bar{S}_{ml}(-\mathrm{i}c_1, \eta)(1-\eta^2)\, \mathrm{d}\eta \,. \end{aligned} \tag{5.324}$$

This equation can be written in the matrix form

$$\Lambda(-\mathrm{i}c_2)\, \Delta(-\mathrm{i}c_2, -\mathrm{i}c_1) - \Delta(-\mathrm{i}c_2, -\mathrm{i}c_1)\, \Lambda(-\mathrm{i}c_1) = (c_2^2 - c_1^2)\, \Omega(-\mathrm{i}c_2, -\mathrm{i}c_1)\,, \tag{5.325}$$

where $\Omega = \{\omega_{nl}^{(m)}\}_{n,l=m}^{\infty}$, Λ is a diagonal matrix whose elements are the eigenvalues λ_{ml}.

Using the equations

$$\begin{aligned} \Delta(-\mathrm{i}c_2, -\mathrm{i}c_1) &= \Delta(-\mathrm{i}c_2, -\mathrm{i}c_1)\, \Omega(-\mathrm{i}c_1, -\mathrm{i}c_1)\,, \\ \Omega(-\mathrm{i}c_1, -\mathrm{i}c_1) &= I - \Gamma^2(-\mathrm{i}c_1, -\mathrm{i}c_1)\,, \end{aligned} \tag{5.326}$$

and Eqs. (5.322)–(5.323) and (5.317), in the first approximation with respect to ξ_0, we have

$$\tilde{R}_2^r - R_0^r = -\xi_0 (c_2^2 - c_1^2) \left[\Gamma^2(-\mathrm{i}c_1, -\mathrm{i}c_1)\right]^r \,, \tag{5.327}$$

where the index r means that we consider the matrix elements for which the sum of the row and column numbers is even.

References

Andrejewski, W., 1953: Die Beugung electromagnetischer Wellen an der leitenden Kreissscheibe und an der kreisformigen Offnung im leitenden ebenen Schirm, *Z. angew. Phys.*, **5**, 178–186.

Asano, S., and G. Yamamoto, 1975: Light scattering by spheroidal particle, *Appl. Opt.*, **14**, 29–49.

Barber, P. W., and S. C. Hill, 1990: *Light Scattering by Particles: Computational Methods*, Singapore: World Scientific.

Barber, P. W., and C. Yeh, 1975: Scattering of electromagnetic waves by arbitrarily shaped dielectric bodies, *Appl. Opt.*, **14**, 2864–2872.

Bohren, C., and D. Huffman, 1983: *Absorption and Scattering of Light by Small Particles*, New York: John Wiley & Sons.

Colton, D., and R. Kress, 1984: *Integral Methods in Scattering Theory*, New York: John Wiley & Sons.

Farafonov, V. G., 1983: Diffraction of a plane electromagnetic wave by a dielectric spheroid, *(Sov.) Diff. Equat.*, **19**, 1765–1777.

Farafonov, V. G., 1984: Scattering of electromagnetic waves by a perfectly conducting spheroid, *(Sov.) Radiotech. Electron.*, **29**, 1857–1865.

Farafonov, V. G., 2001: New recursive solution of the problem of scattering of electromagnetic radiation by multilayered spheroidal particles, *Opt. Spectr.*, **90**, 743–752.

Farafonov, V. G., 2011: A unified approach using spheroidal functions for solving the problem of light scattering by a axisymmetric particles, *J. Math. Sci.*, **175**, 698–723.
Farafonov, V. G., and V. B. Il'in, 2006: Single light scattering: computational methods, In *Light Scattering Reviews*, A. A. Kokhanovsky (ed.), Berlin: Springer-Praxis, pp. 125–177.
Farafonov, V. G., and S. Y. Slavyanov, 1980: Diffraction of a plane wave by a perfectly conducting spheroid, *(Sov.) Radiotech. Electron.*, **25**, 2056–2065.
Farafonov, V. G., and N. V. Voshchinnikov, 2012: Light scattering by a multilayered spheroidal particle, *Appl. Opt.*, **51**, 1586–1597.
Farafonov, V. G., N. V. Voshchinnikov, and V. V. Somsikov, 1996: Light scattering by a core-mantle spheroidal particle, *Appl. Opt.*, **35**, 5412–5426.
Farafonov, V. G., V. B. Il'in, and A. A. Vinokurov, 2007: On use of the field expansions in terms of spheroidal functions, *J. Quant. Spectr. Rad. Transfer*, **106**, 33–41.
Farafonov, V. G., V. B. Il'in, and A. A. Vinokurov, 2010: Near- and far-field light scattering by nonspherical particles: applicability of methods that involve a spherical basis, *Opt. Spectr.*, **109**, 432–443.
Flammer, C., 1953: The vector wave function solution of electromagnetic wave by circular discs and apertures. II. The diffraction problems, *J. Appl. Phys.*, **24**, 1224–1231.
Flammer, C., 1957: *Spheroidal Wave Functions*, Stanford: Stanford University Press.
Hovenier, J. W., K. Lumme, M. I. Mishchenko, N. V. Voshchinnikov, D. W. Mackowski, and J. Rahola, 1996: Computations of scattering matrices of four types of nonspherical particles using diverse methods, *J. Quant. Spectr. Rad. Transfer*, **55**, 695–205.
van de Hulst, H. C., 1957: *Light Scattering by Small Particles*, New York: John Wiley & Sons.
Il'in, V.B., N. V. Voshchinnikov, V. G. Farafonov, Th. Henning, and A. Ya. Perelman, 2002: Light scattering tools for cosmic dust modeling, in Videen, G., and M. Kocifaj (eds), *Optics of Cosmic Dust*, Kluwer Academic Publishers, pp. 71–88.
Il'in, V. B., V. G. Farafonov, and E. V. Farafonov, 2007: Extended boundary condition method in combination with field expansions in terms of spheroidal functions, *Opt. Spcctr.*, **102**, 278–289.
Jackson, J. D., 1975: *Classical Electrodynamics*, New York: John Wiley & Sons.
Kahnert, F. M., 2003a: Surface-integral formulation for electromagnetic scattering in spheroidal coordinates, *J. Quant. Spcctr. Rad. Transfer*, **77**, 61–78.
Kahnert, F. M., 2003b: Numerical methods in electromagnetic scattering theory, *J. Quant. Spectr. Rad. Transfer*, **79-80**, 775–824.
Kantorovich, L. V., and V. I. Krylov, 1964: *Approximate Methods of Higher Analysis*, New York: John Wiley & Sons.
Komarov, I. V., L. I. Ponomarev, and S. Yu. Slavyanov, 1976: *Spheroidal and Coulomb Spheroidal Functions*, Moscow: Nauka.
Lopatin, V. N., and F. Ya. Sid'ko, 1988: *Introduction in Optics of Cell's Suspension*, Novosibirsk: Nauka.
Meixner, J., 1950: Theorie der Beugung electromagnetischer Wellen in der vollkommen leitenden Kreissscheibe und verwandte Problemen, *Ann. Phys.*, **6(7)**, 157–168.
Meixner, J., and W. Andrejewski, 1953: Strenge Theorie der Beugung ebener electromagnetischer Wellen an der vollkommen leitenden Kreissscheibe und an der kreisformigen Offnung im vollkommen leitenden ebenen Schirm. *Ann. Phys.*, **6**, 227–236.
Meixner, J., and F. W. Schefke, 1954: *Mathieusche Funktionen und Scheroid Funktionen*, Berlin: Springer Verlag.
Meixner, J., F. W. Schefke, and G. Wolf, 1980: Mathieu functions and spheroidal functions and their mathematical foundations. *Lect. Notes in Mathem.*, 837–886.

Mishchenko, M. I., L. D. Travis, and D. W. Mackowski, 1996: T-matrix computations of light scattering by nonspherical particles: a review, *J. Quant. Spectr. Rad. Transfer*, **55**, 535–575.

Mishchenko, M. I., J. W. Hovenier, and L. D. Travis, 2000: *Light Scattering by Nonspherical Particles*, San Diego: Academic Press.

Mishchenko, M. I., L. D. Travis, and A. Lacis, 2002: *Scattering, Absorption and Emission of Light by Small Particles*, Cambridge, UK: Cambridge University Press.

Morse, P. M., and H. Feshbach, 1953: *Methods of Theoretical Physics*, New York: McGraw-Hill.

Onaka, T., 1980: Light scattering by spheroidal grains, *Ann. Tokyo Astron. Observ.*, **18**, 1–54.

Schultz, F. M., K. Stamnes, and J. J. Stamnes, 1998: Scattering of electromagnetic waves by spheroidal particles: a novel approach exploiting the T-matrix computed in spheroidal coordinates, *Appl. Opt.*, **37**, 7875–7896.

Seker, S. S., 1986: Radar cross-section of thin dielectric bodies, *IEEE Proc.*, **H-133**, 305–309.

Sinha, B. P., and R. H. McPhie, 1977: Electromagnetic scattering by prolate spheroids for a plane waves with arbitrary polarization and angle of incidence, *Radio Sci.*, **12**, 171–184.

Stratton, J. A., 1941: *Electromagnetic Theory*, New York: McGrow-Hill.

Voshchinnikov, N. V., 1996: Electromagnetic scattering by homogeneous and coated spheroids: calculations using the separation of variables method, *J. Quant. Spectr. Rad. Transfer*, **55**, 627–636.

Voshchinnikov, N.V., and V. G. Farafonov, 1988: Characteristics of radiation scattered by prolate and oblate perfectly conducting spheroids, *(Sov.) Radiotech. Electron.*, **33**, 1364–1373.

Voshchinnikov, N. V., and V. G. Farafonov, 1993: Optical properties of spheroidal particles, *Astrophys. Space Sci.*, **204**, 19–86.

Voshchinnikov, N. V., and V. G. Farafonov, 2000: Applicability of quasi-static and Rayleigh approximations for spheroidal particles, *Opt. Spectr.*, **88**, 71–75.

Voshchinnikov, N. V., and V. G. Farafonov, 2003: Computation of radial prolate spheroidal wave functions using Jáffe's series expansions, *Comp. Math. Math. Phys.*, **43**, 1299–1309.

Voshchinnikov, N. V., V. B. Il'in, Th. Henning, B. Michel, and V. G. Farafonov, 2000: Extinction and polarization of radiation by absorbing spheroids: shape/size effects and some benchmarks, *J. Quant. Spectr. Rad. Transfer*, **65**, 877–893.

Zakharova, N. T., and M. I. Mishchenko, 2000: Scattering properties of needlelike and platelike ice spheroids with moderate size parameters, *Appl. Opt.*, **39**, 5052–5057.

Part II

Radiative Transfer

6 Radiative transfer and optical imaging in biological media by low-order transport approximations: the simplified spherical harmonics (SP_N) approach

Jorge Bouza Domínguez and Yves Bérubé-Lauzière

6.1 Introduction

Radiative transfer theory (RTT) is a valuable theoretical framework for describing the propagation of optical radiation in turbid media (Ishimaru, 1978; Wang and Wu, 2007). RTT has succeeded in fields such as astronomy and astrophysics (Chandrasekhar, 1960), remote sensing of the earth surface and atmosphere (Melnikova et al., 2012), heat transfer (Howell et al., 2010; Modest, 2003; Atalay, 2006), and, particularly, in biomedical optics (Wang and Wu, 2007; Hielscher et al., 2011; Klose, 2010a). The fundamental equation in RTT is the radiative transfer equation (RTE) (Wang and Wu, 2007). The RTE is the most accurate model for describing light propagation in biological tissue, with no approximation regarding the angular, spatial or temporal dependences (Hielscher et al., 2011). The RTE is an integro-differential equation that depends on a set of optical parameters (index of refraction, absorption, scattering and scattering function) that describe the medium (Ishimaru, 1978). The validity limits of the RTE rest on the model conceived to describe light propagation, and should be established for each physical situation (Martí López et al., 2003). Analytical solutions of the RTE are only known for simple geometries (Ishimaru, 1978; Liemert and Kienle, 2011b). Thus, numerical techniques are used in practical situations where complex geometries and/or heterogeneous optical property distributions need to be considered (Tarvainen, 2006). Solving the RTE for biological media carries a considerable numerical burden (Tarvainen, 2006; Klose and Larsen, 2006). In imaging applications, the RTE needs to be solved anew at each iteration step of an optimization algorithm in order to determine optimal optical parameters (Dehghani et al., 2009b; Arridge and Schotland, 2009). This is an implicit limitation of RTE-based image reconstruction algorithms in pre-clinical and clinical imaging and therapeutics, where the diagnosis time matters.

To reduce computation time, the diffusion equation (DE) is frequently used instead of the RTE (Wang and Wu, 2007; Dehghani et al., 2009b). The DE is derived from the RTE using the diffusion approximation which assumes that the field appearing in the RTE is almost isotropic at each point (Ishimaru, 1978; Wang and Wu, 2007). Unfortunately, there are several practical situations where the DE fails, as in the vicinity of sources (Martí López et al., 2004) and in the case of small geometries, low scattering, or high absorption (Hielscher et al., 1998; Chen et al.,

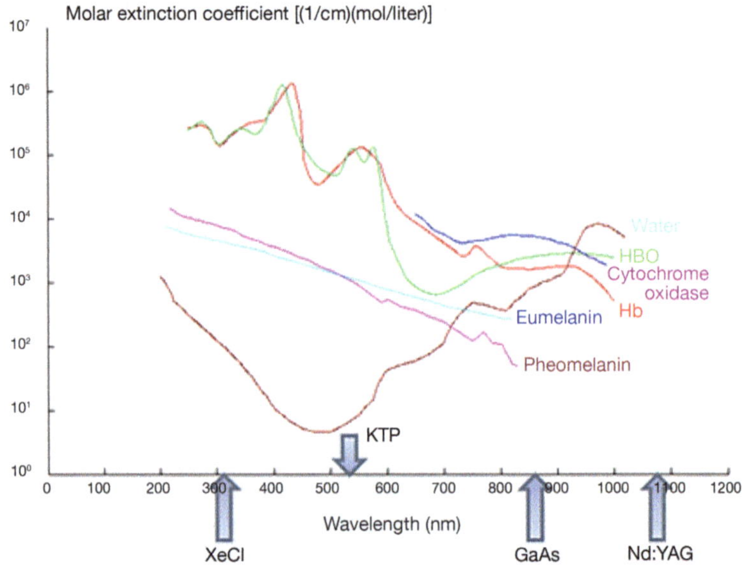

Fig. 6.1. Molar extinction coefficient (absorption/concentration) of several chromophores. Arrows at the bottom indicate the emission wavelength of common lasers.

2001). For example, in the so-called therapeutic window (600 to 1000 nm), vascularized tissues present high absorption because of chromophore absorption spectra, see Fig. 6.1 for the molar extinction coefficient values of common chromophores.

In such circumstances, the radiative field cannot be accurately described by the DE. Then, DE-based radiation dose calculations will yield wrong estimates and spatial resolution and quantitativeness of retrieved optical coefficients maps can be seriously affected. To overcome the drawbacks of the DE and avoid the RTE's computational burden, low-order transport models with simplified angular dependences were recently brought to biomedical optics (Klose and Larsen, 2006; Chu et al., 2009; Bouza Domínguez and Bérubé-Lauzière, 2010; Bouza Domínguez and Bérubé-Lauzière, 2011a). Some of these models are derived from the RTE using the simplified spherical harmonics approximation (SP_N) (Klose and Larsen, 2006). SP_N models have been developed for steady-state – or continuous-wave (CW) (Klose and Larsen, 2006), frequency-domain (FD) (Chu et al., 2009) and time-domain (TD) problems (Bouza Domínguez and Bérubé-Lauzière, 2010; Bouza Domínguez and Bérubé-Lauzière, 2011a), opening new possibilities in treatment and imaging applications of biomedical optics.

We herein review the use of SP_N models in describing radiative transfer in biological media. We also survey the outcomes of using SP_N models in optical imaging. With this, we hope to motivate further developments and applications of SP_N models in therapeutics and optical imaging of biological media.

6.2 Light transport in biological media

In tissue optics, RTT describes the emission, propagation, scattering and absorption of radiation. It provides a macroscopic view of the propagation of light (Ishimaru, 1978). In RTT, only the flow of energy through the medium is considered. Interference and diffraction effects are neglected and magnitudes such as the radiance bear the physical meaning. Through the use of the Stokes vector, polarization can, however, be accounted for in RTT. Additionally, concepts such as ray, ray congruence and ray divergence can be extrapolated from geometrical optics and employed to elaborate a mathematical model of light propagation (Martí López et al., 2003). Applying the law of the conservation of energy in a differential volume element, it is possible to derive an expression for the radiance variation in terms of the optical properties of the medium, leading to the RTE. Corresponding boundary conditions for the radiance can be obtained as well.

In the next section, we write down the expressions for the RTE and its boundary conditions. We introduce the reduced incident and the diffused radiance its corresponding components and review two well-known approximations: the spherical harmonics approximation leading to the so-called P_N equations, and the diffusion approximation (DA) leading to the DE.

6.2.1 The radiative transfer equation

The standard way of deriving the RTE[1] leads to the following expression (Wang and Wu, 2007)

$$\frac{\eta}{c}\frac{\partial}{\partial t}L(\mathbf{r},\widehat{\mathbf{s}},t)+\widehat{\mathbf{s}}\cdot\nabla L(\mathbf{r},\widehat{\mathbf{s}},t)$$
$$=-\left[\mu_{\mathrm{a}}(\mathbf{r})+\mu_{\mathrm{s}}(\mathbf{r})\right]L\left(\mathbf{r},\widehat{\mathbf{s}},t\right)+\mu_{\mathrm{s}}(\mathbf{r})\int_{4\pi}p(\mathbf{r},\widehat{\mathbf{s}},\widehat{\mathbf{s}}')L(\mathbf{r},\widehat{\mathbf{s}}',t)\,\mathrm{d}\Omega'+q\left(\mathbf{r},\widehat{\mathbf{s}},t\right), \tag{6.1}$$

where $L(\mathbf{r},\widehat{\mathbf{s}},t)$ is the radiance at point $\mathbf{r}$ in the direction specified by the unit vector $\widehat{\mathbf{s}}$, η is the refractive index, c is the speed of light in vacuum, $\mu_{\mathrm{a}}(\mathbf{r})$ and $\mu_{\mathrm{s}}(\mathbf{r})$ are respectively the absorption and scattering coefficients, $p(\mathbf{r},\widehat{\mathbf{s}},\widehat{\mathbf{s}}')$ is the normalized scattering function (also customarily called the 'phase function') which represents the probability of a photon being scattered in direction $\widehat{\mathbf{s}}$ when coming from direction $\widehat{\mathbf{s}}'$, $\mathrm{d}\Omega'$ is a differential element of solid angle, $q(\mathbf{r},\widehat{\mathbf{s}},t)$ is a source distribution per unit volume and ∇ denotes the gradient operator with respect to the $\mathbf{r}$ coordinates.

Frequently, the radiance is decomposed into collimated and diffuse components and Eq. (6.1) is posed in terms of the latter (Ishimaru, 1978). This approach is quite useful when applying low-order transport approximations since the angular dependence of the diffuse radiance is less pronounced than the total radiance. Hence, low-order transport approximations better reproduce (with the addition of the col-

[1]In the standard derivation of the RTE, the refractive index is a piecewise constant function not a continuous function. Besides, ray divergence effects are neglected in the discussion here (Martí López et al., 2003).

limated component) the radiative field than full RTE-based low-order transport approximations. In addition, we point out the fact that measurements in diffuse optical tomography (DOT) are done in terms of the diffused component only.

The solution of Eq. (6.1) requires the specification of initial and boundary conditions. For a smooth tissue–air interface ∂V and in the presence of specular Fresnel reflection, the vacuum boundary condition is substituted by the partly-reflecting boundary condition (Case and Zweifel, 1967; Duderstadt and Martin, 1979; Ishimaru, 1978)

$$L(\mathbf{r}',\widehat{\mathbf{s}},t) = B_T(\mathbf{r}',\widehat{\mathbf{s}},t) + R_F(\widehat{\mathbf{n}}\cdot\widehat{\mathbf{s}}')L(\mathbf{r}',\widehat{\mathbf{s}}',t)\,, \quad \mathbf{r}'\in\partial V,\ \widehat{\mathbf{s}}\cdot\widehat{\mathbf{n}}<0\,, \tag{6.2}$$

where R_F is the angle dependent Fresnel coefficient (Born and Wolf, 2003), $B_T(\mathbf{r},\widehat{\mathbf{s}},t)$ is the radiance of the exterior source transmitted inside the medium, $\widehat{\mathbf{n}}$ is the outer normal to the surface ∂V and $\widehat{\mathbf{s}}' = \widehat{\mathbf{s}} - 2(\widehat{\mathbf{n}}\cdot\widehat{\mathbf{s}})\widehat{\mathbf{n}}$. Here, $\widehat{\mathbf{s}}'$ is a vector that points outward and is the specular reflection of vector $\widehat{\mathbf{s}}$. For a smooth tissue–tissue interface, the flux balance can be expressed as Marshak conditions (Marshak, 1947; Davidson and Sykes, 1957; Faris, 2002)

$$\begin{aligned}
\int_{\widehat{\mathbf{s}}\cdot\widehat{\mathbf{n}}>0} L_1(\mathbf{r},\widehat{\mathbf{s}},t)\,(\widehat{\mathbf{s}}\cdot\widehat{\mathbf{n}})\;\mathrm{d}\Omega &= \int_{\widehat{\mathbf{s}}\cdot\widehat{\mathbf{n}}<0} R_{F,1}L_1(\mathbf{r},\widehat{\mathbf{s}},t)\,(-\widehat{\mathbf{s}}\cdot\widehat{\mathbf{n}})\;\mathrm{d}\Omega \\
&\quad + \int_{\widehat{\mathbf{s}}\cdot\widehat{\mathbf{n}}>0} \left[1-R_{F,2}\right] L_2(\mathbf{r},\widehat{\mathbf{s}},t)\,(\widehat{\mathbf{s}}\cdot\widehat{\mathbf{n}})\;\mathrm{d}\Omega \\
\int_{\widehat{\mathbf{s}}\cdot\widehat{\mathbf{n}}<0} L_2(\mathbf{r},\widehat{\mathbf{s}},t)\,(-\widehat{\mathbf{s}}\cdot\widehat{\mathbf{n}})\;\mathrm{d}\Omega &= \int_{\widehat{\mathbf{s}}\cdot\widehat{\mathbf{n}}>0} R_{F,2}L_2(\mathbf{r},\widehat{\mathbf{s}},t)\,(\widehat{\mathbf{s}}\cdot\widehat{\mathbf{n}})\;\mathrm{d}\Omega \\
&\quad + \int_{\widehat{\mathbf{s}}\cdot\widehat{\mathbf{n}}<0} \left[1-R_{F,1}\right] L_1(\mathbf{r},\widehat{\mathbf{s}},t)\,(-\widehat{\mathbf{s}}\cdot\widehat{\mathbf{n}})\;\mathrm{d}\Omega
\end{aligned} \tag{6.3}$$

where $L_i(\mathbf{r},\widehat{\mathbf{s}},t)$ denotes the radiance in medium $i = 1,2$ and $R_{\mathrm{F},i}$ is the angle dependent Fresnel reflection coefficient for medium i.

6.2.2 Spherical harmonics expansion and the P_N approximation

In the spherical harmonics expansion, the transport equation is reduced to a system of coupled partial differential equations (PDEs) with no angular-dependence (Case and Zweifel, 1967; Davidson and Sykes, 1957). The angular dependent functions appearing in Eq. (6.1), such as the radiance $L(\mathbf{r},\widehat{\mathbf{s}},t)$ and the source distribution $q(\mathbf{r},\widehat{\mathbf{s}},t)$, are expanded along spherical harmonics $Y_{l,m}(\widehat{\mathbf{s}}) \equiv Y_{l,m}(\theta,\phi)$ (θ and ϕ being respectively the polar and the azimuthal angles of spherical coordinates) as

$$L(\mathbf{r},\widehat{\mathbf{s}},t) = \sum_{l=0}^{\infty}\sum_{m=-l}^{l}\left(\frac{2l+1}{4\pi}\right)^{\frac{1}{2}}\psi_{l,m}(\mathbf{r},t)Y_{l,m}(\widehat{\mathbf{s}})\,, \tag{6.4}$$

$$q(\mathbf{r},\widehat{\mathbf{s}},t) = \sum_{l=0}^{\infty}\sum_{m=-l}^{l}\left(\frac{2l+1}{4\pi}\right)^{\frac{1}{2}}q_{l,m}(\mathbf{r},t)Y_{l,m}(\widehat{\mathbf{s}})\,, \tag{6.5}$$

where, following (Case-Zweifel, 1967), the normalization factor $\left[(2l+1)/4\pi\right]^{\frac{1}{2}}$ is introduced for the convenience that it results in simpler final expressions. Properties of the spherical harmonics can be found in (Abramowitz and Stegun, 1965).

For the scattering function $p(\mathbf{r}, \widehat{\mathbf{s}}, \widehat{\mathbf{s}}')$, the reasonable assumption is made that it only depends on the angular change between the incident and the scattered directions of a photon, i.e. $p(\mathbf{r}, \widehat{\mathbf{s}}, \widehat{\mathbf{s}}') = p(\mathbf{r}, \widehat{\mathbf{s}} \cdot \widehat{\mathbf{s}}')$. In this case, the phase function can be expanded along Legendre polynomials as

$$p(\mathbf{r}, \widehat{\mathbf{s}} \cdot \widehat{\mathbf{s}}') = \sum_{l=0}^{\infty} \sum_{m=-l}^{l} \left(\frac{2l+1}{4\pi}\right) p_l(\mathbf{r}, t) P_l(\widehat{\mathbf{s}} \cdot \widehat{\mathbf{s}}') \,. \tag{6.6}$$

Making use of the addition theorem for spherical harmonics, this can be rewritten as

$$p(\mathbf{r}, \widehat{\mathbf{s}} \cdot \widehat{\mathbf{s}}') = \sum_{l=0}^{\infty} \sum_{m=-l}^{l} p_l(\mathbf{r}) Y_{l,m}(\widehat{\mathbf{s}}) Y_{l,m}(\widehat{\mathbf{s}}') \,. \tag{6.7}$$

Inserting Eqs. (6.4), (6.5) and (6.7) into Eq. (6.1) and after some algebra (see recurrence relations for $Y_{l,m}(\widehat{\mathbf{s}})$ in (Abramowitz and Stegun, 1965), an infinite set of coupled PDEs is obtained

$$\begin{aligned}
&\left(\frac{\eta}{c}\frac{\partial}{\partial t} + \mu_{tr}(\mathbf{r})\right) \psi_{l,m}(\mathbf{r}, t) \\
&\quad + \frac{1}{2l+1}\left(\frac{\partial}{\partial z}\left[(l+1-m)^{\frac{1}{2}}(l+1+m)^{\frac{1}{2}}\psi_{l+1,m}(\mathbf{r}, t) + (l-m)^{\frac{1}{2}}(l+m)^{\frac{1}{2}}\psi_{l-1,m}(\mathbf{r}, t)\right]\right. \\
&\quad - \frac{1}{2}\left(\frac{\partial}{\partial x} - i\frac{\partial}{\partial y}\right)\left[(l+m)^{\frac{1}{2}}(l+m-1)^{\frac{1}{2}}\psi_{l-1,m-1}(\mathbf{r}, t)\right. \\
&\qquad\qquad \left. -(l-m+2)^{\frac{1}{2}}(l-m+1)^{\frac{1}{2}}\psi_{l+1,m-1}(\mathbf{r}, t)\right] \\
&\quad - \frac{1}{2}\left(\frac{\partial}{\partial x} + i\frac{\partial}{\partial y}\right)\left[-(l-m)^{\frac{1}{2}}(l-m-1)^{\frac{1}{2}}\psi_{l-1,m+1}(\mathbf{r}, t)\right. \\
&\qquad\qquad \left.\left. +(l+m+1)^{\frac{1}{2}}(l+m+2)^{\frac{1}{2}}\psi_{l+1,m+1}(\mathbf{r}, t)\right]\right) \\
&= \mu_s(\mathbf{r}) p_l \psi_{l,m}(\mathbf{r}, t) + q_{l,m}(\mathbf{r}, t) \,,
\end{aligned} \tag{6.8}$$

where $\mu_{tr}(\mathbf{r}) = \mu_a(\mathbf{r}) + \mu_s(\mathbf{r})$ is the transport coefficient and $i = \sqrt{-1}$. Truncating the series in Eqs. (6.4)–(6.7) at $l = N$ (this is the so-called P_N *approximation*), a system of $(N+1)^2$ coupled first-order PDEs is obtained. These are known as the P_N equations, which have been used as the forward model for imaging the scattering and absorption properties of biological media (Wright et al., 2007). Using the finite element method (FEM) for discretizing the forward model, initial results show an improvement over reconstructions based on the diffusion equation (Wright et al., 2007).

6.2.3 P_1 and the diffusion approximation

Truncating the expansions in Eqs. (6.4)–(6.7) at $l = N = 1$ (P_1 approximation), leads to four equations that can be grouped in vector form as (Wang and Wu, 2007)

$$\frac{\eta}{c}\frac{\partial \psi_0(\mathbf{r},t)}{\partial t} + \nabla \cdot \mathbf{J}(\mathbf{r},t) + \mu_a(\mathbf{r})\psi_0(\mathbf{r},t) = \varepsilon_0(\mathbf{r},t)\,, \tag{6.9}$$

$$\frac{\eta}{c}\frac{\partial \mathbf{J}(\mathbf{r},t)}{\partial t} + \frac{1}{3}\nabla\psi_0(\mathbf{r},t) + \left[\mu_a(\mathbf{r}) + {\mu'}_{\mathrm{s}}(\mathbf{r})\right]\mathbf{J}(\mathbf{r},t) = \boldsymbol{\varepsilon}_1(\mathbf{r},t)\,, \tag{6.10}$$

where $\mu_{\mathrm{s}}' = \mu_{\mathrm{s}}(1-g)$ is the so-called reduced scattering coefficient and g is known as the anisotropy coefficient (Ishimaru, 1978; Wang and Wu, 2007), $\psi_0(\mathbf{r},t)$ and $\mathbf{J}(\mathbf{r},t)$ are the fluence rate and the radiant current density vector, respectively, with $\mathbf{J}$ given by

$$\mathbf{J}(\mathbf{r},t) = \int_{4\pi} L(\mathbf{r},\hat{\mathbf{s}},t)\,\hat{\mathbf{s}}\;\mathrm{d}\Omega = \sqrt{\frac{2\pi}{3}}\left(\psi_{1,-1} - \psi_{1,-1}, -i(\psi_{1,-1} + \psi_{1,-1}), \sqrt{2}\psi_{1,0}\right)\,. \tag{6.11}$$

The function $\varepsilon_0(\mathbf{r},t)$ and the vector $\boldsymbol{\varepsilon}_1(\mathbf{r},t)$ embody the first two orders of the expansion of the source term (Eq. 6.5), similarly to ψ and $\mathbf{J}$ for L. Eqs. (6.9) and (6.10) are known as the P_1 equations and constitute the starting point to derive the diffusion equation (DE). The DE is considered valid at macroscopic length scales[2] (Van Rossum and Nieuwenhuizen, 1999) and derived under the assumption that the radiance has a weak angular dependence, originated by a high albedo scattering medium, i.e. $\mu_a \ll \mu_s$ (Wang and Wu, 2007). To derive the DE, $\mathbf{J}(\mathbf{r},t)$ is algebraically eliminated from Eq. (6.9) by using Eq. (6.10) under the condition known as the diffusion approximation (DA) (Wang and Wu, 2007)

$$\tau_0\left|\frac{\partial}{\partial t}\mathbf{J}(\mathbf{r},t)\right| \ll |\mathbf{J}(\mathbf{r},t)|\;, \quad \tau_0 = \frac{\eta}{c\left[\mu_a(\mathbf{r}) + {\mu'}_{\mathrm{s}}(\mathbf{r})\right]}\,. \tag{6.12}$$

The DA imposes constraints on the relative time variation of $\mathbf{J}(\mathbf{r},t)$, which contains the odd-order first terms of the series in Eq. (6.4). If in addition to the DA, we have an isotropic source distribution, then $\boldsymbol{\varepsilon}_1(\mathbf{r},t) = \mathbf{0}$, and the classical DE is obtained (Wang and Wu, 2007)

$$\frac{\eta}{c}\frac{\partial \phi_0(\mathbf{r},t)}{\partial t} + \nabla\cdot\left[D(\mathbf{r})\nabla\phi_0(\mathbf{r},t)\right] + \mu_a(\mathbf{r})\phi_0(\mathbf{r},t) = \varepsilon_0(\mathbf{r},t)\,, \tag{6.13}$$

where $D(\mathbf{r})$ is the standard diffusion coefficient

$$D(\mathbf{r}) = \frac{1}{3\left[\mu_a(\mathbf{r}) + {\mu'}_{\mathrm{s}}(\mathbf{r})\right]} \approx \frac{1}{3{\mu'}_{\mathrm{s}}(\mathbf{r})}\,, \tag{6.14}$$

and the inequality $\mu_a \ll \mu_s$ has been used to approximate the diffusion coefficient. Next, we introduce an approximation to the RTE similar to the P_N approximation: the simplified spherical harmonics approximation.

6.3 The simplified spherical harmonics approximation

Additionally to the time variable (or modulation frequency if FD methods are used), the radiation field is position and direction-dependent. Thus, elaborating numerical schemes for solving the RTE may involve a discretization method for

[2]Length scales such as $\lambda \ll l'_{tr} \ll l_s$, where λ is the wavelength, l'_{tr} the transport mean free path and l_s the sample size.

up to a six-dimensional space. For this reason, RTE approximations that preserve enough accuracy while reducing computation time are highly desirable for practical applications. That is the purpose of low-order transport models such as the P_N approximation (Arridge, 1999; Wright et al. 2007). However, the large number and complexity of the P_N equations (as an extra feature, mixed spatial derivatives are contained in the equations) limit the applicability of the P_N approximation. An alternative is the simplified P_N (SP_N) approximation which transforms the RTE into a system of coupled diffusion equations (elliptic in the steady-state case or parabolic in the time-dependent case) that depend solely on space and time (Klose and Larsen, 2006). The SP_N equations display a significantly reduced complexity compared to the P_N equations, and allow the application of DE-like numerical schemes and solvers (Klose and Larsen, 2006; Chu et al., 2009; Bouza Domínguez and Bérubé-Lauzière, 2010, Montejo et al., 2011).

Applying the SP_N approximation, a methodology coined the SP_N method, originated in the field of nuclear reactor theory (Fletcher, 1983; de Oliveira, 1986). In its early days, the SP_N method lacked firm theoretical foundations, which hampered its use. Further developments allowed resolving this issue and expanding the applications of the SP_N equations to other fields such as heat transport (Larsen et al., 2002), coupled electron–photon transport problems (Kotiluoto et al., 2007), and biomedical optics (Klose and Larsen, 2006).

The SP_N equations have been derived in three ways: (*i*) as a multidimensional generalization of the P_N equations for geometries with planar symmetry – so-called the formal or heuristic derivation (Gelbard, 1960), (*ii*) as an asymptotic correction to the diffusion approximation (Larsen et al., 1996) and (*iii*) using a rigorous variational analysis approach (Tomasevic and Larsen, 1996; Brantley and Larsen, 2000). In biomedical optics, it has been demonstrated that SP_N equations provide transport-like solutions for modeling visible and near-infrared light propagation in small tissue geometries and specially, in the presence of high absorption ($\sim$1 cm^{-1}) (Klose and Larsen, 2006). In addition, these results are achieved with only a fraction of the computational cost of a transport calculation and a minimum of twice the cost of DE calculations (Klose and Larsen, 2006). Moreover, SP_N equations have been introduced in luminescence imaging and provided the inherent advantages of transport-like solutions in model-based image reconstruction algorithms (Klose, 2009; Klose, 2012).

Subsequently, we first present the heuristic derivation of the SP_N equations and corresponding boundary conditions based on the planar symmetry assumption for time-independent problems arising in biomedical optics (the more rigorous derivation based on variational analysis is not discussed here as it would require too much space, and is beyond the scope of the present work). Following this, we will discuss the SP_N equations for the frequency and time domains.

6.3.1 The steady-state SP_N equations

For deriving the SP_N equations, we will assume that the optical properties of the medium vary only along a given axis, and not along directions perpendicular to this axis (i.e., we have a medium with planar symmetry). This also assumes that there is azimuthal symmetry (i.e. no dependence on the spherical angle ϕ, see Fig. 6.2). The

z axis will then be called the axis of symmetry. For a medium with such symmetry, the time-independent RTE (or the equivalent, in terms of the diffuse component of the radiance) has the following form (Klose and Larsen, 2006)

$$\begin{aligned} \varpi \frac{\partial}{\partial z}\psi(z,\varpi) = & -\left[\mu_{\mathrm{a}}(z)+\mu_{\mathrm{s}}(z)\right]\psi(z,\varpi) \\ & +\int_{-1}^{1}\widetilde{\mu}_{\mathrm{s}}(z,\varpi,\varpi')\psi(z,\varpi')\,\mathrm{d}\varpi' + \frac{Q(z)}{2}, \end{aligned} \tag{6.15}$$

where $\psi(z,\varpi)$ can be either the radiance $L(z,\varpi)$ for a medium with embedded isotropic sources or its diffuse radiance component, z is the coordinate along the axis of symmetry oriented along the unit vector $\widehat{\mathbf{k}}$ and $\varpi = \widehat{\mathbf{s}}\cdot\widehat{\mathbf{k}}$ is the cosine of the angle between a given direction of propagation $\widehat{\mathbf{s}}$ and $\widehat{\mathbf{k}}$, $\widetilde{\mu}_{\mathrm{s}}(z,\varpi,\varpi') = \mu_{\mathrm{s}}p(z,\varpi,\varpi')$ is the differential scattering coefficient (or modified phase function) and $Q(z)$ represents a time-independent isotropic source.

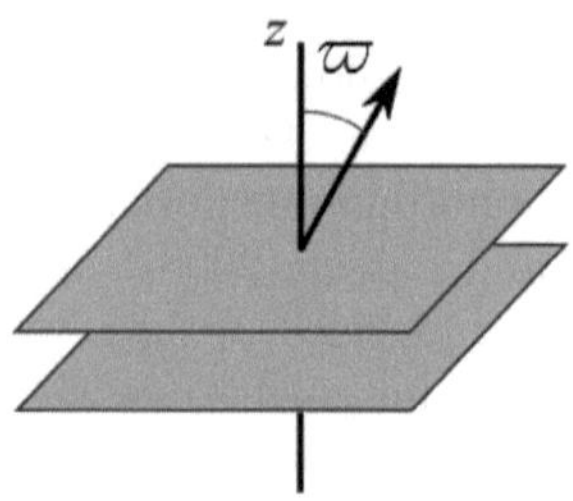

Fig. 6.2. Planar symmetry in a medium.

The corresponding boundary condition is

$$\psi(z,\varpi) = B_T(z,\varpi) + R_F(\varpi)\psi(z,-\varpi)\,, \quad z\in S,\ 0<\varpi<1\,, \tag{6.16}$$

where for the diffuse component of the radiance, we can simply assume that $B_T(z,\varpi) = 0$. If we integrate Eq. (6.15) over the interval $[-1,1]$, we obtain the exact equation

$$\frac{\mathrm{d}\psi_1(z)}{\mathrm{d}z} = -\mu_{\mathrm{a}}(z)\psi_0(z) + \frac{Q(z)}{2}, \tag{6.17}$$

where $\psi_0(z)$ and $\psi_1(z)$ are the fluence and the radiant current density for a medium with planar symmetry. In addition, they are the zeroth and first moments of the Legendre expansion of $\psi(z,\varpi)$ (see below the radiance Legendre expansion).

We may develop the modified phase function $\widetilde{\mu}_{\mathrm{s}}(z,\varpi,\varpi')$ and $\psi(z,\varpi)$ along Legendre polynomials as follows (we assume that $\widetilde{\mu}_{\mathrm{s}}(z,\varpi,\varpi') = \widetilde{\mu}_{\mathrm{s}}(z,\varpi\cdot\varpi')$, i.e. that scattering only depends on the angle between the incident and scattered directions)

$$\widetilde{\mu}_{\mathrm{s}}(z,\varpi,\varpi') = \sum_{n=0}^{\infty}\left(\frac{2n+1}{2}\right)\mu_s(z)g_n(z)P_n(\varpi)P_n(\varpi')\,, \tag{6.18}$$

$$\psi(z,\varpi) = \sum_{n=0}^{\infty}\left(\frac{2n+1}{2}\right)\psi_n(z)P_n(\varpi)\,. \tag{6.19}$$

where $g_n(z)$ and $\psi_n(z)$ are the Legendre moments of the phase function and radiance, respectively. In the case of the Henyey–Greenstein phase function, $g_n(z) = [g(z)]^n$, where g describes the possibly space-dependent degree of anisotropy of the scattering (the anisotropy parameter).

Substituting the Legendre expansions into the main equation for $\psi(z, \varpi)$, multiplying both sides by $P_{n'}(\varpi)$, $n \neq 0$, and integrating over the interval $[-1, 1]$, we obtain

$$\mu_n(z)\psi_n(z) + \frac{\mathrm{d}}{\mathrm{d}z}\left[\frac{n+1}{2n+1}\psi_{n+1}(z) + \frac{n}{2n+1}\psi_{n-1}(z)\right] = 0\,, \quad n = 1, \ldots, \infty\,, \tag{6.20}$$

where the orthogonality property and the recurrence relation of the Legendre polynomials have been used (Abramowitz and Stegun, 1965). Here, we defined the nth-order transport coefficients as $\mu_n(z) = \mu_a(z) + \mu_s(z)[1 - g_n(z)]$. The former equation allows expressing the moments $\psi_n(z)$ as

$$\psi_n(z) = -\frac{1}{\mu_n(z)}\frac{\mathrm{d}}{\mathrm{d}z}\left[\frac{n+1}{2n+1}\psi_{n+1}(z) + \frac{n}{2n+1}\psi_{n-1}(z)\right]\,. \tag{6.21}$$

If the Legendre expansion of the radiance given in Eq. (6.19) is truncated at a given order N, which can be selected to be odd, the *odd-order P_N equations* for planar geometries are obtained. Next, we employ Eqs. (6.21) to algebraically eliminate the odd-order moments. After some algebra, we obtain final equations for the even-order moments

$$\begin{aligned}
&\mu_n(z)\psi_n(z) - \frac{n+1}{2n+1}\frac{\mathrm{d}}{\mathrm{d}z}\left\{\frac{1}{\mu_{n+1}(z)}\frac{d}{dz}\left[\frac{n+2}{2n+3}\psi_{n+2}(z) + \frac{n+1}{2n+3}\psi_n(z)\right]\right\}\\
&- \frac{n}{2n+1}\frac{\mathrm{d}}{\mathrm{d}z}\left\{\frac{1}{\mu_{n-1}(z)}\frac{\mathrm{d}}{\mathrm{d}z}\left[\frac{n}{2n-1}\psi_n(z) + \frac{n-1}{2n-1}\psi_{n-2}(z)\right]\right\} = \delta_{n,0}Q(z)\,,\\
&n = 0, 2, \ldots, N-1\,.
\end{aligned} \tag{6.22}$$

Eq. (6.22) is a system of $K = (N+1)/2$ coupled one-dimensional elliptic equations with K unknowns $\{\psi_n(z)\}_{i=0,2,\ldots,N-1}$, K being even. As mentioned before, odd-order moments can be obtained from the solution of Eq. (6.22) and back-substitution in Eq. (6.21), up to the truncated order N.

The extension of Eq. (6.22) to the three-dimensional case is obtained by replacing z by $\mathbf{r}$, and substituting each operator by its 3-D counterpart in 3-D, i.e. the partial derivative $\partial/\partial z$ becomes the gradient operator $\nabla \equiv [\partial/\partial x,\ \partial/\partial y,\ \partial/\partial z]$. Here, the second angular dependence of the spherical harmonics expansion (the generalization of the P_N approximation for 3-D) is neglected. The final simplified P_N equations (SP_N equations) for steady-state 3-D problems are thus

$$\begin{aligned}
&\mu_n(\mathbf{r})\psi_n(\mathbf{r}) - \frac{n+1}{2n+1}\nabla\cdot\left\{\frac{1}{\mu_{n+1}(\mathbf{r})}\nabla\left[\frac{n+2}{2n+3}\psi_{n+2}(\mathbf{r}) + \frac{n+1}{2n+3}\psi_n(\mathbf{r})\right]\right\}\\
&- \frac{n}{2n+1}\nabla\left\{\frac{1}{\mu_{n-1}(\mathbf{r})}\nabla\left[\frac{n}{2n-1}\psi_n(\mathbf{r}) + \frac{n-1}{2n-1}\psi_{n-2}(\mathbf{r})\right]\right\} = \delta_{n,0}Q(\mathbf{r}),\\
&n = 0, 2, \ldots, N-1\,.
\end{aligned} \tag{6.23}$$

Because of the approximations performed in deriving Eq. (6.23), the solution of the SP_N equations Eq. (6.23) does not converge to the exact transport solution (see Eqs. (6.15)) as $N \to \infty$. Instead, the solution of Eq. (6.23) converges asymptotically to the transport solution so there is an optimal order N for each physical situation (Klose and Larsen, 2006).

To simplify the notation, it is convenient to rewrite Eq. (6.23) in matrix format. For this, we introduce the column vector of even-order moments $\mathbf{\Psi}(\mathbf{r}) = [\psi_{k'}(\mathbf{r})]^T$, $k' = 0, 2, 4, \ldots, N-1$, which is in turn rewritten in terms of the vector of composite moments $\mathbf{\Phi}(\mathbf{r}) = [\varphi_k(\mathbf{r})]^T$, $k = 1, \ldots, K$. The relationship between the even-order moments and the composite moments, and its inverse, can be expressed in a convenient matrix notation. Up to $N = 7$ (higher orders can be obtained as well from Eq. (6.23)), this relationship is given by

$$\mathbf{\Psi}(\mathbf{r}) = \mathbf{T}\mathbf{\Phi}(\mathbf{r}), \quad \mathbf{\Phi}(\mathbf{r}) = \mathbf{T}^{-1}\mathbf{\Psi}(\mathbf{r}), \quad \text{where } \mathbf{T}^{-1} = \begin{bmatrix} 1 & 2 & 0 & 0 \\ 0 & 3 & 4 & 0 \\ 0 & 0 & 5 & 6 \\ 0 & 0 & 0 & 7 \end{bmatrix}.$$

The composite moments allow diagonalizing the 'diffusive operator' containing the differential operators having the form $-\nabla \cdot (D_K \nabla)$. This leads to the matrix form of the steady-state SP_N equations (or CW-SP_N model)

$$(D_r + \mathbf{C})\mathbf{\Phi}(\mathbf{r}) = \mathbf{Q}(\mathbf{r}). \tag{6.24}$$

The term D_r is a diagonal $K \times K$ matrix operator whose elements are all on the main diagonal and given by

$$\operatorname{diag}_{(0)}(D_r) = [-\nabla \cdot (D_1 \nabla) \quad -\nabla \cdot (D_2 \nabla) \quad \cdots \quad -\nabla \cdot (D_K \nabla)], \quad k = 1, \ldots, K, \tag{6.25}$$

where $D_k = 1/[(4k-1)\mu_{2k-1}]$ and the expression $\operatorname{diag}_{(0)}(\ )$ denotes the list of the main diagonal elements. Note that we use index (0) for the main diagonal and positive (negative) values for diagonals located under (below) the main diagonal (such notation will be used again in the sequel). The components of the (symmetric) matrix $\mathbf{C}$ are linear combinations of the transport coefficients μ_n. The explicit expressions for the columns of $\mathbf{C}$ in Matlab notation (up to $N = 7$) are given by

$$\mathbf{C}(:,1) = \begin{bmatrix} \mu_0(\mathbf{r}) \\ -\frac{2}{3}\mu_0(\mathbf{r}) \\ \frac{8}{15}\mu_0(\mathbf{r}) \\ -\frac{16}{35}\mu_0(\mathbf{r}) \end{bmatrix}, \quad \mathbf{C}(:,2) = \begin{bmatrix} -\frac{2}{3}\mu_0(\mathbf{r}) \\ \frac{4}{9}\mu_0(\mathbf{r}) + \frac{5}{9}\mu_2(\mathbf{r}) \\ -\frac{16}{45}\mu_0(\mathbf{r}) - \frac{4}{9}\mu_2(\mathbf{r}) \\ \frac{32}{105}\mu_0(\mathbf{r}) + \frac{8}{21}\mu_2(\mathbf{r}) \end{bmatrix},$$

$$\mathbf{C}(:,3)=\begin{bmatrix} \frac{8}{15}\mu_0(\mathbf{r}) \\ -\frac{16}{45}\mu_0(\mathbf{r})-\frac{4}{9}\mu_2(\mathbf{r}) \\ \frac{64}{225}\mu_0(\mathbf{r})+\frac{16}{45}\mu_2(\mathbf{r})+\frac{9}{25}\mu_4(\mathbf{r}) \\ -\frac{128}{525}\mu_0(\mathbf{r})-\frac{32}{105}\mu_2(\mathbf{r})-\frac{54}{175}\mu_4(\mathbf{r}) \end{bmatrix},$$

$$\mathbf{C}(:,4)=\begin{bmatrix} -\frac{16}{35}\mu_0(\mathbf{r}) \\ \frac{32}{105}\mu_0(\mathbf{r})+\frac{8}{21}\mu_2(\mathbf{r}) \\ -\frac{128}{525}\mu_0(\mathbf{r})-\frac{32}{105}\mu_2(\mathbf{r})-\frac{54}{175}\mu_4(\mathbf{r}) \\ \frac{256}{1225}\mu_0(\mathbf{r})+\frac{64}{245}\mu_2(\mathbf{r})+\frac{324}{1225}\mu_4(\mathbf{r})+\frac{13}{49}\mu_6(\mathbf{r}) \end{bmatrix}. \tag{6.26}$$

The column vector $\mathbf{Q}(\mathbf{r})$ contains the information about the source; it is given by

$$\mathbf{Q}(\mathbf{r})=Q(\mathbf{r})\begin{bmatrix} 1 & -\frac{2}{3} & \frac{8}{15} & -\frac{16}{35} \end{bmatrix}^T. \tag{6.27}$$

6.3.2 SP_N boundary conditions and measurement modeling

The boundary conditions (BCs) associated with Eq. (6.24) can be obtained by inserting Eq. (6.19) into Eq. (6.16) and carrying out integrations similarly to what was done previously. In a convenient matrix form, these BCs are

$$\mathbf{A}\boldsymbol{\Phi}(\mathbf{r})+\mathbf{B}\frac{\partial}{\partial\hat{\mathbf{n}}}\boldsymbol{\Phi}(\mathbf{r})=\mathbf{S}(\mathbf{r}), \quad \mathbf{r}\in\partial V, \tag{6.28}$$

where $\partial/\partial\hat{\mathbf{n}}$ denotes the derivative along the outward-pointing normal $\hat{\mathbf{n}}$ to the boundary. The boundary matrices $\mathbf{A}$, $\mathbf{B}$ and vector $\mathbf{S}$ (external source vector) depend on the reflectivity properties of the boundary and the optical coefficients of the medium. We assume $\mathbf{S}=\mathbf{0}$ for SP_N equations originated from the RTE in terms of the diffuse component of the radiance, since it is related to the exterior source. The mentioned terms have the following form (up to $N=7$)

$$\mathbf{A}=\begin{bmatrix} 1/2+A_1 & -1/8-C_1 & 1/16-E_1 & -5/128-G_1 \\ -1/8-C_2 & 7/24+A_2 & -41/384-E_2 & 1/16-G_2 \\ 1/16-C_3 & -41/384-E_3 & 407/1920+A_3 & -233/2560-G_3 \\ -5/128-C_4 & 1/16-E_4 & -233/2560-G_4 & 3023/17920+A_4 \end{bmatrix}, \tag{6.29}$$

$$\mathbf{B}=\begin{bmatrix} (1+B_1)/3\mu_1 & -D_1/\mu_3 & -F_1/\mu_5 & -H_1/\mu_7 \\ -D_2/3\mu_1 & (1+B_2)/7\mu_3 & -F_2/\mu_5 & -H_2/\mu_7 \\ -D_3/3\mu_1 & -F_3/\mu_3 & (1+B_3)/11\mu_5 & -H_3/\mu_7 \\ -D_4/3\mu_1 & -F_4/\mu_3 & -H_4/\mu_5 & (1+B_4)/15\mu_7 \end{bmatrix}. \tag{6.30}$$

$$\mathbf{S} = \begin{bmatrix} \int_{\widehat{\mathbf{s}}\cdot\widehat{\mathbf{n}}>0} B_T(\mathbf{r},\widehat{\mathbf{s}}) 2\left|\widehat{\mathbf{s}}\cdot\widehat{\mathbf{n}}\right| \,\mathrm{d}\Omega \\ \int_{\widehat{\mathbf{s}}\cdot\widehat{\mathbf{n}}>0} B_T(\mathbf{r},\widehat{\mathbf{s}}) \left[5\left|\widehat{\mathbf{s}}\cdot\widehat{\mathbf{n}}\right|^3 - 2\left|\widehat{\mathbf{s}}\cdot\widehat{\mathbf{n}}\right|\right] \mathrm{d}\Omega \\ \int_{\widehat{\mathbf{s}}\cdot\widehat{\mathbf{n}}>0} B_T(\mathbf{r},\widehat{\mathbf{s}}) \left[\frac{63}{4}\left|\widehat{\mathbf{s}}\cdot\widehat{\mathbf{n}}\right|^5 - \frac{35}{2}\left|\widehat{\mathbf{s}}\cdot\widehat{\mathbf{n}}\right|^3 + \frac{15}{4}\left|\widehat{\mathbf{s}}\cdot\widehat{\mathbf{n}}\right|\right] \mathrm{d}\Omega \\ \int_{\widehat{\mathbf{s}}\cdot\widehat{\mathbf{n}}>0} B_T(\mathbf{r},\widehat{\mathbf{s}}) \left[\frac{429}{8}\left|\widehat{\mathbf{s}}\cdot\widehat{\mathbf{n}}\right|^7 - \frac{693}{8}\left|\widehat{\mathbf{s}}\cdot\widehat{\mathbf{n}}\right|^5 + \frac{315}{8}\left|\widehat{\mathbf{s}}\cdot\widehat{\mathbf{n}}\right|^3 - \frac{35}{8}\left|\widehat{\mathbf{s}}\cdot\widehat{\mathbf{n}}\right|\right] \mathrm{d}\Omega \end{bmatrix}, \tag{6.31}$$

where the coefficients $(A_1, \ldots, H_1,\ A_4, \ldots, H_4)$ (see Appendix A of Klose and Larsen (2006) for their explicit expressions) are linear combinations of the angular moments $R_{F,n}$ of the angle-dependent Fresnel coefficient R_F

$$R_{F,n} = \int_0^1 x^n R_F(x)\,\mathrm{d}x\,. \tag{6.32}$$

The Robin-type boundary condition Eq. (6.28) can be considered as the Marshak condition (Davidson and Sykes, 1957) for the vector of composite moments at a tissue–air interface, where Fresnel reflections occurs. For $N = 1$, Eq. (6.28) becomes the common Robin (or Marshak) boundary condition for the DE (Marshak, 1947; Ishimaru, 1978). At tissue–tissue interfaces, the normal component of the radiant current density vector $J_{no} = \mathbf{J}(\mathbf{r},t)\cdot\widehat{\mathbf{n}}$ is continuous (we can just add both equations appearing in Eq. (6.3)). Then, corresponding boundary conditions can be found by mere substitution of the expansion Eq. (6.19) in this condition and grouping of similar terms.

To end up with the derivation of the SP_N equations as a forward model in inverse problems, we need an expression for relating the outgoing light to the vector of composite moments $\mathbf{\Phi}(\mathbf{r})$. If we take as measurements a finite collection of exitance values (outgoing normal component of the radiant current density vector) $J_{no}^{(out)}$, then

$$J_{no}^{(out)} = \left[\mathbf{j}_1 - \mathbf{j}_2(\mathbf{B})^{-1}\mathbf{A}\right] \mathbf{\Phi}\left(\mathbf{r}_d, t\right) = \mathbf{V}_\mu \mathbf{\Phi}\left(\mathbf{r}_d, t\right)\,, \tag{6.33}$$

where $\mathbf{V}_\mu$ is the measurement operator (vector) that depends on the optical properties of the medium, $\mathbf{r}_d$ is a position where a measurement is made ('detector position'), and the vectors $\mathbf{j}_1$ and $\mathbf{j}_2$ have the following expressions (up to $N = 7$)

$$\mathbf{j}_1 = \begin{bmatrix} 1/4 + J_0 \\ (1/4 + J_0)(-2/3) + (5/16 + J_2)(1/3) \\ (1/4 + J_0)(8/15) + (5/16 + J_2)(-4/15) + (-3/32 + J_4)(1/5) \\ (1/4{+}J_0)(-15/35) + (5/16{+}J_2)(8/35) + (-3/32{+}J_4)(-6/35) + (13/256{+}J_6)(1/7) \end{bmatrix}^T, \tag{6.34}$$

$$\mathbf{j}_2 = \left[-\left(\frac{0.5 + J_1}{3\mu_1}\right),\ \left(-\frac{J_3}{7\mu_3}\right),\ \left(-\frac{J_5}{11\mu_5}\right),\ \left(-\frac{J_7}{15\mu_7}\right)\right]. \tag{6.35}$$

The coefficients $(J_0, J_1, \ldots)$ depend on the angular moments of the angle-dependent Fresnel reflection coefficient $R_{F,n}$ (Eq. (6.32)), and can be found in (Klose and Larsen, 2006).

The SP_N equations Eqs. (6.24) along with their boundary conditions – Eq. (6.28) have been numerically implemented and compared with DE-based and transport (RTE-based) calculations (Klose and Larsen, 2006). The calculations are performed in small homogeneous geometries: $1 \times 1\,\mathrm{cm}$ and $2 \times 2\,\mathrm{cm}$, that mimic a tomographic slice of a small animal. The absorption coefficient values are taken from 0.01 to $2\,\mathrm{cm}^{-1}$ (high absorption). The reduced scattering coefficient is kept constant as $10\,\mathrm{cm}^{-1}$ but considering a variation of the scattering coefficient from 10 to $50\,\mathrm{cm}^{-1}$ and the anisotropy factor g from 0 to 0.8. In every case, a medium with a refractive index value of $\eta = 1.37$ is considered as surrounded by air. In a second round, numerical experiments are carried out in a $2 \times 2\,\mathrm{cm}$ diffusive medium with $\mu_a = 0.01\,\mathrm{cm}^{-1}$, $\mu_s = 10\,\mathrm{cm}^{-1}$, $g = 0$ and non-reentry boundary conditions $\eta = 1$. Highly absorbing inclusions ($\mu_a = 2\,\mathrm{cm}^{-1}$) are embedded in the diffusive medium. Following an analysis of the experiment results, see (Klose and Larsen, 2006) for details, the authors concluded that the SP_N equations Eqs. (6.24) (*i*) can accurately model light propagation in small tissue geometries at visible and near-infrared wavelengths, (*ii*) provide transport-like solutions with a considerably reduced computational cost in comparison with RTE-based calculations and (*iii*) improve DE solutions in transport-like domains with high absorption and small geometries.

6.3.3 Analytical solutions

Analytical solutions are essential for experiments with simple geometries and validation of numerical approaches. Recently, steady-state analytical solutions have been found for infinite (SP_3 and SP_5 equations) and semi-infinite (SP_3 equations) homogeneous media (Liemert and Kienle, 2010; Liemert and Kienle, 2011a). In addition, a methodology for the generalization of the results to the frequency- and time-domain cases is suggested. The final expressions for the composite moments are set out as linear combinations of DE free space Green's functions. Next, we write down the main results for the aforementioned geometries.

6.3.3.1 Infinite homogeneous medium

An infinite homogeneous medium with an isotropic point source located at the origin of coordinates $Q(\mathbf{r}) = \delta(r)/4\pi r^2$ (r is the distance from the source location) has an inherent spherical symmetry. This symmetry allows the following spherical wave expansion of the composite moments and the source

$$\varphi_i(r) = \frac{1}{2\pi^2 r} \int_0^\infty p\,\widehat{\varphi}_i(p)\,\sin(pr)\,\mathrm{d}p\,, \quad i = 1, 2 \tag{6.36}$$

$$Q(r) = \frac{1}{2\pi^2 r} \int_0^\infty p\,\sin(pr)\,\mathrm{d}p\,, \quad i = 1, 2 \tag{6.37}$$

where the hat over a quantity means the transformed quantity in the p-space. Introducing Eqs. (6.36) and (6.37) into Eq. (6.24) leads to the following system of linear equations

$$L(\mu_n, p)\widehat{\mathbf{\Phi}}(p) = \widehat{\mathbf{Q}}\,, \tag{6.38}$$

where $\widehat{\mathbf{Q}} = [1 - 2/3]^T$ and $L(\mu_n, p)$ is a matrix whose coefficients depend on the optical properties of the medium (μ_n) and p. For $N = 3$ (hereon, following (Liemert and Kienle, 2010), we show results for $N = 3$ and 5 only) we get

$$L(\mu_n, p) = \begin{bmatrix} p^2/3\mu_1 + \mu_a & -2\mu_a/3 \\ -2\mu_a/3 & p^2/7\mu_3 + 4\mu_a/9 + 5\mu_2/9 \end{bmatrix}. \tag{6.39}$$

From Eq. (6.38), the composite moment functions $\widehat{\varphi}_i(p)$, $i = 1, 2$ are determined as the ratio of even-order polynomials in p

$$\widehat{\varphi}_i(p) = \begin{cases} \dfrac{F_i^{(1)}(p^2)}{p^4 + \alpha p^2 + \beta}, & i = 1, 2, \text{ for } N = 3 \\ \dfrac{F_i^{(2)}(p^2)}{p^6 + \alpha p^4 + \beta p^2 + \gamma}, & i = 1, 2, 3 \text{ for } N = 5 \end{cases}, \quad F_i^{(m)}(x) = \sum_{j=i-1}^{m} a_{ij} x^j\,, \tag{6.40}$$

where the coefficients a_{ij} appearing in the definition of the polynomial $F_i^{(m)}(x)$ are given by

$$a_{10} = \frac{35}{3}\mu_1\mu_2\mu_3\,, \quad a_{11} = 3\mu_1\,, \quad a_{21} = -\frac{14}{3}\mu_3\,, \quad \text{for } N = 3\,, \tag{6.41}$$

$$\begin{aligned} a_{10} &= \frac{231}{5}\mu_1\mu_2\mu_3\mu_4\mu_5\,, \quad a_{11} = \frac{35}{3}\mu_1\mu_2\mu_3 + 33\mu_1\mu_5\left(\frac{16}{45}\mu_2 + \frac{9}{25}\mu_4\right), \quad a_{12} = 3\mu_1\,, \\ a_{21} &= -\frac{462}{25}\mu_3\mu_4\mu_5\,, \quad a_{22} = -\frac{14}{3}\mu_3\,, \quad a_{32} = \frac{88}{15}\mu_5\,, \qquad \text{for } N = 5\,. \end{aligned} \tag{6.42}$$

The coefficients α, β and γ for the polynomials of the denominator are real positive numbers that depend on the transport coefficients

$$\alpha = 3\mu_a\mu_1 + \frac{28}{9}\mu_a\mu_3 + \frac{35}{9}\mu_2\mu_3\,, \quad \beta = \frac{35}{3}\mu_a\mu_1\mu_2\mu_3\,, \quad \text{for } N = 3\,, \tag{6.43}$$

$$\begin{aligned} \alpha &= 3\mu_a\mu_1 + \frac{28}{9}\mu_a\mu_3 + \frac{35}{9}\mu_2\mu_3 + 11\mu_5\left(\frac{64}{225}\mu_a + \frac{16}{45}\mu_2 + \frac{9}{25}\mu_4\right), \\ \beta &= \mu_a\mu_1\left(\frac{35}{3}\mu_2\mu_3 + \frac{176}{15}\mu_2\mu_5 + \frac{297}{25}\mu_4\mu_5\right) + \mu_3\mu_4\mu_5\left(\frac{308}{25}\mu_a + \frac{77}{5}\mu_2\right), \\ \gamma &= \frac{231}{5}\mu_a\mu_1\mu_2\mu_3\mu_4\mu_5\,, \quad \text{for } N = 5\,. \end{aligned} \tag{6.44}$$

For $N = 3$, using the expressions for α and β given in Eq. (6.49), it can be shown that the polynomial discriminant $\sqrt{\alpha^2 - 4\beta}$ of the denominator in Eq. (6.40) is always positive. Therefore, a partial fractions expansion using the polynomial zeros $\lambda_{1,2} = \left(-\alpha \pm \sqrt{\alpha^2 - 4\beta}\right)/2$ is possible. Thus, the composite moments can be written as

$$\widehat{\varphi}_i(p) = \frac{A_i}{p^2 + p_1^2} + \frac{B_i}{p^2 + p_2^2}, \quad p_j = \sqrt{-\lambda_j}, \quad j = 1, 2, \tag{6.45}$$

where

$$A_i = \frac{F_i^{(1)}(\lambda_1)}{k_2^2 - k_1^2}, \quad B_i = -\frac{F_i^{(1)}(\lambda_2)}{k_2^2 - k_1^2}. \tag{6.46}$$

In the case of an infinite medium, it is known that the Green's function $G(r)$ of the steady-state DE is given by

$$G(r) = \frac{e^{-\mu_{\mathit{eff}}\, r}}{4\pi D r} = \frac{1}{2\pi^2 r} \int_0^\infty \frac{p \sin(pr)}{p^2 + \mu_{\mathit{eff}}}\, \mathrm{d}p,$$
$$D = \frac{1}{3\left(\mu_a + \mu'_{\mathrm{s}}\right)}, \quad \mu_{\mathit{eff}} = \sqrt{3\mu_a(\mu_a + \mu'_{\mathrm{s}})}. \tag{6.47}$$

A comparison of Eqs. (6.45) and (6.47) shows that the composite moment functions $\varphi_i(r)$, $i = 1, 2$ can be written as a superposition of two DE free space Green's functions $G(r)$ as follows

$$\varphi_i(r) = A_i \frac{e^{-p_1 r}}{4\pi r} + B_i \frac{e^{-p_2 r}}{4\pi r}, \quad p_j = \sqrt{-\lambda_j}, \quad j = 1, 2. \tag{6.48}$$

Similarly, for $N = 5$ the polynomial appearing in the denominator of Eq. (6.40) can be decomposed into three partial fractions. It is thus possible to demonstrate that each composite moment function can be expanded into three DE free space Green's functions $G(r)$ as

$$\varphi_i(r) = A_i \frac{e^{-p_1 r}}{4\pi r} + B_i \frac{e^{-p_2 r}}{4\pi r} + C_i \frac{e^{-p_3 r}}{4\pi r}, \quad p_j - \sqrt{-\lambda_j}, \quad j = 1, 2, 3, \tag{6.49}$$

where the expressions for the expansion coefficients are

$$A_i = \frac{F_i^{(2)}(\lambda_1)}{\left(p_2^2 - p_1^2\right)\left(p_3^2 - p_1^2\right)}, \quad B_i = -\frac{F_i^{(2)}(\lambda_2)}{\left(p_2^2 - p_1^2\right)\left(p_3^2 - p_2^2\right)},$$
$$C_i = \frac{F_i^{(2)}(\lambda_3)}{\left(p_3^2 - p_1^2\right)\left(p_3^2 - p_2^2\right)}, \quad p_j = \sqrt{-\lambda_j}, \ j = 1, 2, 3. \tag{6.50}$$

The zeros of polynomials λ_j appearing in Eq. (6.50) can be calculated from Viète's trigonometric method for obtaining roots of third-degree polynomials as (see Liemert and Kienle, 2010)

$$\lambda_j = 2\sqrt{\frac{\xi}{3}} \cos\left[\frac{\nu + 2(j-1)\pi}{3}\right], \quad \xi = \frac{1}{3}\alpha^2 - \beta,$$
$$\nu = \arccos\left[-\frac{3}{2\xi}\sqrt{\frac{3}{p}}\left(\frac{2}{27}\alpha^3 - \frac{1}{3}\alpha\beta + \gamma\right)\right], \quad j = 1, 2, 3. \tag{6.51}$$

The analytical solutions provided by Eqs. (6.48) and (6.49) are easy to implement. The analog expressions for frequency-domain problems can be calculated by setting $\mu_n(z) = \mu_a(z) + \mu_s(z)\,[1 - g_n(z)] + i(\eta\omega/c)$, where η is the refractive index of the medium, ω is the angular frequency of the intensity modulated source, c is the speed of light in the vacuum and $i = \sqrt{-1}$. By performing the inverse Fourier transform, we can also obtain analytical formulas in time-domain. In the case of the SP_1 (the DE), an analytical formula for time-domain problems can be directly derived (Wang and Wu, 2007).

In (Liemert and Kienle, 2010), Eqs. (6.48) and (6.49) are compared with Monte Carlo simulations (Wang and Wu, 2007) in the steady-state and time domains and with DE solutions. For the steady-state, the numerical experiments are carried out with an infinite homogeneous medium. The optical properties of the medium are $\mu_s' = 1\,\mathrm{mm}^{-1}$, $g = 0.9$ and values of 0.2 and $2\,\mathrm{mm}^{-1}$ are used for the absorption coefficient. An isotropic point source is placed at the origin of coordinates. Then, the steady-state fluence rate versus distance from the isotropic source is calculated using the SP_N and the corresponding DE solutions and simulated using the Monte Carlo method. A comparison of the results showed that the SP_N solutions are in much better agreement that the DE-based solutions with the Monte Carlo simulations. Particularly, the SP_N solutions accurately reproduce Monte Carlo simulations at all distances from the source including both far and very close (<0.5 mm) to the source. In a second set of experiments, the time-resolved reflectance from a semi-infinite scattering medium, with a perpendicular incident pencil beam, is calculated using the SP_N and the DE solutions. The optical properties of the medium are $\mu_s' = 1\,\mathrm{mm}^{-1}$, $g = 0.9$, $\mu_a = 0.1\,\mathrm{mm}^{-1}$ and $\eta = 1.4$. The medium is considered as surrounded by air. The reflectance time-dependence is calculated at distances of 6.5, 9.5, and 12.5 mm from the position where the beam impinges. The results showed that SP_N solutions describe light propagation even for very short time (<100 ps) values, where the DE fails.

6.3.3.2 Semi-infinite homogeneous medium

In this subsection, we present the analytical solution of the SP_N equations for a semi-infinite geometry with an embedded isotropic point source. This solution is due to Liemert and Kienle (Liemert and Kienle, 2011a). Contrary to their approach, we do not use the formalism of bras and kets (i.e. the Dirac formalism of quantum mechanics), which we find less accessible and can be cumbersome to the non-initiated, it is also not absolutely necessary to reach the solution as elementary linear algebra means are sufficient. Although the exposition of the results is carried out for $N = 3$, the methodology presented can be used to achieve similar results for higher orders.

In the present case, the physical situation has an inherent cylindrical symmetry. Thus, we can expand the composite moments and the δ-source distribution using the zero-order Hankel transform

$$\varphi_i(\mathbf{r}) = \frac{1}{2\pi}\int_0^\infty \varphi_i\,(q,z)\,J_0\,(q\rho)\,q\,\mathrm{d}q\,, \quad \delta\,(\mathbf{r}-\mathbf{r}') = \frac{\delta\,(z-z')}{2\pi}\int_0^\infty J_0\,(q\rho)\,q\,\mathrm{d}q\,, \tag{6.52}$$

where $J_0(x)$ is the zeroth-order Bessel function of the first kind. Substituting Eq. (6.52) into Eq. (6.24) yields a system of second order differential equations for $\mathbf{\Phi}(\mathbf{r})$

$$\frac{\mathrm{d}^2\mathbf{\Phi}}{\mathrm{d}z^2} = \left(\mathbf{M}_\mu + q^2\mathbf{I}_{2\times 2}\right)\mathbf{\Phi} + \delta\left(z - z'\right)\boldsymbol{\varepsilon}\,, \tag{6.53}$$

where the coefficient matrix $\mathbf{M}_\mu$ and the vector $\boldsymbol{\varepsilon}$ are

$$\mathbf{M}_\mu = \begin{bmatrix} 3\mu_a\mu_1 & -2\mu_a\mu_1 \\ -14\mu_a\mu_3 & \frac{28}{9}\mu_a\mu_3 + \frac{35}{9}\mu_2\mu_3 \end{bmatrix}, \quad \boldsymbol{\varepsilon} = \frac{1}{3}\begin{bmatrix} -9\mu_1 \\ 14\mu_3 \end{bmatrix}, \tag{6.54}$$

and $\mathbf{I}_{2\times 2}$ is the 2×2 identity matrix.

The solution of the boundary value problem $\mathbf{\Phi}(q,z)$ posed by Eqs. (6.53) and (6.28) can be obtained by using the superposition principle

$$\mathbf{\Phi}\left(q, z; z'\right) = \mathbf{\Phi}^{(h)}\left(q, z\right) + \mathbf{\Phi}^{(p)}\left(q, z; z'\right)\,, \tag{6.55}$$

where $\mathbf{\Phi}^{(h)}\left(q, z\right)$ is the solution to the source-free problem (homogeneous component) and $\mathbf{\Phi}^{(p)}\left(q, z; z'\right)$ is a particular solution of Eq. (6.53).

The solution to the source-free problem

$$\frac{\mathrm{d}^2\mathbf{\Phi}^{(h)}}{\mathrm{d}z^2} - \left(\mathbf{M}_\mu + q^2\mathbf{I}_{2\times 2}\right)\mathbf{\Phi}^{(h)} = \mathbf{0}\,, \tag{6.56}$$

will be sought in a form with exponential dependence as follows (similarly to scalar ODEs with constant coefficients)

$$\mathbf{\Phi}(q, z) = e^{\lambda(q)z}\mathbf{w}\,, \tag{6.57}$$

with $\mathbf{w}$ a two-component vector independent of q. Inserting this solution into Eq. (6.56) leads to the following

$$\left[\mathbf{M}_\mu - \left(\lambda^2 - q^2\right)\mathbf{I}_{2\times 2}\right]\mathbf{w} = \mathbf{0}\,. \tag{6.58}$$

Hence, for the proposed vector given in Eq. (6.57) to be a solution of the homogeneous equation, $\mathbf{w}$ must be an eigenvector of $\mathbf{M}_\mu$, and $\lambda^2 - q^2$ must be equal to an eigenvalue. At this point, the eigenvalues of $\mathbf{M}_\mu$ must thus be calculated. After some algebra, these are found to be positive (hence they will be denoted by ς_1^2 and ς_2^2), and given by

$$\varsigma_{1/2}^2 = \alpha \pm \sqrt{\alpha^2 - \beta}\,, \tag{6.59}$$

with

$$\alpha = \frac{3}{2}\mu_a\mu_1 + \frac{28}{18}\mu_a\mu_3 + \frac{35}{18}\mu_2\mu_3\,, \quad \beta = \frac{35}{3}\mu_a\mu_1\mu_2\mu_3\,. \tag{6.60}$$

Now that the eigenvalues are found, the associated eigenvectors $\mathbf{w}_1$ and $\mathbf{w}_2$ can be calculated. Let $\mathbf{w}_i = [\xi_i, \eta_i]$, $i = 1, 2$, then ξ_i and η_i must satisfy

$$\left(3\mu_a\mu_1 - \varsigma_i^2\right)\xi_i - \left(2\mu_a\mu_1\right)\eta_i = 0\,. \tag{6.61}$$

Hence $\mathbf{w}_i$ can be taken as follows

$$\mathbf{w}_i = \begin{bmatrix} 2\mu_a\mu_1 \\ 3\mu_a\mu_1 - \varsigma_i^2 \end{bmatrix} . \tag{6.62}$$

The spectral decomposition of $\mathbf{M}_\mu$ will be written as

$$\mathbf{M}_\mu = \mathbf{LDL}^{-1} , \tag{6.63}$$

where

$$\mathbf{D} = \mathrm{diag}\left(\varsigma_1^2, \varsigma_2^2\right) , \tag{6.64}$$

and $\mathbf{L}$ has as its columns the eigenvectors $\mathbf{w}_1$ and $\mathbf{w}_2$. These eigenvectors being column vectors, we may thus write

$$\mathbf{L} = [\mathbf{w}_1\mathbf{w}_2] = \begin{bmatrix} w_{1;1} & w_{2;1} \\ w_{1;2} & w_{2;2} \end{bmatrix} = \begin{bmatrix} 2\mu_a\mu_1 & 2\mu_a\mu_1 \\ 3\mu_a\mu_1 - \varsigma_1^2 & 3\mu_a\mu_1 - \varsigma_2^2 \end{bmatrix} . \tag{6.65}$$

Here $w_{i;j}$ denotes the jth component of vector $\mathbf{w}_i$. Reverting back to λ, we have that $\lambda^2 - q^2$ can be either equal to ς_1^2 or ς_2^2. Hence, there will be 4 possible values for λ, these being $\pm\lambda_i(q)$ with

$$\lambda_i(q) = \sqrt{q^2 + \varsigma_i^2} , \quad i = 1, 2 . \tag{6.66}$$

Now, since we must have $\mathbf{\Phi}^{(h)}(q, z) \to \mathbf{0}$ when $z \to \infty$, we can only retain the possible values of λ, that are negative. Thus, the homogeneous solution can be expressed as the following superposition

$$\mathbf{\Phi}^{(h)}(q, z) = c_1(q)\, e^{-\lambda_1(q)z}\mathbf{w}_1 + c_2(q)\, e^{-\lambda_2(q)z}\mathbf{w}_2 , \tag{6.67}$$

where $c_1(q)$ and $c_2(q)$ will be determined later on using the BCs.

Now, to find the particular solution $\mathbf{\Phi}^{(p)}(q, z; z')$, the following Fourier (or plane wave) decomposition of the particular vector of composite moments, and of the Dirac delta function are used

$$\mathbf{\Phi}^{(p)}(q, z; z') = \frac{1}{2\pi}\int_{-\infty}^{\infty} \mathbf{\Phi}^{(p)}(q, k)\, e^{ik(z'-z)}\, \mathrm{d}k , \quad \delta(z - z') = \frac{1}{2\pi}\int_{-\infty}^{\infty} e^{ik(z'-z)}\, \mathrm{d}k . \tag{6.68}$$

Now, recall that $\mathbf{\Phi}^{(p)}(q, z; z')$ must satisfy the following vector differential equation

$$\frac{d^2\mathbf{\Phi}^{(p)}}{dz^2} - \left(\mathbf{M}_\mu + q^2\mathbf{I}_{2\times 2}\right)\mathbf{\Phi}^{(p)} = \delta(z - z')\,\boldsymbol{\varepsilon} . \tag{6.69}$$

Inserting the Fourier decompositions into this last equation gives the following linear vector equation

$$\left[\mathbf{M}_\mu + \left(k^2 + p^2\right)\mathbf{I}_{2\times 2}\right]\mathbf{\Phi}^{(p)}(q, k) = -\boldsymbol{\varepsilon} . \tag{6.70}$$

This is similar to the eigenvalue equation encountered before in Eq. (6.58), except that here the left-hand side is not zero. To solve this equation, we use the spectral decomposition of $\mathbf{M}_\mu$, giving as solution

$$\mathbf{\Phi}^{(p)}(q,k) = \mathbf{L}\left\{\left[\mathbf{D} + \left(k^2+p^2\right)\mathbf{I}_{2\times 2}\right]^{-1}\mathbf{L}^{-1}(-\boldsymbol{\varepsilon})\right\} . \tag{6.71}$$

The quantity in braces is a vector column, and the result on the left-hand side of the last equation is seen to be a linear superposition (or combination) of the columns of $\mathbf{L}$, which are the eigenvectors of $\mathbf{M}_\mu$.[3] $\mathbf{L}$ being a 2×2 matrix, its inverse is easily calculated to be

$$\mathbf{L}^{-1} = \frac{1}{\det(\mathbf{L})}\begin{bmatrix} w_{2;2} & -w_{2;1} \\ -w_{1;2} & w_{1;1} \end{bmatrix} , \tag{6.72}$$

where

$$\det(\mathbf{L}) = 2\mu_a\mu_1\left(\varsigma_1^2 - \varsigma_2^2\right) . \tag{6.73}$$

With these last results, the particular solution given in Eq. (6.16) can be more explicitly written as

$$\mathbf{\Phi}^{(p)}(q,k) = \frac{1}{\det(\mathbf{L})}\left(\frac{14\mu_3 w_{2;1} + 9\mu_1 w_{2;2}}{k^2+\lambda_1^2(q)}\mathbf{w}_1 - \frac{14\mu_3 w_{1;1} + 9\mu_1 w_{1;2}}{k^2+\lambda_2^2(q)}\mathbf{w}_2\right) . \tag{6.74}$$

Introducing the constants

$$h_1 = \frac{14\mu_3 w_{2;1} + 9\mu_1 w_{2;2}}{3\det(\mathbf{L})} , \qquad h_2 = \frac{14\mu_3 w_{1;1} + 9\mu_1 w_{1;2}}{3\det(\mathbf{L})} , \tag{6.75}$$

the last expression for $\mathbf{\Phi}^{(p)}(q,k)$ can more succinctly be written as

$$\mathbf{\Phi}^{(p)}(q,k) = \frac{h_1}{k^2+\lambda_1^2(q)}\mathbf{w}_1 - \frac{h_2}{k^2+\lambda_2^2(q)}\mathbf{w}_2 . \tag{6.76}$$

Taking the inverse Fourier transform of the last expression, we get[4]

$$\mathbf{\Phi}^{(p)}(q,z) = \frac{h_1}{2}\frac{e^{-\lambda_1(q)|z-z'|}}{\lambda_1(q)}\mathbf{w}_1 - \frac{h_2}{2}\frac{e^{-\lambda_2(q)|z-z'|}}{\lambda_2(q)}\mathbf{w}_2 . \tag{6.77}$$

All the pieces to obtain the complete solution $\mathbf{\Phi}$ decomposed as in Eq. (6.55) have now been found. The use of the BC given in Eq. (6.28) (using that $\widehat{\mathbf{n}} = -\widehat{\mathbf{z}}$) allows determining the constant coefficients $c_1(q)$ and $c_2(q)$ appearing in Eq. (6.67) by solving two equations for these yet two unknowns (we shall not do this explicitly here). As a final step, inverting the Hankel transform leads to the following final expression for the composite moments

[3]That the product of a matrix with a vector on its right is a linear combination of the columns of the matrix is called the 'column point of view of matrix multiplication', see standard modern texts on linear algebra such as Strang (2005) or Lay (2011).

[4]Using the $1/2\pi$ normalization conventions of the Fourier transforms given in Eq. (6.68), the inverse Fourier transform of $1/(k^2+a^2)$ is $e^{-a|z-z'|}/2a$.

$$\begin{aligned}\boldsymbol{\Phi}(\mathbf{r}) &= \boldsymbol{\Phi}(\rho, z)\\ &= \frac{h_1}{4\pi}\frac{\mathrm{e}^{-\varsigma_1\sqrt{\rho^2+(z-z')^2}}}{\sqrt{\rho^2+(z-z')^2}}\mathbf{w}_1 - \frac{h_2}{4\pi}\frac{\mathrm{e}^{-\varsigma_2\sqrt{\rho^2+(z-z')^2}}}{\sqrt{\rho^2+(z-z')^2}}\mathbf{w}_2\\ &+ \frac{1}{2\pi}\left(\int_0^\infty c_1(q)\, e^{-\lambda_1(q)z} J_0(q\rho)\, q\, \mathrm{d}q\right)\mathbf{w}_1\\ &+ \frac{1}{2\pi}\left(\int_0^\infty c_2(q)\, e^{-\lambda_2(q)z} J_0(q\rho)\, q\, \mathrm{d}q\right)\mathbf{w}_2 .\end{aligned} \tag{6.78}$$

Finally, the fluence $\psi_0(z)$ and the reflectance $R(\rho)$ at the boundary $z = 0$ can be calculated for $N = 3$ as

$$\psi_0(z) = \begin{bmatrix} 1 & -2/3 \end{bmatrix} \boldsymbol{\Phi} = \phi_1 - \frac{2}{3}\phi_2 , \tag{6.79}$$

$$R(\rho) = \left[\left(\frac{1}{4}+J_0\right),\ -\frac{2}{3}\left(\frac{1}{4}+J_0\right)+\frac{1}{3}\left(\frac{5}{16}+J_2\right)\right]\boldsymbol{\Phi} + \left[\frac{1+2J_1}{6\mu_1}\quad \frac{J_3}{7\mu_3}\right]\frac{\mathrm{d}\boldsymbol{\Phi}}{\mathrm{d}z} . \tag{6.80}$$

These results for the SP_3 equations (derived from the RTE) have been used to compare reflectance values given in Eq. (6.80) with Monte Carlo simulations (Wang and Wu, 2007) and DE solutions (Liemert and Kienle, 2011a). The numerical experiments considered an isotropic point source located at one transport mean free path $l'_{tr} = 1/\mu_1$ inside a semi-infinite homogeneous medium (refractive index 1.4) surrounded by air. For typical values in the near infrared (NIR) ($\mu_a = 0.01\,\mathrm{mm}^{-1}$, $\mu_s = 10\,\mathrm{mm}^{-1}$) and blue or green wavelengths ($\mu_a = 1\,\mathrm{mm}^{-1}$, $\mu_s = 10\,\mathrm{mm}^{-1}$) ranges, the SP_3 equations were shown to give results that better agree with those of Monte Carlo simulations than the DE for distances to the source >1 mm. However, at small distances (<1 mm) to the isotropic point source, the SP_3 displayed no improvements compared to the DE (Liemert and Kienle, 2011a). A comparison of the SP_3 solution (and the DE solution) for an isotropic point source as above with Monte Carlo simulations for an infinitely narrow beam shows that the isotropic solutions do not match well the Monte Carlo results in this case. This means that the approximation of such a beam by an isotropic point source is not a good approximation, contrary to what is pervasively assumed in biomedical optics.

The numerical experiments described in (Liemert and Kienle, 2010; Liemert and Kienle, 2011a) lead naturally to the questions of which order N to employ in a given practical situation, and how accurate it can be near sources. First, searching for an optimal N while exploring higher orders ($N = 5, 7$) should be attempted (Klose and Larsen, 2006). In addition, since the radiative field is more anisotropic near sources, a better accuracy can be achieved with the RTE_d as the starting point to apply the SP_N approximation. Despite these recommendations, there is another problem not covered yet. At short distances to the source, the radiative field is not modified to much extent by scattering and absorption events, and the source emission pattern prevails. In the case of an isotropic point source embedded in the medium, the radiative field propagates along divergent rays starting from the location of the source. Such a situation is not considered by standard radiative transfer models where the divergence of rays always has a cylindrical form (Martí López et al.,

2003). Thus, sources are considered as located at the infinity, presumably a fact inherited from RTT applications in astronomy and astrophysics. Finally, we want to stress that in a comparison with real experimental data, modeling the source is a decisive step, see Ducros (2009) and references therein.

6.3.4 Frequency-domain simplified spherical harmonics equations

Lately, frequency-domain SP_N equations (FD-SP_N) have been derived in biomedical optics (Chu et al., 2009). To obtain the FD-SP_N, as proposed in the literature, the Fourier transform is applied to the time-dependent RTE Eq. (6.1). Hence, the result resembles the time-independent RTE and the previous steps on deriving the SP_N equations Eq. (6.24) are pursued. Finally, the FD-SP_N equations have the same form than that of Eq. (6.24), but with the introduction of the complex-valued nth-order transport coefficients

$$\mu_n^*(\mathbf{r}) = \mu_a(\mathbf{r}) + \frac{i\eta\omega}{c} + \mu_s(\mathbf{r})\left[1 - g_n(\mathbf{r})\right] , \tag{6.81}$$

The FD-SP_N equations and its corresponding BC have thus the same form as Eqs. (6.24) and by considering μ_n^* instead of μ_n and working in the frequency-domain (complex magnitudes). Measurements are then related to the quantization of amplitude, phase and even direct-current exitance (Chu and Dehghani, 2009; Xu et al., 2010; Xu et al., 2011), as it is common in frequency-domain systems (Wang et al., 2008).

A finite element method implementation of the FD-SP_N equations is available in the literature (Chu et al., 2009). In that work, several numerical experiments are performed with a 3-D slab of dimensions $40 \times 20 \times 30\,\text{mm}$. Three different cases are considered (1) an homogeneous medium where $\mu_a = 0.001\,\text{mm}^{-1}$, $\mu_s = 2\,\text{mm}^{-1}$, $g = 0.5$ and $\eta = 1.37$; (2) a similar homogeneous medium with the same optical properties but a different absorption coefficient of $0.01\,\text{mm}^{-1}$ and (3) a three-layer slab where the upper and the bottom sections have identical optical properties of $\mu_a = 0.001\,\text{mm}^{-1}$, $\mu_s = 1\,\text{mm}^{-1}$, $g = 0$ and $\eta = 1.37$. The middle layer has the following optical properties $\mu_a = 0.2\,\text{mm}^{-1}$, $\mu_s = 2\,\text{mm}^{-1}$, $g = 0.5$ and $\eta = 1.37$. A comparison of FD-SP_N-based and DE-based calculations with Monte Carlo simulations demonstrate that for $N > 1$, the FD-SP_N model shows increased accuracy compared with the DE in both the phase and amplitude of boundary data. Also, a high difference was found between the predicted light distribution by the DE and the SP_7 in regions near the source (modeled as an isotropic point source) and regions with high absorption ($0.2\,\text{mm}^{-1}$).

6.3.5 Time-domain simplified spherical harmonics equations

Time-domain SP_N equations (TD-SP_N) have been obtained, implemented and validated to solve problems in radiative transfer and biomedical optics. TD-SP_N models have been derived in three different ways: (i) via formal asymptotic analysis (Frank et al., 2007), (ii) by direct forward and back-substitution of the moment functions, leading to an integro-differential final form with temporal convolution operators (Bérubé-Lauzière et al., 2009) and (iii) imposing diffusive-type conditions over odd-order moment functions, similarly to the diffusion approximation (Bouza Domínguez and Bérubé-Lauzière, 2010). These three different approaches lead to different equations. The last approach has been to the most studied of the three for problems in biomedical optics. We will thus present the derivation pertaining to the last approach, leading to the so-called the time-dependent parabolic SP_N equations (TD-pSP_N) (Bouza Domínguez and Bérubé-Lauzière, 2010).

For a medium with planar symmetry, the time-dependent RTE only differs from Eq. (6.15) in the additional term $\eta\partial\psi(z,\varpi,t)/c\partial t$. Following the same steps as in Section 6.3.1, we arrive to an expression identical to Eq. (6.20) for the Legendre moments of the radiance, except for the additional term $\eta\partial\psi_n(z,t)/c\partial t$. Now, the direct back substitution of the odd-orders leads to equations with mixed terms of spatial and time partial derivatives. These equations are not of the diffusion-type, but contain convolution operators over the Legendre moment functions (Bérubé-Lauzière et al., 2009). To preserve the parabolic nature of the equations, we impose the following diffusive conditions on the time-derivatives of the odd-order moments (compare with the diffusion approximation Eq. (6.12))

$$\tau_n\left|\frac{\partial}{\partial t}\psi_n(z,t)\right| \ll |\psi_n(z,t)|\ , \quad \tau_n = \frac{\eta l_n}{c}\,. \tag{6.82}$$

These last conditions limit the relative time variation of the odd-order moments within the characteristic time τ_n. For $N = 1$, Eq. (6.82) turns out to be the well-known DA (Eq. (6.12)) for a planar geometry. So far, a complete study of Eq. (6.82) and therefore, the TD-pSP_N model validity, in terms of frequency modulation and pulse width values in time-resolved problems is pending for completion. An analysis of the TD-pSP_N model validity in the solution of forward and inverse problems can influence the design of experimental sets specifically built for using this model.

Imposing the diffusive conditions provides an expression similar to Eq. (6.21). Thus, the algebraic elimination of odd-moments in terms of the even-moments becomes possible. The extension of these results to 3-D and the introduction of the time-dependent vector of composite moments $\mathbf{\Phi}(\mathbf{r},t)$ leads to a system of $K = (N+1)/2$ coupled *parabolic* PDEs. This model, known as the *time-dependent parabolic SP_N equations* (TD-pSP_N) (Bouza Domínguez and Bérubé-Lauzière, 2010), has the same form as Eq. (6.24) except for the introduction of the term $[\eta\partial\mathbf{T}\mathbf{\Phi}(\mathbf{r},t)/c\partial t$. For $N = 1$, TD-pSP_N equations become the DE. Also, in this approach, the boundary conditions remain the same as in the steady-state case.

The TD-pSP_N model has been numerically implemented using a combined finite difference – finite element scheme (Bouza Domínguez and Bérubé-Lauzière, 2010). In this work, the model (for $N = 3$) is compared with DE-based numerical solutions and Monte Carlo simulations. The numerical experiments are carried out in a 2×2 cm homogeneous medium (see Fig. 6.3 left) for two different regimes: (1) a diffusive regime where $\mu_a = 0.04\,\mathrm{cm}^{-1}$, $\mu_s' = 20\,\mathrm{cm}^{-1}$ and $\eta = 1$ (no refractive index mismatch) and (2) a near-nondiffusive regime (Hielscher et al., 1998) where $\mu_a = 1\,\mathrm{cm}^{-1}$, $\mu_s' = 10\,\mathrm{cm}^{-1}$ and $\eta = 1$ (also, no refractive index mismatch). An isotropic point source, Dirac delta function in time, is placed at the center of the medium.

In both cases, the time-dependent fluence values are calculated at the boundary using the numerical solution provided by the TD-pSP_3 equations and the DE.

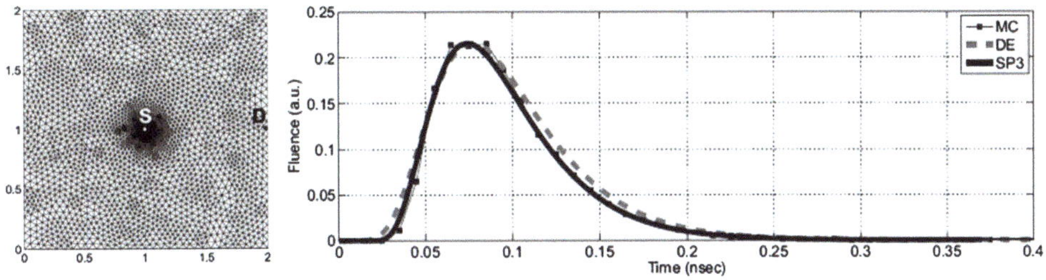

Fig. 6.3. Numerical experiment in a 2-D homogeneous medium (left) with $\eta = 1$, $g = 0.9$ and $\mu_a/\mu_s{}' = 0.1$ (this is a near-nondiffusive regime condition, see Hielscher et al. (1998)) and an isotropic point source, Dirac delta function in time, located at point S. At right, fluence profile at the detector point D calculated using the DE, the TD-pSP_3 equations and the Monte Carlo method. The TD-pSP_3 model better reproduces the Monte Carlo results than the DE, especially for those parts of the curve corresponding to early arriving photons and at long times.

For the diffusive regime, the results showed that TD-pSP_3 equations accurately reproduce the Monte Carlo results. For the near-nondiffusive regime, the TD-pSP_3 solution better reproduces the Monte Carlo results at the early times and at long times than the DE, see Fig. 6.3 at right.

In a second round of experiments, an absorptive inclusion is embedded in a $2 \times 2\,\mathrm{cm}$ homogeneous medium where $\mu_a = 0.01\,\mathrm{cm}^{-1}$, $\mu_s' = 10\,\mathrm{cm}^{-1}$ and $\eta = 1$, see Fig. 6.4. Three increasing values are assumed for the inclusion absorption coefficient $\mu_a = 0.05$, 0.1 and $1\,\mathrm{cm}^{-1}$ (last value corresponding to high absorption). The fluence profile is calculated in the homogeneous medium in the presence and absence of the absorptive inclusion using the TD-pSP_3 equations and the DE. With

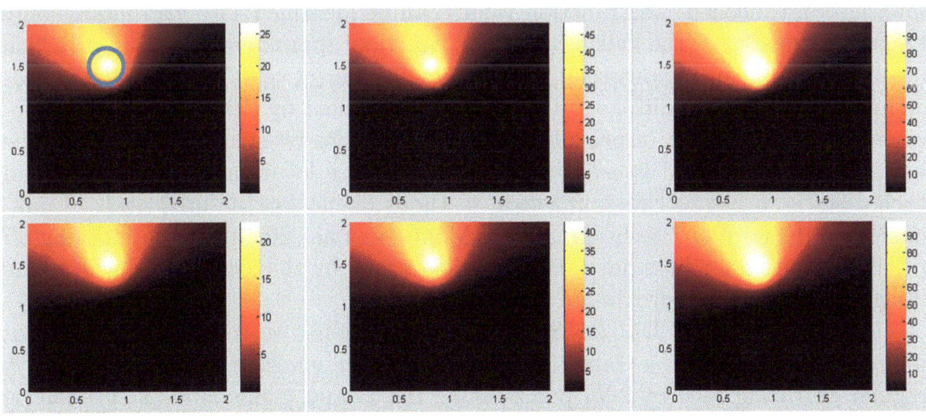

Fig. 6.4. Numerical experiments with a 2-D homogeneous medium ($\eta = 1$, $\mu_a = 0.01\,\mathrm{cm}^{-1}$ and $\mu_s{}' = 10\,\mathrm{cm}^{-1}$) with an isotropic point source (as in Fig. 6.3) and an absorptive inclusion (see small circle in top-left figure) which takes values of 0.05, 0.1 and $1\,\mathrm{cm}^{-1}$ (left to right). Color represents the percentage difference of fluence values with respect to the results in a medium with no inclusion, for the TD-pSP_3 equations (upper row) and the DE (lower row), at 220 ps. The contrast of the fluence fields is higher for the TD-pSP_3 equations than for the DE. This situation is repeated at different times.

these values, the percentage difference of fluence values with respect to the results in a medium with no inclusion is calculated, as a contrast or sensitivity measure, for each model. In all the cases and times, the TD-pSP_3 equations showed a higher sensitivity compared to the DE; see Fig. 6.4. Thus, the TD-pSP_3 equations seem more appropriate for describing light propagation in small geometries in the presence of absorptive inhomogeneities than the DE (Bouza Domínguez and Bérubé-Lauzière, 2010).

6.4 Numerical solutions

In the presence of complex geometries and/or heterogeneous media, it becomes necessary to resort to numerical methods, either implemented on structured or unstructured grids (also called meshes with nodes and elements as components (Jin, 2002)). So far, numerical solutions to SP_N-based equations for boundary problems have been achieved with finite-difference (Klose and Larsen, 2006; Bérubé-Lauzière et al., 2009; Klose and Pöschinger, 2011), finite volume (Montejo et al., 2011) and finite element (Chu et al., 2009; Bouza Domínguez and Bérubé-Lauzière, 2010; Bouza Domínguez and Bérubé-Lauzière, 2011; Lu et al., 2010; Tian et al., 2010; Zhong et al., 2011) methods.

Next, we make a brief exposition of those methods for the solution of SP_N-based equations.

6.4.1 Finite-difference method

Finite difference methods (FDM) rely on structured grids, which confers them several coding advantages as they are less memory demanding (no cell connectivity information is needed) and function values can be identified with grid indices only (Agarwal, 2000). At the cell level, a low function approximation is used, which favors the FDM for regions requiring a large number of cells. On the other hand, complex boundary conditions are difficult to implement by FDM. In addition, representation of irregular (especially curved) geometries by structured grids can be inexact, unless the grid is especially refined at these locations. An alternative in such cases is to use blocking-off region methods or block-structured grids (Talukdar, 2006; Klose and Pöschinger, 2011; Montejo et al., 2010). In the blocking-off method the exterior boundary ∂S is approximated by the junction of grid points lying in S that best approximate ∂S.

To implement the FDM for SP_N—based equations, we consider a regular domain S located in the xy-plane (2-D, for simplicity), enclosed by the curve ∂S. A regular grid composed by $2N_S$ points along the x- and y-axis $\mathbf{r}_i = (x_i, y_i)$ can be defined as

$$x_i = (i-1)\,\Delta x\,, \quad y_i = (j-1)\,\Delta y\,, \quad i,j = 1,\ldots,N_S\,, \tag{6.83}$$

where Δx and Δy are the grid separations along the x- and y-axis, respectively. We order the grid by the values of the i-index first and then by the j-index values. Let the discrete values of each composite moment be denoted by $\overline{\varphi}_{k,i,j} \approx \varphi_k(x_i, y_j)$.

Then, it is convenient to use centered finite difference approximations for the differential operators, see Eq. (6.25), acting over each φ_k as

$$
\begin{aligned}
-\nabla\cdot(D_k\nabla\varphi_k) \approx & -\left(\frac{D_{k,i+1,j}}{\Delta x^2}\right)\overline{\varphi}_{k,i+1,j} \\
& + \left(\frac{D_{k,i+1,j}+D_{k,i-1,j}}{\Delta x^2}+\frac{D_{k,i,j+1}+D_{k,i,j-1}}{\Delta y^2}\right)\overline{\varphi}_{k,i,j} \\
& - \left(\frac{D_{k,i-1,j}}{\Delta x^2}\right)\overline{\varphi}_{k,i-1,j} - \left(\frac{D_{k,i,j+1}}{\Delta y^2}\right)\overline{\varphi}_{k,i,j+1} \\
& - \left(\frac{D_{k,i,j-1}}{\Delta y^2}\right)\overline{\varphi}_{k,i,j-1}\,.
\end{aligned}
\tag{6.84}
$$

With this discrete approximation, the discretized equation for the CW-SP_N model is

$$
\begin{aligned}
&\left(\overline{\mathbf{K}}+\overline{\mathbf{M}}\right)\overline{\mathbf{\Phi}}_{i,j}=\overline{\mathbf{Q}}_{i,j}\,,\quad \overline{\mathbf{\Phi}}_{i,j}\approx\mathbf{\Phi}\left(x_i,y_j\right)\,,\quad \overline{\mathbf{Q}}_{i,j}\approx\mathbf{Q}\left(x_i,y_j\right)\,,\\
&k=1,\ldots,K\,,\quad i,j=1,\ldots,N_S\,,
\end{aligned}
\tag{6.85}
$$

where the vector $\overline{\mathbf{\Phi}}_{i,j}$ is ordered by the values of the indices i, j and k consecutively. Here, we have introduced the terms $\overline{\mathbf{K}}$ (a diagonal block matrix) and $\overline{\mathbf{M}}$ (block matrix). These matrices are composed themselves of block matrices named $\overline{\mathbf{K}}_k$ and $\overline{\mathbf{M}}_{k_1,k_2}$ which are banded diagonal matrices, respectively. Diagonal entries of $\overline{\mathbf{K}}_k$ and $\overline{\mathbf{M}}_{k_1,k_2}$ have the following form

$$
\begin{aligned}
\operatorname{diag}_{(0)}\left(\overline{\mathbf{K}}_k\right)(i,j) &= \frac{D_{k,i+1,j}+D_{k,i-1,j}}{\Delta x^2}+\frac{D_{k,i,j+1}+D_{k,i,j-1}}{\Delta y^2}\,,\\
\operatorname{diag}_{(1)}\left(\overline{\mathbf{K}}_k\right)(i,j) &= -\frac{D_{k,i+1,j}}{\Delta x^2}\,,\\
\operatorname{diag}_{(-1)}\left(\overline{\mathbf{K}}_k\right)(i,j) &= -\frac{D_{k,i-1,j}}{\Delta x^2}\,,\\
\operatorname{diag}_{(N_s-1)}\left(\overline{\mathbf{K}}_k\right)(i,j) &= -\frac{D_{k,i,j+1}}{\Delta y^2}\,,\\
\operatorname{diag}_{-(N_s-1)}\left(\overline{\mathbf{K}}_k\right)(i,j) &= -\frac{D_{k,i,j-1}}{\Delta y^2}\,,
\end{aligned}
\tag{6.86}
$$

$$
\operatorname{diag}_{(0)}\left(\overline{\mathbf{M}}_{k_1,k_2}\right)=\mathbf{C}\left(k_1,k_2\right)\big|_{i,j}\,,\quad k_1,k_2=1,\ldots K\,,
\tag{6.87}
$$

where $\mathbf{C}\left(k_1,k_2\right)\big|_{i,j}$ means that we evaluate at the grid indices (i,j) the entry (k_1,k_2) of the matrix $\mathbf{C}$, see Eq. (6.26). For the FD-SP_N model, we obtain a similar system to Eq. (6.85), but with the complex nth order transport coefficients (Eq. (6.81)) in the matrix entries of Eqs. (6.86) and (6.87).

To derive a discrete formulation for the time-dependent parabolic SP_N equations, the time derivative can be replaced by a finite difference scheme. For this, the total time of study T is divided in regular intervals of size Δt and samples t_m for the time variable are generated as

$$
t_m=m\Delta t\,,\quad \Delta t=\frac{T}{M}\,,\quad m=0,\ldots,M-1\,.
\tag{6.88}
$$

Implicit finite difference schemes, such as the forward Euler or the Crank–Nicholson (CN) schemes (Agarwal, 2000), are preferred to insure unconditional stability. Introducing a control parameter $\theta \in [0,1]$, the different difference schemes for the time-dependence can be written in a unified way as

$$\left(\frac{\eta}{c\Delta t}\overline{\mathbf{T}}+\theta\overline{\mathbf{K}}+\theta\overline{\mathbf{M}}\right)\overline{\boldsymbol{\Phi}}^{m+1}_{i,j} = \left[(1-\theta)\overline{\mathbf{K}}+(1-\theta)\overline{\mathbf{M}}\right]\overline{\boldsymbol{\Phi}}^{m}_{i,j}+\theta\overline{\mathbf{Q}}^{m+1}_{i,j}+(1-\theta)\overline{\mathbf{Q}}^{m}_{i,j}\,, \tag{6.89}$$

where the matrix $\overline{\mathbf{T}}$ has the same type of structure as $\overline{\mathbf{M}}$. The value $\theta = 0$ corresponds to the explicit or backward scheme (conditionally stable), $\theta = 1/2$ is the Crank–Nicholson scheme, and $\theta = 1$ is the full implicit scheme, with the latter two being unconditionally stable.

Solving Eqs. (6.85) and (6.89) leads to high-dimensional sparse linear systems, which means that sparse matrix techniques can be used to save storage requirements, and solutions can be calculated in highly reduced CPU times compared to dense matrix techniques (Saad, 2003). Moreover, direct and iterative methods for solving sparse linear systems are widely available in the literature (Saad, 2003; Davis, 2006; Press et al., 2007).

6.4.2 Finite volume method

The finite volume method (FVM) is a *conservative* discretization method (Versteeg and Malalasekera, 2007). The partial differential equations serving as forward model are transformed into an integral formulation of the underlying conservation laws and discretized directly in physical space. The physical volume V is partitioned into small volumes ΔV called 'cells', with such a partition to be denoted by τ here. The partition τ can be carried in the form of regular or irregular meshes, e.g. a distribution of cubes or a mesh of tetrahedrae (Versteeg and Malalasekera, 2007). The magnitudes of interest in the problem to which the FVM is applied are replaced in the equations by their average values in 'cells'. This step is carried out after integrating the equations over the partition τ. A cell-centered scheme stores the variable values at all cell centers whereas a node-centered scheme stores the variable values at the nodes. The FVM allows the discrete representation of complex volumes without the implicit FDM implementation mesh refinement on irregular boundaries. Furthermore, cell averaging diminishes the problem dimensionality (number of unknowns) which is convenient for large volumes. As a disadvantage, the FVM does not provide accurate results in the case of discontinuous (or widely varying) coefficients that can appear in the forward model. This problem can be avoided if the coefficient discontinuities coincide with cell boundaries, which can be achieved by refining the mesh (at the cost of increasing the problem dimensionality).

To apply the FVM to SP_N—based equations, we make a partition τ of the volume of interest V into non-overlapping control volumes ΔV_i centered at the mesh points p_i. Next, we follow the node-centered scheme as presented in (Montejo et al., 2011) for the CW-SP_N model. The integration of Eqs. (6.24) over a finite volume ΔV centered at the mesh point p yields

$$-\iint_{\partial\Delta V}\overline{D}\nabla\boldsymbol{\Phi}\cdot\hat{\mathbf{n}}\,\mathrm{d}S+[\mathbf{C}]_p[\boldsymbol{\Phi}]_p\Delta V=[\mathbf{Q}]_p\Delta V\,, \tag{6.90}$$

where the Gauss–Ostrogradsky Theorem has been applied to obtain the first term, and where $\mathrm{diag}_{(0)}\left(\overline{D}\right) = [D_1 \quad D_2 \quad \cdots \quad D_K]$, $k = 1, \ldots, K$ (this term can be approximated by its value at p), $\partial\Delta V$ denotes the boundary of ΔV composed of a number of faces, $\widehat{\mathbf{n}}_{\partial\Delta V}$ is the outer normal to $\partial\Delta V$, and the notation $[\,]_p$ represents the discrete approximation of the enclosed magnitude at the center p of a finite volume element. The vector $[\mathbf{\Phi}]_p$ is ordered by each mesh node p and by the k index, consecutively, i.e. the values for φ_1 for all the nodes come first, then followed by the values of φ_2, etc.

The first term of Eq. (6.90), i.e.

$$J_{n,\Delta V}^{(\mathbf{\Phi})} = -\iint\limits_{\partial\Delta V} \overline{D}\nabla\mathbf{\Phi}\cdot\hat{\mathbf{n}}_{\partial\Delta V}\,\mathrm{d}S\,, \tag{6.91}$$

represents a flux through $\partial\Delta V$. Eq. (6.91) can be approximated by replacing the gradient operation with finite differences at each face composing $\partial\Delta V$. If such approximations are used for the flux term for all control volumes (including boundary conditions in the same way), Eq. (6.90) generates the following matrix system

$$\left\{\widehat{\mathbf{K}} + [\mathbf{C}]_p\right\}[\mathbf{\Phi}]_p = [\mathbf{Q}]_p\,, \tag{6.92}$$

where $\widehat{\mathbf{K}}$ is a $K \times K$ block *diagonal matrix* composed of A_k sparse banded matrices whose explicit form depends on the chosen finite difference scheme at the faces. Eq. (6.92) is a linear system whose solution can be obtained by the GMRES or matrix decomposition (if it is advantageous, given the problem dimensionality) methods (Press et al., 2007).

For the FD-SP_N model, the resulting equation will have the same form as Eq. (6.92), except for considering the complex nth-order transport coefficients (see Eq. (6.81)) in the matrix entries. For the time-dependent parabolic SP_N equations we obtain a system of differential equations

$$\left\{\widehat{\mathbf{K}} + [\mathbf{C}_p + \frac{\eta}{c}\frac{\mathrm{d}}{\mathrm{d}t}[\mathbf{T}]_p\right\}[\mathbf{\Phi}(t)]_p = [\mathbf{q}]_p\,, \tag{6.93}$$

which again can be solved using finite differences with a control parameter $\theta \in [0, 1]$, (compare with Eq. (6.89)

$$\begin{aligned}\left(\frac{\eta}{c\Delta t}[\mathbf{T}]_p + \theta\widehat{\mathbf{K}} + \theta[\mathbf{C}]_p\right)[\mathbf{\Phi}]_p^{m+1}\\ = \left[(1-\theta)\widehat{\mathbf{K}} + (1-\theta)[\mathbf{C}]_p\right][\mathbf{\Phi}]_p^m + \theta[\mathbf{Q}]_p^{m+1} + (1-\theta)[\mathbf{Q}]_p^m\,.\end{aligned} \tag{6.94}$$

The structure of Eq. (6.94) suggests the use of matrix decomposition methods to accelerate the iterative process of finding the solution. Otherwise, the generalized minimal residual method (GMRES) can be employed. Alternatively for the time variable, Runge–Kutta techniques can be used.

6.4.3 Finite element method

The finite element method (FEM) is a highly versatile approach for dealing with media with intricate geometries and heterogeneous distributions of material (here optical) properties (Jin, 2002). Boundary conditions are added to the formulation naturally, no matter the boundarie's complexity. As with the FVM, the FEM starts with a partition of the volume of interest into non-overlapping elements. The information on the partition or mesh takes the form of nodes and elements that are related by a connectivity matrix. In the FEM, functions representing optical properties or the light profile in the medium are approximated by piecewise linear functions or polynomials within each element. Hence, a highly refined mesh is not needed in regions with spatially slowly varying functions. Compared with the FVM, the FEM is usually more computationally intensive in terms of the problem dimensionality. The FEM exclusively deals with functions evaluated at nodes, while in the FVM it is possible to only deal with point-averaged information.

To implement the FEM for the CW-SP_N model, the volume of interest is partitioned into l non-overlapping elements τ_j, $j = 1, \ldots, l$, such that $V = \bigcup_{j=1}^{l} \tau_j$. The elements are defined via d vertex nodes $\widetilde{\mathbf{N}}_i$, $i = 1, \ldots, d$. The nodes can be separated into d_1 internal nodes and d_2 boundary nodes where the boundary conditions are satisfied. Thus, $d = d_1 + d_2$ and the solution $\mathbf{\Phi}(\mathbf{r})$ to Eq. (6.24) can be approximated by the piecewise polynomial and continuous function $\mathbf{\Phi}^h(\mathbf{r})$ as

$$\mathbf{\Phi}(\mathbf{r}) \approx \mathbf{\Phi}^h(\mathbf{r}) = \sum_{i=1}^{d} \mathbf{\Phi}_i u_i(\mathbf{r}), \quad u_i(\mathbf{r}) \in \Omega^h, \tag{6.95}$$

where Ω^h is a finite-dimensional subspace spanned by the basis functions $u_i(\mathbf{r})$, $i = 1, \ldots, d$. Hence, we can find $\widetilde{\mathbf{\Phi}} = \{\mathbf{\Phi}_i\}$, $i = 1, \ldots, d$ from which the solution can be obtained everywhere through the interpolation rule given in Eq. (6.95). Using the Galerkin method (Jin, 2002; Gockenbach, 2006), we can calculate the equivalent numerical solution of Eq. (6.24) $\widetilde{\mathbf{\Phi}}$ as

$$\left[\widetilde{\mathbf{K}} + \widetilde{\mathbf{M}} + \widetilde{\mathbf{\Pi}}\right] \widetilde{\mathbf{\Phi}} = \widetilde{\mathbf{F}} + \widetilde{\mathbf{\Gamma}}. \tag{6.96}$$

Here, $\widetilde{\mathbf{K}}$ represents a 'compound' stiffness matrix and can be described as a diagonal block matrix composed of 'elemental stiffness matrices' $\widetilde{\mathbf{K}}_k$, $k = 1, \ldots, K$, with entries (i, j) given by the expressions

$$\widetilde{\mathbf{K}}_k(i,j) = \int_V \frac{1}{(4k-1)\,\mu_{2k-1}} \nabla u_i(\mathbf{r}) \cdot \nabla u_j(\mathbf{r}) \mathrm{dV}, \quad k = 1, \ldots, K, \quad i, j = 1, \ldots, d. \tag{6.97}$$

The structure of the 'compound' mass matrix $\widetilde{\mathbf{M}}$ is similar to the matrix $\overline{\mathbf{M}}$ discussed for the FDM method and it is composed of 'elemental mass matrices' $\widetilde{\mathbf{M}}_{k_1,k_2}$, $k_1, k_2 = 1, \ldots, K$ with the following entries

$$\widetilde{\mathbf{M}}_{k_1,k_2}(i,j) = \int_V C(k_1, k_2)\, u_i(\mathbf{r}) u_j(\mathbf{r}) \mathrm{dV}, \quad k_1, k_2 = 1, \ldots, K, \quad i, j = 1, \ldots, d, \tag{6.98}$$

where $\mathrm{C}(k_1, k_2)$ are the elements of the matrix $\mathbf{C}$ (Eq. (6.26)). The structure of matrix $\widetilde{\mathbf{\Pi}}$ is also similar to $\overline{\mathbf{M}}$ and it is composed of matrices $\widetilde{\mathbf{\Pi}}_{k_1,k_2}$, $k_1, k_2 = 1, \ldots, K$, with the following entries

$$\widetilde{\mathbf{\Pi}}_{k_1,k_2}(i,j) = \int_{\partial V} \frac{\Theta(k_1,k_2)}{(4k_1-1)\,\mu_{2k_1-1}} u_i(\mathbf{r}) u_j(\mathbf{r})\,\mathrm{d}\sigma\,, \\ k_1, k_2 = 1, \ldots, K\,, \quad i,j = 1, \ldots, d \tag{6.99}$$

where $\mathrm{d}\sigma$ is an element of area on the boundary ∂V and $\Theta(k_1, k_2)$ are the elements of the matrix $\mathbf{\Theta} = \mathbf{B}^{-1}\mathbf{A}$, see the boundary matrices $\mathbf{A}$ and $\mathbf{B}$ in Eq. (6.28).

The 'compound' force load vector $\widetilde{\mathbf{F}}$ is composed of terms $\widetilde{\mathbf{F}}_k$, $k = 1, \ldots, K$, ('elemental force load vectors') which are column vectors with the following entries

$$\widetilde{\mathbf{F}}_k = \int_V Q(k) u_i(\mathbf{r})\mathrm{dV}\,, \quad k = 1, \ldots, K\,, \quad i = 1, \ldots, d\,, \tag{6.100}$$

where $Q(k)$ are the components of column vector $\mathbf{Q}$.

The column vector $\widetilde{\mathbf{\Gamma}}$ is similar to $\widetilde{\mathbf{F}}$ and originates from the external source distribution $\mathbf{S}$ at the boundary. This vector is composed of terms $\widetilde{\mathbf{\Gamma}}_k$, $k = 1, \ldots, K$, which are column vectors of length d_1 given by

$$\widetilde{\mathbf{\Gamma}}_k(i) = \int_{\partial V} \frac{G(k)}{(4k-1)\,\mu_{2k-1}} u_i(\mathbf{r}) u_j(\mathbf{r})\,\mathrm{d}\sigma\,, \quad k = 1, \ldots, K\,, \quad i = 1, \ldots, d\,, \tag{6.101}$$

where $G(k)$ are the elements of the vector $\mathbf{G} = \mathbf{B}^{-1}\mathbf{S}$.

As for the previous numerical methods discussed, the discretized equations for the FD-SP_N model have the same form as for the steady-state situation, but substituting the transport coefficient by the complex transport coefficients (Eq. (6.81)). For the TD-pSP_N equations, we can write directly the FEM-discretized equations as

$$\left[\widetilde{\mathbf{K}} + \widetilde{\mathbf{M}} + \widetilde{\mathbf{\Pi}} + \frac{\eta \widetilde{\mathbf{T}}}{c}\frac{\mathrm{d}}{\mathrm{d}t}\right]\widetilde{\mathbf{\Phi}}(t) = \widetilde{\mathbf{F}} + \widetilde{\mathbf{\Gamma}}\,, \tag{6.102}$$

where the matrix $\widetilde{\mathbf{T}}$ has an expression similar to that of $\widetilde{\mathbf{M}}$, with analogous entries

$$\widetilde{\mathbf{T}}_{k_1,k_2}(i,j) = \int_V T(k_1,k_2)\, u_i(\mathbf{r}) u_j(\mathbf{r})\mathrm{dV}\,, \quad k_1, k_2 = 1, \ldots, K\,, \quad i,j = 1, \ldots, d\,, \tag{6.103}$$

The solution of Eq. (6.102) can be achieved by a finite difference scheme or Runge–Kutta methods as previously for the FDM and FVM approaches (Eqs. (6.89) and (6.94)).

6.5 Diffuse optical tomography based on SP_N models

Diffuse optical tomography is an imaging technique that aims to recover interior maps based on the transillumination of a biological body and the (generally exterior) measurement of the light that has propagated through the body. DOT involves the recording of the exiting light and digital data processing, to obtain representative images of the relevant internal properties of the biological body (Wang and Wu, 2007). Model-based iterative image reconstruction algorithms in DOT rely on the accuracy of a discretized forward model to reproduce collected measurements (Klose and Hielscher, 2008; Dehghani et al., 2009; Arridge and Schotland, 2009). Lately, model-based DOT has been attempted with the FD-SP_N (Chu and Dehghani, 2009) and the TD-pSP_N equations (Bouza Domínguez and Bérubé-Lauzière, 2011b; Bouza Domínguez and Bérubé-Lauzière, 2011c). We now proceed to describe the main features and results of the implemented DOT algorithms with these models.

6.5.1 DOT based on the FD-SP_N model

For the FEM-discretized FD-SP_N model, the inverse problem has been posed as an unconstrained optimization problem with a regularization term (Chu and Dehghani, 2009; Wang et al., 2011)

$$\boldsymbol{\mu} = \underset{\{\boldsymbol{\mu}^h\}}{\operatorname{argmin}} \sum_{s=1}^{S} \sum_{d=1}^{D} (M_{s,d} - P_{s,d})^2 + \lambda \left(\boldsymbol{\mu}^h - \boldsymbol{\mu}_0\right)^2, \tag{6.104}$$

where 'argmin' stands for argument of the minimum. In the last equation, the vector $\boldsymbol{\mu}$ represents the nodal values of an optical parameter set (e.g. absorption and scattering coefficients); the summation is over the total number of configuration sources S (modeled as isotropic point sources) and detector positions D, and the terms $M_{s,d}$ and $P_{s,d}$ represent the measurements and the forward model predictions, respectively. Phase and amplitude data are considered in Eq. (6.104). The Tikhonov regularization parameter λ appears multiplying the L_2-regularization term, where $\boldsymbol{\mu}_0$ represents the *a priori* estimate of $\boldsymbol{\mu}$. The solution to the optimization problem cast in Eq. (6.104) is found by the Levenberg–Marquardt method (Press et al., 2007) which employs Jacobian calculations of phase and amplitude with respect to $\boldsymbol{\mu}$. The Jacobian calculations are performed using the perturbation method and the reciprocity approach, see (Arridge and Schotland, 2009) for details.

In (Chu and Dehghani, 2009), several numerical experiments are conceived to test the FD-SP_N model performance on retrieving the absorption and scatter properties. The experiments involve small geometries and tissue typical optical coefficient values in the NIR spectrum, see the article for details. A distinction in the retrieved image accuracy, artefact presence (significant near the boundary) and cross-talk effects is found for different orders N. Particularly, for $N = 3$ and 5 the reconstructions performed acceptably well. Errors in the reconstructed values are within 24% of the expected values and the worse results are obtained by the DE (SP_1) in the absorption coefficient reconstruction. The reported results support the use of the FD-SP_N model (orders $N = 3$ and 5) in DOT. The authors explained that further improvements in the image reconstructions can be expected with the optimization of the regularization parameter and selection of stopping criteria.

6.5.2 DOT based on the TD-pSP_N model

Recently, DOT based on the FDM-FEM discretized TD-pSP_N equations has been carried out (Bouza Domínguez and Bérubé-Lauzière, 2011b; Bouza Domínguez and Bérubé-Lauzière, 2011c). This time, the inverse problem is posed as the following constrained optimization problem

$$\boldsymbol{\mu} = \underset{\{\boldsymbol{\mu}^h\}}{\operatorname{argmin}} \frac{1}{2} \sum_{s=1}^{S} \sum_{d=1}^{D} \sum_{m=1}^{M} \left(\frac{M_{s,d}^{(m)} - P_{s,d}^{(m)}}{\sigma_{s,d}} \right)^2 , \quad P_{s,d}^{(m)} = \widehat{M} \widetilde{\boldsymbol{\Phi}}_s^{(m)}$$
$$\text{subject to} \begin{cases} \widetilde{\mathbf{W}} \widetilde{\boldsymbol{\Phi}}_s^{(m)} = \left(\frac{\eta}{c}\right) \widetilde{\mathbf{T}} \widetilde{\boldsymbol{\Phi}}_s^{(m-1)} + \widetilde{\boldsymbol{\Upsilon}}_s^{(m)} , & m = 1, \ldots, M , \quad s = 1, \ldots, S \\ \boldsymbol{\mu}_l \le \boldsymbol{\mu} \le \boldsymbol{\mu}_u , \end{cases} , \tag{6.105}$$

The same notation as in Eq. (6.104) is used. An additional summation over the time steps appears here in the objective function compared to that in Eq. (6.104) to account for the time dependence of the light field. The quantities $\sigma_{s,d}^{(m)}$ are the standard deviations of the measurements; which are mainly determined by shot noise. $\widehat{M}$ is the measurement operator which acts over the time-dependent vector of composite moments $\widetilde{\boldsymbol{\Phi}}_s^{(m)}$. The matrix $\widetilde{\mathbf{W}} = \Delta t\big(\widetilde{\mathbf{K}} + \widetilde{\mathbf{M}} + \widetilde{\mathbf{\Pi}}\big) + (\eta/c)\widetilde{\mathbf{T}}$ and vector $\widetilde{\boldsymbol{\Upsilon}}^{(m)} = \Delta t\big[\widetilde{\mathbf{F}}^{(m)} + \widetilde{\boldsymbol{\Gamma}}^{(m)}\big]$ result from the FDM-FEM numerical discretization scheme; see Section 6.4 (the Euler finite difference scheme is employed). The vectors $\boldsymbol{\mu}_l$ and $\boldsymbol{\mu}_u$ are lower and upper bounds over the set of optical coefficients $\boldsymbol{\mu}$ to be recovered.

The inverse problem cast in Eq. (6.105) contains the forward model and bounds over the optical coefficient values as constraints. In addition, the time-dependence of the forward model increases the complexity of the optimization problem mainly because of the necessary time-stepping and the increased dimensionality of the problem compared to the CW and FD cases. In (Bouza Domínguez and Bérubé-Lauzière, 2011b; Bouza Domínguez and Bérubé-Lauzière, 2011c), the authors decided to employ a 'nested analysis and design' (NAND) method (Hazra, 2010). Basically, in the NAND method the implicit dependence of the state constraints (or $\widetilde{\boldsymbol{\Phi}}_s^{(m)}$; we employ terminology of constrained optimization theory) with the design variables (or $\boldsymbol{\mu}$) is considered. Then, constraints posed by the forward model are eliminated. The solution to the optimization problem given in Eq. (6.105) is obtained through a Sequential Quadratic Programming (SQP) algorithm (Nocedal, and Wright, 2006). SQP uses the gradient of the objective function in the iteration process and a Hessian approximation by the damped BFGS method to avoid computing second derivatives. Finally, a time-dependent adjoint differentiation scheme (see Arridge and Schotland (2009) for the topic of adjoint variables) is utilized to calculate the gradient and reduce the computation time.

To investigate the performance of the TD-pSP_N model in recovering optical properties of biological media, several numerical experiments are conducted in (Bouza Domínguez and Bérubé-Lauzière, 2011b; Bouza Domínguez and Bérubé-Lauzière, 2011c). The experiments involve a circular two-dimensional medium (background medium) with a 1.5 cm of radius. The medium is homogeneous with optical properties $\mu_a = 0.01\,\text{cm}^{-1}$, $\mu_s = 80\,\text{cm}^{-1}$, $g = 0.9$ and $\eta = 1.4$ and it is considered as surrounded by air. In the multi-parameter reconstructions, absorptive and scattering inclusions are embedded in the background medium; see Fig. 6.5. Increasing values of μ_a for the absorptive inclusion are considered: 0.05, 0.1, and $1\,\text{cm}^{-1}$ (high-absorption case) which correspond to diffusion coefficient values of 0.0414, 0.0412 and 0.037 cm. For the scattering inclusion, the value of $120\,\text{cm}^{-1}$ is assumed for its scattering coefficient. Multi-parameter reconstructions (absorption and diffusion coefficient maps) were performed with the DE and the TD-pSP_N equations as the forward models and the results compared, see (Bouza Domínguez and Bérubé-Lauzière, 2011b; Bouza Domínguez and Bérubé-Lauzière, 2011c) for details.

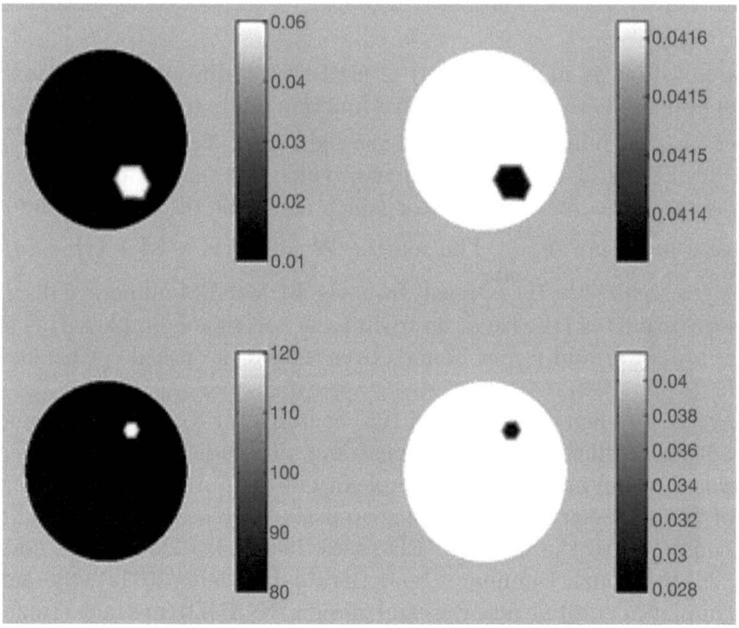

Fig. 6.5. Numerical experiments for the multi-parametric inverse problem. Absorption (only one value of $\mu_a = 0.05\,\text{cm}^{-1}$ is represented) and scattering coefficient ($120\,\text{cm}^{-1}$) distribution (top and bottom, left column) and diffusion coefficient distribution for each type of inclusion (top and bottom, right column).

In all the experiments, the TD-pSP_N model ($N > 1$) recovered accurately the absorptive and scattering inclusion values; see Bouza Domínguez and Bérubé-Lauzière (2011b) and Bouza Domínguez and Bérubé-Lauzière (2011c) for the comparison details. Particularly, the results obtained with $N = 3$ outperformed the DE. For the reconstructed absorption maps, the errors with respect to the original values (in percent, taking the maximum of the reconstructed values) for the DE are of 19%, 16% and 8% ($\mu_a = 0.05$, 0.1, and $1\,\text{cm}^{-1}$). The corresponding errors for the TD-pSP_3 equations are of 0.1%, 8% and below 1%; see Figs. 6.6 and 6.7 and compare with Fig. 6.5, to partly appreciate these results. For the reconstructed dif-

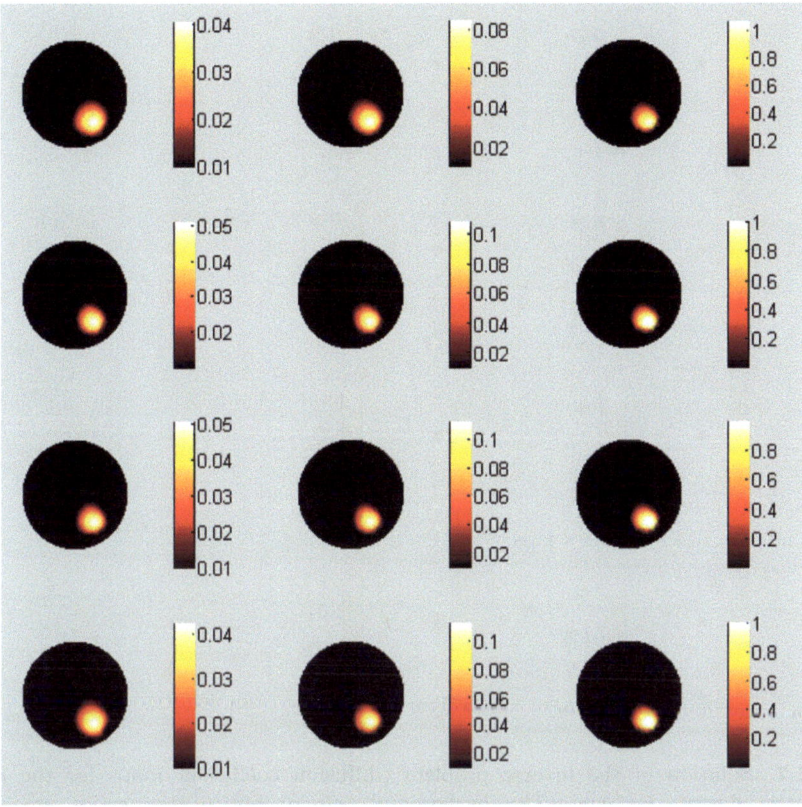

Fig. 6.6. Solution of the inverse problem (absorption coefficient map) for the multi-parametric inverse problem. The background (circle) optical properties are $\mu_a = 0.01\,\text{cm}^{-1}$, $\mu_s = 80\,\text{cm}^{-1}$, $g = 0.9$ and $\eta = 1.4$ and the medium is considered as surrounded by air. Values of the absorption coefficient for the absorptive inclusion are: 0.05, 0.1, and $1\,\text{cm}^{-1}$ (left to right). Images are plotted for the orders $N = 1$, 3, 5 and 7 (first, second, third and fourth rows).

fusion maps, the DE and the TD-pSP_3 equations presented approximately the same errors in the reconstruction of the scattering heterogeneity in the cases $\mu_a = 0.05$ and $0.1\,\text{cm}^{-1}$. For the case $\mu_a = 1\,\text{cm}^{-1}$, the DE error is greater than 40% while in the case of the TD-pSP_3 equations it is only 6%. A similar behaviour is observed for the same reconstructed diffusion maps but at the position of the absorptive inclusion. In addition, reconstructed images presented artifacts (almost negligible spots, at the boundary) and cross-talk effects which vary with the order N, with the DE-based reconstructions delivering the worst results. In this work, the authors concluded that the DOT algorithm based on the TD-pSP_N model ($N > 1$) can accurately replace DE-based algorithms, especially in the physical situations where the DE fails.

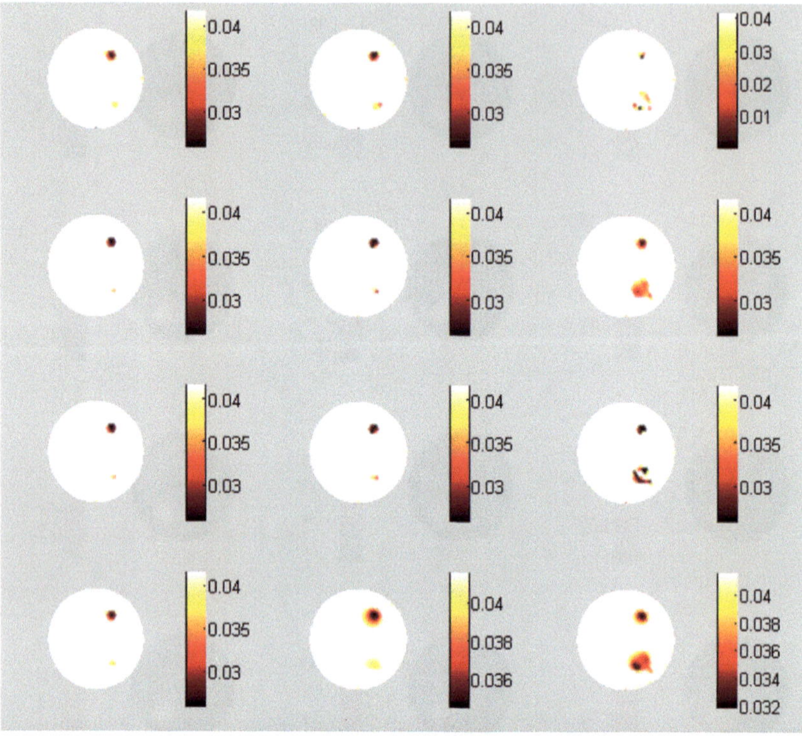

Fig. 6.7. Solution of the inverse problem (diffusion coefficient map) for the multi-parametric inverse problem. The background (circle) optical properties are $\mu_a = 0.01\,\mathrm{cm}^{-1}$, $\mu_s = 80\,\mathrm{cm}^{-1}$, $g = 0.9$ and $\eta = 1.4$ and the medium is considered as surrounded by air. Images are plotted for the orders $N = 1$, 3, 5 and 7 (first, second, third and fourth rows) where the absorptive heterogeneity takes the following values: 0.05, 0.1, and $1\,\mathrm{cm}^{-1}$ (left to right). At the absorptive heterogeneity (see Fig. 6.5), the diffusion coefficient takes the following values $D = 0.0414$, 0.0412 and 0.037 cm. At the scattering heterogeneity (see Fig. 6.5), the diffusion coefficient has the value $D = 0.0277\,\mathrm{cm}$.

6.6 Molecular imaging of luminescence sources based on SP_N models

Optical molecular imaging of luminescence sources (OMI) is a promising discipline of biomedical optics. OMI allows the study of biological processes and medical treatment, as well as the diagnosis and follow-up of diseases (Weissleder and Ntziachristos, 2003; Hielscher, 2005; Ntziachristos, 2006; Rao et al., 2007; Willmann et al., 2008; Klose, 2009; Mitchell et al., 2011; Elwell and Cooper, 2011). Compared to intrinsic imaging (or DOT), luminescent light increases measurement sensitivity of experimental systems to specific targets or physiological processes occurring in biological tissues (Weissleder and Ntziachristos, 2003; Hielscher, 2005; Ntziachristos, 2006). Applications of OMI are mainly focused in small animal imaging, although clinical imaging has been lately targeted (Burgess et al., 2010; Pleijhuis et al., 2011). Tomographic methods in OMI would supply researchers and physicians with three-dimensional visualization of tissue structure and functions. Current modalities are bioluminescence and fluorescence imaging and, more recently, Cerenkov luminescence imaging (Mitchell et al., 2011). Tomographic imaging by these methods involves the solution of an inverse source problem. For solving an inverse source problem, the optical properties of the medium are needed. Optical properties of luminescence emitting media (or substrate media) can be supplemented by additional anatomical information from CT/MRI scans and tabulated optical coefficient values (Alexandrakis et al., 2005; Alexandrakis et al., 2006; Klose et al., 2010). If no *a priori* information exists about the substrate medium, or the information is insufficient, a complementary DOT reconstruction can be carried.

Inverse problems based on deterministic models frequently use the DE (Naser and Patterson, 2011; Larusson et al., 2011; Zhu et al., 2011). However, in many practical situations luminescence sources are located deep into small geometries of tissue, in the presence of high absorption, such as of internal organs. Under those conditions, the DE fails as a model of light propagation in tissues and transport calculations are mandatory to gain in accuracy (Hielscher et al., 1998). In this context, deterministic models based on the SP_N approximation are preferred and used to perform OMI. In this section, we review recent results on the use of SP_N-based models in bioluminescence and fluorescence DOT. We also include the latest applications of SP_N-based models in Cerenkov optical imaging.

6.6.1 Bioluminescence imaging

Bioluminescence originates in chemical reactions and does not require external excitation sources (Klose, 2009; Contag and Bachmann, 2002; Welsh and Kay, 2005; Vo-Dinh, 2003). The chemical reactions involve the interaction of an administered light-producing substrate (usually luciferine) and a transfected enzyme (luciferase from firefly, Renilla, or Aequorin). Luciferase catalyzes the oxidation of luciferine. causing light emission. The bioluminescent source density changes slowly with time and the source can be assumed to be steady. Although the DE is frequently used as the forward model, at emission wavelengths less than <650 nm (Renilla or Gaussia luciferase) light is strongly absorbed by tissues, violating the limits of DE validity. In addition, the ratio μ_s'/μ_a at visible and NIR wavelengths varies over a wide range for some organs such as bone, and for others such as heart and liver, this ratio does not go beyond 10 (Vo-Dinh, 2003). Hence, physical situations where transport calculations are necessary can occur (Hielscher et al., 1998). These difficulties are aggravated in the presence of small geometries and isotropic point-like bioluminescence sources, circumstances where the DE is out of its comfort zone (Martí López et al., 2004; Hielscher et al., 1998; Klose and Larsen, 2006).

To overcome the drawbacks of the DE in such situations, the CW-SP_N model (Eq. (6.24)) has been used as the forward model in bioluminescence imaging (Klose et al., 2010; Lu et al., 2009; Tian et al., 2010; Klose, 2012). Particularly, the CW-SP_3 equations are frequently chosen, since they can provide transport-like solutions with low computational cost (Klose et al., 2010). Reconstruction techniques with the CW-SP_3 equations also employ spectrally resolved information in order to reduce the inherent ill-posedness of inverse source problems (Lu et al., 2009). In the literature, the following reconstruction techniques for small animal imaging have been attempted: (*i*) a gradient-based optimization method with regularization (Lu et al., 2009), (*ii*) the algebraic reconstruction method, with *a priori* estimation of the absorption distribution by an evolution strategy (ES) algorithm (Klose et al., 2010) and a generalized graph cuts optimization method (Tian et al., 2010). Next, we provide some details on the mentioned source reconstruction techniques.

In (Lu et al., 2009), a FEM discretization of the CW-SP_3 equations is used to calculate the model predictions as the exitance values $J_{no}^{(out)}$. Then, the following bound-constrained least-squares problem is posed

$$\min_{\mathbf{0}<\mathbf{Q}<\mathbf{Q}^{\mathrm{sup}}} \left\| \mathbf{A}_{(FEM)}\mathbf{Q} - \mathbf{M} \right\| + \lambda\eta(\mathbf{Q}) \,, \tag{6.106}$$

where $\| \ \|$ represents the L_2 or Euclidean norm, $\mathbf{Q}^{sup}$ is the upper bound of the source density distribution $\mathbf{Q}$, the matrix $\mathbf{A}_{(FEM)}$ appears in the FEM discretization process and includes the measurement operator $\widehat{M}$ (see Eqs. (6.33) and (6.96)), $\mathbf{M}$ is the vector of measurements, λ is the regularization parameter and $\eta(\mathbf{Q})$ is a penalty function. The minimization of Eq. (6.106) is performed by the limited memory variable metric-bound constrained quasi-Newton method (BLMVM) (Benson and Moré, 2001). In the BLMVM, an approximate Hessian is calculated by vector-vector multiplications, which assures easy matrix inversion and reduces memory and computation time. An implementation of the BLMVM is available in the Toolkit for Advance Optimization (TAO) (Website TAO, 2012). A fully parallel version of the reconstruction algorithm including FEM assembly is also provided in (Lu et al., 2009).

In Klose et al. (2010), an ES algorithm minimizes an objective function similar to Eq. (6.106) (no regularization term is included this time) to estimate the average absorption coefficients at each wavelength. The goal is to diminish the inaccuracy on the determination of the optical parameters which could lead to mislocation of the source position. The ES is an iterative method that searches for an optimal selection of parameters by probing the global search parameter space (Beyer and Schwefel, 2002; Dirk, 2002). This method uses selection and mutation as natural-resembling operations and it is comparatively faster than gradient-based approaches (Dirk, 2002). Average absorption distributions are then used in the inverse source problem. To find the solution of the inverse source problem, the CW-SP_3 equations are solved by the FDM. A linear relation between model predictions and source density distribution is derived, $\mathbf{P} = \mathbf{A}_{(FDM)}\mathbf{Q}$ where $\mathbf{A}_{(FDM)}$ is an $m \times n$ matrix which appears similarly to $\mathbf{A}_{(FEM)}$, see Eq. (6.106). To speed-up the calculations, the reciprocity principle is used (Dehghani et al., 2008). The inverse source problem posed by the linear system of equations $\mathbf{A}_{(FDM)}\mathbf{Q} = \mathbf{M}$ is solved by the algebraic reconstruction technique (ART) (Natterer, 2001). The ART (or Kaczmarz method) is a method for solving linear systems of equations that exploit sparseness (Natterer, 2001; Nikazad, 2008). Finally, the ART iteratively computes the solution using the following formula

$$Q_{k+1} = Q_k + \xi_k \left(\frac{M_i - \langle a_i, Q_k \rangle}{\left\| a_i \right\|^2} \right) a_i \,, \tag{6.107}$$

where $\langle \ , \ \rangle$ represents the scalar product, ξ_k is a relaxation parameter, $i = k \bmod m + 1$, M_i is the ith component of $\mathbf{M}$ and a_i is the ith row of the matrix $\mathbf{A}_{(FDM)}$.

In (Tian et al., 2010), the CW-SP_3 equations are discretized using the FEM. A linear relation between the exitance values and the source density distribution is derived. The solution to the inverse source problem is determined by minimizing the energy function $E(\mathbf{Q})$

$$E(\mathbf{Q}) = \left\|\mathbf{A}_{(FEM)}\mathbf{Q} - \mathbf{M}\right\| + \lambda \left\|\mathbf{Q}\right\| , \tag{6.108}$$

which is a particular case of the objective function appearing in Eq. (6.106), considering the penalty function as the L_2 norm of $\mathbf{Q}$. To optimize the energy function given in Eq. (6.108), a gradient-free optimization method called generalized graph cuts (GGC) is employed. GGC is an efficient optimization tool that is applied in computer vision and graphics (Boykov and Kolmogorov, 2004; Kolmogorov and Zabih, 2004; Kolmogorov and Rother, 2007). Lately, GGC has been rediscovered in other disciplines including bioluminescence imaging (Tian et al., 2010; Liu et al., 2010). As described in (Tian et al., 2010; Liu et al., 2011), a graph containing the FEM mesh is built and an equivalent graph expression for Eq. (6.108) is found. Then, the energy function Eq. (6.108) is minimized using a quadratic pseudo-boolean optimization method (Kolmogorov and Rother, 2007).

6.6.2 Fluorescence imaging

Fluorescence imaging by direct methods relies on active, or activatable, probes which are excited by external sources or specific enzymes (Rao et al., 2007; Klose, 2009, 2012). Indirect methods are used in gene activation and regulation with the introduction of transgenes, which induce the production of fluorescence proteins (Rao et al., 2007). Fluorescent probes possess their specific properties in terms of converting excitation light into emitted (fluoresced) light. These are the molar extinction coefficient ε and the quantum yield ς. In TD methods, the fluorescence lifetime τ, which characterizes the fluorescence emission dynamics is also included in the studies. The fluorescence lifetime is sensitive to local metabolite concentrations or environmental conditions within tissues (Nothdurft et al., 2009), and thus provides information about such factors. When distributed into biological tissues, fluorescent probes contribute to the overall absorption (absorption of a fluorophore being equal to ε times the concentration C). Fluorescence imaging requires a forward model that maps fluorophore distribution to fluorescence data, as the straightforwardly used DE (Ntziachristos, 2006; Zacharopoulos et al., 2010; Zhu et al., 2011). However, in the presence of high absorption, see for example (Comsa et al., 2008) and references in (Bouza Domínguez and Bérubé-Lauzière, 2011a), the DE cannot compete with quantitativeness of biomarkers offered by nuclear imaging techniques.

With this perspective, fluorescence tomography has been recently attempted with the CW-SP_N model as the forward model for describing both the excitation (ex) and the fluorescence (fl) light propagation (Klose, 2010b; Klose and

Pöschinger, 2011; Klose et al., 2011; Klose, 2012). Thus, the model consists in a system of two SP_N equations as

$$\left[D_r^{(i)} + \mathbf{C}^{(i)}\right] \boldsymbol{\Phi}^{(i)}(\mathbf{r}) = \mathbf{Q}^{(i)}(\mathbf{r}) \,, \quad i = ex, fl \,, \tag{6.109}$$

where the components of the source vector $\mathbf{Q}^{(fl)}(\mathbf{r})$ are proportional to the fluorescence source term $Q^{fl}(\mathbf{r})$, which quantify the interaction between the excitation light and the fluorescent response

$$Q^{fl}(\mathbf{r}) = \varsigma \varepsilon C \psi_0^{ex}(\mathbf{r}) \,. \tag{6.110}$$

Eq. (6.109) is discretized using the FDM and the resulting algebraic system of equations is solved by the successive over-relaxation (SOR) method (Klose et al., 2011; Saad, 2003). Then, the inverse problem is posed similarly to Eq. (6.106) and iteratively solved for C using an expectation-maximization (EM) method (Dempster et al., 1977; Wernick and Aarsvold, 2004). Based on the abovementioned works, notable improvements have been achieved in the area of hyperspectral excitation-resolved fluorescence tomography. Here, fluorophores with broad molar extinction spectra are used as probes, and allow exploiting the spectral properties of tissue oxy- and deoxy-hemoglobin components in ranges where their molar extinction varies widely (Klose and Pöschinger, 2011). In another work, the authors employ the FEM as the discretization method for Eqs. (6.109) (Han et al., 2010). Also, the usually sparse/spatially-reduced properties of fluorophore distributions are used in a regularization scheme as *a priori* information. The solution of the inverse problem is searched by an iteratively reweighted scheme which approximates the L_1-norm regularization (Han et al., 2010; Wang et al., 2011). A sampling procedure (visual inspection) is chosen to determine the optimal value for the regularization parameter.

In FD, a similar formulation to Eq. (6.109) has been derived (Lu et al., 2010), this time using the FD-SP_N equations and introducing the complex-valued nth-order transport coefficients Eq. (6.81). The resulting equations have been discretized through the FEM and a parallel adaptive FEM is used. Finally, the quest for lifetime imaging has recently triggered the development of new TD forward models based on the TD-pSP_N equations (Bouza Domínguez and Bérubé-Lauzière, 2011a). In this work, a set of TD-pSP_N equations has been obtained for describing the time-dependent propagation of the excitation light and the ensuing fluorescent response. This time, the time-dependent fluorescence source term $Q^{fl}(\mathbf{r}, t)$ quantifies the temporal interaction between the excitation field and the fluorescence emission. The coupling between excitation and fluorescence emission can be described through a convolution operation as

$$Q^{fl}(\mathbf{r}, t) = \frac{\varsigma \varepsilon C(\mathbf{r})}{\tau} \int_{t'=0}^{t'=t} \psi^{ex}(\mathbf{r}, t') \exp\left(\frac{t' - t}{\tau}\right) \mathrm{d}t' \,. \tag{6.111}$$

A FEM/FDM numerical implementation is described in the same work. Numerical simulations with three-dimensional biological media provide new information on the influence of fluorophore distribution on the TD curves, see Fig. 6.8. This approach should lead, in a near future, to the solution of a nonlinear inverse problem for recovering lifetime spatial maps τ.

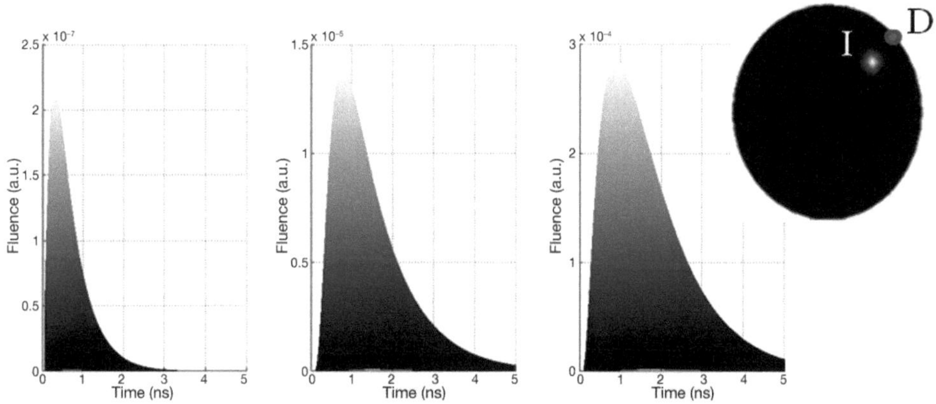

Fig. 6.8. Numerical simulations for a cylindrical homogeneous medium ($\eta = 1.4$, $\mu_a = 0.13\,\text{cm}^{-1}$, $\mu_s = 100\,\text{cm}^{-1}$ and $g = 0.9$.) with an absorptive fluorescent inclusion ($\mu_a = 3\,\text{cm}^{-1}$ and $\tau = 0.56\,\text{ns}$) located at point I; see tomographic cut of the cylinder at right. Plots represent the fluorescence TD curves at point D, the closest point to I. We consider three different fluorophore distributions (left to right): a point inclusion, a small spherical inclusion and a Gaussian distributed inclusion. There is a noteworthy change in the shape of the curve.

6.6.3 Cerenkov luminescence imaging

Cerenkov luminescence imaging (CLI) is an evolving technology that uses optical photons generated by positron emission tomography (PET) radiotracers (Robertson et al., 2009; Liu et al., 2010b; Boschi et al., 2009; Spinelli et al., 2010, Dothager et al., 2010). Cerenkov radiation is created by high-energy charged particles that momentarily exceed the speed of light in the medium in which they propagate (Robertson et al., 2009). PET radionuclides and most of β-emitting radionuclides with biomedical applications produce measurable Cerenkov radiation in water or in tissue (Boschi et al., 2009). The Cerenkov light spectrum is continuous, in contrast to fluorescence or emission spectra that have characteristic spectral peaks. The relative intensity is proportional to frequency thus: higher frequencies (ultraviolet/blue) are most intense. At ultraviolet/blue wavelengths, Cerenkov radiation is highly absorbed by tissue components (water, hemoglobin, cytochromes, etc.). Large absorption coefficients make the DE less accurate and transport calculations are required (Hielscher et al., 1998).

Cerenkov radiation can be detected by current optical imaging methods. Retrieving the distribution of Cerenkov optical sources becomes an inverse luminescence source problem, as in bioluminescence tomography. Moreover, radionuclide activity levels which are necessary to inject and produce detectable optical signals are typical of small animal imaging (Spinelli et al., 2010; Li et al., 2010). Therefore, CLI provides considerable advantages regarding drug discovery and in general, biomedical research. Cerenkov luminescence tomography for small animal imaging has been attempted with success and even a multispectral approach has been developed (Spinelli et al., 2011). However, the reconstruction results are limited by the use of the DE as the forward model (discretized using the FEM) in media where large absorption occurs in small geometries. In addition, the inverse source problem is posed as an unconstrained optimization problem which accentuates the ill-posedness. Typically, a linear least-squares type, objective function with a Tikhonov regularization term is employed (Spinelli et al., 2011; Zhong et al., 2011). Then, the solution can be iteratively retrieved by the preconditioned conjugate gradient (PCG) method (Li et al., 2010) or a non-negative least square optimization algorithm (Spinelli et al., 2011).

Lately, the FEM-discretized CW-SP_3 equations have been used in a model-based reconstruction algorithm to perform whole-body Cerenkov luminescence tomography (Zhong et al., 2011b). The inverse source problem is posed as a linear least-squares objective function with a regularization term or penalty function similarly as in Eq. (6.106). The penalty function is set as a linear combination of L_2 (ridge-regression penalty) and L_1 (lasso-regression penalty) norms of the source density distribution Q. This type of regularization is known as *elastic net regularization* and is used for moderating both smoothing and sparsity effects in the reconstruction (Friedman et al., 2010; Van der Kooij, 2007). The components of the vector $\mathbf{Q}$ are computed by first applying a soft-threshold operation, to account for lasso penalty, and consequently a proportional shrinkage, to account for the ridge penalty. Details of the algorithm and its derivation can be found in (Friedman et al., 2010). In (Zhong et al., 2011b), a number of experiments concerning small animal imaging are performed. A comparison between DE and SP_3-based reconstructions using the mentioned algorithm is carried out. The impact of the high-absorption tissues ($\sim$1 cm^{-1}) is evaluated. There is a substantial reduction in the source localization error (more than an order of magnitude) when the CW-SP_3 equations serve as the forward model. Thus, the work strongly supports the use of the CW-SP_N model in CLI preclinical studies and opens a pathway to clinics, where the use radioactive contrast agents is widely accepted. More recently, Klose has discussed the use of the SP_3 equations for Cerenkov light tomography in a multi-spectral framework (Klose, 2012).

6.7 Summary

Light propagation models based on the SP_N approximation have been derived, implemented, and used to solve problems in biomedical optics during the last decade. In the literature, both the standard RTE and the source-divergence RTE are employed to obtain low-order transport models by introducing the SP_N approximation. In particular, the equations derived from the source-divergence RTE are able to correctly describe light propagation near point sources, a common physical situation in biomedical optics. Applying the SP_N approximation to the diffuse component of the radiance results in a better description of radiative transport in tissues since angular dependencies are attenuated with the reduced diffuse components formulation. SP_N equations have been obtained for steady state, frequency and time domains. Steady-state and frequency-domain SP_N equations are equivalent under a simple transformation of the transport coefficients into complex coefficients. Time-domain SP_N models have different forms in dependence of the derivation. In particular, the time-domain parabolic SP_N equations constitute the light propagation model that has been most studied thus far for applications in biomedical optics. SP_N models have been extended to describe luminescent light transfer in biological tissues.

Recently, SP_N models have been employed in DOT and luminescence imaging with significant success. Image reconstructions of absorption and scattering inhomogeneities show an appreciable improvement in localization and quantitativeness in comparison to DE results. Furthermore, the presence of artifacts and cross-talk effects is reduced by the use of SP_N-based DOT algorithms. In inverse source problems (see references cited in Section 6.6), DE-based results have been improved through the use of SP_N-based algorithms. By accurately modeling Cerenkov light propagation in biological tissues, the SP_N equations have also opened a way to satisfactory radionuclide and optical images co-registration. Additionally, SP_N—based CLI offers an alternative (with both functional and anatomical information) to costly PET instrumentation, with no limitations regarding clinically approved targeted agents as in other luminescence imaging modalities. As a further step in CLI, optical signals can be reinforced by spectrally coupling Cerenkov radiation at ultraviolet/blue wavelengths to far-red and near-infrared emitting quantum nanoparticles or fluorophores, resulting in an improvement of reconstructed images (Dothager et al., 2010).

These results demonstrate that the SP_N models are an alternative to computationally costly transport calculations (calculations are speeded-up by near to two orders of magnitude), and a solution to DE failures in a considerable number of experimental situations. In general, the reconstructions algorithms that employ SP_N-based forward models have been evolving by including spectral information and constrained optimization features. Further efforts should be addressed on (*i*) improving the numerical schemes for calculating the model predictions, (*ii*) reducing the ill-posedness of the inverse problem by imposing constraints in both the parameter space (optical coefficients) and the forward model, and (*iii*) augmenting the robustness of the inverse problem formulation, e.g. in the choice of the objective function.

Acknowledgments

J. Bouza-Domínguez acknowledges financial support from the FQRNT (Québec – Programme de bourses d'excellence pour étudiants étrangers – PBEEE). Y. Bérubé-Lauzière acknowledges financial support from an NSERC Discovery Grant (Canada) for the present work. Yves Bérubé-Lauzière is member of the FRQ-S-funded Centre de recherche clinique Étienne-Le Bel.

References

Abramowitz, M., and Stegun I. A., 1965: *Handbook of Mathematical Functions: with Formulas, Graphs, and Mathematical Tables*, Dover Publications: Mineola, U.S.A.

Agarwal, R. P., 2000: *Difference Equations and Inequality: Theory, Methods, and Applications*, CRC Press: New York.

Alexandrakis, G., Rannou, F. R., Chatziioannou, A. F., 2005: Tomographic bioluminescence imaging by use of a combined optical-PET (OPET) system: a computer simulation feasibility study, *Phys. Med. Biol.*, **50**, 4225–4241.

Alexandrakis, G., Rannou, F. R., Chatziioannou, A. F., 2006: Effect of optical property estimation accuracy on tomographic bioluminescence imaging: simulation of a combined optical–PET (OPET) system, *Phys. Med. Biol.*, **51**, 2045–2053.

Arridge, S. R., 1999: Optical tomography in medical imaging, *Inverse Probl.*, **15**, R41–R93.

Arridge, S. R., and Schotland, J. C., 2009: Optical tomography: forward and inverse problems, *Inverse Probl.*, **25**, 123010.

Atalay, M. A., 2006: P_N solutions of radiative heat transfer in a slab with reflective boundaries, *J. Quant. Spectr. Rad. Transfer*, **101**, 100–108.

Benson, S. J., and Moré, J. A. 2001: Limited-memory variable-metric algorithm for bound-constrained minimization, *Technical Report ANL/MCS-P909-0901*, Argonne National Laboratory.

Bérubé-Lauzière, Y., Issa, V., Bouza Dominguez, J., 2009: Simplified spherical harmonics approximation of the time-dependent equation of radiative transfer for the forward problem in time-domain diffuse optical tomography, *Proc. of SPIE*, **7174**, 717403-11.

Beyer, H. G. and Schwefel, H.-P., 2002: Evolution strategies, *Nat. Comput.*, **1**, 3–52.

Boschi, F., Spinelli, A. E., D'Ambrosio, D., Calderan, L., Marengo, M., Sbarbati, A., 2009: Combined optical and single photon emission imaging: preliminary results, *Phys. Med. Biol.*, **54**, L57–L62.

Bouza Domínguez, J., and Bérubé-Lauzière, Y., 2010: Diffuse light propagation in biological media by a time-domain parabolic simplified spherical harmonics approximation with ray-divergence effects, *Appl. Opt.*, **49**, 1414–1429.

Bouza Domínguez, J., and Bérubé-Lauzière, Y., 2011a: Light propagation from fluorescent probes in biological tissues by coupled time-dependent parabolic simplified spherical harmonics equations, *Biomed. Opt. Express*, **2**, 817–837.

Bouza Domínguez, J., and Bérubé-Lauzière, Y., 2011b: Diffuse optical tomography of absorption in biological media using time-dependent parabolic simplified spherical harmonics equations, *Proc. SPIE*, **8088**, 80880S.

Bouza Domínguez, J., and Y. Bérubé-Lauzière Y., 2012: Diffuse optical tomographic imaging of biological media by time-dependent parabolic SPN equations: a two-dimensional study. *J. Biomed. Optics*, 17(8):086012-1-086012-14.

Born, M., and Wolf, E., 2003: Principles of Optics, 7th edn, Cambridge University Press: Cambridge, UK.

Boykov, Y., and Kolmogorov, V., 2004: An experimental comparison of min-cut/max-flow algorithms for energy minimization in vision, *IEEE Trans. Patt. Anal. and Mach. Intell.*, **26**, 1124–1137.

Brantley, P. S., and Larsen, E. W., 2000: The simplified P3 approximation, *Nucl. Sci. Eng.*, **134**, 1–21.

Burgess, S. A., Ratner, D., Chen, B. R., Hillman, E. M. C., 2010: Fiber-optic and articulating arm implementations of laminar optical tomography for clinical applications, *Biomed. Opt. Express*, **1**, 780–790.

Case, K. M., and Zweifel, P. F., 1967: Linear Transport Theory, Addison-Wesley.

Chandrasekhar, S., 1960: Radiative Transfer, Dover Publications: New York.

Chen, B., Stamnes, K., Stamnes, J. J., 2001: Validity of the Diffusion Approximation in Bio-Optical Imaging, *Appl. Opt.*, **40**, 6356–6366.

Chu, M., Vishwanath, K., Klose, A. D., Dehghani, H., 2009: Light transport in biological tissue using three-dimensional frequency-domain simplified spherical harmonics equations, *Phys. Med. Biol.*, **54**, 2493–2500.

Chu, M., and Dehghani, H., 2009: Image reconstruction in diffuse optical tomography based on simplified spherical harmonics approximation, *Opt. Express*, **16**, 17780–17791.

Contag, C. H., and Bachmann, M. H., 2002: Advances in In Vivo bioluminescence imaging of gene expression. *Annu. Rev. Biomed. Eng.*, **4**, 235–260.

Comsa, D. C., Farrell, T. J., Patterson, M. S., 2008: Quantitative fluorescence imaging of point-like sources in small animals, *Phys. Med. Biol.*, **53**, 5797–5814.

Davidson, B., and Sykes, J. B., 1957: Neutron Transport Theory, Oxford University Press: Oxford.

Davis, T. A., 2006: Direct Methods for Sparse Linear Systems (Fundamentals of Algorithms), SIAM: Philadelphia.

Dehghani, H., Davis, S. C., Pogue, B. W., 2008: Spectrally resolved bioluminescence tomography using the reciprocity approach, *Med. Phys.*, **35**, 4863–4871.

Dehghani, H., Eames, M. E., Yalavarthy, P. K., Davis, S. C., Srinivasan, S., Carpenter, C. M., Pogue, B. W., Paulsen, K. D., 2009a: Near infrared optical tomography using NIRFAST: Algorithm for numerical model and image reconstruction, *Commun. Numer. Meth. Eng.*, **25**, 711–732.

Dehghani, H., Srinivasan, S., Pogue, B. W., Gibson, A., 2009b: Numerical modelling and image reconstruction in diffuse optical tomography, *Phil. Trans. Royal. Soc. A*, **367**, 3073–3093.

Dempster, A. P., Laird, N. M., Rubin, D. B., 1977: Maximum likelihood from incomplete data via the EM algorithm, *J. R. Stat. Soc. Ser. B Stat. Methodol.*, **39**, 1–38.

de Oliveira, C. R. E., 1986: An arbitrary geometry finite element method for multigroup neutron transport with anisotropic scattering, *Prog. Nucl. Energ.*, **18**, 227–236.

Dirk, V. A., 2002: *Noisy Optimization with Evolution Strategies (Genetic Algorithms and Evolutionary Computation)*, Springer: New York.

Dothager, R. S., Goiffon, R., Jackson, E., Harpstrite, S., Piwnica-Worms, D., 2010: Cerenkov radiation energy transfer (CRET) imaging: a novel method for optical imaging of PET isotopes in biological systems, PLoS ONE, 5, e13300.

Ducross, N., 2009: *Tomographie optique de fluorescence dans les milieux diffusants: apport de l'information temporelle*, Ph.D. thesis, Université Claude Bernard – Lyon I.

Duderstadt, J. J., and Martin, W. R., 1979: *Transport Theory*, John Wiley and Sons: New York.

Elwell, C. E., and Cooper, C. E., 2011: Making light work: illuminating the future of biomedical optics, *Phil. Trans. Royal Soc. A*, **369**, 4358–4379.

Faris, G. W., 2002: Diffusion equation boundary conditions for the interface between turbid media: a comment, *J. Opt. Soc. Am. A*, **19**, 519–520.

Fletcher, J., 1983: Solution of the multigroup neutron transport equation using spherical harmonics, *Nucl. Sci. Eng.*, **84**, 33–46.

Frank, M., Klar, A., Larsen, E., Yasuda, S., 2007: Time-dependent simplified P_N approximation to the equations of radiative transfer, *J. Comput. Phys.*, **226**, 2289–305.

Friedman, J. H., Hastie, T., Tibshirani, R., 2010: Regularization paths for generalized linear models via coordinate descent, *J. Stat. Softw.*, **33**, 1–22.

Gelbard, E. M., 1960: Application of spherical harmonics methods to reactor problems, WAPD-BT-20, Bettis Atomic Power Laboratory.

Gockenbach, M. S., 2006: *Understanding and Implementing the Finite Element Method*, SIAM: Philadelphia.

Han, D., Tian, J, Liu, K., Feng, J., Zhang, B., Ma, X., Qin, C., 2010: Sparsity-promoting tomographic fluorescence imaging with simplified spherical harmonics approximation, *IEEE Trans. Biomed. Eng.*, **57**, 2564–2567.

Hazra, S. B., 2010: PDE-constrained optimization methods, in *Large-Scale PDE-Constrained Optimization in Applications*, Springer-Verlag: New York.

Hielscher, A. H., Alcouffe, R. E., Barbour, R. L., 1998: Comparison of finite-difference transport and diffusion calculations for photon migration in homogeneous and heterogeneous tissues, *Phys. Med. Biol.*, **43**, 1285–1302.

Hielscher, A. H., 2005: Optical tomography imaging of small animals, *Curr. Opin. Biotechnol.*, **16**, 79–88.

Hielscher, A. H., Kim, H. K., Klose, A. D., 2011: Forward models of light transport in biological tissue, in *Handbook of Biomedical Optics*, CRC Press: Boca Raton.

Howell, J. R., Siegel, R., Menguc, M. P., 2010, *Thermal Radiation Heat Transfer*, 5th edn, CRC Press: New York.

Ishimaru, A., 1978: *Wave Propagation and Scattering in Random Media*, Academic Press: New York.

Jin, J., 2002: *The Finite Element Method in Electromagnetics*, John Wiley and Sons Inc.: New York.

Klose, A. D., and Larsen, E., 2006: Light transport in biological tissue based on the simplified spherical harmonics equations, *J. Comput. Phys.*, **220**, 441–70.

Klose, A. D., and Hielscher, A. H., 2008: Optical tomography with the equation of radiative transfer, *Int. J. Numer. Method Heat Fluid Flow.*, **18**, 443–464.

Klose, A. D., 2009: Radiative transfer of luminescence light in biological tissue, in *Light Scattering Reviews 4*, A. A. Kokhanovsky, ed., Springer – Praxis: Chichester, pp. 293–345.

Klose, A. D., Beattie, B. J., Dehghani, H., Vider, L., Le, C., Ponomarev, V., Blasberg, R., 2010: In vivo bioluminescence tomography with a blocking-off finite-difference SP_3 method and MRI/CT co-registration, *Med. Phys.*, **37**, 329–338.

Klose, A. D., 2010a: The forward and inverse problem in tissue optics based on the radiative transfer equation: A brief review, *J. Quant. Spectrosc. Radiat. Transfer*, **111**, 1852–1853.

Klose, A. D., 2010b: Hyperspectral excitation-resolved fluorescence tomography with the SP_3 equations, Proc. OSA-BIOMED Conf., BSuD51.

Klose A. D., and Pöschinger, T., 2011: Excitation-resolved fluorescence tomography with simplified spherical harmonics equations, *Phys. Med. Biol.*, **56**, 1443–1469.

Klose, A. D., Tekabe, Y., Johnson, L., 2011: Hyperspectral fluorescence tomography Of quantum dots using the simplified spherical harmonics equations, *Proc. SPIE*, 8088, 80880V.

Klose, A. D., 2012: Multispectral optical luminescence tomography with the simplified spherical harmonics equations, in *Light Scattering Reviews 7*, A. A. Kokhanovsky, ed., Springer – Praxis, Chichester, UK.

Kolmogorov, V., and Zabih, R., 2004: What energy functions can be minimized via graph cuts?, *IEEE Trans. Patt. Anal. and Mach. Intell.*, **26**, 147–159.

Kolmogorov, V., and Rother, C., 2007: Minimizing non-submodular functions with graph cuts-a review, *IEEE Trans. Patt. Anal. and Mach. Intell.*, **29**, 1274–1279.

Kotiluoto, P., Pyyry, J., Helminen, H., 2007: Multitrans SP_3 code in coupled photon-electron transport problems, *Radiat. Phys. Chem.*, **76**, 9–14.

Larusson, F., Fantini, S., Miller, E. L., 2011: Hyperspectral image reconstruction for diffuse optical tomography, *Biomed. Opt. Express*, **2**, 946–965.

Larsen, E. W., Morel, J. E., McGhee, J. M., 1996: Asymptotic derivation of the multigroup P_1 and simplified P_N equations with anisotropic scattering, *Nucl. Sci. Eng.*, **123**, 328.

Larsen, E. W., Thömmes, G., Klar, A., Seaïd, M., Götz T., 2002: Simplified P_N approximations to the equations of radiative heat transfer and applications, *J. Comp. Phys.*, **183**, 652–675.

Lay, D. C., 2011: *Linear Algebra and Its Applications*, 4th edn, Addison-Wesley, Boston.

Li, C., Mitchell, G., Cherry, S., 2010: Cerenkov luminescence tomography for small-animal imaging, Opt. Lett., **35**, 1109–1111.

Liemert, A., and Kienle, A., 2010: Analytical solutions of the simplified spherical harmonics equations, *Opt. Lett.*, **35**, 3507–3509.

Liemert, A., and Kienle, A., 2011a: Comparison between radiative transfer theory and the simplified spherical harmonics approximation for a semi-infinite geometry, *Opt. Lett.*, **36**, 4041–4043.

Liemert, A., and Kienle, A., 2011b: Analytical solution of the radiative transfer equation for infinite-space fluence, *Phys. Rev. A*, **83**, 039903.

Liu, K., Tian, J., Lu, Y-J., Qin, C-H., Zhu, S-P., Zhang, X., 2010: A fast bioluminescent source localization method based on generalized graph cuts with mouse model validations, *Opt. Express*, **18**, 3732–3745.

Liu, H., Ren, G., Miao, Z., Zhang, X., Tang, X., Han, P., Ghambir, S. S. Cheng Z., 2010b: Molecular optical imaging with radioactive probes, *PLoS ONE*, **5**, e9470.

Liu, K. Tian, J., Sr., Qin, C., Yang, X., Zhu, S., Han, D., Ping Wu, Xiaoqian Dai, 2011: In vivo heterogeneous tomographic bioluminescence imaging via a higher-order approximation forward model, *Proc. SPIE*, **7965**, 79650C.

Lu, Y., Douraghy, A., Machado, H. B., Stout, D., Jie, T., Herschman, H., Chatziioannou, A. F., 2009: Spectrally resolved bioluminescence tomography with the third-order simplified spherical harmonics approximation, *Phys. Med. Biol.*, **54**, 6477–6493.

Lu, Y., Zuh, B., Shen, H., Rasmussen, J. C., Wang G., Sevick-Muraca, E. M., 2010: A parallel adaptive finite element simplified spherical harmonics approximation solver for frequency domain fluorescence molecular imaging, *Phys. Med. Biol.*, **55**, 4625-4645.

Martí López, L., Bouza Domínguez, J., Hebden, J. C., Arridge, S. R., Martínez Celorio, R. A., 2003: Validity conditions for the radiative transfer equation, *J. Opt. Soc. Am. A*, **20**, 2046–2056.

Marshak, R. E., 1947: Note on the spherical harmonic method as applied to the Milne problem for a sphere, *Phys. Rev.*, **71**, 443–446.

Melnikova, I., Kuznetsov, A., Seroukhova, O., Pozdnyakov, D., Vasilyev, A., 2012: Remote Sensing of the Environment and Radiation Transfer. An Introductory Survey, Springer Verlag, New York.

Mitchell, G. S., Gill, R. K., Boucher, D. L., Li C., Cherry, S. R., 2011: In vivo Cerenkov luminescence imaging: a new tool for molecular imaging, *Phil. Trans. Royal Soc. A*, **28**, 369, 4605–4619.

Modest, M. F., 2003, *Radiative Heat Transfer*, 2nd edn, Academic Press, New York.

Montejo, L. D., Klose, A. D., Hielscher, A. H., 2010: Implementation of the equation of radiative transfer on block-structured grids for modeling light propagation in tissue, *Biomed. Opt. Express*, **1**, 861-878.

Montejo, L. D., Kim, H. K., Hielscher, A. H., 2011: A finite volume algorithm for modeling light transport with the time-independent simplified spherical harmonics approximation to the equation of radiative transfer, *Proc. SPIE*, **7896**, 78960J.

Naser, M. A., and Patterson, M. S., 2011: Improved bioluminescence and fluorescence reconstruction algorithms using diffuse optical tomography, normalized data, and optimized selection of the permissible source region, *Biomed. Opt. Express*, **2**, 169–184.

Natterer, F., 2001: *The Mathematics of Computerized Tomography*, SIAM, Philadephia.

Nikazad, T., 2008: *Algebraic reconstruction methods*, Ph.D. thesis, Linköping University.

Nocedal, J., and Wright, S., 2006: Sequential Quadratic Programming, in *Numerical Optimization*, Springer Verlag, New York.

Nothdurft, R. E., Patwardhan, S. V., Akers, W., Yunpeng, Y., Achilefu, S., Culver, J. P., 2009: In vivo fluorescence lifetime tomography, *J. Biomed. Opt.*, **14**, 024004.

Ntziachristos, V., 2006: Fluorescence molecular imaging, *Annu. Rev. Biomed. Eng.*, **8**, 1–33.

Pleijhuis, R. G. Langhout, G. C., Helfrich, W., Themelis, G., Sarantopoulos, A.,, Crane, L. M., Harlaar, N. J., de Jong, J. S., Ntziachristos, V., van Dam, G. M., 2011: Near-infrared fluorescence (NIRF) imaging in breast-conserving surgery: assessing intra-operative techniques in tissue-simulating breast phantoms, *Eur. J. Surg. Oncol.*, **37**, 32–39.

Press, W. H., Teukolsky, S. A., Vetterling, W. T., Flannery, B. P., 2007: Numerical Recipes 3rd Edition: The Art of Scientific Computing, Cambridge University Press, Cambridge, UK.

Rao, J., Dragulescu-Andrasi, A., Yao, H., 2007: Fluorescence imaging in vivo: recent advances, *Curr. Opin. Biotechnol.*, **18**, 17–25.

Ripoll, J., 2000: Light diffusion in turbid media with biomedical application, Ph.D. thesis, Universidad Autónoma de Madrid.

Robertson, R., Germanos, M. S., Li, C., Mitchel, G. S., Cherry, S. R., Silva, M. D., 2009: Optical imaging of Cerenkov light generation from positron-emitting radiotracers, *Phys. Med. Biol.*, **54**, N355–365.

Saad, Y., 2003: Iterative Methods for Sparse Linear Systems, SIAM, Philadelphia.

Spinelli, A. E., D'Ambrosio, D., Calderan, L., Marengo, M., Sbarbati, A., Boschi, F., 2010: Cerenkov radiation allows in vivo optical imaging of positron emitting radiotracers, *Phys. Med. Biol.*, **55**, 483–495.

Spinelli, A. E., Kuo, C., Rice, B. W., Calandrino, R., Marzola, P., Sbarbati, A., Boschi, F., 2011: Multispectral Cerenkov luminescence tomography for small animal optical imaging, *Opt. Express*, **19**, 12605–12618.

Strang, G., 2005: *Linear Algebra and its Applications*, 4th edn, Brooks Cole, Stamford.

Talukdar, P., 2006: Discrete transfer method with the concept of blocked-off region for irregular geometries, *J. Quant. Spectr. Rad. Transfer*, **98**, 238–248.

Tarvainen, T., 2006: *Computational methods for light transport in optical tomography*, Ph.D. thesis, University of Kuopio.

Tian, J., Liu, K., Lu, Y., Qin, C., Yang, X., Zhu, S., Han, D., Feng, J., Ma, X., Chang, Z., 2010: Evaluation of the simplified spherical harmonics approximation in bioluminescence tomography through heterogeneous mouse models, *Opt. Express*, **18**, 20988–21002.

Tomasevic, D. I., and Larsen, E. W., 1996: The simplified P2 approximation, *Nucl. Sci. Eng.*, **122**, 309–325.

Van der Kooij, A., 2007: *Prediction accuracy and stability of regression with optimal scaling transformations*, Ph.D. thesis, University of Leiden.

Van Rossum, M. C. W., and Nieuwenhuizen, Th. M., 1999: Multiple scattering of classical waves: microscopy, mesoscopy, and diffusion, *Rev. Mod. Phys.*, **71**, 313–371.

Versteeg, H., and Malalasekera, W., 2007: An Introduction to Computational Fluid Dynamics: The Finite Volume Method, Prentice Hall: New Jersey.

Vo-Dinh T. (ed.), 2003: Biomedical Photonics Handbook, CRC Press: Boca Raton.

Wang, L., and Wu, H., 2007, Biomedical Optics: Principles and Imaging, Wiley-Interscience, Hoboken, NJ.

Wang, J., Davis, S. C., Srinivasan, S., Jiang, S., Pogue, B. W., Paulsen, K. D., 2008: Spectral tomography with diffuse near-infrared light: inclusion of broadband frequency domain spectral data, *J. Biomed. Opt.*, **13**, 041305-041305.

Wang, Y., Yagola, A. G., Yang, C., 2011: Optimization and Regularization for Computational Inverse Problems and Applications, Springer Verlag: New York.

Website of Toolkit for Advanced Optimization (TAO). Consulted January, 10th, 2012. http://www.mcs.anl.gov/research/projects/tao/.

Weissleder, R., and Ntziachristos, V., 2003: Shedding light onto live molecular targets, *Nat. Med.*, **9**, 123–128.

Welsh, D. K., and Kay S. A., 2005: Bioluminescence imaging in living organisms, *Curr. Opin. Biotechnol.*, **16**, 73–78.

Wernick, M. N., and Aarsvold, J. N., 2004: Emission Tomography: The Fundamentals of PET and SPECT, Academic Press: San Diego.

Willmann, J. K., van Bruggen, N., Dinkelborg, L. M., Gambhir, S. S., 2008: Molecular imaging in drug development, *Nat. Rev. Drug Discov.*, **7**, 591–607.

Wright, S., Schweiger, M., Arridge, S. R., 2007: Reconstruction in optical tomography using the P_N approximations, *Meas. Sci. Technol.*, **18**, 79–86.

Xu, G., Piao, D., Bunting, C. F., Dehghani, H., 2010: The pain and gain of DC-based diffuse optical tomography reconstruction – New insights into an old-like problem, OSA-BIOMED Conf., BSuD54.

Xu, G., Piao, D., Dehghani, H., 2011: The utility of direct-current as compared to frequency domain measurements in spectrally-constrained diffuse optical tomography toward cancer imaging technology, *Tech. Cancer Res. Treat.*, **10**, 403–416.

Zacharopoulos, A., Garofalakis, A., Ripoll, J., Arridge, S. R., 2010: Development of i nvivo fluorescence imaging with the matrix-free method, *J. Phys.*, **255**, 012006.

Zhong, J., Tian, J., Yang, X., Qin, C., 2011: Whole-body Cerenkov luminescence tomography with the finite element SP_3 method, *Annals of Biomed. Eng.*, **39**, 1728–1735.

Zhong, J., Qin, C., Yang, X., Zhu, S., Zhang, X., Tian, J., 2011b: Cerenkov luminescence tomography for in vivo radiopharmaceutical imaging, *Int. J. Biom. Imag.*, 641618.

Zhu, Q., Dehghani, H., Leblond, F., El-Ghussein, F., Pogue, B. W., 2011: Development and evaluation of a time-resolved near-infrared fluorescence finite element model, *Proc. SPIE*, **7896**, 78960T.

Zhu, Q., Dehghani, H., Tichauer, K. M., Holt, R. W., Vishwanath, K., Leblond, F., Pogue, B. W., 2011: A three-dimensional finite element model and image reconstruction algorithm for time-domain fluorescence imaging in highly scattering media, *Phys. Med. Biol.*, **56**, 7419.

7 Transillumination of highly scattering media by polarized light

Evgenii E. Gorodnichev, Sergei V. Ivliev, Alexander I. Kuzovlev,
and Dmitrii B. Rogozkin

7.1 Introduction

Visualization of various objects hidden in highly scattering turbid media is one of the most general and important problems of modern statistical and biomedical optics. Significant interest in this field in the last two decades was stimulated by diagnostic applications [1–8]. The major difficulty encountered in imaging through these media is due to the multiple scattering effect. Multiple scattering leads to loss of directionality of the incident beam, resulting in the image blurring. There are several approaches to the problem of imaging through highly scattering media using infrared and visible light [1–8]. Some of them are concentrated on selecting 'image bearing' photons. Such quasi-straightforward propagating photons were called 'snake photons' and are expected to carry information necessary for the image reconstruction [9].

Among the techniques based on separating quasi-straightforward propagating photons, the polarization-gated method, owing to its instrumental simplicity, holds a particular position (see, e.g., [3, 5, 8, 10–25]). This method relies on the fact that strongly scattered (i.e. diffusive) photons get depolarized and therefore can be removed by detecting only the polarized component of transmitted light.

The degree of polarization depends on the number of scattering events and, correspondingly, on the photon path length in the medium. Therefore the photons propagating along nearly straight lines and passing the shortest path are the least depolarized. Polarization of such early arrival photons (the ballistic and snake photons) remain close to the initial one. Therefore polarization can be used for time-gating the ballistic and snake components of the output signal with a temporal gate of the order of the polarization decay time (in accordance with [10, 11] the degree of polarization, regardless of the sample thickness, is different from zero over the initial 100 ps after the arrival of the ballistic component).

Depolarization of a linearly polarized pulse was first studied in [26]. From the results [26] obtained with a Monte Carlo code it follows that the photons transmitted through a thick slab ($\sigma_{tr} z \gg 1$, σ_{tr} is the transport (or reduced) scattering coefficient, z is the slab thickness) with delays $\Delta = ct - z > z$ are completely depolarized. More recently, conclusions drawn in [26] were confirmed by experimental [10, 11, 27] and numerical [27, 28] results.

The fact that the polarization of snake photons differs little from the initial one has already been used for optical imaging of objects hidden inside turbid media [3, 5, 12–25]. From published experimental data it follows that the polarization-gated method enables one to improve the image contrast and resolution and can be considered as one of the most promising approaches being currently studied for transillumination of highly scattering biological tissues.

In the polarization-difference techniques (see, e.g., [13–16,19,23–25]) the 'image-bearing' component of light is extracted by subtracting the detected cross-polarized signal from the co-polarized one. As diffusive photons with completely randomized polarization make the same contribution to both polarized components of scattered radiation, subtraction of cross-polarized component from the co-polarized one filters out the diffusive photons from the ballistic and snake photons. The feasibility of these techniques depends on the depolarization characteristics of the medium, which, in their turn, are influenced by concentration, size and shape of scattering particles, their refractive index (the influence of these parameters on depolarization is experimentally studied with the use of the well-characterized tissue phantoms, e.g. aqueous suspensions of polystyrene microspheres and Intralipid [5, 10, 11, 13, 24]).

Transillumination of tissues with visible and near-infrared light relies on the differences in optical parameters of an embedded object and surrounding tissues. The ultimate goal is to image and characterize millimeter-sized objects in the sample up to several centimeters in thickness (see, e.g., [1, 5, 29–31]). As a rule, the embedded object is characterized by higher absorption [29–33]. The difference in the transport scattering coefficient between the embedded object and the surrounding tissue is less significant.

For surrounding tissues, the absorption coefficient is of the order of $\sigma_a = (1 \div 3) \cdot 10^{-3}\,\mathrm{mm}^{-1}$. The absorption coefficient of an object embedded in tissue may range up to $0.1\,\mathrm{mm}^{-1}$ [29–33]. The typical values of the transport scattering coefficient σ_{tr} fall in the range between $0.4\,\mathrm{mm}^{-1}$ and $1.6\,\mathrm{mm}^{-1}$ [29–33]. The transport optical thickness $\sigma_{tr} z$ of the test samples may vary from a few tenths up to several tens. The probable values of coefficient σ_{tr} for an object usually differ from those for surrounding tissues by $20 \div 30\%$ (see, e.g., [29–31]).

For different types of tissues, the mean cosine of single-scattering angle $\langle \cos\gamma \rangle$ (i.e. the anisotropy factor of scattering) varies in the range between 0.75 and 0.95. High anisotropy of single scattering, $1 - \langle \cos\gamma \rangle \ll 1$, is common to all types of tissues.

As it is difficult to quantify all the morphological parameters (concentration, size, shape and refractive index of scatterers) of a complex scattering medium such as biological tissue, the influence of these parameters on propagation and depolarization of light is usually studied on well-characterized tissue phantoms.

Samples of monodisperse polystyrene microspheres diluted with water to the desired concentration are the most frequently used tissue phantoms. The typical particle diameter varies in the range $0.1 \div 2\,\mu\mathrm{m}$ (see, e.g., [5, 11, 13, 24–28]). The relative refractive index of polystyrene particles in water is equal to $n = 1.19$.

The refractive index of scatterers in actual tissues is, as a rule, lower than that of polystyrene microspheres. Therefore aqueous suspension of silica microspheres, the refractive index $n = 1.03$ of which is comparable to the refractive index fluctuations in actual biological tissues, appears to be of interest for modeling tissue properties [24,25].

Transport optical coefficients of tissue phantoms are commonly calculated with the Mie theory [34–37].

In what follows, polarization-difference optical imaging of objects hidden in highly scattering media is studied. Previous theoretical works devoted to this problem dealt with numerical simulations (see, e.g., [16,18–21,23]). Contrary to these works, we present a simple analytical theory that makes possible semi-quantitative describing the image characteristics and gives an insight into their dependence on optical parameters of the medium. Our approach is based on the basic mode approximation [38–41] in the vector radiative transfer equation. The results obtained take into account the contribution from the polarized snake-photons of the output signal, and enable us to reproduce experimental data on the polarization-gated transillumination of tissue-like media. To develop the model of depolarization we consider propagation of an ultrashort pulse of polarized light through a turbid sample. The temporal profiles of the degree of polarization are studied. For linearly polarized light, the degree of polarization is shown to be different from zero over the initial temporal interval of the order of the transport mean free time $(\sigma_{tr} \cdot c)^{-1}$ (c is the speed of light). The results of our calculations correlate well with the results of numerical simulations and experimental data. According to our results the values of the polarization modes differ from the intensity only by the factors that describe additional attenuation in the domain of temporal delays. Within our approach, it takes only some quantities, namely, the transport scattering, absorption and depolarization coefficients to describe depolarization of light in the medium. For both the intensity of the transmitted pulse and the basic modes of polarization, the edge-spread functions are calculated within the Fokker–Planck model which allows for highly forward scattering of snake-photons. The results obtained make it possible to determine the dependence of the polarization-difference image characteristics (e.g., spatial resolution and contrast) on the time-gate duration. Our calculations are employed to simulate the images of simple-shaped objects hidden in a highly scattering medium under conditions of CW illumination. The contributions of the photons passing by the object and passing through it are taken into account. For the photons passing through the object, the depth-average scattering and depolarization coefficients are substituted for the depth-dependent ones. Examples of 1-D and 2-D images are presented. Sensitivity of the image contrast to the polarization state of light and variations in optical properties of the medium and the object is discussed.

7.2 General relations

Consider a beam of polarized light incident on a medium normally to its surface. The medium is assumed to be a statistically isotropic disordered ensemble of large-scale scatterers (size a is larger than wavelength λ). The polarization state of scattered light is generally described by the Stokes column vector [5,34–37]

$$\hat{S} = \begin{pmatrix} I \\ Q \\ U \\ V \end{pmatrix} \tag{7.1}$$

where the four Stokes parameters are defined by the following relations

$$\begin{aligned} I &= \langle E_{\|} E_{\|}^* + E_{\perp} E_{\perp}^* \rangle \\ Q &= \langle E_{\|} E_{\|}^* - E_{\perp} E_{\perp}^* \rangle \\ U &= \langle E_{\|} E_{\perp}^* + E_{\|}^* E_{\perp} \rangle \\ V &= \mathrm{i}\langle E_{\|} E_{\perp}^* - E_{\|}^* E_{\perp} \rangle \end{aligned} \tag{7.2}$$

The Stokes parameters and the components $E_{\|}$ and $E_{\perp}$

$$\mathbf{E} = \mathbf{e}_{\|} E_{\|} + \mathbf{e}_{\perp} E_{\perp} \tag{7.3}$$

of the electric field appearing in Eq. (7.2) are defined in the system of unit vectors $\mathbf{e}_{\|} = \partial\mathbf{n}/\partial\theta$, $\mathbf{e}_{\perp} = [\mathbf{e}_{\|}, \mathbf{n}]$, and $\mathbf{n}$. The unit vector $\mathbf{n} = (\sin\theta\cos\varphi, \sin\theta\sin\varphi, \cos\theta)$ is the direction of propagation of the transverse electromagnetic wave, the vector $\mathbf{e}_{\|}$ lies in the plane formed by the vectors $\mathbf{n}_0$ and $\mathbf{n}$ (where $\mathbf{n}_0$ is the internal normal to the surface), the vector $\mathbf{e}_{\perp}$ is perpendicular to this plane. The brackets $\langle \ldots \rangle$ denote statistical averaging.

As shown in [42], to describe multiple scattering of light in turbid media, it is more convenient to go from linear basis (7.3) to the circular basis,

$$\mathbf{E} = \mathbf{e}_{+} E_{+} + \mathbf{e}_{-} E_{-} \tag{7.4}$$

where

$$\mathbf{e}_{\pm} = \frac{1}{\sqrt{2}} (\mathbf{e}_{\|} \mp \mathrm{i}\mathbf{e}_{\perp}) \tag{7.5}$$

In this basis, the electric field is the superposition of waves with left (+) and right (−) circular polarizations.

For the circular representation, the following column vector [36,42]

$$\hat{I} = \begin{pmatrix} \langle E_{-} E_{+}^* \rangle \\ \langle |E_{+}|^2 \rangle \\ \langle |E_{-}|^2 \rangle \\ \langle E_{-}^* E_{+} \rangle \end{pmatrix} = \frac{1}{2} \begin{pmatrix} Q - \mathrm{i}U \\ I - V \\ I + V \\ Q + \mathrm{i}U \end{pmatrix} = \frac{1}{\sqrt{2}} \begin{pmatrix} I_2 \\ I_0 \\ I_{-0} \\ I_{-2} \end{pmatrix} \tag{7.6}$$

is an analog of Stokes vector (7.1).

The nonstationary (time-dependent) vector transfer equation for the parameters I_m $(m = \pm 0, \pm 2)$ has the form [36,42]

$$\left\{ \frac{1}{c}\frac{\partial}{\partial t} + \mathbf{n}\frac{\partial}{\partial \mathbf{r}} + \sigma_{tot} \right\} I_m(\mathbf{r}, \mathbf{n}, t) = \sigma \int d\mathbf{n}' \, d_{mk}(\mathbf{n}, \mathbf{n}') I_k(\mathbf{r}, \mathbf{n}', t) \tag{7.7}$$

where $\sigma_{tot} = \sigma + \sigma_a$ is the total extinction coefficient, σ and σ_a are the coefficients of elastic scattering and absorption, respectively, c is the speed of light. The phase matrix appearing in Eq. (7.7) is given by [36, 42]

$$d_{mn}(\mathbf{n},\mathbf{n}') =$$
$$\begin{pmatrix} a_+^{(2,3)}\exp(2i\chi_+) & b_+\exp(-2i\beta) & b_+\exp(-2i\beta) & a_-^{(2,3)}\exp(2i\chi_-) \\ b_+\exp(-2i\beta') & a_+^{(1,4)} & a_-^{(1,4)} & b_-\exp(2i\beta') \\ b_-\exp(-2i\beta') & a_-^{(1,4)} & a_+^{(1,4)} & b_+\exp(2i\beta') \\ a_-^{(2,3)}(-2i\chi_-) & b_-\exp(2i\beta) & b_-\exp(2i\beta) & a_+^{(2,3)}\exp(-2i\chi_+) \end{pmatrix} \tag{7.8}$$

where $\chi_\pm$ and β, β' are defined as

$$\chi_\pm = \pi - (\beta \pm \beta') \tag{7.9}$$

$$\cos 2\beta = 1 - \frac{2(1-\mu'^2)(1-\cos^2\psi)}{1-(\mathbf{nn}')^2}$$
$$\sin 2\beta = \frac{2\sqrt{1-\mu'^2}(\mu'\sqrt{1-\mu^2} - \mu\sqrt{1-\mu'^2}\cos\psi)\sin\psi}{1-(\mathbf{nn}')^2} \tag{7.10}$$

$$\mathbf{nn}' = \mu\mu' + \sqrt{(1-\mu^2)(1-\mu'^2)}\cos\psi,$$
$$\mu = \mathbf{nn_0} = \cos\theta, \quad \mu' = \mathbf{n}'\mathbf{n_0} = \cos\theta', \quad \psi = \varphi - \varphi' \tag{7.11}$$

Functions $\cos 2\beta'$ and $\sin 2\beta'$ are obtained from functions $\cos 2\beta$ and $\sin 2\beta$, respectively, by interchanging μ and μ'.

Functions $a_\pm^{(i,j)}(\mathbf{nn}')$ $(i,j = 1,\ldots,4)$ and $b_\pm(\mathbf{nn}')$ entering into Eq. (7.8) are expressed in terms of the elements of the scattering matrix within the standard linear representation [34–37],

$$\begin{pmatrix} a_1(\mathbf{nn}') & b_1(\mathbf{nn}') & 0 & 0 \\ b_1(\mathbf{nn}') & a_2(\mathbf{nn}') & 0 & 0 \\ 0 & 0 & a_3(\mathbf{nn}') & b_2(\mathbf{nn}') \\ 0 & 0 & -b_2(\mathbf{nn}') & a_4(\mathbf{nn}') \end{pmatrix} \tag{7.12}$$

by the following relations:

$$a_\pm^{(i,j)} = \frac{a_i \pm a_j}{2}, \quad b_\pm = \frac{b_1 \pm ib_2}{\sqrt{2}} \tag{7.13}$$

For the forward scattering, $\mathbf{n} = \mathbf{n}'$, equalities $a_2(1) = a_3(1)$, $b_1(1) = 0$, $b_2(1) = 0$ [36] are valid and therefore $a_-^{(2,3)}(1) = 0$, $b_\pm(1) = 0$.

The matrix element a_1 appearing in matrix (7.12) is the phase function. It is normalized by relation

$$\int d\mathbf{n}' a_1(\mathbf{nn}') = 1 \tag{7.14}$$

In the case of spherical scatterers the diagonal elements of matrix (7.12) satisfy equalities $a_1(\mathbf{nn}') = a_2(\mathbf{nn}')$ and $a_3(\mathbf{nn}') = a_4(\mathbf{nn}')$. For particles of given radius and refractive index, matrix elements $a_{\pm}^{(i,j)}$ and $b_{\pm}$ entering into Eq. (7.8) can be expressed in terms of the scattering amplitudes [35, 36]

$$a_{\pm}^{(1,4)}(\mathbf{nn}') = a_{\pm}^{(2,3)}(\mathbf{nn}') = \frac{n_0}{4\sigma}|A_{\parallel}(\mathbf{nn}') \pm A_{\perp}(\mathbf{nn}')|^2 \qquad (7.15)$$

$$b_{\pm}(\mathbf{nn}') = \frac{n_0}{2\sqrt{2}\sigma}(A_{\parallel}(\mathbf{nn}') \pm A_{\perp}(\mathbf{nn}'))(A_{\parallel}(\mathbf{nn}') \mp A_{\perp}(\mathbf{nn}'))^* \qquad (7.16)$$

where $A_{\parallel}$ and $A_{\perp}$ are the scattering amplitudes of the waves polarized, respectively, parallel and perpendicularly to the scattering plane, n_0 is the number of scatterers per unit volume. The values of $A_{\parallel}$ and $A_{\perp}$ can be calculated with the Mie theory [34–37]. In the case of the Born spherical scatterers ($ka|n-1| \ll 1$, where $k = 2\pi/\lambda$, a and n are the radius and the relative refractive index of the scatterers), amplitudes $A_{\parallel}$ and $A_{\perp}$ are related by the following equation [34, 35]:

$$A_{\parallel}(\mathbf{nn}') = (\mathbf{nn}')A_{\perp}(\mathbf{nn}') \qquad (7.17)$$

For the forward scattering $A_{\parallel}(1) = A_{\perp}(1)$.

7.3 Basic mode approximation

Properties of the scattering matrix have been discussed in many publications (see, e.g., [36, 37]). The interest in this problem is caused by wide applications of optical methods for studying various scattering media (aerosols, seawater, biological tissues, colloidal suspensions, etc.).

As single scattering by large inhomogeneities occurs predominantly through small angles ($1 - \langle\cos\gamma\rangle \ll 1$) [34–37], the off-diagonal elements in Eq. (7.8), that are proportional to $a_{-}^{(2,3)}$ and $b_{\pm}$, appears to be small as compared to the other elements of the corresponding phase matrix.

The relationship between the elements of matrix (7.8) allows us to develop an iterative procedure for solving the vector radiative transfer equation (see Eq. (7.7)). In the first approximation, neglecting the elements proportional to $a_{-}^{(2,3)}$ and $b_{\pm}$, we obtain three independent equations. These equations describe the propagation of the basic polarization modes [38–41].

The scalar mode, the intensity I, obeys the scalar radiative transfer equation,

$$\left\{\frac{1}{c}\frac{\partial}{\partial t} + \mathbf{n}\frac{\partial}{\partial \mathbf{r}} + \sigma_{tot}\right\} I(\mathbf{r},\mathbf{n},t) = \sigma \int d\mathbf{n}' a_1(\mathbf{nn}') I(\mathbf{r},\mathbf{n}',t) \qquad (7.18)$$

The fourth Stokes parameter V that corresponds to the basic mode of circular polarization is governed by the following equation:

$$\left\{\frac{1}{c}\frac{\partial}{\partial t} + \mathbf{n}\frac{\partial}{\partial \mathbf{r}} + \sigma_{tot}\right\} V(\mathbf{r},\mathbf{n},t) = \sigma \int d\mathbf{n}' a_4(\mathbf{nn}') V(\mathbf{r},\mathbf{n}',t) \qquad (7.19)$$

Factoring out the azimuth-dependent harmonics $\exp(\pm 2i\varphi)$ in $I_{\pm 2}$ (these functions are responsible for the transformation of $I_{\pm 2}$ under rotations),

$$I_{\pm 2} = \frac{1}{\sqrt{2}} W(\mathbf{r}, \mathbf{n}, t) \exp(\pm 2i\varphi)$$

we arrive at the following transfer equation for the basic mode of linear polarization W [38–41]:

$$\left\{ \frac{1}{c}\frac{\partial}{\partial t} + \mathbf{n}\frac{\partial}{\partial \mathbf{r}} + \sigma_{tot} \right\} W(\mathbf{r}, \mathbf{n}, t) = \sigma \int d\mathbf{n}' a_+^{(2,3)}(\mathbf{n}\mathbf{n}') \exp\left(2i(\chi_+ - \psi)\right) W(\mathbf{r}, \mathbf{n}', t) \tag{7.20}$$

where χ_+ and ψ are defined by Eqs. (7.9), (7.11).

Without resorting to the circular representation, equations (7.19) and (7.20) were derived also in [43].

The equations for V and W (see Eqs. (7.19), (7.20)) differ from the scalar transfer equation (Eq. (7.18)) by the form of the phase functions. The phase functions appearing in Eqs. (7.19) and (7.20) are a_4 and $a_+^{(2,3)} \exp(2i(\chi_+ - \psi))/2$, respectively. The difference between these phase functions and the phase function a_1 entering into Eq. (7.18) gives rise to nonzero effective 'absorption' in Eqs. (7.19) and (7.20) (even in the absence of true absorption). The effective 'absorption' in Eqs. (7.19) and (7.20) is responsible for the additional attenuation of V and W as compared to the intensity I and describes the effect of depolarization of circularly and linearly polarized light .

There are two different, 'geometrical' and 'dynamical', mechanisms of depolarization of electromagnetic waves in propagation through a random medium. These mechanisms were first pointed out within the framework of the study of wave propagation through a turbulent atmosphere [44, 45].

The 'geometrical' mechanism of depolarization is due to Rytov's rotation of the polarization plane. The plane of polarization turns, as the ray of light propagates along a nonplanar curve. Depolarization occurring in multiple scattering of linearly polarized light is a result of superposition of randomly oriented polarizations of waves propagating along different random paths. Therefore, the 'geometrical' depolarization occurs at depths of the order of transport mean free path l_{tr} ($l_{tr} = \sigma_{tr}^{-1}$), simultaneously with isotropization of the beam of light over the directions of propagation [46]. The situation is different in the case of circularly polarized light . Circularly polarized light can be presented as a superposition of two linearly cross-polarized waves shifted in phase by $\pi/2$. In multiple scattering, the Rytov effect results in the turn of the polarization plane of each linearly polarized wave, but has no effect on the phase shift between them. Therefore, a circularly polarized wave propagating along any random trajectory is unaffected by the Rytov rotation (or, what is the same, by the 'geometrical' mechanism of depolarization).

The pure geometrical depolarization can be obtained in the limit $a_1 = a_2 = a_3$ and $b_1 = b_2 = 0$ (or, for spherical particles, $A_{\|} = A_{\perp}$).

The difference between elements a_i, $i = 1, \ldots, 4$ (or, for spherical particles, the difference between amplitudes $A_{\|}$ and $A_{\perp}$) are responsible for the 'dynamical' mechanism of depolarization. Physically, the 'dynamical' mechanism is due to the difference in amplitudes between two components of the single-scattered wave that are polarized, respectively, parallel and perpendicularly to the scattering plane. By this mechanism, multiply scattered light depolarizes as spread in amplitudes increases. The 'dynamical' depolarization occurs independently of the initial polarization of light. In particular, circularly polarized light depolarizes only due to the 'dynamical' mechanism [46]. For linearly polarized light , the role of one or the other mechanism depends on the relative refractive index of the scatterers, their size and shape. As a rule, the geometrical mechanism can be either dominant or as important as the dynamical mechanism of depolarization [38–41, 46].

Consider transmission of a wide stationary beam through thick layers of a turbid medium.

Solutions to Eqs. (7.18)–(7.20) for the basic modes should be sought in the form of series expansions in generalized spherical harmonics. In the scalar transfer theory, this approach is known as P_l-approximation [47]. The expansion in generalized spherical harmonics for the vector transfer equation, originally proposed in [42], was used in a number of studies as a basis for numerical integration and in analytical calculations of the Stokes parameters of un polarized light propagating through optically thick layers (see, e.g., [36, 48]). Recently, this approach was applied to evaluate spatial moments of the photon distribution from a pulsed source of light [49].

Functions $I(z, \mu)$ and $V(z, \mu)$ can be represented as series expansions in Legendre polynomials $P_l(\mu)$:

$$I(z,\mu) = \sum_{l=0} \frac{2l+1}{4\pi} I(z,l) P_l(\mu), \quad V(z,\mu) = \sum_{l=0} \frac{2l+1}{4\pi} V(z,l) P_l(\mu) \tag{7.21}$$

The coefficients $I(z, l)$ and $V(z, l)$ appearing in expressions (7.21) satisfy equations

$$\frac{l}{(2l+1)} \frac{\partial I(z,l-1)}{\partial z} + \frac{(l+1)}{(2l+1)} \frac{\partial I(z,l+1)}{\partial z} + [\sigma + \sigma_a - \sigma a_1(l)] I(z,l) = 0 \tag{7.22}$$

$$\frac{l}{(2l+1)} \frac{\partial V(z,l-1)}{\partial z} + \frac{(l+1)}{(2l+1)} \frac{\partial V(z,l+1)}{\partial z} + [\sigma + \sigma_a - \sigma a_4(l)] V(z,l) = 0 \tag{7.23}$$

where

$$a_{1,4}(l) = 2\pi \int_{-1}^{1} d\mu a_{1,4}(\mu) P_l(\mu) \tag{7.24}$$

The angular dependence of the integral term in Eq. (7.20) (see, e.g., [36]), tells that a solution to Eq. (7.20) should be sought as an expansion in generalized spherical harmonics $P_{22}^l(\mu)$:

$$W(z,\mu) = \sum_{l=2} \frac{2l+1}{4\pi} W(z,l) P_{22}^l(\mu) \tag{7.25}$$

Detailed definitions and properties of the generalized spherical harmonics can be found in [36, 50]. Substituting Eq. (7.25) into Eq. (7.20), we obtain [39–41]

$$\frac{l^2-4}{l(2l+1)} \frac{\partial W(z,l-1)}{\partial z} + \frac{4}{l(l+1)} \frac{\partial W(z,l)}{\partial z} + \frac{(l+1)^2-4}{(l+1)(2l+1)} \frac{\partial W(z,l+1)}{\partial z} + [\sigma + \sigma_a - \sigma a_+^{(2,3)}(l)] W(z,l) = 0 \tag{7.26}$$

where coefficients $a_{+}^{(2,3)}(l)$ are expressed as

$$a_{+}^{(2,3)}(l) = 2\pi \int_{-1}^{1} d\mu \left[\frac{a_2(\mu) + a_3(\mu)}{2}\right] P_{22}^{l}(\mu) \tag{7.27}$$

We can take advantage of Eqs. (7.23) and (7.26), to evaluate the basic polarization modes in the asymptotic state of propagation at large z ($z \gg l_{tr}$). Analysis of the asymptotic limit is important for various practical applications, because it highlights the difference in propagation between linearly and circularly polarized beams (e.g., see [51–62]).

In the asymptotic solution of the vector radiation transfer equation, we can factor out the angular dependence from the expression describing the decay of polarization with increasing z.

Substituting $V(z,l)$ and $W(z,l)$ approximated by exponentials $V(z,l) = V(l)\exp(-\epsilon_V z)$ and $W(z,l) = W(l)\exp(-\epsilon_W z)$ into Eqs. (7.23) and (7.26) we arrive at eigenvalue problems. The attenuation coefficients ϵ_V and ϵ_W are the smallest eigenvalues. The corresponding sets of coefficients $V(l)$ and $W(l)$ give the eigenvectors.

The asymptotic solutions for basic polarization modes have the form

$$V^{(as)}(z,\mu) = \exp(-\epsilon_V z) \sum_{l=0} \frac{2l+1}{4\pi} V(l) P_l(\mu) = C_V \Phi_V(\mu) \exp(-\epsilon_V z) \tag{7.28}$$

$$W^{(as)}(z,\mu) = \exp(-\epsilon_W z) \sum_{l=2} \frac{2l+1}{4\pi} W(l) P_{22}^{l}(\mu) = C_W \Phi_W(\mu) \exp(-\epsilon_W z) \tag{7.29}$$

where coefficients C_V and C_W are defined as $C_V = V(l=0)$, $C_W = W(l=2)$.

Functions $\Phi_V(\mu)$ and $\Phi_W(\mu)$ appearing in Eqs. (7.28) and (7.29) describe the asymptotic angular profile of basic modes V and W. In accordance with Eqs. (7.28) and (7.29), they are normalized by conditions

$$2\pi \int_{-1}^{1} d\mu \Phi_V(\mu) = 1 \tag{7.30}$$

$$2\pi \int_{-1}^{1} d\mu P_{22}^{l=2}(\mu) \Phi_W(\mu) = 1 \tag{7.31}$$

The attenuation coefficients of the circularly and linearly polarized modes are determined by the smallest roots of equation

$$\det\left([\sigma(1 - a_4(l)) + \sigma_a]\delta_{l,m} - \frac{\epsilon_V}{(2l+1)} (l\delta_{l-1,m} + (l+1)\delta_{l+1,m}) \right) = 0 \tag{7.32}$$

$$\det\left([\sigma(1 - a_{+}^{(2,3)}(l)) + \sigma_a]\delta_{l,m} - \epsilon_W \left(\frac{l^2-4}{l(2l+1)}\delta_{l-1,m} + \frac{4}{l(l+1)}\delta_{l,m} + \frac{(l+1)^2-4}{(l+1)(2l+1)}\delta_{l+1,m} \right)\right) = 0 \tag{7.33}$$

When the optical properties of the medium are known, solutions to Eqs. (7.32), (7.33) can be found numerically with the use of the determinants truncated at some l_{max}.

The attenuation coefficients of basic modes V and W calculated for suspensions of polystyrene and silica particles in water are shown in Fig. 7.1. Absorption was neglected. The values of $a_4(l)$ and $a_+^{(2,3)}(l)$ entering into Eqs. (7.32) and (7.33) were calculated with the Mie theory. The insignificant difference between the results obtained within the l-polynomial approximation for $l = 2$ and $l = 9$ illustrates good convergence of expansions (7.28) and (7.29).

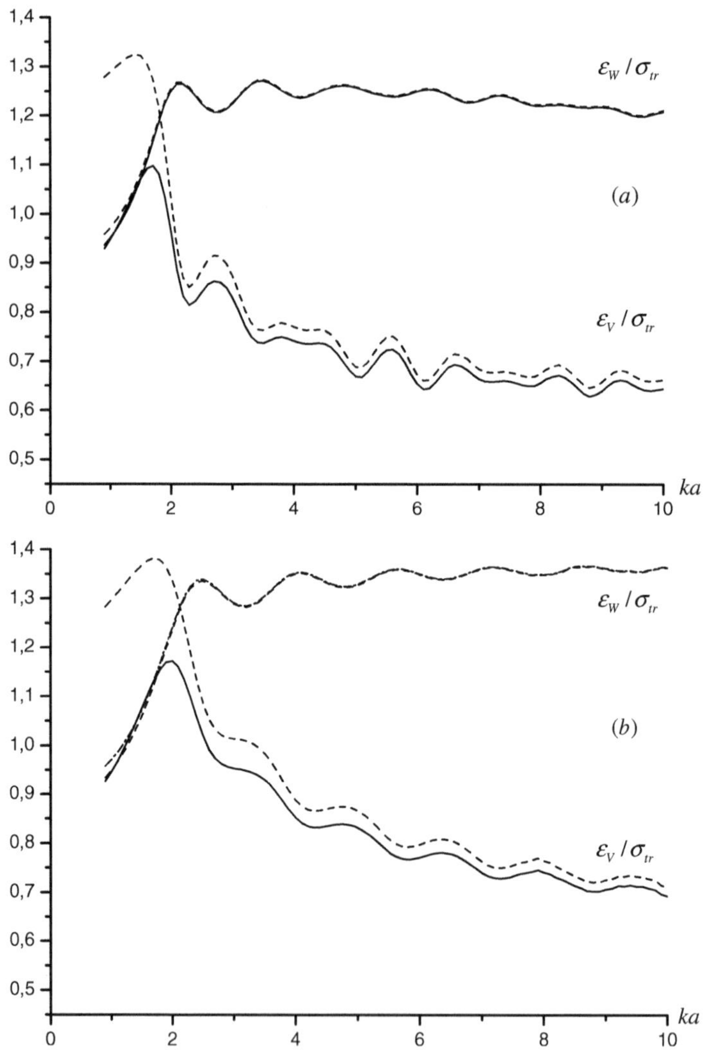

Fig. 7.1. Attenuation coefficients ϵ_W (upper curves) and ϵ_V (lower curves) as a function of radius a of polystyrene (a) and silica (b) microspheres in water. The l-polynomial approximation (solid and dashed curves correspond to 9 and 2 polynomials, respectively, k is the wave number of light in water).

Figure 7.2 illustrates anisotropy of the angular profile of the linearly polarized mode in the asymptotic state. These results were obtained by truncating the expansion in spherical harmonics at $l_{max} = 3$ and 10.

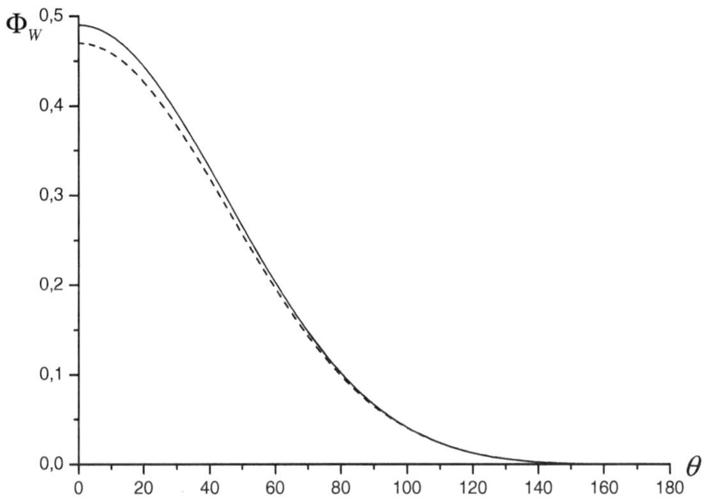

Fig. 7.2. Angular profile of the linearly polarized mode in the asymptotic state. Suspension of polystyrene microspheres in water, $ka = 5$, the l-polynomial approximation (solid and dashed curves correspond to 9 and 2 polynomials, respectively).

7.4 Pulse propagation

An ultrashort pulse of light propagating in a scattering medium experiences multiple scattering events, resulting in broadening of its temporal profile. Early arrival photons of the output pulse propagate along nearly straight lines and can be used to image objects hidden inside the medium (see, e.g., [9–11]). The polarization of such photons is close to the initial one.

From experimental [10, 11] and numerical [26–28] results it follows that polarization of transmitted light is retained over a small interval near $ct = z$. This fact can be used for polarization-gating the ballistic and snake photons with a gate of the order of polarization decay time.

It what follows the propagation of an ultrashort pulse of polarized light through a turbid medium is considered within the basic mode approximation. A simple theoretical model is developed to calculate temporal profiles of the degree of polarization and the depolarization ratio for the pulse transmitted through a turbid slab. It is demonstrated that the degree of polarization of linearly polarized light differs from zero over the initial interval that is of the order of the transport mean free time. In the medium with large inhomogeneities (size a is larger than wavelength λ), the circular polarization exhibits slower decrease with time. The results of calculations are shown to agree with the experimental data and the results of numerical simulation.

Consider the normal incidence of an ultrashort pulse of polarized light on the surface of the medium. The beam spatial width is assumed to be greater than transverse broadening of the beam in the medium.

For early arrival photons of the output pulse that propagate along nearly straight lines, the forward peaked angular distribution is to be expected. Therefore the equations for the basic modes can be transformed within the small-angle approximation (see, e.g., [38, 39, 41]).

The small-angle transfer equations for the intensity has the form [63]

$$\left(\frac{\partial}{\partial z}+\frac{\theta^2}{2}\frac{\partial}{\partial \Delta}\right)\tilde{I}(z,\boldsymbol{\theta},\Delta)=\tilde{I}_{st} \tag{7.34}$$

$$\tilde{I}_{st}=\sigma\int d\boldsymbol{\theta}' a_1(|\boldsymbol{\theta}-\boldsymbol{\theta}'|)\left(\tilde{I}(z,\boldsymbol{\theta}',\Delta)-\tilde{I}(z,\boldsymbol{\theta},\Delta)\right) \tag{7.35}$$

where $\tilde{I}(z,\boldsymbol{\theta},\Delta)=c^{-1}\exp(\sigma_a ct)I(z,\boldsymbol{\theta},\Delta)$, vector $\boldsymbol{\theta}$ denotes the angle between vectors $\mathbf{n}$ and $\mathbf{n}_0$, and $\Delta = ct - z$ is the difference between path length ct and depth z.

The boundary condition for Eq. (7.34) is written as

$$I(z=0,\boldsymbol{\theta},\Delta)=I_0\delta(\Delta)\delta(\boldsymbol{\theta}) \tag{7.36}$$

The small-angle version of Eq. (7.19) for the basic mode of circular polarization can be represented as [64–66]

$$\left(\frac{\partial}{\partial z}+\frac{\theta^2}{2}\frac{\partial}{\partial \Delta}\right)\tilde{V}(z,\boldsymbol{\theta},\Delta)=\tilde{V}_{st}-\frac{\sigma_{dep}^{(V)}}{4\pi}\int d\boldsymbol{\theta}'\tilde{V}(z,\boldsymbol{\theta}',\Delta) \tag{7.37}$$

where $\tilde{V}(z,\boldsymbol{\theta},\Delta)=c^{-1}\exp(\sigma_a ct)V(z,\boldsymbol{\theta},\Delta)$ and

$$\sigma_{dep}^{(V)}=\sigma\int d\mathbf{n}'(a_1(\mathbf{nn}')-a_4(\mathbf{nn}')) \tag{7.38}$$

The expression for $\tilde{V}_{st}$ is obtained from (7.35) by substitution of $\tilde{V}$ for $\tilde{I}$.

The last term on the right-hand side of Eq. (7.37) is responsible for depolarization of circularly polarized light . Such a form of this term results from the assumption that $a_1(\mathbf{nn}')-a_4(\mathbf{nn}')$ is a nearly constant function of the angular variable as compared to V [67].

The boundary condition for Eq. (7.37) is similar to boundary condition (7.36). When the incident light is circularly polarized, $V_0 = I_0$.

Propagation of the basic mode of linear polarization is described by Eq. (7.20). The small-angle approximation can be applied to Eq. (7.20) in the following way. Expansion of the angle-dependent coefficients on the left- and right-hand sides of Eq. (7.20) in terms of small angle θ with allowance for the first nonvanishing terms yields the following equation [64–66]:

$$\left(\frac{\partial}{\partial z}+\frac{\theta^2}{2}\frac{\partial}{\partial \Delta}\right)\tilde{W}(z,\boldsymbol{\theta},\Delta)=$$
$$\tilde{W}_{st}-\frac{\sigma_{tr}\theta^2}{2}\tilde{W}(z,\boldsymbol{\theta},\Delta)-\frac{\sigma_{dep}^{(W)}}{4\pi}\int d\boldsymbol{\theta}'\tilde{W}(z,\theta',\Delta) \tag{7.39}$$

where $\tilde{W}(z, \boldsymbol{\theta}, \Delta) = c^{-1} \exp(\sigma_a ct) W(z, \boldsymbol{\theta}, \Delta)$ and

$$\sigma_{dep}^{(W)} = \sigma \int d\mathbf{n}' (a_1(\mathbf{nn}') - a_+^{(2,3)}(\mathbf{nn}')) \tag{7.40}$$

The expression for $\tilde{W}_{st}$ is derived from Eq. (7.35) by substitution of $\tilde{W}$ for $\tilde{I}$.

Two additional terms on the right-hand side of Eq. (7.39) result from the difference between functions $a_+^{(2,3)}(\mathbf{nn}') \exp(2i\chi_+)$ and $a_1(\mathbf{nn}')$, entering into Eqs. (7.20) and (7.18), respectively, and are responsible for depolarization of linearly polarized light .

In accordance with results [40, 41], we can write

$$\begin{aligned} a_+^{(2,3)} \exp(2i(\chi_+ - \psi)) &= a_1 - a_1[1 - \exp(2i(\chi_+ - \psi))] - (a_1 - a_+^{(2,3)}) \\ &\quad + (a_1 - a_+^{(2,3)})[1 - \exp(2i(\chi_+ - \psi))] \end{aligned} \tag{7.41}$$

Each term in Eq. (7.41) has a certain physical meaning. In particular, the second and third terms are responsible for geometrical and dynamical depolarization, respectively. The fourth term describes the combined effect of both types of depolarization. In highly forward scattering media, the second and third terms in Eq. (7.41) are less than the first one. The fourth term is the least in Eq. (7.41) as the term of higher order in the single-scattering angle. Neglecting the last term in Eq. (7.41), we obtain

$$\begin{aligned} &\sigma \int d\mathbf{n}' a_+^{(2,3)}(\mathbf{nn}') \exp(2i(\chi_+ - \psi)) \tilde{W}(z, \mathbf{n}', t) \\ &\approx \sigma \int d\mathbf{n}' a_1(\mathbf{nn}') \tilde{W}(z, \mathbf{n}', t) - 2\sigma_{tr} \frac{1 - \cos\theta}{1 + \cos\theta} \tilde{W}(\mathbf{r}, \mathbf{n}, t) \\ &\quad - \frac{\sigma_{dep}^{(W)}}{4\pi} \int d\mathbf{n}' \tilde{W}(z, \mathbf{n}', t) \end{aligned} \tag{7.42}$$

The second term in the right-hand side of Eq. (7.42) is derived under the assumption that the scattering phase function is highly forward peaked, and

$$\sigma \int d\mathbf{n}' a_1(\mathbf{nn}')(1 - \exp(2i(\chi_+ - \psi))) \approx 2\sigma_{tr} \frac{1 - \cos\theta}{1 + \cos\theta} \tag{7.43}$$

Within the small-angle approximation, Eq. (7.43) takes the form

$$2\sigma_{tr} \frac{1 - \cos\theta}{1 + \cos\theta} \cdot \tilde{W} \approx \frac{\sigma_{tr}\theta^2}{2} \cdot \tilde{W}$$

Figure 7.3 illustrates the accuracy of approximation (7.43). The term under consideration describes depolarization of light due to the Rytov effect [38–41].

The last term in Eqs. (7.39) and (7.42) describes the dynamic depolarization and results from the following approximation:

$$\sigma \int d\mathbf{n}' (a_1(\mathbf{nn}') - a_+(\mathbf{nn}')) \tilde{W}(z, \mathbf{n}', t) \approx \frac{\sigma_{dep}^{(W)}}{4\pi} \int d\mathbf{n}' \tilde{W}(z, \mathbf{n}', t) \tag{7.44}$$

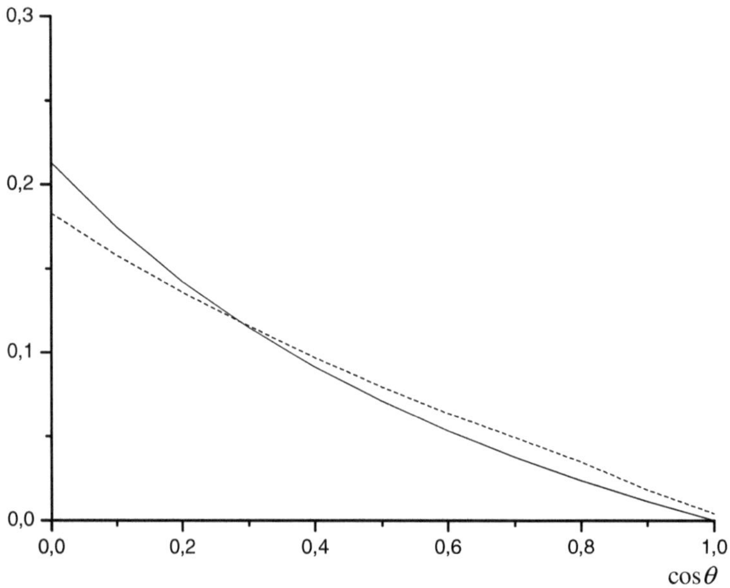

Fig. 7.3. Coefficient of geometrical depolarization for aqueous suspension of polystyrene microspheres ($ka = 5$, a is the radius of scattering particles and k is the wave number of light in water) calculated by expression $2\frac{\sigma_{tr}}{\sigma}\frac{1-\cos\theta}{1+\cos\theta}$ (solid line) and by formula $\int d\mathbf{n}' a_1(\mathbf{nn}')(1-\exp(2\mathrm{i}(\chi_+ - \psi)))$ (dashed line).

where coefficient $\sigma_{dep}^{(W)}$ is defined by Eq. (7.40). Equality (7.44) is based on the assumption that the angular dependence of $a_1(\mathbf{nn}') - a_+(\mathbf{nn}')$ is virtually isotropic as compared with $\tilde{W}(\mathbf{r}, \mathbf{n}', t)$.

The expressions for the additional terms in Eqs. (7.39) and (7.42) should be considered as a reasonable approximation for modeling multiple scattering of polarized light in tissue phantoms (e.g., in aqueous suspension of polystyrene particles).

Equations (7.34), (7.37) and (7.39) enable us to express approximately basic polarization modes V and W in terms of the intensity [64–66].

To model scattering in tissue-like media, we take advantage of $\tilde{I}_{st}$ within the Fokker–Planck approximation [47] (see, also [68, 69])

$$\tilde{I}_{st} = \frac{\sigma_{tr}}{2} \cdot \frac{1}{\theta}\frac{\partial}{\partial\theta}\theta\frac{\partial}{\partial\theta}\tilde{I} \tag{7.45}$$

where σ_{tr} is the transport scattering coefficient. This simple approximation is best suited to analytical treatment and semi-quantitative description of wave propagation and depolarization in media with highly forward scattering [70, 71].

Within the framework of Eq. (7.45) the relationships between the polarization modes and the intensity can be written as [64]

$$V(z, \theta, \Delta) = \exp\left(-2\sigma_{dep}^{(V)}\Delta\right) I(z, \theta, \Delta) \tag{7.46}$$

$$W(z, \theta, \Delta) = \exp\left(-\left(\sigma_{tr} + 2\sigma_{dep}^{(W)}\right)\Delta\right) I(z, \theta, \Delta) \tag{7.47}$$

In accordance with these relationships, photons that propagate along a straight line ($ct = z$ and $\Delta = 0$) are completely polarized.

The linear polarization is retained only over a short initial interval that is no longer than the transport mean free time l_{tr}/c ($l_{tr} = \sigma_{tr}^{-1}$ is the transport mean free path). In the case of the propagation through thick samples ($z \gg l_{tr}$), the initial polarization is retained only for the early arrival photons that are scattered through small angles, the mean square of the multiple-scattering angle at depth z and time instant t [72], $\langle\theta^2\rangle_{z,\Delta} = 4\Delta/z < (\sigma_{tr}z)^{-1} \ll 1$. The spatial diffusion of radiation ($\Delta \gg z$) is accompanied by complete depolarization.

Depolarization of the circularly polarized light as a rule occurs slower. The depolarization time is of the order of $l_{dep}^{(V)}/c$, where $l_{dep}^{(V)} = (\sigma_{dep}^{(V)})^{-1}$ is the mean free path between the depolarizing collisions. For media with large inhomogeneities, as a rule, inequality $\sigma_{dep}^{(V)} \ll \sigma_{tr}$ is valid (see, e.g., [46]). Thus, even the diffusive radiation can be polarized. This effect is manifested in both transmission and reflection (see, e.g., [23, 46, 51–54]).

In the diffusive limit ($\Delta \gg z$) Eq. (7.46) should be replaced by

$$V(z,\theta,\Delta) = \exp\left(-\sigma_{dep}^{(V)}ct\right) I(z,\theta,\Delta) \tag{7.48}$$

Eqs. (7.46) and (7.48) can be combined into a single formula,

$$\tilde{V}(z,\theta,\Delta) = \exp\left(-\sigma_{dep}^{(V)} \cdot ct(1 - \langle\cos\theta\rangle_{z,t})\right) \tilde{I}(z,\theta,\Delta) \tag{7.49}$$

where $\langle\cos\theta\rangle_{z,t}$ is mean cosine of multiple-scattering angle, which can be approximated by

$$\langle\cos\theta\rangle_{z,t} = \exp\left(-\frac{2\Delta}{z}\right) \tag{7.50}$$

From Eqs. (7.47) and (7.49) it follows that the degree of polarization of the linearly and circularly polarized beams can be described by

$$P_L = \frac{W}{I} = \exp\left(-\left(\sigma_{tr} + 2\sigma_{dep}^{(W)}\right)\Delta\right) \tag{7.51}$$

$$P_C = \frac{V}{I} = \exp\left(-\sigma_{dep}^{(V)} \cdot ct(1 - \langle\cos\theta\rangle_{z,t})\right) \tag{7.52}$$

The depolarization ratios can be easily found from these equations. For example, the ratio of two cross-polarized components of intensity is equal to

$$D = \frac{I_\perp}{I_\parallel} = \frac{I - W}{I + W} = \tanh\left[\frac{1}{2}\left(\sigma_{tr} + 2\sigma_{dep}^{(W)}\right)\Delta\right] \tag{7.53}$$

Coefficients σ_{tr}, $\sigma_{dep}^{(W)}$ and $\sigma_{dep}^{(V)}$, entering into Eqs. (7.51) and (7.52), depend on the relative refractive index, size and shape of scattering inhomogeneities. For spherical particles, coefficients σ_{tr} and $\sigma_{dep}^{(W)} = \sigma_{dep}^{(V)}/2$ can be calculated with the Mie theory or can be taken from experimental data (coefficients σ_{tr}, $\sigma_{dep}^{(W)}$ and $\sigma_{dep}^{(V)}$ for several media can be found in [40, 41], see also Fig. 7.4). Coefficients σ_{tr} and $\sigma_{dep}^{(W)}$ satisfy inequality $\sigma_{dep}^{(W)} < \sigma_{tr}/2$.

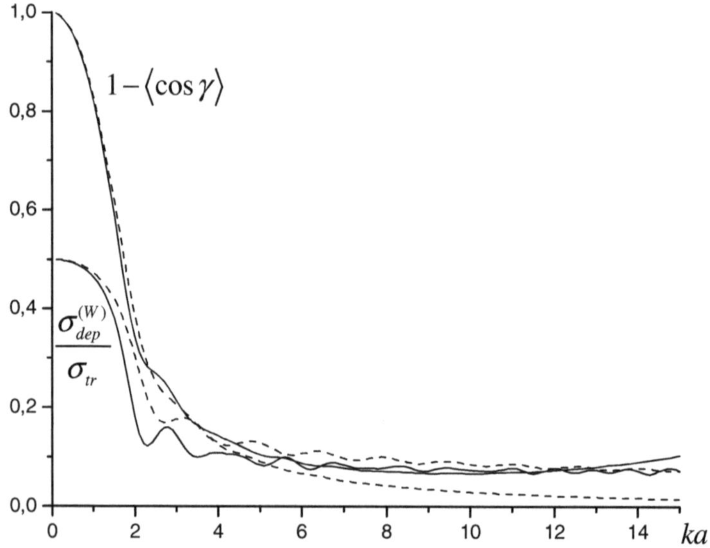

Fig. 7.4. Dependence of $\sigma_{dep}^{(W)}/\sigma_{tr}$ and $1 - \langle\cos\gamma\rangle$ on parameter $x = ka$ (a is the radius of scattering particles, k is the wave number in the medium) for water suspensions of polystyrene (solid curves) and silica (dashed curves) microspheres.

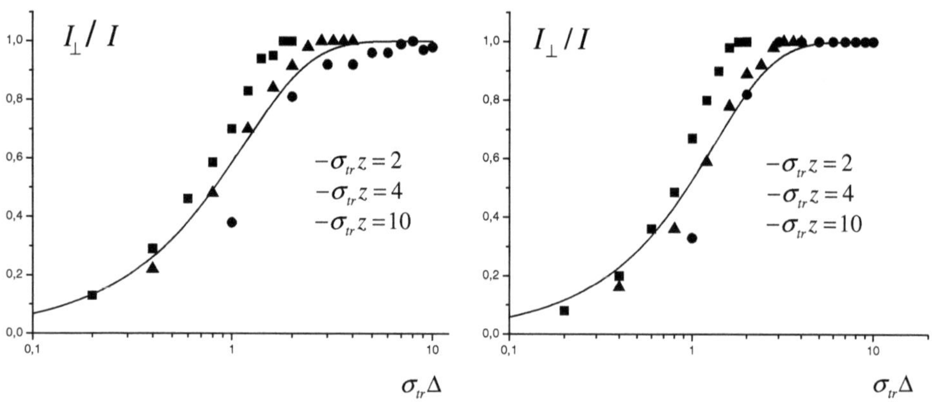

Fig. 7.5. Depolarization ratio $I_\perp/I_\parallel$ as a function of 'transport optical delay' $\sigma_{tr}\Delta$ for water suspension of 0.3- (a) and 0.99- (b) μm diameter polystyrene microspheres, $\lambda = 633$ nm. Solid lines are the results of our calculations with Eq. (7.53). Symbols are the results of Monte Carlo simulation [26]. The values of $\sigma_{dep}^{(W)}$ are determined with the Mie formulas, $\sigma_{dep}^{(W)}/\sigma_{tr} = 0.17$ (a) and 0.086 (b).

The theoretical results presented above can be compared to the experimental and numerical data obtained in [10, 11, 26–28] which describe transmission of polarized pulses through optically thick turbid samples.

According to Eq. (7.53) depolarization of the linearly polarized wave depends only on delay Δ and is unaffected by the slab thickness z. This conclusion is confirmed by comparison with data of Monte Carlo simulation for aqueous suspension of polystyrene microspheres [26]. The depolarization ratio was calculated in [26] for different values of the transport optical thickness and presented as a function of the normalized delay Δ/z. When going from Δ/z to delay Δ, the data of simulation take the form shown in Fig. 7.5. Within the accuracy of the analysis, data [26] tend to a universal function of delay Δ.

The dependence of the depolarization ratios $I_\perp/I_\parallel$ and I_+/I_- on delay Δ, as applied to linearly and circularly polarized waves, is illustrated in Fig. 7.6, where our results and the results of numerical integration of the vector transfer equation with the discrete ordinate method [28] are presented. The figure clearly shows the difference between the linearly and circularly polarized waves in the depolarization rates.

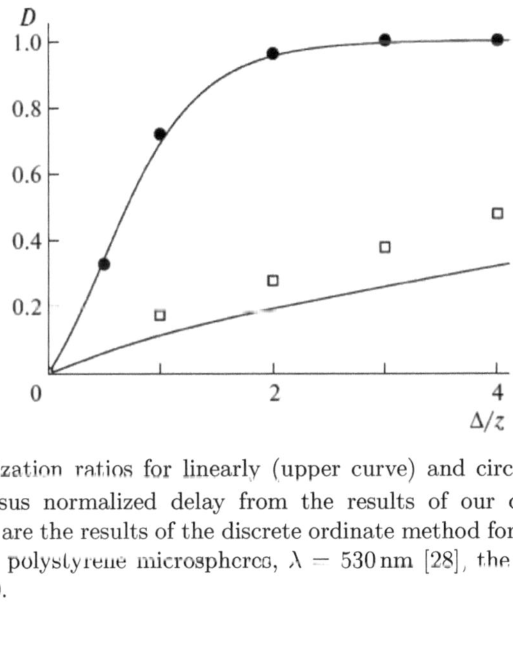

Fig. 7.6. Depolarization ratios for linearly (upper curve) and circularly (lower curve) polarized light versus normalized delay from the results of our calculation ($\sigma_{dep}^{(V)} = 0.15 \cdot \sigma_{tr}$). Symbols are the results of the discrete ordinate method for aqueous suspension of 2.0 μm diameter polystyrene microspheres, $\lambda = 530$ nm [28], the optical thickness of the sample $\sigma z = 10$.

For ratio I_+/I_-, the difference between our results and the data of numerical calculations [28] may be due to a low accuracy of calculations [28]. As analysis shows the numerical results [28] contradict equality $I_\parallel + I_\perp = I_+ + I_-$. Inequality $I_\parallel + I_\perp > I_+ + I_-$ gets more clearly defined as Δ increases. This is evidently demonstrated by the time-dependence of P_C at large delays, $\Delta > z$, where the exponential decay of P_C is changed by the anomalous Gaussian dependence (see Fig. 6 in [28]).

The results of our calculations are also in agreement with data of experimental measurements [11]. Data [11] were obtained for the different samples identical in thickness and transport mean free path l_{tr}. Thus the values of the characteristic time of depolarization in the samples should to be appear of the same order of magnitude. This is illustrated in Fig. 7.7, where data [11] for the degree of linear polarization are compared to our calculations with Eq. (7.51).

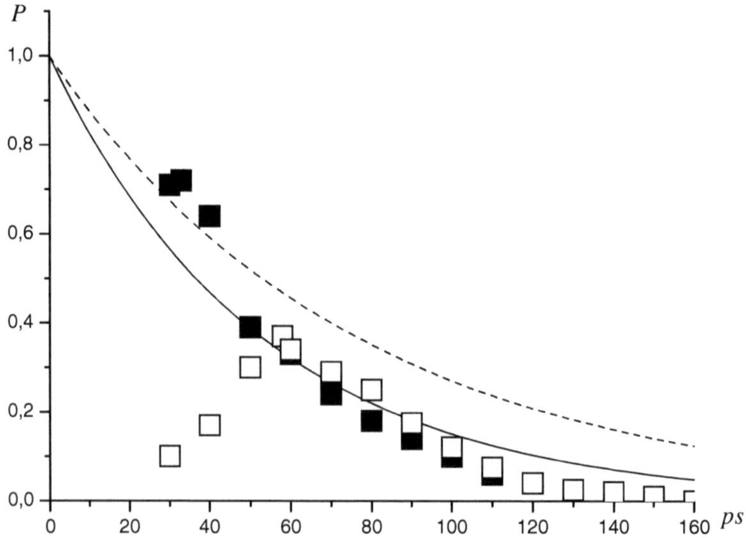

Fig. 7.7. Temporal profiles of the degree of linear polarization for aqueous suspensions of $0.2\,\mu$m (closed squares) and $1.07\,\mu$m (open squares) polystyrene microspheres for a 50 mm thick sample, transport mean free path $l_{tr} = 20\,$mm, wavelength $\lambda = 532\,$nm [11]. Curves are the results of our calculations for $\sigma_{dep}^{(W)}/\sigma_{tr} = 0.34$ and 0.08, respectively.

7.5 Model of depolarization

Thus, we arrive at the following model of depolarization within the basic mode approximation. The polarization state of circularly and linearly polarized radiation can be described by basic modes I, V, and W. The values of polarization modes V and W differ from intensity I only by factors that are responsible for attenuation in the domain of delays $\Delta = ct - z$. The corresponding attenuation factors are determined by expressions (7.47) and (7.46). In going to continuous (CW) illumination, the expressions for the basic modes should be integrated with respect to delay Δ:

$$I(z) = \int_0^\infty d\Delta\; I(z,\Delta) = \exp\left(-\sigma_a z\right) \tilde{I}(z,\sigma_a) \tag{7.54}$$

$$V(z) = \int_0^\infty d\Delta\; V(z,\Delta) = \exp\left(-\sigma_a z\right) \tilde{I}(z,\sigma_a + 2\sigma_{dep}^{(V)}) \tag{7.55}$$

$$W(z) = \int_0^\infty d\Delta\; W(z,\Delta) = \exp\left(-\sigma_a z\right) \tilde{I}(z,\sigma_a + \sigma_{tr} + 2\sigma_{dep}^{(W)}) \tag{7.56}$$

Here, we factor out the attenuation due to absorption of the straightforward propagating photons and introduce the modified intensity $\tilde{I} = \exp\left(\sigma_a z\right) I$.

It is instructive to compare Eqs. (7.55) and (7.56) with the phenomenological model of depolarization proposed in [20].

According to [20], the linear polarization is described by the quantity that is similar to W. This quantity is obtained by integrating the product of the path distribution $P(s)$ and the attenuation factor $\exp(-\sigma_L s)$ over paths s. Distribution $P(s)$ is actually the transmitted pulse profile. Coefficient σ_L is assumed to describe depolarization per unit path length. Thus, our results (7.55), (7.56) and the result of [20] differ essentially in the attenuation factors. In contrast to the corresponding result of [20], our factor $\exp(-(\sigma_{tr} + 2\sigma_{dep}^{(W)})\Delta)$ describes the attenuation as a function of delay $\Delta = ct - z$ and not of path ct, and equals unity at $ct = z$ (i.e., the straightforward propagating photons retain their initial polarization).

Let us apply our model of depolarization to studying the degree of polarization under conditions of CW illumination. To determine the intensity entering into Eqs. (7.54)–(7.56) we take advantage of the results obtained within the small-angle Fokker-Planck approximation [69]. This approximation takes into account highly forward scattering ($1 - \langle\cos\gamma\rangle \ll 1$) of light in biological tissues and tissue-like media [70, 71] and enables us to elaborate simple semi-quantitative model of propagation of polarized light .

In accordance with [69] (see also [39]), the intensity of a narrow beam can be presented in the analytical form

$$\tilde{I}(z,\boldsymbol{\rho},\boldsymbol{\theta}) = \frac{1}{\pi^2 A_0(z)\Delta^2(z)} \exp\left\{-\frac{1}{\Delta^2(z)}(A_1(z)\rho^2 - 2A_2(z)\boldsymbol{\rho}\boldsymbol{\theta} + A_3(z)\theta^2)\right\} \tag{7.57}$$

where vectors $\boldsymbol{\theta} = (\theta_x, \theta_y)$ and $\boldsymbol{\rho} = (x, y)$ characterize the direction of propagation of scattered photons and their transverse displacement from the beam axis, respectively,

$$A_0(z) = \cosh(z\sqrt{\sigma_a\sigma_{tr}}), \quad A_1(z) = 2\sqrt{\frac{\sigma_{tr}}{\sigma_a}}\tanh(z\sqrt{\sigma_a\sigma_{tr}}) \tag{7.58}$$

$$A_2(z) = \frac{2}{\sigma_a}\left(1 - \frac{1}{\cosh(z\sqrt{\sigma_a\sigma_{tr}})}\right), \quad A_3(z) = \frac{2z}{\sigma_a}\left(1 - \frac{\tanh(z\sqrt{\sigma_a\sigma_{tr}})}{z\sqrt{\sigma_a\sigma_{tr}}}\right) \tag{7.59}$$

$$\Delta^2(z) = A_1(z)A_3(z) - (A_2(z))^2, \tag{7.60}$$

Substituting Eq. (7.57) into Eqs. (7.55) and (7.56) we can obtain the expressions for the degree of polarization.

Under the conditions that are typical for many experiments (see, e.g., [1,9,20]) the sample surface is illuminated by a narrow beam of light, the beam axis is perpendicular to both boundaries of the sample. The transmitted light is collected behind the sample by the on-axis detector with a narrow field of view. In the case of 'a narrow beam geometry' (see, e.g., [20,27]), the area of the detector is assumed to be much less than the cross-section of the transmitted beam. By contrast, the case where the transmitted light is collected from the whole cross-section of the beam corresponds to 'a wide beam geometry' (see, e.g., [13, 24]).

As applied to transmission through a medium with no absorption ($\sigma_a = 0$), the degree of polarization can be written in the form

$$P = \frac{\xi}{\sinh\xi} \tag{7.61}$$

where ξ is equal to

$$\xi_L = z\sqrt{(\sigma_{tr} + 2\sigma_{dep}^{(W)})\sigma_{tr}} \tag{7.62}$$

$$\xi_C = z\sqrt{2\sigma_{dep}^{(V)}\sigma_{tr}} \tag{7.63}$$

for linearly and circularly polarized light , respectively. Eq. (7.61) is valid for a wide beam geometry and corresponds to the conditions of experiments [53–56].

For a narrow beam geometry, the depth dependence of the degree of polarization is described by expression

$$P = \frac{1}{12}\frac{\xi^4}{\xi\sinh\xi - 2\cosh\xi + 2} \tag{7.64}$$

where ξ is determined by Eqs. (7.62) or (7.63).

The validity of our theoretical approach can be tested by comparison with experimental data [20, 53–56].

The results of measurements [20, 53, 54] of the degree of linear polarization in aqueous suspensions of polystyrene microspheres and the results of our calculations are illustrated in Figs. 7.8 and 7.9. The coefficients entering into Eq. (7.62) were calculated with the Mie theory. For the conditions of experiments [53, 54] (see Fig. 7.8), the values of ratio $\sigma_{dep}^{(W)}/\sigma_{tr}$ are equal to 0.076 and 0.08, respectively. The theoretical curves corresponding to such close values of $\sigma_{dep}^{(W)}/\sigma_{tr}$ appear to be indistinguishable. For the same reason, it is hard to discriminate the theoretical curves shown in Fig. 7.9.

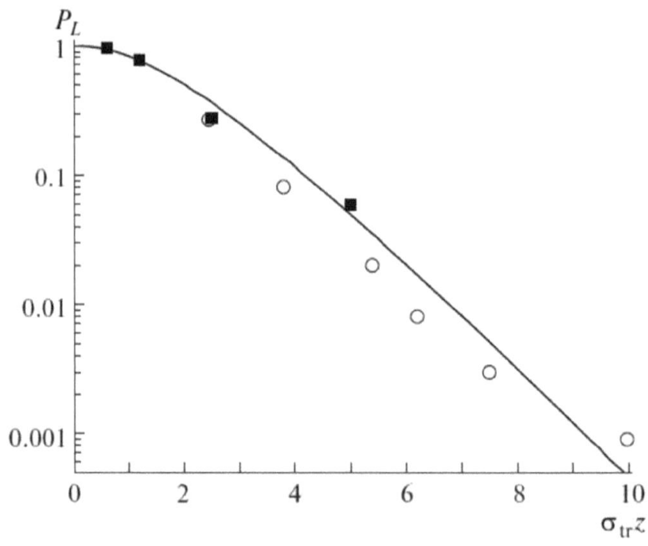

Fig. 7.8. Degree of linear polarization in aqueous suspension of polystyrene particles as a function of the transport optical thickness. Solid line is the result of calculations with Eqs. (7.61) and (7.62). Symbols ○ and ■ are the experimental data for 1.05 μm diameter particles ($\lambda = 670$ nm) [53] and 1.072 μm diameter particles ($\lambda = 633$ nm) [54], respectively. A wide beam geometry.

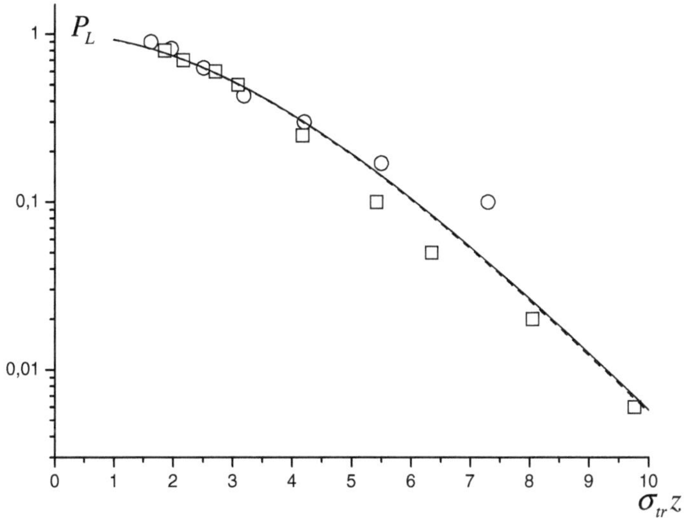

Fig. 7.9. Degree of linear polarization in aqueous suspension of polystyrene microspheres as a function of the transport optical thickness of the sample. Squares and circles are experimental data [20] for 2.19 μm ($\lambda = 532$ nm, $\sigma_{dep}^{(W)} = 0.070 \cdot \sigma_{tr}$) and 1.07 μm ($\lambda = 514$ nm, $\sigma_{dep}^{(W)} = 0.075 \cdot \sigma_{tr}$) diameter microspheres, respectively. Solid and dashed lines are the results of our calculations for a narrow beam geometry (see Eq. (7.64)).

As follows from Figs. 7.8 and 7.9 our theoretical formulae (7.61), (7.64) are in good agreement with data [20, 53, 54].

Matching values of $\sigma_{dep}^{(W)}/\sigma_{tr}$, we can fit the theoretical curves to the corresponding experimental data. The results of application of such a procedure to data of measurements [54–56] are shown in Fig. 7.10. Experimental data [54–56] describe the depth-dependence of the degree of polarization in biological tissues and tissue phantoms and enable us to estimate the values of the corresponding depolarization coefficients. As follows from the results of the fitting procedure, the depolarization ratio $\sigma_{dep}^{(W)}/\sigma_{tr}$ ranges from 0.1 (polystyrene microsphere suspension, arterial tissue) to 0.75 (Intralipid, myocardial tissue). For the polystyrene microspheres, the restored value agrees well with the calculated one (see Figs. 7.8 and 7.9).

For circularly polarized light , simple relations (7.61) and (7.63) are also in agreement with experimental data. Comparison between the measured [53] and calculated degree of circular polarization is illustrated in Fig. 7.11. Coefficients σ_{tr} and $\sigma_{dep}^{(V)}$ entering into Eq. (7.63) were determined with the Mie theory. The experimental setup of [53] corresponds to a wide beam geometry. Aqueous suspension of polystyrene microspheres were used as a scattering sample.

In tissue phantoms (suspensions of polystyrene and silica microspheres [24, 25, 53, 54], Intralipid [54]) circularly polarized light depolarizes more slowly than light with linear polarization (compare, e.g., Figs. 7.8 and 7.11). By contrast the rate of depolarization in actual tissues can be higher for circularly polarized light . This is illustrated in Fig. 7.12 where experimental data [54–56] and the corresponding fitting curves are presented.

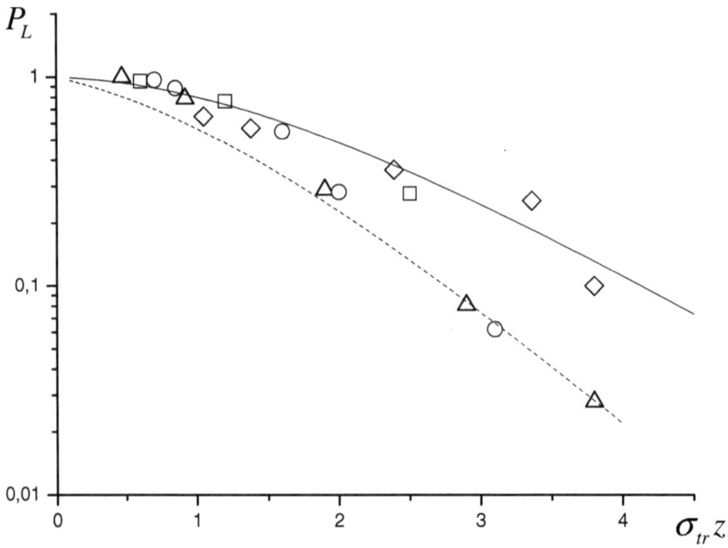

Fig. 7.10. Degree of linear polarization in various media as a function of the transport optical thickness of the sample. Symbols (□ – suspension of 1.07 μm diameter polystyrene microspheres in water, ◇ – arterial tissue, ○ – Intralipid suspension, △ – myocardial tissue) are experimental data ($\lambda = 633\,\text{nm}$) [54–56]. Fitting curves are the results of our calculations with Eqs. (7.61) and (7.62) for $\sigma_{dep}^{(W)} = 0.1 \cdot \sigma_{tr}$ and $\sigma_{dep}^{(W)} = 0.75 \cdot \sigma_{tr}$.

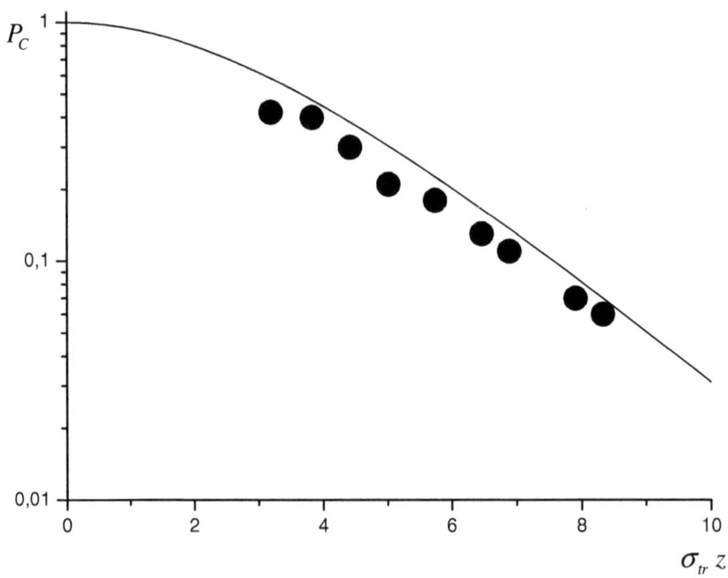

Fig. 7.11. Degree of circular polarization in aqueous suspension of polystyrene microspheres as a function of the transport optical thickness. Symbols are experimental data [53] for $ka = 6.43$. Solid line is the result of our calculations for a wide beam geometry (see Eqs. (7.61) and (7.63), $\sigma_{dep}^{(V)} = 0.176 \cdot \sigma_{tr}$).

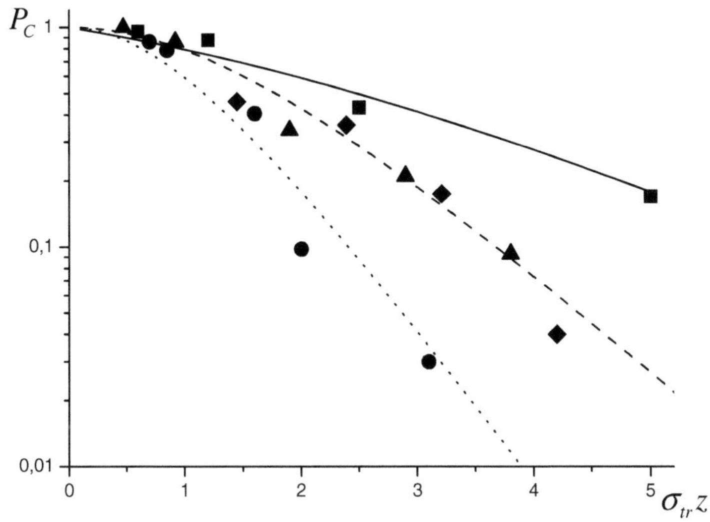

Fig. 7.12. Degree of circular polarization in various media as a function of the transport optical thickness of the sample. Symbols (■ – suspension of 1.07 μm diameter polystyrene microspheres in water, ♦ – arterial tissue, • – Intralipid suspension, ▲ – myocardial tissue) are experimental data ($\lambda = 633$ nm) [54–56], fitting curves are the results of our calculations with Eqs. (7.61) and (7.63) for $\sigma_{dep}^{(V)} = 0.2 \cdot \sigma_{tr}$ (polystyrene microspheres), for $\sigma_{dep}^{(V)} = 0.75 \cdot \sigma_{tr}$ (arterial tissue and Intralipid suspension) and $\sigma_{dep}^{(V)} = 1.75 \cdot \sigma_{tr}$ (myocardial tissue).

7.6 Polarization-difference imaging through highly scattering media

The degree of polarization depends on the photon path length in the medium. The quasi-straightforward propagating photons pass the shortest path and are the least depolarized. This feature underlies polarization-difference imaging of objects hidden inside turbid media (see, e.g., [13–16,19,23–25]). In this technique the image-bearing component of light is extracted by subtracting the detected cross-polarized signal from the co-polarized one. The feasibility of such a technique depends on the depolarization characteristics of the medium, which, in their turn, are influenced by concentration, sizes and shape of scattering inhomogeneities, their refractive index.

To investigate potentialities of the polarization-gated imaging of objects hidden in highly scattering media, we take advantage of the model of depolarization developed above within the basic mode approximation. This model has been validated by comparison with data of experiments and numerical simulation and can be applied to describe the results of polarization-difference transillumination of tissue-like phantoms.

As before, we restrict our consideration by the Fokker–Planck model of scattering. Within the framework of this model, it takes only a few quantities (transport scattering coefficient σ_{tr} , absorption coefficient σ_a and depolarization coefficients $\sigma_{dep}^{(W)}$, $\sigma_{dep}^{(V)}$) to calculate the image of an object embedded in the scattering sample.

7.6.1 General relations. Edge spread function

The spatial resolution and the image contrast available by polarization-gated transillumination of a tissue-equivalent phantom (a turbid medium with an inserted obstacle) can be evaluated from the line scans across the tested sample. We start our analysis with a completely absorbing object embedded in a highly scattering sample.

To describe the shadow from the object, the so-called edge-spread function is convenient to use. In accordance with definition [13,20,29], the edge-spread function describes the spatial distribution of radiation passing by the absorbing half-plane edge versus the position of the source-detector axis (see Fig. 7.13).

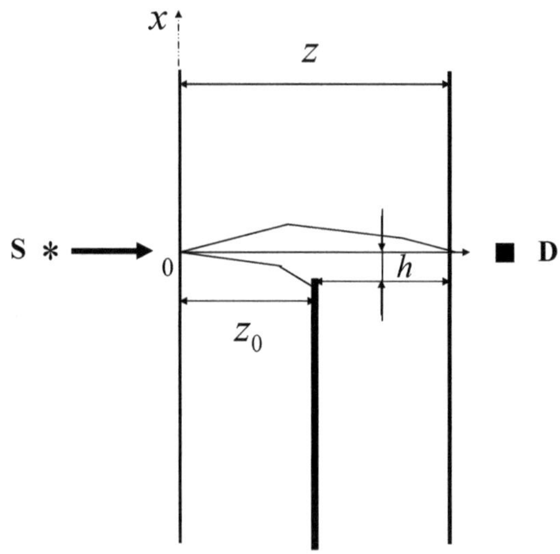

Fig. 7.13. Source-detector geometry for optical scanning of a scattering sample containing an absorbing half-plane.

Let us consider a scattering slab with an absorbing half-plane ($x < 0$). The slab is assumed to be illuminated by a collimated narrow beam. The source-detector axis is perpendicular to both boundaries of the slab and the half-plane. The transmitted light is collected behind the sample by the on-axis detector with a narrow field of view. Such conditions are typical for many experiments on imaging through turbid tissues (see, e.g., [1, 9, 13, 20, 24, 25]).

For a δ-pulsed source, the intensity of light passing by the inhomogeneity (the distance between the absorbing half-plane and the input boundary of the slab is assumed to be equal z_0) can be written in the form [65, 66]

$$E(z,h,t) = \int_0^t dt' \int_h^\infty dx' \int_{-\infty}^\infty dy' \int d\boldsymbol{\theta}'$$
$$G(z,\boldsymbol{\rho}=0,\boldsymbol{\theta}=\boldsymbol{\theta}_0,t|z_0,\boldsymbol{\rho}',\boldsymbol{\theta}',t')I(z_0,\boldsymbol{\rho}',\boldsymbol{\theta}',t') \qquad (7.65)$$

where $G(z, \boldsymbol{\rho}, \boldsymbol{\theta}, t|z_0, \boldsymbol{\rho}', \boldsymbol{\theta}', t')$ and $I(z, \boldsymbol{\rho}, \boldsymbol{\theta}, t)$ are the Green function of the time-dependent (or nonstationary) radiative transfer equation and the intensity of light in the medium with no inhomogeneity, $\boldsymbol{\theta}_0$ is the initial direction of the beam propagation (in our case $\boldsymbol{\theta}_0 = 0$), $\boldsymbol{\rho} = (x, y)$ is the transverse displacement from the beam axis, h is the distance between the source-detector axis and the half-plane edge ($h > 0$ corresponds to the shadow region). The intensity $I(z, \boldsymbol{\rho}, \boldsymbol{\theta}, t)$ entering into Eq. (7.65) can be expressed in terms of the Green function by relation

$$I(z, \boldsymbol{\rho}, \boldsymbol{\theta}, t) = I_0 G(z, \boldsymbol{\rho}, \boldsymbol{\theta}, t|z' = 0, \boldsymbol{\rho}' = 0, \boldsymbol{\theta}' = \boldsymbol{\theta}_0, t' = 0) \tag{7.66}$$

where I_0 is total flux of the incident radiation. Eq. (7.65) is valid under the assumption that the projection of the photon velocity onto z-axis does not change its sign. Therefore, Eq. (7.65) is fully suitable for calculating the contribution from the snake photons to the intensity.

With generalization of Eq. (7.65), the image of an arbitrary-shaped screen can be described by the following formula:

$$E(z, \mathbf{h}, t) = \int_0^t dt' \left(\int_{-\infty}^{\infty} dx' \int_{-\infty}^{\infty} dy' - \int_S dx' dy' \right)$$
$$\int d\boldsymbol{\theta}' G(z, \boldsymbol{\rho} = 0, \boldsymbol{\theta} = \boldsymbol{\theta}_0, t|z_0, \boldsymbol{\rho}', \boldsymbol{\theta}', t') I(z_0, \boldsymbol{\rho}', \boldsymbol{\theta}', t') \tag{7.67}$$

where integration is carried out over the whole plane excepting the obstacle area, vector $\mathbf{h}$ denotes the position of the source-detector axis in the (x, y)-plane.

As follows from Eqs. (7.65) and (7.67) calculations of the edge-spread function reduce to searching for the Green function of the radiative transfer equation for a homogeneous scattering medium.

Within the small-angle Fokker-Planck approximation, the explicit analytical expression for the Green function is given by [65, 66]

$$G(z, \boldsymbol{\rho}, \boldsymbol{\theta}, \Delta|z, \boldsymbol{\rho}', \boldsymbol{\theta}', \Delta') = \frac{\exp(-\sigma_a(z - z'))}{2\pi \mathrm{i}}$$
$$\cdot \int_{-\mathrm{i}\infty}^{\mathrm{i}\infty} dp \frac{\exp(p(ct - (z - z')))}{\pi a(z - z')} \int \frac{d\mathbf{q}}{(2\pi)^2} \int \frac{d\mathbf{k}}{(2\pi)^2} \int \frac{d\mathbf{k}'}{(2\pi)^2}$$
$$\cdot \exp\Big(\mathrm{i}\mathbf{q}(\boldsymbol{\rho} - \boldsymbol{\rho}') + \mathrm{i}\mathbf{k}\boldsymbol{\theta} + \mathrm{i}\mathbf{k}'\boldsymbol{\theta}'$$
$$- \frac{q^2(z - z')}{2\sigma_a} - \frac{\mathbf{q}(\mathbf{k} + \mathbf{k}')}{\sigma_a} - \frac{1}{4}((\mathbf{k}^2 + \mathbf{k}'^2) b(z, z') + 2\mathbf{k}\mathbf{k}' c(z, z'))\Big) \tag{7.68}$$

where

$$a(z, z') = \sinh\left(\sqrt{\sigma_{tr} p} (z - z')\right) \tag{7.69}$$

$$b(z, z') = 2\sqrt{\frac{\sigma_{tr}}{p}} \cosh\left(\sqrt{\sigma_{tr} p} (z - z')\right) \tag{7.70}$$

$$c(z, z') = 2\sqrt{\frac{\sigma_{tr}}{p}} \frac{1}{\cosh\left(\sqrt{\sigma_{tr} p} (z - z')\right)} \tag{7.71}$$

The results (7.65), (7.67), (7.68) can be readily generalized to case of a polarized beam. Within the model put forward above (see Eqs. (7.55) and (7.56)) the edge-spread functions for circularly and linearly polarized light are derived from Eqs. (7.68)–(7.71) by substitution of the corresponding 'effective' absorption coefficients.

Expression (7.68) for the Green function and formulas (7.55) and (7.56) for modes V and W, respectively, are a basis for calculations of the degree of polarization and the intensity of light in a multiply scattering medium.

The results obtained in the small-angle approximation (see Eqs. (7.65), (7.67), and (7.68)) are valid provided that $1 - \langle\cos\theta\rangle_{z,t} < 1$, where $\langle\cos\theta\rangle_{z,t}$ is the mean cosine of the multiple scattering angle at depth z at instant t. Within the Fokker–Planck approximation [72] $1 - \langle\cos\theta\rangle_{z,t} \approx \langle\theta^2\rangle_{z,t}/2 = 2\Delta/z$ and therefore the above-mentioned inequality takes the form $\Delta < z/2$.

7.6.2 Time-resolved polarization imaging

According to [5, 10–12], polarization can be used for time-gating the ballistic and snake photons of the beam transmitted through the scattering sample. However, a duration of the corresponding time-gate has not been determined until recently. To evaluate this quantity, we compare the time-gate dependence of the image profiles obtained, respectively, with the intensity of light and with the difference between the polarized components.

As applied to a pulsed beam, the edge-spread function can be approximated by the expression [66]

$$E(z,h,t) = \frac{1}{2} I(z,t) \left\{ 1 - \operatorname{erf}\left(\sqrt{\frac{z^2}{z_0(z-z_0)\langle\rho^2\rangle_{z,\Delta}}}\, h \right) \right\} \tag{7.72}$$

where $I(z,t)$ is the intensity of radiation at large distances from the absorbing half-plane (or, in the absence of it) and $\langle\rho^2\rangle_{z,\Delta}$ is the mean square of the transverse displacement of the photons, $\operatorname{erf}(x)$ is the error function [73]. Quantities $I(z,t)$ and $\langle\rho^2\rangle_{z,\Delta}$ are given by [72]

$$I(z,t) = \frac{cI_0\exp(-\sigma_a ct)}{(4\pi)^{5/2}(\sigma_{tr}z^2)^3} \cdot \left(\frac{\sigma_{tr}z^2}{\Delta}\right)^{9/2} \cdot \exp\left(-\frac{\sigma_{tr}z^2}{4\Delta}\right) \tag{7.73}$$

$$\langle\rho^2\rangle_{z,\Delta} = 8 \cdot \frac{\Delta^2}{\sigma_{tr}z} \tag{7.74}$$

Eqs. (7.72) and (7.73) can be derived from the general formula (7.68) in the limit $\Delta < \sigma_{tr}z^2$. In this case, we can use the asymptotic expressions for the functions entering into Eq. (7.68) at great p and perform integrating in Eq. (7.68) explicitly (see, e.g., [63,72]). In transmission through an optically thick slab ($\sigma_{tr}z > 1$), results (7.72) and (7.73) are always valid for the snake component, which in accordance with [3,4,9] (see, also [74]) can be defined by condition $\Delta < z/2$.

Eqs. (7.72) and (7.73) are valid provided that the field of view and the radius of the detector are small as compared with $\sqrt{\langle\theta^2\rangle_{z,\Delta}}$ and $\sqrt{\langle\rho^2\rangle_{z,\Delta}}$, respectively. These equations are normalized to unit solid angle and area of the detector.

Eq. (7.72) can be generalized to an obstacle of more complex shape. In particular, the profile of the shadow from absorbing stripe $-a/2 < x < a/2$ (a is the stripe width) is described by

$$\begin{aligned} E(z,h,t) &= \frac{1}{2} I(z,t) \\ &\cdot \left\{ 2 - \operatorname{erf}\left(\sqrt{\frac{z^2}{z_0(z-z_0)\langle \rho^2 \rangle_{z,\Delta}}} \left(\frac{a}{2} + h \right) \right) \right. \\ &\left. - \operatorname{erf}\left(\sqrt{\frac{z^2}{z_0(z-z_0)\langle \rho^2 \rangle_{z,\Delta}}} \left(\frac{a}{2} - h \right) \right) \right\} \end{aligned} \tag{7.75}$$

For a given time-gate, the energy of the detected signal is obtained by integrating Eqs. (7.72) and (7.75) over the corresponding temporal interval [4, 9, 29]:

$$\varepsilon = \int_{t_0}^{t_0+\Delta t} E \, dt \tag{7.76}$$

where t_0 is the onset of detecting and Δt is the time-gating interval. The image contrast is defined as

$$C = \frac{\varepsilon(h \to \infty) - \varepsilon(h = 0)}{\varepsilon(h \to \infty) + \varepsilon(h = 0)} \tag{7.77}$$

where $\varepsilon(h \to \infty)$ is the background signal that is determined from the measurements at large distances from the obstacle and $\varepsilon(h = 0)$ is the signal at the center of the image.

Results (7.65)–(7.75) are generalized to the case of a polarized beam in the following way. For polarized light , we should introduce the edge-spread functions for the polarization modes V and W. Within the framework of the model put forward above, the edge-spread functions for the circularly and linearly polarized light can be derived from Eqs. (7.72) and (7.73) by substitution of the basic polarization modes $V(z,t)$ and $W(z,t)$ for intensity $I(z,t)$.

As follows from Eqs. (7.55) and (7.56), propagation of polarized light is characterized by the effective attenuation coefficient in the domain of temporal delays Δ. Therefore, the edge-spread functions for the polarization modes differ from Eqs. (7.72) and (7.73) only by the corresponding Δ-dependent factors (see Eqs. (7.46) and (7.47)).

In many experiments, detection of the polarization difference is employed (see, e.g., [13–16, 19, 21–25]). This method is based on subtracting of the intensity of the cross-polarized component from the intensity of radiation with the initial polarization. The substraction of one component from the other is assumed to cancel the contribution from the diffusive (or, strongly scattered) photons. Thus, the difference $I_{\parallel} - I_{\perp}$ is governed only by the contribution of the quasi-straightforward propagating photons (i.e., the ballistic and snake photons).

The difference between the intensities of linearly polarized components coincides with the basic mode W [38–41] and therefore, the polarization-difference image of an object immersed in a scattering medium is described by the spatial profile (in particular, by the edge-spread function) of this mode. Correspondingly, the polarization-difference image for circularly polarized light is determined by the spatial profile of mode V.

The edge-spread functions for intensity I and polarization mode W are illustrated in Fig. 7.14. The calculations are performed within the framework of the above-proposed model. From the figure it follows that the edge-spread function for the intensity blurs with increasing Δ. At the same time, the edge-spread function for the linearly polarized mode exhibits minor variations. Therefore, the polarization difference, as compared to the intensity, provides higher spatial resolution (or, sharpness).

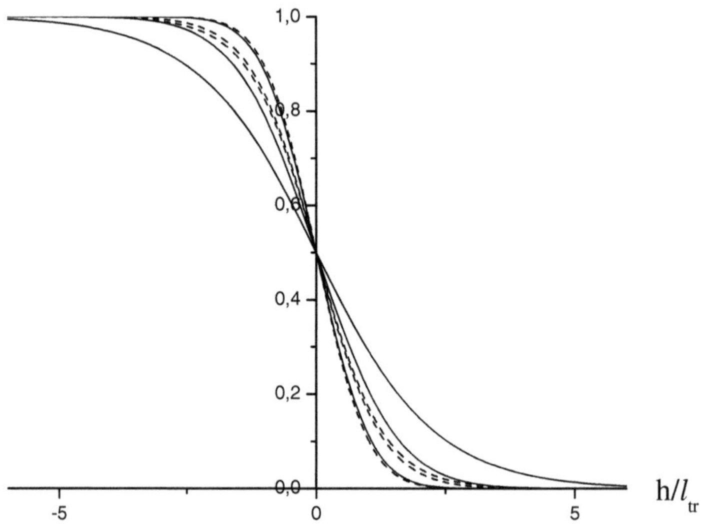

Fig. 7.14. Edge-spread function for intensity I (solid lines) and for linearly polarized mode W (dashed lines). Transport optical thickness $\sigma_{tr}z = 10$, $\sigma_{dep}^{(W)} = 0.1 \cdot \sigma_{tr}$, $z_0 = z/2$. The normalized time-gate $\Delta/z = 0.3,\ 0.5$, and 1.0 (from upper to lower curves at $h < 0$), $t_0 = z/c$.

Combining the corresponding edge-spread functions (see Eq. (7.75)) we can calculate the shadow profile for an absorbing stripe. The image of the stripe is illustrated in Fig. 7.15.

Figs. 7.14 and 7.15 show that the difference between the images obtained with the usual intensity and the polarization difference gets more noticeable as the normalized time-gating interval Δ/z increases.

The effect of polarization on the image contrast (see definition (7.77)) is shown in Fig. 7.16. From the presented results it follows that the polarization enhances distinctly the image contrast at time-gates $\Delta/z > 0.4$–0.5. For shorter time-gating intervals, the image contrast for the intensity does not differ from that obtained with the polarization-difference method. As the time-gate increases, the contrast of the polarization-difference image tends to the gate-independent value that corresponds to the CW illumination.

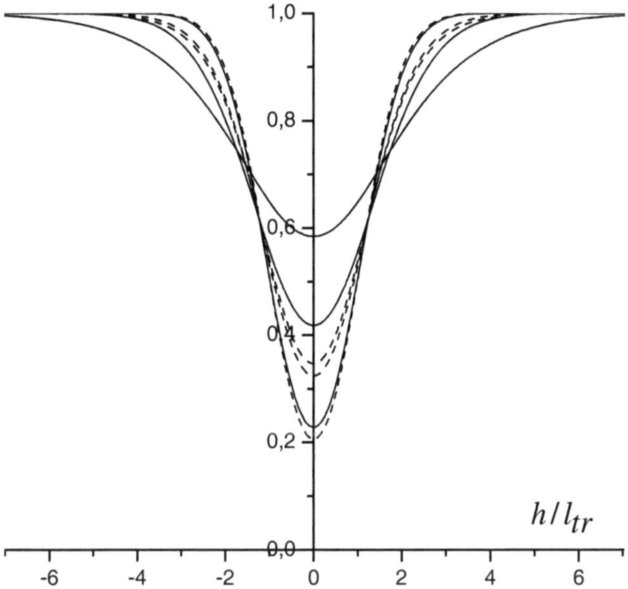

Fig. 7.15. Image of an absorbing stripe. The solid and dashed lines correspond to the intensity and the polarization difference profiles, respectively. The normalized time-gate $\Delta/z = 0.3$, 0.5, and 1.0 (from lower to upper curves). The transport optical thickness $\sigma_{tr}z = 10$, $\sigma_{dep}^{(W)} = 0.1 \cdot \sigma_{tr}$, stripe width $a = 0.2z$, and stripe position $z_0 = z/2$.

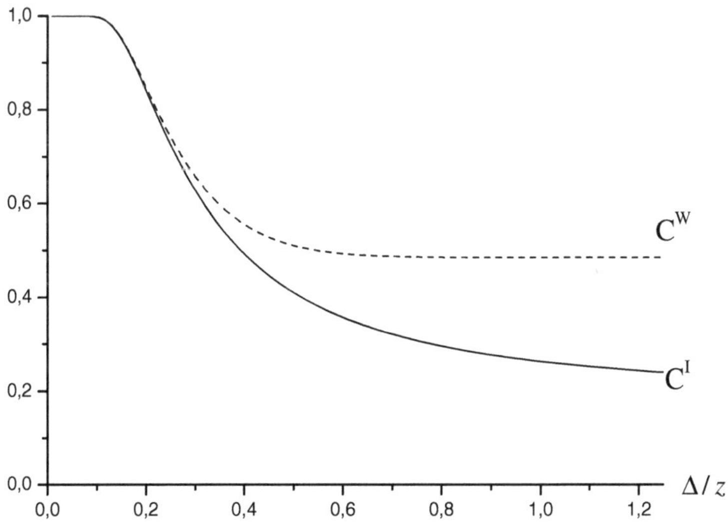

Fig. 7.16. Image contrast of an absorbing stripe versus the normalized time-gate. Solid and dashed curves describe the contrast of intensity and polarization difference profiles, respectively. Transport optical thickness $\sigma_{tr}z = 10$, $\sigma_{dep}^{(W)} = 0.1 \cdot \sigma_{tr}$, stripe width $a = 0.2 \cdot z$, and stripe position $z_0 = z/2$.

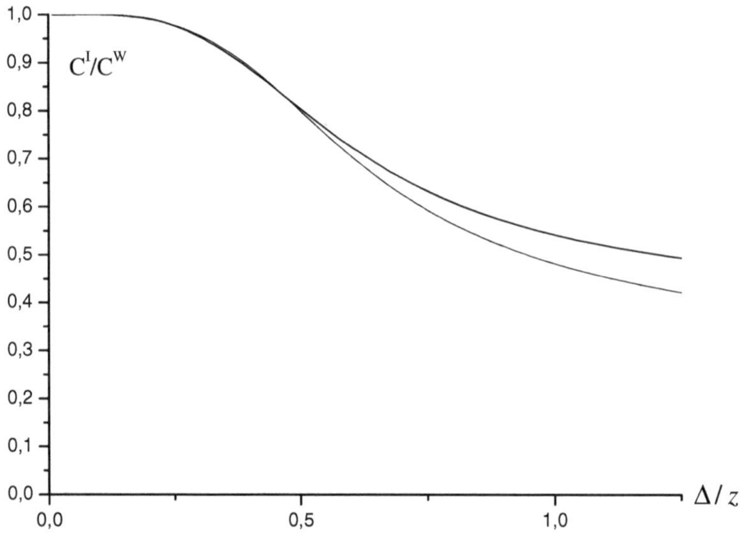

Fig. 7.17. Ratio C^I/C^W versus the normalized time-gate ($\sigma_{tr}z = 10$ (upper curve), $\sigma_{tr}z = 15$ (lower curve), $\sigma_{dep}^{(W)} = 0.1 \cdot \sigma_{tr}$).

This conclusion is confirmed additionally by the results presented in Fig. 7.17. From the figure it is evident that ratio C^I/C^W exhibits virtually universal behavior in the range $\Delta/z < 0.5$ and depends weakly on $\sigma_{tr}z$ at relatively large values of the normalized time-gate Δ/z.

Based on the above-presented results, we conclude that the polarization difference is equivalent to time-gating the snake component of the output signal with the interval of detection of the order of $\Delta \approx 0.5\ z$. For time-gate $\Delta < 0.5\ z$ there is no difference between the images obtained with the polarization-sensitive method and without it. For time-gate $\Delta > 0.5\ z$, the polarization-difference image does not depend on the detection interval and corresponds to the conditions of the CW illumination. Note that inequality $\Delta < 0.5\ z$ virtually coincides with the selection rule for the snake photons (see, e.g., [9, 29, 74]).

7.6.3 Polarization-difference imaging under CW illumination

Experimental and numerical studies [12–25] of polarization-difference imaging through scattering media deal with the CW illumination. Therefore, to compare with data [12–25] our results (see Eqs. (7.65) and (7.67)) should be integrated over time.

For the symmetric position of the inhomogeneity $z_0 = z/2$ (see, e.g., [12, 13, 20, 24]) the edge-spread function takes the form

$$E(z,h) = \frac{1}{2\pi^2} \cdot \frac{\exp(-\sigma_a z)}{A_0(z)\Delta^2(z)} \left\{1 - \mathrm{erf}\left(\frac{h}{\delta}\right)\right\}, \tag{7.78}$$

where

$$\delta = \left(\frac{\Delta^2(z_0)}{2A_1(z_0)} \right)^{1/2} = (A_3(z_0/2))^{1/2} \tag{7.79}$$

and functions $A_i(z)$, $(i = 0 \dots 3)$ and $\Delta^2(z)$ are defined by Eqs. (7.58)–(7.60), respectively. Eq. (7.78) corresponds to a narrow beam geometry.

For a non-absorbing medium ($\sigma_a = 0$), the quantities entering into Eq. (7.78) are equal to

$$A_0(z) = 1, \quad \Delta^2(z) = \frac{1}{3}\sigma_{tr}^2 z^4, \quad \delta = \left(\frac{1}{12}\sigma_{tr} z_0 \right)^{1/2} \cdot z_0 \tag{7.80}$$

To obtain the edge-spread function as applied to a wide beam geometry, we should integrate the Green function appearing in Eqs. (7.65) and (7.67) over transverse displacement $\boldsymbol{\rho}$. The corresponding result is given by

$$E(z,h) = \frac{1}{2\pi} \cdot \frac{\exp(-\sigma_a z)}{A_0(z)A_1(z)} \left\{ 1 - \operatorname{erf}\left(\frac{h}{\delta} \right) \right\} \tag{7.81}$$

where

$$\delta = \left(\frac{1}{2}A_3(z_0) + A_3(z_0/2) \right)^{1/2} \tag{7.82}$$

In Eqs. (7.81) and (7.82), as before, the obstacle position is assumed to be symmetric, $z_0 = z/2$. For a non-absorbing medium ($\sigma_a = 0$), we have

$$A_1(z) = 2\sigma_{tr} z, \quad \delta = \left(\frac{5}{12}\sigma_{tr} z_0 \right)^{1/2} \cdot z_0 \tag{7.83}$$

The image profile for stripe $-a/2 < x < a/2$ is given by superposition of two edge-spread functions and obtained from Eqs. (7.78) and (7.81) by substitution of

$$2 - \operatorname{erf}\left(\frac{a/2 + h}{\delta} \right) - \operatorname{erf}\left(\frac{a/2 - h}{\delta} \right) \tag{7.84}$$

for

$$\left(1 - \operatorname{erf}\left(\frac{h}{\delta} \right) \right)$$

The results presented above are easily generalized to the case of a polarized beam. For polarized light, in addition to the intensity, the edge-spread functions for polarization modes V and W should be introduced. Within the framework of the Fokker–Planck model the corresponding edge-spread functions can be derived from Eqs. (7.55) and (7.56) by substitution of novel 'effective' absorption coefficients for σ_a (see Eqs. (7.55) and (7.56))

The effect of polarization on the image contrast is illustrated in Figs. 7.18 and 7.19 where the shadow profile of an absorbing stripe and the dependence of the image contrast on the sample thickness are shown. From the presented results

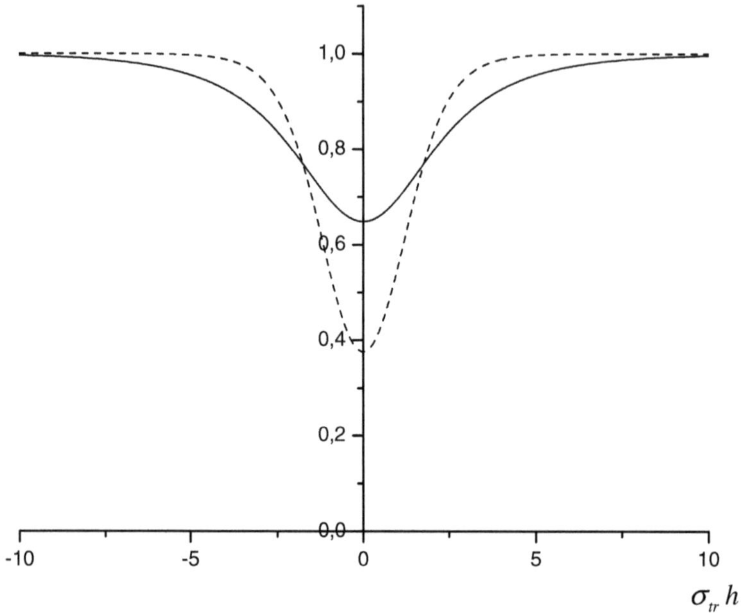

Fig. 7.18. Intensity (solid curve) and polarization-difference (dashed curve) profiles of the image of an absorbing strip. Transport optical thickness $\sigma_{tr} z = 10$, $\sigma_{dep}^{(W)} = 0.1 \cdot \sigma_{tr}$. The stripe of width $a = 2 \cdot l_{tr}$ is positioned in the center plane of the sample, $z_0 = z/2$. A narrow beam geometry.

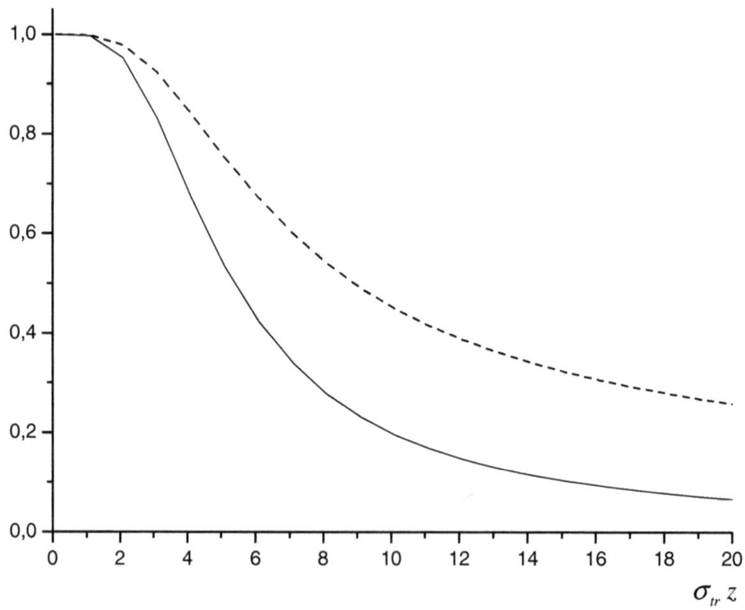

Fig. 7.19. Image contrast versus transport optical thickness $\sigma_{tr} z$. The objects are same as in Fig. 7.18. Solid and dashed curves correspond to the intensity and polarization-difference profiles, respectively.

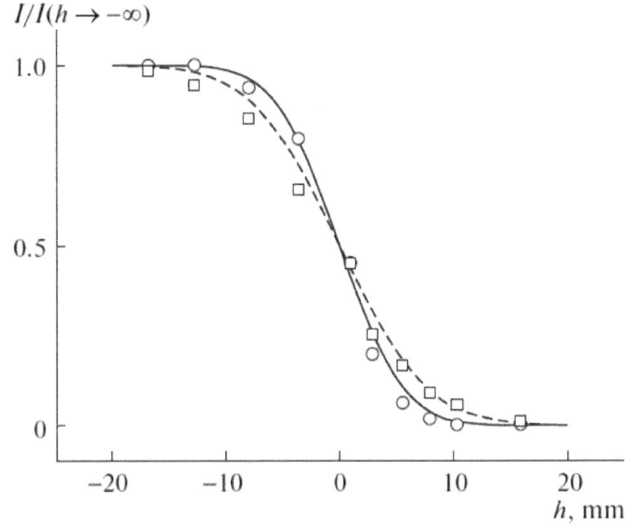

Fig. 7.20. Edge-spread function for the normalized intensity. Solid ($\sigma_{tr}z = 2.4, z = 40$ mm) and dashed ($\sigma_{tr}z = 4.5, z = 40$ mm) curves are the results of our calculations with Eqs. (7.78) and (7.79). Symbols are experimental data [20] for the samples of the corresponding optical thickness.

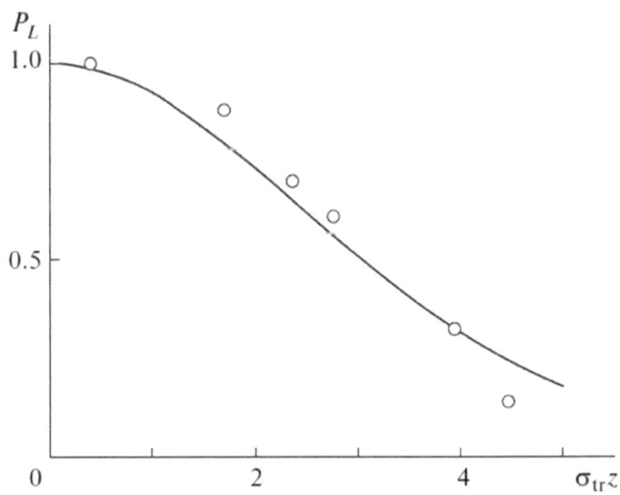

Fig. 7.21. Degree of linear polarization in diluted milk as a function of the transport optical thickness of the sample. Solid curve is the result of our theoretical calculations with Eq. (7.64) for $\sigma_{dep}^{(V)} = 0.1 \cdot \sigma_{tr}$, symbols are experimental data [20], $\lambda = 514$ nm.

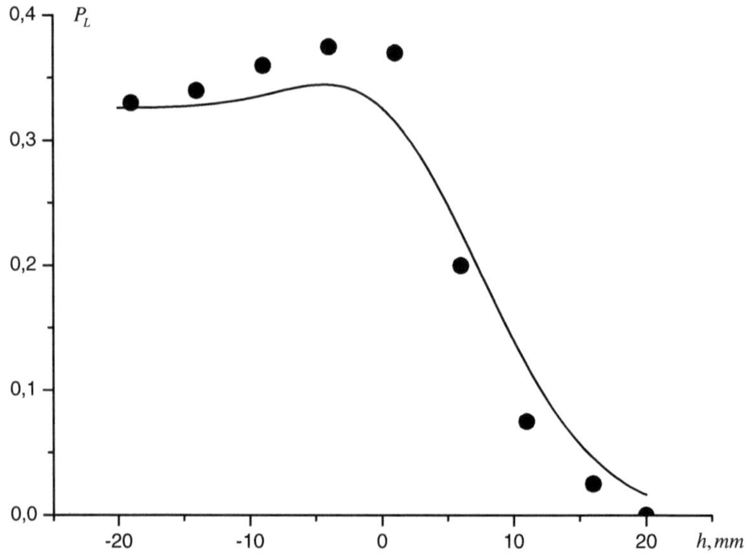

Fig. 7.22. Edge-spread function for the degree of linear polarization $P_L = W/I$. Solid curve is the result of our calculations for $\sigma_{tr} z = 4.0$, $z = 40\,\text{mm}$ and $\sigma_{dep}^{(W)} = 0.1 \cdot \sigma_{tr}$. Symbols are experimental data [20].

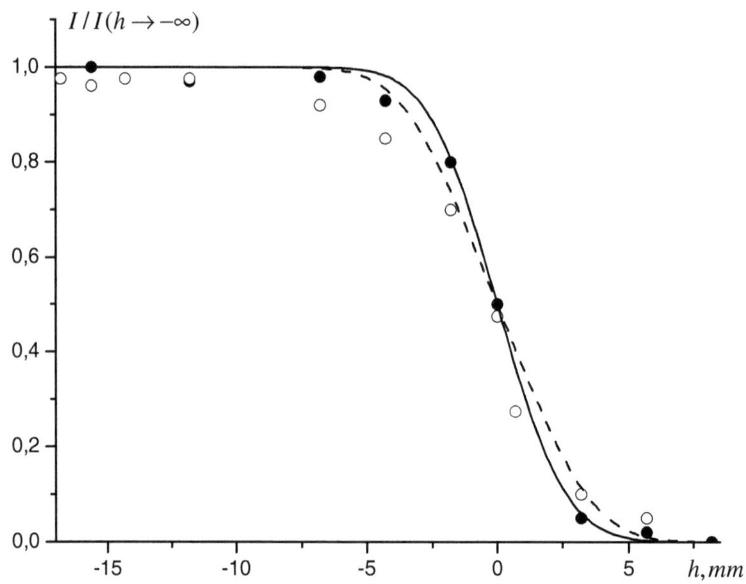

Fig. 7.23. Edge-spread function for the co- and cross- polarized light (suspension of $1.05\,\mu\text{m}$ diameter polystyrene microspheres in water, $\sigma_{tr} z = 1.54$, $z = 10\,\text{mm}$, $\lambda = 633\,\text{nm}$). Our calculations of $(I + W)/2$ and $(I - W)/2$ are shown by solid and dashed lines, respectively, $\sigma_{dep}^{(W)} = 0.074\ \sigma_{tr}$. Symbols are data of experiment [13].

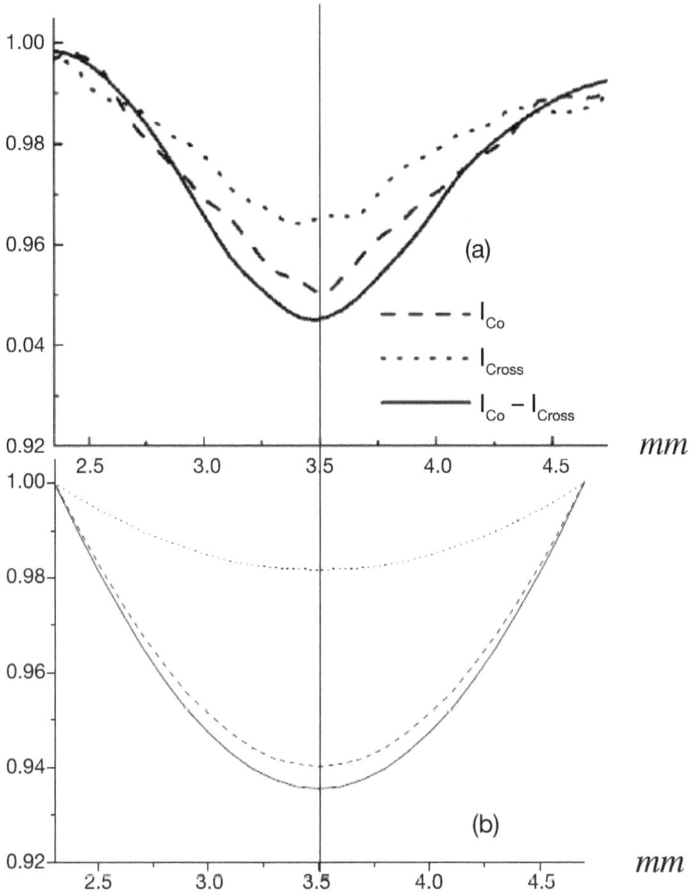

Fig. 7.24. Image of an absorbing obstacle from data of experiment [24] (a) and our calculations (b). The solid curve describes the polarization-difference profile V, the dashed and dotted curves correspond, respectively, to the intensity of light with the initial polarization $(I+V)/2$, and to the intensity of de polarized light , $(I-V)/2$. Aqueous suspension of $1.08\,\mu$m diameter polystyrene microspheres, $z = 10\,$mm, $\sigma_{tr}z = 2$, $\sigma_{dep}^{(V)} = 0.16 \cdot \sigma_{tr}$, $\lambda = 633\,$nm. A 1.8 mm diameter absorbing wire is placed in the sample center, $z_0 = z/2$. A wide beam geometry.

it follows that the polarization-difference method distinctly enhances the image contrast in transillumination of optically thick samples, $\sigma_{tr}z \gg 1$.

The results obtained above for the edge-spread function (see Eqs. (7.78),(7.81)) enable us to describe experimental data [13, 20, 24] on the polarization-gated imaging through scattering media. For the intensity of light transmitted through the medium with no absorption ($\sigma_a = 0$) this statement is illustrated in Fig. 7.20.

To describe data [20] for the degree of linear polarization, first we match $\sigma_{dep}^{(W)}/\sigma_{tr}$ from comparison of our theoretical dependence Eq. (7.64) with the corresponding data [20] far away from the obstacle edge (see Fig. 7.21). For the matched

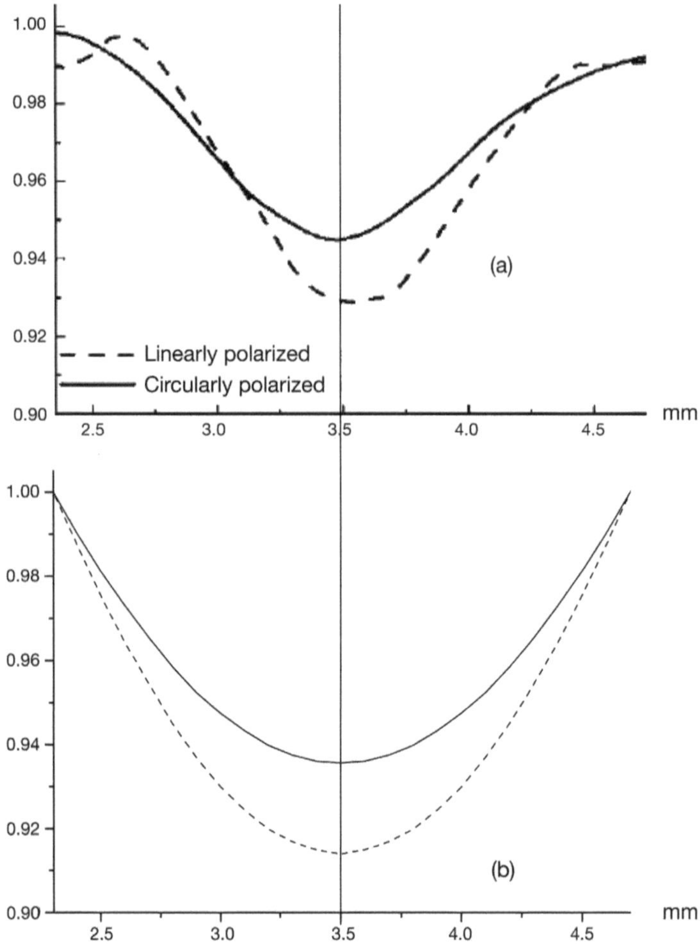

Fig. 7.25. Comparison of the polarization-difference profiles (measured (a) [24] and calculated (b)) for circularly (solid curve) and linearly (dashed curve) polarized light . The experimental conditions are the same as in Fig. 7.24.

value of $\sigma_{dep}^{(W)}/\sigma_{tr}$ the theoretical edge spread for the degree of polarization correlates well with experimental data [20] (see Fig. 7.22).

The results presented in Fig. 7.23 show how the spatial resolution of the image depends on the polarization state of light. In Fig. 7.23, the normalized edge spread functions for the intensity of the co- $((I + W)/2)$ and cross-polarized $((I - W)/2)$ components are shown. Ratio $\sigma_{dep}^{(W)}/\sigma_{tr}$ was calculated with the Mie theory. As follows from the figure, the results of our calculations are in good agreement with experimental data [13]. Our results for I and W underlying Fig. 7.23 are obtained from Eq. (7.81) with no additional fitting parameters.

Our theoretical results correlate also with measurements of the polarization-difference images of finite-sized objects. Examples of the corresponding profiles for the intensity of polarized components and the polarization-difference are shown in Figs. 7.24 and 7.25. The profiles presented in the figures were measured in experiment [24] and also calculated within the framework of our approach. The analytical results obtained for an absorbing stripe were employed. No adjusted parameters to give a good fit to experimental data [24] were used in our calculations.

7.7 Image simulation

Once the validity of our theoretical approach has been tested by comparison with experimental and numerical data, let us apply our results to the image simulation under more realistic conditions.

In the presence of a partially absorbing and scattering inhomogeneity, the intensity of radiation transmitted through the sample can be represented as the sum of the intensity of radiation passing by the inhomogeneity and the intensity of radiation passing through it. The first quantity has been calculated above in the context of studying completely absorbing obstacles.

To calculate the intensity of radiation passing through the inhomogeneity we take advantage of the model of the depth-average optical properties. Within the framework of the model [75], the heterogeneous medium is changed for the homogeneous medium but with the same values of the total optical thickness with respect to scattering and absorption,

$$\sigma_{tr}^{eff} = \frac{1}{z}\int\limits_0^z dz'\ \sigma_{tr}(z'),\ \sigma_a^{eff} = \frac{1}{z}\int\limits_0^z dz'\ \sigma_a(z') \tag{7.85}$$

Within this approximation, the intensity of radiation transmitted, for example, through a strip-like inhomogeneity takes the form

$$E^{(t)}(z,h) = \frac{1}{2}\tilde{I}\left\{\mathrm{erf}\left(\frac{a/2+h}{\tilde{\delta}}\right) - \mathrm{erf}\left(\frac{a/2-h}{\tilde{\delta}}\right)\right\} \tag{7.86}$$

where quantities $\tilde{I}$ and $\tilde{\delta}$ are calculated for the medium with optical properties given by Eq. (7.85). We can obtain the expressions for $\tilde{I}$ and $\tilde{\delta}$ from Eqs. (7.78) or (7.81) by substituting $(\sigma_{tr}(z-\Delta z)+\sigma_{tr2}\Delta z)/z$ and $(\sigma_a(z-\Delta z)+\sigma_{a2}\Delta z)/z$ for σ_{tr} and σ_a, respectively. Here σ_{tr2} and σ_{a2} are the optical coefficients of the inhomogeneity, Δz is its thickness.

To calculate the corresponding contribution for linearly and circularly polarized modes, W and V, the model of the depth-average optical properties should be extended with allowance for depolarization coefficients $\sigma_{dep}^{(W)}$ and $\sigma_{dep}^{(V)}$.

The results obtained above can be used as a basis for simulating the image of an arbitrary-shaped obstacle.

The cross-section of such an inhomogeneity can be marked off into squares (or rectangles). The sum of the contributions from each element to the detected intensity can be written as

$$E = I + \frac{1}{4} \sum_i \left(\tilde{I}_i \cdot \tilde{f}_i - I \cdot f_i \right) \tag{7.87}$$

where the sum is taken over all elements. Quantity $\tilde{I}_i$ is calculated with the depth-average absorption and transport scattering coefficients which are calculated along a straight line passing through ith element. Quantity f_i is given by

$$f_i = \left\{ \mathrm{erf}\left(\frac{h_{x,i}^-}{\delta}\right) + \mathrm{erf}\left(\frac{h_{x,i}^+}{\delta}\right) \right\} \cdot \left\{ \mathrm{erf}\left(\frac{h_{y,i}^-}{\delta}\right) + \mathrm{erf}\left(\frac{h_{y,i}^+}{\delta}\right) \right\} \tag{7.88}$$

where $h_{x,i}^\pm = a_i/2 \pm (h_x - x_i)$, $h_{y,i}^\pm = b_i/2 \pm (h_y - y_i)$, vector (x_i, y_i) determines the position of ith element center. Quantity $\tilde{f}_i$ differs from f_i only by substitution of the corresponding depth-average optical coefficients for the coefficients of the surrounding medium. The E-functions for linearly and circularly polarized modes, $E^{(W)}$ and $E^{(V)}$, respectively, are obtained from Eq. (7.87) with allowance for Eqs. (7.55) and (7.56).

From Eqs. (7.87) and (7.88) it follows that simulation of 2-D images in polarization-gated transillumination can be reduced to elementary procedures. The cross-section of an object is to be represented as a set of squares (or rectangles). Next we calculate the depth-average optical coefficients for each element. Finally, summing over all contributions (see Eq. (7.87)), we find the spatial distribution of the usual intensity or the corresponding differential polarization intensity in 2-D image.

As an illustration of the approach described above, let us give a number of examples of 2D-image simulation.

Fig. 7.26 demonstrates an ability to distinguish two objects from their images with different polarizations. The conditions are similar to those occurring in the experiments. The depolarization coefficients were chosen equal to $\sigma_{dep}^{(W)} = \sigma_{dep}^{(V)}/2 = 0.1 \cdot \sigma_{tr}$. These values are typical for the phantoms composed of polystyrene and silica microspheres.

Examples of 2D-images of an absorbing screen and a low-contrast inhomogeneity are presented in Figs. 7.27–7.29. The case of low-contrast objects is more frequent in actual diagnostic conditions (see, e.g., [29–33]). According to [29–33], the absorption coefficient in the objects is $2 \div 3$ times greater than that in the surrounding tissue. Variations in the transport scattering coefficient do not exceed 20%.

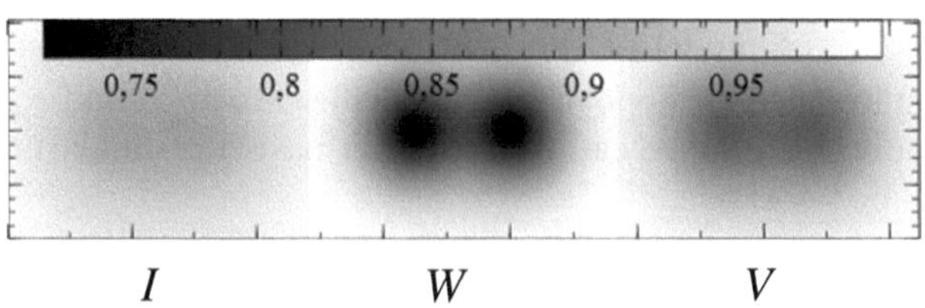

Fig. 7.26. 2-D image of two absorbing screens. Screens 2 mm × 2 mm in size are separated by a 2 mm gap and positioned in the center plane of the sample, $z_0 = z/2$. The thickness and the transport optical thickness of the sample are 10 mm and 10, respectively.

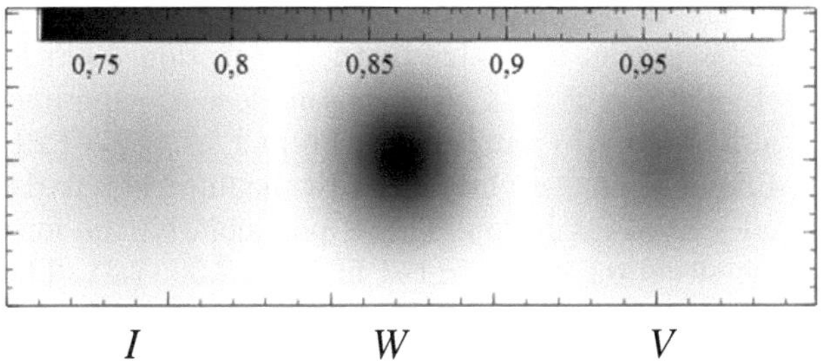

Fig. 7.27. 2-D image of a completely absorbing screen. Thickness of the sample $z = 10\,\text{mm}$. A $2\,\text{mm} \times 2\,\text{mm}$ screen is positioned in the center plane of the sample, $z_0 = z/2$. The transport scattering and absorption coefficients of the sample are $\sigma_{tr} = 1.0\,\text{mm}^{-1}$, $\sigma_a = 0$, respectively. The depolarization coefficients of the object and the surrounding medium equal $\sigma_{dep}^{(W)} = 0.5 \cdot \sigma_{dep}^{(V)} = 0.1 \cdot \sigma_{tr}$.

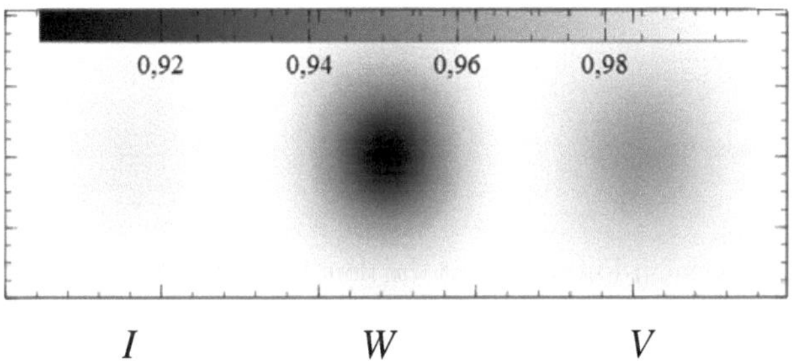

Fig. 7.28. 2-D image of a low contrast object. Thickness of the sample $z = 10\,\text{mm}$. A $2\,\text{mm} \times 2\,\text{mm} \times 2\,\text{mm}$ cube is positioned in the center plane of the sample, $z_0 = z/2$. The transport scattering and absorption coefficients of the sample are $\sigma_{tr} = 1.0\,\text{mm}^{-1}$, $\sigma_a = 0$, respectively. The optical coefficients of the object are $\sigma_{tr2} = 1.2\,\text{mm}^{-1}$, $\sigma_{a2} = 0.01\,\text{mm}^{-1}$. The depolarization coefficients of the object and the surrounding medium equal $\sigma_{dep}^{(W)} = 0.5 \cdot \sigma_{dep}^{(V)} = 0.1 \cdot \sigma_{tr}$.

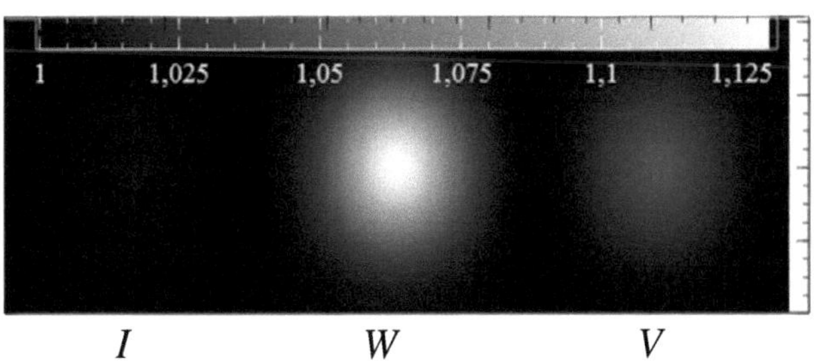

Fig. 7.29. The same as in Fig. 7.28, but $\sigma_{tr2} = 0.8\,\text{mm}^{-1}$.

However the difference in σ_{tr} between the object and the surrounding medium exceeds essentially, as a rule, the value of the absorption coefficient. In our calculations, parameters of the objects were chosen close to experimental conditions [30, 31], $\sigma_{tr} = 1\,\text{mm}^{-1}$, $\delta\sigma_{tr} = \pm 0.2\,\text{mm}^{-1}$. From Figs. 7.27–7.29 it follows that the polarization-difference images obtained with the linearly polarized light exhibit better contrast and spatial resolution. This is particularly true for the phantoms.

A more realistic situation is illustrated in Figs. 7.30, 7.31. The coefficients of depolarization in the object and the surrounding medium were taken close to those in biological tissues, $\sigma_{dep}^{(W)} = 0.35 \cdot \sigma_{tr}$, $\sigma_{dep}^{(V)} = 0.7 \cdot \sigma_{tr}$. In this case the difference between the images obtained with linearly and circularly polarized light is less distinctive.

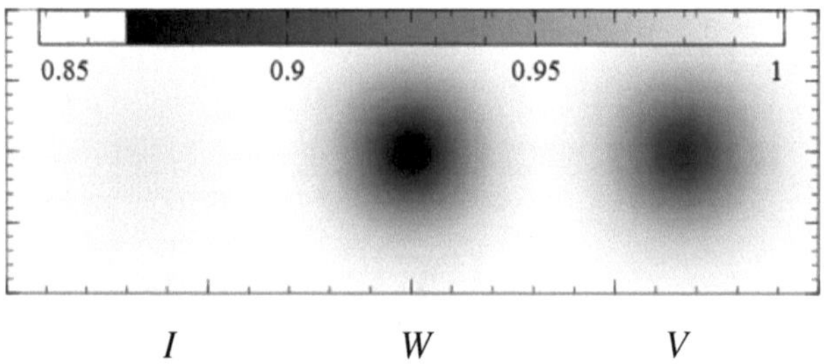

Fig. 7.30. The same as in Fig. 7.28, but the depolarization coefficients of the object and the surrounding medium equal $\sigma_{dep}^{(W)} = 0.5 \cdot \sigma_{dep}^{(V)} = 0.35 \cdot \sigma_{tr}$.

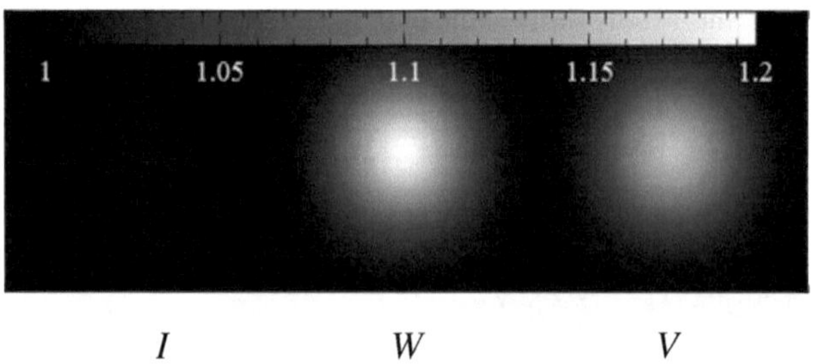

Fig. 7.31. The same as in Fig. 7.30, but $\sigma_{tr2} = 0.8\,\text{mm}^{-1}$.

From the results presented above it follows that the polarization-difference technique provides rather good contrast and spatial resolution. More contrast image is obtained for more attenuated polarization mode. The above-mentioned feature has already been noted in a number of experiments (see, e.g., [25]).

Thus, the applicability of the polarization-difference technique is primarily governed by the threshold of detecting the polarization modes. The threshold value of the degree of polarization depends both on the polarization sensitivity of the detector and on the contribution from the polarized component of the diffusive background (the latter can be estimated as $\sigma_a\theta^2/3\sigma_{tr}$, where angle θ is counted from the forward direction [39–41]). These conditions restrict the applicability of the polarization-difference approach by relatively thin layers. The currently available experimental data relate to the values of the transport optical thickness no greater than $\sigma_{tr} z \sim 10$.

As has been pointed out above the difference in σ_{tr} (and, consequently, in $\sigma_{dep}^{(W)}$ and $\sigma_{dep}^{(V)}$) between objects and surrounding tissues exceeds the typical values of the absorption coefficients. Therefore, the polarization-gating techniques are suited for measurements of the difference in scattering properties rather than in absorption.

7.8 Conclusions

We have described an approach for calculating the polarization-difference image of an object embedded in a highly scattering medium. This approach rests on the basic mode approximation in the vector radiative transfer equation. For additional simplifications we have taken advantage of the small-angle Fokker-Planck approximation. As a result we have first advanced an analytical model of propagation of polarized light through tissue-like media. Our model gives an insight into depolarization of light in the medium and is suited to semi-quantitative calculations of the depth-dependence of the degree of polarization and the image profiles. According to this model, the polarization state of radiation can be determined by the basic modes, namely, by intensity I, circularly and linearly polarized modes, V and W, respectively. The values of V and W differ from the intensity only by the factors that are responsible for depolarization and describe attenuation in domain of temporal delays $\Delta = ct - z$. To go to the case of continuous illumination, the corresponding expressions should be integrated over delay Δ. Polarization-difference imaging has been shown to be equivalent to transillumination with the time-gating interval of the order of $\Delta \sim 0.5\ z$ (i.e. virtually with the interval that separates the quasi-straightforward propagating (or snake-) photons from the diffusive ones). Our results relate the image characteristics to the optical properties of the medium and reproduce the available experimental data for tissue-like phantoms with no adjusting parameters. Within the framework of our approach a numerical procedure has been proposed to simulate polarization-difference images of millimeter-sized inhomogeneities immersed in highly scattering media. Examples of simulation of 2-D images have been presented for a number of cases which closely resemble actual conditions in transillumination of biological tissues.

Acknowledgments

This work was supported by the Russian Presidential Grants for the Support of the Leading Scientific Schools (project NSh-5992.2012.2) and the International Science and Technology Center (project no. 3691).

References

1. J.C. Hebden, S.R. Arridge, and D.T. Delpy, Optical imaging in medicine: experimental techniques, *Phys. Med. Biol.*, **42**, pp. 825–840, 1997.
2. S.R. Arridge and J.C. Hebden, Optical imaging in medicine: modeling and reconstruction, *Phys. Med. Biol.*, **42**, pp. 841–853, 1997.
3. V.V. Tuchin, Coherence-domain methods in tissue and cell optics, *Laser Phys.*, **8**, pp. 807–849, 1998.
4. K. Michielsen, H. De Raedt, J. Przeslawski, and N. Garcia, Computer simulation of time-resolved optical imaging of objects hidden in turbid media, *Phys. Rep.*, **304**, pp. 89–144, 1998.
5. V. Tuchin, L. Wang, and D. Zimnyakov, *Optical Polarization in Biomedical Applications*, Springer (2006).
6. S.R. Arridge and J.C. Schotland, Optical tomography: forward and inverse problems, *Inverse Problems*, **25**, pp. 123010/1-59, 2009.
7. A.D. Klose, Radiative transfer of luminescence light in biological tissue, in *Light Scattering Reviews*, ed. A.A.Kokhanovsky, Springer-Praxis, pp. 293–345, 2009.
8. N. Ghosh and I.A. Vitkin, Tissue polarimetry: concepts, challenges, applications, and outlook, *J. Biomed. Opt.*, **16**, pp. 110801/1-29, 2011
9. B.B. Das and R.R. Alfano, Ultrafast time-gated approach in optical biomedical image, *Current Science, Special Section: Biomedical Applications of Lasers*, **77**, pp. 885–892, 1999.
10. H. Horinaka, K. Hashimoto, K. Wada, and Y. Cho, Extraction of quasi-straight-forward-propagating photons from diffused light transmitting through a scattering medium by polarization modulation, *Opt. Lett.*, **20**, pp. 1501–1503, 1995.
11. S.G. Demos and R.R. Alfano, Temporal gating in highly scattering media by the degree of optical polarization, *Opt. Lett.*, **21**, pp. 161–163, 1996.
12. O. Emile, F. Bretenaker, and A.L. Floch, Rotating polarization imaging in turbid media, *Opt. Lett.*, **21**, pp. 1706–1708, 1996.
13. S.P. Morgan, M.P. Khong, and M.G. Somekh, Effects of polarization state and scatterer concentration on optical imaging through scattering media, *Appl. Opt.*, **36**, pp. 1560–1565, 1997.
14. S.P. Schilders, X.S. Gan, and M. Gu, Resolution improvement in microscopic imaging through turbid media based on differential polarization gating, *Appl. Opt.*, **37**, pp. 4300–4302, 1997.
15. S.P. Schilders, X.S. Gan, and M. Gu, Effect of scatterer size on microscopic imaging through turbid media based on differential polarization gating, *Opt. Commun.*, **157**, pp. 238–248, 1998.
16. X.S. Gan, S.P. Schilders, and M. Gu, Image enhancement through turbid media under a microscope by use of polarization gating methods, *JOSA*, **A16**, pp. 2177–2184, 1999.
17. H. Ramachandran and A. Narayanan, Two-dimensional imaging through turbid media using a continuous wave light source, *Opt. Commun.*, **154**, pp. 255–260, 1998.
18. V. Gopal, S. Mujumdar, H. Ramachandran, and A.K. Sood, Imaging in turbid media using quasi-ballistic photons, *Opt. Commun.*, **170**, pp. 331–345, 1999.
19. J.S. Tyo, Enhancement of the point-spread function for imaging in scattering media by use of polarization-difference imaging, *JOSA*, **A17**, pp. 1–10, 2000.
20. D. Zimnyakov and Y.P. Sinichkin, A study of polarization decay as applied to improved imaging in scattering media, *J. Opt. A: Pure and Appl. Opt.*, **2**, pp. 200–208, 2000.
21. M. Moscoso, J.B. Keller, and G. Papanicolaou, Depolarization and blurring of optical images by biological tissue, *JOSA*, **A18**, pp. 948–960, 2001.

22. S. Mujumdar and H. Ramachandran, Imaging through turbid media using polarization modulation: dependence on scattering anisotropy, *Opt. Commun.*, **241**, pp. 1–9, 2004.
23. G. Yao, Differential optical polarization imaging in turbid media with different embedded objects, *Opt. Commun.*, **241**, pp. 255–261, 2004.
24. P. Shukla, R. Sumathi, S. Gupta, and A. Pradhan, Influence of size parameter and refractive index on the scatterer on polarization-gated optical imaging through turbid media, *JOSA*, **A24**, pp. 1704–1713, 2007.
25. P. Shukla and A. Pradhan, Polarization-gated imaging in tissue phantoms: effect of size distribution, *Appl. Opt.*, **48**, pp. 6099–6104, 2009.
26. P. Bruscaglioni, G. Zaccanti, and Q. Wei, Transmission of a pulsed polarized light beam through thick turbid media: numerical results, *Appl. Opt.*, **32**, pp. 6142–6150, 1993.
27. X. Wang, L.V. Wang, C.W. Sun, and C.C. Yang, Polarized light propagation through scattering media: time-resolved Monte Carlo simulations and experiments, *J. Biomed. Opt.*, **8**, pp. 608–617, 2003.
28. A. Ishimaru, S. Jaruwatanadilok, and Y. Kuga, Polarized pulse waves in random duscrete scatterers, *Appl. Opt.*, **40**, pp. 5495–5502, 2001.
29. G. Mitic, J. Kolzer, J. Otto, E. Plies, G. Solkner, and W. Zinth, Time-gated transillumination of biological tissues and tissuelike phantoms, *Appl. Opt.*, **33**, pp. 6699–6710, 1994.
30. D.J. Hall, J.C. Hebden, and D.T. Deply, Imaging very-low-contrast objects in breast-like scattering media with a time-resolved method, *Appl. Opt.*, **36**, pp. 7270–7276, 1997.
31. L.C. Enfield, A.P. Gibson, N.L. Everdell, D.T. Delpy, M. Schweiger, S.R. Arridge, C. Richardson, M. Keshtgar, M. Douek, and J.C. Hebden, Three-dimensional time-resolved optical mammography of the uncompressed breast, *Appl. Opt.*, **46**, pp. 3628–3638, 2007.
32. V.G. Peters, D.R. Wyman, M.S. Patterson, and G.L. Frank, Optical properties of normal and diseased human breast tissues in visible and near infrared, *Phys. Mod. Biol.*, **35**, pp. 1317-1334, 1990.
33. W.-F. Cheong, Summary of optical properties, in *Optical-Thermal Response of Laser-Irradiated Tissue*, A.J.Welch and M.J.C. van Gemert, eds., Plenum, New York, pp. 275–303, 1995.
34. A. Ishimaru, *Wave Propagation and Scattering in Random Media*, Academic Press, New York, 1978.
35. C.F. Bohren and D.R. Hoffman, *Absorption and Scattering of Light by Small Particles*, Wiley, 1983.
36. M.I. Mishchenko, L.D. Travis, and A.A. Lacis, *Multiple Scattering of Light by Particles*, Cambridge University Press, Cambridge, 2006.
37. A.A. Kokhanovsky, *Polarization Optics of Random Media*, Praxis Publishing, 2003.
38. E.E. Gorodnichev, A.I. Kuzovlev, and D.B. Rogozkin, Depolarization of light in small-angle multiple scatteringin random media, *Laser Phys.*, **9**, pp. 1210–1227, 1999.
39. E.E. Gorodnichev, A.I. Kuzovlev, and D.B. Rogozkin, Multiple scattering of polarized light in turbid media with large particles, in *Light Scattering Reviews*, ed. A.A.Kokhanovsky, Springer-Praxis, pp. 291–337, 2006.
40. E.E. Gorodnichev, A.I. Kuzovlev, and D.B. Rogozkin, Depolarization of multiply scattered light in transmission through a turbid medium with large particles, *Opt. Commun.*, **260**, pp. 30–45, 2006.
41. E.E. Gorodnichev, A.I. Kuzovlev, and D.B. Rogozkin, Multiple scattering of polarized light in a turbid medium, *JETP*, **104**, pp. 319–341, 2007.

42. I. Kuscer and M. Ribaric, Matrix formalism in the theory of diffusion of light, *Opt. Acta*, **6**, pp. 42–51, 1959.
43. E.P. Zege and L.I. Chaikovskaya, Approximate theory of linearly polarized light propagation through a scattering medium, *JQSRT*, **66**, pp. 413–435, 2000.
44. Yu.A. Kravtsov, Geometrical depolarization of light in a turbulent atmosphere, *Izv. Vyssh. Uchebn. Zaved., Radiofiz.*, **13**, pp. 281–285, 1970.
45. Yu.A. Kravtsov and Yu.I. Orlov, *Geometrical Optics of Inhomogeneous Media*, Nauka, Moscow, 1980 [in Russian].
46. E.E. Gorodnichev, A.I. Kuzovlev, and D.B. Rogozkin, Diffusion of circularly polarized light in a disordered medium with large-scale inhomogeneities, *JETP Lett.*, **68**, pp. 22–28, 1998.
47. V.V. Sobolev, *Light Scattering in Planetary Atmospheres*, Pergamon, Oxford, 1975.
48. R.D.M. Garcia and C.E. Siewert, A generalized spherical harmonics solution for radiative transfer models that include polarization effects, *JQSRT*, **36**, pp. 401–423, 1986.
49. W. Cai, M. Lax, and R.R. Alfano, Analytical solution of the polarized photon transport equation in an infinite uniform medium using cumulant expansion, *Phys. Rev.*, **E63**, pp. 016606/1-10, 2000.
50. I.M. Gelfand, R.A. Minlos, and Z.Ya. Shapiro, *Representations of the Rotation and Lorentz Groups and their Applications*, Oxford, 1963.
51. F.C. MacKintosh, J.X. Zhu, D.J. Pine, and D.A. Weitz, Polarization memory of multiple scattering light, *Phys. Rev.*, **B40**, pp. 9342–9345, 1989.
52. J.M. Schmitt, A.H. Gandjbakhche, and R.F. Bonner, Use of polarized light to discriminate short-path photons in a multiply scattering medium, *Appl. Opt.*, **31**, pp. 6535–6546, 1992.
53. D.Bicout, C.Brosseau, A.S.Martinez, and J.M.Schmitt, Depolarization of multiply scattered waves by spherical diffusers: influence of the size parameter, *Phys. Rev.*, **E49**, pp. 1767–1770, 1994.
54. V. Sankaran, K. Schenenberger, J.T. Walsh, and D.J. Maitland, Polarization discrimination of coherently propagating light in turbid media, *Appl. Opt.*, **38**, pp. 4252–4261, 1999.
55. V. Sankaran, M.J. Everett, D.J. Maitland, and J.T. Walsh, Comparison of polarized-light propagation in biological tissue and phantoms, *Opt. Lett.*, **24**, pp. 1044–1046, 1999.
56. V. Sankaran, J.T. Walsh, Jr., and J.D. Maitland, Comparative study of polarized light propagation in biological tissues, *J. Biomed. Opt.*, **7**, pp. 300–306, 2002.
57. N. Ghosh, P.K. Gupta, H.S. Patel, B. Jain, and B.N. Singh, Depolarization of light in tissue phantoms-effect of collection geometry, *Opt. Commun.*, **222**, pp. 93–100, 2003.
58. N. Ghosh, A. Pradhan, P.K. Gupta, S. Gupta, V. Jaiswal, and R.P. Singh, Depolarization of light in a multiply scattering medium: Effect of the refractive index of a scatterer, *Phys. Rev.*, **E70**, pp. 066607/1-7, 2004.
59. X. Ni and R.R. Alfano, Time-resolved backscattering of circularly and linearly polarized light in a turbid medium, *Opt. Lett.*, **29**, pp. 2773–2775, 2004.
60. M. Xu and R.R. Alfano, Circular polarization memory of light, *Phys. Rev.*, **E72**, pp. 065601/1-4, 2005.
61. M. Xu and R.R. Alfano, Random walk of polarized light in turbid media, *Phys. Rev. Lett.*, **95**, pp. 213901/1-4, 2005.
62. N. Ghosh, P.K. Gupta, A. Pradhan, and S.K. Majumdar, Anomalous behaviour of depolarization of light in a turbid medium, *Phys. Lett.*, **A354**, pp. 236–242, 2006.
63. D.B. Rogozkin, Light pulse beam propagation in anisotropically scattering medium, *Izv. Acad. Sci. USSR, Atmos. Ocean. Phys.*, **23**, pp. 275–281, 1987.

64. E.E. Gorodnichev, S.V. Ivliev, A.I. Kuzovlev, and D.B. Rogozkin, Depolarization of light in the pulse transmission through the medium with large inhomogeneities, *Laser Phys.*, **20**, pp. 2021–2028, 2010.
65. E.E. Gorodnichev, S.V. Ivliev, A.I. Kuzovlev, and D.B. Rogozkin, Transmission of polarized light through turbid media, *Optics and Spectroscopy*, **110**, pp. 586–594, 2011.
66. E.E. Gorodnichev, S.V. Ivliev, A.I. Kuzovlev, and D.B. Rogozkin, Imaging through turbid media by polarized light , *Laser Phys.*, **22**, pp. 566–574, 2012.
67. E.E. Gorodnichev, A.I. Kuzovlev, and D.B. Rogozkin, Propagation of circularly polarized light in media with large-scale inhomogeneities, *JETP*, **88**, pp. 421–432, 1999.
68. V.S. Remizovich, D.B. Rogozkin, and M.I. Ryazanov, Propagation of a light signal in a substance with large-scale random inhomogeneities with photon path-length fluctuations due to multiple scattering taken into account, *Izv. Acad. Sci. USSR, Atmos. Ocean. Phys.*, **18**, pp. 480–485, 1982.
69. V.S. Remizovich, D.B. Rogozkin, and M.I. Ryazanov, Propagation of a modulated narrow light beam in a scattering medium with allowance for photon-path fluctuations during multiple scattering, *Radiophys. Quantum Electron.*, **25**, pp. 639–645, 1982.
70. A.D. Kim, Light propagation in biological tissues containing an absorbing plate, *Appl. Opt.*, **43**, pp. 555–563, 2004.
71. P. Gonzalez-Rodriguez and A.D. Kim, Light propagation in tissues with forward-peaked and large-angle scattering, *Appl. Opt.*, pp. 2599–2609, 2008.
72. V.S. Remizovich, D.B. Rogozkin, and M.I. Ryazanov, Propagation of a light-flash signal in a turbid medium, *Izv. Acad. Sci. USSR, Atmos. Ocean. Phys.*, **19**, pp. 796–801, 1983.
73. *Handbook of Mathematical Functions*, Abramowitz, M. and Stegun, I.A., Eds., New York: Dover, 1965.
74. N.G. Chen, J. Bai, Estimation of quasi-straightforward propagating light in tissues, *Phys. Med. Biol.*, **44**, pp. 1669–1676, 1999.
75. D. Grosenick, H. Wabnitz, and H. Rinneberg, Time-resolved imaging of solid phantoms for optical mammography, *Appl. Opt.*, **36**, pp. 221–231, 1997.

8 On the application of the invariant embedding method and the radiative transfer equation codes for surface state analysis

Victor P. Afanas'ev, Dmitry S. Efremenko and Alexander V. Lubenchenko

8.1 Introduction

The equivalence of equations, describing the various physical phenomena, always provides a mutual enrichment of theories. For example, the analogy of the force lines of the electric field and the current lines of a viscous incompressible fluid created the Ostrogradsky–Gauss theorem. The authors of this chapter have background in the theory of electron transfer [1–3] and light ion transfer [4]. Atmospheric remote sensing and electron spectroscopy have much in common in the principle of their methodology. In both cases the inverse problem is solved on the base of spectra of the reflected beam of photons or particles. The spectra are formed due to the interaction between the beam and the investigated medium.

In this chapter our attention is focused on scattering with a highly anisotropic scattering phase function $\omega(\theta)$, where θ is the scattering angle. We introduce a degree of elongation $k = \omega(0)/\omega(\pi)$. For considered problems the parameter k lies between 10^3 and 10^{11}. Minimum values of the degree of elongation appear in optical scattering problems with aerosols of a small fraction. Maximum values of k occur in the description of photon scattering by a coarse fraction aerosol and light ion scattering with energies of several MeV ($1\,\text{eV} = 1.6 \times 10^{-19}\,\text{J}$). Such high energies are common for Rutherford backscattering spectroscopy (RBS). For electron scattering the elongation has intermediate values: $k \simeq 10^4$–10^8. The small-angle approximation is efficiently used for the highest values of k, for instance, in RBS [5]. Many computational problems have been discovered and solved during the numerical solution of optical problems. This fact has stimulated the development of transport equation numerical solutions, excluding as many approximations to describe the multiple scattering as possible.

Electrons with energies in units of keV, scattered by heavy elements (such as gold), have $k \approx 10^4$. It is known that calculations based on small-angle approximation in this situation leads to large errors. Nevertheless small-angle approximation is acceptable for the description of electron scattering with energy $\sim 10\,\text{keV}$ in the samples of light elements (such as beryllium or carbon).

The verification of all, even the most reliable, solutions proved by all the theorems of existence and uniqueness should be based on comparison with experimental data. Ideally, the structure and the composition of the sample component should be investigated by independent methods. All the above conditions are fulfilled in the study of angular distributions of elastically scattered electrons. Elastic peak electron spectroscopy (hereafter EPES) is an actively developed method for sample composition analysis in near-surface layers. The investigated depth is from 0.5 nm to 50 nm. The phenomenon of electron elastic reflection has been known since the Germer/Devison experiment on low-energy electron diffraction – LEED [6]. The possibility of using the phenomenon of electron elastic reflection was predicted in [7]. The classical works [8–11] should be referred in the context of the development of EPES for the surface analysis of solids. However, the EPES method requires the elastic energy loss spectra measured with high-energy resolution (< 0.5 eV for the initial energy of some keV). That is why it has been developed only in the twenty-first century. The EPES method is often called 'the electron Rutherford backscattering'. The name indicates the prototype – Rutherford backscattering (RBS) [12], in which the sample is irradiated not with electrons, but with light ions. There are less strict requirements on the accuracy of the spectra measurements in the RBS method, but a proton accelerator to MeV energies is required. Despite this, the RBS method was widely used in the twentieth century, mainly due to a simple and intuitive classical small-angle model used to interpret the measured spectra. The main difficulty in the interpretation of reflected electron spectra is the consideration of multiple scattering.

The main goal of our research is to create an analytical tool to interpret the EPES spectra. EPES deals with the spectra measured for a reflection and a transmission. The energy of the probing electron beam lies between 10^3 eV and 10^5 eV. For these energies, the interaction between fast electrons and solids can be separated into two independent types of scattering according to the Fermi hypothesis. The following nomenclature is adopted. The electron interaction with the electronic structure of solids is called 'an inelastic scattering'. It can be local (ionization) and nonlocal (the excitation of the plasmons). The interaction of an electron beam with a nucleus is called 'an elastic scattering'. In the range 10^3–10^5 eV the 'bremsstrahlung' can be neglected. The approximation of a broad beam is used – a monoenergetic, monodirectional flux of the electrons falls on the sample. The flux of the reflected electrons will be denoted by the function R.

EPES can be used not only for the diagnosis of a solid surface, but also for the verification of the computational models. For example, the surface excitation parameters (SEP) [13, 14] can be extracted from EPES spectra. Accurate description of the elastic peak shape is used as well for calculations of the spectra measured on the wider energy loss range 0–100 eV (called 'the reflection electron energy loss spectroscopy' – REELS) [15].

Among the modern methods of the theoretical description of the optical radiation and the particle multiple scattering, there are both analytical and numerical methods based on solution of the radiative transfer equation and statistical modeling methods (Monte Carlo (MC) methods) [16, 17], providing the numerical solutions of the transfer equation [18].

The radiative transfer equation (RTE) considers the radiation itself as the energy flow regardless of its nature. The main characteristic is the radiation intensity. The RTE can be successfully used not only for optical radiation [19, 20], but also for particles [21, 22]: electrons, ions, neutrons. The study of optical radiative transfer in turbid media is based on photometrical representation and formulated in terms of ray optics, while the study of electron flow transfer in solids is based on the classical description of electron–solid interaction. The classical description is only applicable if the quantum effects can be neglected. This is valid if the de Broglie wavelength of the moving particle is less than the characteristic dimension of an 'elementary' scattering volume. The distance between atoms of the sample is about a few angstroms. For electrons scattered in a solid, this distance is considered as a characteristic dimension. The electron wavelength approaches this limit by the energies less than 150 eV.

The methods of statistical modeling enable us to solve the problem of electron multiple scattering in inhomogeneous media that approach the real samples as closely as possible [23–26]. Still for an adequate computing accuracy of the various characteristics, it is necessary to analyze about 10^7–10^9 trajectories. This takes too much time, even using modern computing machinery. In addition, as a rule, the problem of the experimental data interpretation includes the inverse radiative transfer problem. In this case, the statistical modeling appears to be inefficient. The analytical methods of RTE solving enable us to analyze the experimental data in terms of simple physical parameters and to solve the ill-posed and inverse problems of radiative transfer theory.

A detailed description of multiple scattering is required to develop the precise analytical methods for RTE computations. These methods should be based on an exact solution of the boundary-value problem of the RTE. The solution complexity of the boundary-value problem is determined, first of all, by the boundary conditions: the radiation of the source impinges upon the upper bound, while the lower bound is not illuminated. As shown in [27], such inhomogeneous boundary-value problem results in that the RTE solution is not a simple function, but it belongs to the class of generalized functions. This fact complicates the RTE solution significantly even in the simplest case – radiation backscattering from a homogeneous semi-infinite medium with an isotropic single scattering law.

The propagation of radiation in real media (the ocean and atmospheric aerosol) and the propagation of electrons in solids are characterized by the high anisotropy of single elastic scattering. The total scattering cross-section exceeds the transport scattering cross-section considerably. In this case, for the analytical and numerical solution of RTE boundary-value problems, some special approximate methods based on small-angle approximation are used [28–30].

The backscattered radiation is traditionally computed in the quasi-single scattering model based on the small-angle approximation [31–35]. The main problem of the quasi-single scattering approximation is an ambiguity in the separation of the single scattering phase function into a 'sharp' small-angle part describing a forward motion and a 'blunt' part describing a backscattering. The solution of this problem may not be formalized since it is a subjective decision. A backscattered signal can be interpreted with good precision by a successful separation. On the other hand, the study of the ray trajectories in turbid media using statistical modeling programs shows that the small-angle approximation does not describe the backscattered radiation flow from the real media in some cases. Therefore the applicability range problem of the small-angle approximation, specifically of the quasi-single scattering approximation, turns out to be urgent.

Recently, many numerical methods for optical RTE solution based on the solution discretization or its expansion by an orthonormal basis into special functions have been developed [36–39]. Using these methods for the problem of radiation propagation in media with an anisotropic single scattering law leads to practically insuperable computation difficulties as the solution instability increases considerably. To overcome this instability, the number of solved equations increases significantly. Therefore, the computational time increases as well and it can approach the computational times of the MC method. It is more effective to solve the problem of instability using the special methods based on the selection of a singular part, of an *a priori* known type, and a regular part. For the regular part, an equation is deduced from the RTE that is solved numerically thereupon. Such a solution is more stable and converges more quickly. As a singular part in [40], the Dirac δ-function was used; in [41] it was the small-angle solution.

Most of the solution methods of the RTE boundary-value problem are developed for optical radiation scattering. Optical radiation transfer in turbid media and electron transfer in solids are described by equivalent equations. But the parameters of the scattering medium are of a different physical nature. Optical radiation and electron propagation are similar due to the purely elastic scattering. Still by the multiple scattering, the photons are absorbed and the electrons are stopped. The absorption and stopping are absolutely different physical processes. Whereas photon absorption supposes just its instantaneous disappearance, electron stopping never leads to disappearance: the electrons lose energy, slow down and finally stop. The equations of optical radiation and electron flow transfer differ deeply though being similar. Therefore, the development of a rapid and precise method for computing the characteristics of multiply scattered electrons with an anisotropic single scattering law is of current interest.

This research is focused on the development of an analytical tool to interpret EPES spectra taking into account the methodology experience from radiative transfer theory.

This chapter is organized as follows. The structure of elastically scattered electron spectra is discussed in the Section 8.2. The radiative transfer models used for electron spectra interpretation are considered in Section 8.3 and in Section 8.4. Section 8.5 is devoted to the case of the semi-infinite medium and the formulation of the synthetic algorithm to solve the Ambartsumyan equation. The verification of the Rubin–Everhart model and the small-angle approximation on the base of DISORT, MDOM, NMSS code is shown in Section 8.6. In Section 8.7 the developed models are applied for surface analysis. The chapter concludes with a summary, Section 8.8.

8.2 The structure of the elastic peak

8.2.1 The energy shift of elastic peaks

The idea of EPES is based on the term *elastic energy loss* (also referred as *recoil energy*). Consider an electron with initial energy E_0, velocity V_0 and mass m, that is incident on the motionless nucleus M. Then the elastic scattering occurs on the

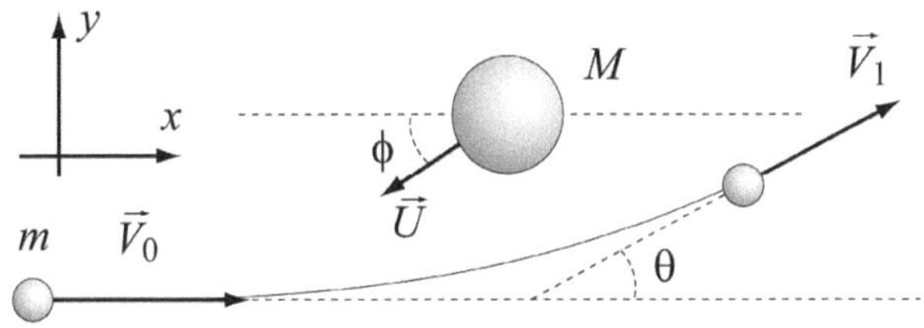

Fig. 8.1. The scattering of the electron with the mass m on the nuclear with mass M.

electrostatic potential of a nucleus. As a result, the trajectory of the electron is rotated by the angle θ. The nucleus achieves the velocity U by angle ϕ (Fig. 8.1) while the electron gets the velocity V_1.

In the case of elastic scattering the energy and momentum conservation laws can be applied.

$$\begin{cases} \frac{mV_0^2}{2} = \frac{mV_1^2}{2} + \frac{MU^2}{2} \\ (x)\colon mV_0 = mV_1 \cos\theta + MU\cos\phi \\ (y)\colon 0 = mV_1 \sin\theta - MU\sin\phi \end{cases} \tag{8.1}$$

From this system the ratio of the incident electron energy and scattered electron energy is determined as following:

$$\frac{E_1}{E_0} = \left(\frac{\sqrt{M^2 - m^2 \sin^2\theta} + m\cos\theta}{m + M} \right)^2 \tag{8.2}$$

The right side is called the 'kinematic factor' and depends on the angle of scattering and the ratio m/M. From (8.2) the relation for energy losses can be achieved $\Delta E = E_0 - E_1$. Since $M \gg m$, the energy loss reads as:

$$\Delta E = \frac{2m}{M}(1 - \cos\theta)E_0 \tag{8.3}$$

Thus, the energy loss depends on the mass of the nucleus, which elastically scatters the incident electron. Therefore, for instance, the energy spectrum of the elastically reflected electrons by a two-component system consisting of atoms with the masses M_1 and M_2 will be formed by the electrons with energy losses

$$\Delta E_1 = \frac{2m}{M_1}E_0\,(1 - \cos\theta) \quad \text{and} \quad \Delta E_2 = \frac{2m}{M_2}E_0\,(1 - \cos\theta)$$

The reader can notice that Eq. (8.3) is similar to the expression $\Delta\lambda = h(mc)^{-1}(1-\cos\theta)$, describing the *Compton* scattering [42]. That is why the elastic scattering in the paper [43] is referred as the Compton scattering.

8.2.2 The broadening of elastic peaks

The spectrum of energy losses will be generated not only by electrons with the energy loss corresponding to (8.3), but also with energies close to it. With a good degree of accuracy the energy distribution of the electrons elastically scattered from the nucleus mass M can be described by a Gaussian distribution

$$G\left(\Delta, \Delta E, \sigma_{\text{total}}\right) = \frac{1}{\sqrt{2\pi}\sigma_{\text{total}}} \exp\left(-\frac{(\Delta - \Delta E)^2}{2\sigma_{\text{total}}^2}\right) \tag{8.4}$$

Here Δ is the energy loss, ΔE is the elastic energy loss, σ_{total} is the standard deviation of the normal distribution. The elastic peak broadening is defined by four factors:

1. ***The energy spread of electron guns:*** the flux of electrons is not strictly monoenergetic. The energy distribution of the electrons emitted by an electron gun is described by the function $G\left(\Delta, 0, \sigma_G\right)$.
2. ***The Doppler effect due to the thermal motion of the target nucleus:*** Eq. (8.3) was obtained under the assumption that the nucleus is motionless. If the electron is moving, then the kinetic energy with respect to the nucleus E_0^{eff} can be larger or smaller than E_0. Thus, the recoil energy reads as:

   $$\Delta E' = \frac{2m}{M} E_0^{\text{eff}} \left(1 - \cos\theta\right) \tag{8.5}$$

   The Doppler effect leads to the additional broadening of the elastic peak, which is equivalent to the convolution of the spectrum with the function $G\left(\Delta, 0, \sigma_D\right)$. The value σ_D depends on the sample temperature and can be reduced by sample cooling. The molecular-kinetic theory provides an estimation for σ_D:

   $$\sigma_D = \sqrt{2E_0 R^G T m / M} \tag{8.6}$$

   here RG is the universal gas constant. The quantum-mechanical theory that takes into account the influence of phonon excitations on the broadening of the peaks of elastically scattered electrons, is presented in [44].
3. ***The resolution of the energy analyzer:*** the energy analyzer introduces an additional broadeningσ_A into measured spectra.
4. ***The multiple elastic scattering.*** The theoretical calculations and Monte Carlo simulations [2] indicate that the multiple elastic scattering leads to a broadening of the elastic peak and its shift. Usually these effects can be neglected.

The impact of each factor is described by the convolution of the measured spectra with Gaussian function with a broadening σ_{total}. Due to the normal distribution properties, the total broadening σ_{total} can be calculated as

$$\sigma_{\text{total}} = \sqrt{\sigma_G^2 + \sigma_D^2 + \sigma_A^2} \tag{8.7}$$

A criterion of EPES applicability is similar to the Rayleigh criterion in optics. Consider a sample with two the elements (the masses are M_1 and M_2), forming the peaks of elastically scattered electrons with broadenings $\sigma 1$ and $\sigma 2$. According to Eq. (8.3), the position of maxima are determined by relations

$$\Delta E_1 = \frac{2m}{M_1}(1-\cos\theta)\,E_0\,, \qquad \Delta E_2 = \frac{2m}{M_2}(1-\cos\theta)\,E_0$$

The distance between the elastic peaks on the energy axis is $|\Delta E_2 - \Delta E_1|$. The peaks can be resolved if the distance between them exceeds the half-width $|\Delta E_2 - \Delta E_1| > \sigma_1 + \sigma_2$. So we have:

$$E_0 > \frac{\sigma_1 + \sigma_2}{2m\,(1-\cos\theta)}\frac{M_1 M_2}{|M_2 - M_1|} \tag{8.8}$$

Thus, for the implementation of EPES one needs:

- the high-energy of an electron probing beam (10–40 keV);
- a large difference between the masses of the sample elements (for instance, hydrogen in metals);
- a large scattering angle (the scattering angle in the equipment where EPES spectra were resolved [43] for the first time was 120 degrees);
- a measurement equipment with high-energy resolution (state-of-the-art energy analyzers have an apparatus function with the broadening ~0.01 eV).

The way how to extract the information about the sample composition with the condition (8.8) violated will be described in the next subsection.

8.2.3 Qualitative analysis of the experimental spectra of elastically scattered electrons

Let us make a qualitative analysis of the spectra measured at the Australian National University by M. Vos. Figure 8.2 shows the spectrum of the electrons elastically scattered from a three-layer system Au/(Si + N)/Si. The positions of the elastic peaks in the experiment coincide with the calculations by Eq. (8.3):

$$\Delta E_{\mathrm{Au}} = 0.33\,\mathrm{eV}, \quad \Delta E_{\mathrm{Si}} = 2.3\,\mathrm{eV}, \quad \Delta E_{\mathrm{N}} = 4.6\,\mathrm{eV}.$$

The reader can note that the width of the elastic peak is smaller for greater mass of the nucleus, which corresponds to the Eq. (8.6).

The EPES method is efficient, particularly for the detection of the bound hydrogen and its isotopes. Figure 8.3 shows the experimental data from [43]. The hydrogen elastic peak is located far enough from the second elastic peak, which makes it easy to resolve them. However, the hydrogen elastic peak is located in an area in which the inelastic loss background becomes significant. In fact, this background is formed by the electrons that suffer the inelastic collisions.

The spectrum of elastically reflected electrons $R(\Delta)$ can be represented in the form of Gaussian sum:

$$R(\Delta) = \sum_k S_k G(\Delta, \Delta E_k, \sigma_k) \tag{8.9}$$

where S_k is the area under the peak of kth element, ΔE_k is the recoil energy for the kth element, σ_k is the broadening of kth element peak. The position of the peaks

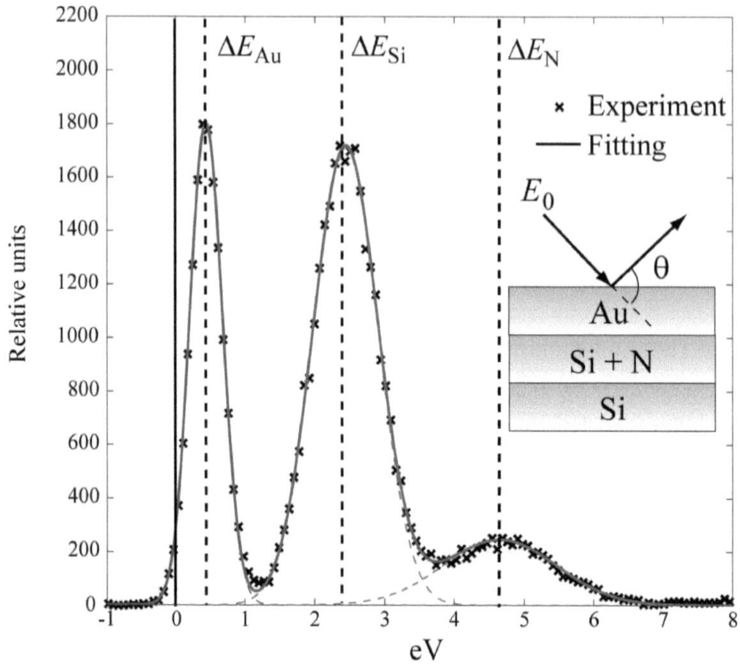

Fig. 8.2. The elastically reflected energy spectrum for the three-layer system (Au)/(Si + N)/(Si). The scattering angle is $\theta = 120°$, the probing beam energy is $E_0 = 40\,\text{keV}$.

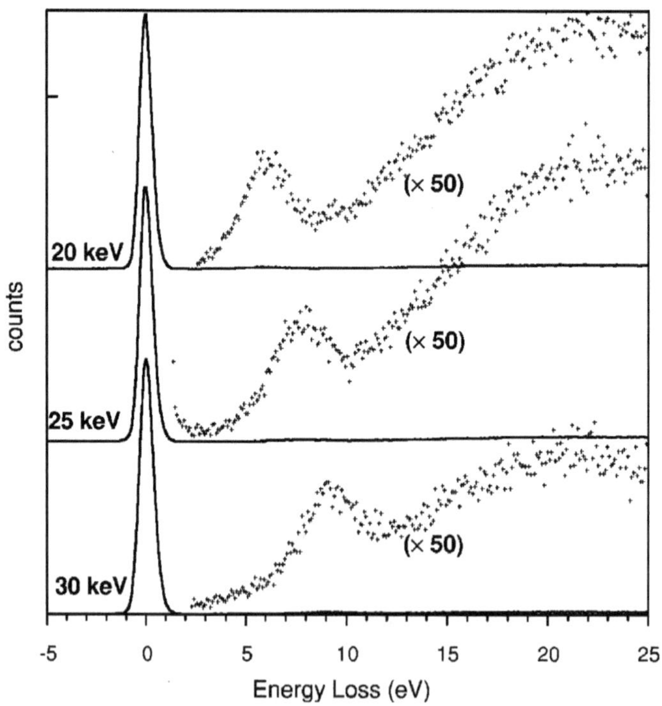

Fig. 8.3. The spectrum of the electrons elastically scattered by the formvar film containing hydrogen [43].

depends on the sample composition (in other words, it depends on the presence of specific chemical elements). However, the information about the structure sample is contained in the coefficients *Sk*. The quantitative EPES requires binding between coefficients *Sk* and the scattering medium properties. It can be established by radiative transfer models. The next section is devoted to the transport models for electrons in solids.

8.3 Models of elastic electron transport in solids

8.3.1 Review of electron transport models in solids

The calculation of elastically scattered electron intensity can be carried out by two independent methods. The first method is based on the calculation of the pass-length distributions taking into account only the elastic scattering and then applying the Bouguer law for removing particles being scattered inelastically. This method has been implemented in [45–47]. The disadvantage of this approach is the need to compute the pass-length distributions.

Monte Carlo simulations (MC) are widely used for theoretical calculations. However, the MC simulations of the properties of the reflected or transmitted beam always contain a stochastic error. The efficiency of the local estimation method [16] is low for electrons since they not only change the direction of motion, but also lose their energy. The dispersion of the MC method increases drastically in the case of a strong anisotropy of the single scattering phase function (the case of electron energy of some kiloelectronvolts). For instance, 10^7 trajectories are simulated in [48, 49] to compute the angular distributions of the reflected 1-keV electrons. The average error of MC simulations is 2%. For 10-keV electrons the error is 15%. The higher number of trajectories is used to reduce the dispersion. The stochastic error is proportional to $N^{-0.5}$, where N is the number of simulated particles. Thus, the computations appear to be time-consuming. This fact can eliminate all advances of the MC method.

Analytical solutions are especially valuable for inverse problems. The term *inverse problem* for the electron spectroscopy means a determination of scattering medium properties from the experimentally measured spectral and spatial characteristics of the scattered electrons. High computational speed can be achieved if the spectra are determined on the basis of analytical expressions.

We have already mentioned that the method '*e*-Rutherford scattering' [50, 51] implies the prototype – the Rutherford backscattering of light ions [52]. This method needs the accelerator of fast ions. Nevertheless one of the main advantages of RBS is the possibility of using a simple classical model of the small-angle deflection of Rubin [33]. In this theory, the particle moves straight ahead before and after the act of the strong scattering. The energy loss occurs during the elastic scattering. The process of the straight forward motion is accompanied by inelastic energy losses. This model was used by Everhart [32] for the total reflection coefficient calculations. It was shown that most adequately the 'straight forward' model describes the reflection coefficients of the sample, whenever the nuclear charge of sample atoms does not exceed the value of $Z < 30$ [33].

The development of the 'straight ahead' approach in the elastic scattering channel is the small-angle approximation, which takes into account the multiple elastic scattering of the electrons during their motion in the sample. It leads to the isotropization of the flux. This approach is proposed in the classical paper of Goudsmit and Saunderson [53]. In the small-angle model used in [50, 54, 55], the process of the straight-line motion is not accompanied by inelastic energy losses. Also, it is assumed that all particles move along the initial direction. At least one 'strong' scattering is required for the electrons to lose energy. The expression 'strong scattering' means scattering by a large angle. It changes the trajectory of motion significantly. The multiple small-angle scattering with a 'strong' collision called 'quasi-single scattering approximation' [56]. The discussion about alternative analytical methods for calculating the characteristics of the spectra of the elastically scattered electrons can be found in [57].

Let us show that the energy losses during the small-angle movement are negligibly small comparing to the losses due to the strong single scattering. Consider a movement along the half-circle due to the multiple elastic scattering (Fig. 8.4(a)). The electron suffers N collusions during the movement and it changes the direction by the angle θ/N. Then the total energy loss will be:

$$\Delta\bar{E}' = E_0 N \int_0^{\gamma} \frac{2m}{M} \sin^2\frac{\theta}{2N}\, d\theta \tag{8.10}$$

For the simplicity, the case of the backscattering ($\gamma = \pi$) is considered. Since $N \gg 1$,

$$\sin^2\frac{\theta}{2N} \approx \frac{\theta^2}{4N^2}$$

and Eq. (8.10) is simplified

$$\Delta\bar{E}' = E_0 \frac{m\,\pi^3}{6MN}$$

The energy loss value for a single backscattering (see Fig. 8.4(b)) is

$$\Delta\bar{E} = E_0 \frac{4m}{M}$$

And the ratio is expressed by

$$\frac{\Delta\bar{E}'}{\Delta\bar{E}} = \frac{\pi^3}{24N} \tag{8.11}$$

So $\Delta\bar{E}' \ll \Delta\bar{E}$ whenever $N \gg 1$.

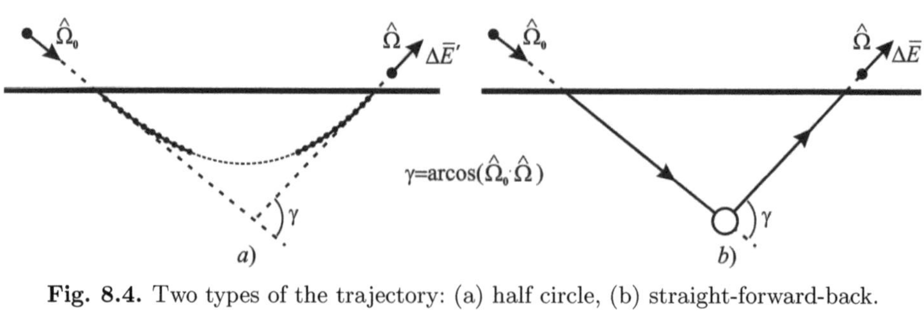

Fig. 8.4. Two types of the trajectory: (a) half circle, (b) straight-forward-back.

8.3.2 The model of elastic electron scattering by a single plane layer

Let us consider a theoretical model, which is used for the investigation of multiple electron scattering in solids. The standard assumption of a flat layer is used. The sample is considered to be a plane-parallel layer with the thickness d and with constant scattering properties in the whole volume. Internal sources are absent. The axis OZ is perpendicular to the upper border inside a medium. Zero of the coordinate system is placed at the upper border. Then z is the coordinate on the axis OZ. A mono-energetic and mono-directional electron beam impinges on the surface along the direction $\hat{\Omega}_0$ with density F_0 and initial energy E_0. The unit vectors along the corresponding directions are indicated by '^'. The direction of the particle movement $\hat{\Omega} = (\cos\theta \sin\varphi \sin\theta \sin\varphi \cos\varphi)$ is defined by the polar angle θ and the azimuth angle φ.

The electrons change direction and lose the energy Δ as a consequence of the scattering in the medium. Only the elastic energy losses are of interest. Nothing but the 'strong' elastic scattering event changes the propagation direction significantly. The probability that no inelastic processes occur over a path-length u is given by $\exp(-n_0\sigma_{\text{in}}u)$, where σ_{in} is the inelastic scattering cross-section and n_0 is the atomic concentration. In the following we use:

$$\exp\left(-n_0\,\sigma_{\text{in}}u\right) = \exp\left(-u/l_{\text{in}}\right)$$

where l_{in} is the inelastic mean free path.

As the probability of the large angle scattering ('strong' scattering) for electrons with an energy of a few kiloelectronvols is much smaller than for the small-angle scattering, the electron movement is well described by the single scattering approximation. Moreover, the electron reflection and transition functions can be separated by the energy and angular variables.

Given the path-length distribution of the electrons reflected by the layer $A_R(u, d, \hat{\Omega}_0, \hat{\Omega})$ or transmitted through the layer $A_T(u, d, \hat{\Omega}_0, \hat{\Omega})$, the spectra of elastically reflected electrons are described by the reflection function $R(d, E_0, \Delta, \hat{\Omega}_0, \hat{\Omega})$, while the spectra of the elastically-transmitted particles are described by the transmission function $T(d, E_0, \Delta, \hat{\Omega}_0, \hat{\Omega})$:

$$\begin{aligned}
R(d,\Delta,\hat{\Omega}_0,\hat{\Omega}) &= G(\Delta,\hat{\Omega}_0\cdot\hat{\Omega})R_{\text{el}}(d,\hat{\Omega}_0,\hat{\Omega}) \\
&= G(\Delta,\hat{\Omega}_0\cdot\hat{\Omega})\int_0^\infty A_R(u,d,\hat{\Omega}_0,\hat{\Omega})\,e^{-n_0\,\sigma_{\text{in}}u}\,du \\
T(d,\Delta,\hat{\Omega}_0,\hat{\Omega}) &= G(\Delta,\hat{\Omega}_0\cdot\hat{\Omega})T_{\text{el}}(d,\hat{\Omega}_0,\hat{\Omega}) \\
&= G(\Delta,\hat{\Omega}_0\cdot\hat{\Omega})\int_0^\infty A_T(u,d,\hat{\Omega}_0,\hat{\Omega})\,e^{-n_0\,\sigma_{\text{in}}u}\,du
\end{aligned} \tag{8.12}$$

with G being a Gaussian function which maxima and broadenings are depends on the geometry, as it has been described in the previous paragraphs.

8.3.3 The optical similarity

The functions $A_R(.)$ and $A_T(.)$ are the solutions of the boundary value problem for the transfer of the elastically scattered electrons in a layer. The description of the multiple elastic interaction between the electron beam and the scattering medium is based on the Lewis–Spencer equation [58]:

$$\mu\frac{\partial}{\partial z}A\big(u,z,\hat{\Omega}_0,\hat{\Omega}\big)+\frac{\partial}{\partial u}A\big(u,z,\hat{\Omega}_0,\hat{\Omega}\big)+n_0\,\sigma_{\mathrm{el}}\,A\big(u,z,\hat{\Omega}_0,\hat{\Omega}\big)$$
$$=n_0\int_{4\pi}A\big(u,z,\hat{\Omega}_0,\hat{\Omega}'\big)\omega_{\mathrm{el}}\big(\hat{\Omega}\cdot\hat{\Omega}'\big)\,d\Omega' \tag{8.13}$$

with the following boundary conditions

$$\begin{cases} A\big(0,z,\hat{\Omega}_0,\hat{\Omega}\big)=0,\, z>0,\\ A\big(u,0,\hat{\Omega}_0,\hat{\Omega}\big)=\delta\big(\hat{\Omega}_0-\hat{\Omega}\big)\,,\mu>0,\\ A\big(u,d,\hat{\Omega}_0,\hat{\Omega}\big)=0,\mu<0, \end{cases} \tag{8.14}$$

where u is a path-length, $A\big(u,z,\hat{\Omega}_0,\hat{\Omega}\big)$ is a path-length distribution of electrons with the angle of incidence Ω_0 and moving in the direction $\hat{\Omega}$ at the point z, $\mu=z\hat{\Omega}$ is cosine of the angle between the direction of motion $\hat{\Omega}$ and the axis OZ, ω_{el} is the differential cross-section for the elastic scattering of the electrons in solids. The particles that escape the inelastic interactions are of interest. The probability of escaping the inelastic interactions is determined by the exponential law of Bouguer. Thus, a new function is introduced:

$$L\big(z,\hat{\Omega}_0,\hat{\Omega}\big)=\int_0^{\infty}A\big(u,z,\hat{\Omega}_0,\hat{\Omega}\big)\,\exp\big(-n_0\,\sigma_{\mathrm{in}}u\big)\,du\,, \tag{8.15}$$

which describes the elastically scattered electron flux. Applying this transformation to (8.13) and (8.14), one can get the following equation:

$$\mu\frac{\partial}{\partial\tau}L\big(\tau,\hat{\Omega}_0,\hat{\Omega}\big)+L\big(\tau,\hat{\Omega}_0,\hat{\Omega}\big)=\frac{\lambda}{4\pi}\int_{4\pi}L\big(\tau,\hat{\Omega}_0,\hat{\Omega}'\big)x\big(\hat{\Omega}\cdot\hat{\Omega}'\big)\,d\Omega' \tag{8.16}$$

with the following boundary condition

$$\begin{cases} L\big(0,\hat{\Omega}_0,\hat{\Omega}\big)=\delta\big(\hat{\Omega}_0-\hat{\Omega}\big),\;\mu>0,\\ L\big(\tau,\hat{\Omega}_0,\hat{\Omega}\big)=0,\;\mu<0. \end{cases} \tag{8.17}$$

Equation (8.16) is the same as the radiative transfer equation, which was rigorously studied by Chandrasekhar [19] and Sobolev [20]. In optics, the function L is called the 'radiance'. The transformation (8.15) reveals the connection between two branches of radiative transfer: the transfer of the photons in the turbid medium and the transfer of the elastically scattered electrons. Many theories that describe the light scattering for all wavelengths ([59–61]) have been developed. However, the atmospheric parameters (such as the particle size distribution, the composition of the atmosphere) are poorly known. In the case of electron spectroscopy it is possible to create the sample with known properties. The elastic scattering cross-sections are also well-known from quantum mechanics theory and experiments. Therefore, the radiative transfer models can be verified by means of the electron spectroscopy [62].

8.3.4 Equations for elastically reflected and elastically transmitted electrons derived by the invariant-embedding method

The radiative transfer equation has excessive information about the radiance behavior inside the medium. Usually one is interested in the scattered radiance which escapes from the medium. Ambartsumian [63] obtained a nonlinear integral equation for the radiance factor in the case of a semi-infinite medium on the basis of the invariant embedding method. Then, Chandrasekhar [19] developed the ideas of Ambartsumian and derived four integral-differential equations for the reflectance and transmittance for a finite layer. These equations are nonlinear and heterogeneous. But they are more convenient for the boundary value problem solution from the numerical point of view.

The general solution of the invariant embedding leads to a system of four equations for the transmission and the reflection. The equations are derived in the following way: a layer of the thickness $d\tau$ is added at the top (Fig. 8.5) or at the bottom (Fig. 8.6) of the medium. The thickness of the medium is τ. The added layer is thin enough for the multiple scattering to be neglected. Furthermore, the difference of the transmittance and the reflectance due to this addition is derived. Bearing in mind that only a single scattering is possible in the added layer, one can get a system of equations, obtained by Chandrasekhar [19]:

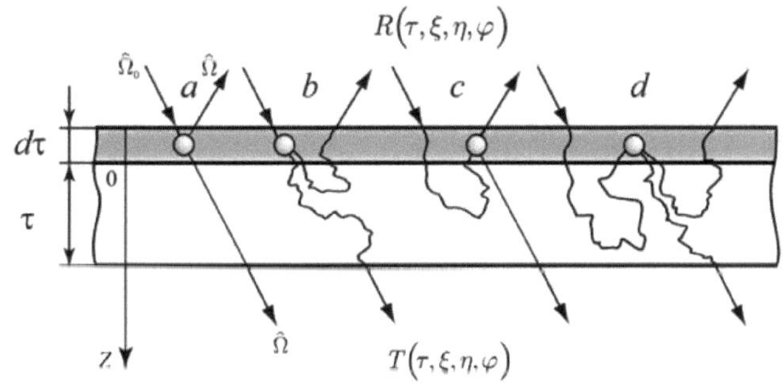

Fig. 8.5. The processes that change the transmittance and the reflectance in the layer of thickness $d\tau$ above the medium.

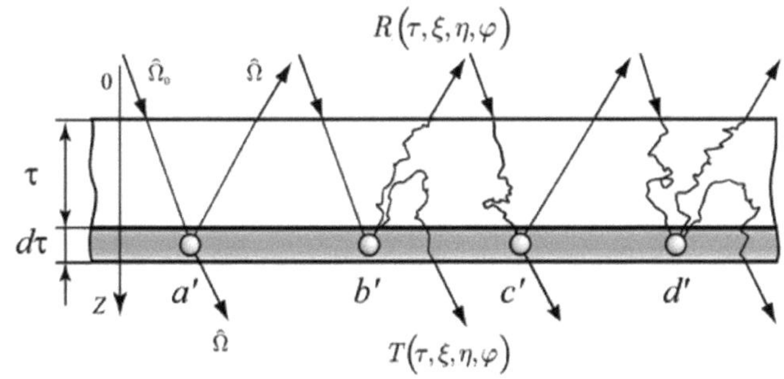

Fig. 8.6. The processes that change the transmittance and the reflectance in the layer of thickness $d\tau$ below the medium.

$$
\begin{aligned}
&\frac{\partial}{\partial\tau}R(\tau,\xi,\eta,\varphi)+\left(\frac{1}{\xi}+\frac{1}{\eta}\right)R(\tau,\xi,\eta,\varphi)=\overbrace{\lambda x(\xi,-\eta,\varphi)}^{a}\\
&+\overbrace{\frac{\lambda}{4\pi}\int_0^{2\pi}\int_0^1 x(\xi,\eta',\varphi')\,R(\tau,\eta',\eta,\varphi-\varphi')\,\frac{d\eta'}{\eta'}\,d\varphi'}^{b}\\
&+\overbrace{\frac{\lambda}{4\pi}\int_0^{2\pi}\int_0^1 R(\tau,\xi,\eta',\varphi')\,x(\eta',\eta,\varphi-\varphi')\,\frac{d\eta'}{\eta'}\,d\varphi'}^{c}\\
&+\overbrace{\frac{\lambda}{16\pi^2}\int_0^{2\pi}\int_0^{2\pi}\int_0^1\int_0^1 R(\tau,\xi,\eta',\varphi')\,x(-\eta',\eta',\varphi'-\varphi'')\,R(\tau,\eta',\eta,\varphi-\varphi'')\,\frac{d\eta''}{\eta''}\,\frac{d\eta'}{\eta'}\,d\varphi''\,d\varphi'}^{d}
\end{aligned}
\tag{8.18}
$$

$$
\begin{aligned}
&\frac{\partial}{\partial\tau}T(\tau,\xi,\eta,\varphi)+\frac{1}{\xi}T(\tau,\xi,\eta,\varphi)=\overbrace{\lambda\, e^{-\frac{\tau}{\eta}}x(\xi,\eta,\varphi)}^{a}\\
&+\overbrace{\frac{\lambda}{4\pi}\int_0^{2\pi}\int_0^1 x(\xi,\eta',\varphi')\,T(\tau,\eta',\eta,\varphi-\varphi')\,\frac{d\eta'}{\eta'}\,d\varphi'}^{b}\\
&+\overbrace{\frac{\lambda}{4\pi}\,e^{-\frac{\tau}{\eta}}\int_0^{2\pi}\int_0^1 R(\tau,\xi,\eta',\varphi')\,x(-\eta',\eta,\varphi-\varphi')\,\frac{d\eta'}{\eta'}\,d\varphi'}^{c}\\
&+\overbrace{\frac{\lambda}{16\pi^2}\int_0^{2\pi}\int_0^{2\pi}\int_0^1\int_0^1 R(\tau,\xi,\eta',\varphi')\,x(-\eta',\eta',\varphi'-\varphi'')\,T(\tau,\eta',\eta,\varphi-\varphi'')\,\frac{d\eta''}{\eta''}\,\frac{d\eta'}{\eta'}\,d\varphi''\,d\varphi'}^{d}
\end{aligned}
\tag{8.19}
$$

$$
\begin{aligned}
&\frac{\partial}{\partial\tau}R(\tau,\xi,\eta,\varphi)=\overbrace{\lambda\, e^{-\left(\frac{\tau}{\xi}+\frac{\tau}{\eta}\right)}x(\xi,-\eta,\varphi)}^{a'}\\
&+\overbrace{\frac{\lambda}{4\pi}\,e^{-\frac{\tau}{\xi}}\int_0^{2\pi}\int_0^1 x(\xi,-\eta',\varphi')\,T(\tau,\eta',\eta,\varphi-\varphi')\,\frac{d\eta'}{\eta'}\,d\varphi'}^{b'}\\
&+\overbrace{\frac{\lambda}{4\pi}\,e^{-\frac{\tau}{\eta}}\int_0^{2\pi}\int_0^1 T(\tau,\xi,\eta',\varphi')\,x(\eta',-\eta,\varphi-\varphi')\,\frac{d\eta'}{\eta'}\,d\varphi'}^{c'}\\
&+\overbrace{\frac{\lambda}{16\pi^2}\int_0^{2\pi}\int_0^{2\pi}\int_0^1\int_0^1 T(\tau,\xi,\eta',\varphi')\,x(\eta',-\eta',\varphi'-\varphi'')\,T(\tau,\eta',\eta,\varphi-\varphi'')\,\frac{d\eta''}{\eta''}\,\frac{d\eta'}{\eta'}\,d\varphi''\,d\varphi'}^{d'}
\end{aligned}
\tag{8.20}
$$

$$\frac{\partial}{\partial\tau}T(\tau,\xi,\eta,\varphi)+\frac{1}{\eta}T(\tau,\xi,\eta,\varphi)=\overbrace{\lambda\, e^{-\frac{\tau}{\xi}}x(\xi,\eta,\varphi)}^{a'}$$
$$+\overbrace{\frac{\lambda}{4\pi}\,e^{-\frac{\tau}{\xi}}\int_0^{2\pi}\int_0^1 x(\xi,-\eta',\varphi')\,R(\tau,\eta',\eta,\varphi-\varphi')\,\frac{d\eta'}{\eta'}\,d\varphi'}^{b'}$$
$$+\overbrace{\frac{\lambda}{4\pi}\int_0^{2\pi}\int_0^1 T(\tau,\xi,\eta',\varphi')\,x(\eta',\eta,\varphi-\varphi')\,\frac{d\eta'}{\eta'}\,d\varphi'}^{c'}$$
$$+\overbrace{\frac{\lambda}{16\pi^2}\int_0^{2\pi}\int_0^{2\pi}\int_0^1\int_0^1 T(\tau,\xi,\eta',\varphi')\,x(\eta',-\eta',\varphi'-\varphi'')\,R(\tau,\eta',\eta,\varphi-\varphi'')\,\frac{d\eta''}{\eta''}\,\frac{d\eta'}{\eta'}\,d\varphi''\,d\varphi'}^{d'}. \tag{8.21}$$

The scattering results in changing the direction of movement and losing the energy. As has been mentioned above, the reflected particles which suffer only the elastic scatterings are taken into account. They must suffer at least one 'strong' collision, considerably altering the direction of motion, to be reflected. Since the probability of scattering at large angles (the 'strong' scattering) for the keV energy electrons is orders of magnitude smaller than at small angles, the electron transfer inside the medium is properly described in a framework of the quasi-single scattering model.

8.4 The quasi-single scattering approximation and the quasi-multiple scattering approximation

To solve the derived equations, some approximations are used, which in fact are the variances of the quasi-single scattering approximation. A quasi-single scattering approximation [31, 34, 35] applied to Eq. (8.16) leads to a well-known problem. The single scattering phase function has to be split into the sharp 'small-angle' part for the scattering in a forward direction, and into a smooth part for the scattering in a backward direction. This separation is not unique, has no rigorous justification and, hence, is often sophisticated. For Eqs. (8.18) to (8.21) such a separation is unnecessary. Instead of this, two kinds of multiple scattering are considered. The first kind ('strong multiple scattering') converts a descending flux of the particles into an ascending flux, while the second kind does not. Such separation is already contained in Eqs. (8.18) to (8.21). Obviously, for the reflection the number of processes of the first kind should be odd. In the quasi-single scattering approximation only one strong scattering happens. If the absorption before and after the strong scattering is assumed, then we get the 'classical quasi-single scattering approximation'. If a small-angle scattering happens before and after the strong scattering, then such model is called 'the small-angle quasi-single scattering approximation'. Now we derive the analytical solutions for Eqs. (8.18) to (8.21) implying the described models.

8.4.1 The single scattering model

Excluding the processes b, c and d in (8.18) and (8.19), we get the system of equations describing the single scattering model – the particles suffer one and only one elastic scattering:

$$\begin{cases} \frac{d}{d\tau}R^1(\tau,\xi,\eta,\varphi) + \left(\frac{1}{\xi}+\frac{1}{\eta}\right) R^1(\tau,\xi,\eta,\varphi) = \lambda\; x(\xi,-\eta,\varphi) \\ \frac{d}{d\tau}T^1(\tau,\xi,\eta,\varphi) + \frac{1}{\xi}T^1(\tau,\xi,\eta,\varphi) = \lambda \exp\left(-\frac{\tau}{\eta}\right) x(\xi,\eta,\varphi) \end{cases} \tag{8.22}$$

The solution of the system of equations (8.22) for flat, mono-directional source reads as:

$$\begin{cases} R^1(\tau,\xi,\eta,\varphi) = \lambda\,\frac{\eta\xi}{\eta+\xi}x(\xi,-\eta,\varphi)\left(1-\exp\left(-\left(\frac{1}{\eta}+\frac{1}{\xi}\right)\tau\right)\right) \\ T^1(\tau,\xi,\eta,\varphi) = \lambda\frac{\eta\xi}{\eta-\xi}\,x(\xi,\eta,\varphi)\left(\exp\left(-\frac{1}{\xi}\tau\right)-\exp\left(-\frac{1}{\eta}\tau\right)\right) \end{cases} \tag{8.23}$$

8.4.2 Linearization of the system of equations in a model with one strong collision

If we neglect the process d in Eq. (8.18), and the processes c and d in (8.19) and b' and d' in (8.21), we obtain the linearized system, which describes the transfer with one strong collusion:

$$\begin{aligned} &\frac{\partial}{\partial\tau}R(\tau,\xi,\eta,\varphi) + \left(\frac{1}{\xi}+\frac{1}{\eta}\right) R(\tau,\xi,\eta,\varphi) = \lambda\; x(\xi,-\eta,\varphi) \\ &\quad+\frac{\lambda}{4\pi}\int_0^{2\pi}\int_0^1 x(\xi,\eta',\varphi')R(\tau,\eta',\eta,\varphi-\varphi')\frac{d\eta'}{\eta'}\,d\varphi' \\ &\quad+\frac{\lambda}{4\pi}\int_0^{2\pi}\int_0^1 \tilde{R}(\tau,\xi,\eta',\varphi')x(\eta',\eta,\varphi-\varphi')\,\frac{d\eta'}{\eta'}\,d\varphi' \end{aligned} \tag{8.24}$$

$$\begin{aligned} &\frac{\partial}{\partial\tau}T_1(\tau,\xi,\eta,\varphi) + \frac{1}{\xi}T_1(\tau,\xi,\eta,\varphi) = \lambda \exp\left(-\frac{\tau}{\eta}\right) x(\xi,\eta,\varphi) \\ &\quad+\frac{\lambda}{4\pi}\int_0^{2\pi}\int_0^1 x(\xi,\eta',\varphi')T_1(\tau,\eta',\eta,\varphi-\varphi')\,\frac{d\eta'}{\eta'}\,d\varphi' \end{aligned} \tag{8.25}$$

$$\begin{aligned} &\frac{\partial}{\partial\tau}T^1(\tau,\xi,\eta,\varphi) + \frac{1}{\eta}T^1(\tau,\xi,\eta,\varphi) = \lambda \exp\left(-\frac{\tau}{\xi}\right) x(\xi,\eta,\varphi) \\ &\quad+\frac{\lambda}{4\pi}\int_0^{2\pi}\int_0^1 T^1(\tau,\xi,\eta',\varphi')x(\eta',\eta,\varphi-\varphi')\,\frac{d\eta'}{\eta'}\,d\varphi' \end{aligned} \tag{8.26}$$

This model (8.24) was investigated in works [3, 4] for the reflectance.

8.4.3 The classical quasi-single scattering approximation

For the keV energy electron scattering in a solid, the probability of small-angle scattering is the orders of magnitude higher than that for large-angle scattering. In other words, the elongation parameter k, presented in Section 8.1, is large. For 40 keV for gold $k = 10^5$, for carbon $k = 10^7$. This fact allows us to use the classical quasi-single scattering approximation, also referred as the Rubin–Everhart model [32, 33]. This model is extensively used for the interpretation of RBS spectra of fast light ions [30, 31], and for an estimation of the elastically scattered electron intensity [54, 51]. Since the scattering phase function has a strong forward peak, it can be replaced by the Dirac function:

$$x(\hat{\Omega} \cdot \hat{\Omega}_0) = 2\delta(\hat{\Omega}_0 - \hat{\Omega}) \tag{8.27}$$

Substituting the approximation (8.27) in the integrals in (8.24), (8.25), (8.26), one can get the system of ordinary differential equations:

$$\begin{cases} \dfrac{\partial}{\partial \tau} R(\tau,\xi,\eta,\varphi) + \left(\dfrac{1}{\xi} + \dfrac{1}{\eta}\right) R(\tau,\xi,\eta,\varphi) \\ \quad = \lambda\, x(\xi,-\eta,\varphi) + \lambda \left(\dfrac{1}{\xi} + \dfrac{1}{\eta}\right) R(\tau,\xi,\eta,\varphi) \\ \dfrac{\partial}{\partial \tau} T_1(\tau,\xi,\eta,\varphi) + \dfrac{1}{\xi} T_1(\tau,\xi,\eta,\varphi) \\ \quad = \lambda \exp\left(-\dfrac{\tau}{\eta}\right) x(\xi,\eta,\varphi) + \dfrac{\lambda}{\xi} T_1(\tau,\xi,\eta,\varphi) \\ \dfrac{\partial}{\partial \tau} T^1(\tau,\xi,\eta,\varphi) + \dfrac{1}{\eta} T^1(\tau,\xi,\eta,\varphi) \\ \quad = \lambda \exp\left(-\dfrac{\tau}{\xi}\right) x(\xi,\eta,\varphi) + \dfrac{\lambda}{\eta} T^1(\tau,\xi,\eta,\varphi) \end{cases} \tag{8.28}$$

The functions T^1 and T_1 describe the symmetric processes: a strong scattering then the movement without scattering; on the other hand: the movement without scattering – then a strong scattering. However, the movement may be accompanied by inelastic scatterings. This is the difference between the classical quasi-scattering model and single scattering approximation, where an inelastic scattering is absent. Since the processes have equal probability, the resulting transmission reads as:

$$T(\tau,\xi,\eta,\varphi) = \frac{1}{2}\left(T_1(\tau,\xi,\eta,\varphi) + T^1(\tau,\xi,\eta,\varphi)\right) \tag{8.29}$$

Solving (8.28) using (8.29), one can get the expressions for the transmittance and the reflectance:

$$\begin{aligned} T(\tau,\xi,\eta,\varphi) &= \frac{\lambda}{2}\eta\xi \left[\frac{x(\xi,\eta,\varphi)}{\eta(1-\lambda)-\xi} \left(\exp\left(-\frac{1}{\eta}\tau\right) - \exp\left(-\frac{1-\lambda}{\xi}\tau\right)\right) \right. \\ &\quad \left. + \frac{x(\xi,\eta,\varphi)}{\xi(1-\lambda)-\eta} \left(\exp\left(-\frac{1}{\xi}\tau\right) - \exp\left(-\frac{1-\lambda}{\eta}\tau\right)\right) \right] \end{aligned} \tag{8.30}$$

$$R(\tau,\xi,\eta,\varphi) = \frac{\xi\eta}{\xi+\eta}\frac{\lambda}{1-\lambda}\, x(\xi,-\eta,\varphi) \left(1 - \exp\left(-(1-\lambda)\left(\frac{1}{\eta}+\frac{1}{\xi}\right)\tau\right)\right) \tag{8.31}$$

8.4.4 The small-angle quasi-single scattering approximation

The equations (8.24), (8.25), (8.26) can be solved by the spherical harmonics method [64, 65] taking into account multiple scatterings during the movement before and after the strong scattering. The scattering phase function $x(\varsigma, \eta', \varphi')$ has a sharp peak for ς near to η' on the interval $-1 \leq \eta' \leq 1$, in other words, for the small-angle scattering. Therefore, we can extend the limits of integration over η' on the interval $[-1, 1]$ and use the saddle-point method:

$$\begin{aligned} &\int_0^{2\pi}\int_0^1 x(\varsigma, \eta', \varphi') f(\tau, \eta', \eta, \varphi - \varphi') \frac{d\eta'}{\eta'}\, d\varphi' \\ &\qquad \approx \frac{1}{\varsigma}\int_0^{2\pi}\int_{-1}^1 x(\varsigma, \eta', \varphi') f(\tau, \eta', \eta, \varphi - \varphi')\, d\eta'\, d\varphi' \end{aligned} \tag{8.32}$$

The operation (8.32) is the calculation of scattering integrals in a small-angle approximation. However, the processes a and a' in Eqs. (8.18) to (8.21) are defined only on the interval $0 < \eta' \leq 1$. Let us define new functions, which are equal to zero on the interval $[-1, 0]$:

$$x^+(\xi, \eta, \varphi) = \begin{cases} x(\xi, \eta, \varphi), & \eta > 0 \\ 0, & \eta < 0 \end{cases}$$

$$x^-(\xi, \eta, \varphi) = \begin{cases} x(\xi, -\eta, \varphi), & \eta > 0 \\ 0, & \eta < 0 \end{cases} \tag{8.33}$$

Then we expand the scattering phase function x, the reflection and transmission functions R and T by the spherical function series:

$$x(\xi, \eta, \varphi) = x(\hat{\Omega}_0 \cdot \hat{\Omega}) = \sum_{l,m} x_l^m\, Y_l^m(\hat{\Omega}_0) Y_l^m(\hat{\Omega}) \tag{8.34}$$

$$R(\xi, \eta, \varphi) = \sum_{l,m} r_l^m\, Y_l^m(\hat{\Omega}_0) Y_l^m(\hat{\Omega}), \quad T(\xi, \eta, \varphi) = \sum_{l,m} t_l^m\, Y_l^m(\hat{\Omega}_0) Y_l^m(\hat{\Omega}) \tag{8.35}$$

where x_l^m, r_l^m, t_l^m are the expansion coefficients of the single scattering phase function, the reflectance and the transmittance respectively. If a phase function is highly anisotropic, then there is a good approximation for the functions x^+ and x^- in the form of series:

$$x_l^{m+}(\xi, \eta, \varphi) = \sum_{l,m} x_l^m{+}Y_l^m(\hat{\Omega}_0) Y_l^m(\hat{\Omega}), \quad x_l^{m-}(\xi, \eta, \varphi) = \sum_{l,m} x_l^m{-}Y_l^m(\hat{\Omega}_0) Y_l^m(-\hat{\Omega}) \tag{8.36}$$

In [3, 66], the following representation of the highly anisotropic single scattering phase functions has been proposed:

$$x^-(\hat{\Omega}_0 \cdot \hat{\Omega}) \approx x(\hat{\Omega}_0 \cdot \hat{\Omega}) - 2\delta(\hat{\Omega}_0 \cdot \hat{\Omega})$$

Thus, the expansion coefficients by spherical functions read as:

$$x_l^{m\,-} = \left(2 - x_l^m\right), \quad x_l^{m\,+} = x_l^m \tag{8.37}$$

Substituting (8.32) to (8.36), one can get a linear system of the ordinary differential equations with zero boundary conditions for the expansion coefficients T and R. Using (8.32) for the transmittance, the solution in the small-angle quasi-single scattering approximation is derived:

$$\begin{aligned} R(\tau,\xi,\eta,\varphi) &= \lambda\frac{\xi\eta}{\xi+\eta}\sum_{l,m}\frac{(2-x_l^m)}{(1-\lambda x_l^m/2)} \\ &\times\left(1-\exp\left(-\left(1-\frac{\lambda}{2}x_l^m\right)\left(\frac{1}{\xi}+\frac{1}{\eta}\right)\tau\right)\right)Y_l^m(\hat{\Omega}_0)Y_l^m(-\hat{\Omega}) \end{aligned} \tag{8.38}$$

$$\begin{aligned} T(\tau,\xi,\eta,\varphi) = \frac{\lambda}{2}\xi\eta\sum_{l,m}\Bigg(&\frac{\lambda x_l^m}{\eta(1-\lambda x_l^m/2)-\xi}\left(\exp\left(-\frac{\tau}{\eta}\right)-\exp\left(-\frac{1-\lambda x_l^m/2}{\xi}\tau\right)\right) \\ &+\frac{\lambda x_l^m}{\xi(1-\lambda x_l^m/2)-\eta}\left(\exp\left(-\frac{\tau}{\xi}\right)-\exp\left(-\frac{1-\lambda x_l^m/2}{\eta}\tau\right)\right)\Bigg)Y_l^m(\hat{\Omega}_0)Y_l^m(\hat{\Omega}) \end{aligned} \tag{8.39}$$

In the obtained solutions of the boundary problem the singular term (Dirac function) is included. It does not describes scattered particles. Since the Dirac function cannot be represented in the form of finite series, it should be analytically subtracted for the numerical computations, implying that T and R expansion coefficients turn to zero whenever l goes to infinity.

8.4.5 The quasi-multiple small-angle approximation. The nonlinear term in the radiative transfer equation

In the quasi-single small-angle approximation it is assumed that the particle suffers only one 'strong' scattering. The more the single scattering albedo and the less the single scattering phase function anisotropy is, the higher is the probability of 'strong' scatterings. In this case the quasi-multiple small-angle approximation has to be used [67]. There the particle suffers some 'strong' scatterings, but the movement between strong collisions is a small-angle one. There is no limitation on the number of 'strong' scatterings. This model is applicable for particles with a path-length both less and more than the transport path length.

Let us derive the formulas for this model.

Further the substitute

$$r:\, R = \frac{\mu\mu_0}{\mu+\mu_0}r$$

is used. Then, the last nonlinear term in Eq. (8.18) can be rewritten as

$$\begin{aligned} &\frac{\lambda\eta\xi}{16\pi^2}\int_0^{2\pi}\int_0^{2\pi}\int_0^1\int_0^1\frac{1}{(\xi+\eta')(\eta''+\eta)}r(\tau,\xi,\eta',\varphi')\,x(-\eta',\eta',\varphi'-\varphi'') \\ &\times r(\tau,\eta',\eta,\varphi-\varphi'')\,d\eta''\,\eta'\,d\varphi''\,d\varphi' \end{aligned}$$

In the integrand, the function $x\left(-\eta', \eta', \varphi' - \varphi''\right)$ has a sharp peak at grazing scattering angles (the flux of electrons moving almost parallel to the upper boundary layer). In this case the maximum is attained, if $\eta' \to 0$ and $\eta \to 0$, $\varphi'' \to \varphi'$. In the region near maximum, one can write following expressions:

$$\begin{aligned}\int_0^{2\pi}\int_0^1 & r\left(\tau, \xi, \eta' \to 0, \varphi'\right) x\left(-\eta' \to 0, \eta', \varphi' - \varphi''\right) d\eta'\, d\varphi' \\ &= \int_0^{2\pi}\int_{-1}^0 r\left(\tau, \xi, \eta' \to 0, \varphi'\right) x\left(-\eta' \to 0, \eta', \varphi' - \varphi''\right) d\eta'\, d\varphi'\end{aligned} \tag{8.40}$$

The equality (8.40) allows us to extend the region of integration by the variable η' in the nonlinear term:

$$\begin{aligned}\int_0^{2\pi}\int_0^1 & r\left(\tau, \xi, \eta' \to 0, \varphi'\right) x\left(-\eta' \to 0, \eta', \varphi' - \varphi''\right) d\eta'\, d\varphi' \\ &= \frac{1}{2}\int_0^{2\pi}\int_{-1}^1 r\left(\tau, \xi, \eta' \to 0, \varphi'\right) x\left(-\eta' \to 0, \eta', \varphi' - \varphi''\right) d\eta'\, d\varphi'\end{aligned} \tag{8.41}$$

Taking into account (8.41), the nonlinear term reads as:

$$\frac{1}{4}\frac{\lambda}{16\pi^2}\int_0^{2\pi}\int_0^{2\pi}\int_{-1}^1\int_{-1}^1 r(\tau,\xi,\eta',\varphi')x(-\eta',\eta',\varphi'-\varphi'')r(\tau,\eta',\eta,\varphi-\varphi'')\,d\eta''\,d\eta'\,d\varphi''\,d\varphi'$$

Expanding the functions $r\left(\tau, \xi, \eta', \varphi'\right)$ and $r\left(\tau, \eta', \eta, \varphi - \varphi''\right)$ into the spherical harmonics series and using the orthogonality properties for expansion coefficients, one can get the following equation:

$$r_l^m\left(\tau'\right) + \frac{d}{d\tau'} r_l^m\left(\tau'\right) = \frac{\lambda}{4} x_l^{m\,-} + \frac{\lambda}{2} x_l^{m\,+} r_l^m\left(\tau'\right) + \frac{\lambda}{4}\left(r_l^m\left(\tau'\right)\right)^2 x_l^{m\,-} \tag{8.42}$$

where

$$\tau' = \tau\left(\frac{\eta + \xi}{\eta\xi}\right) \tag{8.43}$$

with a boundary condition:

$$r_l^m(0) = 0$$

After some derivations, one can find the solution of (8.42):

$$r_l^m\left(\tau'\right) = r_l^m(\infty)\left(1 - \frac{1 - f_l^m}{\exp\left(\tau' y_l^m/2\right) - f_l^m}\right) \tag{8.44}$$

where

$$r_l^m(\infty) = 2\frac{\lambda x_l^{m\,-}}{1 - \lambda x_l^{m\,+}/2}\left[1 + \sqrt{1 - \frac{\left(x_l^{m\,-}\lambda\right)^2}{\left(2 - \lambda x_l^{m\,+}\right)^2}}\right]^{-1} \tag{8.45}$$

are the expansion coefficients for the reflection function in the case of the semi-infinite medium,

$$y_l^m = \sqrt{\left(2-\lambda\, x_l^{m\,+}\right)^2 - \left(\lambda^2\, x_l^{m\,-}\right)^2}, \quad f_l^m = \frac{2-\lambda\, x_l^{m\,+} - y_l^m}{2-\lambda\, x_l^{m\,+} + y_l^m}$$

The final solution reads as

$$\tilde{R}_M(\tau,\xi,\eta,\varphi) = \frac{\xi\eta}{\xi+\eta}\sum_{l,m} r_l^m(\infty)\left(1-\frac{1-f_l^m}{\exp\left(\frac{y_l^m}{2}\left(\frac{1}{\eta}+\frac{1}{\xi}\right)\tau\right)-f_l^m}\right)Y_l^m(\hat{\Omega}_0)Y_l^m(-\hat{\Omega}) \tag{8.46}$$

The expression (8.46) comes to the expression (8.38), if $f_l^m \to 0$, $y_l^m \to (2-\lambda\, x_l^{m\,+})$, $\sqrt{1-\left(x_l^{m\,-}\lambda/\left(2-\lambda x_l^{m\,+}\right)\right)^2} \to 1$. It leads us to the following condition:

$$\frac{x_l^{m\,-}\lambda}{2-\lambda x_l^{m\,+}} < \frac{x_0^{0-}\lambda}{2-\lambda x_0^{0\,+}} < \frac{b\lambda}{1-\lambda} < \varepsilon \ll 1 \tag{8.47}$$

where $b = x_0^{0-}/2 = \int_{-1}^{0} x(\eta)\, d\eta/2$ is the probability of the 'strong' scattering, ε is the value much less than 1. From (8.47) one can derive the estimation of the single scattering albedo when the quasi-single model can be used:

$$\lambda < \lambda_g \approx \frac{\varepsilon}{\varepsilon + b} \tag{8.48}$$

For Henyey–Greenstein single scattering phase function $b \approx (1-q)^{3/2}/2$. From (8.48) one can get the condition for applicability of the quasi-single scattering model:

$$\lambda < \lambda_g \approx \frac{2\varepsilon}{2\varepsilon + (1-q)^{3/2}} \tag{8.49}$$

For instance, for $\varepsilon = 0.1$ and $q = 0.9$ the parameter λ_g is 0.86, while for $q = 0.95$ λ_g is 0.95. The single scattering albedo for electrons of keV energies in solids varies from 0.5 till 0.8. The scattering is rather anisotropic ($q = 1 - \sigma_{\mathrm{tr}}/\sigma_{\mathrm{el}} > 0.9$, σ_{tr} is the transport cross-section, σ_{el} is the elastic cross-section). Thus, the inequality (8.49) holds true for the case of EPES spectroscopy and the impact of the nonlinear term derived in the small-angle approximation is not significant. This conclusion is illustrated in the Fig. 8.7, where the angular distributions of electrons elastically scattered by a gold sample is considered. The incident angle is normal. The energy of the incident electrons is 1 keV. Also we apply the code DISORT, which solves the Eq. (8.16), to compute the 'etalon' angular distribution. The single scattering approximation has a significant error. The quasi-single small-angle model also has an error due to the strong influence of the multiple scattering and the fast isotropization of the electron beam in the gold. The nonlinear term slightly improves the small-angle solution, but still the error is significant.

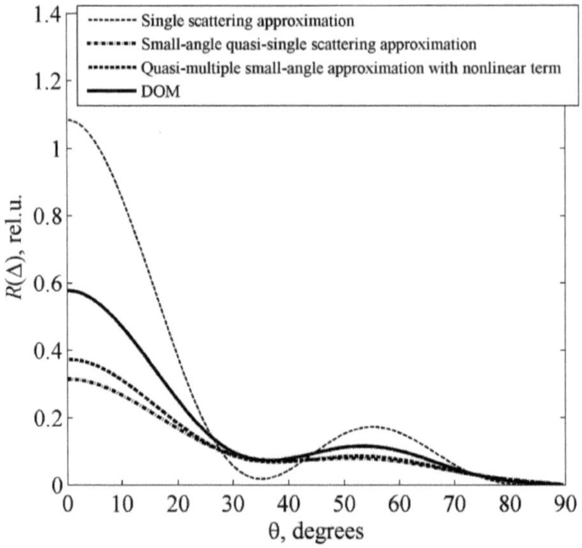

Fig. 8.7. The angular distributions of elastically scattered electrons. The reflection from the gold sample. The angle of incidence is normal. The initial energy is 1 keV.

8.4.6 Scattering by two-layer systems

Let us consider the scattering by a two-layer sample. The thickness of the first (upper) layer is τ_1 while the thickness of the second (bottom) layer is τ_2. In the quasi-single scattering approximation, the reflectance for the upper layer does not depend on the properties of lower layers. Thus, all the expressions derived in the previous paragraphs are suitable for the upper layer. The bottom index '1' is used whenever we refer to the single-layer solution. Now we are interested in the reflection function $R_2(\tau_1, \tau_2, \xi, \eta, \varphi)$ and in the transmission function $T_2(\tau_1, \tau_2, \xi, \eta, \varphi)$ for the bottom layer. The boundary conditions for the reflection and transmission functions of the second layer satisfy the following boundary conditions:

$$R_2\left(0, \tau_2, \xi, \eta, \varphi\right) = R_1\left(\tau_2, \xi, \eta, \varphi\right) \tag{8.50}$$

$$T_2\left(0, \tau_2, \xi, \eta, \varphi\right) = T_1\left(\tau_2, \xi, \eta, \varphi\right) \tag{8.51}$$

With the boundary conditions (8.50) and (8.51) the solution of the system of equations (8.24) to (8.26) in the small-angle approximation through the spherical harmonics method (8.34) to (8.36) has a following form:

$$\begin{aligned} R_1\left(\tau_1, \hat{\Omega}_0, \hat{\Omega}\right) &= \frac{\xi\eta}{\xi+\eta} \sum_{l,m} \frac{\lambda_1\left(2 - x_{1\,l}^{m}\right)}{\left(1 - \lambda_1\, x_{1\,l}^{m}/2\right)} \\ &\times \left(1 - \exp\left(-\left(1 - \frac{\lambda_1}{2}\, x_{1\,l}^{m}\right)\left(\frac{1}{\xi} + \frac{1}{\eta}\right)\tau_1\right)\right) Y_l^m(\hat{\Omega}_0) Y_l^m(-\hat{\Omega}) \end{aligned} \tag{8.52}$$

$$R_2(\tau_1,\tau_2,\hat{\Omega}_0,\hat{\Omega}) = \frac{\xi\eta}{\xi+\eta}\sum_{l,m}\frac{\lambda_2\,(2-x_{2\,l}^m)}{(1-\lambda_2\,x_{2\,l}^m/2)} \times\left(1-\exp\left(-\left(1-\frac{\lambda_2}{2}\,x_{2\,l}^m\right)\left(\frac{1}{\xi}+\frac{1}{\eta}\right)\tau_2\right)\right) \times\exp\left(-\left(1-\frac{\lambda_1}{2}\,x_{1\,l}^m\right)\left(\frac{1}{\xi}+\frac{1}{\eta}\right)\tau_1\right)Y_l^m(\hat{\Omega}_0)Y_l^m(-\hat{\Omega}) \tag{8.53}$$

$$T_1(\tau_1,\Delta,\hat{\Omega}_0,\hat{\Omega}) = \frac{\xi\eta}{2}\sum_{l,m}\left\{\frac{\lambda_1 x_{1l}^m}{\eta(1-\lambda_1 x_{1l}^m/2)-\xi}\left(\exp\left(-\frac{\tau_1}{\eta}\right)-\exp\left(-(1-\lambda_1 x_{1l}^m/2)\frac{\tau_1}{\xi}\right)\right)\exp\left(-\frac{\tau_2}{\eta}\right) + \frac{\lambda_1 x_{1l}^m}{\xi(1-\lambda_1 x_{1l}^m/2)-\eta}\left(\exp\left(-\frac{\tau_1}{\xi}\right)-\exp\left(-(1-\lambda_1 x_{1l}^m/2)\frac{\tau_1}{\eta}\right)\right)\exp\left(-\left(1-\frac{\lambda_2}{2}x_{2l}^m\right)\frac{\tau_2}{\eta}\right)\right\} \times Y_l^m(\hat{\Omega}_0)Y_l^m(\hat{\Omega}) \tag{8.54}$$

$$T_2(\tau_1,\tau_2,\Delta,\hat{\Omega}_0,\hat{\Omega}) = \frac{\xi\eta}{2}\sum_{l,m}\left\{\frac{\lambda_2 x_{2l}^m}{\eta(1-\lambda_2 x_{2l}^m/2)-\xi}\left(\exp\left(-\frac{\tau_2}{\eta}\right)-\exp\left(-(1-\lambda_2 x_{2l}^m/2)\frac{\tau_2}{\xi}\right)\right)\exp\left(-\frac{\tau_1}{\xi}\right) + \frac{\lambda_2 x_{2l}^m}{\xi(1-\lambda_2 x_{2l}^m/2)-\eta}\left(\exp\left(-\frac{\tau_2}{\xi}\right)-\exp\left(-(1-\lambda_2 x_{2l}^m/2)\frac{\tau_2}{\eta}\right)\right)\exp\left(-\left(1-\frac{\lambda_1}{2}x_{1l}^m\right)\frac{\tau_1}{\xi}\right)\right\} \times Y_l^m(\hat{\Omega}_0)Y_l^m(\hat{\Omega}) \tag{8.55}$$

These solutions contain the singularity as well as (8.38) and (8.39) which should be subtracted.

From the last equations one can derive the solution for the important case from the practical point of view – the homogeneous layer $\tau_1 = \tau$ on the semi-infinite substrate $\tau_2 = \infty$. The reflection functions for the layer R_1 and the substrate R_2 can be expressed as

$$R_1(\tau,\hat{\Omega}_0,\hat{\Omega}) = \frac{\xi\eta}{\xi+\eta}\sum_{l,m}\frac{\lambda_1\,(2-x_{1\,l}^m)}{(1-\lambda_1\,x_{1\,l}^m/2)} \times\left(1-\exp\left(-\left(1-\frac{\lambda_1}{2}\,x_{1\,l}^m\right)\left(\frac{1}{\xi}+\frac{1}{\eta}\right)\tau\right)\right)Y_l^m(\hat{\Omega}_0)Y_l^m(-\hat{\Omega}) \tag{8.56}$$

$$R_2(\infty,\hat{\Omega}_0,\hat{\Omega}) = \frac{\xi\eta}{\xi+\eta}\sum_{l,m}\frac{\lambda_2 x_{2\,l}^m\,(\lambda_2-1)}{(1-\lambda_2\,x_{2\,l}^m/2)} \times\exp\left(-\left(1-\frac{\lambda_1}{2}\,x_{1\,l}^m\right)\left(\frac{1}{\xi}+\frac{1}{\eta}\right)\tau\right)Y_l^m(\hat{\Omega}_0)Y_l^m(-\hat{\Omega}). \tag{8.57}$$

If the spectrum is measured in the relative units, then an informative value is the ratio R_1/R_2. This value is a function of the layer thickness τ. Therefore, the magnitude of the layer thickness can be determined by measuring the reflectance of the elastically scattered electrons.

The reader might be wondering why we use sophisticated approximations instead of rigorous numerical RTE solutions. The reason is that to analyze the surface structure we have to divide the reflection function for the whole system into the reflection functions for each layer. In other words, we have an additional variable – the energy loss. Thus, the main reason for the usage of approximate models is to perform such a split but not to solve the boundary problem.

8.4.7 Scattering by a multi-component sample

To calculate the reflectance of the elastically scattered electrons for a two-component sample, two things should be noticed:

1. The inelastic scattering in the solids with elements A+B is not a superposition of the inelastic scattering in the single-component solid samples with elements A and B. It is caused primarily by the inelastic energy losses on the plasmons.
2. The recoil energy in the small-angle scattering is too small to be measured in the present experiments [2].

That is why the intensity of the elastically scattered electrons for a sample with elements A and B can be calculated as follows. The total intensity $R_{\mathrm{A+B}}$ for a multi-component layer is derived using the effective elastic cross-sections, while the inelastic cross-sections are computed by TPP-2M equation [68]. The effective elastic cross-section is computed in the following way:

$$\sigma^{\mathrm{el}}_{\mathrm{A+B}} = \frac{n_{\mathrm{A}}\sigma^{\mathrm{el}}_{\mathrm{A+B}} + n_{\mathrm{B}}\sigma^{\mathrm{el}}_{\mathrm{A+B}}}{n_{\mathrm{A}} + n_{\mathrm{B}}} \tag{8.58}$$

Then the total intensity is divided proportionally to the value of the single scattering phase function for a certain angle and a concentration of the specific element.

$$R_{\mathrm{A}} = \frac{n_{\mathrm{A}} x_{\mathrm{A}}(\Omega, \Omega_0)}{n_{\mathrm{A}} x_{\mathrm{A}}(\Omega, \Omega_0) + n_{\mathrm{B}} x_{\mathrm{B}}(\Omega, \Omega_0)} R_{\mathrm{A+B}} \tag{8.59}$$

8.5 Backscattering from a semi-infinite sample

This section is intended to summarize the RTE solutions for semi-infinite medium as well as to present a synthetic approach. The elastic electron scattering takes place on atoms of a sample. The potential of those atoms is spherically symmetrical. Therefore, the single scattering phase function $x(\hat{\Omega}_0 \cdot \hat{\Omega})$ depends only on the cosine of the scattering angle $\gamma = \hat{\Omega}_0 \cdot \hat{\Omega}$. It can be expanded into series by azimuthal harmonics, as in the case of the optical radiation scattering [19, 20]:

$$x(\hat{\Omega}_0 \cdot \hat{\Omega}) = \sum_{m=0}^{\infty} x^m(\xi, \eta) \cos(m\varphi) \tag{8.60}$$

where $x^m(\xi,\eta)$ are the azimuthal harmonics of the single scattering phase function. Such representation leads to expansion of the backscattering function into series by azimuthal harmonics $R(\tau,\xi,\eta,\varphi)$:

$$R(\tau,\xi,\eta,\varphi) = \sum_{m=0}^{\infty} R^m(\tau,\xi,\eta)\cos(m\varphi) \tag{8.61}$$

In the case of semi-infinite medium, the derivative over optical depth in (8.18) vanishes. The equation (8.18) implying (8.60) and (8.61) for the reflection by the semi-infinite medium reads as:

$$\begin{aligned} &\frac{1}{\xi}R^m(\xi,\eta) + R^m(\xi,\eta)\frac{1}{\eta} = \frac{\lambda}{4}x^m(\xi,-\eta) \\ &+\frac{\lambda}{2}\int_0^1 x^m(\xi,\eta')R^m(\eta',\eta)\frac{d\eta'}{\eta'} + \frac{\lambda}{2}\int_0^1 R^m(\xi,\eta')x^m(\eta',\eta)\frac{d\eta'}{\eta'} \\ &+\lambda\int_0^1\int_0^1 R^m(\xi,\eta')x^m(-\eta',\eta'')R^m(\eta',\eta)\frac{d\eta''}{\eta''}\frac{d\eta'}{\eta'} \end{aligned} \tag{8.62}$$

Eq. (8.62) will be called the Ambartsumian equation.

8.5.1 The expansion by the number of elastic collisions

Let us consider the most natural solution method of Eq. (8.62) from the physical point of view: the computations of the multiple elastic collisions. This method is based on the neglect of the elastic collisions of the number more than a certain number. In the theory of the optical radiative transfer, this method has also been used [69, 20]. In references [70, 71], the multiple elastic scattering method was used to solve the problem of the elastic electron backscattering.

Now we apply the multiple elastic scattering method to solve Eq. (8.62). If the integral terms in (8.62) are removed, we obtain a solution describing backscattering with the single elastic scattering only. Then, substituting this expression into the right-hand side of Eq. (8.62), we regard the double elastic backscattering only. Repeated use of this iterative scheme results in a solution in an expanded Neumann form:

$$R^m(\xi,\eta) = \frac{\eta\xi}{\eta+\xi}\sum_{k=1}^{\infty}\rho_k^m(\xi,\eta)\lambda^k \tag{8.63}$$

The physical meaning of the expansion by degrees of λ^k is a representation of the universal backscattering function in the expanded series by the number of elastic collisions.

Using (8.63), the recurrent relations for the functions $\rho_k^m(\xi,\eta)$ can be derived:

$$\rho_1^m(\xi,\eta) = \frac{1}{4}x^m(\xi,-\eta)$$

$$\rho_2^m(\xi,\eta) = \frac{\eta}{8}\int_0^1 x^m(\xi,\eta')\,x^m(\eta',-\eta)\,\frac{d\eta'}{\eta'+\eta} + \frac{\xi}{8}\int_0^1 x^m(\xi,-\eta')\,x^m(\eta',\eta)\,\frac{d\eta'}{\eta'+\xi}$$

$$\rho_k^m(\xi,\eta) = \frac{\eta}{2}\int_0^1 x^m(\xi,\eta')\,\rho_{k-1}^m(\eta',\eta)\,\frac{d\eta'}{\eta'+\eta} + \frac{\xi}{2}\int_0^1 \rho_{k-1}^m(\xi,\eta')\,x^m(\eta',\eta)\,\frac{d\eta'}{\eta'+\xi}$$
$$+\eta\xi\sum_{i=1}^{k-2}\int_0^1\int_0^1 \rho_i^m(\xi,\eta')\,x^m(\eta',-\eta'')\,\rho_{k-1-i}^m(\eta',\eta)\,\frac{d\eta'}{\eta'+\xi}\frac{d\eta''}{\eta''+\eta},\quad k>2 \tag{8.64}$$

The functions $\rho_k^m(\xi,\eta)$ determine the probability that the particle changes its moving direction in the medium exactly k times.

With the modern level of computer engineering, it is easy to perform the calculations by Eq. (8.64) as in [72]. The Neumann series (8.63) converges rapidly whenever the single albedo λ is small enough. Then the contribution of collisions of the small numbers is significant while the high orders of scatterings can be neglected. If λ approaches to one, the series (8.63) converges extremely slowly, as the functions $\rho_k^m(\xi,\eta)$(8.64) are slowly decaying by the large numbers of collisions. In this case, the multiple scattering will dominate over the single scattering, and it will be necessary to consider the hundreds of the series terms (8.63) to obtain an acceptable computing accuracy. In addition, if the single scattering phase function is strongly anisotropic, the numerical integration within the given accuracy in the iterative formula (8.64) will require many integration nodes. Such difficulties make computations based on Eqs. (8.63) and (8.64) inefficient.

8.5.2 Expansion by the number of 'strong' elastic scatterings

Let us select a plane which is parallel to the sample surface at any depth. Then we divide the radiation field crossing the plane into upward and downward currents. For the backscattering at least one 'strong' elastic collision takes place. It changes the direction of the downward current onto upward. Besides 'strong' collusions there are the multiple collisions that do not change the radiation line with respect to the surface normal. Such separation is unique and relevant for each collision depending on the scattered radiation direction. For a backscattering, the number of the 'strong' collisions should be odd.

The selection of a 'strong' elastic scattering enables to expand the universal backscattering function by the number of 'strong' collisions:

$$R^m(\xi,\eta) = \sum_{k=0}^{\infty} R_{2k+1}^m(\xi,\eta) \tag{8.65}$$

The series (8.65) converges hundreds times faster than (8.63) (see Fig. 8.8). The reason is that each term of series relates to the multiple elastic scattering that does not change the moving direction with respect to the surface normal. The greater the anisotropy of the single scattering phase function, the less the 'strong' scattering probability and the fewer the terms of Eq. (8.65) that are to be considered to reach the given accuracy.

If the expansion by the number of 'strong' collisions (8.65) is used, the nonlinear integral equation (8.62) breaks up into linear equations with a known right-hand side for each odd number of the 'strong' elastic collisions:

$$\int_0^1 G(\xi,\eta')\, R_{2k+1}^m(\eta',\eta)\,\frac{d\eta'}{\eta'} + \int_0^1 R_{2k+1}^m(\xi,\eta')\, G(\eta',\eta)\,\frac{d\eta'}{\eta'} = B_k(\xi,\eta) \quad (8.66)$$

where

$$B_k(\xi,\eta) = \sum_{i=0}^{k-1}\int_0^1\int_0^1 R_{2i+1}^m(\xi,\eta')\, D(\eta',\eta'')\, R_{2(k-i)-1}^m(\eta',\eta)\,\frac{d\eta''}{\eta''}\frac{d\eta'}{\eta'} \quad \text{for } k>0$$

$$D(\eta',\eta'') = \lambda\, x^m(-\eta',\eta''),\quad G(\xi,\eta) = \delta(\xi-\eta) - \frac{\lambda}{2}x^m(\xi,\eta).$$

8.5.3 The discrete ordinate method

Eqs. (8.62) and (8.66) are solved by the discrete ordinate method. For the first time, this method was used in transfer theory by Schuster [73] and Schwarzschild [74] who studied the RTE with an isotropic single scattering phase function. Separating the radiation field into the upward and downward flows, they replaced the RTE by a system of two equations. This method was developed and reworked by Chandrasekhar [19]. The discrete ordinate method is based on the quadrature formulae application. The quadrature formulae adopted for the various integrands can be found in [75]. To achieve a good accuracy, it is reasonable to choose the quadrature formulae of high precision order. The best results for smooth solutions are given by the Gauss or Gauss–Christoffel quadrature formulae. An elementary trapezoidal formula can be also used by making the grid twice as dense and precising the solution in turn. This also gives a good precision result, but requires considerably greater number of nodes than in the Gauss formulae.

We substitute the integrals in the integral equations by quadrature sums. For example:

$$\int_0^1 x^m(\xi,\eta')\, R^m(\eta',\eta)\,\frac{d\eta'}{\eta'} \approx \sum_{i=1}^{N} x^m(\xi,\psi_i)\, R^m(\psi_i,\eta)\,\frac{s_i}{\psi_i}$$

where ψ_i and s_i are the nodes and the weights of a quadrature formula. The number of nodes N in the discrete ordinates method depends on the anisotropy of the single scattering phase function and the required solution accuracy. The problem of choosing N will be considered below.

Then a grid $\{\xi_i,\eta_j\}$ of variables ξ and η is introduced:

$$\xi_i = \psi_i, i=1,\ldots,N, \eta_j = \psi_j, j=1,\ldots,N.$$

Also for this grid we determine: the matrix $\mathbf{R} = R^m(\xi_i,\eta_j)$ of the scattering function azimuthal harmonics, the matrices $\mathbf{G} = G(\xi_i,\psi_j)\, s_j/\psi_j$, $\mathbf{B}_0 = B_0(\xi_i,\eta_j)$

and $\mathbf{D} = s_i D(\xi_i, \psi_j) s_j / \xi_i \psi_j$. The matrices are square and have a size $N \times N$. The Ambartsumian equation (8.62) in the grid $\{\xi_i, \eta_j\}$ in the matrix form reads as:

$$\mathbf{G}\,\mathbf{R} + \mathbf{R}\,\mathbf{G}' = \mathbf{B}_0 + \mathbf{R}\,\mathbf{D}\,\mathbf{R} \tag{8.67}$$

where the sign '′' denotes a transposed matrix. Eq. (8.67) will be called the discrete Ambartsumian equation (DAE). The equation for the number of elastic collisions (8.66) in the same grid in the matrix form reads as:

$$\mathbf{G}\,\mathbf{R}_{2k+1} + \mathbf{R}_{2k+1}\,\mathbf{G}' = \mathbf{B}_k, \quad \mathbf{B}_k = \sum_{i=0}^{k-1} \mathbf{R}_{2i+1}\,\mathbf{D}\,\mathbf{R}_{2(k-i)-1} \tag{8.68}$$

8.5.4 Solution of the discrete Ambartsumian equation

In control theory, a nonlinear equation of the type (8.67) is called the *continuous-time algebraic Riccati equation* and Eq. (8.68) is called the *Lyapunov equation.* Some various methods for the solution of such matrix equations have been developed. The detailed reviews and the comparisons of those methods can be found in [76–79]. The following solution methods of the Lyapunov equation are assigned: the Bartels–Stewarts method [80], the Hessenberg–Schur method [81], and the solution methods of the Riccati equation: the Newton method [82–84], the sign function method [85], and the Schur vector method [86]. The algorithms, presented in widely distributed mathematical libraries, are based on these methods, for example, SLICOT (Subroutine LIbrary for COntrol Theory) [87].

The methods mentioned above are used, first of all, to solve control theory problems. To use these methods in transfer theory, they have to be adopted. In [88, 89], the non-symmetric algebraic Riccati equation (NARE) was derived for the particle transfer in media with an isotropic single scattering law. This equation is analogous to the DAE with an isotropic single scattering phase function. The NARE was solved using the Newton method in these works. In Ref. [90], to solve the NARE, a special algorithm was designed – the structure-preserving doubling algorithm (SDA).

Computations of scattered particle angular distributions in media with an isotropic single scattering law by the SDA method is several times faster than by the Newton method. As was mentioned above, the isotropic single scattering law is not valid for electron scattering. In this case, as numerical experiments evidence, the best result of the mentioned methods is achieved by the Newton method.

In this chapter, a new effective solution method of the DAE based on expansion by the number of 'strong' collisions is presented. We expand the DAE (8.67) by the number of 'strong' collisions. Then we get the Lyapunov equation (8.68) for each number of collisions. The solution of the Lyapunov equation within the standard Bartels–Stewarts method [80] is based on the Schur decomposition. In the Newton method (8.68) the Schur decomposition is performed for each iteration to solve the Lyapunov equation. For a highly anisotropic single scattering phase function, only a few iterations are needed. In this case, the standard Bartels–Stewarts method becomes inefficient. To solve Eq. (8.68), an expansion by the eigenvectors instead of the Schur decomposition is used. Such expansion has to be performed only once, so it increases the computation speed greatly. We will call it the multiple strong scattering method (MSS).

As [41] shows, the computational speed may increase by the given accuracy if the solution is divided into a small-angle part and a regular part. The code developed by V.P. Budak on the basis of this idea is called MDOM. The computational speed and efficiency can be increased if a combination of different methods is used. The first step is to calculate the initial approximation using Eq. (8.44). Then MSS is applied for some number of the strong scattering K_1 and a further J iterations are performed by the Newton method. Then we calculate K_2 strong calculations by MSS. This synthetic method is denoted by 'Newton and multiple strong scattering method (NMSS)'.

8.5.5 The computation accuracy and time

The algorithms and the solution methods of the Ambartsumian equation studied above are based on the discrete ordinates method. The accuracy of the Ambartsumian equation solution will be determined by the approximation accuracy of the integral equation (8.62) by the DAE (8.67). The approximation accuracy depends on the number of nodes N in the discrete ordinates method. The greater is N, the less is the approximation error. However, the greater is N, the longer is the computation time t, bearing in mind that $t \sim N^3$. For practical applications the computational speed cannot be divorced from accuracy considerations. Thus, a method of choosing the optimal number of nodes, N_{opt}, is needed to achieve the given accuracy within the minimal computing time. N_{opt} depends on the initial energy of the incident particle, various parameters of the scattering medium, the quadrature formula and the method of solution of the Ambartsumian equation.

The problem of estimating the accuracy of a nonlinear integral equation, such as the Ambartsumian equation, seems rather intricate. One solution of this problem is the following: the derived solution $\mathbf{R}$ is compared to an 'ideal' solution $\mathbf{R}_{\text{id}}$ in the grid $\{\xi_i, \eta_j\}$with the given integration nodes number N. Then the mean relative error is computed:

$$\delta R = \frac{\|\mathbf{R} - \mathbf{R}_{\text{id}}\|}{\|\mathbf{R}_{\text{id}}\|} \tag{8.69}$$

where $\|\ldots\|$ are the norms of the vectors and of the matrices. The 'ideal' solution is found by two different methods. In this chapter, the MDOM [41] and the NMSS ($J = 2$, $K_1 = 10$, $K_2 = 4$) methods are used for this purpose. The integration nodes number, N_{id}, was set so that the mean relative deviation of solutions R_{id} derived by these methods was the smallest. The mean relative deviation of the solutions obtained by two methods is $\delta R_{\text{id}} \cong 10^{-6}$. It has been achieved by $N_{\text{id}} = 500$. δR_{id} is limited by the computing accuracy: of the nodes ψ_i and the weights s_i of the quadrature formula, of the phase function expansion coefficients by the Legendre polynomials, and of the azimuthal harmonics of the phase function.

Since the solution method of the discrete equation (8.67) does not influence on the solution accuracy of the integral equation (8.62), the numerical integration error of the DAE should be considerably less than the approximation error of the integral Ambartsumian equation. The solution accuracy of the DAE (8.67) is estimated by the mean relative error using the expression approved in the control theory for the Riccati equation accuracy:

$$\delta R_{DAE} = \frac{\|\mathbf{GR} + \mathbf{RG}' - \mathbf{B}_0 - \mathbf{RDR}\|}{\|\mathbf{GR}\| + \|\mathbf{RG}'\| + \|\mathbf{B}_0\| + \|\mathbf{RDR}\|} \tag{8.70}$$

Given the number of the integration nodes N, the relative computing error of the phase function in the same grid using the normalizing condition $\int_{-1}^{1} x(\mu)\, d\mu = 2$ is estimated:

$$\delta x_N = \left| 1 - \sum_{i=1}^{N} \left(x(\psi_i) + x(-\psi_i) \right) s_i/2 \right| \tag{8.71}$$

The relative error δx_N can be used to estimate the quality of the computing method. If δx_N is less than the relative error δR for the considered method, the upcoming errors will be caused by the solution method itself. Thus, many more integration nodes than necessary will have to be taken to achieve the given accuracy.

To choose an optimal number of integration nodes, N_{opt}, the upper limit of the relative error ε has to be specified. For practical needs, an accuracy of about 1% is enough. For our calculations, let us set the upper limit of the relative error of 0.1%, so $\varepsilon \approx 10^{-3}$. Summarizing all above, let us write the necessary condition for choosing N_{opt}:

$$\delta R_{DAE} \ll \delta R \leq \delta x_N \leq \varepsilon \approx 10^{-3} \tag{8.72}$$

Let us pass to the numerical experiments. The experiments were performed for the backscattered and for the elastically backscattered electrons angular distributions. The single scattering phase functions are computed on the basis of the ELSEPA code [91], the mean inelastic path lengths are computed by the formula TPP-2M [92]. The programming environment is MATLAB 7.9, which is convenient for matrix multiplications. For our numerical experiments, the computer was used with the following features: the processor was Intel(R) Core(TM) 2 Duo CPU E6750 @ 2.66 GHz, the RAM size was 2~GB, the OS was Windows XP.

In Fig. 8.8(a), the mean relative error δR_{DAE}(at the left side) and the computation time t in seconds (at the right side) as functions of the iteration number (or the highest number of collisions in the calculated sum), K, for the various computation methods are plotted. The calculations were performed for the electrons of the initial energy $E_0 = 5\,\text{keV}$ impinging normally to the surface (in this case, just the zero harmonic is enough) onto a sample of Al. The number of nodes in the discrete ordinates method was fixed: $N = 100$. In Fig. 8.8(a), the computing results of the Newton method [84] are shown as asterisks, the results of the SDA method [90] are shown as five-point stars, and the results of the calculations with the Neumann series (8.63) and (8.64) are shown by the triangles. In the latter case, K indicates the number of elastic collisions. In Fig. 8.8(a), the results for the MSS method are shown (rhombs) and as well as for two versions of the NMSS method: (1) one Newton iteration $J = 1$ (hollow circles), (2) two Newton iterations $J = 2$, $K_2 = 4$ (solid circles). $K_1 = K$ shows the number of the 'strong' collisions the calculation was performed up to.

On the basis of the numerical experiments performed for various sample materials and initial electron energies, it can be concluded that the most effective (shortest computing time and best accuracy) solution method for the DAE is the

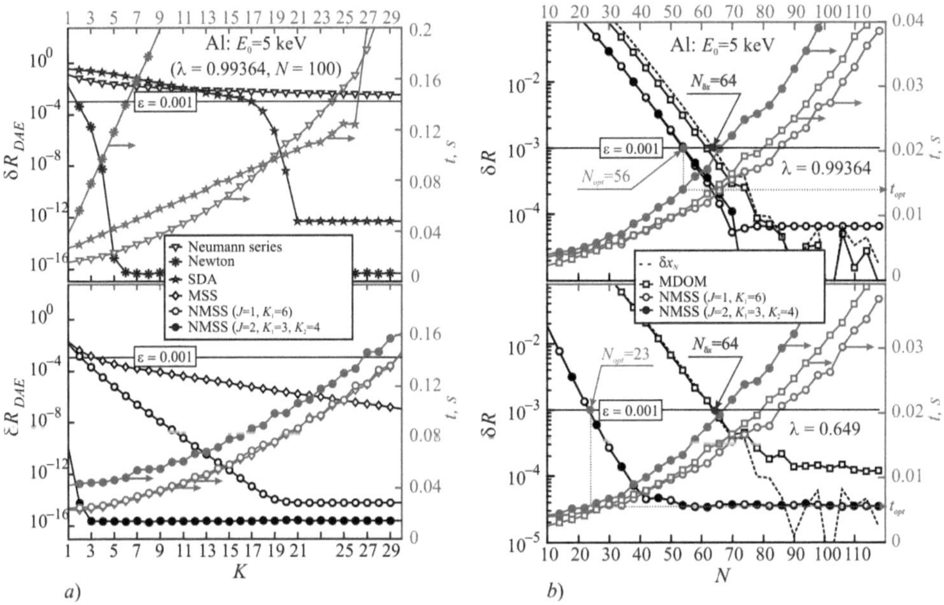

Fig. 8.8. (a) The comparison of the computation results by various methods: the mean relative error of the DAE solution (the log vertical scale on the left) and the computing time in seconds (the linear vertical scale on the right) over the iteration number K and the integration nodes number N. The computing results obtained by the following methods are plotted: the Newton method (asterisks), the SDA (five-pointed stars), the Neumann series (triangles), the MSS method (rhombuses), the NMSS with the parameters $J = 1$, $K_1 = 6$ (hollow circles), the NMSS with $J = 2$, $K_1 = 3$, $K_2 = 4$ (solid circles). (b) The comparison of the computation results by the various methods: the mean relative error (the log vertical scale on the left) and the computation time in seconds (the linear vertical scale on the right) depending on the iteration number K and the integration nodes number N. The computing results obtained by the following methods are plotted: the MDOM (squares), the NMSS with the parameters $J = 1$, $K_1 = 6$ (hollow circles), the NMSS with $J = 2$, $K_1 = 3$, $K_2 = 4$ (solid circles); the mean relative error δx_N is plotted by a dashed line.

NMSS method. The Newton method is not worse than the NMSS as regards accuracy, but the computation time is few times longer.

There is no need for a relative error $\delta R_{DAE} \approx 10^{-15}$ from the practical point of view. The accuracy $\delta R_{DAE} \approx 10^{-6}$ is enough not to increase the relative error δR. In this case, less K is used by calculation and therefore the computing time decreases. For example, for the NMSS method, parameters may be the following: $J = 1$ and $K_1 = 6$.

In Fig. 8.8(b), the mean relative error δR and the computing time versus the integration nodes number N for the various computation methods are plotted. The calculations were performed for the electrons with the initial energy $E_0 = 5$ keV impinging normally onto the surface of a sample made of Al. The results of the calculations by the MDOM method [41] are plotted by squares; calculations by the NMSS method with the parameters $J = 1$ and $K_1 = 6$ are plotted by the hollow circles; the calculations by the NMSS method with the parameters $J = 2$, $K_1 = 3$, $K_2 = 4$ are plotted by the solid circles. For the MDOM method $N_{\text{opt}} \approx N_{\delta x}$, for the NMSS method $N_{\text{opt}} < N_{\delta x}$.

Fig. 8.8(b) shows the optimal integration nodes number N_{opt} for the NMSS method with the parameters $J = 2$, $K_1 = 3$, $K_2 = 4$. Using N_{opt} the computing time is: $t_{\text{opt}} = 0.014\,\text{s}$ for $\lambda = 0.99364$, $t_{\text{opt}} = 0.006\,\text{s}$ for $\lambda = 0.649$.

8.5.6 Angular distributions of the elastically scattered electrons

For the graphical representation of the angular distribution computation results, it is convenient to use the planes of the incidence and pickup (see Fig. 8.9). Further we shall study the angular distributions in the pickup plane. In this case, to indicate the pickup angles, one variable θ is enough. The magnitude of θ is equal to the polar pickup angle and the sign of θ shows the azimuthal angle. If $\theta > 0$ $(+\theta)$, the azimuthal angle equals φ. If $\theta < 0$ $(-\theta)$, the azimuthal angle equals $\varphi + \pi$.

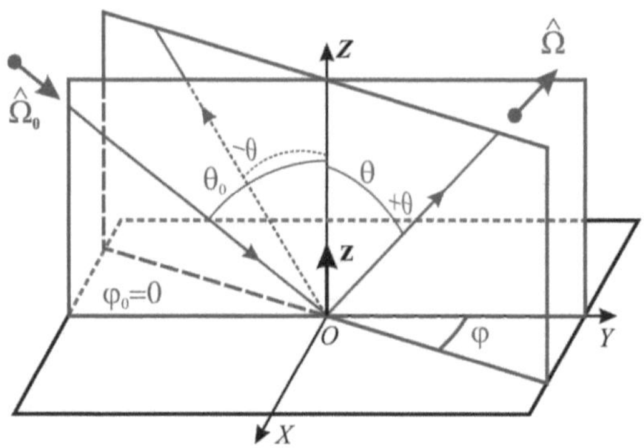

Fig. 8.9. The incidence and pickup planes

In Fig. 10, the angular distributions obtained by various methods are shown. The results of the NMSS method are plotted by a solid line; the results of the MDOM [41] are plotted by a dashed line; the results of the DISORT [40] are plotted by a dash-dot line. The DISORT is used in many programs designed for atmosphere radiation propagation modeling (Streamer, MODTRAN, SBDART). The DISORT may be downloaded from an FTP server [93]. In the figures, the angular distributions and the angular distribution absolute error depending on the pickup angle are shown. The absolute error was computed as the difference of the solution derived by the method and the 'ideal' solution. The solution by the MDOM method with the integration nodes number $N_{\text{id}} = 500$ was considered 'ideal'. For the comparison the results obtained by the MDOM and DISORT are multiplied by $\pi \cos\theta \cos\theta_0$. Fig. 8.9 shows the following items for each method: the integration node number N, the mean relative error δR, and the computation time t in seconds.

The calculations were performed for a pickup plane coinciding with the incidence plane $\varphi = 0$. In Fig. 8.10, the elastically backscattered electron angular dis-

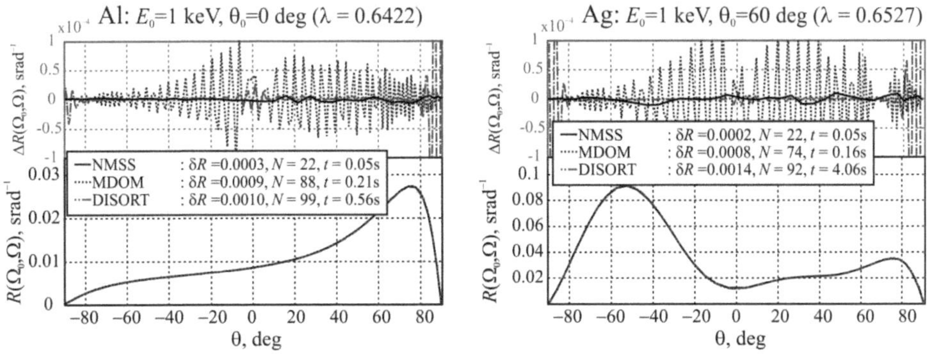

Fig. 8.10. The electron angular distributions and the absolute angular distribution error depending on the pickup angle. The computing results by the following methods are indicated: the NMSS (solid line), the MDOM (dashed line), the DISORT method (dash-and-dot line). The mean relative error δR, the integration nodes number N and the computing time t in seconds are given for each method.

tributions $R(\xi, \eta, \varphi)$ in the pickup plane and the elastically backscattered electrons angular distribution absolute error $\Delta R(\xi, \eta, \varphi)$ depending on the pickup angle are shown.

The efficiency of a method can be estimated by the product $T_{\text{eff}} = \delta R \cdot t$ of the mean relative error and the computation time. For the NMSS method $10^{-5} < T_{\text{eff}} < 10^{-4}$, for the MDOM $10^{-4} < T_{\text{eff}} < 10^{-3}$, for the DISORT $10^{-2} < T_{\text{eff}} < 10^{0}$. The method NMSS is synthetic. The NMSS method possesses the best features of the methods it is based on. This is an explanation for its high efficiency. In this case, computation time does not exceed a second for a wide range of the sample materials, for the initial energies from 0.2 to 50 keV and for any angle of incidence. At the same time, the mean relative error is less than 0.1%.

The NMSS method as any other solution method of the discrete Ambartsumian equation (Neumann series, MSS, Newton, SDA) computes the matrix of the scattering function values on the initial and on the sighting angle grids directly. The numerical experiments vividly demonstrate the high efficiency of the NMSS method among all solution methods of the discrete Ambartsumian equation. The codes MDOM and DISORT provide the vector on grid by the sighting angles for a certain initial angle. Thus, it is not quite correct to compare the calculation efficiency using the MDOM, DISORT and NMSS methods. Moreover, the codes can be optimized for the semi-infinite medium. The rigorous comparison will be the subject of complementary investigations exceeding the limits of this work.

8.6 Approbation of the theoretical models based on the discrete ordinate method (DISORT, MDOM, NMSS)

8.6.1 The comparison of DISORT, MDOM, NMSS calculations with Bronstein and Pronin experiments

The angular distributions of the elastically scattered electrons are useful for the verification of transport models in solids. The measurements of angular distributions of the elastically scattered electrons in the case of a single-component sample do not require a high-energy resolution. The problem of the angular distribution of the elastically scattered electrons is equivalent to the radiative transfer in the turbid media. Consequently, the single scattering albedo and the single scattering phase function provide a complete description of this problem. The single scattering model provides the following expression for the reflection function in the case of the semi-infinite medium:

$$R(\hat{\Omega}_0, \hat{\Omega}) = \lambda \frac{\xi\eta}{\xi + \eta}\, x(\xi, -\eta, \varphi) \tag{8.73}$$

In the Rubin–Everhart model the reflectance is described by the expression:

$$R(\hat{\Omega}_0, \hat{\Omega}) = \lambda \frac{\xi\eta}{\xi + \eta} \frac{1}{1-\lambda}\, x(\xi, -\eta, \varphi) \tag{8.74}$$

The expression for the angular distribution of the elastically scattered electrons in the small-angle approximation has the form:

$$R(\hat{\Omega}_0, \hat{\Omega}) = \frac{\xi\eta}{\xi + \eta} \sum_{l,m} \frac{x_l^m \lambda\,(\lambda - 1)}{(1 - \lambda\, x_l^m/2)} Y_l^m(\hat{\Omega}_0) Y_l^m(-\hat{\Omega}) \tag{8.75}$$

The reader can observe that in the first two cases the shape of the angular distribution follows the shape of the phase function in a backward direction. That does not agree with the experiment data. However, for $\lambda^k k1$ the expression (8.74) becomes (8.73). This case corresponds to a strong absorption (the probability of inelastic scattering is much higher than the probability of elastic scattering). Consequently, the particles that have experienced only one single elastic scattering dominate in the signal of the elastically scattered particles. The angular dependences of the single scattering approximation and Rubin–Everhart model are identical. However, the absolute values of intensities are different by an order of magnitude.

We give an analytical treatment of the pioneering experiments of Bronstein and Pronin [8] on the basis of the models presented in this chapter. The elastic scattering cross-sections are taken from [94], the values required to calculate the single scattering albedo are defined on the basis of a formula TPP-2M [68]. Table 8.1 provides the background information on the key parameters for the angular distributions of the elastically scattered electrons – the transport mean free path, the inelastic mean free path (IMFP), the elastic mean free path (EMFP), and the single scattering albedo.

A detailed description of the experimental setup is given in [9]. The energy of a probe beam is 1000 eV; the incidence angle is normal. Also, our calculations are compared to Monte Carlo simulations, performed by W. S. M. Werner [48].

Table 8.1.

$E0$,	Be ($Z=4$)				Cu ($Z=29$)				Ag ($Z=47$)				Au ($Z=79$)			
keV	l_{in}, nm	l_{el}, nm	l_{tr}, nm	λ nm	l_{in}, nm	l_{el}, nm	l_{tr}, nm	λ nm	l_{in}, nm	l_{el}, nm	l_{tr}, nm	λ nm	l_{in}, nm	l_{el}, nm	l_{tr}, nm	λ nm
0.5	1.22	1.87	14.8	0.39	1.09	0.45	1.64	0.71	0.91	0.51	1.72	0.64	0.84	0.39	1.57	0.68
1	2.06	3.51	47.9	0.37	1.74	0.64	3.75	0.73	1.47	0.71	3.37	0.67	1.33	0.55	2.36	0.71
2	3.59	6.78	160	0.35	2.93	0.95	9.59	0.76	2.45	1.00	7.80	0.71	2.21	0.75	4.63	0.75
5	7.75	16.5	820	0.32	6.11	1.68	37.4	0.78	5.09	1.66	27.5	0.75	4.58	1.18	14.2	0.80
10	14.0	32.2	2854	0.30	10.91	2.73	111.9	0.80	9.08	2.54	77.4	0.78	8.12	1.71	36.9	0.83
20	25.7	62.1	9964	0.29	19.75	4.60	350.8	0.81	16.4	4.00	228	0.80	14.63	2.55	100	0.85
40	47.4	117.2	32970	0.29	36.20	7.94	1087	0.82	30.1	6.50	667	0.2	26.7	3.89	277	0.87

Figures 8.11–8.14 show the angular distribution for Be, Cu, Ag and Au, respectively. For Be (Fig. 8.11) all angular distributions are in a good agreement. Here we can also observe that the error of the approximate models is higher for the heavy elements (Ag, Au). However, the experiment data, the simulations and DISORT, MDOM, NMSS solutions are in good coincidence. Since the deviance between these codes is negligibly small, we will not distinguish the solutions of these methods on the plots by different notations. The common notation 'DOM' is used for all of them.

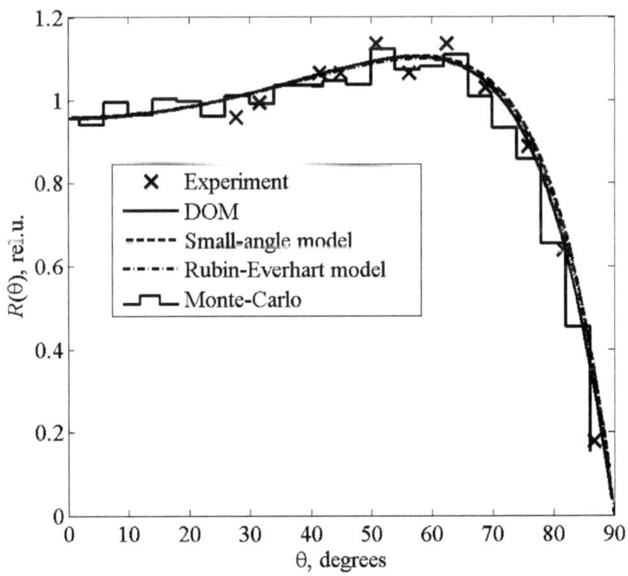

Fig. 8.11. The angular distributions of the elastically reflected electrons for Be – the experiment and the calculations; the probing energy is 1000 eV.

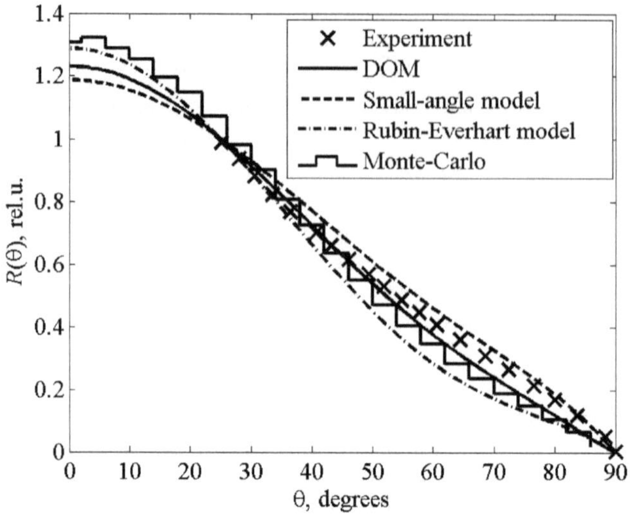

Fig. 8.12. The same as for Fig. 8.11, but for Cu.

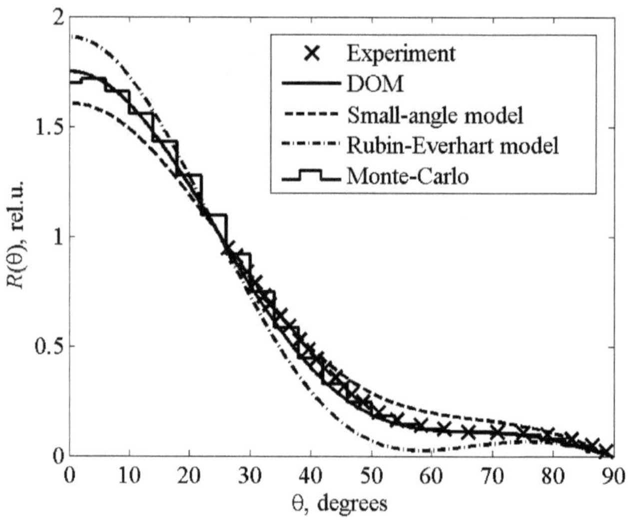

Fig. 8.13. The same as for Fig. 8.11, but for Ag.

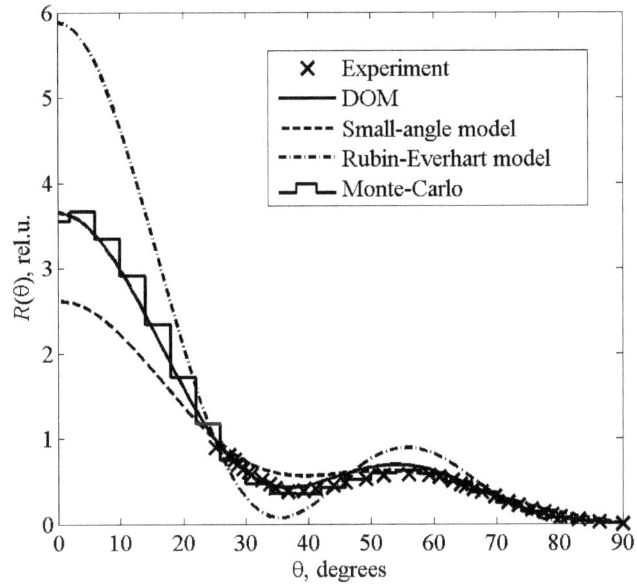

Fig. 8.14. The same as for Fig. 8.11, but for Au.

The DOM solution coincides with the experimental data and Monte Carlo simulations (Monte Carlo simulations are taken from [48]). This fact reveals the possibility of using the discrete ordinate method for the calculations of the elastic electron reflectance. The solution can be obtained numerically with the accuracy limited by the accuracy of the phase function and the single scattering albedo. From another point of view, the discrete ordinate method can be used for the verification of the approximate transport models.

8.6.2 Influence of multiple scattering on the form of angular distributions of the elastically scattered electrons

In the 1–10 keV energy range there is a lack of the experimental data on the angular distributions of the elastically scattered electrons. DOM calculations can be compared to the results of Monte Carlo simulations. Figures 8.15–8.17 show the angular distributions of the elastically scattered electrons at 2–10 keV. The exact solution and Monte Carlo simulations agree while the calculations in the quasi-single model and in the small-angle approximation describe the angular dependence with a significant error.

For energies exceeding 10 keV there is no experimental data for angular distributions of the elastically scattered electrons. Monte Carlo simulations can be problematic, as the probability of the backscattering due to elastic processes is low. However, in some works [50, 54] the experiments were carried out for energy 40 keV. And the Rubin–Everhart approximation is used to describe the experimental data [33]. Now we evaluate the error of the Rubin–Everhart model in this case.

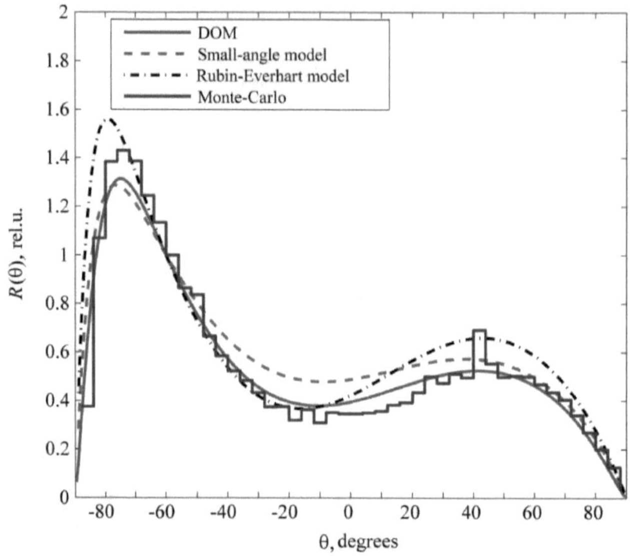

Fig. 8.15. The angular distributions of the elastically reflected electrons for Au – the calculations; the probing energy is 2000 eV, the angle of incidence is 60 degrees to the surface normal.

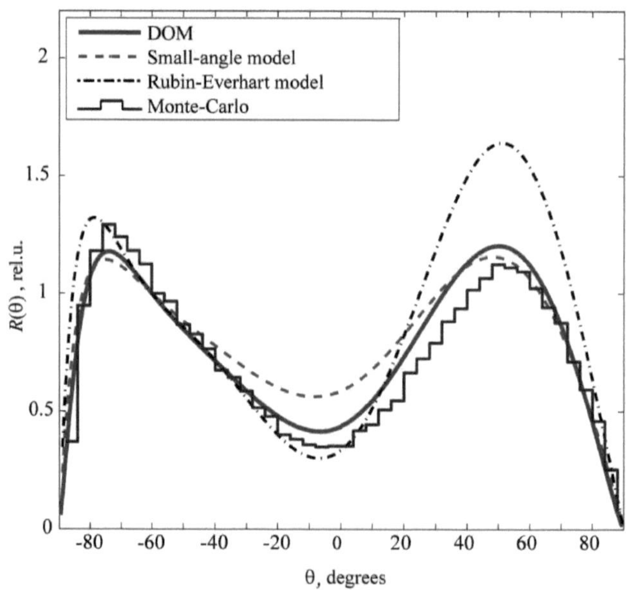

Fig. 8.16. The same as Fig. 8.15, but for probing energy 5000 eV.

Figures 18 and 19 show the evolution of the angular distributions with increasing energy. At 40 keV the small-angle approximation and Rubin–Everhart model almost coincide with the solution DOM.

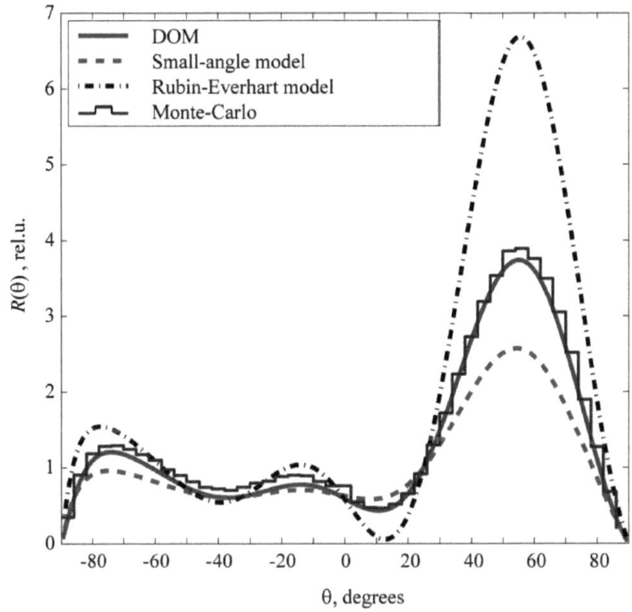

Fig. 8.17. The same as Fig. 8.15, but for 10 000 eV.

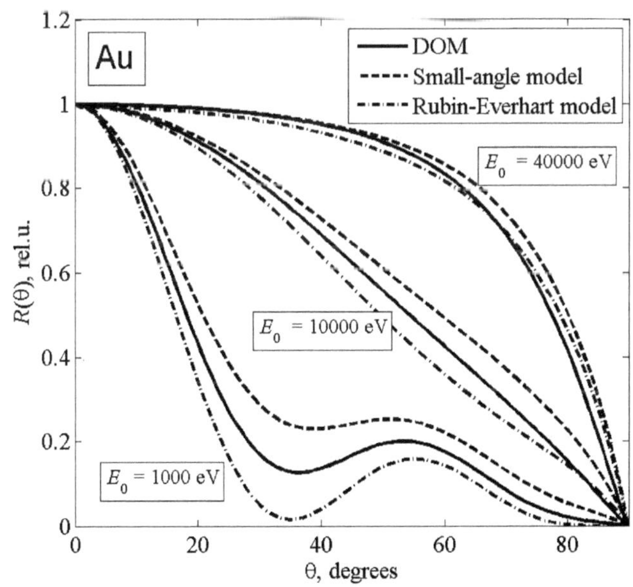

Fig. 8.18. The angular distributions of the elastically scattered electrons for the different probing energies – Au, the normal angle of incidence.

Note that Rubin–Everhart model gives the solution, which is near to DOM solution. For other elements and the finite thicknesses, the error of the approximate models is even lower. Thus, Figs. 8.18 and 8.19 can be considered as the justification

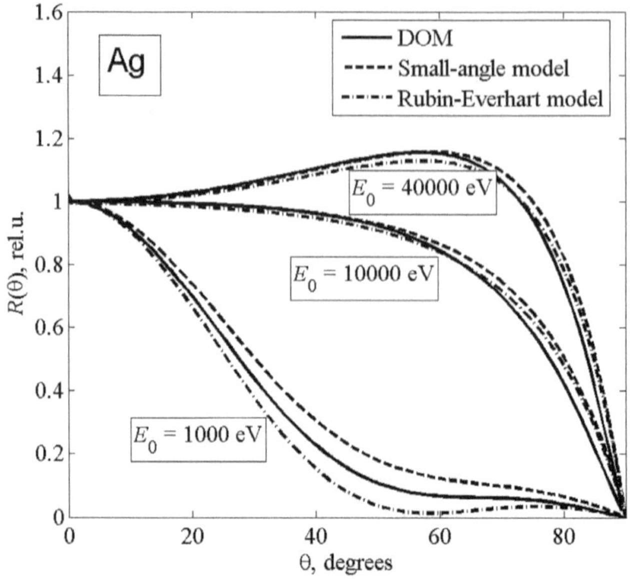

Fig. 8.19. The angular distributions of the elastically scattered electrons for the different probing energies – Ag, the angle of incidence is normal.

of the usage of Rubin–Everhart model for EPES spectra calculations at rather high energies.

8.6.3 The asymptotic formula for angular distributions of the elastically scattered electrons

In the previous paragraphs one can see that the error of the approximate models is decreasing with the increasing energy. Let us give an explanation. The reflection function of the elastically reflected electrons is determined only by the albedo and the scattering phase function. We will analyze these two quantities. The reason for errors of the approximate models is an inaccurate account of the multiple scatterings. Figure 8.20 shows the single scattering albedo for Be, Ag, Cu, Au as function of the probing energy. In the energy range of interest 10–100 keV, the single scattering albedo changes slightly.

Let us analyze the change of the elastic scattering phase function with increasing energy. The degree of elongation of the phase function is determined by the most important parameter of particle transport in solids – the transport mean free path l_{tr} – the distance covered on average by a collimated beam before it has lost the memory of its original direction. If the energy increases, then the degree of anisotropy of the phase function also increases (it is becoming more stretched-forward). This means that the multiple scattering will become dominantly 'small-angle'. The probability of large-angle scattering is reduced and negligible as compared to the small-angle scattering. In addition, it is necessary to take into account the inelastic scattering (the absorption of the particles). The main quantitative parameter that determines the contribution of the multiple scattered particles in the intensity of the scattered particles is the ratio $l_{\text{in}}/l_{\text{tr}}$, which decreases with the increasing energy. It is shown in Fig. 8.21.

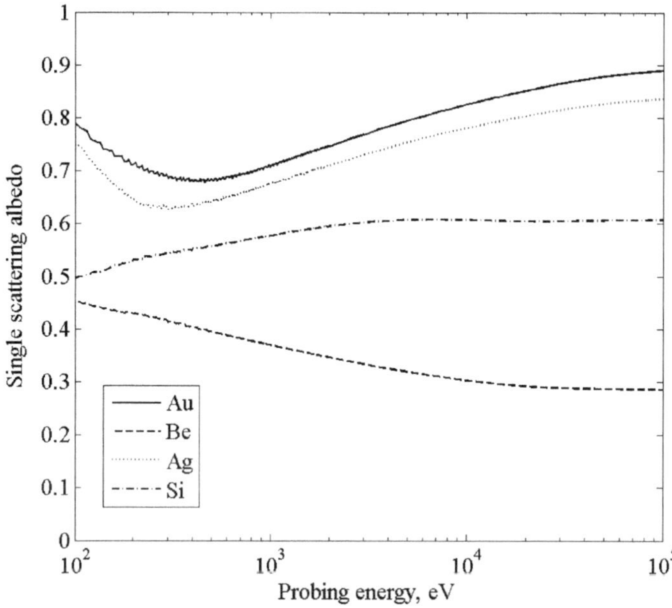

Fig. 8.20. The dependence of the single scattering albedo on the probing energy.

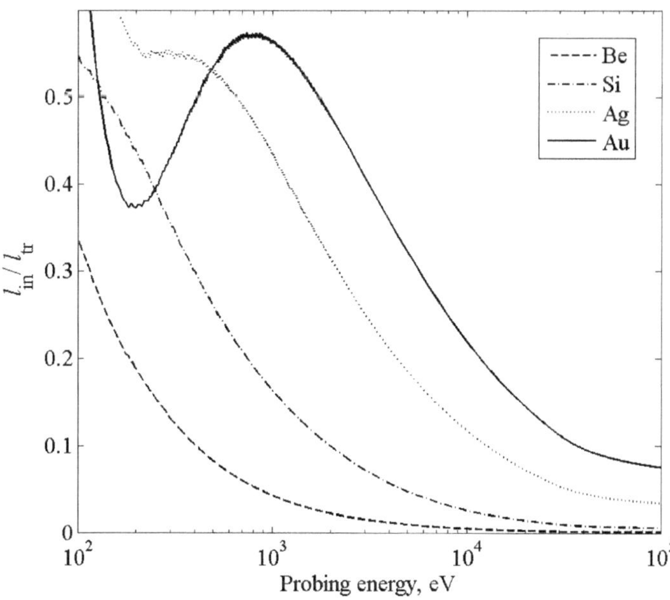

Fig. 8.21. The dependence of ratio $l_{\text{in}}/l_{\text{tr}}$ versus the probing energy.

In other words, the pass-length of the elastically reflected electrons is much smaller than the transport path. Therefore, at high probing energies, the shape of the angular distributions of the elastically scattered electrons can be described by the Rubin–Everhart model and by the single scattering model. However, in the calculations of the intensity absolute values, the single scattering model makes an error by factor $1-\lambda$, as it ignores the particles that are scattered in a 'straight-forward' direction. If $\lambda \ll 1$ then the number of such particles is negligibly small, and both models give the same results. In the single-scattering model, the shape of the angular distribution follows the shape of the scattering phase function:

$$R(\xi,-\eta,\varphi) = x(\xi,-\eta,\varphi)\,\frac{\xi\eta}{\xi+|\eta|} \tag{8.76}$$

With increasing energy, the differential elastic scattering cross-section of all elements (and hence their scattering phase functions) are getting closer to the Rutherford cross-section (as shown in Fig. 8.22):

$$x_R(\theta) = \left(\frac{Ze^2}{16\pi E\varepsilon_0}\right)^2 \left(\sin\left(\frac{\theta}{2}\right)^2 + \delta^2\right)^{-2} \tag{8.77}$$

where δ is the screening parameter of the Coulomb potential. Consequently, at sufficiently high energies the angular distributions are described by a universal dependence for all elements. Since we are interested in the relative values, the rate

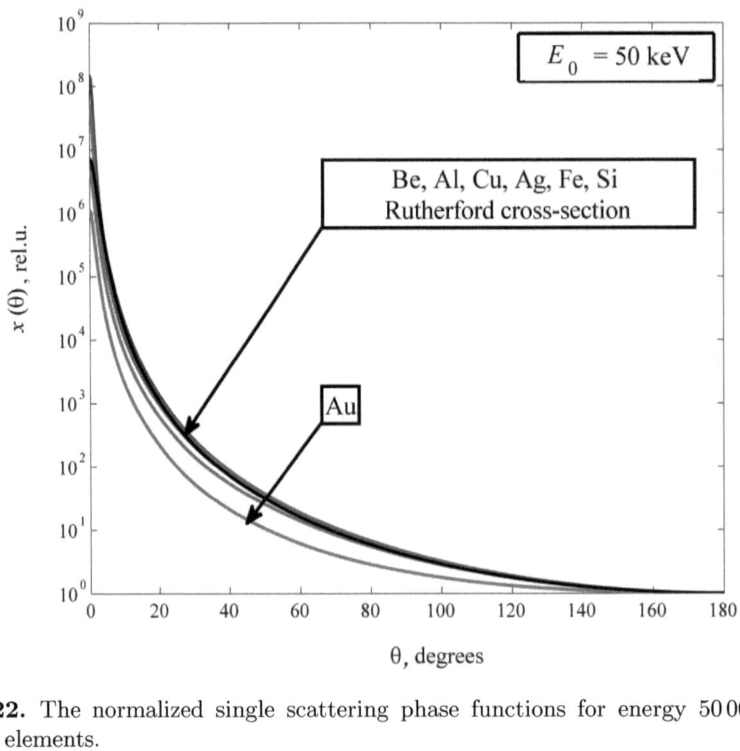

Fig. 8.22. The normalized single scattering phase functions for energy 50 000 eV for different elements.

$\left[(Ze^2)/(16\pi E\varepsilon_0)\right]^2$ is discarded. Then the expression (8.77) is substituted in (8.76) for the case of the normal incidence ($\xi = 1$, $\cos\theta = \eta$). Also the screening parameter can be ignored, since it affects the small-angle scattering and does not affect the reflection. After some simplifications one can derive an expression which describes the asymptotic behavior of the angular distribution of the elastically scattered electrons for the normal angle of incidence:

$$R_a(\theta) = \frac{\lambda}{1-\lambda}\frac{\cos\theta}{\cos^6\frac{\theta}{2}} \tag{8.78}$$

The angular distribution for 50 keV is plotted on Fig. 8.23. With the exception of Au the angular distribution for all elements come to the function (8.78). For gold the angular distribution does not come to (8.78) also at greater energy. The energy of the electrons in the Au inner shells is comparable to the energy of the probing electrons [95]. That is why the differential cross-section of Au is not described by the Rutherford cross-section with enough accuracy. In addition, the shape of the differential cross-section is affected by the relativistic effects.

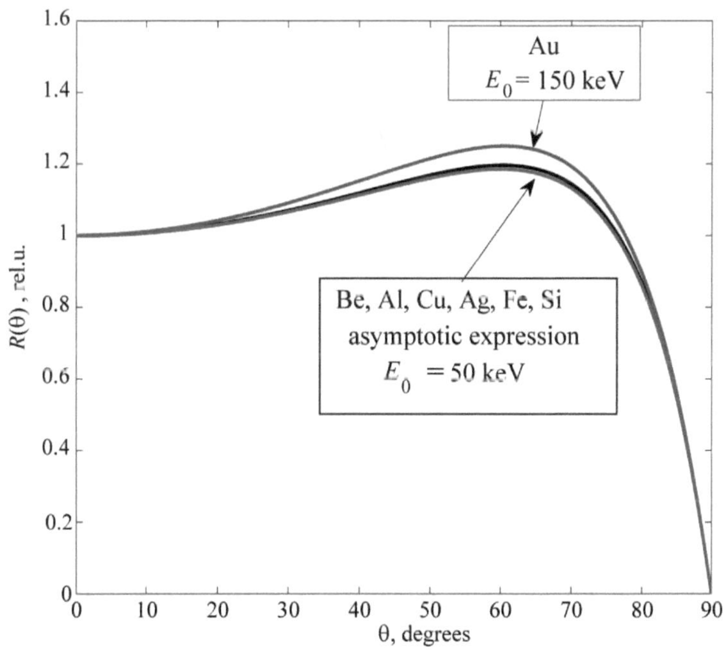

Fig. 8.23. The angular distributions of electrons – asymptotic behavior.

8.6.4 Effects of the multiple scattering on the total elastic reflection coefficient

Until now, we have been interested in the shape of angular distributions. Since the spectra are usually measured in relative units, their normalization could be arbitrary. However, it is technically difficult, but possible, to measure angular distributions in absolute values – for a fixed current of the probing electrons for the different samples. The total reflection coefficient depends on the number Z. Then the question of the absolute values of the angular distribution becomes relevant.

The simulations are performed for the semi-infinite medium of Cu, Ag and Au for the normal angle of incidence. We compute the flux of the elastically reflected electrons using the Rubin–Everhart model and DOM. The error of the Rubin–Everhart model as the function of the probing energy is shown in Fig. 8.24. The simulations evidence that the Rubin–Everhart model yields an error level below 5% at the energy 5–15 keV for medium elements (Cu, Ag) and at the energy 30 keV for heavy elements (Au).

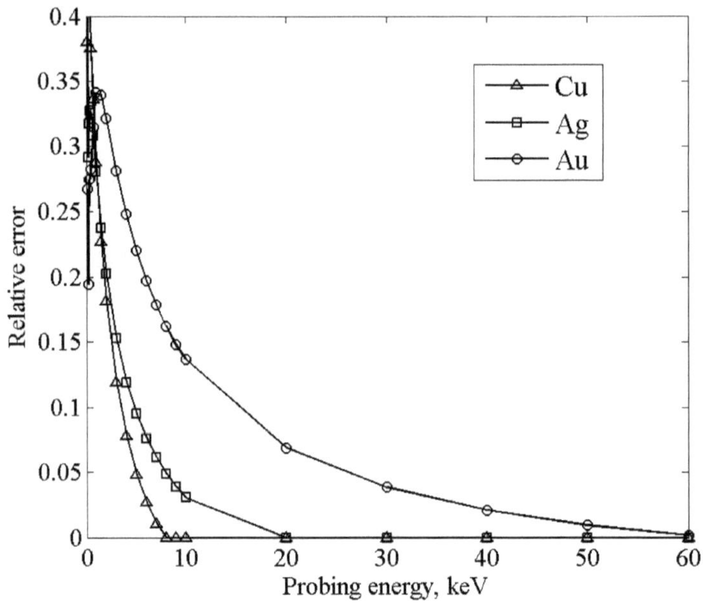

Fig. 8.24. The error of Rubin-Everhart model for the semi-infinite medium as a function of the probing energy.

8.6.5 The influence of surface plasmons on the angular distribution of elastically scattered electrons

Plasmons are the excitations of the electron system of solids. Due to the influence of the surface plasmons, the effective single scattering albedo can be changed. Indeed, when the electron approaches the surface of a solid, it loses the energy for the excitation of plasmons. In this case elastic scattering is not possible, because of the screening by the electron gas. That is why the single scattering albedo $\lambda = \sigma_{\rm el}/(\sigma_{\rm el}+\sigma_{\rm in})$ at the surface should be less than at the depth of the sample. In other words, the inelastic mean free path-length depends on the depth. This conclusion was also reached by the authors of paper [96], but from the pure quantum-mechanics approach. To verify this idea, the angular distributions from the paper [97] are used. They are interpreted in the three-layer model (Fig. 8.25) with different albedo values in each layer.

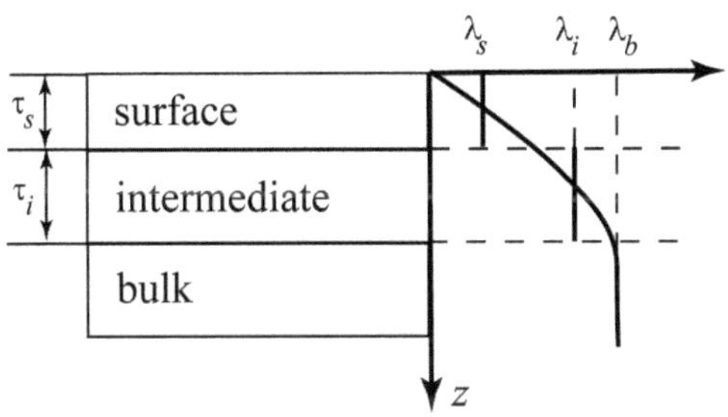

Fig. 8.25. Three-layer model of a sample with different values of the single scattering albedo.

It is difficult to determine the thickness of the layers in nanometers, as the calculation gives only the relative values (the thickness over the inelastic mean free path). The 'absolute values' of the inelastic and elastic mean free paths are sometimes unknown. Table 8.2 provides the recovered albedo dependence on depth. The code DISORT has been used to compute the angular distributions. The distributions are plotted in Figs. 8.26 and 8.27.

It is interesting to note that for heavy elements, the angular distributions do not need 'the multilayer correction'. They are well described on the basis of the tabulated values of the single scattering albedo.

Table 8.2. The dependence of the albedo on the depth of the sample.

	Si	**Fe**
Layer 1 (surface)	$\lambda_s = \lambda_b/2.5$, $\tau_s = l_{\rm in}/5$	$\lambda_s = \lambda_b/2$, $\tau_s = l_{\rm in}/5$
Layer 2 (intermediate)	$\lambda_i = \lambda_b/1.2$, $\tau_i = l_{\rm in}/3$	$\lambda_i = \lambda_b/1.4$, $\tau_i = l_{\rm in}/4$
Layer 3 (bulk)	$\lambda_b = 0.57$	$\lambda_b = 0.71$

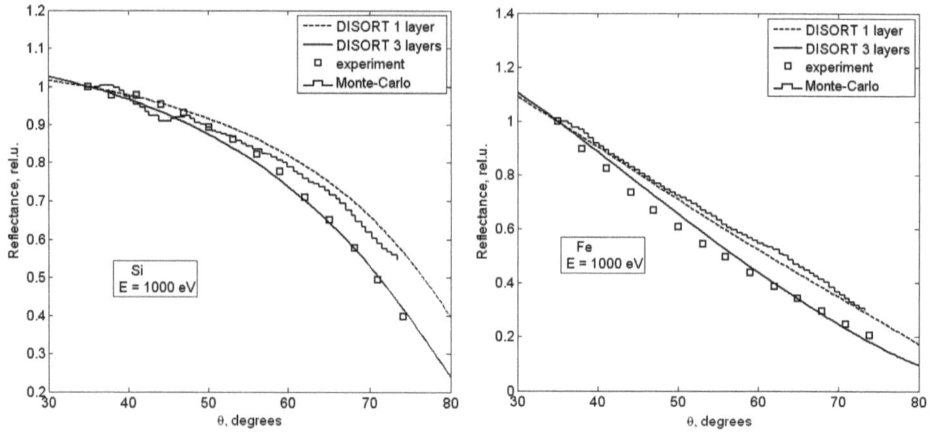

Fig. 8.26. The angular distributions of the elastically scattered electrons for Si and Fe – 1000 eV.

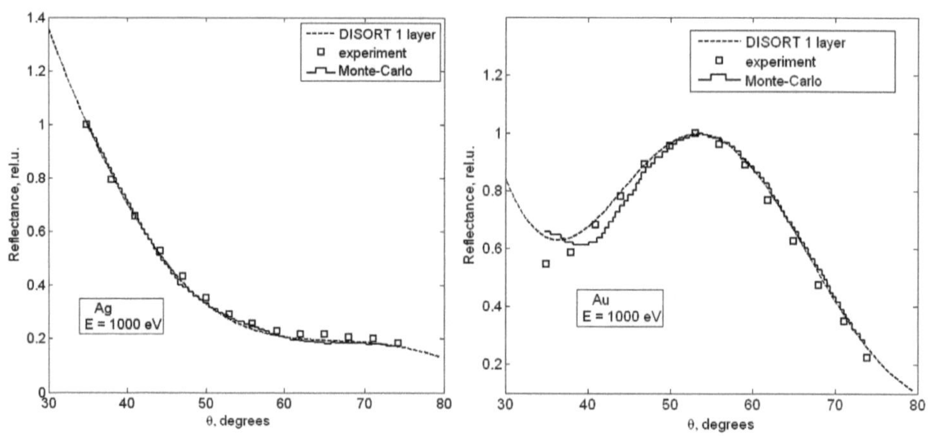

Fig. 8.27. The same as in Fig. 8.26, but for Ag and Au.

8.7 The practical applications of small-angle models

8.7.1 The comparison with the Monte Carlo simulations for Au+Si sample

Let us investigate the reliability of the Rubin–Everhart model for determining the thicknesses of deposited layers. A two-layer model Au/Si for 8 keV is considered. The purpose of the calculations is to determine the ratio of the number of electrons scattered on Au to the total number of the elastically scattered electrons $R_{\mathrm{Au}}/R_{Au+Si}$. The parameters of the simulation model are the follows: a layer of Au: $\lambda = 0.8$, $\tau = 1.0$; a substrate Si: $\lambda = 0.6$, $\tau = \infty$. The scattering phase functions are taken from [98]. 109 particles are simulated. The output is the energy spectrum of the elastically scattered electrons and the angular distribution of the elastically scattered electrons. The normal angle of incidence is chosen to avoid the azimuth differentiation of the spectrum. The spectrum is formed by the particles with the angle of reflection 45 ± 1 with respect to the normal. The simulated spectrum is shown in Fig. 8.28. The position of the peak maximum is well described by (8.3). The elastic peak is asymmetric due to the influence of multiple elastic scattering. So the spectrum cannot be presented as the sum of Gaussian functions.

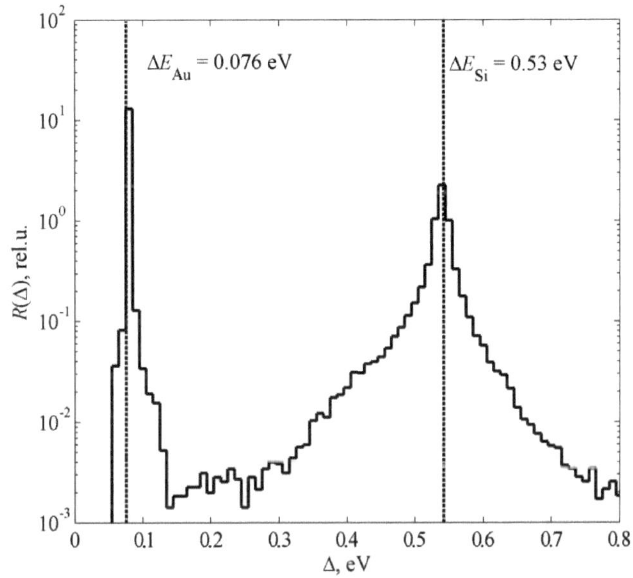

Fig. 8.28. The spectrum of the elastically reflected electrons – the Monte Carlo simulations.

Using this plot one can calculate precisely the area under the first peak corresponding to Au, and the sum of the areas under the two peaks.

$$\frac{R_{\mathrm{Au}}}{R_{\mathrm{Au+Si}}} = \frac{13.43}{20.03} = 0.67$$

To determine the ratio $R_{\mathrm{Au}}/R_{Au+Si}$ it is convenient to use such a function (Fig. 8.29):

$$H(\Delta) = \int_0^{\Delta} R(\varepsilon)\, d\varepsilon \tag{8.79}$$

The physical meaning of function $H(\Delta)$ is the number of electrons with energy losses less than or equal to Δ.

The calculations in the Rubin–Everhart model (8.31), in the small-angle approximation (8.38) and by the code DISORT differ by less than 3%. Similarly, the angular dependence of the relations $R_{\mathrm{Au}}/R_{\mathrm{Au+Si}}$ and $R_{\mathrm{Au}}/R_{\mathrm{Si}}$ are calculated (Fig. 8.30). Having DOM as a reference, it is possible to suggest the optimal geometry of the experiment with the minimal error of the simple models. For instance, for the considered case the deviation is minimal for the angle of scattering 135

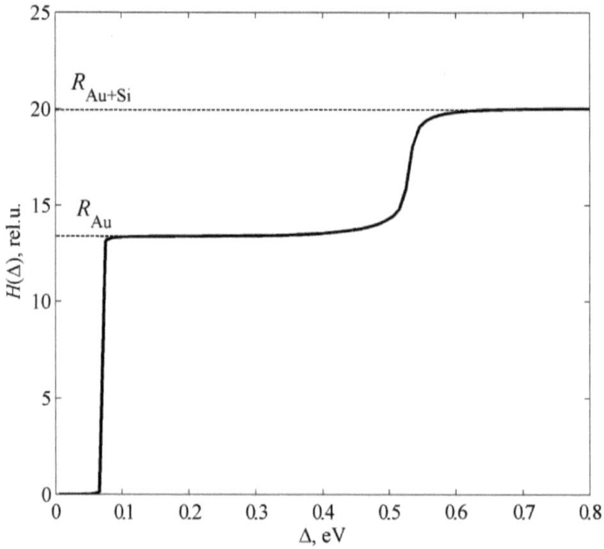

Fig. 8.29. Function $H(\Delta)$ for two-layer system (Au+Si)/Au.

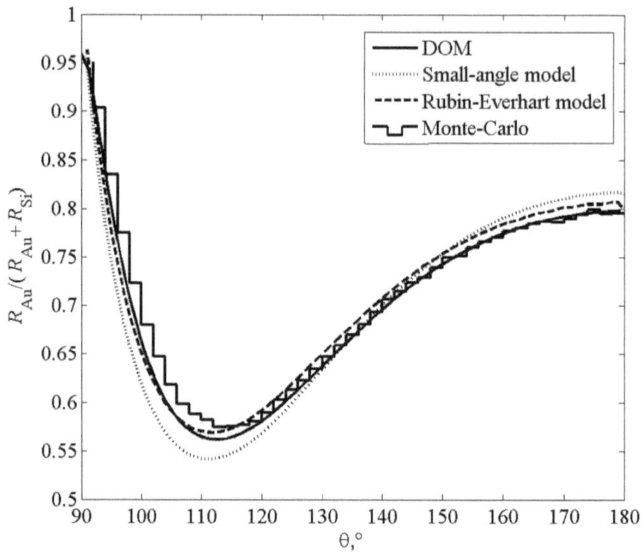

Fig. 8.30. The angular dependence of the ratio $R_{Au}/(R_{Au}+R_{Si})$.

degrees, which corresponds to the reflection angle 45 degrees with respect to the normal.

8.7.2 The stratified analysis of the samples by means of EPES

Now we apply the Rubin–Everhart approximation for the interpretation of EPES spectra for the systems 'carbon on silicon' and 'silicon nitride on silicon'. The experiments were performed by the team of M. Vos at the Australian National University.

The interpretation is implemented in a two-layer model since it is possible to extract reliably only one parameter from one spectrum. Fig. 8.31 shows the spectrum of the electrons elastically scattered from a two-layer sample – C layer/substrate Si. The spectrum was approximated by the sum of two Gaussian functions. The area ratio is $S_{\mathrm{C}}/S_{\mathrm{Si}} = 0.14$. The analytical formulas (8.56) and (8.57) yields the ratio of the area for an array of the layer thicknesses. The thickness can be easily found by the fitting procedure, which is very effective due to analytical solutions. The obtained thickness is 52 nm.

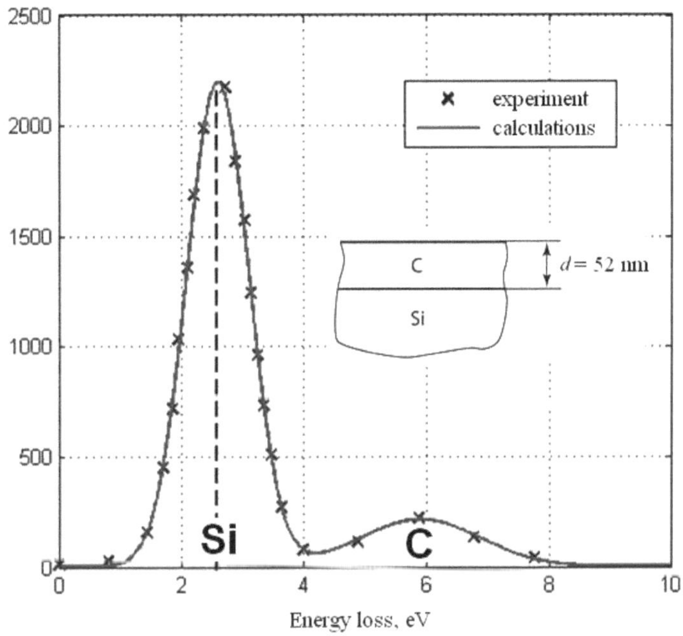

Fig. 8.31. The EPES spectrum for the carbon-silicon sample; the initial energy is 40 keV.

Figure 8.32 shows the experimental spectra and the fitting of the EPES spectrum from the system 'silicon nitride layer on silicon substrate'. From the experiment data, the ratio value is $S_{\mathrm{Si}}/S_{\mathrm{N}} = 5.5$ which corresponds to the thickness of Si_3N_4 30 nm.

It is possible to interpret N spectra in the $N+1$-layer model without additional assumptions. The spectra can be measured in the different geometries and for different probing beam energies. In both cases the effective probing depth lp of the sample is changed [99, 100]:

$$l_p \sim \frac{l_{\mathrm{in}}\left(E_0\right)}{\mu + \mu_0}\mu\mu_0 \tag{8.80}$$

where μ_0 and μ are the cosines of the incidence and of the reflection. Physically, the probing depth is a depth from which the elastic scattered electrons come to the energy analyzer. Thus, this is the sample area investigated by EPES.

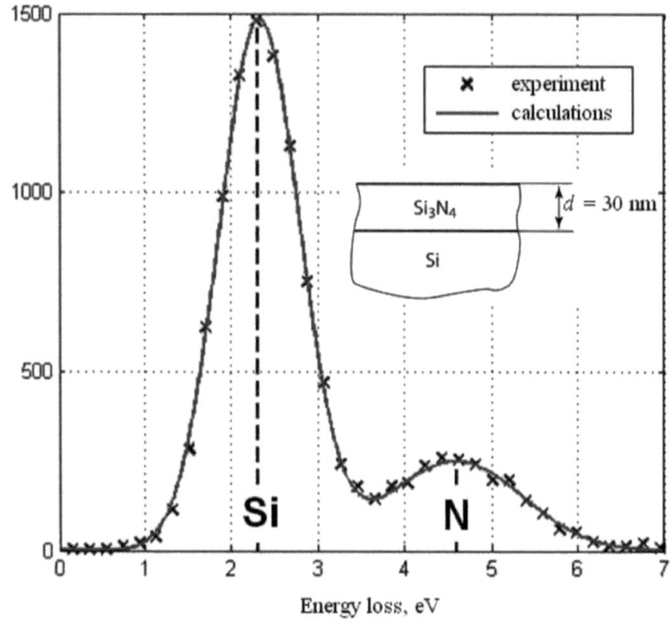

Fig. 8.32. The EPES spectra for two-layer sample Si_3N_4–Si; the initial energy is 40 keV.

With high accuracy one can assume the inelastic mean free path to be directly proportional to the probing energy. Thus, increasing the energy, we explore the deeper layers of the sample – the energy scanning method. And vice versa – the high-resolution depth is equivalent to the small value of l_p (8.80). This fact imposes an upper limit on the probing beam energy while the relation (8.8) defines the lower constraint.

The composition of the sample is reconstructed from the EPES spectra based on the condition

$$\left(\frac{R^1}{R^2}\right)^{\exp} = \left(\frac{R^1}{R^2}\right)^{\text{teor}} \tag{8.81}$$

here R_1 is a reflection function for a layer, R_2 is a reflection function for a substrate. However, not just one, but a number of samples, can satisfy the relation (8.81). Let us consider a model problem: the ratio for Au and Si peaks is

$$R_{\text{Au}}/R_{\text{Si}} = 1 \tag{8.82}$$

We need to find the distribution of Au atoms in a Si substrate. The first layer is determined by the thickness t_1 and the relative concentration r – the number of gold atoms per one atom of the silicon (Fig. 8.33). By changing both parameters,

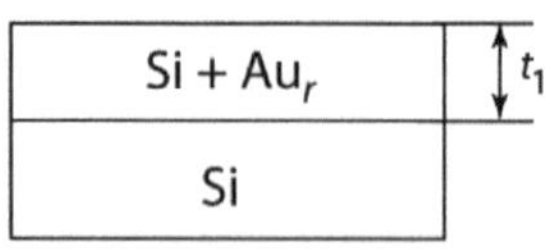

Fig. 8.33. The sample 'layer Si+Aur – substrate Si'.

Table 8.3. The examples t_1 and r, satisfying the condition $R_{Au}/R_{Si} = 1$.

r	t_1 (8 keV), nm	t_1 (15 keV), nm	$\frac{\lvert T_1(8\,\text{keV}) - T_1(15\,\text{keV})\rvert}{T_1(8\,\text{keV})} \cdot 100\%$
0.05	4.2	3.3	21%
0.1	1.61	1.45	9.93%
0.2	0.72	0.68	5.53%
0.4	0.34	0.33	2.9%
1	0.133	0.130	0%

the condition (8.82) can be satisfied. The examples of possible combinations t_1 and r are given in Table 8.3.

The relation (8.82) can also be reached in the model Si-Au-Si. Some examples are shown in Fig. 8.34.

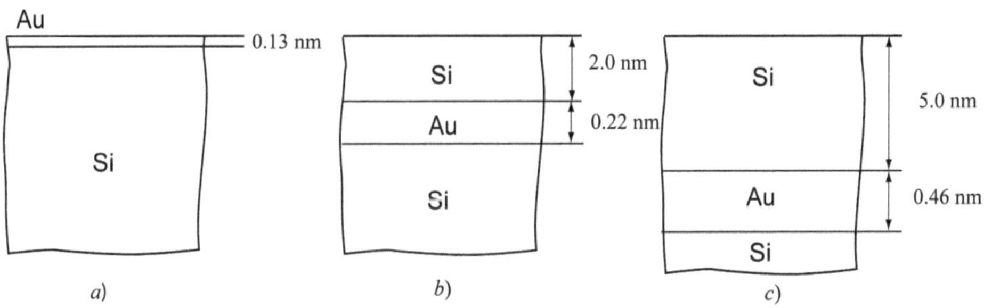

Fig. 8.34. The sample 'layer Si + layer Au – substrate Si'.

The values of ratio R_{Au}/R_{Si} are given in Table 8.4 for different probing beam energies. The difference in ratios R_{Au}/R_{Si} is due to the different probing depth (8.80).

Thus, the restoration of the multilayer solid structure based on one spectrum of the elastically scattered electrons is ambiguous – there is an endless number of options that satisfy condition (8.81). In order to exclude the redundant solutions of the inverse problem, some additional conditions are needed. These conditions can

Table 8.4. The values of ratio R_{Au}/R_{Si} and thicknesses for different energies.

Values of ratio R_{Au}/R_{Si}				
Scheme on Fig. 8.30	8 keV	10 keV	12 keV	15 keV
a	**1.00**	1.02	1.15	1.02
b	**1.00**	1.16	1.31	1.33
c	**1.00**	1.27	1.44	1.59
$\tau_{\text{lim}}^{\text{Si}}$, nm	5.83	7.08	8.28	10.0
$\tau_{\text{lim}}^{\text{Au}}$, nm	2.82	3.38	3.95	4.77

be formulated, for example, from the theoretical distribution functions of particles in the sample caused by the ion bombardment or diffusing atoms. If there is no prior information, the additional spectra measured at the different energies or geometry have to be used. In the case of N spectra the inverse problem is solved on the base of the following functional minimization:

$$F = \sum_{i=1}^{N} \left[\left(\frac{R^1}{R^2} \right)_i^{\text{exp}} - \left(\frac{R^1}{R^2} \right)_i^{\text{teor}} \right] \to \min \tag{8.83}$$

In the research-and-educational center 'Nanotechnology' in the Moscow Power Engineering institute the spectra of elastically reflected electrons were measured for the sample 'layer Au + substrate Si' [101]. First, the Au layer was deposited on the Si substrate, and then the sample was subjected to argon ion bombardment. Such ion mixing changes the initial distribution of Au atoms in the sample. The problem was to restore the Au atom distribution by means of EPES spectroscopy. EPES spectra were measured for the energy of incident beam 8 keV, 10 keV, 12 keV and 15 keV.

To interpret the experiment results, we use the Rubin–Everhart model. The calculations are performed for a sample with the number of layers from 1 to 5. The functional minimum is reached in a three-layer model. A further increase in the number of layers does not decrease the functional (8.83). The calculated and experimental ratios are plotted in Fig. 8.35. The extracted profile of the Au relative concentration is presented in Table 8.5.

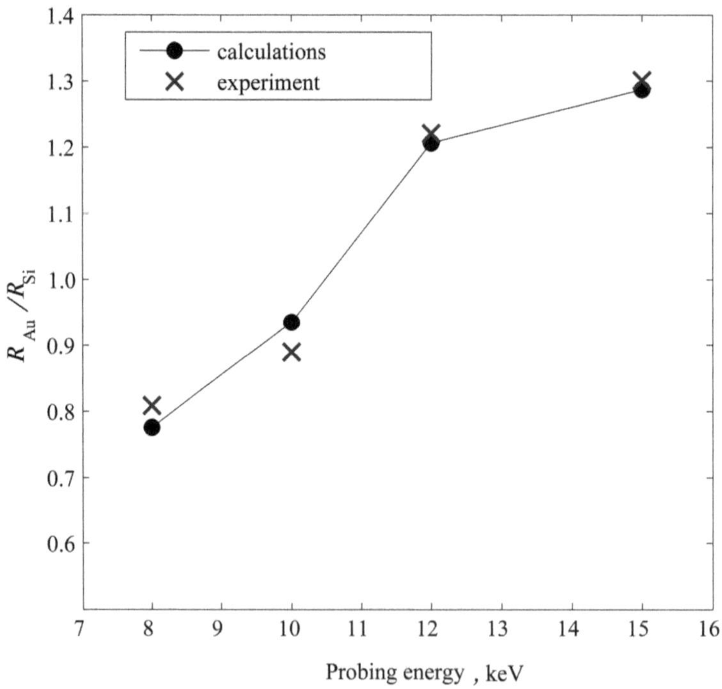

Fig. 8.35. The ratio of Au and Si for the energy scanning method – the experiment data and calculations.

Table 8.5. The dependence of the relative concentration $r(z)$ over the sample depth, extracted from EPES spectra.

	Thickness, nm	Relative concentration r
Layer 1	0.41	0.181
Layer 2	0.20	0.077
Substrate	8	0

8.7.3 Determination of the thickness of the deposited layer in the case of a low-energy resolution

Consider a two-layer system 'layer Au + substrate Ni' from [102], when the energy resolution is low, so that it is not sufficient to separate the peaks with the widths σ_{Au}, σ_{Ni} of the elastically scattered electrons by the nuclei with the masses M_{Au}, M_{Ni}. In this case the condition (8.8) is violated:

$$E_0 < \frac{\sigma_{\mathrm{Au}} + \sigma_{\mathrm{Ni}}}{2m\,(1 - \cos\theta)} \frac{M_{\mathrm{Au}} M_{\mathrm{Ni}}}{|M_{\mathrm{Au}} - M_{\mathrm{Ni}}|} \tag{8.84}$$

The elastic peaks formed by the electrons scattered from various elements are united. Just a single peak of the elastically scattered electrons is observed in the spectrum. The area beneath the united peak is given by the sum:

$$R_{\mathrm{Au+Ni}} = R_{\mathrm{Au}} + R_{\mathrm{Ni}} \tag{8.85}$$

where $R_{\mathrm{Au+Ni}}$ is the area under the peak; R_{Au} and R_{Ni} is the area under the peaks if they could be separated. In this case an analysis like that described in the previous paragraphs cannot be performed. However, the thickness of the deposited layer can be determined if the absolute values of the elastic peak area is known for the clean substrate. In this case, the following relation contains the information about the layer thickness:

$$R_{\mathrm{Au+Ni}}/R_{\mathrm{Ni}} = \left(R_{\mathrm{Au}} + R_{\mathrm{Ni}}\right)/R_{\mathrm{Ni}} \tag{8.86}$$

The ratio (8.86) is a function of the Au layer thickness and can be calculated in the small-angle approximation and Monte Carlo. Also DOM provides the exact solution for $R_{\mathrm{Au+Ni}}$.

In [102] the Au layer thickness was controlled by X-ray Photo-emission Spectroscopy (XPS). The experimental and theoretical values of the ratio (8.86) are plotted in Figs. 8.36 and 8.37. They evidence a good agreement between the experimental data and the theoretical models.

Note that the behavior of ratio (8.86) is different for energies 500 eV and 2000 eV. At 500 eV the differential elastic cross-section for Au is smaller than the differential elastic cross-section for Ni by some orders of magnitude for the specific angle of scattering. This is why the ratio is decreasing. For 2000 eV the situation is opposite. The differential cross-sections are shown in the Figs. 8.38 and 8.39.

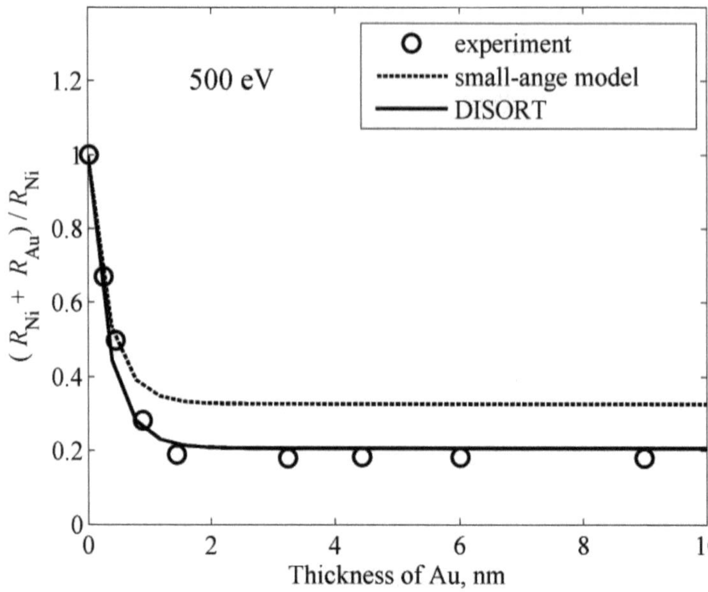

Fig. 8.36. The dependence of ratio (8.86) over layer thickness. Probing energy is 500 eV.

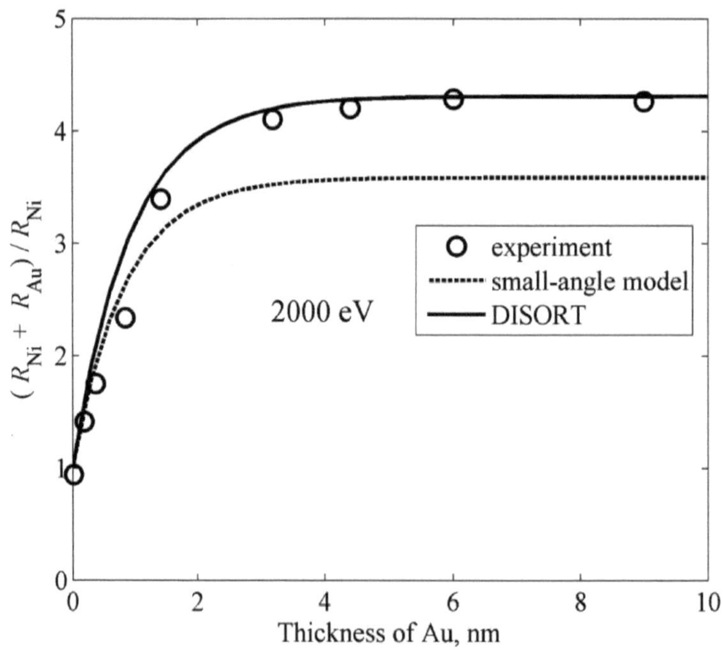

Fig. 8.37. The same as in Fig. 8.36, but for probing energy 2000 eV.

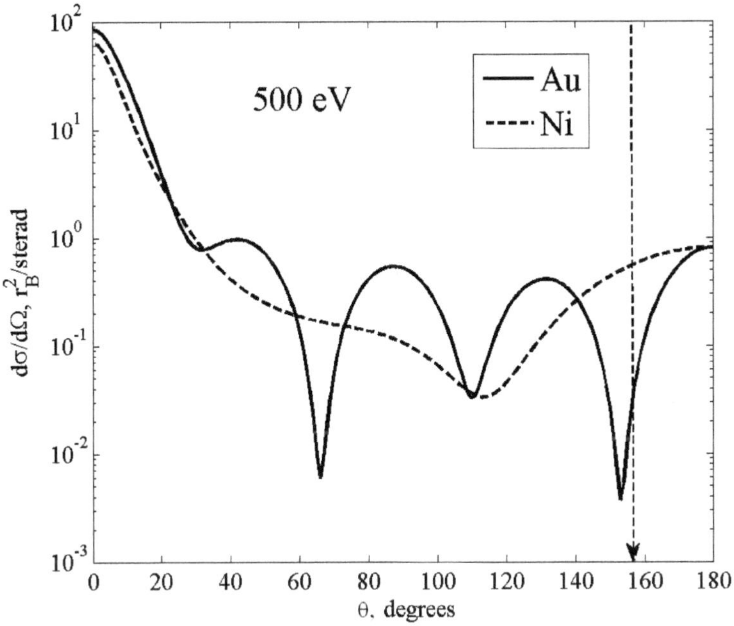

Fig. 8.38. The differential cross-section for Au and Ni: 500 eV. The vertical arrow denotes the scattering angle in [102].

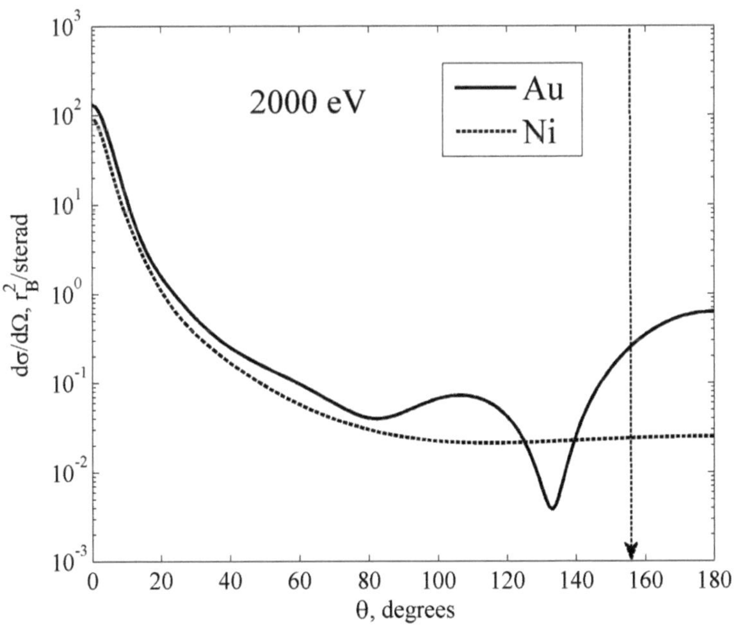

Fig. 8.39. The same as in Fig. 8.38, but for the probing energy 2000 eV.

8.8 Conclusions

The main motivation of this chapter was to reveal the similarity between the remote sensing and the elastic peak electron spectroscopy. The radiative transfer codes based on the discrete ordinate method (such as DISORT, MDOM) are used for verification of the approximate solutions of the radiative transfer equation for electrons.

The problem of computations of the angular distributions of the particles backscattered from a semi-infinite sample is considered. This problem is numerically solved with the accuracy limited by the computer representation of numbers with a floating point by various methods and codes: Neumann series, NMSS, Newton, SDA, MDOM, DISORT. It is shown that the most effective of the considered methods is the NMSS method for electron energy range 0.2 keV to 50 keV, which is synthetic and inherits the advances of other methods.

The electron transfer in solids with a highly anisotropic single scattering phase function is considered. The boundary problem for transmission and reflection functions of elastically scattered electrons is solved on the basis of the invariant embedding method. Four integral-differential equations with nonlinear term are derived. They are linearized and solved analytically due to the high anisotropy in the single scattering phase-function.

Quantitative electron spectroscopy requires the computations of the reflection and transmission functions not for the whole system, but for each layer separately. To derive the analytical relations, some approximations have been used. We analyzed the following models: the single scattering model, the classical quasi-single scattering model (Rubin–Everhart model) and the small-angle quasi-single scattering model. The analytical expressions for the last model are derived on the basis of the spherical harmonics method. The error of the approximate models is estimated. It depends on the incident electron energy and the nuclear charge of the sample. Having the discrete ordinate method solution as reference we found the cases where the Rubin–Everhart model could be used for elastic peak electron spectra interpretation.

In the future, the authors plan also to interpret X-ray photo-emission experiments on the basis of the invariant embedding method.

Acknowledgments

Authors would like to thank the participants of the seminar 'Solid-state electronics' in Moscow State University for the discussion of the results and critical notes, and Sergey Korlin, PhD (NASA), for valuable hints. Also the authors express their gratitude to two anonymous reviewers whose valuable comments helped to improve the chapter.

References

1. Afanas'ev V.P., Efremenko D.S. and Lubenchenko A.V. Spectroscopy of charged particles elastically scattered by plane-parallel solid layers, *J. Surf. Invest. X-ray, Synchrotron and Neutron Techniques.* 2008, **2**, 371–375.

2. Afanas'ev V.P., Afanas'ev M.V., Lubenchenko A.V., Batrakov A.A., Efremenko D.S. and Vos M. Influence of multiple elastic scattering on the shape of the elastically scattered electron peak, *J. Electron Sperctr. Relat. Phenomen.* 2010, **177**, 35–41.
3. Afanas'ev V.P. and Naujoks D. Backscattering of fast electrons, *Phys. Stat. Sol.* 1990, **164**, 133–140.
4. Afanas'ev V.P. and Naujoks D. Backscattering of light ions, *Z. Phys. B.* 1991, **86**, 39–47.
5. Rubin S., Passell T.O. and Bailey L.E. Chemical analysis of surfaces by nuclear methods, *Anal. Chem.* 1957, **29**, 736–743.
6. Landau L.D. and Lifshitz E.M. *Quantum Mechanics: Non-relativistic Theory*, 1977.
7. Boersch H., Wolter R. and Schoenebeck H.Z. Elastische energieverluste kristallgestreuter elektronen, *Zeitschrift fur Physik.* 1967, **199**, 124–134.
8. Bronstein I.M. and Pronin V.P. Elastic reflection of moderate energy electrons from solids, *Soviet Physics-Solid State.* 1975, **7**, 1363–1364.
9. Bronshtein I.M., Pronin V.P. and Stozharov V.M. Elastic reflection of electrons from Au, *Soviet Physics-Solid State.* 1975, **16**, 1374–1377.
10. Bronshtein I.M. and Pronin V.P. Elastic scattering of electrons during the adsorption of Au on Be and Be on Au, *Fiz. Tverd. Tela.* 1975, **17**, 2502–2504.
11. Artemyev V.P., Makarov V.V. and Petrov N.N. On the possibilities of applying the method of electron Rutherford backscattering to study the surface layers of solids, *Bulletin of the Russian Academy of Sciences: Physics.* 1985, **49**, 1765–1769.
12. Borisov A.M., Virgil'ev Y.S. and Mashkova E.S. Structure modification and emission properties of carbon-based materials under high-dose irradiation, *J. Surf. Invest.: X-ray, Synchrotron and Neutron Techniques.* 2008, **2**, 52–67.
13. Tanuma S., Ichimura S. and Goto K. Estimation of surface excitation correction factor for 200–5000 eV in Ni from absolute elastic scattering electron spectroscopy, *Surf. Interf. Anal.* 2000, **30**, 212–216.
14. Gurban S. et al. Ag, Ge and Sn reference samples for elastic peak electron spectroscopy (EPES), used for experimental determination of the inelastic mean free path and the surface excitation parameter, *Surf. Interf. Anal.* 2002, **34**, 206–210.
15. Tougaard S. et al. Intercomparison of methods for separation of REELS elastic peak intensities for determination of IMFP, *Surf. Interf. Anal.* 2001, **31**, 1–10.
16. Sobol I.M. *A Primer for the Monte Carlo Method*, CRC Press, 1994.
17. Marchuk G.I., Mikhailov G.A., Nazaraliev M.A., Darbinjan R.A., Kargin B.A. and Elepov B.S. *The Monte Carlo Methods in Atmospheric Optics*, Springer, 1980.
18. Mikhailov G.A. *New Monte Carlo Methods with Estimating Derivatives*, Brill Academic Publishers, 1995.
19. Chandrasekhar S. *Radiative Transfer*, Dover Pubns, 1960.
20. Sobolev V.V. *Light Scattering in Planetary Atmospheres*, Pergamon Press, 1975.
21. Kalashnikov N., P., Remizovich V.S. and M.I. Ryazanov *Collisions of Fast Charged Particles in Solids*, Gordon & Breach, 1985.
22. Davison B. *Neutron Transport Theory*, Clarendon Press, 1957.
23. Liljequist D. Simplified models for the Monte Carlo simulation of energy distributions of keV electrons transmitted or back-scattered in various solids, *Journal of Physics D: Appl. Phys.* 1978, **11**, 839.
24. Jablonski A. Elastic backscattering of electrons from surfaces, *Surf. Sci.* 1985, **151**, 166–182.
25. Werner W.S.M. Electron transport in solids for quantitative surface analysis, *Surf. Interface Anal.* 2001, **31**, 141–176.
26. Ali E.S.M. and Rogers D.W.O. Energy spectra and angular distributions of charged particles backscattered from solid targets, *J. Phys. D: Appl. Phys.* 2008, **41**, 055505.

27. Case K.M. and Zweifel P.F. *Linear Transport Theory*, Addison-Wesley Reading, 1967.
28. Goudsmit S. and Saunderson J.L. Multiple scattering of electrons, *Phys. Rev.* 1940, **58**, 36–42.
29. Lewis H.W. Multiple scattering in an infinite medium, *Phys. Rev.* 1950, **78**, 526–529.
30. Bethe H.A. Moliere's theory of multiple scattering, *Phys. Rev.* 1953, **89**, 1256–1266.
31. Gordon H.R. Simple calculation of the diffuse reflectance of the ocean, *Appl. Opt.* 1973, **12**, 2803–2804.
32. Everhart T.E. Simple theory concerning the reflection of electron from solids, *J. Appl. Phys.* 1960, **31**, 1483–1490.
33. Rubin S. Surface analysis by charged particle spectroscopy, *Nucl. Instr. and Methods.* 1959, **5**, 177–183.
34. Oswald R., Kasper E. and Gaukler K.H. A multiple scattering theory of elastic electron backscattering from amorphous surfaces, *J. Electron Spectrosc. Relat. Phenom.* 1993, **61**, 251–274.
35. Tombuyses B. and Dubus A. Contribution to the study of elastic electron backscattering from polycrystalline metallic targets by Monte Carlo simulation, *Math. Comput. Simul.*. 1998, **47**, 483–491.
36. Dave J.V. A Direct solution of the spherical harmonies approximation to the radiative transfer equation for an arbitrary solar elevation. Part II: Results, *J. Atmospheric Sciences.* 1975, **32**, 1463–1474.
37. Barichello L.B., Garcia R.D.M. and Siewert C.E. A spherical-harmonics solution for radiative-transfer problems with reflecting boundaries and internal sources, *J. Quant. Spectrosc. Radiat. Transfer.* 1998, **60**, 247–260.
38. Muldashev T.Z., Lyapustin A.I. and Sultangazin U. Spherical harmonics method in the problem of radiative transfer in the atmosphere-surface system, *J. Quant. Spectrosc. Radiat. Transfer.* 1999, **61**, 393–404.
39. Chalhoub E.S. Discrete-ordinates solution for radiative-transfer problems, *J. Quant. Spectrosc. Radiat. Transfer.* 2003, **76**, 193–206.
40. Stamnes K., Tsay S.C., Wiscombe W. and Jayaweera K. Numerically stable algorithm for discrete-ordinate-method radiative transfer in multiple scattering and emitting layered media, *Appl. Opt.*. 1988, **27**, 2502–2509.
41. Budak V.P. and Korkin S.V. On the solution of a vectorial radiative transfer equation in an arbitrary three-dimensional turbid medium with anisotropic scattering, *J. Quant. Spectrosc. Radiat. Transfer.* 2008, **109**, 220—234.
42. Compton A.H. A quantum theory of the scattering of X-rays by light elements, *Phys. Rev.* 1923, **21**, 483–502.
43. Vos M. Observing atom motion by electron-atom Compton scattering, *Phys. Rev. A.* 2001, **65**, 012703–012708.
44. Kover L., Fujikawa T. and Suzuki R. Theory of recoil effects of elastically scattered electrons and of photoelectrons, *JESRP.* 2006, **151**, 170–177.
45. Afanas' ev V.P. *Electron and Ion Spectroscopy of Solids*, Publishing House of MPEI [in Russian], 2005.
46. Afanas'ev V.P., Efremenko D.S. and Lubenchenko A.V. Spectroscopy of charged particles elastically scattered by plane-parallel solid layers, *J. Surf. Invest.: X-ray, Synchrotron and Neutron Techniques.* 2008, **2**, 371–375.
47. Borodyansky S. Effects of elastic scattering on energy spectra of emitted and backscattered electrons, *Surf. Interf. Anal.* 1993, **20**, 811–814.
48. http://eapclu.iap.tuwien.ac.at/~werner/.
49. Jablonski A. and Zemek J. Angle-resolved elastic-peak electron spectroscopy: Solid-state effects, *Surf. Sci.* 2006, **600**, 4464–4474.

50. Vos M. Detection of hydrogen by electron Rutherford backscattering, *Ultramicroscopy.* 2002, **92**, 143–149.
51. Li Y.G., Ding Z.J., Zhang Z.M. and Tokesi K. Monte Carlo calculation of electron Rutherford backscattering spectra and high-energy reflection electron energy loss spectra, *Nuclear Instruments and Methods in Physics Research Section B.* 2009, **267**, 215–220.
52. Alford T.L., Feldman L.C. and Mayer J.W. *Fundamentals of Nanoscale Film Analysis*, Springer Science, 2007.
53. Goudsmit S.A. and Saunderson J.L. Multiple scattering of electrons, *Phys. Rev.* 1940, **58**, 36–42.
54. Went M.R. and Vos M. High-resolution study of quasi-elastic electron scattering from a two-layer system, *Surf. Sci.* 2006, **600**, 2070–2078.
55. Vos M., Chatzidimitriou-Dreismann C.A., Abdul-Redah T. and Mayers J. Electron and neutron scattering from polymer films at high momentum transfer, *Nuclear Instruments and Methods in Physics Research Section B.* 2005, **227**, 233–250.
56. Budak V.P. and Lubenchenko A.V. Precision and application range of the quasi-single scattering approximation in the backscattering signal calculation, *Atmospheric and Oceanic Optics.* 2007, **20**, 577–582.
57. Jablonski A. Modeling of elastic and inelastic electron backscattering from surfaces, *Prog. in Surf. Sci.* 2005, **79**, 3–27.
58. Scott W.T. The theory of small-angle multiple scatteringof charged particles, *Rev. Mod. Phys.* 1963, **35**, 231–241.
59. Mie G. Beitrage zur Optic truber Medien, speziell kolloidaler Metallosungen, *Ann. Phys..* 1908, **25**, 377–445.
60. van de Hulst H.C. *Light Scattering by Small Particles*, John Wiley & Sons, 1957.
61. Deirmendjian D. *Electromagnetic Scattering on Spherical Polydispersions*, Elsevier, 1969.
62. Afanas'ev V.P., Budak V.P., Efremenko D.S. and Lubenchenko A.V. Angular distributions of electrons and light ions elastically reflected from a solid surface, *J. Surf. Invest.: X-ray, Synchrotron and Neutron Techniques.* 2010, **4**, 488–493.
63. Ambartsumian V.A. A new method of calculation of the light scattering in turbid medium, *Izvestiya Acad. Sci. USSR, Ser. Geograph. and Geophys. Sci.* 1942, **3**, 97–103 [in Russian].
64. Nikiforov A.F., Uvarov V.B. and Boas R.P. *Special Functions of Mathematical Physics*, Birkhäuser, 1988.
65. Muldashev T.Z. and Sultangazin U.M. The method of spherical harmonics for solving the radiation transfer problem in a plane-parallel atmosphere, *USSR Computational Mathematics and Mathematical Physics.* 1986, **26**, 151–159.
66. Landau L.D. and Lifschitz E.M. *Lehrbuch der Theoretischen Physik, Quantenmechanik*, Akademie-Verlag, 1979.
67. Lubenchenko A.V. Angular distributions of charged particles and radiation, *J. Surf. Invest. X-ray, Synchrotron and Neutron Techniques.* 2003, **4**, 22–31.
68. Tanuma S., Powell C.J. and Penn D.R. Calculation of electron inelastic mean free paths (IMFPs) VII. Reliability of the TPP-2M IMFP predictive equation, *Surf. Interf. Anal.* 2003, **35**, 268–275.
69. Irvin W.M. Multiple scattering by large particles, *The Astrophysical Journal.* 1965, **142**, 1563–1575.
70. Vicanek M. Electron transport processes in reflection electron energy loss spectroscopy (REELS) and X-ray photoelectron spectroscopy (XPS), *Surf. Sci..* 1999, **440**, 1–40.

71. Glazov L.G. and Tougaard S. Electron backscattering from surfaces: The invariant-embedding approach, *Phys. Rev. B.* 2003, **68**, 155409.
72. Glazov L.G. and Tougaard S. Electron backscattering from surfaces: Azimuth-resolved distributions, *Phys. Rev. B.* 2005, **72**, 85406.
73. Schuster A. The influence of radiation of the transmission of heat, *Phil. Mag.* 1903, **5(6)**, 243–257.
74. Schwarzschild K. Ueber das Gleichgewicht der Sonnenatmosphaere, *Nachr. Konig. Gesel. der Wiss.*. 1906, **1**, 41–53.
75. Krylov V.I. and Stroud A. *Approximate Calculation of Integrals*, Dover Publications, 2006.
76. Barnett S. and Storey C. Some applications of the Lyapunov matrix equation, *J. Inst. Maths Applies.* 1968, **4**, 33–42.
77. Hammarling S.J. Numerical solution of the stable, non-negative definite Lyapunov equation, *IMA Journal of Numerical Analysis.* 1982, **2**, 303–323.
78. Sorensen D.C. and Zhou Y. Direct methods for matrix sylvester and Lyapunov equations, *Journal of Applied Mathematics.* 2003, **6**, 277—303.
79. Benner P., Li J.R. and Penzl P. Numerical solution of large-scale Lyapunov equations, Riccati equations, and linear-quadratic optimal control problems, *Numer. Linear Algebra Appl.*. 2008, **15**, 755–777.
80. Bartels R. and Stewart G. Solution of the matrix equation AX + XB = C: Algorithm 432, *Comm. ACM.* 1972, **15**, 820—826.
81. Golub G.H., Nash S. and Van Loan C.F. A Hessenberg–Schur method for the problem AX + XB = C, *IEEE Trans. Automat. Control.* 1979, **AC-24**, 909–913.
82. Kleinman D. On an iterative technique for Riccati equation computations, *IEEE Trans. Automat. Control.* 1968, **AC-13**, 114–115.
83. Lancaster P. and Rodman L. *The Algebraic Riccati Equation*, Oxford University Press, 1995.
84. Benner P. and Byers R. Newton's metod with exact line search for solving the algebraic Riccati equations, *IEEE Trans. Automat. Control.* 1998, **43**, 101–107.
85. Byers R. Solving the algebraic Riccati equation with the matrix sign function, *Linear Algebra Appl.*. 1987, **85**, 267–279.
86. Laub A. A Schur method for solving algebraic Riccati equations, *IEEE Trans. Automat. Control.* 1979, **AC-24**, 913–921.
87. Benner P., Mehrmann V., Sima V., Van Huffel S. and Varga A. SLICOT – a subroutine library in systems and control theory, *Applied and Computational Control, Signals, and Circuits.* 1999, **1**, 499–539.
88. Juang J. and Chen I.D. Iterative solution for a certain class of algebraic matrix Riccati equations arising in transport theory, *Transport Theory Statist. Phys.* 1993, **22**, 65 – 80.
89. Juang J. and Lin W.W. Nonsymmetric algebraic Riccati equations and Hamiltonian-like matrices, *SIAM J. Matrix Anal. Appl.*. 1999, **20**, 228–243.
90. Guo X.X., Lin W.W. and Xu S.F. A structure-preserving doubling algorithm for nonsymmetric algebraic Riccati equation, *Numer. Math.* 2006, **103**, 393–412.
91. Salvat F., Jablonski A. and Powell C.J. ELSEPA–Dirac partial-wave calculation of elastic scattering of electrons and positrons by atoms, positive ions and molecules, *Comput. Phys. Commun.* 2005, **165**, 157–190.
92. Tanuma S., Powell C.J. and Penn D.R. Calculation of electron inelastic mean free paths (IMFPs) VII. Reliability of the TPP-2M IMFP predictive equation, *Surf. Interface Anal.*. 2003, **35**, 268–275.
93. ftp://climate1.gsfc.nasa.gov/wiscombe/Multiple_Scatt.

94. Salvat F., Jablonski A. and Powel C.J. Elastic electron-atom scattering in amplitude-phase representation with application to electron diffraction and spectroscopy, *Comput. Phys. Commun.* 2005, **165**, 157–190.
95. Harrison W.A. *Solid State Theory*, Dover Publications, 1970.
96. Graat P.C.J., Somers M.A.J. and Bottger A. Generalization of the Tougaard method for inelastic-background estimation in electron spectroscopy: Incorporation of a depth-dependent inelastic mean free path, *Surf. Interf. Anal.* 1995, **23**, 44–49.
97. Zemek J., Jiricek P., Werner W.S.M., Lesiak B. and Jablonski A. Angular-resolved elastic peak electron spectroscopy: experiment and Monte Carlo calculations, *Surf. Interf. Anal.* 2006, **38**, 615–619.
98. Salvat F., Jablonski A. and Powell C.J. Dirac partial-wave calculation of elastic scattering of electrons and positrons by atoms, positive ions and molecules, *Comput. Phys. Commun.* 2005, **165**, 157–190.
99. Zommer L. Interface effect for EPES sampling depth for overlayer/substrate systems, *Surf. Sci.* 2006, **600**, 4735–4740.
100. Zommer L. and Jablonski A. EPES sampling depth paradox for overlayer/substrate system, *J. of Electron Spectr. and Relat. Phenom.* 2006, **150**, 56–61.
101. Kostanovskiy I.A., Afanas'ev V.P. and Batrakov A.A. Electron Rutherford back scattering spectroscopy with energy and angular scanning for thin oxide films diagnostics, *Nanotechnology (IEEE-NANO), 11th IEEE International Conference on Nanotechnology.* 2011, 760–763.
102. Jablonski A., Hansen H.S., Jansson C. and Tougaard S. Elastic electron backscattering from surfaces with overlayers, *Phys. Rev. B.* 1992, **45**, 3694–3702.

9 On some trends in the progress of astrophysical radiative transfer

Arthur G. Nikoghosian

9.1 Introduction

The objective of this review is to provide some insight into trends in the development of one of the most important fields of theoretical astrophysics – the theory of radiative transfer. The foundation of the field goes back to the pioneering works of Schuster, Schwarzschild, Milne and Eddington appearing at the beginning of the last century in connection with modeling the stellar atmospheres. Since then the approaches proposed in these works have made rapid progress forming a theory usually referred to as 'classical'. Parallel to this an alternative approach was offered by Ambartsumian with his methods based on the principle of invariance and the laws of addition of layers. With their deep physical content the methods have been extremely flexible in applications and efficient in numerical computations. Their major role in the theory is well-known. However, the modern fundamental results in this direction make expedient reviewing the most important of them to elucidate their place in existing theory and significance from the point of view of its further progress.

The study of the equation of radiation transfer and its solutions in various forms occupied an important place in the first works aimed at interpreting the stellar spectra. Of course, the simplest, and therefore rough, models, in which the medium is assumed to be plane-parallel, stationary, homogeneous and purely absorbing were investigated first of all. The last assumption substantially simplifies the problem of finding the field of radiation in the atmosphere, since in this case the state of the radiating gas obeys the equilibrium laws of Saha and Boltzmann with the local values of temperature and density. In this approximation, referred to as the approximation of LTE (local thermodynamic equilibrium), the source function, appearing in the transfer equation, is given by the Kirchhoff–Planck law. The situation changes drastically, when one takes into account the scattering of radiation which is of particular importance in the problem of the spectral line formation. Now the state of the radiating gas depends not only on the local values of thermodynamic parameters, but also on the field of radiation at the particular point, which establishes coupling between different volumes inside the atmosphere. The equation of radiation transfer in this case is integro-differential and its solution, in general, encounters great difficulties.

In spite of the rough assumptions, the first works in this direction in many respects contributed to the physical understanding of the studied processes and stimulated the development of theory. The most important articles in this field were assembled by Menzel [1]. The approach, which was developed in them and became classical, consisted in finding the source function as a function of the depth in the atmosphere, which, in its turn, made it possible to determine the field of radiation in it. In the simplest cases of isotropic and monochromatic scattering the problem mathematically is reduced to the solution of an integral equation of Fredholm's type with the kernel, which is the exponential integral dependent on the modulus of a difference of the arguments. For instance, in the simplest case of isotropic scattering the problem of finding the radiation field in a semi-infinite atmosphere illuminated by a beam of parallel rays at an angle of $\arccos\varsigma$ is reduced, as is well-known, to the solution of the following integral equation

$$S(\tau,\varsigma) = \frac{\lambda}{2}\int_0^\infty \mathrm{Ei}\left(|\tau-\tau'|\right) S(\tau',\varsigma)\, d\tau' + \frac{\lambda}{4} e^{-\tau/\varsigma}, \tag{9.1}$$

where λ is the single-scattering albedo (or the probability of photon re-radiation during the elementary act of scattering) and S is the source function. Its solution allows, in particular, the determination of the intensity of the radiation emerging from the atmosphere, i.e., the quantity, directly measured during the observations.

In contrast to the conventional approach described above, Ambartsumian proposed the new method, named by him 'the principle of invariance ', which allowed us to find the outgoing intensity without the preliminary determination of the light regime at all depths. By the principle of invariance , he implies such transformation of the initial atmosphere, which does not influence the global optical characteristics of a medium [2–5]. It is obvious that during this determination the term ' principle of invariance ' can be used only in the singular. The meaning of this remark will become clear further on. The application of the principle significantly facilitates the solution of the problems of radiation transfer, revealing from the very beginning the structure of the desired solutions which, in its turn, is a great help in determining the field of radiation inside the atmosphere. As was shown subsequently (see below, Section 9.5, and also [6]), the principle of invariance is a special case of the more general variational principle connected with the translational transformation of optical depth. The application of this principle makes it possible to derive for different quantities in different problems the large number of important relationships, which sometimes are possible to identify immediately, on the basis of simple physical and/or probabilistic considerations. Such relationships, which follow from the principle of invariance , can be called *the invariance relations*.

As a result of Ambartsumian's studies in 1941–1947 on the theory of radiative transfer, another rather effective method, named by him the method of addition of layers, was also proposed [7] (see also [8]). It answers the question, how the global optical characteristics of the absorbing and scattering media (coefficients of reflection and transmission) during their joining are added. It is obvious that this sufficiently general posing of the question appears naturally, if we abandon the requirement that the optical properties of a medium should remain unchanged during the addition to it of additional layer. The obtained relationships reveal the functional form of the optical characteristics of the composite atmosphere; moreover, which is important, all parameters and functions, which describe elementary processes, play the role of arbitrary parameters. The method gives a key to the solution of the transfer problems for inhomogeneous atmospheres, and on the other hand, it was basis for the so-called method of 'invariant imbedding' developed afterwards.

9.2 The principle of invariance

The invariance principle was formulated for the first time by Ambartsumian in treating the problem of diffuse reflection of light from a semi-infinite homogeneous atmosphere [3,4]. Considerations underlied the principle based on the apparent fact that the addition to this medium of a layer of infinitesimal optical thickness $\Delta\tau$, possessing the same properties as the original one, must not change its reflectivity. This thesis was referred to by Ambartsumian as the principle of invariance . This implies that the total contribution of processes relevant to the added layer will be equal to zero.

The reflectance of a medium is characterized by the reflection function $\rho(\eta, \varsigma)$, where ς and η are respectively the cosines of the angles of incidence and reflection. The contribution of all the other possible processes is of a higher order of smallness with respect to $\Delta\tau$ and may be ignored. Then in the simplest case of isotropic scattering the condition of invariance of the reflection function for the semi-infinite atmosphere can be written in the form

$$(\eta + \xi)\,\rho(\eta, \xi) = \frac{\lambda}{2}\varphi(\eta)\,\varphi(\xi)\,, \tag{9.2}$$

where the function

$$\varphi(\eta) = 1 + \eta \int_0^1 \rho(\eta, \eta')\, d\eta' \tag{9.3}$$

is referred to as the Ambartsumian φ-function. The last two equations implies that the function φ satisfies the following functional equation

$$\varphi(\eta) = 1 + \frac{\lambda}{2}\eta \int_0^1 \frac{\varphi(\eta)\,\varphi(\eta')}{\eta + \eta'}\, d\eta'\,, \tag{9.4}$$

usually called the Ambartsumian equation. Eq. (9.2) shows that the reflectance $\rho(\eta, \varsigma)$ is expressed through a function of one variable and is a symmetrical function of its arguments. The quantity $\eta\rho(\eta, \varsigma)\, d\eta$ possesses a probabilistic meaning, namely, it gives the probability that the quantum incident on the medium in the direction ς will be reflected by it in the directional interval η, $\eta + d\eta$.

In the same paper [2] the principle of invariance was applied for solving the problem of the diffuse reflection and transmission for the medium of finite optical thickness. In this case the layer of infinitesimal optical thickness $\Delta\tau$ is added to one of boundaries while just such a layer is subtracted from the opposite side. For the reflectance $\rho(\eta, \varsigma)$ and diffuse part of transmittance $\sigma(\eta, \varsigma)$ this results in (for convenience of the further discussion, we adopt here somewhat different notation):

$$\rho(\eta, \varsigma) = \frac{\lambda}{2}\frac{\varphi(\eta)\,\varphi(\varsigma) - \psi(\eta)\,\psi(\varsigma)}{\eta + \varsigma}\,, \qquad \sigma(\eta, \varsigma) = \frac{\lambda}{2}\frac{\psi(\eta)\,\varphi(\varsigma) - \varphi(\eta)\,\psi(\varsigma)}{\eta - \varsigma}\,. \tag{9.5}$$

The auxiliary functions $\varphi(\eta)$ and $\psi(\eta)$ are determined from the following system of functional equations

$$\varphi(\eta) = 1 + \frac{\lambda}{2}\eta \int_0^1 \frac{\varphi(\eta)\,\varphi(\eta') - \psi(\eta)\,\psi(\eta')}{\eta + \eta'}\,d\eta', \tag{9.6}$$

$$\psi(\eta) = e^{-\tau_0/\eta} + \frac{\lambda}{2}\eta \int_0^1 \frac{\psi(\eta)\,\varphi(\eta') - \varphi(\eta)\,\psi(\eta')}{\eta - \eta'}\,d\eta', \tag{9.7}$$

where τ_0 is the optical thickness of the medium. These functions are also called the Ambartsumian functions. It is clear that the reflectance and transmittance as well as the functions $\varphi(\eta)$ and $\psi(\eta)$ depend also on the optical thickness of the medium, nevertheless, for brevity, τ_0 is not indicated explicitly among arguments.

As has already been indicated, the starting point for the determination of the intensity of radiation outgoing from the medium is here not the equation of transfer, which allows us to find the required quantity only after the regime of radiation is found for all depths in the atmosphere. It is obvious that, in view of the linearity of the problem, knowledge of the functions of reflection and transmission makes it possible to determine the intensity of the outgoing radiation for any flux falling on the medium. On the other hand, formulas (9.2) and (9.5) give the solution not just of one particular problem of diffuse reflection (for the semi-infinite medium) or the problem of diffuse reflection and transmission (for the finite medium). In fact they make it possible to reveal the structure of the global optical characteristics of medium, as such, expressing in this case the unknown quantities through the functions of one variable. The approach itself in many respects contributed to the presence of a number of important relations connecting with each other different characteristics of the radiation field, in the problems of radiative transfer theory most frequently encountered in astrophysical applications. Such relationships were obtained at different times by a number of authors (for example, see [6, 9, 10]).

It should be noted that the relations (9.2), (9.4) were obtained by Ambartsumian earlier by another method in considering the scattering of light by the atmospheres of planets [11]. The way, chosen in the mentioned work, consists in the formal differentiation of the initial integral equation for the source function over the optical depth. From a purely mathematical point of view the way proposed is of large importance, since it shows how the solution of the integral equation of the Fredholm type with the difference kernel can be reduced to the solution of a functional equation.

The idea on the invariant property of the global optical characteristics of the scattering and absorbing atmosphere with respect to the layer addition was employed in the problem of radiation diffusion through the optically thick medium [12,13]. This research implies that the function $\varphi(\eta)$ admits a physical interpretation which concerns the angular distribution of the intensity of radiation transmitted through the optically thick atmosphere in the absence of true absorption. Various meanings can be ascribed to this function, of which the limb-darkening law for the Sun is one of the astrophysical examples.

In various astrophysical problems one encounters, as is known, the necessity to find the radiation field within the medium. An important advantage of the invariance principle is that knowledge of the intensities of radiation outgoing from a medium facilitates the solution of this problem essentially (see, e.g., [14–16]). For instance, being applied to Eq. (9.1), the invariance principle makes it possible to reduce the solution of this Fredholm-type integral equation to the solution of following Volterra-type equation for an auxiliary function $\Phi(\tau)$ related with the resolvent function of Eq. (9.1)

$$\Phi(\tau) = L(\tau) + \int_0^{\tau} L(\tau - \tau')\, \Phi(\tau')\, d\tau' , \tag{9.8}$$

where the kernel-function

$$L(\tau) = \frac{\lambda}{2} \int_0^1 \varphi(\varsigma)\, e^{-\tau/\varsigma} \frac{d\varsigma}{\varsigma} \tag{9.9}$$

is known well in the radiative transfer theory [17, 18]. An explicit expression for the function $\Phi(\tau)$ was obtained in [19].

For illustration, we limited our consideration to the simplest case of monochromatic scattering, though the described picture and conclusions remain valid for the much more general statement of the transfer problem. From the pure mathematical point of view, the principle of invariance may be considered as a way of reducing the boundary-value problem usually formulated for the source function to the solution of the initial-value or Cauchy problem.

9.2.1 Anisotropic scattering

The first works on the principle of invariance concern monochromatic and isotropic scattering. It was apparent, however, that the principle may also be applied under much more general assumptions to the elementary act of scattering. That is why as early as in the paper [20] Ambartsumian handles the problem of diffuse reflection of light from a semi-infinite plane-parallel atmosphere for anisotropic scattering with arbitrary phase function. It is noteworthy that here he employs the expansion of the phase function $x(\gamma)$ (γ is the angle of scattering) in a series of Legendre polynomials which was suggested earlier by him in [21, 22]

$$x(\gamma) = \sum_{i=0}^{\infty} x_i P_i(\cos\gamma) . \tag{9.10}$$

or with use of the summation theorem for spherical functions

$$x(\eta, \eta', \varphi - \varphi') = \sum_{m=0}^{\infty} \cos m(\varphi - \varphi') \sum_{i=m}^{\infty} c_{im} P_i^m(\eta)\, P_i^m(\eta') , \tag{9.11}$$

where the constants c_{im} are expressed through the coefficients x_i, and $P_i^m(\eta)$ are the associated Legendre polynomials. Then an expansion similar to (9.11) holds for the reflection function

$$\rho(\eta, \eta', \varphi - \varphi') = \sum_{m=0}^{\infty} \rho_m(\eta, \eta') \cos m(\varphi - \varphi') . \tag{9.12}$$

Now the solution of the problem is written by means of the functions $\varphi_i^m(\eta)$

$$\rho_m(\eta,\eta') = \frac{\lambda}{4}\sum_{i=m}^{\infty}(-1)^{i+m} c_{im}\frac{\varphi_i^m(\eta)\varphi_i^m(\eta')}{\eta+\eta'}, \tag{9.13}$$

which are determined from the following set of functional equations

$$\varphi_i^m(\eta) = P_i^m(\eta) + 2\frac{(-1)^{i+m}}{2-\delta_{0m}}\int_0^1 \rho_m(\eta,\eta')P_i^m(\eta')\,d\eta', \tag{9.14}$$

where δ_{km} is the Kronecker symbol.

In the same paper these results were illustrated in treating two special cases: scattering with two-term and the Rayleigh phase functions, which are of astrophysical importance.

9.2.2 Partial redistribution over frequencies and directions

It is well known that the multiple scattering of the line radiation in various astrophysical media undergoes redistribution over frequencies and directions. The transfer problem that arises in the general case of partial redistribution is similar in many respects to that for anisotropic scattering. This analogy is especially distinct when one uses the bilinear expansions of the redistribution functions. Thus, for example, in the simplest case of a pure Doppler redistribution r_{I} [23,24] it has been shown [25] that the expansion

$$r_{\mathrm{I}}(x',x,\gamma) = \frac{1}{\sqrt{\pi}\sin\gamma}\exp\left\{\frac{x^2+x'^2-2xx'\cos\gamma}{\sin^2\gamma}\right\} = \sum_{k=0}^{\infty}\cos^k\gamma\,\alpha_k(x)\,\alpha_k(x') \tag{9.15}$$

holds, where x and x' are the so-called dimensionless frequencies of the incident and scattered photons measured from the center of the line in units of the Doppler width and $\alpha_k(x) = \left(2^k\sqrt{\pi}k!\right)^{-1/2}\exp(-x^2)H_k(x)$ is an orthonormal system of functions with weight $\exp\left(x^2\right)$ expressed in terms of Hermite polynomials $H_k(x)$. Analogous bilinear expansion was obtained in [25] also for the case of the combined Doppler and damping effects, the latter being due to radiation and collision. Let us consider the directionally averaged redistribution law

$$r_{\mathrm{I}}(x',x) = \int_{\max(|x|,|x'|)}^{\infty}\exp\left(-u^2\right)du = \sum_{k=0}^{\infty}A_k\alpha_{2k}(x')\,\alpha_{2k}(x), \tag{9.16}$$

where $A_k = 1/(2k+1)$. In this case the one-dimensional problem of the spectral line formation reduces to solving the infinite set of functional equations for the functions $\varphi_k(x)$

$$\varphi_k(x) = \alpha_{2k}(x) + \frac{\lambda}{2}\sum_{m=0}^{\infty}A_m\int_{-\infty}^{\infty}\frac{\varphi_m(x)\,\varphi_m(x')}{\alpha(x)+\alpha(x')}\alpha_{2k}(x')\,dx', \tag{9.17}$$

where $\alpha(x)$ is the profile of the absorption coefficient. This is the natural generalization of equation (9.4) to the case of partial redistribution over frequencies. Using expansions of types (9.15) and (9.16) makes it much easier to solve the corresponding transfer problems for the line-radiation [26–29]. Quasi-analytic and numerical methods have been developed for solving these problems [29–32]. It was shown that the accuracy in solving the truncated set of equations (9.17) can be highly increased in view of the fact that to any fixed value of x there corresponds the number n beginning of which the functions $\varphi_k(x)$ may be replaced by the known functions $\alpha_{2k}(x)$. Physically this reflects the fact that the formation of the far wings of the spectral line is due to the single scattering.

The invariance principle has played an important role in the theory of radiative transfer. It has been especially effective as applied to relatively complicated problems in radiative transfer theory. According to the idea suggested in [33], this theory can be constructed so that it is based on the invariance principle, while the radiative transfer equation and the invariance relationships follow directly from it. Subsequently, as the theory developed, the feasibility of this approach became obvious and the direct use of this approach seemed preferable in some cases. The advantage of the approach lies in the profound intuitive content of the invariance principle and the existence of a close connection with the characteristic features of the physical problem under consideration, the symmetry property, and the boundary and initial conditions. In addition, as is well known from physics, there is the relationship between invariance principles and conservation laws. In view of the importance of all these questions, we examine them in more detail in Section 9.5 using radiative transfer in a plane-parallel atmosphere as an example.

9.3 Quadratic and bilinear relations of radiative transfer theory

As early as in 1977 Rybicky derived quadratic integrals of the transfer equation; however, some important points that arose there remained unanswered for a long time [34]. He studied the problems, for which the transfer equation admits integrals that involve quadratic moments of the radiation field. When the boundary conditions (or other constraints) are required to be met, the integrals convert into nonlinear relations of specific form for quantities characterizing the radiation field in the atmosphere. The so-called Q- and R- integrals involve as a particular case a certain class of 'surface' results, some of which are known well in transfer theory.

Further generalization of Rybicki's results for monochromatic, isotropic scattering in the plane-parallel medium was given in [35]. The so-called 'two-point' relations have been found; they couple the intensities of equally directed radiation at two different depths in the atmosphere. The more general concept of 'bilinear integrals' (or relations) was introduced in [36] for quadratic integrals that connect the radiation fields of two separate transfer problems but referred to the same optical depth. Following the given terminology, it was reasonable to introduce also the concept of 'two-point bilinear integrals' for those coupling with each other radiation fields referred to both different transfer problems and diverse optical depths. Evidently this latter type of relations comprises all other types as the specific cases. Regardless of the new results, the main question on the physical nature of existence of quadratic and bilinear relations remained abstruse for a long time until appearing the mentioned paper [36]. The idea that there must exist some relationship to the invariance principle suggests itself and is justified if for no other reason than that the majority of nonlinear equations in transfer theory are associated in some or other way with the invariance technique. Most commonly these equations admit a plain physical interpretation and can be established by means of a direct use of the mathematical model of the physical process. Having this in mind, one may naturally be tempted to derive the requisite bilinear relations in a similar manner, thus making clear the physical significance of these relations. This kind of attempt has been made by Hubeny in [37,38], who provided some intuitive insight into the physical nature of quadratic results of the transfer theory.

The more general and mathematically rigorous derivation of quadratic and bilinear relations was given in [39, 6] on the basis of the principle of invariance . We elucidated the profound connection between invariance principle, quadratic relations and conservation laws resulting from general variational principle.

Let us treat the process of monochromatic, isotropic scattering in a semi-infinite, plane-parallel atmosphere. Suppose also that the atmosphere is homogeneous and does not contain energy sources. Turning to Eqs. (9.2), (9.3) and making use the expression for the zero-moment α_0 of $\varphi(\eta)$

$$\alpha_0 = \int_0^1 \varphi(\eta)\, d\eta = 2\left(1 - \sqrt{1-\lambda}\right)/\lambda, \tag{9.18}$$

one can write

$$\sqrt{1-\lambda}\varphi(\eta) = 1 - \int_0^1 \rho(\eta, \eta')\, \eta'\, d\eta' \tag{9.19}$$

and

$$(1-\lambda)(\eta+\xi)\,\rho(\eta,\xi) = \frac{\lambda}{2}\left(1 - \int_0^1 \rho(\eta,\eta')\,\eta'\,d\eta'\right)\left(1 - \int_0^1 \rho(\xi,\eta')\,\eta'\,d\eta'\right). \tag{9.20}$$

Alongside the reflection coefficient we introduce into consideration the function $Y(\tau, \eta, \mu)$ that characterizes the probability of the photon exit from atmosphere in the direction μ, if originally it was moving at depth t with the directional cosine η. The symmetry property of the Y-function follows from the reciprocity principle and can be represented in the form:

$$|\eta|\, Y(\tau, \eta, \mu) = |\mu|\, Y(\tau, -\mu, -\eta) = |\mu|\, \tilde{Y}(\tau, \mu, \eta), \tag{9.21}$$

where for convenience we introduce the function $\tilde{Y}$ with angular arguments referenced from inner normal direction. This function also admits a probabilistic interpretation, namely, $\tilde{Y}(\tau, \eta, \mu)\, d\eta$ is the probability that a photon incident on the atmosphere with the directional cosine μ will move (in general, as a result of multiple scattering) at depth τ within the directional interval $(\eta, \eta + d\eta)$. It is clear that $\tilde{Y}(0, \mu, \eta) = \eta\rho(\eta, \mu)$ In fact, $\tilde{Y}(\tau, \mu, \eta)/|\eta|$ is none other than the Green function for the source-free problem called also the surface Green function [18]. This quantity completely determines the radiation field throughout the semi-infinite atmosphere that is illuminated by the external monodirectional source of unit intensity.

It is obvious that

$$Y(\tau,-\eta,\mu) = \int_0^1 Y(\tau,\eta',\mu)\,\rho(\eta',\eta)\,\eta'\,d\eta',$$

$$\tilde{Y}(\tau,\mu,-\eta) = \eta\int_0^1 \tilde{Y}(\tau,\mu,\eta')\,\rho(\eta,\eta')\,d\eta'. \tag{9.22}$$

Now, multiplying equation (9.2) by the product $Y(\tau,\xi,\mu)Y(\tau',\eta,\mu')$ and integrating over ξ and η from 0 to 1, we arrive at the first fundamental result

$$\int_{-1}^{1} Y(\tau,\xi,\mu)\,Y(\tau',-\xi,\mu')\,d\xi = \frac{\lambda}{2}\left(\int_{-1}^{1} Y(\tau,\xi,\mu)\,d\xi\right)\left(\int_{-1}^{1} Y(\tau',\xi,\mu')\,d\xi\right) \tag{9.23}$$

where the relations (9.22) and symmetry property of the reflection coefficient are used. We shall see later that the described procedure admits a simple physical interpretation which makes it possible to obtain the same result directly by means of some modification of invariance idea. The probabilistic meaning assigned to the function Y sets one thinking that some statistical explanation may be suggested for equation (9.23) as well. Indeed, this equation implies that $\lambda/2$ can be regarded as the correlation coefficient of two random events so that this result can be stated in the probabilistic language as follows.

Two random events of two photons exit from a semi-infinite atmosphere in certain fixed (diverse, in general) directions, if they were originally moving in opposite directions at some different optical depths, are correlated with the correlation coefficient equal to $\lambda/2$.

The second fundamental result generating quadratic and bilinear R-relations, can be found by a similar manner from equation (9.20). Multiplying this equation by $\tilde{Y}(\tau,\mu,\eta)\tilde{Y}(\tau',\mu',\xi)$, and integrating over η and ξ in the range $(0,1)$, in light of the second of Eqs. (9.22), we obtain

$$(1-\lambda)\int_{-1}^{+1} \tilde{Y}(\tau,\mu,\xi)\,\tilde{Y}(\tau',\mu',-\xi)\,d\xi$$
$$= \frac{\lambda}{2}\left(\int_{-1}^{+1} \tilde{Y}(\tau,\mu,\xi)\,\xi\,\frac{d\xi}{|\xi|}\right)\left(\int_{-1}^{+1} \tilde{Y}(\tau',\mu',\xi)\,\xi\,\frac{d\xi}{|\xi|}\right). \tag{9.24}$$

Utilizing the reversibility property (9.21) in equations (9.23) and (9.24), one can write out another pair of equations for the functions $\tilde{Y}$ and Y

$$\int_{-1}^{+1} \tilde{Y}(\tau,\mu,\xi)\,\tilde{Y}(\tau',\mu',-\xi,)\,\frac{d\xi}{\xi^2}$$
$$= \frac{\lambda}{2}\left(\int_{-1}^{+1} \tilde{Y}(\tau,\mu,\xi)\,\frac{d\xi}{|\xi|}\right)\left(\int_{-1}^{+1} \tilde{Y}(\tau',\mu',\xi)\,\frac{d\xi}{|\xi|}\right), \tag{9.25}$$

and

$$(1-\lambda)\int_{-1}^{+1} Y(\tau,\xi,\mu)\,Y(\tau',-\xi,\mu',)\,\xi^2\,d\xi$$
$$= \frac{\lambda}{2}\left(\int_{-1}^{1} Y(\tau,\xi,\mu)\,\xi\,d\xi\right)\left(\int_{-1}^{1} Y(\tau',\xi,\mu')\,\xi\,d\xi\right). \tag{9.26}$$

Relations (9.23)–(9.26) constitute a wealth of information about radiation fields in the source-free media. Some of their versatile consequences will be examined below. Now we present merely the special result following from these relations, when $t = t'$ and $\mu = \mu'$

$$\int_0^1 Y(\tau,\xi,\mu) Y(\tau,-\xi,\mu)\, d\xi = \frac{\lambda}{4}\left(\int_{-1}^{+1} Y(\tau,\xi,\mu)\, d\xi\right)^2 , \tag{9.27}$$

$$(1-\lambda)\int_0^{+1} Y(\tau,\xi,\mu) Y(\tau,-\xi,\mu,)\, \xi^2\, d\xi = \frac{\lambda}{4}\left(\int_{-1}^{+1} Y(\tau,\xi,\mu)\, \xi\, d\xi\right)^2 . \tag{9.28}$$

Thus, we arrived at quadratic Q- and R- relations representing the prototypes of those obtained in [34]. The derived bilinear and quadratic relations may be applied to different transfer problems of astrophysical interest to give a number of new results.

9.3.1 The problem of diffuse reflection

Let us start, for instance, with the problem of diffuse reflection from a semi-infinite atmosphere, which is illuminated from outside by a parallel beam of radiation of unit intensity with directional cosine μ. Using superscripts '+' and '−' to denote the intensities with angular arguments $+\eta$ and $-\eta$, respectively, by virtue of the probabilistic meaning of function Y given above, one may write

$$I^+(\tau,\eta,\mu) = \tilde{Y}(\tau,\mu,-\eta)/\eta, I^-(\tau,\eta,\mu) = \tilde{Y}(\tau,\mu,\eta)/\eta \tag{9.29}$$

Now equations (9.23) and (9.26) correspondingly yield

$$Q(\tau,\mu;\tau'\mu') = \lambda J(\tau,\mu) J(\tau',\mu'), \quad (1-\lambda) R(\tau,\mu;\tau'\mu') = \lambda H(\tau,\mu) H(\tau',\mu') , \tag{9.30}$$

where

$$J(\tau,\mu) = \frac{1}{2}\int_{-1}^{+1} \left[I^+(\tau,\eta,\mu) + I^-(\tau,\eta,\mu)\right] d\eta ,$$

$$H(\tau,\mu) = \frac{1}{2}\int_{-1}^{+1} \left[I^+(\tau,\eta,\mu) - I^-(\tau,\eta,\mu)\right] \eta\, d\eta \tag{9.31}$$

are the mean intensity and flux, respectively; we introduced also two quadratic moments of the radiation field given by

$$Q(\tau,\mu;\tau'\mu') = \frac{1}{2}\int_{-1}^{+1} \left[I^+(\tau,\eta,\mu) I^-(\tau',\eta,\mu') + I^+(\tau',\eta,\mu') I^-(\tau,\eta,\mu)\right] d\eta , \tag{9.32}$$

$$R(\tau,\mu;\tau'\mu') = \frac{1}{2}\int_{-1}^{+1} \left[I^+(\tau,\eta,\mu) I^-(\tau',\eta,\mu') + I^+(\tau',\eta,\mu') I^-(\tau,\eta,\mu)\right] \eta^2\, d\eta . \tag{9.33}$$

The angular arguments μ and μ', which specify the directions of incidence, enter into Eqs. (9.30) as parameters so that the relations of this type can be written for arbitrary angular distribution of the illuminating radiation. Moreover, they enable one to establish the relationship between radiation fields of two diverse problems with different angular distribution for incident radiation.

To give an important insight into the source-free problem and the group of related problems we decide upon the simple and physically intelligible formula, second of Eqs. (9.22) which was employed in deriving Eqs. (9.30). By virtue of formulas (9.29) it can be rewritten as

$$I^{+}(\tau,\eta,\mu)=\eta\int_0^1\rho(\eta,\varsigma)\,I^{-}(\tau,\varsigma,\mu)\,d\varsigma\,, \tag{9.34}$$

which states the obvious fact that, in the absence of internal sources, the upward radiation (and ultimately the radiation field at any optical depth t), is completely determined by the intensity of inward radiation. Indeed, taking account of that

$$I^{-}(\tau,\eta,\mu)=\lambda\int_0^{\tau}J(\tau,\mu)\,e^{-(\varsigma-t)/\eta}\,\frac{dt}{\eta}+\frac{1}{2}\delta(\eta-\mu)\,e^{-\tau/\mu} \tag{9.35}$$

and the source function $S(\tau,\mu)=\lambda J(\tau,\mu)$, we insert formula (9.35) into Eq. (9.34) to obtain

$$S(\tau,\mu)=\int_0^{\tau}L(\tau-t)\,S(t,\mu)\,dt+\frac{\lambda}{2}\varphi(\mu)\,e^{-\tau/\mu}\,. \tag{9.36}$$

Thus, knowledge of the reflectance of the atmosphere makes it possible to reduce the classical boundary-value problem of determining the internal field of radiation to the Volterra-type Eq. (9.36) associated with an initial value problem. This result is not unexpected, however, and is important in the sense that similar equations may be written (see below) for some special internal-source problems as well while formula (9.34) is no more valid. This possibility stems from appropriate quadratic and bilinear relations to be derived.

9.3.2 Uniformly distributed energy sources

The results obtained above are sufficient to write down easily two-point bilinear relations for atmospheres with uniformly distributed sources. We assume the initial sources of energy are due to thermal emission so that the source-function has a form

$$S(\tau)=\lambda J(\tau)+(1-\lambda)\,B\,, \tag{9.37}$$

where B = *const* is related to the Planck function. As was shown in [27], the transfer problem for uniformly distributed sources is closely connected with that of diffuse reflection. An especially simple relationship exists between the problem with the source-function (9.37) and the diffuse-reflection problem with isotropic incident radiation. The plain probabilistic considerations based on the obvious fact that a photon, moving somewhere in the semi-infinite atmosphere, will either be destroyed or escape it, lead to the following relations

$$\tilde{I}_*^{\pm}(\tau,\eta)=\int_0^1 I_*^{\pm}(\tau,\eta,\mu)\,d\mu=1-I^{\pm}(\tau,\eta,B)/B\,, \tag{9.38}$$

where intensities relevant to the diffuse-reflection problem are supplied by an asterisk. It is customary to mark B explicitly as an argument identifying the problem under consideration. Incorporating formulas (9.38) in equations (9.27) applied to the two separate problems with different values of sources (B and B'), we find

$$Q(\tau, B; \tau', B') = \lambda S(\tau, B) S(\tau', B') - (1-\lambda) BB', \tag{9.39}$$

$$\begin{aligned}(1-\lambda) R(\tau, B; \tau', B') = \lambda H(\tau, B) H(\tau', B') \\ + (1-\lambda)\left[BK(\tau', B') + B'K(\tau, B)\right] \\ - (1-\lambda) BB'/3,\end{aligned} \tag{9.40}$$

where

$$K(\tau, B) = \frac{1}{2}\int_0^1 \left[I^+(\tau, \eta, B) + I^-(\tau, \eta, B)\right]\eta^2 \, d\eta \tag{9.41}$$

is the *K-moment* and other quantities are given by formulas (9.31)–(9.33) with μ and μ' replaced by B and B', respectively. These two-point bilinear equations obtained are the further generalization of existing results. Similar relations connecting the radiation fields for the diffuse-reflection problem, and that for an atmosphere with internal sources, may be readily found. The next question concerns the integral equation for the source function. Starting again with relation (9.34), we see that in this case it takes a form

$$I^+(\tau, \eta, B) = \eta \int_0^1 \rho(\eta, \varsigma) I^-(\tau, \varsigma, B) \, d\varsigma + B\left[2 - \varphi(\eta)\right], \tag{9.42}$$

where the formulas (9.38) and (9.3) are used. Hence for the source-function (9.37) we have

$$S(\tau, B) = B\sqrt{1-\lambda} + \frac{\lambda}{2}\int_0^1 \varphi(\varsigma) I^-(\tau, \varsigma, B) \, d\varsigma. \tag{9.43}$$

Note that this equation could be found from the first of bilinear relations (9.39) on setting $\tau' = 0$ and $B = B'$. This result was obtained also in [35]. Now expressing I^- in terms of S yields

$$S(\tau, B) = \int_0^\tau L(\tau - t) S(t, B) \, dt + B\sqrt{1-\lambda}. \tag{9.44}$$

Thus, in accordance with what is being said above, this special internal-source problem as well is reducible to the Volterra-type equation. As a matter of fact, the mere existence of linking formulas (9.38) indicates that all mathematical results for the diffuse-reflection problem have their counterparts in the transfer under consideration.

Now having the appropriate results for the diffuse-reflection problem, one can with only slightly more effort develop bilinear relations for the problem of exponentially distributed internal energy sources as well as the Milne problem. Referring the interested reader to the paper [36] for details of derivations, here we limit ourselves by presenting the final results.

9.3.3 Exponentially distributed energy sources

For the sources of the form $b(\tau, m) = (1-\lambda)\, B\, e^{-m\tau}$ we will obtain

$$\begin{aligned}\lambda Q(\tau, m; \tau', m') &= S(\tau, m)\, S(\tau', m') \\ &\quad - \left[b(\tau, m)\, I^{+}\left(\tau', m^{-1}, m'\right) + b(\tau', m')\, I^{+}\left(\tau', m'^{-1}, m\right)\right] ,\end{aligned} \tag{9.45}$$

where $S(\tau, m) = \lambda J(\tau, m) + b(\tau, m)$, and

$$\begin{aligned}\lambda(1-\lambda)\, R(\tau, m; \tau', m') &= \left[\lambda H(\tau, m) - b(\tau, m)\right]\left[\lambda H(\tau', m') - b(\tau', m)\right] \\ &\quad - (1-\lambda)\left[m^{-2} b(\tau', m')\, I^{+}\left(\tau, m'^{-1}, m\right) + m^{-2} b(\tau, m)\, I^{+}\left(\tau, m^{-1}, m'\right)\right] .\end{aligned} \tag{9.46}$$

This relation is valid for arbitrary values of m and m'. It is apparent that the procedure described here may be performed to establish a relationship between problems that correspond to exponentially and either diffuse-reflection problem or the problem of uniformly distributed sources: the integral equation for $S(\tau, m)$ may be derived starting again with formulas (9.34) to find

$$I^{+}(\tau, \eta, m) = \eta \int_0^1 \rho(\eta, \varsigma)\, I^{-}(\tau, \varsigma, m)\, d\varsigma + b(\tau, m)\, \eta \rho\left(\eta, m^{-1}\right) , \tag{9.47}$$

which implies the following integral equation

$$S(\tau, m) = \int_0^\tau L(\tau - t)\, S(t, m)\, dt + (1-\lambda)\, B \varphi\left(m^{-1}\right) e^{-m\tau} . \tag{9.48}$$

9.3.4 The Milne problem

Two-point quadratic relations for the Milne problem have been derived by Ivanov in [35], so we limit the discussion to a brief description of our approach to obtaining the bilinear relations with only the final results being presented. Besides, the solution of the Milne problem is determined to within a constant factor, therefore bilinear relations connecting two separate Milne problems only trivially differ from two-point quadratic relations. For simplicity the conservative case ($\lambda = 1$) will be treated.

The starting point is the obvious relation

$$I^{+}(\tau, \eta) = I^{+}(0, \eta) + \eta \int_0^1 \rho(\eta, \varsigma)\, I^{-}(\tau, \varsigma)\, d\varsigma , \tag{9.49}$$

which can be viewed as a counterpart of Eqs. (9.34), (9.42) and (9.47) for the Milne problem. Now, as is well known (see, e.g., [9,24]).

$$I^{+}(0, \eta) = \left(\sqrt{3}/4\right) F \varphi(\eta) , \tag{9.50}$$

where F is the normalizing factor determined by the emergent flux.

Multiplying invariance equation (9.2) by $I^{+}(\tau,\eta,F)\,I^{-}(\tau,\eta,F')$ (parameters F and F' as an additional argument are introduced to distinguish the problem) and taking account of formulas (9.49) and (9.50), one derives

$$Q(\tau,F;\tau',F') = J(\tau,F)\,J(\tau',F') - (3/16)\,FF' , \tag{9.51}$$

where Q and J are given by formulas (9.32) and the first of (9.31), respectively, with F, F' as parameters. In a similar manner one can obviously write out two-point bilinear relations linking the Milne problem with any of the problems treated above. An especially simple relation may be established with the diffuse-reflection problem (for $\lambda = 1$)

$$Q(\tau,\mu;\tau',F) = J(\tau,\mu)\,J(\tau',F) . \tag{9.52}$$

The integral equation for the source function of the Milne problem follows from both Eqs. (9.49) and (9.51) on setting in the latter $\tau = 0$

$$S(\tau,F) = \int_0^\tau L(\tau - t)\,S(t,F)\,dt + \left(\sqrt{3}/4\right)F . \tag{9.53}$$

Thus, again, as in the preceding paragraphs, we arrive at the Volterra-type integral equation. Scrutinizing the results obtained, we observe that all treated problems possess some common features, of which the possibility of reduction to the Volterra-type equation is an example. An alternative important characteristic is that the radiation fields corresponding to these special distributions of sources are connected with each other by means of bilinear relations. We shall see below that this class of problems may be supplemented.

9.4 The modified principle of invariance

In this section, we shall give an important insight into the physical nature of this kind of nonlinear relations. The preceding considerations suggest the idea that there must exist a close interconnection between bilinear relations and the invariance principle. For this reason, let us return to the original formulation of the invariance principle sketched at the outset of the review. We shall consider a semi-infinite, plane-parallel and source-free homogeneous atmosphere. For simplicity of exposition, the probabilistic approach to the problem will be adopted.

Let a photon be incident upon the boundary plane $\tau = 0$ of the atmosphere at an angle $\cos^{-1}\mu$ to the inner normal, and we are interested in the probability of the photon exit at some angle $\cos^{-1}\mu'$ to the upper normal direction. The classical formulation of the invariance principle assumes the addition (or removal) of a thin layer to (from) the surface of the atmosphere, presuming the optical properties of the layer and atmosphere to be the same. It is completely clear, however, that physically there is no difference whether we add (or remove) the layer to (from) the top of the atmosphere or do it somewhere within it with some selected layer $(\tau, \tau+\Delta)$. In the latter case we are interested in the probability of reflection subject to condition that the level τ was intersected. Recalling the probabilistic meaning of functions Y and $\tilde{Y}$ introduced in Section 9.3, it is readily seen that the probability of the mentioned event is

$$d\mu \int_0^1 Y(\tau, -\varsigma, \mu')\, \tilde{Y}(\tau, \mu, \varsigma)\, d\varsigma\,. \tag{9.54}$$

Further, we must determine the same probability accounting of the elementary processes of interaction with the selected layer $(\tau, \tau + \Delta\tau)$, and equate to that of the formula (9.54) in accordance with what has been said. One may distinguish three types of processes of the first order of $\Delta\tau$ associated with the layer.

(i) The scattering photon passes the selected layer with no interaction. The probability associated with this possibility is

$$\begin{aligned} d\mu \int_0^1 Y(\tau, -\varsigma, \mu')\, \tilde{Y}(\tau, \mu, \varsigma)\, d\varsigma - \Delta\tau d\mu \int_0^1 Y(\tau, -\varsigma, \mu')\, \tilde{Y}(\tau, \mu, \varsigma)\, \frac{d\varsigma}{\varsigma} \\ - \Delta\tau\, d\mu \int_0^1 Y(\tau, \eta, \mu')\, d\eta \int_0^1 \rho(\eta, \varsigma)\, \tilde{Y}(\tau, \mu, \varsigma)\, d\varsigma \end{aligned} \tag{9.55}$$

(ii) The photon enters the selected layer by crossing the plane τ and is scattered in the layer in some direction specified by η. For this process we have

$$\Delta\tau\, d\mu' \frac{\lambda}{2} \left(\int_{-1}^{+1} Y(\tau, \eta, \mu')\, d\eta \right) \left(\int_{-1}^{+1} \tilde{Y}(\tau, \mu, \varsigma)\, \frac{d\varsigma}{\varsigma} \right). \tag{9.56}$$

(iii) The photon enters the selected layer from below, i.e. by crossing the plane $\tau + \Delta\tau$, and is scattered in the layer. The associated probability is

$$\Delta\tau\, d\mu' \frac{\lambda}{2} \left(\int_{-1}^{1} Y(\tau, \eta, \mu')\, d\eta \right) \left(\int_0^1 d\eta' \int_0^1 \rho(\eta', \varsigma)\, \tilde{Y}(\tau, \mu, \varsigma)\, d\varsigma \right). \tag{9.57}$$

Now adding up expressions (9.55)–(9.57) and equating the result to that of (9.54), by virtue of formulas (9.21) and the second of Eqs. (9.22) finally we obtain

$$\int_{-1}^{+1} Y(\tau, \varsigma, \mu)\, Y(\tau, -\varsigma, \mu')\, d\varsigma = \frac{\lambda}{2} \left(\int_{-1}^{+1} Y(\tau, \varsigma, \mu)\, d\varsigma \right) \left(\int_{-1}^{+1} Y(\tau, \varsigma, \mu')\, d\varsigma \right). \tag{9.58}$$

Thus we have arrived at the specific version of the bilinear Q-relation (see Eq. (9.23)) written for $\tau = \tau'$. On setting $\tau = 0$ in (9.58) and taking account of the boundary condition $Y(0, \zeta, \mu) = \delta(\zeta - \mu)$, we are led to invariance equation (9.2) for the reflection coefficient. We see that equation (9.58) is more informative as compared to Eq. (9.2) and can be regarded as the extension of Ambartsumian's equation to all depths in the atmosphere.

9.5 The variational formalism

As was above said, the theory of radiative transfer can be made to rest on the principles of invariance. We saw that the invariance equation, being combined with some simple physical reasoning, makes it possible to derive the more informative quadratic and bilinear relations, for which the former equation is a special (surface) result. It turned out that the resulting Q-relations may be envisioned as a manifestation of a somewhat generalized version of the classical principle of invariance and can be obtained immediately. These facts indicate the fundamental nature of the invariance property of transfer problems and set one thinking that there must exist some general formulation of problems that implies both the equation of transfer and the principles of invariance as kinds of laws.

Now we shall see that the problems of transfer of radiation in the plane-parallel homogeneous atmosphere admit a variational formulation, the equation of transfer then being the Euler–Lagrange equation and the bilinear Q-relation being the conservation law due to form-invariance of the suitable Lagrangian. In fact, a single functional comprises all the information on features of the problem and allows a systematic connection between symmetries and conservation laws. Being the first integrals of the Euler–Lagrange equation, the conservation laws may facilitate the solution of the problem under consideration and assist in its interpretation. Two salient problems, encountered in having recourse to the variational principle, are the existence of the principle for a given problem and the derivation of appropriate conservation laws. The former of these problems for systems of partial differential equations was solved by Vainberg [40], who showed that this problem is equivalent to determining whether an operator is potential or not. The derivation of the conservation laws is based on Noether's theorem [41], which suggests a systematic procedure for establishing these laws from a direct study of the variational integral. An important generalization of Noether's theorem to encompass the integro-differential equations was given by Tavel [42].

While the variational approach is widely used in various branches of theoretical physics, this is not the case in the field of the radiative transfer theory, with the only exception being the paper of Anderson [43] who employs Tavel's results to establish the conservation law suitable for the case of non-isotropic scattering. Krikorian and Nikoghossian used the results of the rigorous mathematical theory in applying the Lagrangian formalism to the one-dimensional transfer problem [39].

Let us start with the transfer equations for the function Y

$$\pm\eta\frac{dY(\tau,\pm\eta,\mu)}{d\tau} = -Y(\tau,\pm\eta,\mu) + \frac{\lambda}{2}\int_{-1}^{1} Y(\tau,\eta',\mu)\,d\eta'\,. \tag{9.59}$$

From these equations one can easily obtain

$$\eta^2\frac{d^2\Phi}{d\tau^2} = -\Phi(\tau,\eta,\mu) - \lambda\int_0^1 \Phi(\tau,\eta',\mu)\,d\eta'\,, \tag{9.60}$$

where we introduced notation $\Phi(\tau,\eta,\mu) = Y(\tau,+\eta,\mu) + Y(\tau,-\eta,\mu)$.

One may readily check the self-adjointness of this equation so that the variational formulation is admitted. The Lagrangian density L corresponding to

Eq. (9.59) was obtained in [43]

$$L\left(\Phi, \Phi', \tau, \eta, \mu\right) = \Phi^2 + \left(\eta \Phi'\right)^2 - 2\Phi U\,, \tag{9.61}$$

where

$$U\left(\tau, \mu\right) = \frac{\lambda}{2} \int_0^1 \Phi\left(\tau, \eta', \mu\right)\, d\eta'\,. \tag{9.62}$$

In accordance with [43], the Euler–Lagrange equation has a form

$$\frac{\partial L}{\partial \Phi} - \frac{d}{d\tau}\frac{\partial L}{\partial \Phi'} + \lambda \int_0^1 \frac{\partial L}{\partial U}\, d\eta' = 0\,. \tag{9.63}$$

One will make sure that insertion of the Lagrangian (9.61) into (9.63) yields the transfer equation (9.60). It is important that both the transfer equation (9.60) and the Lagrangian density (9.61) do not depend explicitly on τ, or stated differently, they are form-invariant under infinitesimal trans-formation

$$\tau \to \tau' = \tau + \delta\tau\,, \quad \eta = \eta'\,, \quad \mu = \mu'\,, \tag{9.64}$$

where the quantity $\delta\tau$ is allowed to be an arbitrary infinitesimal function of τ. This implies that the transformation (9.64), i.e. translation of the optical depth, is the symmetry transformation for the system (9.59) and suggests a certain conservation law as follows

$$\int_0^1 \left[L - \frac{\partial L}{\partial \Phi}\Phi'\right] d\eta = \mathrm{const}\,, \tag{9.65}$$

which, in view of (9.61), takes the form

$$\int_0^1 \left[\Phi^2\left(\tau, \eta, \mu\right) \quad \eta^2 \Phi'^2\left(\tau, \eta, \mu\right) - 2U\left(\tau, \mu\right) \Phi\left(\tau, \eta, \mu\right)\right] d\eta = \mathrm{const}\,, \tag{9.66}$$

or

$$\int_0^1 Y\left(\tau, \varsigma, \mu\right) Y\left(\tau, -\varsigma, \mu\right) d\varsigma = \frac{\lambda}{4}\left(\int_{-1}^{+1} Y\left(\tau, \varsigma, \mu\right)\, d\varsigma\right)^2 + \mathrm{const}\,. \tag{9.67}$$

This relation is, in essence, a prototype of the Q-integral obtained by Rybicki in [34]. The above considerations imply that by its content the integral (9.67) is an analog of the momentum conservation law in mechanics and is due to the axes translation transformation.

For semi-infinite atmosphere $Y\left(\tau, \pm\varsigma, \mu\right) \to 0$ as $\tau \to \infty$ so that, const $= 0$. More general relation for this case may be derived if we treat problems differing with each other by the value of the parameter μ

$$\int_0^1 Y\left(\tau, \varsigma, \mu\right) Y\left(\tau, -\varsigma, \mu'\right) d\varsigma = \frac{\lambda}{2}\left(\int_{-1}^{+1} Y\left(\tau, \varsigma, \mu\right)\, d\varsigma\right)\left(\int_{-1}^{+1} Y\left(\tau, \varsigma, \mu'\right)\, d\varsigma\right)\,. \tag{9.68}$$

This equation was obtained in two different ways, particularly on the basis of non-complicated physical reasoning. It holds everywhere where λ does not vary with depth.

The variational formalism not only allows us to elucidate the physical meaning of the invariance principle but also enables us to derive along with many known results a great number of new relations of great importance for the theory and applications. It allows us also to find out some statistical characteristics of the diffusion process in the atmosphere [36, 44]. Some of the known nonlinear relations possess a fairly obvious physical or/and probabilistic meaning and can be written immediately on the basis of simple arguments. However, as we shall show below, by no means all of them follow from the variational principle, so they cannot be recognized as invariance relations or, even less, as invariance principles.

9.5.1 The polynomial distribution of sources

An important question to be answered is the applicability of the Lagrangian formalism to the transfer problems for atmospheres containing energy sources. Inasmuch as both the diffuse reflection and Milne problems obey the homogeneous integro-differential equation (9.60), the suitable quadratic and bilinear integrals are possible to write directly. In the case of the source-containing atmospheres we are led to the inhomogeneous integro-differential equations with appropriate source terms. Generally, the self-adjointness of the transfer equations may be violated so that the direct application of the variational principle becomes impossible. Nevertheless, it was shown above that in two cases of the uniformly and exponentially distributed sources this difficulty was surmounted by using the apparent connection between these two problems and the source-free problem. Now an additional possibility based on the conservation law (9.66) is appeared to reduce the source-containing problems to the source-free problem. As an example we shall briefly consider here the case of polynomial distribution of sources [6]. The first attempt in this direction was made in [34]. The approach proposed there was successful, however, for polynomials of the degree not higher than second, whereas, as we show now, the Q-integrals exist for polynomials for arbitrary high order.

Let us have the transfer equation of the form

$$\eta^2 \frac{d^2 \Phi}{d\tau^2} = -\Phi(\tau, \eta) - \lambda \int_0^1 \Phi(\tau, \eta')\, d\eta' - 2g(\tau)\,, \tag{9.69}$$

where

$$g(\tau) = (1 - \lambda) B(\tau)\,, \quad B(\tau) = \sum_{i=0}^{N} B_i \tau^i\,. \tag{9.70}$$

Let us introduce the function

$$F_N(\tau, \eta) = 2 \sum_{k=0}^{[N/2]} B^{(2k)}(\tau)\, p_{2k}(\eta) \tag{9.71}$$

where N is the degree of polynomials, $B^{(m)} = d^m B/d\tau^m$, and brackets mean the integer part. It can be checked by direct substitution that F_N is a partial solution of Eq. (9.69) if only $p_{2k}(\eta)$ are polynomials of the order $2k$ ($k = 1, 2, \dots$) defined as follows

$$p_{2k}(\eta) = \sum_{i=0}^{k} p_{2i}(0)\, \eta^{2(k-i)}, \tag{9.72}$$

and

$$p_{2k}(0) = \frac{\lambda}{1-\lambda} \sum_{i=1}^{k} \frac{p_{2(k-i)}(0)}{2k+1}, \quad p_0(\eta) = 1. \tag{9.73}$$

Having the rule for construction the function (9.71), one can derive the requisite Q-integrals for any source term with polynomial depth-dependence. Some results for small values of N are presented in [6].

9.6 The group of RSF (reducible to the source-free) problems

Thus, we saw that there exists a group of different frequently occurring radiation transfer problems of astrophysical interest which admit quadratic and bilinear integrals. They can be reduced to the source-free problem. This group includes the Milne problem, the problem of diffuse reflection (and transmission in the case of the atmosphere of finite optical thickness) as well as the problems with exponential and polynomial laws for the distribution of internal energy sources. An important special case of the last type of problem is the problem of radiative transfer in an isothermal atmosphere (i.e. in atmosphere with homogeneous distribution of internal sources). This group of problems is characterized at least by three features. First of all, the invariance principle implies bilinear relations connecting the solutions of the listed problems. It has been recently shown [45] that the group of the RSF-problems admits a class of integrals involving quadratic and bilinear moments of the intensity of arbitrarily high order. Secondly, if the problem can be formulated for a finite atmosphere then the principle allows us to connect its solution with that of the proper problem for a semi-infinite atmosphere. Finally, knowledge of the φ-function reduces their solutions to the Volterra-type equation for the source function with the kernel-function (9.9). We refer to these problems as the group of RSF-problems.

9.7 Arbitrarily varying sources

For further insight into the point, we shall give the general treatment of the problem of sources. Consider the transfer of monochromatic radiation through a semi-infinite homogeneous and plane-parallel atmosphere that contains energy sources varied arbitrarily with depth $g(\tau) = (1-\lambda)B(\tau)$. The source function of this problem, as is known, satisfies integral equation

$$S(\tau) = \frac{\lambda}{2} \int_0^{\infty} \mathrm{Ei}(|\tau - t|)\, S(t)\, dt + g(\tau). \tag{9.74}$$

As was shown in [46, 47], the solution of this equation is equivalent to the solution of the following coupled pair of the Volterra-type equations

$$S(\tau) = \omega(\tau) + \int_0^{\tau} L(\tau - t)\, S(t)\, dt\,, \tag{9.75}$$

$$\omega(\tau) = g(\tau) + \int_{\tau}^{\infty} L(\tau - t)\, \omega(t)\, dt\,. \tag{9.76}$$

As a matter of fact, the set of equations (9.75), (9.76) represents a kind of splitting of the so-called Λ-operator [48] and may become useful in solving the transfer problems. Note that in being applied to this system, the escape probability approach [49] leads, as could be expected, to sufficiently accurate results. For instance, when $B = const$, one may easily find for $\tau = 0$ that $\omega = g/(1-\lambda)$ and $S(0) = \omega = B\sqrt{1-\lambda}$. For deep interiors of the atmosphere we have

$$\lim_{t\to\infty} S(\tau) = \lim_{t\to\infty} \frac{\omega}{L(\tau) + \sqrt{1-\lambda}} = B\,. \tag{9.77}$$

Thus in both cases we get the exact values of the source function, whereas the same approach applied to Eq. (9.74) gives a rather crude estimate for the surface value $S(0) = 2(1-\lambda)B/(2-\lambda)$. Moreover, it easy to see that we obtain correct values of $S(0)$ for all the treated special RSF-problems. This follows from the fact that in these cases Eq. (9.76) allows exact analytical solution so that the problems resolve themselves into the solution of a single Volterra-type equation (9.75) for the source function.

9.8 Finite medium

The approach described for the semi-infinite atmosphere is easy to apply to an atmosphere of finite optical thickness. In fact, the generalization of previous results for such an atmosphere is attained trivially and resolves itself into determining the proper values of the integration constants. To proceed, we introduce by analogy to the semi-infinite atmosphere the quantities $Y(\tau, \pm\eta, \mu; \tau_0)$ with similar probabilistic meaning, characterizing the photon's exit from the boundary $\tau = 0$ of an atmosphere for the photon moving in direction $\pm\eta$ at depth τ (angular arguments are referenced with respect to the outward normal to the surface $\tau = 0$). This implies that

$$\begin{aligned} Y(0, \eta, \mu; \tau_0) &= \delta(\eta - \mu)\,, & Y(\tau_0, \eta, \mu; \tau_0) &= \mu q(\mu, \eta, \tau_0)\,, \\ Y(0, -\eta, \mu; \tau_0) &= \mu\rho(\mu, \eta, \tau_0)\,, & Y(\tau_0, -\eta, \mu; \tau_0) &= 0\,, \end{aligned} \tag{9.78}$$

where

$$q(\mu, \eta, \tau_0) = \eta^{-1}\delta(\eta - \mu)\, e^{-\frac{\tau_0}{\eta}} + \sigma(\mu, \eta, \tau_0)\,, \tag{9.79}$$

is the transmittance of the medium. Since the functions $Y(\tau, \pm\eta, \mu; \tau_0)$ satisfy the partition transfer equations (9.59), the variational principle leads to the quadratic Q-integral (9.67) with the value $(\lambda/4)\,\psi^2(\mu, \tau_0)$ for the constant. It is apparent that with no more effort one may derive many different kinds of bilinear and two-point bilinear relations linking with each other different quantities describing the field of radiation in different depths of the finite media of different optical thickness. Such relations for some of the RSF-problems for finite atmosphere are given in [6]. Of special interest are the nonlinear relations which establish connection between characteristics of the radiation transfer in finite and semi-infinite atmospheres. We limit ourselves to presenting here only two of them which will be mentioned later on

$$\begin{aligned} \eta' \rho_\infty (\eta', \eta) = & -\eta\rho(\eta, \eta', \tau) + \int_0^1 q(\eta, \varsigma, \tau) Y_\infty (\tau, -\varsigma, \eta')\, d\varsigma \\ & - \frac{\lambda}{2}\psi(\eta, \tau) \int_0^1 Y_\infty (\tau, \varsigma, \eta')\, d\varsigma - \frac{\lambda}{2}\varphi_\infty (\eta') \varphi(\eta, \tau)\,, \end{aligned} \tag{9.80}$$

$$\begin{aligned} Y_\infty (\tau, \eta, \eta') = & \, q(\eta, \eta', \tau) - \eta \int_0^1 \rho(\eta, \varsigma, \tau) Y_\infty (\tau, -\varsigma, \eta')\, d\varsigma \\ & - \frac{\lambda}{2}\varphi(\eta, \tau) \int_0^1 Y_\infty (\tau, \varsigma, \eta')\, d\varsigma - \frac{\lambda}{2}\varphi_\infty (\eta') \psi(\eta, \tau)\,, \end{aligned} \tag{9.81}$$

where quantities pertaining to the semi-infinite atmosphere are marked by the sign of infinity.

9.9 Statistical description of the radiation diffusion process

The problem usually posed in a study of radiation transport in a medium is to find the field of radiation at any point in it depending on direction, frequency and so on. But for many reasons, quantities that give a statistical description of the scattering process are of great interest. Its importance is due in the first place to the fact that it facilitates better understanding of the physical essence of a number of effects predicted by the mathematical solution of the problem. Secondly, it makes it possible to determine a number of important physical characteristics of an atmosphere such as the mean radiation intensity, the mean degree of excitation of atoms and so forth. Note also that the problem of finding the radiation field in a medium can ultimately also be regarded as a stochastic problem requiring the determination of the statistical mean of some random variables.

Among the various statistical mean quantities most attention in the literature is devoted to the mean number of scattering events (MNSE) undergone by photons diffusing in a medium. Pioneering here is Ambartsumian's work [3], in which he proposed for this quantity the formula $N = \lambda \partial \ln I / \partial \lambda$, where I is the intensity of radiation in the beam. This quantity was subsequently estimated by many authors for different special cases, though the general treatment of the problem was given by Sobolev in a series of papers [50–52]. It was in these papers that the MNSE was calculated separately for photons that escape the medium as a result of diffusion and the photons that are trapped (i.e. undergo true absorption) in it. However, relations obtained by him, like the physical arguments which provide their basis, cease to hold when allowance is made for absorption and emission in the continuum.

Nor has there yet been a comprehensive study of the more complicated cases when the scattering is anisotropic or subject to partial redistribution over frequencies. Very important also is the statistical description of scattering in its dependence on the initial characteristics of the photon, for instance, the frequency, direction of motion, etc.

The problem of finding the MNSE in its general formulations was treated in a series of our papers [53–56]. A new approach was elaborated for determining any statistical average characterizing a diffusion process in an atmosphere. It is based on the invariance principle and extensive use of generating functions. It was shown that the method is applicable to determining the statistical averages of continuously distributed random variables as well, though it is then necessary to use appropriate characteristic functions. As an illustration, the problem of the average time of a photon travel (ATT) in an atmosphere under the assumption that it spends time only in traveling between scatterings was treated [57]. In the general case, when the photons are thermalizing not only in scattering but also in flight, this average makes it possible to gauge the relative importance of energy dissipation in the medium and of energy flow through a boundary. Another important application of this average is associated with the problems, frequently encountered in astrophysical applications, of the radiation of an atmosphere subject to non-stationary energy sources. In these problems, knowledge of the ATT makes it possible to ascertain whether radiative equilibrium in the medium is established. Some special cases of the ATT problem and problems related with it were treated by a number of authors [8, 58–61].

Under general assumptions concerning the elementary scattering event, equations were obtained in [53–57] for determining the MNSE and ATT in a plane-parallel semi-infinite atmosphere. It was shown that, for moving photons (i.e. not for those being trapped) the problem simplifies and is reduced to differentiation over proper parameter. The dependence of these quantities on the initial characteristics of the photon was established. Here we limit ourselves by presenting a few results obtained on the basis of the invariance principle. Let us consider a photon with frequency x moving at the optical depth τ in a semi-infinite atmosphere in some direction specified by $\pm\eta$ referenced from the direction of outward normal to the surface. We designate $\langle N(\tau,x,\eta)\rangle$ the MNSE for this photon irrespective of whether it exits the medium or undergoes true absorption there. As was shown, this quantity satisfies the partitioned set of equations

$$\begin{aligned}\pm\eta\frac{\partial\langle N(\tau,x,\pm\eta)\rangle}{\partial\tau} &= -v(x)\langle N(\tau,x,\pm\eta)\rangle\\ &+\frac{\lambda}{2}\int_{-1}^{+1}d\eta'\int_{-\infty}^{\infty}\gamma(x,\eta;x',\eta')\langle N(\tau,x,\eta')\rangle\,dx'+v(x)\,,\end{aligned}\tag{9.82}$$

where $v(x)=\alpha(x)+\beta$, β is the ratio of the absorption coefficient in continuum to that in the line center and $\gamma(x,\eta;x',\eta')$ is the function of redistribution over frequency and direction. The boundary condition is $\langle N(0,x,+\eta)\rangle=0$. Other two quantities of interest are $\langle\Omega(\tau,x,\pm\eta)\rangle$ and $R_0(\tau,x,\pm\eta)$. The first is the dimensionless ATT measured in units of the average time of travel between two successive scattering events. The second is the probability of a given photon to thermalize somewhere in the atmosphere. It was shown that these two quantities satisfy the system of equations analogous to those of (9.82) with similar boundary condition with the only difference being in the free term which correspondingly are 1, and $u(x) = (1-\lambda)\,\alpha(x) + \beta$. This implies

$$(1-\lambda)\langle N(\tau, x, \pm\eta)\rangle + \lambda\beta\langle \Omega(\tau, x, \pm\eta)\rangle = R_0(\tau, x, \pm\eta)\,. \tag{9.83}$$

Some results obtained for a semi-infinite atmosphere were generalized in [62] to encompass the finite media as well. The form of equations (9.82) and those for the functions $\langle \Omega(\tau, x, \pm\eta)\rangle$ and $R_0(\tau, x, \pm\eta)$, shows that variational formulation can be applied also to these quantities, making possible to derive quadratic and bilinear relations for them. These kinds of relations are obtained in [63].

9.10 The layers adding method

Because of importance Ambartsumian's arguments in deriving the laws of addition of global optical characteristics of the absorbing and scattering media (coefficients of reflection and transmission) for further discussion, we renew our acquaintance with the method and present together with an explanatory figure the starting auxiliary equations written for the 1-D homogeneous media. Figure 9.1 shows a medium of optical thickness τ_0 divided into two parts with thicknesses τ_1 and τ_2. Each of the media is characterized by reflection r and transmission q coefficients. Based on some simple physical and probability arguments, one can write

$$I_1 = r(\tau_2)\,I_0 + q(\tau_2)\,I_3\,, \tag{9.84}$$
$$I_2 = q(\tau_1)\,I_4\,, \tag{9.85}$$
$$I_3 = r(\tau_1)\,I_4\,, \tag{9.86}$$
$$I_4 = q(\tau_2)\,I_0 + r(\tau_2)\,I_3\,. \tag{9.87}$$

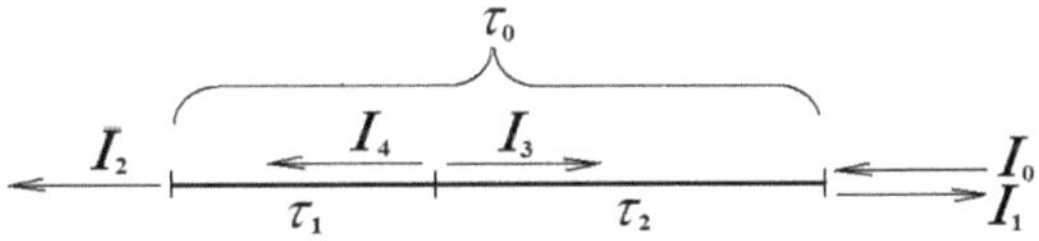

Fig. 9.1. Illustrating the method of adding layers.

In view of the fact that $I_1 = r(\tau_1+\tau_2)\,I_0$, and $I_2 = q(\tau_1+\tau_2)\,I_0$, it is easy to obtain the requisite addition laws for the reflection and transmission coefficients of scattering and absorbing media

$$q(\tau_1+\tau_2) = \frac{q(\tau_1)\,q(\tau_2)}{1 - r(\tau_1)\,r(\tau_2)}\,, \tag{9.88}$$

$$r(\tau_1+\tau_2) = r(\tau_2) + \frac{r(\tau_1)\,q^2(\tau_2)}{1 - r(\tau_1)\,r(\tau_2)}\,. \tag{9.89}$$

These relations are referred to as the laws of addition for the transmission and reflection coefficients. The quantities q, r have a probabilistic meaning and may be correspondingly interpreted as the probabilities of the transmission and reflection of a photon incident on the medium.

Replacing τ_2 by the infinitesimal Δ and passing to the limit when $\Delta \to 0$, we will have

$$\frac{dq}{d\tau_0} = -\left(1 - \frac{\lambda}{2}\right) q(\tau_0) + \frac{\lambda}{2} q(\tau_0) r(\tau_0), \tag{9.90}$$

$$\frac{dr}{d\tau_0} = \frac{\lambda}{2} - (2 - \lambda) r(\tau_0) + \frac{\lambda}{2} r^2(\tau_0). \tag{9.91}$$

The system of nonlinear differential equations we obtain satisfies the initial conditions $q(0) = 1$, $r(0) = 0$. Its solution is:

$$r(\tau_0) = r_0 \frac{1 - e^{-2k\tau_0}}{1 - r_0^2 e^{-2k\tau_0}}, \quad q(\tau_0) = \left(1 - r_0^2\right) \frac{e^{-k\tau_0}}{1 - r_0^2 e^{-2k\tau_0}}, \tag{9.92}$$

where $k = (\lambda/4)\left(1 - r_0^2\right)/r_0$, and r_0 is the coefficient of reflection from a semi-infinite atmosphere.

Thanks to their generality, these relations became a base for various modifications and stimulated the development of new methods in radiative transfer theory. Some results obtained in the field are, in essence, nothing but elaboration of some special cases of the law of the addition of layers. For instance, taking as one of the layers a semi-infinite atmosphere and for another a layer of infinitesimal optical thickness, we are led to the problem considered in the previous section, for which the invariance principle was formulated. When the infinitesimal layer is added to a finite layer, the requisite optical characteristics are found as the functions of optical thickness. Thus, the problem becomes 'imbedded' in a family of similar problems differing by the value of optical thickness. The generalization of this approach to the three-dimensional case was given by Chandrasekhar in [9]. It underlies the method of 'invariant imbedding' developed by Bellman and his co-authors [64, 65]).

Finally, the method of addition of layers plays an important role in solving the problems of radiation transfer in inhomogeneous atmospheres. In this case the medium can be divided into a number of layers in such a way as each of them can be regarded as homogeneous and the addition formulas (9.88), (9.89) are repeatedly applied (see, e.g., [66–68]). It is noteworthy that in the course of derivation of requisite optical parameters the fluxes appearing at the interfaces between adjacent layers are eliminated. Accordingly, in each application of the addition formulas one deals only with the intensities at the boundaries of the composite atmosphere. This is pointed out in some papers [69–71] in which the addition formulas are treated in the case when the component layers are allowed to be inhomogeneous. In the last two works the law of addition of layers is called 'the star product'.

9.10.1 The nature of some nonlinear relations of the radiation transfer theory

We saw that invariant properties of the problems of radiation transfer in homogeneous atmospheres lead to a variety of nonlinear relations. They are widely used and are especially efficient when they are combined with other methods of solution. In the light of these comparatively new results, the question arises to what extent certain nonlinear relations in radiative transfer theory are connected with the variational principles resulting from invariance with respect to the translational transformation of the optical depth. Here we are mainly speaking of the relations used in connection with the invariance principle or with the method of adding layers,

We begin with noting that Eqs. (9.84)–(9.87) have been written without invoking the invariance principle, to which they are, of course, not related. In fact, these formulas remain valid in general cases, when it is inappropriate to speak of invariance properties of the problem. For example, these equations remain in force even when the media are inhomogeneous, with, of course, their polarity property taken into account (see below). We now examine how Eqs. (9.84)–(9.87) are rewritten for the plane-parallel atmosphere. We transform the above problem in the following way: let a parallel beam of radiation with intensity I_0 be incident on the boundary $\tau = 0$ of a medium at an angle of $\arccos \varsigma$ to its inner normal (Fig. 9.2). The intensity of the reflected radiation is related to the reflection function $\rho(\eta, \varsigma, \tau_0)$ by $I(0, \eta, \varsigma) = I_0 \rho(\eta, \varsigma, \tau_0) \varsigma$. In analogous fashion for transmitted radiation we have $I(\tau_0, \eta, \varsigma) = I_0 q(\eta, \varsigma, \tau_0) \varsigma$, in which we have retained the customary notation σ for the diffuse part of the transmitted radiation. For brevity, the dependence of the intensities on the optical thickness of the medium is left out of the argument. Since the medium can now be regarded as consisting of two parts with respective thicknesses τ and $\tau_0 - \tau$, relations analogous to Eqs. (9.84)–(9.87) can be written in the form (I_0 can be taken equal to unity without loss of generality):

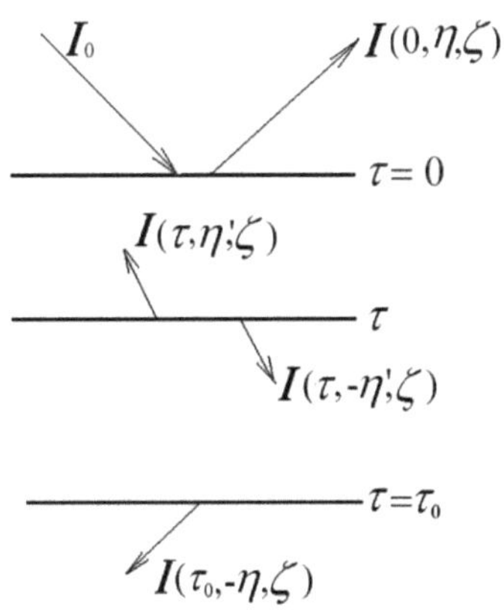

Fig. 9.2. Diffuse reflection of light from a medium of finite thickness.

$$\rho(\eta,\varsigma,\tau_0)\,\varsigma=\rho(\eta,\varsigma,\tau)\,\varsigma+I(\tau,\eta,\varsigma)\;e^{-\tau/\eta}+\int_0^1\sigma(\eta,\eta',\tau)\,I(\tau,\eta',\varsigma)\,\eta'\,d\eta'\,, \quad (9.93)$$

$$\sigma(\eta,\varsigma,\tau_0)\,\varsigma=\sigma(\eta,\varsigma,\tau_0-\tau)\,\varsigma\,e^{-\tau/\eta}+I^*(\tau,-\eta,\varsigma)\;e^{-(\tau_0-\tau)/\eta}$$
$$+\int_0^1\sigma(\eta,\eta',\tau_0-\tau)\,I^*(\tau,-\eta',\varsigma)\,\eta'\,d\eta'\,, \quad (9.94)$$

$$I(\tau,\eta,\varsigma)=\rho(\eta,\varsigma,\tau_0-\tau)\,\varsigma\,e^{-\tau/\varsigma}+\int_0^1\rho(\eta,\eta',\tau_0-\tau)\,I^*(\tau,-\eta',\varsigma)\,\eta'\,d\eta'\,, \quad (9.95)$$

$$I^*(\tau,-\eta,\varsigma)=\sigma(\eta,\varsigma,\tau)\,\varsigma+\int_0^1\rho(\eta,\eta',\tau)\,I(\tau,\eta',\varsigma)\,\eta'\,d\eta'\,, \quad (9.96)$$

where the diffuse part of the intensity of the downward-moving radiation is indicated by an asterisk. The limiting case of Eqs. (9.93), (9.95) and (9.96) for $\tau_0\to\infty$ are of special interest:

$$I_\infty(0,\eta,\varsigma)=I(0,\eta,\varsigma)+\int_0^1 q(\eta,\eta',\tau)\,I_\infty(\tau,\eta',\varsigma)\,\eta'\,d\eta'\,, \quad (9.97)$$

$$I(\tau,\eta,\varsigma)=\rho_\infty(\eta,\varsigma)\,\varsigma\,e^{-\tau/\varsigma}+\int_0^1\rho_\infty(\eta,\eta')\,I^*(\tau,-\eta',\varsigma)\,\eta'\,d\eta'\,, \quad (9.98)$$

$$I^*_\infty(\tau,-\eta,\varsigma)=\sigma(\eta,\varsigma,\tau)\,\varsigma+\int_0^1\rho(\eta,\eta',\tau)\,I_\infty(\tau,\eta,\eta')\,\eta'\,d\eta'\,, \quad (9.99)$$

These equations establish the relationship between the characteristics of the radiation fields in semi-infinite and finite media. We see that all the equations given above (9.93)–(9.99) have a fairly simple physical significance and can be written down at once without invoking the invariance principle.

9.10.2 The Chandrasekhar relations

The nonlinear relations written down by Chandrasekhar in the 1950s and called invariance principles by him [9] are well known in radiative transfer theory. Here we shall examine these relations from the standpoint of their possible connection with the invariance principle.

The problem of diffuse reflection and transmission of radiation by a medium of finite optical depth examined by Chandrasekhar was stated as follows: on a medium of optical thickness τ_0 in direction $(-\mu_0,\varphi_0)$, falls a parallel beam of radiation with flux πF per unit area perpendicular to the direction of incidence. The reflection function S and the transmission function T are introduced so that

$$I(0,\mu,\varphi)=\frac{F}{4\mu}S(\tau_0,\mu,\varphi;\mu_0,\varphi_0)\,, \quad (9.100)$$

$$I^*(\tau_0,-\mu,\varphi)=\frac{F}{4\mu}T(\tau_0,\mu,\varphi;\mu_0,\varphi_0)\,. \quad (9.101)$$

Chandrasekhar wrote down the following four nonlinear equations:

$$I(\tau,\mu,\varphi) = \frac{F}{4\mu} e^{-\tau/\mu_0} S(\tau_0-\tau,\mu,\varphi;\mu_0,\varphi_0) + \frac{1}{4\pi\mu}\int_0^{2\pi} d\varphi \int_0^1 S(\tau_0-\tau,\mu,\varphi;\mu',\varphi')\, I^*(\tau,-\mu',\varphi')\, d\mu', \tag{9.102}$$

$$I^*(\tau,-\mu,\varphi) = \frac{F}{4\mu} T(\tau,\mu,\varphi;\mu_0,\varphi_0) + \frac{1}{4\pi\mu}\int_0^{2\pi} d\varphi \int_0^1 S(\tau,\mu,\varphi;\mu',\varphi')\, I(\tau,\mu',\varphi')\, d\mu', \tag{9.103}$$

$$\frac{F}{4\mu} S(\tau_0,\mu,\varphi;\mu_0,\varphi_0) = \frac{F}{4\mu} S(\tau,\mu,\varphi;\mu_0,\varphi_0) + e^{-\tau/\mu} I(\tau,\mu,\varphi) + \frac{1}{4\pi\mu}\int_0^{2\pi} d\varphi \int_0^1 T(\tau,\mu,\varphi;\mu',\varphi')\, I(\tau,\mu',\varphi')\, d\mu', \tag{9.104}$$

$$\frac{F}{4\mu} T(\tau_0,\mu,\varphi;\mu_0,\varphi_0) = \frac{F}{4\mu} e^{-\tau/\mu_0} T(\tau_0-\tau,\mu,\varphi;\mu_0,\varphi_0) + e^{-(\tau_0-\tau)/\mu} I(\tau,-\mu,\varphi) + \frac{1}{4\pi\mu}\int_0^{2\pi} d\varphi \int_0^1 T(\tau_0-\tau,\mu,\varphi;\mu',\varphi')\, I^*(\tau,-\mu',\varphi')\, d\mu', \tag{9.105}$$

An equation for the limiting case of a semi-infinite atmosphere was derived from Eq. (9.106) with $\tau_0 \to \infty$

$$I(\tau,\mu,\varphi) = \frac{F}{4\mu} e^{-\tau/\mu_0} S(\mu,\varphi;\mu_0,\varphi_0) + \frac{1}{4\pi\mu}\int_0^{2\pi} d\varphi \int_0^1 S(\mu,\varphi;\mu',\varphi')\, I^*(\tau,-\mu',\varphi')\, d\mu'. \tag{9.106}$$

We now set ourselves the task of comparing Eqs. (9.102)–(9.105) with Eqs. (9.93)–(9.96) of the preceding subsection. To do this we neglect the azimuthal dependence in Eqs. (9.102)–(9.105), set $F = 1$, switch to the previous notation, and include the following easily verified relation between the corresponding coefficients of reflection and transmission:

$$(1/2)\, S(\tau_0,\eta,\varsigma) = \eta\varsigma\rho(\eta,\varsigma,\tau_0), \tag{9.107}$$

$$(1/2)\, T(\tau_0,\eta,\varsigma) = \eta\varsigma\sigma(\eta,\varsigma,\tau_0). \tag{9.108}$$

Then it is easy to show that Eqs. (9.102)–(9.105) are the same as Eqs. (9.95), (9.96), (9.93), and (9.94), respectively, of the preceding subsection. Thus, neither these, nor the previous equations have a direct connection to the invariance principle associated with a translational transformation of the optical depth, and so they cannot be derived from the corresponding conservation law. As for Eq. (9.106), it, in turn, is the same as Eq. (9.98). As noted above, besides Eq. (9.98), Eqs. (9.97) and (9.99) also specify a connection between the radiation fields in semi-infinite and finite media. This type of relation can also be written down for a number of other quantities which describe the radiation field in semi-infinite and finite media. For example, based on Eqs. (9.94) and (9.96) we can write

$$\eta\rho_\infty(\eta,\eta') = \eta\rho(\eta,\eta',\tau) + \int_0^1 q(\eta',\varsigma,\tau)\, Y_\infty(\tau,-\varsigma,\eta)\,\varsigma\, d\varsigma\,. \tag{9.109}$$

$$Y_\infty(\tau,\eta',\eta) = q(\eta,\eta',\tau) + \int_0^1 \rho(\eta',\varsigma,\tau)\, Y_\infty(\tau,-\varsigma,\eta)\,\varsigma\, d\varsigma\,, \tag{9.110}$$

These relations in the operator form are given also in [10]. Similar formulas that follow from the invariance principle are much more complicated and cannot be reduced to the above equations. Two of them, that were derived in [6], we gave in Section 9.8 (see Eqs. (9.80), (9.81)), We can see that the two quantities are related to one another by equations of considerably different kinds. These and the other examples given in this chapter show that the nonlinear relations discussed here can be divided into two classes. The first includes those formulas which characterize only the radiative transfer process, itself, in a plane-parallel atmosphere; these have great generality. Equations (9.84)–(9.87), (9.93)–(9.99), (9.102)–(9.106), (9.109), and (9.110) belong to this class. The second, narrower class includes equations that are a consequence of the invariance properties of the specific transfer problem at hand. Many equations of this type for groups of the RSF-problems have been derived in the first part of this review.

9.11 Inhomogeneous atmosphere

When interpreting the radiation from objects in space it is usually necessary to apply various simplifying assumptions regarding their geometry and physical properties. For example, it is often assumed that a radiating medium is homogeneous and stationary, although it clearly has a rather complex structure and is subject to variation in time. This simplifies the problems to a great extent and makes it possible to estimate some characteristics of the radiating medium averaged in some sense. However, the high-resolution observational data available nowadays afford an opportunity for a more detailed investigation of astrophysical objects and analysis of their radiation. This leads to pressure to develop a suitable theory of radiation transfer trough an inhomogeneous atmosphere, providing new efficient methods of computations. Such attempts were made by a number of authors (see, e.g., [72–76]). In connection with solar prominences, in [77–80] we treated the effect of physical inhomogeneities related to the distribution of internal energy sources and to geometrical factors. In an analysis of multiple scattering at the line frequencies, the scattering coefficient was usually assumed to be constant inside the radiating volume. It is, however, evident that such an assumption may be rather crude for the interpretation of radiation in optically thick lines. Henceforth under inhomogeneous atmosphere we will mean atmosphere with the scattering coefficient arbitrarily varying with depth, though methods we develop remain in force also in the case of variation of other parameters which determine an elementary scattering event and the distribution of primary energy sources.

The complexity of the boundary-value transfer problems makes it necessary to develop appropriate analytical techniques in order to make it somehow easier to get a numerical solution. In each individual case, depending on the initial assumptions about the properties of the medium, of elementary scattering events, etc., specialized methods have been developed. One of the first methods of this type was the method based on the invariance principle, i.e., on the symmetry properties of the problem, which avoided the above difficulties in the case of homogeneous atmospheres and allowed us to determine the intensity of the emerging radiation without prior knowledge of the radiation field in the entire atmosphere. There has also been a natural drive to find an alternative statement of the classical problems of radiative transfer theory with the aim of reducing them to initial-value problems (so-called Cauchy problems). With the development of high-speed electronic computers, research in this area has become especially important in connection with the fact that solving this kind of problem is more suited to the computers' capabilities. Among the first papers in this area, we note those of Bellman [81] and Sobolev [15, 16, 82], who developed a method based on extensive use of the 'surface' resolvent function. The idea of this approach goes back to Krein's paper [83].

As was said above, the method of invariant imbedding enables one to readdress standard problems in a way such as to reduce them to the initial value problems. Of the extensive literature in this area, besides the above cited works [64, 65], we note also monographs [84, 85].

The method we propose includes a simple, but, at the same time, universal computational scheme that can be used to determine the radiation field and various characteristics of the scattering process as solutions of corresponding initial-value problems. The idea behind this method developed in [68, 86–89] for solving a given linear problem of radiation transfer involves a preliminary determination of the global optical properties of an atmosphere – the reflection and transmission coefficients, as well as some other related quantities, for a family of atmospheres with different optical thicknesses. This makes it possible to determine the radiation field inside the 1-D inhomogeneous medium without solving any new equations. However, as we shall see below, there exists another route that allows us to reduce the computational process to ordinary matrix multiplication. Regardless of the initial assumptions, the calculations are easily carried out on modern computers and, most importantly, are numerically stable.

For illustration of the basic idea we begin with the simplest scalar case involving the transfer of monochromatic radiation in a one-dimensional inhomogeneous atmosphere. As was shown in [68], the inhomogeneous atmosphere exhibits the so-called polarity property with respect to the sense of the incoming illumination, i.e., its optical properties are described by three parameters: two reflection coefficients and one transmission coefficient.

Figure 9.3 is a schematic illustration of two cases where a photon is incident from outside on a composite medium formed as the result of adding two inhomogeneous scattering and absorbing media with optical thicknesses τ_1 and τ_2. Here we denote the reflection coefficient of each medium when illuminated from the left by an overhead bar.

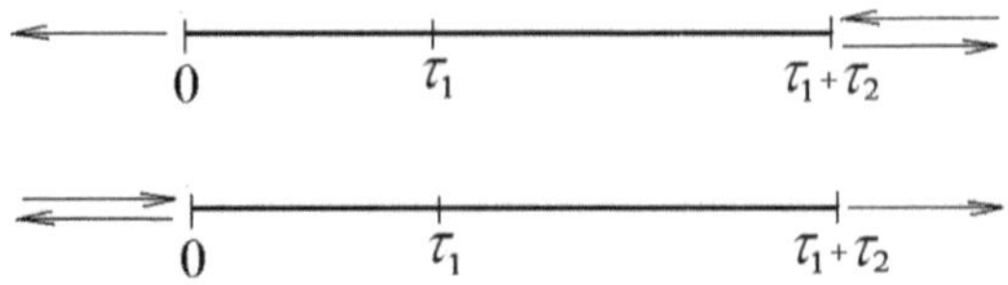

Fig. 9.3. Reflection and transmission of a composite atmosphere.

The formulas for addition in the case shown in the upper part of Fig. 9.3 are

$$q(\tau_1+\tau_2)=\frac{q(\tau_1)\,q(\tau_2)}{1-r(\tau_1)\,\bar{r}(\tau_2)}\,,\quad r(\tau_1+\tau_2)=r(\tau_2)+\frac{r(\tau_1)\,q^2(\tau_2)}{1-r(\tau_1)\,\bar{r}(\tau_2)}\,. \tag{9.111}$$

For the reflection coefficient of the composite atmosphere on the left (lower part of Fig. 9.3) we can write

$$\bar{r}(\tau_1+\tau_2)=\bar{r}(\tau_1)+\frac{r(\tau_1)\,q^2(\tau_2)}{1-r(\tau_1)\,\bar{r}(\tau_2)}\,. \tag{9.112}$$

It should be noted that, in general, the two components of the composite medium differ one from another not only by optical thickness, but also by the form of the scattering coefficient. Even in the case of the scattering coefficient common for two components, their optical characteristics may differ by the range of variation of this coefficient. But this should not cause confusion in further discussion since henceforth τ_2 will be replaced by an infinitely thin layer. The layers addition laws were generalized to the case of inhomogeneous media in [68, 86]. It was also shown that if a medium is illuminated on the side of the boundary $\tau_0=\tau_1+\tau_2$, then the ordinary procedure of taking the limit when τ_2 tends to 0 yields the differential equations which coincide by their form with those in the case of homogeneous atmosphere (see Eqs. (9.90), (9.91)). It is important that $r(\tau_0)$ satisfies a separate equation. This is a Riccati equation and can be solved by one of the standard numerical methods. Note that a fairly high accuracy is obtained even in the simplest case of solving Eq. (9.91) by the Euler method. It is important that this algorithm is numerically stable. In fact, for an arbitrary right-hand side of Eq. (9.91), we have.

$$\lambda(\tau_0)\left[1-r(\tau_0)\right]-2\le 0\,, \tag{9.113}$$

which implies that partial loss of stability is possible only in the case where $\lambda(\tau_0)$ approaches unity asymptotically, so that $r(\tau_0)\to 1$, as well. Greater accuracy can be obtained by using the fourth-order Runge–Kutta procedure in its various modifications (e.g., by Gill [85]). It is obvious that in the course of solving the initial-value problem for some fixed values of τ_0, we determine the reflectance and transmittance of a family of atmospheres with intermediate values of optical thicknesses.

After finding the reflectance, $r(\tau_0)$, the transmission coefficient is determined explicitly using the formula

$$q(\tau_0)=\exp\left[-\int_0^{\tau_0}\varpi(\tau)\,d\tau\right]\,, \tag{9.114}$$

where

$$\varpi(\tau_0) = 1 - \frac{\lambda(\tau_0)}{2}\left[1 + r(\tau_0)\right] . \tag{9.115}$$

It follows from Eqs. (9.90) and (9.91) that as $\tau_0 \to \infty$, $r(\tau_0)$ and $q(\tau_0)$, respectively, approach $\left(2 - \lambda_\infty - 2\sqrt{1-\lambda_\infty}\right)/\lambda_\infty$ and zero, where λ_∞ is the limiting value of the scattering coefficient.

To proceed, let us introduce the functions $P(\tau_0) = q^{-1}(\tau_0)$ and $S(\tau_0) = r(\tau_0)/q(\tau_0) = r(\tau_0)P(\tau_0)$. From Eqs. (9.90) and (9.91), it is not difficult to obtain (see [68, 86])

$$\frac{dP}{d\tau_0} = \left[1 - \frac{\lambda(\tau_0)}{2}\right]P(\tau_0) - \frac{\lambda(\tau_0)}{2}S(\tau_0) , \tag{9.116}$$

$$\frac{dS}{d\tau_0} = \frac{\lambda(\tau_0)}{2}P(\tau_0) - \left[1 - \frac{\lambda(\tau_0)}{2}\right]S(\tau_0) , \tag{9.117}$$

This system of linear equations with the initial conditions $P(0) = 1$, $S(0) = 0$ can be written in vector-matrix notation as

$$\frac{d\mathbf{Y}}{d\tau_0} = \mathbf{A}(\tau_0)\,\mathbf{Y}(\tau_0) , \tag{9.118}$$

where we have introduced notation

$$\mathbf{Y}(\tau_0) = \begin{pmatrix} P(\tau_0) \\ S(\tau_0) \end{pmatrix}, \quad \mathbf{A}(\tau_0) = \begin{pmatrix} a(\tau_0) & -b(\tau_0) \\ b(\tau_0) & -a(\tau_0) \end{pmatrix}, \tag{9.119}$$

$$a(\tau_0) = 1 - \frac{\lambda(\tau_0)}{2}, \quad b(\tau_0) = \frac{\lambda(\tau_0)}{2} . \tag{9.120}$$

The matrix $\mathbf{A}$ has a particular property

$$\mathbf{A}^2(\tau_0) = (1 - \lambda(\tau_0))\,\mathbf{I}, \tag{9.121}$$

where $\mathbf{I}$ is the unit matrix. This implies

$$\mathbf{A}^{-1}(\tau_0) = \left[1 - \lambda(\tau_0)\right]^{-1}\mathbf{A}(\tau_0) . \tag{9.122}$$

Another property of $\mathbf{A}$ is that the related commutator,

$$\mathbf{A}(\tau_1)\,\mathbf{A}(\tau_2) - \mathbf{A}(\tau_2)\,\mathbf{A}(\tau_1) = \left[\lambda(\tau_1) - \lambda(\tau_2)\right]\begin{pmatrix} 0 & 1 \\ 1 & 0 \end{pmatrix}, \tag{9.123}$$

is nonzero. This means that if we seek a solution of Eq. (9.118) in the form of a matrix exponential, then the corresponding Magnus series [90] is infinite. The situation is simpler for homogeneous atmosphere, for which

$$\mathbf{Y}(\tau_0) = \exp(\mathbf{A}\tau_0)\,\mathbf{Y}(0) , \tag{9.124}$$

so that, given Eq. (9.121),

$$\mathbf{Y}(\tau_0) = \left[\mathbf{I}\,\mathrm{ch}(k\tau_0) + \mathbf{A}\,\mathrm{sh}(k\tau_0)\right]\mathbf{Y}(0) , \tag{9.125}$$

where $k = \sqrt{1-\lambda}$. With Eq. (9.120), Eq. (9.125) leads to the results consistent with the standard expressions for the reflection and transmission coefficients [66, 8],

$$P(\tau_0) = \mathrm{ch}(k\tau_0) + \frac{1+k^2}{2k}\mathrm{sh}(k\tau_0), \tag{9.126}$$

$$S(\tau_0) = \frac{1-k^2}{2k}\mathrm{sh}(k\tau_0). \tag{9.127}$$

In conclusion, we note also that $P(\tau_0)$ and $S(\tau_0)$, as shown in [66], satisfy linear equations

$$\frac{d^2P}{d\tau_0^2} - \frac{\lambda'}{\lambda}\frac{dP}{d\tau} - \left(1-\lambda-\frac{\lambda'}{\lambda}\right)P(\tau_0) = 0, \tag{9.128}$$

$$\frac{d^2S}{d\tau_0^2} - \frac{\lambda'}{\lambda}\frac{dS}{d\tau} - \left(1-\lambda+\frac{\lambda'}{\lambda}\right)S(\tau_0) = 0, \tag{9.129}$$

respectively, for the initial conditions

$$P(0) = 1, \quad P'(0) = 1 - \frac{\lambda(0)}{2}, \quad S(0) = 0, \quad S'(0) = \frac{\lambda(0)}{2}.$$

Examples of explicit solutions of these equations in terms of elementary functions are presented in [68, 89],

9.11.1 The radiative transfer equations

One criticism sometimes raised about the applicability of the methods of adding layers and invariant imbedding is that the latter are supposedly not effective for determining the radiation field inside a medium. Here we show that, in fact, these methods make it easy to determine this field as well as a number of other quantities which describe the process of multiple scattering inside a medium.

We now write the transfer equations in terms of $U(\tau,\tau_0)$ and $V(\tau,\tau_0)$, which represent the probabilities that photon will move at the optical depth τ in the direction of decreasing and increasing optical depths, respectively (see Fig. 9.4):

$$\frac{dU}{d\tau} = \left[1-\frac{\lambda(\tau)}{2}\right]U(\tau,\tau_0) - \frac{\lambda(\tau)}{2}V(\tau,\tau_0), \tag{9.130}$$

$$\frac{dV}{d\tau} = \frac{\lambda(\tau)}{2}U(\tau,\tau_0) - \left[1-\frac{\lambda(\tau)}{2}\right]V(\tau,\tau_0), \tag{9.131}$$

These equations satisfy the boundary conditions $U(\tau_0,\tau_0) = 1$, $V(0,\tau_0) = 0$. In classical astrophysical problems, the transfer equations are usually reduced to just these kind of boundary-value problems. It is easy to see on comparing the system of Eqs. (9.130) and (9.131) with Eqs. (9.116) and (9.117) that, as functions of optical depth, the functions $U \to V$ satisfy the same equations as the functions $P \to S$ of the optical thickness. However, in the first case, we have to deal with a boundary-value problem, and in the second, with an initial-value problem. If we introduce a vector with components U and V then the system of Eqs. (9.130) and (9.131) can also be written in the vector-matrix form.

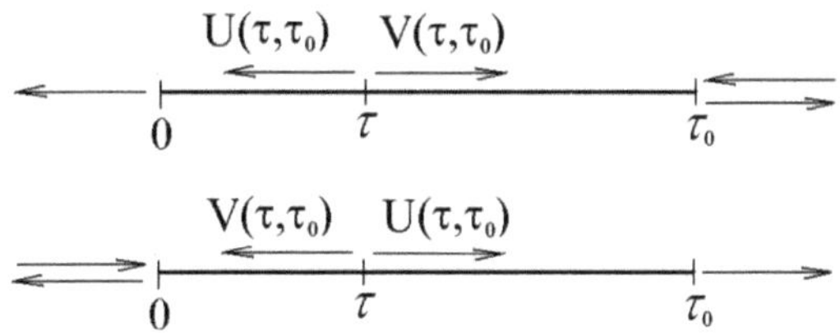

Fig. 9.4. Schematic illustration of radiative transfer in a one-dimensional medium.

Based on some simple physical arguments it is easy to show that

$$q(\tau_0) = q(\tau) U(\tau, \tau_0) \, , \quad V(\tau, \tau_0) = r(\tau) U(\tau, \tau_0) \, . \tag{9.132}$$

These equations follow mathematically from a comparison of the conditions at the boundary $\tau = 0$, $U(0, \tau_0) = q(\tau_0)$ and $V(0, \tau_0) = 0$, with the same conditions for the functions P and S. Strictly speaking, these equations need no proof, since the addition rules for the reflection and transmission coefficients have been derived from them [7].

Using Eq. (9.114), we obtain

$$U(\tau, \tau_0) = \exp\left[-\int_{\tau}^{\tau_0} \varpi(\tau')\, d\tau'\right] , \tag{9.133}$$

and for V

$$V(\tau, \tau_0) = r(\tau) \exp\left[-\int_{\tau}^{\tau_0} \varpi(\tau')\, d\tau'\right] . \tag{9.134}$$

Thus, for determining the radiation field inside a medium it is sufficient to first determine the reflectances of a family of atmospheres by solving Eq. (9.91). It should be noted that, as is clear from the above formulas, the variables τ and τ_0 in expressions for U and V are separated:

$$U(\tau, \tau_0) = q(\tau_0) P(\tau) , V(\tau, \tau_0) = q(\tau_0) S(\tau) \, . \tag{9.135}$$

9.11.2 Determination of some other quantities

Knowledge of r makes it possible to find explicit solutions for a whole series of problems that are frequently encountered in astrophysical applications. Here we consider a few of them.

(i) *Internal energy sources.* Suppose that an inhomogeneous atmosphere contains energy sources of power $B(\tau)$ and it is required to determine the intensity of the radiation emerging from the medium and the radiation field within the medium. If we denote the intensities emerging from the medium through the boundaries $\tau = \tau_0$ and $\tau = 0$ by $I_1(\tau_0)$ and $I_2(\tau_0)$, then, after some simple reasoning, we can write

$$I_1(\tau_0) = \frac{1}{2} q(\tau_0) \int_0^{\tau_0} [1 + r(\tau)] B(\tau)\, d\tau / q(\tau) \, , \tag{9.136}$$

$$I_2(\tau_0) = \frac{1}{2}\int_0^{\tau_0} [\lambda(\tau) I_1(\tau) + B(\tau)] q(\tau)\, d\tau\,. \tag{9.137}$$

Knowledge of the latter is sufficient for determining the intensities $I^+(\tau,\tau_0)$, $I^-(\tau,\tau_0)$ of the radiation directed, respectively, toward decreasing and increasing depths, i.e.,

$$I^+(\tau,\tau_0) = [I_2(\tau_0) - I_2(\tau)]/q(\tau)\,, \tag{9.138}$$

$$I^-(\tau,\tau_0) = r(\tau) I^+(\tau,\tau_0) + I_1(\tau)\,. \tag{9.139}$$

(ii) *Average number of scattering events.* We introduce the notations $N_r(\tau_0)$, $N_q(\tau_0)$ for the MNSE, respectively, for two categories of photons: reflected and transmitted. It has been shown [68] that these quantities can be completely expressed in terms of the above reflection and transmission coefficients for the family of atmospheres by

$$N_r(\tau_0) = \frac{1}{2}\frac{q^2(\tau_0)}{r(\tau_0)}\int_0^{\tau_0} \lambda(\tau)\left[1 + r^2(\tau)\right]\frac{d\tau}{q^2(\tau)}\,, \tag{9.140}$$

$$N_q(\tau_0) = \frac{1}{2}\int_0^{\tau_0} \lambda(\tau)\{1 + r(\tau)[1 + N_r(\tau)]\}\, d\tau\,. \tag{9.141}$$

(iii) *Reflection from the opposite boundary.* For completeness, here we also give the formula for the reflection coefficient of this medium when it is illuminated from the side of the boundary $\tau = 0$ in Fig. 9.4. We have shown [68, 86], that

$$\bar{r}(\tau_0) = \frac{1}{2}\int_0^{\tau_0} \lambda(\tau) q^2(\tau)\, d\tau = \frac{1}{2}\int_0^{\tau_0} \lambda(\tau)\, e^{-2\int_0^{\tau}\varpi(\tau')d\tau'}\, d\tau\,. \tag{9.142}$$

Thus, we can note the major conclusion of this section: *in order to solve the scalar problem of radiation transfer in a homogeneous atmosphere it is enough to know its reflectivity, since all the other quantities of interest are found explicitly.*

9.12 The group theoretical description of the radiation transfer

In this section group theory is used to describe a procedure for adding inhomogeneous absorbing and scattering atmospheres in a one-dimensional approximation. The group representations are derived for the composition of media in three different cases inhomogeneous atmospheres in which the scattering coefficient varies continuously with depth, composite or multicomponent atmospheres, and the special case of homogeneous atmospheres [91].

There are at least two reasons for the importance of radiative transfer problems in one-dimensional atmospheres. First, these problems are usually easier to solve and are extremely convenient to apply. In addition, despite the approximation, they can sometimes provide a satisfactory accuracy for estimating one or another characteristic of a radiation field in a three-dimensional medium with plane-parallel symmetry. Second, solutions to problems in this approximation serve an irreplaceable aid in understanding effects associated with the transfer process, itself, in cases where they are of primary importance, rather than problems associated.

We introduce the concept of a composition or transformation of scattering and absorbing atmospheres, which involves the adding of an additional atmosphere to an initial one in the general case of an inhomogeneous atmosphere (or removing some part from an initial atmosphere). It is assumed that the added or subtracted parts do not contain primary energy sources. The transformations induced in this way form a group if the group product is taken to mean the resultant of two successive transformations. It is easy to see that the other conditions for formation of a group are satisfied. In particular, the role of a unit element is the identity transformation, which leaves the initial atmosphere unchanged, and the inverse elements are transformations which reverse the effect of one or another already performed transformation. The associativity of the group product is evident. We refer to this group of transformations as the GN2 group. It is easy to see that it is not commutative. Among this type of groups, an important role is played by groups associated with the formation of a multicomponent atmosphere. In that case, the transformation is taken to mean the addition (or removal) of a homogeneous medium characterized by an optical thickness and a scattering coefficient. This group (we refer to it as GNH2) is a two-parameter, non-commutative group. The special case where λ is the same for all the added or removed media (group GH2) yields the case of homogeneous atmospheres. In this case, the group is obviously commutative, i.e., is Abelian [92]. It is also a one-parameter, infinite, and continuous group.

Let us now to go back to the composite medium shown in Fig. 9.3 and resultant addition equations (9.111) and (9.112). Eq. (9.111) yields

$$P\left(\tau_1 + \tau_2\right) = P\left(\tau_1\right) P\left(\tau_2\right) - S\left(\tau_1\right) \bar{S}\left(\tau_2\right) , \tag{9.143}$$

where $\bar{S} = \bar{r} q^{-1}$. Further, dividing the second of Eq. (9.111) by the first one and making a series of simple transformations we obtain

$$S(\tau_1 + \tau_2) = P\left(\tau_1\right) S\left(\tau_2\right) + S\left(\tau_1\right) M\left(\tau_2\right) , \tag{9.144}$$

where $M\left(\tau\right) = \left[1 - S\left(\tau\right) \bar{S}\left(\tau\right)\right] / P\left(\tau\right)$. Similar transformations using the addition formulas (9.112) yield

$$\bar{S}(\tau_1 + \tau_2) = P\left(\tau_2\right) \bar{S}\left(\tau_1\right) + \bar{S}\left(\tau_2\right) M\left(\tau_1\right) , \tag{9.145}$$

It can be confirmed by direct testing that there is also an addition law for $M\left(\tau\right)$

$$M\left(\tau_1 + \tau_2\right) = M\left(\tau_1\right) M\left(\tau_2\right) - S\left(\tau_1\right) \bar{S}\left(\tau_2\right) . \tag{9.146}$$

On introducing the matrices

$$A\left(\tau\right) = \begin{pmatrix} P\left(\tau\right) & -\bar{S}\left(\tau\right) \\ S\left(\tau\right) & M\left(\tau\right) \end{pmatrix} , \tag{9.147}$$

it is easy to confirm that they also constitute a group and form a representation of the group GN2. In fact, each element of the group GN2 corresponds to a transformation $T(g)$,

$$\begin{pmatrix} P(\tau_1+\tau_2) \\ S(\tau_1+\tau_2) \end{pmatrix} = \begin{pmatrix} P(\tau_2) & -\bar{S}(\tau_2) \\ S(\tau_2) & M(\tau_2) \end{pmatrix} \begin{pmatrix} P(\tau_1) \\ S(\tau_1) \end{pmatrix}, \tag{9.148}$$

if the medium is illuminated from the right, and $T'(g)$,

$$\begin{pmatrix} P(\tau_1+\tau_2) \\ \bar{S}(\tau_1+\tau_2) \end{pmatrix} = \begin{pmatrix} P(\tau_1) & -S(\tau_1) \\ \bar{S}(\tau_1) & M(\tau_1) \end{pmatrix} \begin{pmatrix} P(\tau_2) \\ \bar{S}(\tau_2) \end{pmatrix}, \tag{9.149}$$

in the opposite case. In addition, the product $g_1 \otimes g_2$ corresponds to the matrix product $\mathbf{A}(\tau_1+\tau_2) = \mathbf{A}(\tau_2)\mathbf{A}(\tau_1)$, i.e., $T(g_1 \otimes g_2) = T(g_2)T(g_1)$ for illumination from the right, and $\tilde{\mathbf{A}}(\tau_1+\tau_2) = \tilde{\mathbf{A}}(\tau_1)\tilde{\mathbf{A}}(\tau_2)$, i.e. $T'(g_1 \otimes g_2) = T'(g_1)T'(g_2)$ for illumination from the left. (Here the tilde denotes the transposed matrix.) The identity transformation obviously corresponds to the unit matrix: $T(e) = \mathbf{E}$ and $T'(e) = \mathbf{E}$. The matrix $\mathbf{A}(\tau)$ is nonsingular (its determinant equals 1), so inverse matrices exist, with

$$\mathbf{A}^{-1}(\tau) = \begin{pmatrix} M(\tau) & S(\tau) \\ -\bar{S}(\tau) & P(\tau) \end{pmatrix}, \quad \tilde{\mathbf{A}}^{-1}(\tau) = \begin{pmatrix} M(\tau) & \bar{S}(\tau) \\ -S(\tau) & P(\tau) \end{pmatrix}. \tag{9.150}$$

These two group representations are isomorphic, since the correspondence between the groups GN2 and $T(g)$, as well as between GN2 and $T'(g)$, are mutually unique. For the infinitesimal operator of the group $T(g)$ we have

$$\Xi = \underbrace{\lim}_{\tau\to 0} \frac{\mathbf{A}(\tau)}{\tau} = \begin{pmatrix} 1-\dfrac{\lambda}{2} & -\dfrac{\lambda}{2} \\ \dfrac{\lambda}{2} & -\left(1-\dfrac{\lambda}{2}\right) \end{pmatrix}. \tag{9.151}$$

Then Eqs. (9.143) and (9.144) can be rewritten in differential form to give Eqs. (9.116) and (9.117).

9.12.1 Radiation field inside a medium

As was shown in the preceding section, knowledge of the reflection coefficient for 1-D inhomogeneous media with different optical thicknesses makes it extremely simple to determine the radiation field inside an atmosphere with a fixed optical thickness. We now consider radiative transfer in a medium of optical thickness τ_0 as indicated in the upper part of Fig. 9.4. Let us suppose that the scattering coefficient varies continuously in the medium. It is evident that if we specify the incident flux, then knowledge of the functions $U(\tau, \tau_0)$ and $V(\tau, \tau_0)$ allows us to find the corresponding intensities.

By analogy with the groups introduced above, we introduce the concept of an optical depth translation group, which involves a transition from one optical depth to another. It is easy to verify that the necessary conditions for formation of a group are satisfied here. Everything said above about the properties of the group GN2 for the optical thickness apply equally to compositions (translations) of the optical depth. The only limitation is that the value of the total thickness obtained as a result of a translation must not exceed the optical thickness of the medium. Given the probabilistic meaning of the quantities introduced above, we can write (cf., Eqs. (9.132))

$$U(\tau, \tau_0) = q(\tau_0) P(\tau) , \quad V(\tau, \tau_0) = q(\tau_0) S(\tau) . \tag{9.152}$$

It is easy to understand from Eqs. (9.152) that the group $T(g)$ simultaneously is a representation of the optical depth translation group. In fact, based on Eq. (9.150), we can write

$$\begin{pmatrix} U(\tau + \tau', \tau_0) \\ V(\tau + \tau', \tau_0) \end{pmatrix} = \begin{pmatrix} P(\tau') & -\bar{S}(\tau') \\ S(\tau') & M(\tau') \end{pmatrix} \begin{pmatrix} U(\tau, \tau_0) \\ V(\tau, \tau_0) \end{pmatrix} . \tag{9.153}$$

Formulas derived from Eq. (9.153) show how the radiation intensities at different depths in the atmosphere are related to one another. The infinitesimal operator for the optical thickness translation group representation is evidently the same as Eq. (9.151) and yields the customary transfer equations, (9.130), (9.131). If the atmosphere is homogeneous, then there is a conservation law which can be written directly from these equations as

$$[U(\tau, \tau_0) - V(\tau, \tau_0)]^2 - (1 - \lambda)[U(\tau, \tau_0) + V(\tau, \tau_0)]^2 = \lambda q^2(\tau_0) . \tag{9.154}$$

Although it is obvious, for a long time this relationship was unknown unlike its special surface manifestation.

We now consider the case when the medium is illuminated from the side of boundary $\tau = 0$ which is illustrated schematically in the lower part of Fig. 9.4. Arguments similar to those used to derive Eqs. (9.152) imply that

$$U(\tau, \tau_0) = q(\tau_0) P(\tau_0 - \tau) , \quad V(\tau, \tau_0) = q(\tau_0) \bar{S}(\tau_0 - \tau) , \tag{9.155}$$

where it is understood that in the quantities corresponding to the thickness $\tau_0 - \tau$, the scattering coefficient varies within the interval $[\tau, \tau_0]$. Based on Eq. (9.149), we have

$$\begin{pmatrix} P(\tau_0 - \tau) \\ \bar{S}(\tau_0 - \tau) \end{pmatrix} = \begin{pmatrix} P(\tau') & -S(\tau') \\ \bar{S}(\tau') & M(\tau') \end{pmatrix} \begin{pmatrix} P(\tau_0 - \tau - \tau') \\ \bar{S}(\tau_0 - \tau - \tau') \end{pmatrix} , \tag{9.156}$$

whence

$$\begin{pmatrix} U(\tau + \tau', \tau_0) \\ V(\tau + \tau', \tau_0) \end{pmatrix} = \begin{pmatrix} M(\tau') & \bar{S}(\tau') \\ -S(\tau') & P(\tau') \end{pmatrix} \begin{pmatrix} U(\tau, \tau_0) \\ V(\tau, \tau_0) \end{pmatrix} . \tag{9.157}$$

In its differential form this formula transforms to the usual transfer equation under the condition $U(0, \tau_0) = 1$ and $V(\tau_0, \tau_0) = 0$. They are needed to examine the frequently encountered practical problem of radiative transfer in semi-infinite atmospheres.

9.12.2 Semi-infinite medium

We take the limit $\tau_0 \to \infty$ in Eqs. (9.157). For the limiting values of U and V which evidently do exist and now depend only on the optical depth, we retain the previous notation. Given this, we have

$$U(\tau) = P(\tau')U(\tau+\tau') - S(\tau')V(\tau+\tau'), V(\tau)$$
$$= \bar{S}(\tau')U(\tau+\tau') + M(\tau')V(\tau+\tau'). \tag{9.158}$$

From this one can obtain an entire series of different formulas, some of which are known from the theory of radiative transfer in homogeneous media. Thus, for example, the first of Eqs. (9.158) can be written as

$$U(\tau+\tau') = q(\tau')U(\tau) + r(\tau')V(\tau+\tau'). \tag{9.159}$$

Given the second of Eqs. (9.155), we write it in the form $V(\tau) = r_\infty U(\tau)$, where r_∞ is the reflection coefficient of a semi-infinite atmosphere. We then find

$$U(\tau+\tau') = q(\tau')U(\tau)/[1 - r_\infty r(\tau')], \quad \text{or} \quad U(\tau+\tau') = U(\tau)U(\tau'). \tag{9.160}$$

The last equation reflects the semi-group property of the function $U(\tau)$. According to the principle of reversibility, the same property possesses the probability of emerging from a semi-infinite atmosphere for a photon moving at some depth in the direction of the surface.

9.12.3 Multicomponent atmosphere

As an example of an inhomogeneous atmosphere we examine a multicomponent medium consisting of a number of homogeneous media which differ from one another in optical thickness and in the value of the scattering coefficient λ. Besides its intrinsic interest, the problem of determining the reflection and transmission properties of this kind of atmosphere is important for developing schemes for numerical calculation of these quantities in media with continuously varying λ.

As was noted, here we are concerned, in general, with the two-parameter noncommutative group GNH2. Since here a transformation is understood to be the addition (or removal) of a homogeneous medium, the group representation can be written in the form

$$A(\tau,\lambda) = \begin{pmatrix} P(\tau,\lambda) & -S(\tau,\lambda) \\ S(\tau,\lambda) & M(\tau,\lambda) \end{pmatrix}. \tag{9.161}$$

Consider a multicomponent atmosphere containing N layers with optical thickness $\tau_0^{(N)}$. Suppose the medium is illuminated from the side of the boundary $\tau_0^{(N)}$, as shown in Fig. 9.5. Then according to Eq. (9.148), for finding the optical characteristics of this medium we will have

$$\begin{pmatrix} P(\tau_0^{(N+1)}) \\ S(\tau_0^{(N+1)}) \end{pmatrix} = \begin{pmatrix} P(\tau_N,\lambda_N) & -S(\tau_N,\lambda_N) \\ S(\tau_N,\lambda_N) & M(\tau_N,\lambda_N) \end{pmatrix} \begin{pmatrix} P(\tau_0^{(N)}) \\ S(\tau_0^{(N)}) \end{pmatrix}, \tag{9.162}$$

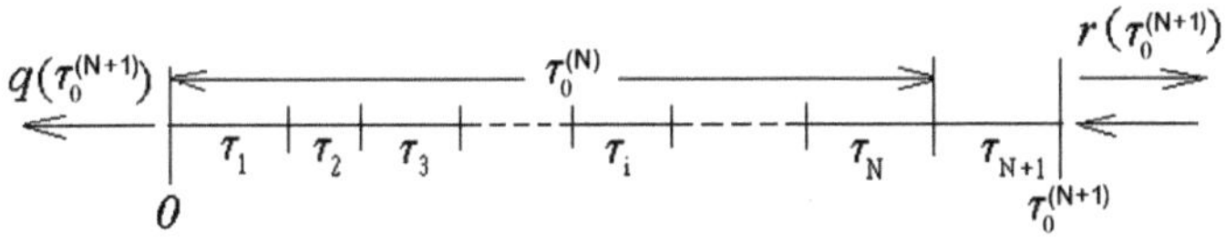

Fig. 9.5. Radiative transfer in a multicomponent atmosphere.

where τ_N and λ_N refer to the added homogeneous layer, so that for the corresponding values of P, S and M there are the following explicit expressions [8,86]

$$P(\tau_N, \lambda_N) = \frac{1}{4k_N}\left[(1+k_N)^2 e^{k_N\lambda_N} - (1-k_N)^2 e^{-k_N\lambda_N}\right], \tag{9.163}$$

$$S(\tau_N, \lambda_N) = \frac{1-k_N^2}{2k_N}\,\mathrm{sh}(k_N\tau_N), \tag{9.164}$$

$$M(\tau_N, \lambda_N) = \frac{1}{4k_N}\left[(1+k_N)^2 e^{-k_N\lambda_N} - (1-k_N)^2 e^{k_N\lambda_N}\right], \tag{9.165}$$

where $k_N = \sqrt{1-\lambda_N}$. Thus, by accumulation subject to the initial conditions $P(\tau_0^{(0)}) = 1$, $S(\tau_0^{(0)}) = 0$, Eq. (9.162) makes it possible to determine the desired optical characteristics of a multicomponent atmosphere. For completeness, we also give a formula obtained in [88] for finding the reflectivity of this atmosphere from its boundary 0.

$$\bar{S}\left(\tau_0^{(N)}\right) = \frac{1}{P\left(\tau_0^{(N-1)}\right)}\left[P\left(\tau_0^{(N)}\right)\bar{S}\left(\tau_0^{(N-1)}\right) + S(\tau_N, \lambda_N)\right]. \tag{9.166}$$

In order to fully evaluate the importance of these formulas, we should note that they can serve as a starting point in developing an alternative algorithm for finding the optical characteristics of an atmosphere in which the reflection coefficient varies continuously in the medium. In particular, taking the optical thickness of the added homogeneous layers to be the same and sufficiently small, we arrive at an extremely effective method for numerical solution of the problem. As the calculations show, in practice it is always possible to provide an accuracy that is entirely sufficient for applied problems.

The group representations introduced here express the symmetry properties of the radiative transfer problem with respect to variations in the optical thickness, as well as in the optical depth when the radiation field inside the medium is considered. At the same time, they make it possible easily to construct a solution to transfer problems for an arbitrary variation in the scattering coefficient in an atmosphere. This allows us to obtain immediately the solution of a family of problems with high accuracy and over a fairly wide range of variation in the optical thickness. An examination of the diffusion of radiation in a multicomponent atmosphere is of special importance for practical problems. The representation of the GNH2 group in this case essentially involves a simple algorithm for adding the optical properties of an arbitrary number of absorbing and scattering media.

9.13 The plane-parallel atmosphere

We devote this section to the problem of radiation transfer in the plane-parallel inhomogeneous medium of finite optical thickness. Instead of extension of the approach described in Section 9.11 over this case we propose an alternative way that allows us to address an initial-value problem. Once again, the central place in the approach occupies the right-hand side reflectance of the medium (see the upper picture in Fig. 9.4). For convenience, under reflectance we shall mean here $r(\eta,\xi,\tau_0)$ simply related to that introduced at the outset of the review $r(\eta,\xi,\tau_0) = \eta\xi\rho(\eta,\xi,\tau_0)$. Simple imbedding arguments yield

$$\frac{dr}{d\tau_0} = -\left(\frac{1}{\eta}+\frac{1}{\xi}\right) r(\eta,\xi,\tau_0) + \frac{\lambda(\tau_0)}{2}\varphi(\eta,\tau_0)\varphi(\xi,\tau_0)\,, \tag{9.167}$$

with $r(\eta,\xi,0) = 0$. By analogy, we deal with transmittance $\tilde{q}(\eta,\xi,\tau_0) = \eta\xi q(\eta,\xi,\tau_0)$, so that

$$\tilde{q}(\eta,\xi,\tau_0) = \xi\delta(\eta-\xi)\exp\left[-\tau_0/\xi\right] + \tilde{\sigma}(\eta,\xi,\tau_0)\,, \tag{9.168}$$

where the diffuse component, $\tilde{\sigma}(\eta,\xi,\tau_0) = \eta\xi\sigma(\eta,\xi,\tau_0)$, as usual, is separated. As it is known [8, 9], the latter is found from

$$\frac{d\tilde{\sigma}}{d\tau_0} = -\frac{1}{\xi}\tilde{\sigma}(\eta,\xi,\tau_0) + \frac{\lambda(\tau_0)}{2}\psi(\eta,\tau_0)\varphi(\xi,\tau_0)\,, \tag{9.169}$$

where the function

$$\psi(\eta,\tau_0) = \int_0^1 \tilde{q}(\eta,\eta',\tau_0)\frac{d\eta'}{\eta'} = \exp\left(-\frac{\tau_0}{\eta}\right) + \int_0^1 \tilde{\sigma}(\eta,\eta',\tau_0)\frac{d\eta'}{\eta'} \tag{9.170}$$

satisfies the set of equations (9.6), (9.7). The initial condition is obvious: $\tilde{\sigma}(\eta,\xi,0) = 0$.

In fact, having found the transmittance, one can use it as an initial condition in solving the classical transfer equations [93]. However, there is an alternative and simpler way of finding the internal field of radiation, of which the transmittance is a special value. To this end let us introduce the function $U(\eta,\tau;\xi,\tau_0)$ similar, in some sense, to transmittance but with more general probabilistic meaning, namely, it describes the probability to find the incident photon moving in direction η at optical depth τ.Here again the diffuse component of U must be separated denoted henceforth by u. By analogy with (9.168) one may write

$$U(\eta,\tau;\xi,\tau_0) = \xi\delta(\eta-\xi)\exp\left[-(\tau_0-\tau)/\xi\right] + u(\eta,\tau;\xi,\tau_0)\,. \tag{9.171}$$

The invariant imbedding approach leads to

$$\frac{du}{d\tau_0} = -\frac{1}{\xi}u(\eta,\tau;\xi,\tau_0) + \frac{\lambda(\tau_0)}{2}\Psi(\eta,\tau,\tau_0)\varphi(\xi,\tau_0)\,, \tag{9.172}$$

where

$$\Psi(\eta,\tau,\tau_0)=\int_0^1 U(\eta,\tau;\eta',\tau_0)\,\frac{d\eta'}{\eta'}=\exp\left(-\frac{\tau_0-\tau}{\eta}\right)+\int_0^1 u(\eta,\tau;\eta',\tau_0)\,\frac{d\eta'}{\eta'}\,, \tag{9.173}$$

and the initial condition, $u(\eta,\tau;\xi,\tau)=\delta(\eta-\xi)$.

It is obvious that $u(\eta,0;\xi,\tau_0)=\tilde{\sigma}(\eta,\xi,\tau_0)$ and $\Psi(\eta,0,\tau_0)=\psi(\eta,\tau_0)$. Thus, there is no need to solve Eq. (9.169) since Eq. (9.172) allows us to find the rightward-directed radiation at any depth τ ($\tau=0$, in particular) for a family of atmospheres with $\tau_0\geq\tau$. Having found the function u, it is easy to determine the counterpart of U for the opposite direction. Simple physical reasoning allows to write

$$V(\eta,\tau;\eta,\tau_0)=\rho(\eta,\xi,\tau)\exp\left(-\frac{\tau_0-\tau}{\xi}\right)+\int_0^1 \rho(\eta,\eta',\tau)\,u(\eta',\tau;\xi,\tau_0)\,\frac{d\eta'}{\eta'}\,. \tag{9.174}$$

Now let us envision the case when the medium is illuminated from the opposite side (the lower picture in Fig. 9.4). The quantities similar to those above, we supply by bars. Reasoning carried out in deriving the requisite quantities uses again the imbedding procedure of augmentation of the optical thickness from the right-hand side of the medium ($\tau=\tau_0$). So we limit ourselves with presenting the final results [93].

The reflectance $\bar{\rho}(\eta,\xi,\tau_0)$ satisfies

$$\frac{d\bar{r}}{d\tau_0}=\frac{\lambda(\tau_0)}{2}\psi(\eta,\tau_0)\,\varphi(\xi,\tau_0)\,, \tag{9.175}$$

subject to the initial condition $\bar{r}(\eta,\xi,0)=0$. Since the right-hand side of this equation can be envisaged as known then its solution is equivalent to calculation of a finite integral. As for the transmission coefficient, the polarity property yields $\bar{\tilde{q}}(\eta,\xi,\tau_0)=\tilde{q}(\xi,\eta,\tau_0)$. The quantities which describe the internal field of radiation are found from

$$\frac{d\bar{V}}{d\tau_0}=\frac{\lambda(\tau_0)}{2}\Psi(\tau,\eta,\tau_0)\,\psi(\xi,\tau_0) \tag{9.176}$$

with condition $\bar{V}(\eta,\tau;\xi,\tau)=0$ and from evident relation

$$\bar{U}(\eta,\tau;\eta,\tau_0)=\psi(\xi,\tau)+\int_0^1 \rho(\eta,\eta',\tau)\,\bar{V}(\eta',\tau;\xi,\tau_0)\,\frac{d\eta'}{\eta'}\,. \tag{9.177}$$

Note that the solution of Eq. (9.176) numerically is also equivalent to calculations of a finite integral. The interested reader is referred to [93] for illustration of numerical results.

Thus, solution of two equations (9.167) and (9.172) is sufficient not only to find the global optical characteristics of an inhomogeneous atmosphere but also to reproduce the internal field of radiation independent of what boundary of it is illuminated. Note also that having information on the external and internal radiation field for one of these two cases is enough to determine the same quantities for the opposite case. As is shown in [93], only one additional differential equation of the first order must be treated to solve an important and frequently encountered in astrophysical applications transfer problem in an atmosphere containing the energy sources. The advantage of the approach is that we deal only with the easily solvable initial-value problems. They have a rather plain physical implication and is preferred compared to the Riccati transformation method [94]. In contrast to the scalar 1-D problem, the property of numerical stability of the proper equations in the vector-matrix case is an important and noticeably difficult point to study.

9.14 Line formation in mesoturbulent atmosphere

The imbedding approach is efficient also in treating the problem of the spectral line formation in a turbulent atmosphere with a spatially correlated velocity field [95]. Because of the high occurrence of turbulent phenomena in the universe, the problem is of extreme importance for astrophysics. Originally, the non-thermal mechanism due to hydrodynamic motions was invoked in order to achieve satisfactory agreement between the theoretical and observed profiles and equivalent widths of spectral lines originating in stellar atmospheres, although no directly observable proofs existed for the hydrodynamic nature (in the customary sense) of this phenomenon. However, the phenomenon of granulation, which is directly observable in the Sun's photosphere, as well as motion on different scales in solar prominences, suggests that this kind of phenomenon is to be expected in other stars as well. This leads to the question of how random variations in the velocity field within a radiating atmosphere affect the observed spectra.

As is known, the hydrodynamic characteristic of turbulent motions is the correlation coefficient along the direction of propagation of a ray. It depends significantly on the type of turbulence and is determined by the degree of correlation between the variations in the velocity field at different points in the medium. The characteristic parameter which describes this correlation on the average is the correlation length, Λ. In the two limiting cases of $\Lambda \to 0$ and $\Lambda \to \infty$, the problem of spectral line formation is greatly simplified, so these cases have been most often examined by astrophysicists in the course of interpreting observed spectra. For values of smaller than the photon mean free path, random variations in the hydrodynamic velocity at nearby points of the medium are essentially independent of one another. In the limit of $\Lambda \to 0$, the velocities of the motions are independent on an atomic level, so that in calculating the profile of an observed line only the Doppler-shifted absorption coefficient in the line is averaged over the velocity. This limiting case corresponds to microturbulence, In the opposite limiting case of $\Lambda \to \infty$, known as macroturbulence, the hydrodynamic velocities at all points vary in unison, so that the mean profile of the spectrum line is a superposition of profiles that are shifted by different amounts, as happens when the radiating object is rotating.

Evidently, the micro- and macroturbulence models are only approximations and do not provide a clear representation of the effect of turbulent motions in the case of arbitrary average sizes for the turbulence cells. This intermediate case, often referred to as mesoturbulence [96], has been examined by many authors. Here we mention only a few of the papers by French and German theorists during the 1970s and 1980s, when this topic underwent a rapid development. An early paper by Traving [97] (see also [98]) dealt with a discrete model problem in which absorption by atoms was replaced by exponential absorption on the part of solid, independently moving turbulence cell with finite, but fixed, dimensions. This approach was developed further [99, 100] under the assumption that the sizes of the cells can vary randomly, with the interface points distributed in space according to the Poisson law (the so-called Kubo-Anderson process [101]). With some simplifying assumptions, a closed expression was obtained for the statistical mean profile of a spectral line under LTE.

A fundamentally different method has been developed [102–103] as a continuum analog of the problem. It reduces to considering a Uhlenbeck–Ornstein process [104] with a Gaussian velocity distribution. The result is Fokker–Planck-type equations for the joint distribution function of the velocities and the radiated intensity in the line. These equations are rather complicated and are solved numerically. In fact, both of the above approaches are approximate and are not consistent with one another.

Now we briefly reproduce the approach proposed in [95] which uses the method of invariant imbedding. Consider an atmosphere of finite optical thickness τ_0 measured at the center of a spectral line in the absence of hydrodynamic motions. We shall assume that the medium contains energy sources of power, $B(\tau)\,\varpi(x)$, where $\varpi(x) = \omega(x) + \beta$, $\omega(x)$ is the profile of the absorption coefficient and β is the ratio of the absorption coefficient in the continuum to that at the line center. The quantity $B(\tau)$ plays the role of a source function and is related to the Planck function. We shall assume that a homogeneous turbulence has developed in the atmosphere, so that the hydrodynamic velocity vector v is a random function that depends on the depth, while the mean characteristics of the velocity field are independent of the depth in the atmosphere [105]. In addition, let us suppose that the probability law according to which the velocity takes one or another value is also independent of the depth. Finally, we assume that the variations in velocity inside the medium are correlated with one another. We are interested in the mean intensity of the radiation of the medium in the direction of the surface normal. emerging from the boundary $\tau = \tau_0$.

Let the function $\langle I(\tau_0, x, u)\rangle$ represent the mean intensity of the radiation emerging from the atmosphere with frequency x under condition that the hydrodynamic velocity at the boundary of the medium is equal to u, the latter being measured in units of thermal velocity. Effect of the radiation on the velocity field will be neglected. We denote by $G(u, u', \rho(l))\,du$ the probability that if the value of the velocity at depth τ' is u', then at depth τ it will lie within the interval u, $u + du$. Because of the uniformity of the process, the correlation coefficient ρ depends only on the distance between the points, $l = |\tau - \tau'|$. In the simplest case of spectral lines under LTE, one can use the rules of addition for statistical mean intensities developed in [106–108] to write for the mean intensity resulting from augmentation of the initial turbulent atmosphere of thickness $\tau_0$by the small layer Δ possessing the same properties

$$\langle I(\tau_0 + \Delta, x, u)\rangle = e^{-\varpi(x-u)\Delta} \int_{-\infty}^{\infty} G(u, u', \rho(\Delta))\, \langle I(\tau_0, x, u')\rangle\, du' + B(\tau_0)\left[1 - e^{-\varpi(x-u)\Delta}\right]. \tag{9.178}$$

This equation is crucial for deriving the equations for the random function $\langle I(\tau_0, x, u)\rangle$. Once it has been determined, the unknown value of the statistical mean intensity $\langle I(\tau_0, x)\rangle$ can be found by virtue of

$$\langle I(\tau_0, x)\rangle = \int_{-\infty}^{\infty} P(u) \langle I(\tau_0, x, u)\rangle \, du \,, \tag{9.179}$$

where $P(u)$ is the distribution law of turbulent velocities.

One may specify the law governing the variations in the non-thermal velocity at different depths by assuming that the process is Markovian and consider a Gaussian distribution in the plane. Then

$$G(u, u', \rho) = \frac{1}{u_t\sqrt{\pi\,(1-\rho^2)}} \exp\left(-\frac{(u-\rho u')^2}{u_t^2\,(1-\rho^2)}\right) , \tag{9.180}$$

where $u_t = \sqrt{\pi}\,\langle u_{\text{turb}}\rangle$ and $\langle u_{\text{turb}}\rangle$ is the mean hydrodynamic velocity in units of thermal velocity. As was proved in [95], there is the bilinear expansion

$$G(u, u', \rho) = \frac{1}{\alpha_0(u')} \sum_{k=0}^{\infty} \rho^k \alpha_k(u)\, \alpha_k(u') \,, \tag{9.181}$$

where

$$\alpha_k(u) = \left(2^k \pi u_t^2 k!\right)^{-1/2} e^{-(u/u_t)^2} H_k\left(\frac{u}{u_t}\right) , \tag{9.182}$$

and $H_k(u)$ are Hermite polynomials. Note that the functions $\alpha_k(u)$ represent an orthonormal set of functions with weight $\alpha_0(u)^{-1}$. For a uniform Markov process the correlation coefficient varies exponentially with the distance between the depths being $\rho(l) = \exp(-l/\Lambda)$, where $l = |\tau - \tau'|$ and Λ is the mean correlation length. In fact, if we consider three depth such that $\tau_1 > \tau_2 > \tau_3$ and take $l_1 = |\tau_1 - \tau_2|$ and $l_2 = |\tau_2 - \tau_3|$, then, using the expansion (9.181), it is easy to show that

$$G(u, u', \rho(l_1 + l_2)) = \int_{-\infty}^{\infty} G(u, u'', \rho(l_2))\, G(u'', u', \rho(l_1))\, du''. \tag{9.183}$$

This equation is essentially the Kolmogorov-Chapman relation for diffusion Markov processes and expresses the multiplicative property of the transition probability for the process (in this case, the conditional velocity distribution function). Of two ways proposed in [92] for finding $\langle I(\tau_0, x, u)\rangle$, we give here that reducing the problem to the integral equation

$$\langle I(\tau_0, x, u)\rangle = \int_{-\infty}^{\infty} \varpi(x - u')\, du' \int_0^{\tau_0} K(\tau_0 - t, u, u') \left[B(t) - \langle I(t, x, u')\rangle\right] dt \,, \tag{9.184}$$

where

$$K(\tau, u, u') = \sum_{n=0}^{\infty} \frac{\alpha_n(u)\, \alpha_n(u')}{\alpha_0(u)} e^{-(n/\Lambda)\tau} \,. \tag{9.185}$$

In two limiting cases of macro- and microturbulence we arrive at known results. Actually, if $\Lambda \to \infty$, and $K(\tau, u, u') \to \delta(u - u')$, then Eq. (9.184) yields after some simple algebra

$$\langle I(\tau_0, x, u)\rangle = \int_0^{\tau_0} B(t)\, e^{-\varpi(x-u)(\tau_0-t)} \varpi(x-u)\, dt\,, \tag{9.186}$$

and, having in mind that $P(u) = \alpha_0(u)$, the final result

$$\langle I(\tau_0, x)\rangle = \int_{-\infty}^{\infty} \alpha_0(u) \langle I(\tau_0, x, u)\rangle\, du\,. \tag{9.187}$$

In the opposite case of $\Lambda \to 0$ and $K(\tau_0, u, u') \to \alpha_0(u)$ so that $\langle I(\tau_0, x, u)\rangle$ does not depend on u and coincides with the requisite quantity $\langle I(\tau_0, x)\rangle$. Finally, we obtain

$$\langle I(\tau_0, x)\rangle = \int_0^{\tau_0} B(t)\, e^{-\gamma(x)(\tau_0-t)} \gamma(x)\, dt\,, \tag{9.188}$$

where

$$\gamma(x) = \int_{-\infty}^{\infty} \alpha_0(u)\, \varpi(x-u)\, du\,. \tag{9.189}$$

Solution of Eq. (9.184) for intermediate values of Λ allows us to study the dependence of the line profile, integral intensity, and width on the mean correlation length and the average value of the hydrodynamic velocity. Referring the interested reader for details of these results to the mentioned paper [95], here we limit ourselves to noting that the transition from a microturbulent regime to a macroturbulent regime occurs within a comparatively narrow range of variation in the correlation length. It is important that the proposed approach yields a solution to the problem for a family of inhomogeneous atmospheres with different optical thicknesses, which, in its turn, makes it easy to determine the radiation field inside the turbulent medium. The approach can be generalized in various ways, in particular, it can be applied without significant changes to the case where the correlation length depends on position within the atmosphere.

References

1. Menzel, D., 1966: ed., *Selected Papers on the Transfer of Radiation*, New York: Dover.
2. Ambartsumian, V.A., 1943: On diffuse reflection of light by turbid medium, *Dokl. Akad. Nauk SSSR*, **38**, 257–261,.
3. Ambartsumian, V.A., 1960: *Scientific Papers*, Vol. 1, Yerevan: Izd AN ArmSSR (in Russian).
4. Ambartsumian, V.A., 1980: On some trends in the development of astrophysics, *Ann. Rev. Astron. Astrophys.*, **18**, 1–13.
5. Ambartsumian, V.A., 1980: Invariance principle and some of its applications, in: *Proc. Symp. 'The Invariance Principle and its Applications'*, 9–17, Yerevan: Izd. AN ArmSSR (in Russian).

6. Nikoghossian, A.G., 1999: The variational formalism and bilinear relations of the radiative transfer theory, *J. Quantit. Spectrosc. Radiat. Transfer*, **61**, 345–359.
7. Ambartsumian, V.A., 1944: On one-dimensional case of the problem of the scattering and absorbing medium of finite optical thickness, *Izv. AN ArmSSR*, No. 1-2, 31–36.
8. Sobolev, V.V., 1963: *A Treatise on Radiative Transfer*, Princeton: Van Nostrand.
9. Chandrasekhar, S., 1960: *Radiative Transfer*, New York: Dover.
10. Yengibarian, N.B., and M.A. Mnatsakanian, 1974: On the linear transfer problems, *Dokl. Akad. Nauk SSSR*, **217**, 533–535.
11. Ambartsumian, V.A., 1942: On reflection of light by atmospheres of planets, *Astron. Zh.*, **19**, 30 (see also [3], 206–222).
12. Ambartsumian, V.A., 1941: Light scattering and absorption in atmospheres of planets, *Uch. Zap. LGU*, No. 11 (see also [3], 181–205).
13. Ambartsumian, V.A., 1944: The diffusion of light through scattering medium of large optical thickness, *Dokl. Akad. Nauk SSSR*, **43**, 106 (see also [3] 256–262).
14. Sobolev, V.V., 1958: On the theory of radiation transfer, *Izv. AN ArmSSR*, **11**, 39–50.
15. Sobolev, V.V. 1959: On the theory of radiation transfer in stellar atmospheres, *Astron. Zh.*, **36**, 573–578.
16. Sobolev, V.V., 1985: *Course in Theoretical Astrophysics* (in Russian), Moscow, Nauka.
17. Ivanov, V.V., 1969: *Radiative Transfer and the Spectra of Celestial Objects* (in Russian), Moscow, Nauka.
18. Case, K.M., and P.F. Zweifel, 1967: *Linear Transport Theory*, Reading, MA: Addison-Wesley.
19. Minin, I.N., 1958: *Dokl. Akad. Nauk SSSR*, **120**, 63.
20. Ambartsumian, V.A., 1943: On the problem of diffuse reflection, *J. Exp. Teoret. Phys.*, **13**, 323–334.
21. Ambartsumian, V.A., 1941: The light scattering and absorption in atmospheres of planets, *Trudi Astron. Obs. LGU*, **12**, 64 (see also [3], 181–205).
22. Ambartsumian, V.A., 1942: New method for studying the light scattering in the foggy medium, *Izv. AN SSSR*, No.3, 97 (see also [3], 223–231).
23. Hummer, D.G., 1962: Non-coherent scattering: I. The redistribution function with Doppler broadening *Mon. Notic. Roy Astron. Soc.*, **125**, 21–37.
24. Mihalas, D., 1970: *Stellar Atmospheres*, San Francisco: Freeman and Co.
25. Nikoghossian, A.G., 1977: The problem of diffuse reflection for redistribution of radiation over frequencies and directions, *Dokl. Akad. Nauk SSSR*, **235**, 786–789.
26. Nikoghossian, A.G., and H.A. Haruthyunian, 1978: The diffuse reflection of light for an arbitrary phase function, *Astrophysics*, **14**, 223–230.
27. Nikoghossian, A.G., and H.A. Haruthyunian, 1979: The radiation field in a semi-infinite atmosphere containing energy sources, *Astrophys. Space Sci.*, **64**, 285–299.
28. Nikoghossian, A.G., and H.A. Haruthyunian, 1980: Application of the invariance principle to solution of various problems of non-coherent and non-isotropic scattering, in: Proc. Symp. '*The Invariance Principle and its Applications*', 232–241. Yerevan: Izd. AN ArmSSR (in Russian).
29. Haruthyunian, H.A., and A.G. Nikoghossian, 1980: Application of the invariance principle to the compton scattering problem, in: *Proc. Symp. 'The Invariance Principle and its Applications'* 431–439, Yerevan: Izd. AN ArmSSR (in Russian).
30. Nikoghossian, A.G., 2007: Diffuse reflection of the line-radiation from a semi-infinite turbulent atmosphere, *Astrophysics*, **50**, 321–332.
31. Nikoghossian, A.G., and H.A. Haruthyunian, 1976: Spectral lines formation for a general law of redistribution, *Dokl. Akad. Nauk SSSR*, **229**, 583–586.

32. Haruthyunian, H.A., and A.G. Nikoghossian, 1976: Radiation scattering for a general law of frequency redistribution, *J. Quantit. Spectrosc. Radiat. Transfer*, **19**, 135–148.
33. Preisendorfer, R.W., 1976: *Hydrologic Optics*, Vol. IV, Honolulu, Hawaii: U.S. Department of Commerce.
34. Rybicki, G.B., 1977: Integrals of the transfer equations, I, *Astrophys. J.*, **213**, 165–176.
35. Ivanov, V.V., 1978: Nonlinear equations in linear problems of radiation transfer in plane atmospheres, *Astron. Zh.*, **22**, 612–618.
36. Nikoghossian, A.G., 1997: Invariance principle and bilinear relations of the radiative transfer theory, I, *Astrophys. J.*, **483**, 849–856.
37. Hubeny, I., 1987a: Probabilistic interpretation of radiative transfer, I, *Astron. Astrophys*, **185**, 332–335,
38. Hubeny, I., 1987b: Probabilistic interpretation of radiative transfer, II, *Astron. Astrophys*, **185**, 336–342,
39. Krikorian, R.A., and A.G. Nikoghossian, 1996: On the lagrangian formalism in the radiative transfer theory, *J. Quantit. Spectrosc. Radiat. Transfer*, **56**, 465–469.
40. Vainberg, M.M., 1964: *Variational Methods for the Study of Non-Linear Operators*, San Francisco: Holden Day.
41. Gelfand, I.M., and S.V. Fomin, 1965: *Calculus of Variation*, Englewood Cliffs, NJ: Prentice-Hall.
42. Tavel, M., 1971: Application of Noether's theorem to the transport equation, *Transport Theory Statist Phys.*, **1**, 271–285.
43. Anderson, D., 1973: Variational principles for the radiative transfer equation of single scattering, *J. Inst. Math. Applic.*, **12**, 55–62.
44. Nikoghossian, A.G., 2000: Non-linear relations for statistical mean quantities describing the multiple scattering process, *Astrophysics*, **43**, 337–342.
45. Nikoghossian, A.G., 2011: Bilinear integrals of the radiative transfer equation, *Astrophysics*, **54**, 242–249.
46. Yengibarian, N.B., 1972: On factorization of the symmetric integral operators, *Dokl. Akad. Nauk SSSR*, **203**, 4.
47. Yengibarian, N.B., and A.G. Nikoghossian, 1973: Non-coherent scattering, *J. Quantit. Spectrosc. Radiat. Transfer*, **13**, 787–811.
48. Hubeny, I., 1992: in: *Atmospheres of Early-type Stars*, U. Huber, C.S. Jeffery, ed., Lecture Notes in Physics, No. 401, p. 377, Berlin: Springer.
49. Rybicki, G.B., 1984: Escape probability methods, in: *Methods in Radiative Transfer*, W. Kalkofen, ed., 21–64, Cambridge: Cambridge University press.
50. Sobolev, V.V., 1966: The number of scatterings during photon diffusion, I, *Astrophysics*, **2**, 69–73.
51. Sobolev, V.V., 1966: The number of scatterings during photon diffusion, II, *Astrophysics*, **2**, 119–123.
52. Sobolev, V.V., 1967: The number of scatterings during photon diffusion, III, *Astrophysics*, **3**, 1–5.
53. Haruthyunian, H.A.and A.G. Nikoghossian, 1983: The mean number of scatterings for radiative transfer with frequency redistribution, *Dokl. Akad. Nauk SSSR*, **268**, 1342–1346.
54. Nikoghossian, A.G., 1984: Statistical description of radiation field on the basis of the invariance principle, I, *Astrophysics*, **21**, 527–537.
55. Nikoghossian, A.G., 1984: Statistical description of radiation field on the basis of the invariance principle, II, *Astrophysics*, **21**, 685–693.

56. Haruthyunian, H.A., and A.G. Nikoghossian, 1987: Statistical description of radiation field on the basis of the invariance principle, IV, *Astrophysics*, **27**, 536–543.
57. Nikoghossian, A.G., 1986: Statistical description of radiation field on the basis of the invariance principle, III, *Astrophysics*, **24**, 89–99.
58. Irvine, W.M., 1964: The formation of absorption bands and the distribution of photon optical paths in a scattering atmosphere, *Bull. Astron. Inst. Netherl.*, **17**, 266–279.
59. Irvine, W.M., 1966: The Distribution of photon optical paths in a scattering atmosphere, *Astrophys. J.*, **144**, 1140–1147.
60. Ivanov, V.V., 1970: The mean free path of a photon in a scattering medium, *Astrophysics*, **6**, 355–367.
61. Hummer, D.G., and P.B. Kunasz, 1980: Energy loss by resonance line photons in an absorbing medium, *Astrophys. J.*, **236**, 609–618.
62. Nikoghossian, A.G., 2006: Statistical description of the radiation field in a homogeneous medium of finite optical thickness, *Astrophysics*, **49**, 229–239.
63. Nikoghossian, A.G., 2000: Nonlinear equations for statistical averages describing process of multiple scattering, *Astrophysics*, **43**, 337–342.
64. Bellman, R., R. Kalaba, and M. Wing, 1960: Invariant imbedding and mathematical physics, I, Particle processes, *J. Math. Phys.*, **1**, 280–308.
65. Bellman, R., R. Kalaba and M. Prestrud, 1963: *Invariant imbedding and radiative transfer in slabs of finite thickness*, New York: American Elsevier.
66. Yanovitskij, E.G., 1997: *Light Scattering in Inhomogeneous Atmospheres*, New York: Springer.
67. Van de Hulst, H.C., 1980: *Multiple Light Scattering Tables, Formulas and Applications*, Vol. 1, New York, Academic Press.
68. Nikoghossian, A.G., 2004: Radiative transfer in one-dimensional inhomogeneous atmospheres, *Astron. Astrophys*, **422**, 1059–1066.
69. Redheffer, R., 1962: On the relation of transmission-line theory to scattering and transfer, *J. Math. Phys.*, **41**, 1–41.
70. Grant, I.P., and G.E. Hrant, 1968: Solution of radiative transfer problems using the invariant Sn method, *Mon. Not. Roy. Astr. Soc.*, **141**, 27–41.
71. Peraiah, A., 1999, Principles of invariance in radiative transfer, *Sp. Sci. Rev.*, **87**, 465–538.
72. Jefferis J.T., and C. Lindsey, 1988: Radiative transfer in inhomogeneous atmospheres – A statistical approach, *Astrophys. J.*, **335**, 372–382.
73. Gu, Y., C. Lindsey, and J.T. Jefferies, 1995: Radiative transfer in stochastic media, *Astrophys. J.*, **450**, 318–333.
74. Cecchi-Pestellini, C., and L. Barletti, 2001: Radiative transfer in a stochastic universe, I, *New Astron.*, **6**, 151–163.
75. Meinkohn, E., and S. Richling, 2002: Radiative transfer with finite elements, II, Lyalpha line transfer in moving media, *Astron. Astrophys.*, **392**, 827–839.
76. Juvela, M., and P. Padovan, 2003: Dust emission from inhomogeneous interstellar clouds: Radiative transfer in 3D with transiently heated particles, *Astron. Astrophys.*, **397**, 201–212.
77. Nikoghossian, A.G., S. Pojoga, and Z. Mouradian, 1997: On the radiative transfer in atmospheres with randomly distributed inhomogeneities, *Astron. Astrophys.*, **325**, 813–818.
78. Pojoga, S., A.G. Nikoghossian, and Z. Mouradian, 1998: A statistical approach to the investigation of fine structure of solar prominences, *Astron. Astrophys.*, **332**, 325–338.

79. Nikoghossian, A.G., S. Pojoga, and Z. Mouradian, 1999: Statistical characteristics of radiation formed in atmosphere with randomly distributed inhomogeneities, *Astron. Astrophys.*, **342**, 785–798.
80. Nikoghossian, A.G., and Z. Mouradian, 2000: Profiles of the spectral lines formed in stochastic multicomponent atmosphere, *Astron. Astrophys.*, **360**, 1086–1095.
81. Bellman, R., 1957: Functional equations in the theory of dynamic programming, VII, A partial differential equation for the Fredholm resolvent, *Proc. Amer. Math. Soc.*, **8**, 435–440.
82. Sobolev, V.V., 1957: Radiation diffusion in a semi-infinite medium, *Dokl. Akad. Nauk SSSR*, **116**, 45–48.
83. Krein, M.G., 1955: On new method of solution of linear integral equations of the first and second kinds, *Dokl. Akad. Nauk SSSR*, **100**, 413–416.
84. Casti, J., and R. Kalaba, 1976: *Imbedding Methods in Applied Mathematics* [Russian translation], Moscow: Mir.
85. Scott, M., 1973: *Invariant Imbedding and its Applications to Ordinary Differential Equations. An Introduction*, Reading, MA: Addison-Wesley.
86. Nikoghossian, A.G., 2004: Radiative transfer in inhomogeneous atmospheres, I, *Astrophysics*, **47**, 104–116.
87. Nikoghossian, A.G., 2004: Radiative transfer in inhomogeneous atmospheres, II, *Astrophysics*, **47**, 248–259.
88. Nikoghossian, A.G., 2004: Radiative transfer in inhomogeneous atmospheres, III, *Astrophysics*, **47**, 412–421.
89. Nikoghossian, A.G., 2011: Solution of linear radiation transfer problems in plane-parallel atmosphere, I, *Astrophysics*, **54**, 553–567.
90. Magnus, W., 1954: On the exponential solution of differential equations for a linear operator, *Comm. Pure and Appl. Math.*, VII(4), 649–673.
91. Nikoghossian, A.G., 2011: Group-theoretical description of radiative transfer in one-dimensional media, *Astrophysics*, **54**, 126–138.
92. Wigner, E., 1959: *Group Theory*, New York: Academic Press.
93. Nikoghossian, A.G., 2012: Solution of linear radiation transfer problems in plane-parallel atmosphere, II, *Astrophysics*, **55**, 261–274.
94. Bellman, R., and G.M. Wing, 1973: *An Introduction to Invariant Imbedding*, New York: Wiley & Sons.
95. Nikoghossian, A.G., 2007: Spectral lines formation in a mesoturbulent atmosphere, *Astrophysics*, **50**, 175–186.
96. Gray, D.F., 1978: Turbulence in stellar atmospheres, *Sol. Phys.*, **59**, 193–236.
97. Traving, G., 1964: Über die Bildung von Fraunhoferlinen in turbulenten Sternatmosphären, *Z. Astrophys.*, **60**, 167–175.
98. Hundt, E., 1973: Microturbulence in stellar spectroscopy: The turbulent element approach, *Astron. Astrophys.*, **29**, 17–21.
99. Frish, H., 1975: Finite eddy-size effects on centre-to-limb variations: An alternative to anisotropic microturbulence, *Astron. Astrophys.*, **40**, 267–276.
100. Frish, H., and U. Frish, 1976: Non-LTE transfer – II, *Mon. Not. Roy. Astron. Soc.*, **175**, 157–175.
101. Bharucha-Reid, A.T., 1960: Elements of the theory of markov processes and their applications, New York: McGraw-Hill.
102. Gail, H.P., and Sedlmayr, E., 1974: Effects of correlated turbulent velocities on photospheric line formation, *Astron. Astrophys.*, **36**, 17–25.
103. Schmid-Burgk, J., 1975: Line formation in turbulent media: Mathematics of profile computation, *Astron. Astrophys.*, **32**, 73–78.

104. Gail, H.P., Sedlmayr, E., and G. Traving, 1975: Non-LTE line formation in turbulent media, *Astron. Astrophys.*, **44**, 421–429.
105. Batchelor, G., 1970: *The Theory of Homogeneous Turbulence*, Cambridge: Cambridge University Press.
106. Nikoghossian, A.G., 2002: Intensity fluctuations of radiation escaping from a multicomponent stochastic atmosphere, I, *Astrophysics*, **45**, 223–231.
107. Nikoghossian, A.G., 2005: Intensity fluctuations of radiation escaping from a multicomponent stochastic atmosphere, II, *Astrophysics*, **48**, 253–261.
108. Nikoghossian, A.G., 2007: Spectral line formation in a dynamically active multicomponent and stochastic atmosphere, *Astrophysics*, **50**, 94–104.

10 A review of fast radiative transfer techniques

Vijay Natraj

10.1 Introduction

Atmospheric radiative transfer involves gas absorption coupled with molecular Rayleigh scattering, in addition to scattering and absorption by clouds and aerosols. Further, computation of heating rates are dependent on absorption and emission of radiation, processes that have a complex dependence on various quantities. Typically, spectral regions contain several overlapping lines with intensities varying over many orders of magnitude. The most accurate method for computing the radiative terms in a molecular atmosphere involves a detailed line-by-line (LBL) calculation of the absorption coefficient versus wavenumber. However, direct numerical solution of the radiative transfer equation over frequency is in most cases too computationally expensive to be used on a routine basis. Therefore a variety of approximations have been developed to accelerate the computational process. This chapter discusses several of these techniques.

The outline of the chapter is as follows. Section 10.2 discusses the k-distribution method, which involves grouping spectral intervals according to absorption coefficient strength, and the correlated-k method. The latter is an extension to inhomogeneous atmospheres; it is assumed that the ordering of absorption coefficient strengths remains the same at all altitudes. In Section 10.3, we introduce a scheme to fit transmission functions with exponential sums for calculating spectrally integrated radiative fluxes, especially when both line absorption and scattering are important. Section 10.4 describes spectral mapping methods, which gain their efficiency by identifying spectral intervals that have similar optical properties. Section 10.5 describes optimum spectral sampling, which is a fast and accurate transmittance parameterization technique that extends the exponential sum fitting of transmittances and k-distribution techniques to vertically inhomogeneous atmospheres with overlapping absorbing species. Section 10.6 provides an overview of several techniques that separate single and multiple scattering, with the multiple-scattering component treated as a function of the absorption optical thickness and the scattering height. Section 10.7 discusses principal component analysis , which reduces the dimensionality of the optical properties. Neural networks are the topic of Section 10.8. Section 10.9 deals with semi-infinite and optically thick media. Section 10.10 provides a brief discussion of lower-order scattering approximations with and without consideration of polarization effects.

10.2 *k*-distribution and correlated-*k* methods

Ambartsumian (1936) was the first to explore ways to speed up radiative transfer calculations. He was interested in studying the effect of absorption lines on the radiative equilibrium of the outer layers of stars. He recognized that for a homogeneous atmosphere, the transmission within a spectral interval is independent of the LBL variation of the absorption coefficient k with respect to wavenumber ν, but depends only on the distribution of k within the interval. The k-distribution requires far fewer points to represent the spectral absorption than is required for LBL computations. This technique was later used by Arking and Grossman (1972) to study the effects of line shape and band structure on temperature profiles in planetary atmospheres, under conditions of radiative equilibrium.

The correlated-k method is an extension of the k-distribution concept to inhomogeneous atmospheres, and was first proposed by Lacis et al. (1979) and independently by Chou and Arking (1980). The former were interested in studying the effects of cirrus clouds on surface temperature while the latter were working on computing atmospheric cooling rates in infrared water vapor bands. The basic idea is as follows. When the broadening of lines is primarily due to molecular collisions (as is the case in the troposphere and lower stratosphere), the line shape can be approximated by a Lorentzian. In the far wings of the lines (which are important in typical atmospheric conditions), the absorption cross-section k at a specific pressure p and temperature T can be expressed in terms of its value at a reference pressure p_r and temperature T_r as follows:

$$k(p,T) = k(p_r,T_r)\frac{p}{p_r}f(T,T_r)\,, \tag{10.1}$$

where f is only a function of temperature. The benefit of using Eq. (10.1) to represent the absorption cross-section is immediately evident. The first term on the right-hand side is a function only of the wavenumber and the other two terms are functions only of pressure and temperature. This leads to very significant savings in radiative transfer computations since we can now compute $k(p_r,T_r)$ offline for spectral regions of interest.

Figure 10.1 shows a comparison of the monochromatic transmittances with those calculated using the k-distribution with far-wing scaling. The solid curve was computed using LBL calculations, and the dashed curve used the far-wing scaling approximation.

Goody et al. (1989) and Lacis and Oinas (1991) further developed the correlated-k technique to efficiently solve for radiative transfer in vertically inhomogeneous, multiple scattering atmospheres. If we denote the frequency distribution of absorption coefficient strengths in a spectral interval as $f(k)$, the cumulative distribution function $g(k)$, that defines the fraction of the interval for which the absorption coefficient is less than k, can be computed as follows:

$$g(k) = \int_0^k f(k')\,dk'\,. \tag{10.2}$$

The inverse of the cumulative frequency distribution, $k(g)$, is the k-distribution. The frequency distribution is normalized to unity; g is in the range $[0,1]$. For a

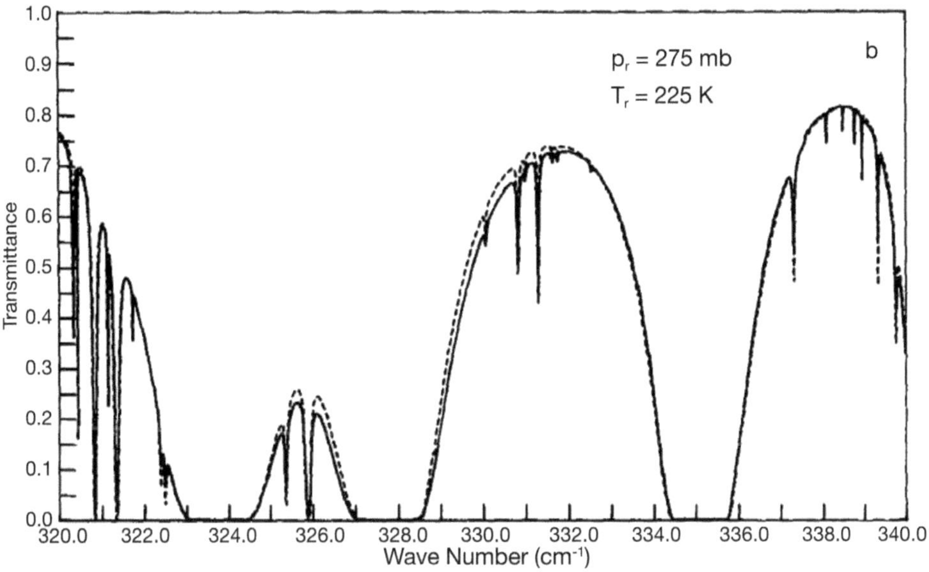

Fig. 10.1. Transmittance comparison between LBL and far-wing scaling calculations. The solid curve represents LBL calculations and the dashed curve represents far-wing scaling calculations [from Chou and Arking, 1980].

homogeneous atmosphere, g is a monotonic function of k and can be uniquely inverted. Hence, integration in wavenumber space can be replaced by one in g-space. However, if the atmosphere is inhomogeneous, the function $g(k)$ varies with altitude and therefore the inversion is not in general unique. Goody et al. (1989) accounted for vertical inhomogeneity by assuming that the k-distributions correlate between different pressure levels. They showed that this assumption is justified in the weak and strong line limits. Goody et al. (1989) and Lacis and Oinas (1991) found that the correlated-k technique was useful to calculate heating and cooling rates in the atmosphere.

Variants of the correlated-k scheme have been utilized by several researchers. Buchwitz et al. (2000) adapted the correlated-k approach to make it applicable for spectral regions containing two strong overlapping line absorbers and arbitrary additional minor or continuum absorbers. Boesche et al. (2008) introduced the k-binning approach. The main difference between k-binning and correlated-k is that in the former the entire absorption band is simulated whereas the latter requires separate calculations for each instrument channel. The radiances for each channel are reconstructed in the k-binning approach from the simulations that represent the entire spectral band. This technique has the advantage that no assumptions about the shape of the sensor weighting function need be made *a priori* for a given spectral interval. Cao et al. (2011) proposed a new technique to optimize the number of k-intervals, the equivalent absorption coefficients and the quadrature weights when using the correlated-k approach for the computation of spectrally integrated three-dimensional (3-D) atmospheric radiance.

10.3 Exponential sum fitting of transmittances

While the correlated-k technique results in a speed increase of two to three orders of magnitude over LBL calculations, further speed increases could be obtained by optimizing the number of k values. Hunt and Grant (1969) devised a procedure called 'exponential sum fitting of transmissions' (ESFT) to perform such an optimization, and Wiscombe and Evans (1977) provided a detailed mathematical basis. In ESFT, the spectral mean transmission $T(u)$ is approximated by a sum of exponentials $E(u)$ at the M monochromatic wavelengths:

$$T(u) \simeq E(u) = \sum_{i=1}^{M} w_i \exp[-k, u] , \tag{10.3}$$

where u is the absorber amount, and the weights w_i satisfy the following constraint:

$$\sum_{i=1}^{M} w_i = 1 . \tag{10.4}$$

The weights are determined as follows. A set $u_n = n\Delta u$, $n = 0, \ldots, N$ of equally spaced values is used to compute the least-squares residual R:

$$R = \sum_{n=0}^{N} w_n [T(u_n) - E(u_n)]^2 . \tag{10.5}$$

The best fit is then defined as that which minimizes R.

Kratz (1995) used the ESFT technique to perform calculations for the Advanced Very High Resolution Radiometer (AVHRR). Kratz et al. (1998) applied the technique to perform minor trace gas radiative forcing calculations. Chou et al. (1993) divided the infrared water vapor and CO_2 absorption spectrum into 10 bands and used ESFT to compute cooling rates in the troposphere and lower stratosphere. Mano (1995) developed a modified procedure to make ESFT suitable for atmospheric radiation calculations, and applied it to H_2O absorption bands in the infrared and near-infrared. Armbruster and Fischer (1996) made further improvements by considering the pressure- and temperature-dependence of absorption by atmospheric gases.

Figure 10.2(a) illustrates the cooling rate in the 800–1380 cm^{-1} spectral region for the minor trace gases (CO_2, N_2O, CH_4, CFC-11, CFC-12, and HCFC-22) in an atmosphere without any other absorbers. Both exact LBL and approximate ESFT results are shown. At the surface, the trace gases have a net cooling effect. Near the tropopause they warm the atmosphere. Higher in the stratosphere, there is again a net cooling effect. Figure 10.2(b) illustrates the change in the cooling rates due to the introduction of the minor trace gases into an atmosphere already containing H_2O and O_3. Addition of water vapor reduces the ability of the trace gases to cool the lower troposphere. This effect is smaller higher up in the atmosphere where the water vapor concentration is lower. It can be seen that the ESFT technique captures these features with high accuracy compared to the reference LBL calculations.

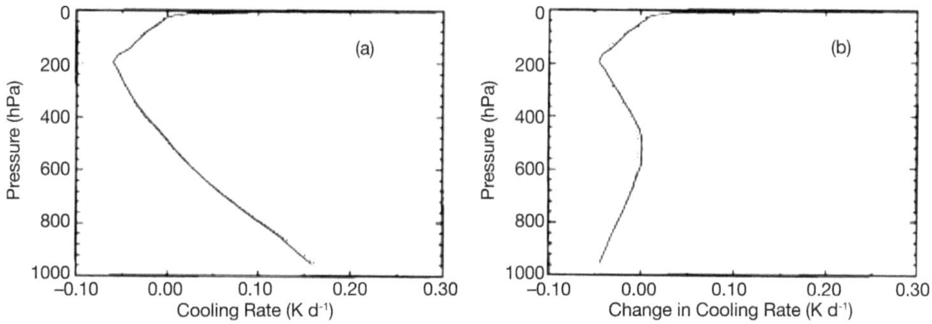

Fig. 10.2. Cooling rate profiles for the 800 to 1380 cm^{-1} spectral range calculated using LBL (solid curve) and ESFT (dashed curve) calculations. (a) Cooling rates resulting from the introduction of the minor trace gases into an atmosphere where no other absorbers are considered. (b) Change in cooling rates resulting from the introduction of the minor trace gases into an atmosphere already containing H_2O and O_3 [from Kratz et al., 1998].

10.4 Spectral mapping

Although spectral properties of a gas are highly correlated from one level to another, the correlation is not strict. While the correlated-k method works reasonably well for computing heating and cooling rates, it does not generate very accurate radiances. Besides, there is no way to tune the model to obtain a specified accuracy for the radiance calculation. To deal with this problem, West et al. (1990) showed how the correlated-k method can be understood in terms of more general spectral mapping transformations. They provided a means to calculate mapping transformations that strictly preserve correlation through all layers of the atmosphere. This technique enabled calculations to be made arbitrarily close to those from a LBL calculation by increasing the number of terms. Spectral mapping could also be used for mixtures of gases, and takes into account spectral variation in the incident solar flux, something the correlated-k procedure could not do.

The spectral mapping atmospheric radiative transfer (SMART) model (Meadows and Crisp, 1996; Crisp, 1997) was the first numerical implementation of mapping techniques. SMART adopts the following procedure. First, it generates a LBL description of the optical properties in a vertically inhomogeneous, scattering, absorbing, and emitting atmosphere. Second, it identifies and bins wavelengths that remain spectrally correlated throughout the atmosphere. Third, it performs a monochromatic radiative transfer calculation for each bin. Finally, it maps the radiances computed for each bin back to the original high-resolution spectral grid.

Figure 10.3 shows a synthetic spectrum of the 1.14–1.2 μm region generated using the spectral mapping technique, compared to a high-resolution spectrum of the Venus atmosphere taken by the Fourier Transform Spectrometer (FTS) on the Canada France Hawaii Telescope (CFHT) (Meadows and Crisp, 1996). Clearly, the radiance fit is very reasonable.

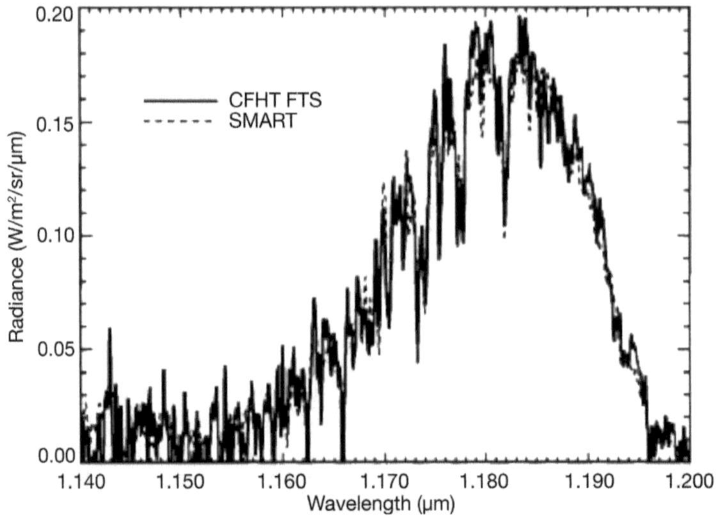

Fig. 10.3. A synthetic spectrum generated using spectral mapping, compared to a high-resolution spectrum of the Venus atmosphere [from Meadows and Crisp, 1996].

10.5 Optimal spectral sampling

There are two main difficulties associated with the correlated-k and ESFT techniques. First, the assumption that the absorption coefficients at different altitudes are perfectly correlated breaks down (1) for a single gas, when there are lines with different strengths or when the line strengths are highly temperature-dependent, and (2) for gas mixtures, when the relative concentrations of absorbers in the mixture vary with altitude.

Moncet et al. (2008) proposed a method called Optimal Spectral Sampling (OSS) to solve this problem by selecting a few specific wavelengths in the interval of interest. The OSS method avoids the basic problems of the correlated-k and ESFT techniques by rewriting Eq. (10.3) as follows:

$$T(u) = \sum_{i=1}^{N} w_i \exp[-k(v_i)u] \,. \tag{10.6}$$

In other words, spectral points are explicitly selected rather than the actual absorption cross-sections. This makes the extension to inhomogeneous atmospheres with multiple layers and mixtures of gases easy, because the terms inside the summation in Eq. (10.6) can be replaced a double sum over layers and gases with no loss of generality.

The obvious issue then is to select the *nodes* v_i. Moncet et al. (2008) use the following procedure for node selection. First, M uniformly spaced spectral locations spanning the channel bandwidth are chosen for every sensor channel. Second, a LBL model is used to compute radiances at these spectral locations for a representative set of atmospheric profiles, surface conditions and viewing geometries. Third, channel radiances, $\overline{R}_s$, for each member s of the set of S training scenes are produced by convolving the monochromatic radiances with the instrument response function. Fourth, for any given set of N nodes, the rms difference between $\overline{R}_s$ and the weighted sum of monochromatic radiances, $R_s(v_i)$, associated with the selected nodes, is computed as follows:

$$\varepsilon_N = \sqrt{\frac{1}{S}\sum_{s=1}^{S}\left[\overline{R}_s - \sum_{i=1}^{N} w_i R_s(v_i)\right]^2} \,. \tag{10.7}$$

The optimal weights are obtained using a robust least squares regression technique, with the sum of the weights equal to unity. Finally, an automated search is used to identify the smallest set of N nodes such that the rms difference ε_N is less than a prescribed tolerance.

It is to be noted that a large number of LBL calculations may need to be performed to find the optimal wavelengths and associated weights. The OSS method was tested on the Atmospheric Infrared Sounder spectral channels. Figure 10.4

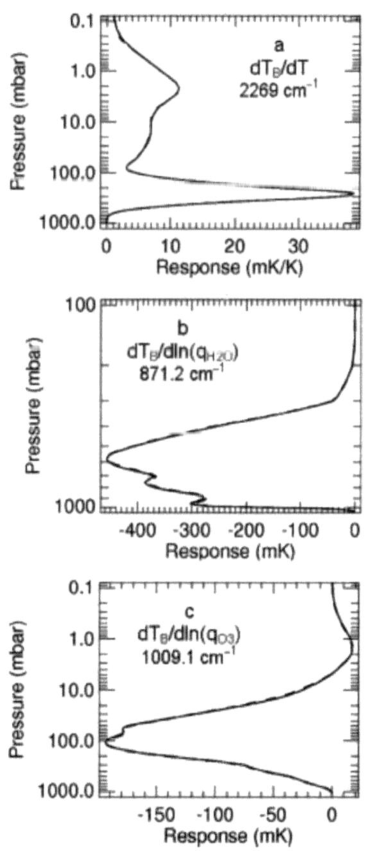

Fig. 10.4. Comparison between LBLRTM (solid) and OSS (dashed) Jacobians for (a) temperature, (b) water vapor, and (c) ozone. The derivatives were converted from radiance to equivalent brightness temperature and, for (b) and (c), they are with respect to the logarithm of the mixing ratio [from Moncet et al., 2008].

shows a comparison of Jacobians (derivatives of radiance with respect to atmospheric or surface parameters) calculated analytically (by direct differentiation of the radiative transfer equation) using OSS with finite difference estimates from the line-by-line LBLRTM model. There is excellent agreement for temperature, water vapor and ozone Jacobians.

10.6 Double-*k*, linear-*k* and low streams interpolation approaches

Duan et al. (2005) introduced a double-k approach to account for the uncorrelated nature of overlapping absorption lines. In this technique, they used, in addition to the total absorption optical thickness k, the absorption optical thickness k^* from the top of the atmosphere to a layer where significant scattering occurs. In this way they accounted not only for the integrated gaseous absorption but also its vertical distribution (see Fig. 10.5 for an example of different vertical profiles of gas absorption) as well as the distribution of scattering material.

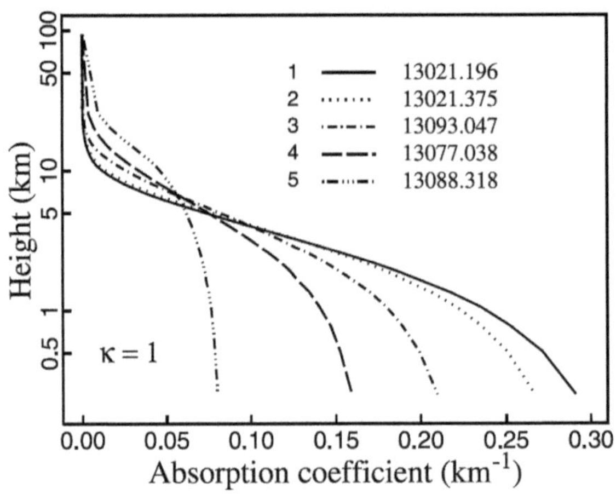

Fig. 10.5. Vertical distributions of oxygen absorption at five wave numbers with the same total absorption optical depth [from Duan et al., 2005].

Further, they took advantage of the fact that the radiative transfer calculations can be split into single and multiple scattering components. Multiple scattering is computationally expensive. However, in general it varies very smoothly with absorption optical thickness. On the other hand, single scattering is not such a smooth function of absorption optical thickness, but can be calculated very quickly, as shown in Fig. 10.6. With this is mind, Duan et al. (2005) computed single scattering exactly at every wavelength using a very fine atmospheric grid. For multiple scattering, they used a smaller number of layers in the discretization and computed the radiance I^{ms} as follows:

$$I^{ms} = g(k) f_k(k^*/k) , \tag{10.8}$$

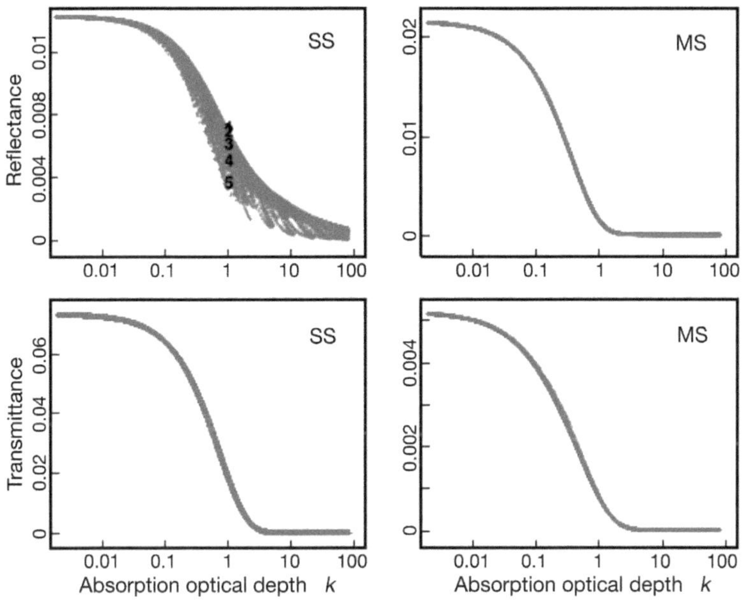

Fig. 10.6. Single-scattering (SS) and multiple-scattering (MS) components of radiances as a function of oxygen absorption optical depths for the oxygen A-band [from Duan et al., 2005].

where piecewise analytical functions are used to define g and f_k, with the coefficients obtained by fitting the calculated radiances at selected k and k^* values.

Hasekamp and Butz (2008) proposed a linear-k approach for efficient and accurate vector multiple scattering calculations in absorption bands. They considered the multiply scattered radiance as a function of total absorption optical thickness and its normalized vertical distribution. The basic principle of the linear-k method is to calculate the multiply scattered radiance for a small set of reference vertical distributions and then compute the radiance for the actual vertical distribution by performing a linear Taylor series expansion around the reference distribution. A second-order polynomial fit is then employed to correct for the total absorption optical thickness compared to the reference value. Clearly, the linear-k approach requires derivatives of the Stokes parameters with respect to the absorption optical thickness in different layers.

O'Dell (2010) introduced the Low Streams Interpolation (LSI) approach, which is fairly similar to Duan et al. (2005). The main differences between the two methods are: (1) LSI parameterizes the difference between multiple scattering calculations made using a low number (2, say) and a high number (24, say) of computational (quadrature) polar cosine directions (streams), as a function of optical quantities very similar to those defined by Duan et al. (2005); (2) polarization is considered by O'Dell (2010) but ignored by Duan et al. (2005). Figure 10.7 shows the errors in Stokes parameters I and Q for 2-stream calculations relative to high-accuracy calculations, as a function of the column-integrated gas optical depth. It is evident that there is a strong relationship between the error and the gas absorption optical

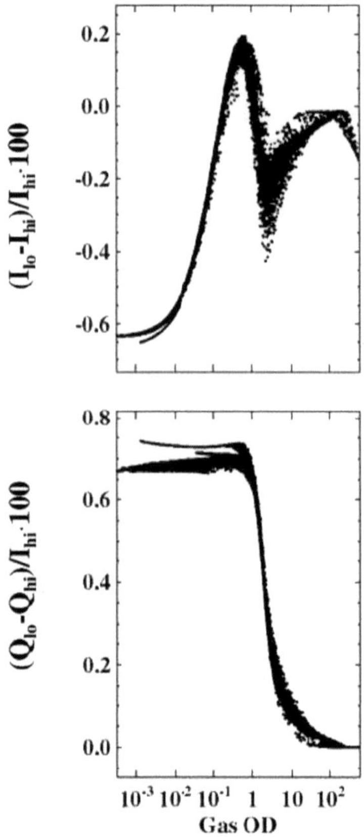

Fig. 10.7. Errors in Stokes parameters I and Q for 2-stream calculations compared to high-accuracy calculations. Errors in I are expressed in percent, while errors in Q are expressed as a percent of the continuum value of I [from O'Dell et al., 2010].

depth τ_g. The idea, then, is to perform 2-stream calculations at all wavelengths and reconstruct the error curve by interpolation using a small number of high-accuracy calculations. The scatter in Fig. 10.7 implies that the vertical structure of gas absorption needs to be accounted for. The basic approach of LSI is to perform a two-dimensional bilinear interpolation of radiance errors in terms of the two variables of τ_g and $\xi^{1/2}$, where ξ is defined similarly to the quantity k^* defined by Duan et al. (2005).

$$\xi = \frac{\tau_g'}{\tau_g}, \tag{10.9}$$

where τ_g' is the cumulative gas absorption optical depth down to the layer where the cumulative scattering optical depth equals a critical value. The critical value is either unity or half the total scattering optical depth (if this is less than unity). Finally, once the error ε_I in Stokes parameter I at a particular wavelength is computed, the corrected monochromatic value of I can be obtained as follows:

$$I = \frac{I_{lo}}{1 + \varepsilon_I}. \tag{10.10}$$

The corresponding expression for Stokes parameter Q is:

$$Q = Q_{lo} - \varepsilon_Q I \,. \tag{10.11}$$

The root mean square (RMS) radiance errors for the LSI technique are typically less than 0.1%.

10.7 Principal component analysis

Natraj et al. (2005) introduced an approach employing principal component analysis (PCA) of optical properties to speed up LBL computations. PCA is a technique that reduces multidimensional data sets to lower dimensions by using orthogonal basis functions. Natraj et al. (2010) improved upon this technique in several ways.

The basic idea is to perform PCA on the optical thickness and single scattering albedo profiles. First, the spectral intervals are grouped into bins based on the cumulative gas absorption optical thickness and the single scattering albedo of the top layer. The gas absorption optical thickness is used rather than the total extinction optical thickness for two primary reasons: (1) gas absorption has greater spectral variation than molecular and particulate scattering, and (2) aerosol and cloud loading can vary substantially from scene to scene; hence, usage of the extinction optical thickness would necessitate changing binning parameters for every scene. The purpose of the single scattering albedo parameter is to account for the vertical structure of gas absorption. Typically, four empirical orthogonal functions (EOFs) are adequate to capture more than 99.99% of the variance in the optical properties. Subsequently, radiance calculations are made for mean bin optical properties and for (positive and negative) perturbations from the mean for the relevant EOFs (typically the first four). From these results, the following quantities can be computed:

$$I_d^0 = \ln\left(I^0/I_2^0\right) ; \tag{10.12a}$$
$$I_{d,k}^{\pm} = \ln\left(I_k^{\pm}/I_{2,k}^{\pm}\right) ; \tag{10.12b}$$
$$Q_d^0 = Q^0 - Q_1^0 ; \tag{10.12c}$$
$$Q_{d,k}^{\pm} = Q_k^{\pm} - Q_{1,k}^{\pm} \,. \tag{10.12d}$$

In Eqs. (10.12), k is the EOF index; subscripts 1 and 2 are the single scatter and two-stream results, respectively (absence of either of these subscripts indicates a benchmark multiple scattering calculation with a high number of streams); superscripts 0, + and − indicate calculations for mean value, positive and negative perturbations, respectively.

Finally, a second-order Taylor series expansion is done to obtain the radiance at the required wavelength.

$$I_l = I_2 \exp\left[I_d^0 + \sum_k \delta I_{d,k} P_{kl} + \frac{1}{2}\sum_k \delta^2 I_{d,k} P_{kl}^2\right] ; \tag{10.13a}$$

$$Q_l = Q_1 + Q_d^0 + \sum_k \delta Q_{d,k} P_{kl} + \frac{1}{2}\sum_k \delta^2 Q_{d,k} P_{kl}^2 \,, \tag{10.13b}$$

where l is the wavelength index; and are, respectively, the first- and second-order central difference operators.

It is to be noted that, like the techniques discussed in Section 10.6, single scattering is computed exactly at every wavelength and PCA is performed only on the multiple scattering component. Also, since the Stokes vector components I and Q are second-order accurate, the Jacobians (with respect to trace gas profiles, for example) are expected to be first-order accurate.

The intensity and polarization are treated differently in this formulation. PCA is performed on the logarithm of the ratio of two intensity calculations, one using a large number of streams and the other using only two streams. Usage of the logarithm avoids negative intensities and takes into account exponential extinction in the absence of scattering. Usage of two-stream calculations reduces the number of terms needed in the PCA expansion for a fixed accuracy. This is because radiative transfer calculations are complicated and nonlinear, for which PCA is not very effective. However, the ratio of n-stream and two-stream intensities is almost linearly dependent on the optical parameters because scattering effects can be considered as a perturbation to gaseous absorption and the two-stream model computes the absorption perfectly in the absence of scattering.

Polarization, on the other hand, is a direct result of scattering (pure absorption does not contribute to polarization if the incident light is unpolarized). Further, single scattering often contributes significantly to the total polarization since multiple scattering is in general depolarizing. Hence, applying PCA directly to the multiply scattered polarized radiance gives excellent results.

Figures 10.8 and 10.9 show the percentage difference between PCA and LBL computations for Stokes parameters I and Q, respectively, for a sample scenario with aerosols, ice and water clouds. The spectral regions considered are the O_2 A-band, the 1.61-μm CO_2 band and the 2.06-μm CO_2 band. There are three interesting features illustrated by these plots. First, there is a clear slope in the results. This is because aerosol/cloud scattering properties are averaged for each bin. The slope goes away when a constant phase matrix is used. Second, the errors are higher for Stokes parameter Q. This is because we use the correlation between two-stream and n-stream intensities for Stokes parameter I, but do not use a similar procedure for Q. The results for Q could be improved using a dedicated four-stream model and using the ratio of n-stream to four-stream values in the analysis. However, it is the degree of polarization that is usually significant and not the exact value of Q. In the continuum, this is usually low. In the line cores it can be high. However, this is a region that can be described by single scattering, which is computed exactly. Third, the largest errors are in the 2.06-μm CO_2 band. This is a region with two significant absorbers (H_2O and CO_2). If we require greater accuracy, we would need to use additional bins for lines with large H_2O absorption, which would result in a mild increase in computation time.

Liu et al. (2006) used PCA to compress channel transmittances or radiances in the infrared into a set of orthogonal eigenvectors called empirical orthogonal functions (EOFs). The channel radiances can be projected onto these EOFs to obtain principal component (PC) scores. Liu et al. (2006) found that 100–250 PCs were sufficient to reproduce channel radiances to 0.01 K accuracy for hyperspectral sensors with thousands of channels. The EOFs capture the spectral variations

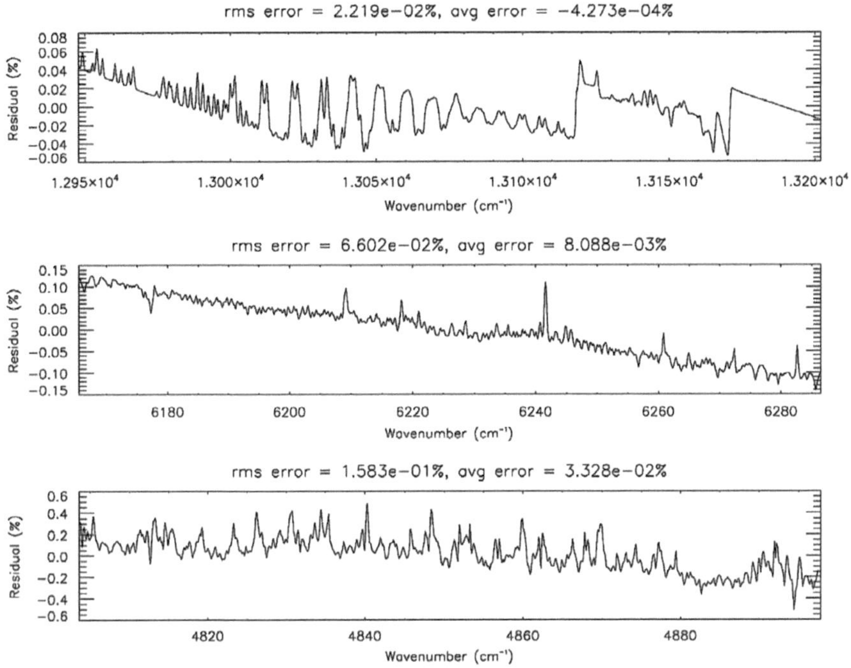

Fig. 10.8. Error in Stokes parameter I for PCA calculations compared to LBL calculations in (top) O_2 A-band; (middle) 1.61-μm CO_2 band; (bottom) 2.06-μm CO_2 band. The error is defined as (PCA–LBL)/LBL×100 [from Natraj et al., 2010].

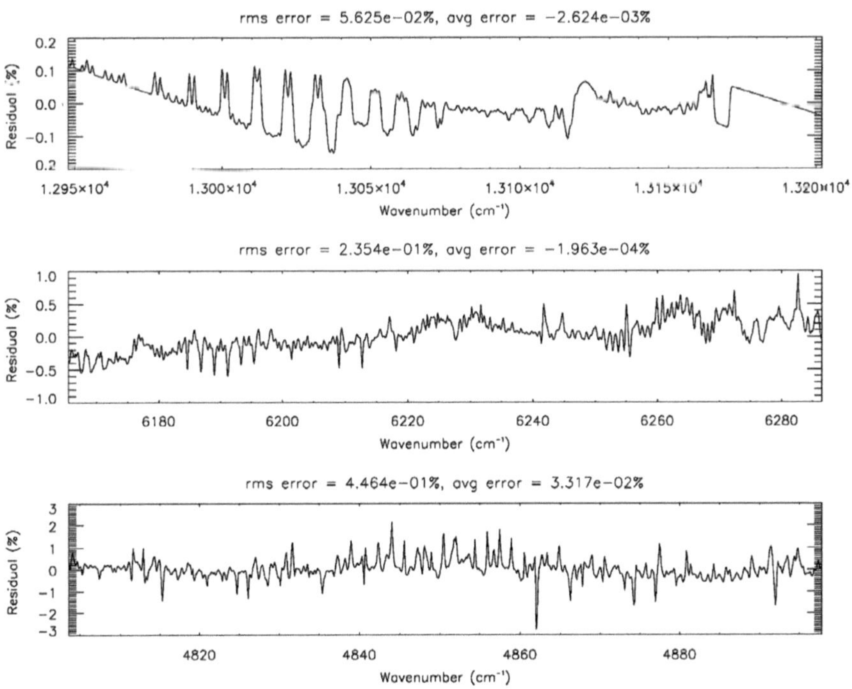

Fig. 10.9. Same as Fig. 10.8 but for Stokes parameter Q [from Natraj et al., 2010].

of the radiances, while the PC scores capture the temperature and composition dependence.

An ensemble of synthetic spectra were used to generate EOFs and PC scores. Since the EOFs and the instrument line shape (ILS) do not depend on the atmospheric state, the PC scores for other scenarios can be obtained as projections of the channel radiance on the EOFs. The channel radiance is a linear combination of monochromatic radiances within the frequency range of that channel, with the weights being the normalized ILS at the frequency grid point. Liu et al. (2006) use a correlation function to select the location of the monochromatic grid points.

10.8 Neural networks

Key and Schweiger (1998) were the first to use neural network techniques to speed up radiative transfer computations. However, their model was limited to broadband calculations since they were mainly interested in radiative heat budgets in the atmosphere and at the surface. Subsequently, Schwander et al. (2001) developed a neural network technique that performed narrowband calculations. A radiative transfer model is used to calculate high-resolution transmittances for a broad range S of scenarios (on the order of 20,000). These calculations are used as training sets to calculate transmittance spectra at a large number N of wavelengths from modeled transmittances at a very small number M of wavelengths. To train the neural network the input data (transmittances at M wavelengths along with auxiliary quantities such as solar zenith angle and total column ozone) is propagated through the network.

The schematic structure of a neural network is shown in Fig. 10.10. The input layer has $M + 2$ neurons, *viz.*, input vector $\mathbf{x}_i$ consisting of transmittances at M wavelengths together with the solar zenith angle and the total ozone amount), and the output layer has N neurons, *viz.*, output vector $\mathbf{o}_k$ comprising transmittances at N wavelengths of interest). One hidden layer with J neurons is used to connect the input and output layers. The neural network is fully interconnected; each neuron in one layer is connected to all neurons in the adjacent layers. There are no connections

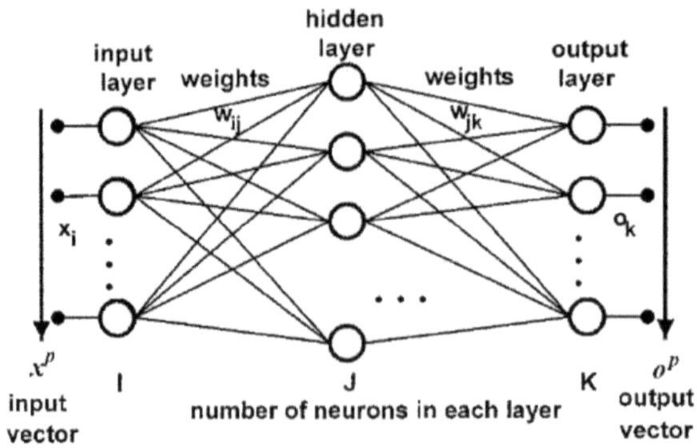

Fig. 10.10. Schematic structure of a neural network used for spectral radiative transfer modeling [from Schwander et al., 2001].

between neurons in the same layer. The signal propagating from one neuron to another is multiplied with a corresponding weight. The weights have to be found using training examples.

To train the neural network, the input data are propagated successively through the network. The resulting output vector is compared with the target vector. From the resulting error between the output and the target vector the weights of the neural network are refined. In this manner, the entire training set is passed through the neural network to complete one epoch. Training continues until the mean difference between the target and output data stops decreasing. Finally, to test the performance of the neural network, a completely different set of spectra is used as a test data set and passed through the network. During this phase, the weights are not adapted.

To test the neural network algorithm, UV indices were calculated on the basis of both the training and the independent test data set. Two sets of model runs were made: (1) global irradiance calculated for 153 wavelengths, and (2) global irradiance calculated for only seven wavelengths between 280 and 700 nm, with the neural network providing the values for the complete set of 153 wavelengths. The deviations between the model runs are presented in Fig. 10.11(a) for the training data set and in Fig. 10.11(b) for the test data set. In both cases, the deviations are within ±5%; in fact, the majority of the UV indices show deviations within ±1%.

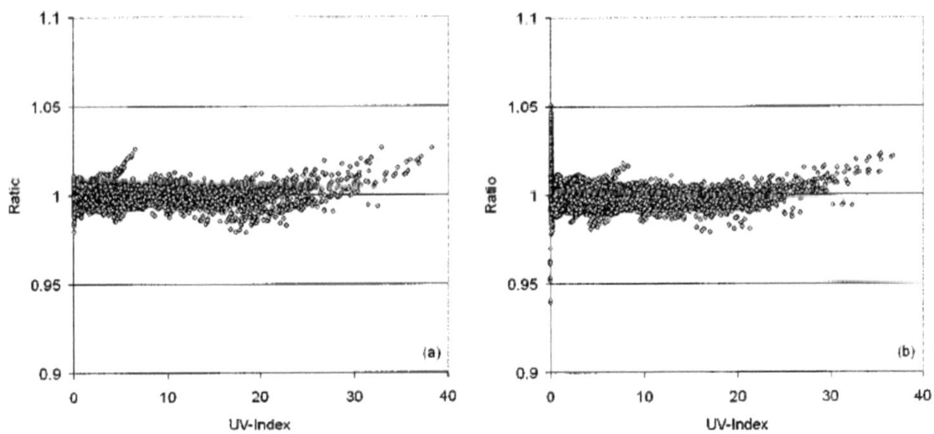

Fig. 10.11. Ratio of UV index calculated with and without use of the neural-network algorithm for (a) training data set and (b) test data set [from Schwander et al., 2001].

Recently, Takenaka et al. (2011) used neural network methods to develop an algorithm for estimating solar radiation from space. The neural network algorithm was applied to data from the Multi-functional Transport Satellite-1 Replacement (MTSAT-1R) geostationary satellite, and estimations were validated against *in situ* observations at four SKYNET sites. The method was also applied to observations from the Advanced Earth Observing Satellite-II/Global Imager (ADEOS-II/GLI).

10.9 Parameterizations for semi-infinite and optically thick media

Ignoring polarization, the reflection function $R_\infty(\mu, \mu_0, \phi)$ for a conservative (non-absorbing) semi-infinite medium can be expressed as:

$$R_\infty(\mu, \mu_0, \phi) = R_{ss}(\mu, \mu_0, \phi) + R_{ms}(\mu, \mu_0, \phi)\,, \tag{10.14}$$

where $R_{ss}(\mu, \mu_0, \phi)$ and $R_{ms}(\mu, \mu_0, \phi)$ are, respectively, the single and multiple scattering contributions, and μ, μ_0 and ϕ are, respectively, the absolute value of the cosine of the incident angle, the cosine of the observation angle and the relative azimuth angle between the observation and incidence directions. The single scattering contribution is given by the following expression (Chandrasekhar, 1950):

$$R_{ss}(\mu, \mu_0, \phi) = \frac{p(\vartheta)}{4(\mu + \mu_0)}\,, \tag{10.15}$$

where $p(\vartheta)$ is the phase function and ϑ is the scattering angle. The multiple scattering contribution can be parameterized in the following form (Kokhanovsky, 2002):

$$R_{ms}(\mu, \mu_0, \phi) = \frac{a + b\mu\mu_0 + c(\mu + \mu_0)}{4(\mu + \mu_0)}\,, \tag{10.16}$$

where a, b and c are constants to be obtained from exact radiative transfer computations.

We can now define the function:

$$D(\mu, \mu_0) = 4(\mu + \mu_0)R_\infty(\mu, \mu_0, \phi) - p(\vartheta)\,. \tag{10.17}$$

Using Eqs. (10.14)–(10.17), it follows that:

$$D(\mu, \mu_0) = a + b\mu\mu_0 + c(\mu + \mu_0)\,. \tag{10.18}$$

For nadir viewing, Eq. (10.18) can be simplified to:

$$D(1, \mu_0) = a + c + (b + c)\mu_0\,. \tag{10.19}$$

Figure 10.12 shows a comparison between exact radiative transfer calculations and a linear fit to Eq. (10.19) for water droplets with an effective radius of 6 μm at a wavelength $\lambda = 0.65\,\mu$m. Assuming that $c = 0$, values for a and b can be derived from the coefficients of the linear fit.

For nadir viewing, Eq. (10.14) can now be simplified as follows:

$$R_\infty(1, \mu_0, \phi) = \frac{a + b\mu_0 + c(1 + \mu_0) + p(\pi - \arccos(\mu_0))}{4(1 + \mu_0)}\,, \tag{10.20}$$

thereby reducing the calculation of the reflection function to that of a phase function.

Figure 10.13(a) shows a comparison of calculations of the reflection function for cloudy media with effective droplet radii of 6 and 16 μm, using Eq. (10.20)

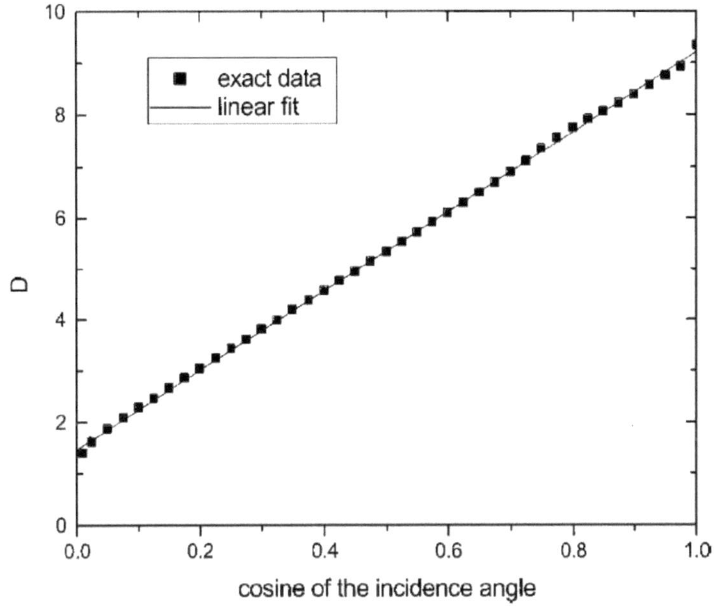

Fig. 10.12. Comparison of the function $D(1, \mu_0)$ as obtained from Eq. (10.17) (symbols) and a linear fit to Eq. (10.19) (solid curve) [from Kokhanovsky, 2002].

and exact radiative transfer calculations. It is clear that the simple approximation given by Eq. (10.20) provides very accurate results for the reflection function of semi-infinite water clouds in the case of nadir observations. Figure 10.13(b) shows that the error is less than 2% for incidence angles less than 85°.

For a thick (but not semi-infinite) layer over a black (nonreflecting) surface, the reflection function can be obtained as follows (Germogenova, 1961, 1963; van de Hulst, 1980; Sobolev, 1984):

$$R(b, \mu, \mu_0, \phi) = R_\infty(\mu, \mu_0, \phi) - tK_0(\mu)K_0(\mu_0)\,, \tag{10.21}$$

where b is the optical thickness of the layer, K_0 is the so-called escape function and t is the global transmittance, which is related to the asymmetry parameter and the escape function.

The escape function describes the angular distribution of photons leaving the semi-infinite non-absorbing layer, and can be computed as follows:

$$K_0(\mu) = \frac{3}{4\pi}\int_0^{2\pi} d\phi \int_0^1 R_\infty(\mu, \mu_0, \phi)(\mu + \mu_0)\mu_0\, d\mu_0\,. \tag{10.22}$$

The global transmittance t can be computed as follows:

$$t = \frac{1}{\alpha + \frac{3}{4}b(1-g)}\,, \tag{10.23}$$

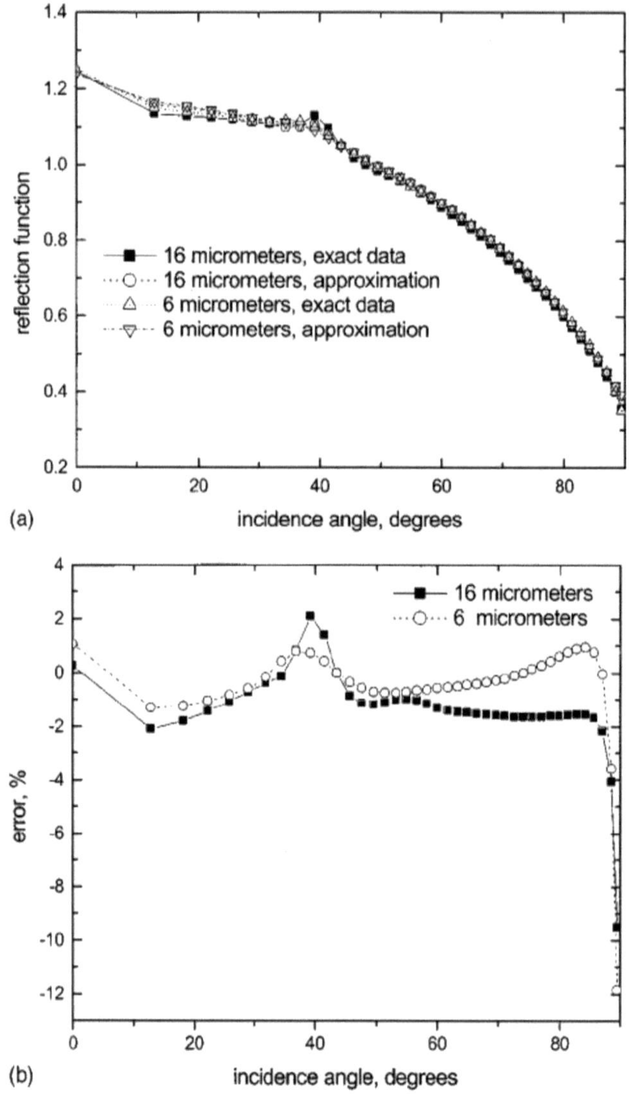

Fig. 10.13. (a) Comparison of calculations of the reflection function for cloudy media with effective droplet radii of 6 and 16 μm, using Equation (10.20) and exact radiative transfer calculations; (b) Difference between calculations using Equation (10.20) and exact calculations [from Kokhanovsky, 2002].

where g is the asymmetry parameter and:

$$\alpha = 3 \int K_0(\mu)\mu^2 \, d\mu \,. \tag{10.24}$$

The escape function can be approximately evaluated as follows:

$$K_0(\mu) = \frac{3}{7}(1 + 2\mu) \,. \tag{10.25}$$

Kokhanovsky and Rozanov (2003) found that further accuracy could be obtained by replacing t in Eq. (10.23) by t^*, where:

$$t^* = t - t_c\,, \tag{10.26}$$

and

$$t_c = \frac{4.86 - 13.08\mu_0 + 12.76\mu_0^2}{b^3}\,. \tag{10.27}$$

Using Eqs. (10.20), (10.21), (10.23)–(10.26), we obtain the following expression for the reflection function for nadir viewing:

$$R(b, 1, \mu_0, \phi) = \frac{a + b\mu_0 + c(1 + \mu_0) + p(\pi - \arccos(\mu_0))}{4(1 + \mu_0)} - \frac{27}{49}\left(\frac{1}{1.07 + \frac{3}{4}b(1-g)} - t_c\right)(1 + 2\mu_0)\,. \tag{10.28}$$

Again, the calculation of the reflection function is reduced to the evaluation of a phase function.

The dependence of the reflection function on the incident angle is shown in Fig. 10.14. The approximation described in Eq. (10.28) works well both for highly reflecting clouds and those with low reflection. Figure 10.15 shows the error due to the approximation as a function of incident angle. The error is typically less than 5% and only a weak function of cloud optical thickness, except for very thin clouds.

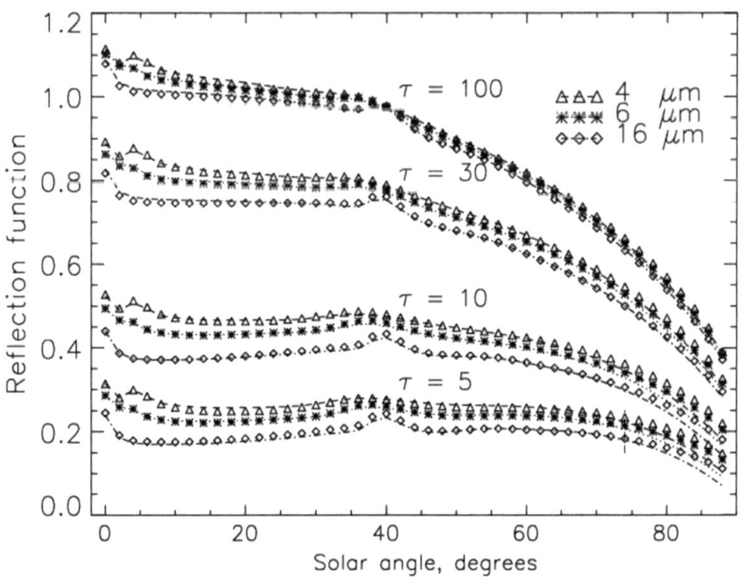

Fig. 10.14. Dependence of the reflection function of a plane–parallel homogeneous water cloud on the incident angle for nadir viewing for different values of the cloud optical thickness and effective droplet radius. Exact calculations are represented by the symbols, and the approximate values calculated using Equation (10.28) by the lines [from Kokhanovsky and Rozanov, 2003].

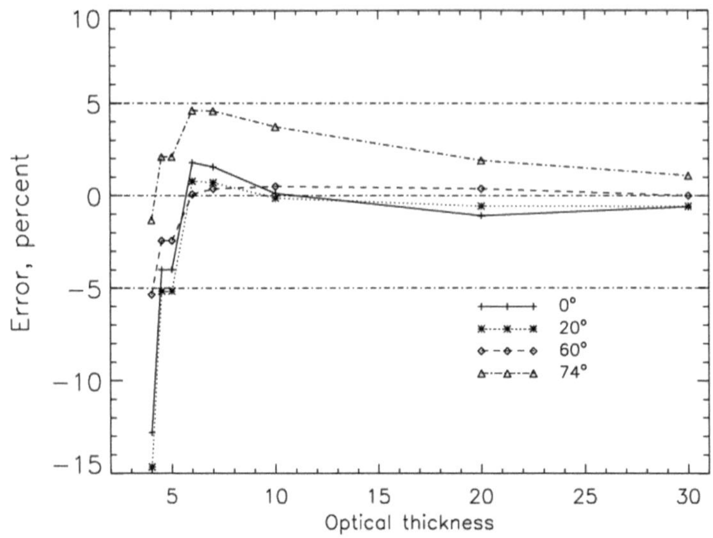

Fig. 10.15. Difference between calculations using Eq. (10.28) and exact calculations as a function of cloud optical thickness for various incident angles [from Kokhanovsky and Rozanov, 2003].

If the surface is Lambertian with albedo A, the reflection function $R_A(b, \mu, \mu_0, \phi)$ can be expressed as follows (van de Hulst, 1980; Kokhanovsky and Nauss, 2006; Nauss and Kokhanovsky, 2011):

$$R_A(b, \mu, \mu_0, \phi) = R(b, \mu, \mu_0, \phi) + \frac{A t_d(\mu) t_d(\mu_0)}{1 - A r_s}, \tag{10.29}$$

where t_d is the diffuse transmittance of the cloud layer and r_s is the spherical albedo. The diffuse transmittance is given by:

$$t_d(\mu) = t K(\mu). \tag{10.30}$$

Similar equations can be derived for weakly absorbing media and for the general case of any single scattering albedo (Germogenova, 1961, 1963; van de Hulst, 1980; Sobolev, 1984; Nauss and Kokhanovsky, 2011).

10.10 Low orders of scattering approximations

Hovenier (1971) provided analytical expressions for single and second-order scattering by homogeneous layers including polarization. The equations for the reflection, $\mathbf{R}_1(\mu, \mu_0, \phi - \phi_0)$, and transmission, $\mathbf{T}_1(\mu, \mu_0, \phi - \phi_0)$, functions for single scattering are particularly simple.

$$\mathbf{R}_1(\mu,\mu_0,\phi-\phi_0)=\frac{1}{4(\mu+\mu_0)}\mathbf{Z}(-\mu,\mu_0,\phi-\phi_0)\left[1-\exp\left(-\frac{b}{\mu}-\frac{b}{\mu_0}\right)\right]; \quad (10.31a)$$

$$\mathbf{T}_1(\mu,\mu_0,\phi-\phi_0)=\frac{1}{4(\mu_0-\mu)}\mathbf{Z}(\mu,\mu_0,\phi-\phi_0)\left[\exp\left(-\frac{b}{\mu_0}\right)-\exp\left(-\frac{b}{\mu}\right)\right];\ \mu\neq\mu_0 \quad (10.31b)$$

$$\mathbf{T}_1(\mu,\mu_0,\phi-\phi_0)=\frac{b}{4\mu_0^2}\mathbf{Z}(\mu,\mu_0,\phi-\phi_0)\exp\left(-\frac{b}{\mu_0}\right). \quad (10.31c)$$

where $\mathbf{Z}$ is the phase matrix, and $\phi-\phi_0$ is the relative azimuth angle between the viewing and incident directions.

The equations for higher orders of scattering can be obtained using the successive orders of scattering technique that involves integration of the results for the previous order of scattering. In particular, the single scattered light acts as a source for the second order of scattering. The expressions for second-order scattering are given below:

$$\mathbf{R}_2(\mu,\mu_0,\phi-\phi_0)=\frac{1}{4\pi}\int_0^1\int_0^{2\pi}\mathbf{Z}(-\mu,-\mu',\phi-\phi')\mathbf{Z}(-\mu',\mu_0,\phi'-\phi_0)g(\mu,\mu_0,\mu')d\mu'd\phi';$$
$$+\frac{1}{4\pi}\int_0^1\int_0^{2\pi}\mathbf{Z}(-\mu,\mu',\phi-\phi')\mathbf{Z}(\mu',\mu_0,\phi'-\phi_0)h(\mu,\mu_0,\mu')d\mu'd\phi'; \quad (10.32a)$$

$$\mathbf{T}_2(\mu,\mu_0,\phi-\phi_0)=\frac{1}{4\pi}\int_0^1\int_0^{2\pi}\mathbf{Z}(\mu,-\mu',\phi-\phi')\mathbf{Z}(-\mu',\mu_0,\phi'-\phi_0)e(\mu,\mu_0,\mu')d\mu'd\phi'$$
$$+\frac{1}{4\pi}\int_0^1\int_0^{2\pi}\mathbf{Z}(\mu,\mu',\phi-\phi')\mathbf{Z}(\mu',\mu_0,\phi'-\phi_0)f(\mu,\mu_0,\mu')d\mu'd\phi'. \quad (10.32b)$$

Expressions for functions e, f, g and h are given in Appendix A.

Kawabata and Ueno (1988) carried out analytic integrations of the invariant imbedding equations over optical thickness to obtain reflection and transmission functions (with polarization neglected) for the first three orders of scattering in vertically inhomogeneous media. Natraj and Spurr (2007) used the same method to provide analytic equations for the first two orders of scattering (2OS) for reflection in inhomogeneous media with polarization effects accounted for.

The azimuth dependence of reflection is expressed by means of a Fourier series expansion:

$$\mathbf{R}(\mu,\mu_0,\phi-\phi_0)=\mathbf{R}_1(\mu,\mu_0,\phi-\phi_0)+\mathbf{R}_{2,c}^0(\mu,\mu_0)$$
$$+2\sum_{m=1}^{M}\left[\mathbf{R}_{2,c}^m(\mu,\mu_0)\cos m(\phi-\phi_0)+\mathbf{R}_{2,s}^m(\mu,\mu_0)\sin m(\phi-\phi_0)\right], \quad (10.33)$$

where the subscripts 1 and 2 refer to the order of scattering, while c and s refer to the cosine and sine components of the Fourier series, respectively. M is the number of Fourier components necessary to achieve Fourier series convergence.

Natraj and Spurr (2007) also defined an intensity correction, I_{corr}, as follows:

$$I_{\text{corr}}(\mu,\mu_0,\phi-\phi_0)\equiv\mathbf{R}(\mu,\mu_0,\phi-\phi_0)\big|_{(1,1)}-R(\mu,\mu_0,\phi-\phi_0), \quad (10.34)$$

where $\mathbf{R}(\mu, \mu_0, \phi - \phi_0)\big|_{(1,1)}$ is the (1,1) element of $\mathbf{R}(\mu, \mu_0, \phi - \phi_0)$ and $R(\mu, \mu_0, \phi - \phi_0)$ is the (scalar) reflection function computed ignoring polarization. Equation (10.34) can also be expanded in a Fourier series:

$$I_{\text{corr}}(\tau; -\mu, \mu_0, \phi - \phi_0) = I^0_{\text{corr}}(\mu, \mu_0) + 2 \sum_{m=1}^{M} I^m_{\text{corr}}(\mu, \mu_0) \cos m(\phi - \phi_0) . \quad (10.35)$$

The sum of the intensity from a scalar calculation and the intensity correction computed above approximates the intensity with polarization included.

If the atmosphere is stratified into n layers, starting from the surface, the single scattering reflection matrix at the top of layer n can be evaluated using the following recurrence relation:

$$\begin{aligned} \mathbf{R}_1(n+1; \Omega_{n+1}) &= \mathbf{R}_1(n; \Omega_n)\Psi(\tau_{n+1}; \mu^{-1} + \lambda_n) \\ &+ \frac{\omega_n \mu^{-1} \lambda_n}{4(\mu^{-1} + \lambda_n)} \left[1 - \Psi(\tau_{n+1}; \mu^{-1} + \lambda_n)\right] \Pi_n(\Omega_n) . \end{aligned} \quad (10.36)$$

$\mathbf{R}_1(n+1; \Omega_{n+1})$ and $\mathbf{R}_1(n; \Omega_n)$ are, respectively, the reflection matrices at levels $n+1$ and n. The phase matrix $\Pi_n(\Omega_n)$ is evaluated using an exact specification of the scattering law based on the use of complete sets of expansion coefficients at the geometry Ω_n. ω_n is the single scattering albedo in layer n, μ is the cosine of the viewing angle, and λ_n is an *average secant* in layer n.

Equation (10.36) is valid in the pseudo-spherical approximation, where all scattering is regarded as taking place in a plane-parallel medium, but the solar beam attenuation is treated for a curved atmosphere. The accuracy of the pseudo-spherical approximation depends on the parameterization used to describe the direct beam attenuation. For most cases, the *average secant* parameterization is sufficient (Spurr, 2002). In a multi-layer atmosphere, slant path transmittances are taken to be exact at layer boundaries, with a simple exponential in optical thickness to approximate the attenuation across layers. For a plane-parallel attenuation, $\lambda_n = 1/\mu_0$. Ψ is defined as follows:

$$\Psi(\tau_{n+1}; y) \equiv \exp\left[-y(\tau_{n+1} - \tau_n)\right] . \quad (10.37)$$

The cosine term for the second order of scattering is given by:

$$\begin{aligned} \mathbf{R}^m_{2,c}(\tau_{n+1}; -\mu, \mu_0) &= \mathbf{R}^m_{2,c}(\tau_n; -\mu, \mu_0)\Psi(\tau_{n+1}; \mu^{-1} + \lambda_n) \\ &+ \frac{\omega_n}{2\mu} \int_0^1 \mathbf{P}^m_c(-\mu, -\mu')\mathbf{V}^m_1(-\mu', \mu_0)\, d\mu' \\ &+ \frac{\omega_n}{2\mu} \int_0^1 \mathbf{P}^m_s(-\mu, -\mu')\mathbf{V}^m_2(-\mu', \mu_0)\, d\mu' \\ &+ \frac{\omega_n \lambda_n}{2} \int_0^1 \mathbf{V}^m_3(-\mu, \mu')\mathbf{P}^m_c(\mu', \mu_0)\, d\mu' \\ &- \frac{\omega_n \lambda_n}{2} \int_0^1 \mathbf{V}^m_4(-\mu, \mu')\mathbf{P}^m_s(\mu', \mu_0)\, d\mu' . \end{aligned} \quad (10.38)$$

Expressions for $\mathbf{P}^m_c$, $\mathbf{P}^m_s$, $\mathbf{V}^m_1$, $\mathbf{V}^m_2$, $\mathbf{V}^m_3$ and $\mathbf{V}^m_4$ are given in Appendix B.

The sine-series contributions to the second-order scattering are:

$$
\begin{aligned}
\mathbf{R}^m_{2,s}(\tau_{n+1}; -\mu, \mu_0) &= \mathbf{R}^m_{2,s}(\tau_n; -\mu, \mu_0)\Psi(\tau_{n+1}; \mu^{-1} + \lambda_n) \\
&+ \frac{\omega_n}{2\mu}\int_0^1 \mathbf{P}^m_c(-\mu, -\mu')\mathbf{V}^m_2(-\mu', \mu_0)\, d\mu' \\
&- \frac{\omega_n}{2\mu}\int_0^1 \mathbf{P}^m_s(-\mu, -\mu')\mathbf{V}^m_1(-\mu', \mu_0)\, d\mu' \\
&+ \frac{\omega_n\lambda_n}{2}\int_0^1 \mathbf{V}^m_3(-\mu, \mu')\mathbf{P}^m_s(\mu', \mu_0)\, d\mu' \\
&- \frac{\omega_n\lambda_n}{2}\int_0^1 \mathbf{V}^m_4(-\mu, \mu')\mathbf{P}^m_c(\mu', \mu_0)\, d\mu' .
\end{aligned}
\tag{10.39}
$$

The Fourier components of the intensity correction can be approximated as:

$$
I^m_{\text{corr}}(\tau; -\mu, \mu_0) = \mathbf{R}^m_{2,c}(\tau; -\mu, \mu_0)\big|_{(1,1)} - R^m_{2,c}(\tau; -\mu, \mu_0) , \tag{10.40}
$$

where $\mathbf{R}^m_{2,c}(\tau; -\mu, \mu_0)\big|_{(1,1)}$ is the (1,1) element of $\mathbf{R}^m_{2,c}(\tau; -\mu, \mu_0)$.

The reflection function for the intensity correction can be evaluated as follows:

$$
\begin{aligned}
R^m_{2,c}(\tau_{n+1}; -\mu, \mu_0) &= R^m_{2,c}(\tau_n; -\mu, \mu_0)\Psi(\tau_{n+1}; \mu^{-1} + \lambda_n) \\
&+ \frac{\omega_n}{2\mu}\int_0^1 P^m_c(-\mu, -\mu')V^m_1(-\mu', \mu_0)\, d\mu' \\
&+ \frac{\omega_n\lambda_n}{2}\int_0^1 V^m_3(-\mu, \mu')P^m_c(\mu', \mu_0)\, d\mu' .
\end{aligned}
\tag{10.41}
$$

Expressions for P^m_c, V^m_1 and V^m_3 are given in Appendix B.

The Fourier components of the intensity correction at the TOA can be finally expressed as:

$$
I^m_{\text{corr}}(\tau_{N+1}) = \mathbf{R}^m_{2,c}(\tau_{N+1}; -\mu, \mu_0)\big|_{(1,1)} - R^m_{2,c}(\tau_{N+1}; -\mu, \mu_0) . \tag{10.42}
$$

The surface boundary condition is the bidirectional reflection distribution function (BRDF) at the surface; reflection functions for the second order of scattering are identically zero.

Figure 10.16 shows the relative errors between the 2OS model and an exact vector model for a sample scenario in the O_2 A-band. The solar, viewing and relative azimuth angles are 50°, 30° and 60°, respectively. The pseudo-spherical approximation was employed for the calculations. The results using the 2OS model are exact in the line cores and most inaccurate ($\sim 30\%$ error in the Stokes parameter Q) in the continuum. However, the continuum is a region dominated by multiple scattering and polarization is least significant there. This suggests that, while the 2OS model may not always provide Stokes parameter Q with sufficiently high accuracy, the degree of polarization ($-Q/I$), or one of its orthogonal components ($I \pm Q$), can be obtained very accurately.

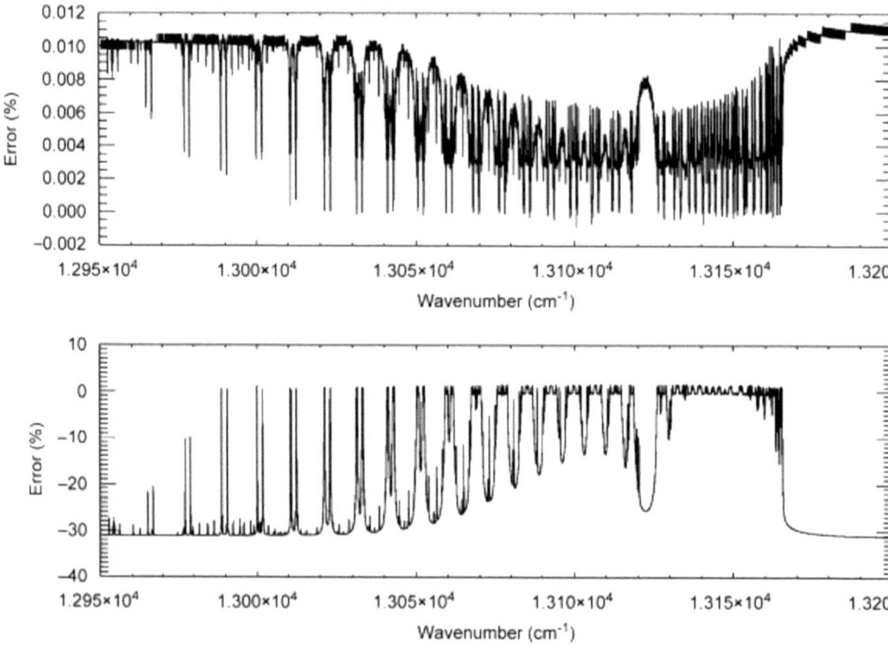

Fig. 10.16. Relative (%) errors between the 2OS model and an exact vector model. (Upper panel) Stokes parameter I; (Lower panel) Stokes parameter Q. The error in the Stokes parameter I is the difference between the sum of the intensity correction from the 2OS model and the scalar intensity, and the intensity from a full vector multiple scattering calculation.

10.11 Conclusions

In this chapter we have summarized several techniques to speed up radiative transfer computations. These include the k-distribution and correlated k-distribution methods, exponential sum fitting of transmittances, spectral mapping methods, optimal spectral sampling double-k, linear-k and low streams interpolation techniques, principal component analysis , neural networks, asymptotic solutions valid for semi-infinite and optically thick layers, and low orders of scattering approximations. The choice of technique depends on the specific application and spectral range under consideration. For example, correlated-k methods do very well for calculation of heating and cooling rates. For sophisticated radiance calculations in the presence of gas absorption; however, one would need to resort to more accurate techniques such as low streams interpolation or principal component analysis .

Acknowledgments

This work was carried out at the Jet Propulsion Laboratory, California Institute of Technology, under a contract with the National Aeronautics and Space Administration. The author acknowledges helpful comments from the anonymous reviewer.

Appendix A: Functions relevant to second order of scattering for homogeneous atmospheres

e, f, g and h can be calculated as follows:

$$e(\mu,\mu_0,\mu') = \frac{1}{4(\mu_0+\mu')}\left\{\begin{array}{l}\dfrac{\mu_0}{\mu_0-\mu}\left[\exp\left(-\dfrac{b}{\mu_0}\right)-\exp\left(-\dfrac{b}{\mu}\right)\right]\\ -\dfrac{\mu'}{\mu'+\mu}\left[\exp\left(-\dfrac{b}{\mu_0}\right)-\exp\left(-\dfrac{b}{\mu'}-\dfrac{b}{\mu_0}-\dfrac{b}{\mu}\right)\right]\end{array}\right\};\quad \mu\neq\mu_0 \quad \text{(A1a)}$$

$$e(\mu,\mu_0,\mu') = \frac{1}{4(\mu_0+\mu')}\left\{\frac{b}{\mu_0}\exp\left(-\frac{b}{\mu_0}\right)-\frac{\mu'}{\mu'+\mu_0}\left[\exp\left(-\frac{b}{\mu_0}\right)-\exp\left(-\frac{b}{\mu'}-\frac{2b}{\mu_0}\right)\right]\right\}.\quad \mu=\mu_0 \quad \text{(A1b)}$$

$$f(\mu,\mu_0,\mu') = \frac{1}{4(\mu_0-\mu')}\left\{\begin{array}{l}\dfrac{\mu_0}{\mu_0-\mu}\left[\exp\left(-\dfrac{b}{\mu_0}\right)-\exp\left(-\dfrac{b}{\mu}\right)\right]\\ -\dfrac{\mu'}{\mu'-\mu}\left[\exp\left(-\dfrac{b}{\mu'}\right)-\exp\left(-\dfrac{b}{\mu}\right)\right]\end{array}\right\};\quad \mu\neq\mu_0\neq\mu' \quad \text{(A2a)}$$

$$f(\mu,\mu_0,\mu') = \frac{1}{4(\mu_0-\mu')}\left\{\frac{b}{\mu_0}\exp\left(-\frac{b}{\mu_0}\right)-\frac{\mu'}{\mu'-\mu_0}\left[\exp\left(-\frac{b}{\mu'}\right)-\exp\left(-\frac{b}{\mu_0}\right)\right]\right\};\quad \mu=\mu_0\neq\mu' \quad \text{(A2b)}$$

$$f(\mu,\mu_0,\mu') = \frac{1}{4(\mu_0-\mu)}\left\{\begin{array}{l}\dfrac{\mu_0}{\mu_0-\mu}\left[\exp\left(-\dfrac{b}{\mu_0}\right)-\exp\left(-\dfrac{b}{\mu}\right)\right]\\ -\dfrac{b}{\mu}\exp\left(-\dfrac{b}{\mu}\right)\end{array}\right\};\quad \mu_0\neq\mu=\mu' \quad \text{(A2c)}$$

$$f(\mu,\mu_0,\mu') = \frac{1}{4(\mu_0-\mu)}\left\{\frac{b}{\mu_0}\exp\left(-\frac{b}{\mu_0}\right)-\frac{\mu}{\mu_0-\mu}\left[\exp\left(-\frac{b}{\mu_0}\right)-\exp\left(-\frac{b}{\mu}\right)\right]\right\};\quad \mu\neq\mu_0=\mu' \quad \text{(A2d)}$$

$$f(\mu,\mu_0,\mu') = \frac{b^2}{8\mu_0^3}\exp\left(-\frac{b}{\mu_0}\right).\quad \mu=\mu_0=\mu' \quad \text{(A2e)}$$

$$g(\mu,\mu_0,\mu') = \frac{1}{4(\mu_0+\mu')}\left\{\begin{array}{l}\dfrac{\mu_0}{\mu_0+\mu}\left[1-\exp\left(-\dfrac{b}{\mu_0}-\dfrac{b}{\mu}\right)\right]\\ +\dfrac{\mu'}{\mu-\mu'}\left[\exp\left(-\dfrac{b}{\mu'}-\dfrac{b}{\mu_0}\right)-\exp\left(-\dfrac{b}{\mu}-\dfrac{b}{\mu_0}\right)\right]\end{array}\right\}; \quad \mu\neq\mu' \tag{A3a}$$

$$g(\mu,\mu_0,\mu') = \frac{1}{4(\mu_0+\mu)}\left\{\frac{\mu_0}{\mu_0+\mu}\left[1-\exp\left(-\frac{b}{\mu_0}-\frac{b}{\mu}\right)\right]-\frac{b}{\mu}\exp\left(-\frac{b}{\mu_0}-\frac{b}{\mu}\right)\right\}. \quad \mu=\mu' \tag{A3b}$$

$$h(\mu,\mu_0,\mu') = \frac{1}{4(\mu_0-\mu')}\left\{\begin{array}{l}\dfrac{\mu_0}{\mu_0+\mu}\left[1-\exp\left(-\dfrac{b}{\mu_0}-\dfrac{b}{\mu}\right)\right]\\ -\dfrac{\mu'}{\mu'+\mu}\left[1-\exp\left(-\dfrac{b}{\mu'}-\dfrac{b}{\mu}\right)\right]\end{array}\right\}; \quad \mu_0\neq\mu' \tag{A4a}$$

$$h(\mu,\mu_0,\mu') = \frac{1}{4(\mu_0+\mu)}\left\{\begin{array}{l}\dfrac{\mu}{\mu_0+\mu}\left[1-\exp\left(-\dfrac{b}{\mu_0}-\dfrac{b}{\mu}\right)\right]\\ -\dfrac{b}{\mu_0}\exp\left(-\dfrac{b}{\mu_0}-\dfrac{b}{\mu}\right)\end{array}\right\}. \quad \mu_0=\mu' \tag{A4b}$$

Appendix B: Functions relevant to second order of scattering for inhomogeneous atmospheres

$\mathbf{V}_1^m$, $\mathbf{V}_2^m$, $\mathbf{V}_3^m$ and $\mathbf{V}_4^m$ can be calculated as follows:

$$\begin{aligned}\mathbf{V}_1^m(-\mu',\mu_0) &= \Phi(\tau_{n+1};\mu^{-1},x',\lambda_n)\mathbf{R}_{1,c}^m(\tau_n;-\mu',\mu_0)\\ &+\frac{\mathbf{P}_c^m(-\mu',\mu_0)\omega_n x'\lambda_n}{4(x'+\lambda_n)(\mu^{-1}+\lambda_n)}\left[1-\Psi(\tau_{n+1};\mu^{-1}+\lambda_n)-(\mu^{-1}+\lambda_n)\Phi(\tau_{n+1};\mu^{-1},x',\lambda_n)\right].\end{aligned} \tag{B1}$$

$$\begin{aligned}\mathbf{V}_2^m(-\mu',\mu_0) &= \Phi(\tau_{n+1};\mu^{-1},x',\lambda_n)\mathbf{R}_{1,s}^m(\tau_n;-\mu',\mu_0)\\ &+\frac{\mathbf{P}_s^m(-\mu',\mu_0)\omega_n x'\lambda_n}{4(x'+\lambda_n)(\mu^{-1}+\lambda_n)}\left[1-\Psi(\tau_{n+1};\mu^{-1}+\lambda_n)-(\mu^{-1}+\lambda_n)\Phi(\tau_{n+1};\mu^{-1},x',\lambda_n)\right].\end{aligned} \tag{B2}$$

$$\begin{aligned}\mathbf{V}_3^m(-\mu,\mu') &= \Phi(\tau_{n+1};x',\lambda_n,\mu^{-1})\mathbf{R}_{1,c}^m(\tau_n;-\mu,\mu')\\ &+\frac{\mathbf{P}_c^m(-\mu,\mu')\omega_n x'\mu^{-1}}{4(\mu^{-1}+\lambda_n)(x'+\mu^{-1})}\left[1-\Psi(\tau_{n+1};\mu^{-1}+\lambda_n)-(\mu^{-1}+\lambda_n)\Phi(\tau_{n+1};x',\lambda_n,\mu^{-1})\right].\end{aligned} \tag{B3}$$

$$\mathbf{V}_4^m(-\mu,\mu') = \Phi(\tau_{n+1}; x', \lambda_n, \mu^{-1})\mathbf{R}_{1,s}^m(\tau_n; -\mu, \mu') + \frac{\mathbf{P}_s^m(-\mu,\mu')\omega_n x' \mu^{-1}}{4(\mu^{-1}+\lambda_n)(x'+\mu^{-1})}\left[1 - \Psi(\tau_{n+1};\mu^{-1}+\lambda_n) - (\mu^{-1}+\lambda_n)\Phi(\tau_{n+1}; x', \lambda_n, \mu^{-1})\right]. \tag{B4}$$

$\mathbf{P}_c^m$ and $\mathbf{P}_s^m$ are, respectively, the cosine and sine components of the mth term in the Fourier expansion of the phase matrix $\mathbf{P}$. Φ and x' are defined as follows:

$$\Phi(\tau_{n+1};\alpha,\beta,\gamma) \equiv \left\{ \begin{array}{l} \dfrac{\Psi(\tau_{n+1};\alpha+\gamma) - \Psi(\tau_{n+1};\beta+\gamma)}{\beta-\alpha}, \ \beta \neq \alpha \\ (\tau_{n+1} - \tau_n)\Psi(\tau_{n+1};\alpha+\gamma), \ \beta = \alpha \end{array} \right\}; \tag{B5a}$$

$$x' \equiv \frac{1}{\mu'}. \tag{B5b}$$

The first order of scattering terms in Eqs. (B1)–(B4) are given below:

$$\mathbf{R}_{1,c}^m(\tau_{n+1}; -\mu, \mu') = \mathbf{R}_{1,c}^m(\tau_n; -\mu, \mu')\Psi(\tau_{n+1}; \mu^{-1} + x') + \frac{\omega_n \mu^{-1} x'}{4(\mu^{-1} + x')}\left[1 - \Psi(\tau_{n+1}; \mu^{-1} + x')\right] \mathbf{P}_c^m(-\mu, \mu'); \tag{B6a}$$

$$\mathbf{R}_{1,c}^m(\tau_{n+1}; -\mu', \mu_0) = \mathbf{R}_{1,c}^m(\tau_n; -\mu', \mu_0)\Psi(\tau_{n+1}; x' + \lambda_n) + \frac{\omega_n x' \lambda_n}{4(x' + \lambda_n)}\left[1 - \Psi(\tau_{n+1}; x' + \lambda_n)\right] \mathbf{P}_c^m(-\mu', \mu_0); \tag{B6b}$$

$$\mathbf{R}_{1,s}^m(\tau_{n+1}; -\mu, \mu') = \mathbf{R}_{1,s}^m(\tau_n; -\mu, \mu')\Psi(\tau_{n+1}; \mu^{-1} + x') + \frac{\omega_n \mu^{-1} x'}{4(\mu^{-1} + x')}\left[1 - \Psi(\tau_{n+1}; \mu^{-1} + x')\right] \mathbf{P}_s^m(-\mu, \mu'); \tag{B6c}$$

$$\mathbf{R}_{1,s}^m(\tau_{n+1}; -\mu', \mu_0) = \mathbf{R}_{1,s}^m(\tau_n; -\mu', \mu_0)\Psi(\tau_{n+1}; x' + \lambda_n) + \frac{\omega_n x' \lambda_n}{4(x' + \lambda_n)}\left[1 - \Psi(\tau_{n+1}; x' + \lambda_n)\right] \mathbf{P}_s^m(-\mu', \mu_0). \tag{B6d}$$

For the intensity correction, we have the following contributions:

$$R_{1,c}^m(\tau_{n+1}; -\mu, \mu') = R_{1,c}^m(\tau_n; -\mu, \mu')\Psi(\tau_{n+1}; \mu^{-1} + x') + \frac{\omega_n \mu^{-1} x'}{4(\mu^{-1} + x')}\left[1 - \Psi(\tau_{n+1}; \mu^{-1} + x')\right] P_c^m(-\mu, \mu'); \tag{B7a}$$

$$R_{1,c}^m(\tau_{n+1}; -\mu', \mu_0) = R_{1,c}^m(\tau_n; -\mu', \mu_0)\Psi(\tau_{n+1}; x' + \lambda_n) + \frac{\omega_n x' \lambda_n}{4(x' + \lambda_n)}\left[1 - \Psi(\tau_{n+1}; x' + \lambda_n)\right] P_c^m(-\mu', \mu_0). \tag{B7b}$$

P_c^m is the mth term in the Fourier expansion of the phase function.

V_1^m and V_3^m can be calculated as follows:

$$V_1^m(-\mu', \mu_0) = \Phi(\tau_{n+1}; \mu^{-1}, x', \lambda_n) R_{1,c}^m(\tau_n; -\mu', \mu_0) \\ + \frac{P_c^m(-\mu', \mu_0)\omega_n x' \lambda_n}{4(x'+\lambda_n)(\mu^{-1}+\lambda_n)} \left[1 - \Psi(\tau_{n+1}; \mu^{-1}+\lambda_n) - (\mu^{-1}+\lambda_n)\Phi(\tau_{n+1}; \mu^{-1}, x', \lambda_n)\right]. \tag{B8}$$

$$V_3^m(-\mu, \mu') = \Phi(\tau_{n+1}; x', \lambda_n, \mu^{-1}) R_{1,c}^m(\tau_n; -\mu, \mu') \\ + \frac{P_c^m(-\mu, \mu')\omega_n x' \mu^{-1}}{4(\mu^{-1}+\lambda_n)(x'+\mu')} \left[1 - \Psi(\tau_{n+1}; \mu^{-1}+\lambda_n) - (\mu^{-1}+\lambda_n)\Phi(\tau_{n+1}; x', \lambda_n, \mu^{-1})\right]. \tag{B9}$$

References

Ambartsumian, V. (1936), The effect of the absorption lines on the radiative equilibrium of the outer layers of the stars, *Publ. Obs. Astron. Univ. Leningrad*, **6**, 7–18.

Arking, A. A., and K. Grossman (1972), The influence of line shape and band structure on temperatures in planetary atmospheres, *J. Atmos. Sci.*, **29**, 937–949.

Armbruster, W., and J. Fischer (1996), Improved method of exponential sum fitting of transmissions to describe the absorption of atmospheric gases, *Appl. Opt.*, **35(12)**, 1931–1941, doi:10.1364/AO.35.001931.

Boesche, E., P. Stammes, R. Preusker, R. Bennartz, W. Knap, and J. Fischer (2008), Polarization of skylight in the O2 A band: Effects of aerosol properties, *Appl. Opt.*, **47(19)**, 3467–3480, doi:10.1364/AO.47.003467.

Buchwitz, M., V. V. Rozanov, and J. P. Burrows (2000), A correlated-k distribution scheme for overlapping gases suitable for retrieval of atmospheric constituents from moderate resolution radiance measurements in the visible/near-infrared spectral region, *J. Geophys. Res.*, **105(D12)**, 15247–15261, doi:10.1029/2000JD900171.

Cao, Y., W. Zhang, Y. Zhang, H. Chang, and M. Cong (2011), A new k-interval selection technique for fast atmospheric radiance calculation in remote sensing applications, *J. Quant. Spectrosc. Radiat. Transfer*, **112(9)**, 1479–1485, 10.1016/j.jqsrt.2011.03.004.

Chandrasekhar, S. (1950), *Radiative Transfer*, Oxford: Clarendon Press.

Chou, M. D., and A. Arking (1980), Computation of infrared cooling rates in the water vapor bands, *J. Atmos. Sci.*, **37**, 855–867.

Chou, M.-D., W. L. Ridgway, and M. M.-H. Yan (1993), One-parameter scaling and exponential-sum fitting for water vapor and CO_2 infrared transmission functions, *J. Atmos. Sci.*, **50(14)**, 2294–2303, doi:10.1175/1520-0469(1993)050<2294:OPSAES>2.0.CO;2.

Crisp, D. (1997), Absorption of sunlight by water vapor in cloudy conditions: A partial explanation for the cloud absorption anomaly, *Geophys. Res. Lett.*, **24(5)**, 571–574, doi:10.1029/97GL50245.

Duan, M., Q. Min, and J. Li (2005), A fast radiative transfer model for simulating high-resolution absorption bands, *J. Geophys. Res.*, **110**, D15201, doi:10.1029/2004JD005590.

Germogenova, T. A. (1961), On the solution of the transfer equation for a plane layer, Zurnal. *Appl. Math. Math. Phys.*, **1**, 1001–1008.

Germogenova, T. A. (1963), Some formulas to solve the transfer equation in a plane layer problem, in *Spectroscopy of Scattering Media*, (ed. B. I. Stepanov), Minsk: AN BSSR, 36–41.

Goody, R., R. West, L. Chen, and D. Crisp (1989), The correlated-k method for radiation calculations in non homogeneous atmospheres, *J. Quant. Spectrosc. Radiat. Transfer*, **42**, 539–550.

Hasekamp, O. P., and A. Butz (2008), Efficient calculation of intensity and polarization spectra in vertically inhomogeneous scattering and absorbing atmospheres, *J. Geophys. Res.*, **113**, D20309, doi:10.1029/2008JD010379.

Hovenier, J. W. (1971), Multiple scattering of polarized light in planetary atmospheres, *Astron. Astrophys.*, **13**, 7–29.

Hunt, G., and I. Grant (1969), Discrete space theory of radiative transfer and its application to problems in planetary atmospheres, *J. Atmos. Sci.*, **26**, 963–972, doi:10.1175/1520-0469(1969)026<0963:DSTORT>2.0.CO;2.

Kawabata, K., and S. Ueno (1988), The first three orders of scattering in vertically inhomogeneous scattering-absorbing media, *Astrophys. Space Sci.*, **150**, 327–344.

Key, J. R., and A. J. Schweiger (1998), Tools for atmospheric radiative transfer: Streamer and FluxNet, *Comp. Geosci.*, **24(5)**, 443–451, doi:10.1016/S0098-3004(97)00130-1.

Kokhanovsky, A. A. (2002), Simple approximate formula for the reflection function of a homogeneous, semi-infinite turbid medium, *J. Opt. Soc. Am. A*, **19(5)**, 957–960.

Kokhanovsky, A. A., and T. Nauss (2006), Reflection and transmission of solar light by clouds: Asymptotic theory, *Atmos. Chem. Phys.*, **6**, 5,537-5,545, doi:10.5194/acp-6-5537-2006.

Kokhanovsky, A. A., and V. V. Rozanov (2003), The reflection function of optically thick weakly absorbing turbid layers: A simple approximation, *J. Quant. Spectrosc. Radiat. Transfer*, **77**, 165–175.

Kratz, D. P. (1995), The correlated k-distribution technique as applied to the AVHRR channels, *J. Quant. Spectrosc. Radiat. Transfer*, **53(5)**, 501–517, doi:10.1016/0022-4073(95)00006-7.

Kratz, D. P., M.-D. Chou, M. M.-H. Yan, and C.-H. Ho (1998), Minor trace gas radiative forcing calculations using the *k* distribution method with one-parameter scaling, *J. Geophys. Res.*, **103(D24)**, 31,647–31,656, doi:10.1029/1998JD200009.

Lacis, A. A., and V. Oinas (1991), A description of the correlated k distribution method for modeling non gray gaseous absorption, thermal emission, and multiple scattering in vertically inhomogeneous atmospheres, *J. Geophys. Res.*, **96**, 9027–9063.

Lacis, A. A., W. C. Wang, and J. E. Hansen (1979), Correlated k-distribution method for radiative transfer in climate models: Application to effect of cirrus clouds on climate, *NASA Conf. Publ.*, **2076**, 309–314.

Liu, X., W. L. Smith, D. K. Zhou, and A. Larar (2006), Principal component-based radiative transfer model for hyperspectral sensors: theoretical concept, *Appl. Opt.*, **45(1)**, 201–209, doi: 10.1364/AO.45.000201.

Mano, Y. (1995), Modified ESFT method for application to atmospheric radiation, *Papers in Meteorology and Geophysics*, **46(1)**, 1–8.

Meadows, V. S., and D. Crisp (1996), Ground-based near-infrared observations of the Venus nightside: The thermal structure and water abundance near the surface, *J. Geophys. Res.*, **101(E2)**, 4595–4622, doi:10.1029/95JE03567.

Moncet, J.-L., G. Uymin, A. E. Lipton, and H. E. Snell (2008), Infrared radiance modeling by Optimal Spectral Sampling, *J. Atmos. Sci.*, **65**, 3917–3934, doi:10.1175/2008JAS2711.1.

Natraj, V., X. Jiang, R.-L. Shia, X. Huang, J. S. Margolis, and Y. L. Yung (2005), Application of principal component analysis to high spectral resolution radiative transfer: A case study of the O2 A band, *J. Quant. Spectrosc. Radiat. Transfer*, **95(4)**, 539–556, doi:10.1016/j.jqsrt.2004.12.024.

Natraj, V., R.-L. Shia, and Y. L. Yung (2010), On the use of principal component analysis to speed up radiative transfer calculations, *J. Quant. Spectrosc. Radiat. Transfer*, **111(5)**, 810–816, doi:10.1016/j.jqsrt.2009.11.004.

Natraj, V., and R. J. D. Spurr (2007), A fast linearized pseudo-spherical two orders of scattering model to account for polarization in vertically inhomogeneous scattering–absorbing media, *J. Quant. Spectrosc. Radiat. Transfer*, **107(2)**, 263–293, doi:10.1016/j.jqsrt.2007.02.011.

Nauss, T., and A. A. Kokhanovsky (2011), Retrieval of warm cloud optical properties using simple approximations, *Remote Sens. Environ.*, **115(6)**, 1,317–1,325, doi:10.1016/j.rse.2011.01.010.

O'Dell, C. W. (2010), Acceleration of multiple-scattering, hyperspectral radiative transfer calculations via low-streams interpolation, *J. Geophys. Res.*, **115**, D10206, doi:10.1029/2009JD012803.

Schwander, H., A. Kaifel, A. Ruggaber, and P. Koepke (2001), Spectral radiative-transfer modeling with minimized computation time by use of a neural-network technique, *Appl. Opt.*, **40(3)**, 331–335, doi:10.1364/AO.40.000331.

Sobolev, V. V. (1984), Integral relations and asymptotic expressions in the theory of radiative transfer, *Astrofizika*, **20**, 123–132.

Spurr, R. J. D. (2002), Simultaneous derivation of intensities and weighting functions in a general pseudo-spherical discrete ordinate radiative transfer treatment, *J. Quant. Spectrosc. Radiat. Transfer*, **75(2)**, 129–175, doi: 10.1016/S0022-4073(02)00014-6.

Takenaka, H., T. Y. Nakajima, A. Higurashi, A. Higuchi, T. Takamura, R. T. Pinker, and T. Nakajima (2011), Estimation of solar radiation using a neural network based on radiative transfer, *J. Geophys. Res.*, **116**, D08215, doi:10.1029/2009JD013337.

van de Hulst, H. C. (1980), *Multiple Light Scattering: Tables, Formulas and Applications*, New York: Academic Press.

West, R, D. Crisp, and L. Chen (1990), Mapping transformations for broadband atmospheric radiation calculations, *J. Quant. Spectrosc. Radiat. Transfer*, **43**, 191–199.

Wiscombe, W. J., and J. W. Evans (1977), Exponential-sum fitting of radiative transmission functions, *J. Comp. Phys.*, **24**, 416–444, doi:10.1016/0021-9991(77)90031-6.

11 Dependence of direct aerosol radiative forcing on the optical properties of atmospheric aerosol and underlying surface

Claudio Tomasi, Christian Lanconelli, Angelo Lupi, and Mauro Mazzola

11.1 Introduction

Airborne aerosol is a suspension of solid particulate matter and/or liquid particles in air, which are often observed as dust, haze and smoke. They present an overall number concentration usually varying between a few hundred per cubic centimeter of air in the remote areas of the planet and more than $10^4\,\text{cm}^{-3}$ in the most polluted urban areas, with sizes ranging mainly between 0.01 and no more than $100\,\mu$m, and therefore varying by more than four orders of magnitude (Heintzenberg, 1994). Aerosol particles are present in the atmosphere as a result of primary emissions or are formed through secondary processes involving both natural and anthropogenic gaseous species. The primary emissions of aerosols are from both natural and anthropogenic sources, of which the former mainly consist of mineral dust mobilized in desert and semi-arid regions, sea-salt from oceans, volcanic dust from violent eruptions of debris and gases, biogenic aerosols, like viruses, bacterial cells, fungi, and spores from plants and animals, and smokes from forest fires. Conversely, anthropogenic aerosols are mainly composed of industrial dust, dust mobilized through agricultural activities, and smokes from fossil fuel combustion and waste and biomass burning associated with various human activities. Secondary aerosols are formed in the atmosphere through chemical (mainly heterogeneous) reactions involving sulfur dioxide, nitrogen oxides, biogenic volatile organic compounds and other chemical species originating from both natural and anthropogenic activities (Tanré et al., 2003; Seinfeld and Pandis, 2006). The sizes of aerosol particles are comparable to the wavelengths of incoming solar radiation, mainly ranging between 0.3 and $4.0\,\mu$m, and are therefore mostly smaller than the terrestrial radiation wavelengths, which mainly vary between about $4.0\,\mu$m and more than $25\,\mu$m. Thus, aerosols interact very strongly with the solar (short-wave) radiation and more weakly with the terrestrial (long-wave) radiation, as clearly stated by the Mie (1908) scattering theory. Because of such interactions, aerosols induce important effects on the radiation budget of the land–atmosphere–ocean system, exerting a powerful influence on the Earth's climate through strong scattering and absorption processes of short-wave radiation and, although less intensively, through extinction (scattering and absorption) of long-wave radiation emitted by both the terrestrial surface and atmosphere toward space.

These radiative effects have been studied exhaustively over the past 40 years through field measurements, satellite-based observations and radiative transfer model simulations (McCormick and Ludwig, 1967; Vonder Haar and Suomi, 1971; Coakley et al., 1983; Charlson et al., 1990, 1991, 1992; Schwartz and Andreae, 1996; Ramanathan et al., 2001a; Kaufman et al., 2002a). The cited studies have shown that the direct aerosol-induced radiative forcing (hereinafter referred to as DARF) effects are particularly intense in the mid-latitude industrial and more densely populated regions of the planet, where the anthropogenic aerosol emissions are particularly strong (Charlson et al., 1991; Kiehl and Briegleb, 1993; Chin et al., 2007). They are considerably weaker in remote oceanic areas (Takemura et al., 2005) and polar regions (Blanchet, 1989; Tomasi et al., 2007), where the background number and mass concentrations of particles are usually much lower than in densely populated regions.

Direct climatic effects are induced by atmospheric aerosols through the abovementioned interactions of the surface–atmosphere system with short- and long-wave radiation:

(1) Scattering and absorption of incident solar radiation generally cause a marked decrease in the flux density of direct solar radiation reaching the surface, leading to an increase in the solar radiation fraction reflected back to space, and an increase in the overall albedo of the climate system (Haywood and Boucher, 2000; Yu et al., 2006; Quaas et al., 2008). At the same time, scattering by aerosols cause an increase in the diffuse solar radiation flux reaching the surface and contribute to the enhancement of the surface–atmosphere system albedo. As mentioned above, such radiative effects can be evaluated by examining ground-based and *in situ* measurements of the columnar aerosol optical parameters and satellite-borne data (Haywood et al., 1999; King et al., 1999; Yu et al. 2004; Anderson et al., 2005; Chung et al., 2005; Remer and Kaufman, 2006; Bates et al., 2006; Bellouin et al., 2005). Absorption of incoming solar radiation is particularly marked in cases where significant contents of soot substances (containing black carbon (BC) and/or elemental carbon (EC)) are present in the airborne particulate matter (Andreae and Gelencsér, 2006), inducing appreciable warming effects in the lower part of the troposphere (Kaufman, 1987; Bellouin et al., 2003; Stott et al., 2006).
(2) Scattering and absorption of long-wave terrestrial radiation (Lubin et al., 2002) can modify the cooling rate of the atmospheric boundary layer, especially in the presence of dense layers containing haze particle polydispersions (Grassl, 1973; Yu et al., 2002). Observations of these radiative effects have been provided by sensors onboard satellite platforms (Zhang and Christopher, 2003). However, aerosols are generally estimated to scatter and absorb the infrared radiation rather weakly on the global scale, only slightly enhancing the greenhouse effect of the atmosphere, which is mainly generated in the cloudless atmosphere by the thermal radiation absorption due to H_2O, CO_2, CH_4, N_2O and many other atmospheric gases (Hansen et al., 1997).

For relatively high number concentrations, airborne aerosols can also produce important indirect effects on the terrestrial climate system, by acting as cloud condensation nuclei and modifying the cloud droplet concentration and size-distribution.

Therefore, they can change significantly the cloud optical properties, by causing substantial increases in cloud albedo and modifying the average cloudiness conditions in various areas of the Earth (Ackerman et al., 1994, 2000, 2004; Kaufman and Fraser, 1997; Feingold et al., 1999; Kruger and Grassl, 2002; Feingold, 2003; Nakajima et al, 2003; Ervens et al., 2005; Kaufman and Koren, 2006; Jiang et al., 2006; Koren et al., 2008). In addition, aerosols can also enhance the liquid water content of clouds, thus altering cloud lifetime, and strongly influencing the heterogeneous chemistry of the atmosphere, causing further relevant indirect effects on the Earth's climate (Schwartz et al., 1995). The IPCC TAR (2001) report (see also Forster et al., 2007) stated that 'the indirect effect is the mechanism by which aerosols modify the microphysical and, hence, the radiative properties, amount and lifetime of clouds'. Key parameters for determining the indirect effect are the effectiveness of an aerosol particle to act as a cloud condensation nucleus, which is a function of the size, chemical composition, mixing state and ambient environment (e.g., Penner et al., 2001). The microphysically induced effect on the cloud droplet number concentration and, hence, the cloud droplet size, with the liquid water content held fixed has been called the *first indirect effect* (Ramaswamy et al., 2001), the *cloud albedo effect* (Lohmann and Feichter, 2005) or the *Twomey effect* (Twomey, 1977). The microphysically induced effect on the liquid water content, cloud height, and lifetime of clouds has been called the *second indirect effect* (Ramaswamy et al., 2001), the *cloud lifetime effect* (Lohmann and Feichter, 2005) or the *Albrecht effect* (Albrecht, 1989). Thus, the IPCC TAR (2001) report classified the indirect effects into two different types, which are denoted as *cloud albedo effect* and *cloud lifetime effect*, respectively, as these terms are more descriptive of the microphysical processes that occur. The cloud albedo effect was considered to be a radiative forcing mechanism because global model calculations could be performed to describe the influence of increased aerosol concentration on the cloud optical properties, while holding the liquid water content of the cloud fixed. It is considered to be a key uncertainty in the radiative forcing of climate, bearing in mind that a best estimate of this radiative forcing effect was not assigned by IPCC TAR (2001), where a range was indicated between 0 and $-2\,\mathrm{W\,m^{-2}}$ for liquid water clouds. The other indirect effects were not considered to constitute radiative forcing processes, because the hydrological cycle is invariably altered through feedback processes in suppressing drizzle, increasing the cloud height and modifying the cloud lifetime in atmospheric models, without evidence of radiative effects. This is the case of the impact of anthropogenic aerosols on the formation and modification of the physical and radiative properties of ice clouds (Penner et al., 2001), although the quantification of a radiative forcing effect from this mechanism was not considered appropriate, given the host of uncertainties and unknowns surrounding ice cloud nucleation and physics. Similarly, the IPCC TAR (2001) report did not include any assessment of the semi-direct effect (Hansen et al., 1997; Ackerman et al., 2000; Jacobson, 2002; Menon et al., 2003; Cook and Highwood, 2004; Johnson et al., 2004), defined as the mechanism by which absorption of short-wave radiation by tropospheric aerosols leads to heating of the troposphere, which in turn changes the relative humidity and stability of the troposphere and thereby influences cloud formation and lifetime. Therefore, the semi-direct effect was not strictly considered a radiative forcing process in the IPCC TAR (2001) document.

In reality, the direct and indirect aerosol-induced effects are accomplished with numerous other important radiation budget changes arising from the variability of sea-surface temperature, selective absorption by atmospheric water vapor and other minor gases, and variations in the cloud coverage characteristics occurring at various tropospheric altitudes as a result of the complex atmospheric circulation processes. The combination of all such effects renders the radiative behavior of the surface–atmosphere system even more difficult to comprehend (Waliser and Graham, 1993; Hansen et al., 1997, 1998; Ramanathan et al., 2001b; Marsden and Valero, 2004; Takemura et al., 2005).

In addition, the irregular distribution of the aerosol radiative effects on the global scale is enhanced by the marked presence of anthropogenic aerosol sources in the industrialized regions of the planet, which tends to render more irregular the spatial distribution of atmospheric heat released by aerosols after their absorption of solar radiation. These discontinuities in time and irregularities in space in turn lead to significant variations in the atmospheric circulation picture (King et al., 1999). In spite of the great spatial and temporal variability in aerosol concentration and composition features, the regional radiative forcing effects induced by aerosols are estimated to exceed or be comparable in magnitude to the greenhouse warming effects occurring in many areas of the Earth, often presenting opposite signs causing warming and cooling effects in different regions (Takemura et al., 2002). Therefore, the interactions between aerosol particles and solar radiation are among the main sources of uncertainty in modeling climate changes within global circulation models (Hansen et al., 1997, 1998). There are still large knowledge gaps on this matter, for instance regarding the dependence of aerosol radiative forcing on surface reflectance for the columnar aerosol polydispersions with different physico-chemical properties (Charlson et al., 1992). The interactions of solar radiation with atmospheric aerosols are estimated to cause direct effects on the radiation budget, modifying the fields of short-wave downwelling radiance and short-wave upwelling radiance at the Top-of-Atmosphere (ToA) level, usually measured in $\mathrm{W\,m^{-2}\,sr^{-1}}$. These changes can vary greatly as a function of numerous parameters, including the solar zenith angle, the shape-parameters of the columnar aerosol particle size-distribution, the complex refractive index of particulate matter, the aerosol single scattering albedo, the intensity of the multiple scattering effects, and the spectral and directional characteristics of surface reflectance (Chylek and Coakley, 1974; Coakley and Chylek, 1975; Grassl and Newiger, 1982). The large variability in the microphysical and chemical composition parameters of aerosols, as well as in the vertical distribution profiles of aerosol mass concentration and radiative parameters, contribute to increase the complexity of the radiative exchange processes taking place between aerosol layers, the terrestrial surface and absorbing minor gases of the atmosphere, which frequently take place through nonlinear mechanisms (Feichter et al., 2004).

The main goal of the present work is to illustrate the dependence features of the instantaneous aerosol direct radiative forcing as a function of the mentioned parameters, by studying the changes in magnitude and the spectral features of these radiative effects occurring at the Top-of-Atmosphere (ToA), at the surface (i.e. at the Bottom-of-Atmosphere, BoA) and inside the atmosphere. Such radiation budget changes depend on (i) the radiative properties of columnar aerosol, as defined in terms of aerosol optical thickness $\tau_a(\lambda)$ and single scattering albedo $\omega(\lambda)$ (both varying as a function of wavelength λ), (ii) characteristics of surface reflectance, and (iii) solar zenith angle θ_o. To provide a general description of aerosol radiative properties, a set of 40 aerosol extinction models of different origin are presented in Section 11.2 for use in the DARF calculations, which are in part chosen among the best-known models proposed in the literature and in part determined here for providing a more complete picture of aerosol radiative parameters (including desert dust, volcanic and biomass burning particle polydispersions). A set of surface reflectance models is presented in Section 11.3 for use in the DARF calculations, including: (i) a sub-set of 16 Bidirectional Reflectance Distribution Function (BRDF) non-lambertian models, and (ii) a corresponding sub-set of 16 lambertian models adapted to yield equivalent broadband albedo features for ocean, vegetation-covered, desert/semi-arid, and polar snow-covered surfaces. The main physical concepts according to which the DARF effects are induced by atmospheric aerosol at the ToA and BoA levels and within the atmosphere are briefly described in Section 11.4, together with the definitions of these radiative quantities and the schematic description of the calculation method adopted to calculate the DARF effects. They are determined for different aerosol compositions and surface reflectance conditions, with the main purpose of investigating the dependence of such instantaneous DARF effects on aerosol optical parameters $\tau_a(\lambda)$ and $\omega(\lambda)$, underlying surface reflectance characteristics and solar zenith angle θ_o, for different surfaces, as described by the two sets of 16 BRDF non-lambertian and lambertian (isotropic) surface reflectance models.

11.2 Aerosol models

The Mie (1908) electromagnetic theory predicts that the volume scattering and absorption coefficients given by an aerosol polydispersion at a certain wavelength λ (and hereinafter indicated with symbols $\beta_{\rm sca}(\lambda)$ and $\beta_{\rm abs}(\lambda)$, respectively) substantially depend on the shape-parameters of the particle size-distribution curve and the complex refractive index of particulate matter, defined in terms of real part $n(\lambda)$ and imaginary part $k(\lambda)$. Similarly, the phase function $P(\Theta)$ of an aerosol polydispersion [which describes the angular distribution of the radiation scattered by aerosol particles in a specific direction (i.e. with a certain scattering angle Θ)] depends on the size-distribution shape parameters and spectral parameters $n(\lambda)$ and $k(\lambda)$. The aerosol optical thickness $\tau_a(\lambda)$ is given at each wavelength by the integral of the volume extinction coefficient $\beta_{\rm ext}(\lambda)$, along the vertical atmospheric path, with $\beta_{\rm ext}(\lambda)$ given by the sum of $\beta_{\rm sca}(\lambda)$ and $\beta_{\rm abs}(\lambda)$. Therefore, this parameter provides a measure of the overall extinction effects due to columnar aerosol. The monochromatic single scattering albedo $\omega(\lambda)$ relative to the atmospheric column is usually expressed in terms of the analytical form $\omega[\lambda, r, n(\lambda) - ik(\lambda)]$ as a function of wavelength λ, radius r and complex refractive index $n(\lambda) - ik(\lambda)$ of particulate matter. Thus, it is calculated over the entire radius range of the particle size-distribution curve for well-defined spectral curves of $n(\lambda)$ and $k(\lambda)$, obtained by means of appropriate aerosol extinction models or retrieved from sky-brightness measurements in almucantar performed with ground-based multi-wavelength sun-sky/radiometers (Nakajima et al., 1996; Holben et al., 1998; Dubovik and King,

2000; Dubovik et al., 2000; Gatebe et al., 2010). This columnar quantity is in practice given by the ratio between the integrals of the volume scattering and extinction coefficients along the vertical path of the atmosphere, both calculated over the entire radius range of the columnar particle size-distribution (for instance by means of appropriate aerosol extinction models) and provides a measure of the fraction of solar radiation extinguished by aerosol particles that is subject to scattering. Conversely, the difference between unity and columnar single scattering albedo gives a measure of the fraction of aerosol-extinguished energy subject to absorption.

For given reflectance characteristics of an underlying surface, the DARF effects can vary largely as a function of aerosol optical thickness $\tau_a(\lambda)$ and columnar single scattering albedo $\omega(\lambda)$. For application studies made by means of aerosol extinction models, $\omega(\lambda)$ is usually calculated at a limited number of wavelengths, properly chosen over the solar radiation spectrum at the wavelengths located in the middle of the so-called atmospheric windows, in such a way to define a limited spectral series of this parameter and evaluate its optical effects throughout the solar spectrum. To investigate the dependence features of the DARF effects (and in particular of the instantaneous direct aerosol-induced radiative forcing at ToA-level ΔF_{ToA}) on the single scattering albedo $\omega(\lambda)$ of the columnar aerosols, it is useful to make use of different sets of aerosol models characterized by various radiative properties and therefore presenting a wide range of $\omega(\lambda)$.

For this purpose, a rather high number of aerosol extinction models of different origins was examined in the present study, partly drawn from the literature and partly defined originally for different number size-distribution functions $N(r)$, and various pairs of refractive index parts $n(\lambda)$ and $k(\lambda)$, suitable for representing atmospheric particle polydispersions of diverse origin and, hence, with different chemical composition features. The radiative characteristics of an overall set of 40 aerosol models of various origins were defined, related to different microphysical, chemical and optical properties. They are:

- the 3 original aerosol models of the 6S code (Vermote et al., 1997a, b) for dry air conditions;
- the 2 supplementary 6S models proposed by Vermote et al. (1997a) to represent the background desert dust (Shettle, 1984), and the El Chichon volcanic stratospheric aerosol model (King et al., 1984), defined here in greater detail and with better accuracy;
- the 14 modified (M-type) aerosol models determined in the present study by using the 6S basic components to simulate a large variety of wet aerosol radiative properties (in terms of their linear combinations defined for air relative humidity RH = 50%), which allowed a complete coverage of the range of aerosol single scattering albedo most commonly observed in reality;
- the 10 OPAC aerosol models defined by Hess et al. (1998) for RH = 50% to represent various wet aerosol polydispersions of different origin;
- the 4 classical aerosol models proposed by Shettle and Fenn (1979) for RH = 50% to represent the Rural, Maritime, Tropospheric and Urban particle polydispersions used in the SBDART code (Ricchiazzi et al., 1998); and
- the 7 additional aerosol models defined in the present study to represent two Saharan dust multimodal polydispersions sampled over northern Italy (Tomasi et al., 1979), three pre- and post-Pinatubo volcanic particle polydispersions

suspended at various stratospheric altitudes (Pueschel et al., 1993; Tomasi et al., 1997), and two biomass burning smoke particle polydispersions sampled by Carr (2005) at Jabiru (Australia) in the free troposphere and atmospheric boundary layer, respectively.

11.2.1 The three 6S original aerosol models

The 6S aerosol models defined by Vermote et al. (1997a, 1997b) are based on the spectral properties and microphysical characteristics of the four basic components established by the International Radiation Commission (WMO, 1983), which consist of dust-like (DL), oceanic (OC), water-soluble (WS) and soot (SO) particulate matter components, the last of which includes both soluble and insoluble organic substances. For the four components, the spectral and geometric patterns of the following main radiative parameters were precisely described in the 6S code (Vermote et al., 1997b): (i) the phase function $P(\Theta)$, defined as a function of scattering angle Θ; (ii) the spectral values of volume extinction, scattering and absorption coefficients $\beta_{\mathrm{ext}}(\lambda)$, $\beta_{\mathrm{sca}}(\lambda)$ and $\beta_{\mathrm{abs}}(\lambda)$; (iii) the spectral values of asymmetry factor $g(\lambda)$; (iv) the particle number size distribution curve $N(r) = dN/d(Log\, r)$; and (v) the spectral values of complex refractive index $n(\lambda) - \mathrm{i}k(\lambda)$. The computations of these parameters were performed at 11 selected wavelengths of the solar spectrum from 0.30 to 3.75 μm, and for (i) the three scattering angles equal to 0°, 90° and 180°, and (ii) 80 supplementary Gaussian angles ranging between 0° and 180°. The number density and volume size-distribution curves of the four 6S basic components are shown in Fig. 11.1 for dry air conditions over the radius range from 10^{-4} to 10^{2} μm. The curves of particle number concentration $N(r)$ are represented as a function of particle radius r in terms of the unimodal log-normal size-distribution curve having the analytical form,

$$N(r) = dN(r)/d(Log\, r) = \frac{N_o}{\sqrt{2\pi}(\ln 10)(Log\, \sigma)} \exp\left[-\frac{1}{2}\left(\frac{Log\, r - Log\, r_c}{Log\, \sigma}\right)^2\right], \quad (11.1)$$

where N_o is the total particle number concentration (measured in cm^{-3}), ln 10 is a constant approximately equal to 2.3026, *Log* is the decadal logarithm (with base = 10), σ is the geometric standard deviation, and r_c is the mode radius (measured in μm). The shape-parameters of the unimodal size-distribution curves used in Eq. (11.1) to represent the DL, OC, WS and SO components for dry-air conditions are given in Table 11.1, together with the corresponding evaluations of the dry-state particulate mass density ρ (measured in g cm^{-3}) and the parameters defining their growth processes, the liquid water uptake due to the particle growth by condensation, and the single scattering albedo properties. Figure 11.1 shows that the WS and SO unimodal components present predominant number concentrations of fine particles, with mode radii equal to 0.005 and 0.0118 μm, respectively, while the OC and DL components consist mainly of coarse particles, with mode radii equal to 0.30 and 0.50 μm, respectively. Conversely, the WS and SO unimodal curves of particle volume concentration exhibit their maxima at radii ranging between 0.1 and 1 μm, whereas those of the OC and DL components describe wide maxima centred at radii of about 5 and 20 μm, respectively, as is typical of polydispersions with high contents of coarse and giant particles.

Table 11.1. Shape-parameters and particulate volume and mass concentration percentages of the four 6S basic components WS (water-soluble), OC (oceanic), DL (dust-like) and SO (soot) used for dry-air and wet-air (RH = 50%) conditions to determine the physico-chemical and radiative properties of the four 6S basic aerosol components used by Vermote et al. (1997a,b) to define the continental (6S-C), maritime (6S-M), and urban (6S-U) particle polydispersions, and the 14 modified (M-type) aerosol models defined in the present study

6S basic compo-nent	Dry particle mode radius r_c (μm)	Geometric standard deviation σ G_r	Average growth radius factor $\Delta V/V$	Percentage volume increase ratio V_1	Volume percentage of dry particles V_2	Volume percentage of liquid water ρ (g cm^{-3})	Dry particle mass density ρ_w (g cm^{-3})	Wet particle mass density Γ_1	Mass percentage of dry particles Γ_2	Mass percentage of liquid water albedo ω	Mean single scattering albedo ω^*	Weighted average single scattering
WS	0.0050	2.990	1.0180	0.0550	0.9479	0.0521	1.86	1.815	0.9713	0.0287	0.923	0.921
OC	0.3000	2.510	1.0694	0.2229	0.8177	0.1823	2.25	2.022	0.9099	0.0901	0.996	0.999
DL	0.5000	2.990	1.0080	0.0232	0.9773	0.0227	2.36	2.329	0.9902	0.0098	0.702	0.690
SO	0.0118	2.000	1.0283	0.0873	0.9197	0.0803	1.62	1.570	0.9489	0.0511	0.149	0.157

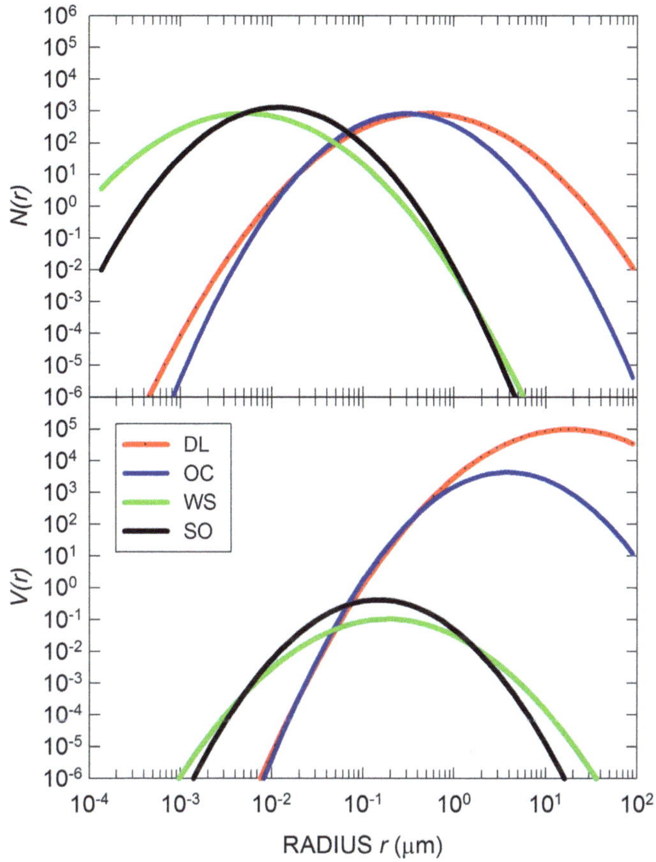

Fig. 11.1. Curves of particle number density size-distribution $N(r)$ (upper part) and volume size-distribution $V(r)$ (lower part) over the radius range from 10^{-4} to 10^{2} μm, for the four dry-air particle polydispersions defined by Vermote et al. (1997b) to represent the 6S basic aerosol components. The four components are all normalized to provide a value of the overall particle number concentration $N_{tot} = 1000\,cm^{-3}$.

The radiative parameters of the four 6S basic components were calculated over the 0.30–3.75 μm wavelength range, using the four monomodal particle size-distribution curves shown in Fig. 11.1. The spectral curves of the following parameters are presented in Fig. 11.2: (i) real part $n(\lambda)$ of dry particulate matter refractive index, (ii) imaginary part $k(\lambda)$ of dry particulate matter refractive index, (iii) volume extinction coefficient $\beta_{ext}(\lambda)$, and (iv) single scattering albedo $\omega(\lambda)$, as given by the ratio $\beta_{sca}(\lambda)/\beta_{ext}(\lambda)$. The comparison shows that the highest spectral values of $n(\lambda)$ in the visible are provided by the SO component, and the lowest ones by the OC component. The SO component also has the highest values of $k(\lambda)$ over the entire spectral range, while the lowest are given by the OC component in the visible and near-infrared range $\lambda < 1.7\,\mu$m, and by the DL component in the infrared range $\lambda > 1.7\,\mu$m. Table 11.2 reports the values of $n(\lambda)$ and $k(\lambda)$ at the 11 selected wavelengths from 0.30 to 3.75 μm, showing that the maritime particles

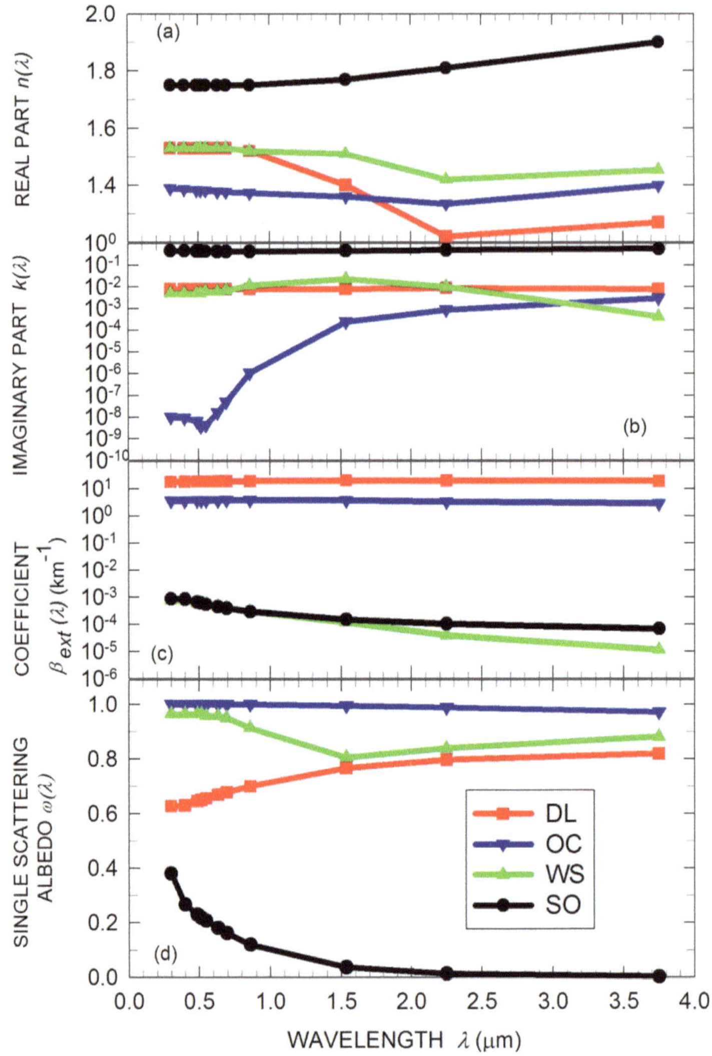

Fig. 11.2. Spectral curves of (a) real part $n(\lambda)$ of dry particulate matter refractive index, (b) imaginary part $k(\lambda)$ of dry particulate matter refractive index, (c) volume extinction coefficient $\beta_{\text{ext}}(\lambda)$, and (d) single scattering albedo $\omega(\lambda)$ over the 0.30–3.75 μm wavelength range of the four 6S basic aerosol components of Vermote et al. (1997b), all normalized to give a value of the overall particle number concentration $N_{\text{tot}} = 1000\,\text{cm}^{-3}$.

represented by the OC component provide very low values of $k(\lambda)$ at all visible wavelengths, clearly indicating that they absorb the visible solar radiation very weakly. The spectral patterns of $\beta_{\text{ext}}(\lambda)$ calculated at the 11 selected wavelengths appear to be nearly constant with wavelength for the DL and OC components and slowly decreasing for the WS and SO components throughout the whole visible and near-infrared range. Consequently, the OC component exhibits values of $\omega(\lambda)$ close to unity at all these wavelengths, whereas the WS component has values of $\omega(\lambda)$

Table 11.2. Values of the real part $n(\lambda)$ and imaginary part $k(\lambda)$ of the dry-air aerosol particles that constitute the four 6S basic aerosol components of Vermote et al. (1997a,b) at 11 selected wavelengths over the 0.30–3.75 μm spectral range

Wavelength	DL		WS		OC		SO	
λ (μm)	$n(\lambda)$	$k(\lambda)$	$n(\lambda)$	$k(\lambda)$	$n(\lambda)$	$k(\lambda)$	$n(\lambda)$	$k(\lambda)$
0.300	1.530	$8.00\ 10^{-3}$	1.530	$5.00\ 10^{-3}$	1.388	$1.00\ 10^{-8}$	1.750	0.470
0.400	1.530	$8.00\ 10^{-3}$	1.530	$5.00\ 10^{-3}$	1.385	$9.90\ 10^{-9}$	1.750	0.460
0.488	1.530	$8.00\ 10^{-3}$	1.530	$5.00\ 10^{-3}$	1.382	$6.41\ 10^{-9}$	1.750	0.450
0.515	1.530	$8.00\ 10^{-3}$	1.530	$5.00\ 10^{-3}$	1.381	$3.70\ 10^{-9}$	1.750	0.450
0.550	1.530	$8.00\ 10^{-3}$	1.530	$6.00\ 10^{-3}$	1.381	$4.26\ 10^{-9}$	1.750	0.440
0.633	1.530	$8.00\ 10^{-3}$	1.530	$6.00\ 10^{-3}$	1.377	$1.62\ 10^{-8}$	1.750	0.430
0.694	1.530	$8.00\ 10^{-3}$	1.530	$7.00\ 10^{-3}$	1.376	$5.04\ 10^{-8}$	1.750	0.430
0.860	1.520	$8.00\ 10^{-3}$	1.520	$1.20\ 10^{-2}$	1.372	$1.09\ 10^{-6}$	1.750	0.430
1.536	1.400	$8.00\ 10^{-3}$	1.510	$2.30\ 10^{-2}$	1.359	$2.43\ 10^{-4}$	1.770	0.460
2.250	1.220	$9.00\ 10^{-3}$	1.420	$1.00\ 10^{-2}$	1.334	$8.50\ 10^{-4}$	1.810	0.500
3.750	1.270	$1.10\ 10^{-2}$	1.452	$4.00\ 10^{-3}$	1.398	$2.90\ 10^{-3}$	1.900	0.570

gradually decreasing from about 0.96 to around 0.80 as the wavelength increases from 0.30 to 3.75 μm. Rather low values of $\omega(\lambda)$ characterize the DL component, which slowly increase with wavelength from nearly 0.6 in the visible to around 0.8 at the two longer infrared wavelengths. The SO component shows very low values of $\omega(\lambda)$ at all wavelengths, decreasing from about 0.3 in the visible to less than 0.05 at wavelengths $\lambda > 1.5$, as a result of its strong absorption properties.

Using different volume percentages of the four 6S basic components, the following three tropospheric aerosol models were defined by Vermote et al. (1997b):

(a) Continental (6S-C) aerosol (trimodal) model, consisting of volume percentages of the three 6S components equal to 70% for DL, 29% for WS and 1% for SO.
(b) Maritime (6S-M) aerosol (bimodal) model, consisting of volume percentages equal to 95% for OC and 5% for WS.
(c) Urban (6S-U) aerosol (trimodal) model, consisting of volume percentages equal to 61% for WS, 22% for SO and 17% for DL.

The shape parameters of the various modes giving form to the size-distribution curves of the three 6S original aerosol models are given in Table 11.3, with their multimodal values of number density concentration N_o normalized to yield an overall particle number concentration of 10^3 cm^{-3}. The mean dry-air particulate mass density ρ of the 6S-C, 6S-M and 6S-U models were calculated as linear combinations of the dry-air mass density values of the 6S basic components, weighted by their mass percentage contents given in Table 11.3. The number size-distributions $N(r)$ and volume size-distributions $V(r)$ of the three 6S dry-air aerosol models are shown in Fig. 11.3 over the $10^{-4}-10^2$ μm radius range, according to the Vermote et al. (1997b) data. The comparison provides evidence of the main differences existing in particle number concentration and volume contributions due to fine particles (with $r \leq 1$ μm) and coarse particles (with $r > 1$ μm), showing that the 6S-C, 6S-U and 6S-M models present comparable number concentrations in the radius range lower than 0.1 μm, and appreciably different values of $N(r)$ within the $0.1 \leq r \leq 1$ μm

Table 11.3. Values of the shape-parameters N_o, σ and r_c used to define the multimodal number size-distributions of the 6S dry-air aerosol models, particle mass density ρ ($\mathrm{g\,cm^{-3}}$), and main radiative parameters (volume extinction, scattering and absorption coefficients at the 0.55 μm wavelength; single scattering albedo ω at the 0.55 μm wavelength, mean single scattering albedo ω, weighted average single scattering albedo ω^*, and mean Ångström's exponent α calculated over the 0.40–0.86 μm wavelength range), as obtained for dry-air conditions according to the Vermote et al. (1997b) assumptions. The unimodal values of N_o are calculated to give a value of the overall particle number concentration $N_{tot} = 1000\,\mathrm{cm^{-3}}$ for all the five 6S dry-air aerosol models: 6S-C (continental, trimodal), 6S-M (maritime, bimodal), 6S-U (urban, trimodal), 6S-D (background desert dust, trimodal), and 6S-V (El Chichon volcanic stratospheric aerosol, trimodal)

6S dry-air aerosol models	First mode			Second mode			Third mode				β_{ext}	β_{sca}	β_{abs}	Monochromatic ω	Mean single scattering albedo	Weighted average	Mean Ångström's exponent
	N_1 ($\mathrm{cm^{-3}}$)	σ	r_c (μm)	N_2 ($\mathrm{cm^{-3}}$)	σ	r_c (μm)	N_3 ($\mathrm{cm^{-3}}$)	σ	r_c (μm)	ρ ($\mathrm{g\,cm^{-3}}$)	(0.55 μm) ($\mathrm{km^{-1}}$)	(0.55 μm) ($\mathrm{km^{-1}}$)	(0.55 μm) ($\mathrm{km^{-1}}$)	(0.55 μm)	ω	ω^*	α
6S-C	741.396	2.990	0.005	258.598	2.000	0.0118	0.006	2.990	0.50	2.208	$5.77\ 10^{-4}$	$5.15\ 10^{-4}$	$6.26\ 10^{-5}$	0.892	0.850	0.883	1.249
6S-M	997.201	2.990	0.005	2.799	2.500	0.300	–	–	–	2.231	$2.06\ 10^{-3}$	$2.04\ 10^{-3}$	$2.20\ 10^{-5}$	0.989	0.986	0.989	0.229
6S-U	967.683	2.990	0.005	32.249	2.000	0.0118	0.068	2.990	0.50	1.892	$5.46\ 10^{-4}$	$3.53\ 10^{-4}$	$1.93\ 10^{-4}$	0.646	0.553	0.632	1.459
6S-D	542.142	2.104	0.001	457.857	3.120	0.0218	0.001	1.860	6.24	2.500 (*)	2.36 100	$2.19\ 10^{0}$	$0.17\ 10^{0}$	0.931	0.938	0.932	0.307
6S-V	592.000	2.500	0.110	400.000	1.500	0.270	8.000	1.100	1.00	1.650 (**)	$6.35\ 10^{-1}$	$6.35\ 10^{-1}$	0.000	1.000	1.000	1.000	0.214

(*) Volz (1973); (**) King et al. (1984).

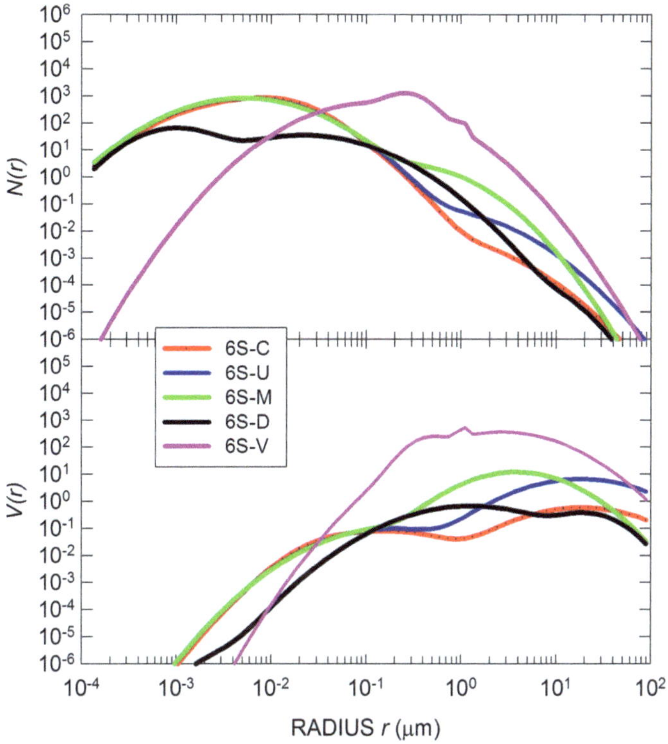

Fig. 11.3. As in Fig. 11.1, for the three original aerosol models of continental (6S-C), maritime (6S-M) and urban (6S-U) origins defined by Vermote et al. (1997a, 1997b) for dry air conditions, and the two supplementary models 6S-D (background desert dust), according to Shettle (1984), and 6S-V (El Chichon volcanic stratospheric aerosol), according to King et al. (1984). The five multimodal size-distribution curves are all normalized to give a value of the overall particle number concentration $N_{tot} = 1000\,\text{cm}^{-3}$.

range, with higher number concentrations for the 6S-M model than those of the 6S-C and 6S-U models. Fig. 11.3 also shows that the 6S-M model provides values of $N(r)$ higher by more than one order of magnitude than those of the 6S-U model, and by nearly two orders of magnitude than those of the 6S-C model in the range typical of coarse particles. The volume size-distribution curves of the 6S-C, 6S-U and 6S-M models exhibit slightly different values over the radius range $r < 0.1\,\mu$m, and clearly higher values of the 6S-M model than those of the 6S-U and 6S-C models over both the fine particle radius range from 0.1 to 1 μm and the coarse particle range from about 1 μm to more than 10 μm.

Figure 11.4 shows the spectral curves of dry particulate matter refractive index parts $n(\lambda)$ and $k(\lambda)$ relative to the 6S-C, 6S-M, and 6S-U aerosol models, both calculated at each wavelength and for each model as a linear combination of the values of the 6S basic component concentration parameters defined by Vermote et al. (1997b). It turns out that the values of $n(\lambda)$ vary in the visible between less than 1.40 (6S-M model) and more than 1.55 (6S-U model), and then all decrease at longer wavelengths, presenting the most marked variations for the 6S-C model,

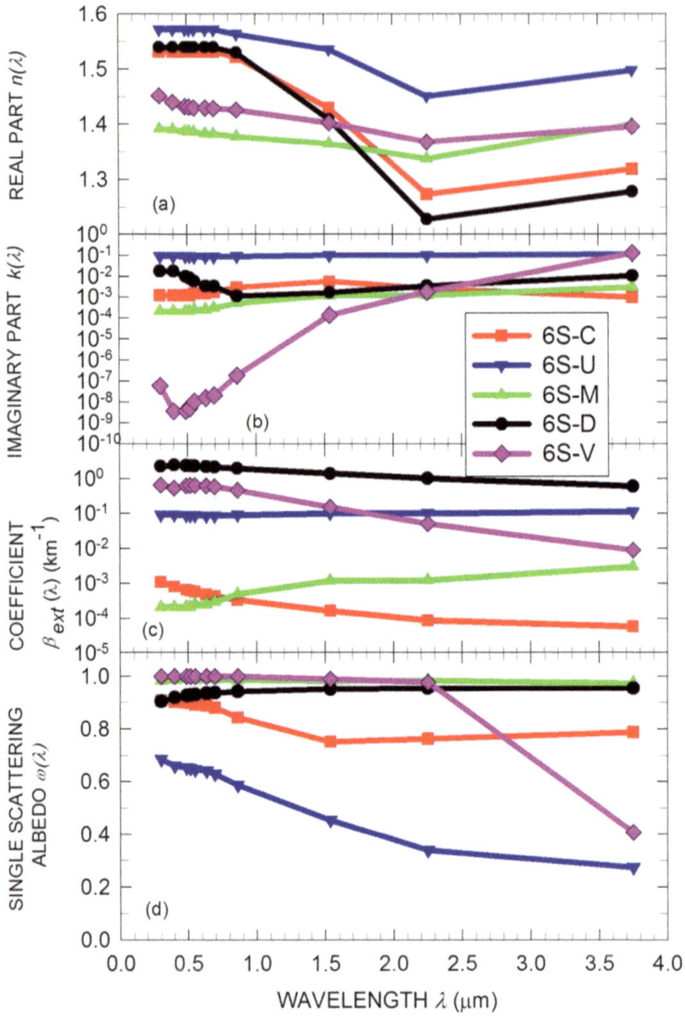

Fig. 11.4. As in Fig. 11.2, for the five 6S aerosol models shown in Fig. 11.3, all normalized to give a value of the overall particle number concentration $N_{tot} = 1000\,cm^{-3}$.

until becoming lower than 1.32 at wavelengths $\lambda > 2\,\mu m$. A large range of $k(\lambda)$ is shown by the 6S original models, with values varying between 10^{-4} (6S-M model) and 10^{-1} (6S-U model) at visible and infrared wavelengths. The spectral curves of coefficient $\beta_{ext}(\lambda)$ and single scattering albedo $\omega(\lambda)$ are also displayed in Fig. 11.4, offering evidence of their more significant variations over the 0.30–3.75 μm wavelength range. It can be seen that $\beta_{ext}(\lambda)$ varies rather slowly as a function of λ in the 6S-M model and more rapidly in the 6S-C and 6S-U models. By contrast, parameter $\omega(\lambda)$ in the 6S-M model assumes values very close to unity at all wavelengths, that of the 6S-C model presents values gradually decreasing from about 0.90 in the visible to less than 0.80 in the infrared, and that of the 6S-U model exhibits values considerably decreasing from around 0.70 in the visible to less than 0.40 at wavelengths $\lambda > 2\,\mu m$. Table 11.3 gives the values of the following radiative parameters of the three original dry-air 6S models defined by Vermote et al. (1997b): (i) the volume extinction, scattering and absorption coefficients calculated at the 0.55 μm wavelength, (ii) the monochromatic single scattering albedo $\omega(0.55\,\mu\text{m})$ given by ratio $\beta_{sca}(0.55\,\mu\text{m})/\beta_{ext}(0.55\,\mu\text{m})$; (iii) the arithmetic mean single scattering albedo ω, calculated over the 0.30 – 3.75 μm spectral range; (iv) the weighted average single scattering albedo ω^*, calculated using as weight-function the spectral curve of direct solar irradiance reaching the surface after its passage through the U. S. Standard Atmosphere (Anderson et al., 1986) for $m = 2$; and (v) the best-fit value of Ångström's exponent α calculated for the spectral series of $\beta_{ext}(\lambda)$ determined at 7 selected wavelengths over the 0.40 – 0.86 μm range (i.e. at wavelengths $\lambda = 0.400$, 0.488, 0.515, 0.550, 0.633, 0.694 and 0.860 μm).

11.2.2 The 6S supplementary aerosol models

Two supplementary aerosol models were proposed by Vermote et al. (1997a) in the 6S code, providing a pair of aerosol extinction models that are difficult to reproduce using a mix of the four 6S basic components. They were defined to represent (i) a desert dust particle polydispersion (6S-D model) based on the Shettle (1984) measurements, and (ii) a volcanic stratospheric particle polydispersion (6S-V model) based on the measurements analyzed by King et al. (1984). Both models were here defined with improved accuracy as follows:

(a) The *background desert dust* aerosol model (hereinafter referred to as 6S-D model) represents a polydispersion of particles, which remain suspended in the atmosphere for days or weeks after their mobilization in arid regions, and can be transported over very long distances by intercontinental winds. Its particle number size-distribution was assumed to consist of three log-normal curves having the form of Eq. (11.1). The values of geometric standard deviation σ and mode radius r_c characterizing each log-normal curve of the 6S-D model are given in Table 11.3, together with the corresponding unimodal values of particle number concentration N_i yielding an overall multimodal particle number density $N_{tot} = 10^3\,\text{cm}^{-3}$. The dry-air particulate mass density ρ was assumed to be equal to 2.50 g cm^{-3}, according to the Volz (1973) evaluations made for Saharan dust transported over the Caribbean region. The size-distribution curves of $N(r)$ and $V(r)$ for the dry-air 6S-D aerosol model are shown in Fig. 11.3 over the 10^{-4}–10^{2} μm radius range, to permit their comparison with the 6S-C, 6S-M and 6S-U curves. The values of $N(r)$ determined for the 6S-D model are seen to be considerably lower than those of the 6S-C and 6S-M models throughout the entire fine particle radius range $r < 0.1\,\mu$m, and to be comparable with those of the 6S-U model within the given size range, while they differ by less than one order of magnitude from those of the three 6S original models over the 0.1–1 μm radius range. For radii varying between 1 μm and more than 20 μm, the 6S-D values of $N(r)$ were in general found to be lower than those of the 6S-M model and comparable with those of the 6S-C model. Correspondingly, the 6S-D size-distribution curve of $V(r)$ yields values that are (i) considerably lower than those of the three 6S original models over the radius range $r < 0.1\,\mu$m, (ii) comparable with them over the 0.1–1 μm radius range, and (iii) lower than those of the 6S-M model and comparable with those of the 6S-C model over the higher radius range.

The complex refractive index of the 6S-D model was determined according to the Shettle (1984) assumptions: (i) the spectral values of $n(\lambda)$ were calculated according to the evaluations made by Carlson and Benjamin (1980) at wavelengths $\lambda < 0.6\,\mu$m, and to those determined by Volz (1973) at longer wavelengths, by integrating these data with those of the 6S dust-like (DL) basic component; and (ii) the spectral values of $k(\lambda)$ were correspondingly determined using the estimates provided by Volz (1973), Carlson and Caverly (1977), Patterson et al. (1977), Patterson (1977, 1981) and Carlson and Benjamin (1980). The spectral curves of $n(\lambda)$ and $k(\lambda)$ obtained for this model are shown in Fig. 11.4 for comparison with those of the three 6S original aerosol models: the values of $n(\lambda)$ are very similar to those of the 6S-C model in the visible and then gradually decrease at infrared wavelengths until becoming appreciably lower than those of the 6S-C model at around 3.00 μm wavelength. Correspondingly, the values of $k(\lambda)$ decrease rapidly with wavelength in the visible and are comparable with those of the 6S-C and 6S-M models for $\lambda > 0.8\,\mu$m. Using the size-distribution shape-parameters given in Table 11.3 and the refractive index data assumed above, calculations of the spectral values of $\beta_{ext}(\lambda)$ and $\omega(\lambda)$ were made for the 6S-D model, obtaining the values shown in Fig. 11.4 over the 0.30–3.75 μm wavelength range. The 6S-D values of $\beta_{ext}(\lambda)$ were found to be higher by nearly three orders of magnitude than those of the three 6S original models, since this multimodal model (assumed to have a total number concentration $N_{tot} = 10^3\,\text{cm}^{-3}$) exhibits considerably more marked relative concentrations of both fine particles with greater sizes and coarse particles than those of the 6S classical models. The values of $\omega(\lambda)$ obtained for the 6S-D model are slightly higher than those of the 6S-C model at visible wavelengths, and gradually increasing with wavelength to become comparable with those of the 6S-M model at the 3.75 μm wavelength. The 6S-D monochromatic values of the volume scattering, absorption and extinction coefficients at the 0.55 μm wavelength are given in Table 11.3, together with those calculated for the three 6S original models.

(b) The *El Chichón stratospheric volcanic* aerosol model (hereinafter referred to as 6S-V model) was defined on the basis of the King et al. (1984) measurements made at the Mauna Loa Observatory to analyze the optical and microphysical parameters of the columnar content of the El Chichon volcanic particles. The size-distribution of the aerosol loading was assumed by King et al. (1984) to consist of three modes: (i) the first was represented in terms of the modified gamma function having the analytical form,

$$dN(r)/d\,(Log\,r) = C_1 r^2 \exp\left(-1.98 r/r_c\right) , \tag{11.2}$$

with columnar particle number $C_1 = 1.674 \times 10^{11}\,\text{cm}^{-2}\,\mu\text{m}^{-2}$, radius r measured in μm, and mode radius $r_c = 0.11\,\mu$m, as defined by McClatchey et al. (1980) for representing the background unperturbed (i.e. without volcanic particles) conditions of the Standard Radiation Atmosphere; and (ii) the second and third were assumed to have analytical forms similar to the log-normal curve given in Eq. (11.1), with columnar particle number constants $C_2 = 50\,C_3$, and $C_3 = 3.869 \times 10^5\,\text{cm}^{-2}\,\mu\text{m}^{-2}$, respectively, and values of σ equal to 1.5 and 1.1, respectively, as derived by King et al. (1984) examining the balloon-borne data recorded by Hofmann and Rosen (1983a,b) for particle samples taken at altitudes of 21.5 to 24.5 km. above the Mauna Loa Observatory, a few months after the El Chichon eruption. However, for reasons of uniformity, we decided here to substitute the first monomodal modified gamma curve with a log-normal curve (as in the second and third modes), which was defined with shape-parameters obtained by applying a best-fit procedure to the original concentration data of McClatchey et al. (1980). The values of the shape-parameters for the three log-normal modes are given in Table 11.3, for the value of the overall multimodal number density concentration $N_{tot} = 10^3\,cm^{-3}$. The corresponding size-distribution curves of $N(r)$ and $V(r)$ calculated for dry-air conditions of the 6S-V trimodal model are presented in Fig. 11.3, where their comparison with the size-distributions of the other 6S models indicates that the particle number concentration $N(r)$ of 6S-V particles is very low at radii smaller than $10^{-2}\,\mu m$ with respect to the three 6S original aerosol models. It becomes comparable with them in the radius range from 10^{-2} to $10^{-1}\,\mu m$, subsequently increasing to predominate markedly on the particle number concentration of the 6S original models over the range from about 0.1 to nearly $70\,\mu m$. The size-distribution of $V(r)$ obtained for the 6S-V model describes a large multimodal maximum, with considerably higher values than those of the other 6S models over the $0.05 \leq r \leq 60\,\mu m$ radius range. This multimodal maximum is substantially given in Fig. 11.3 by the linear combination of the first two modes centered at radii of 0.11 and $0.27\,\mu m$, respectively. The slight peak appearing in both the number and volume size-distribution curves shown in Fig. 11.3 is clearly due to the third mode of volcanic particles centred at radius $r_c = 1\,\mu m$ (see Table 11.3).

The 'long-lived' sulfate aerosols forming in the stratosphere after the El Chichon eruption of spring 1982 were primarily generated through chemical transformation and condensation of SO_2 injected at stratospheric altitudes by the violent volcanic eruption (Turco et al., 1982). These particles were nearly spherical liquid water droplets having a 75% concentration by weight of sulphuric acid, as indicated by the balloon-borne boiling point measurements performed by Hoffman and Rosen (1983) and the residual percentage of liquid water. Therefore, the spectral values of $n(\lambda)$ were directly determined for this aerosol model using the values proposed by Palmer and Williams (1975) for a 75% (by weight) aqueous solution of H_2SO_4, while those of $k(\lambda)$ were assumed according to Palmer and Williams (1975) at wavelengths $\lambda > 0.70\,\mu m$ and taking into account (i) the estimates of Hummel et al. (1988) at wavelenths from 0.35 to $0.70\,\mu m$, and (ii) those of Burley and Johnston (1992) within the $0.25 \leq \lambda \leq 0.34\,\mu m$ wavelength range. Their spectral features are shown in Fig. 11.4 over the 0.30–$3.75\,\mu m$ spectral interval, together with the corresponding spectral curves of radiative parameters $\beta_{ext}(\lambda)$ and $\omega(\lambda)$. As can be seen, the values of $n(\lambda)$ are more similar to those of the 6S-M model, while those of $k(\lambda)$ are smaller than those of the other 6S models by several orders of magnitude at wavelengths $\lambda < 1.5\,\mu m$, due to the negligible absorption properties of these stratospheric particles. Very high values of $\beta_{ext}(\lambda)$ were obtained for the 6S-V model with respect to the three 6S original models, due to the considerably higher content of coarse particles. By contrast, the spectral curve of $\omega(\lambda)$ provides nearly unit values (as in the case of the 6S-M model) over the wavelength range from 0.30 to about $2.5\,\mu m$. The monochromatic values of the main radiative parameters obtained for the 6S-V model are given in Table 11.3.

11.2.3 The 6S modified (M-type) aerosol models

The five 6S models described in the previous two subsections were calculated for dry-air conditions of particulate matter. However, the relative humidity (RH) conditions of air usually observed at tropospheric altitudes for cloudless and clean-air conditions differ appreciably from those observed for dry-air conditions, presenting values of RH often no lower than 50%. Taking into account that moist air conditions with RH $\approx$ 50% favor a very limited growth of particle sizes by condensation, estimated to be equal to a few percentage points only by Hänel (1976), it seems plausible to assume that such wet aerosol models yield slightly higher values of $\beta_{ext}(\lambda)$ and $\omega(\lambda)$ than those shown in Fig. 11.4 for dry-air conditions, due to the moderate uptake of liquid water mass by each particle. To obtain more realistic simulations of the wet aerosol radiative properties, it was decided to calculate also the spectral values of volume extinction, scattering and absorption coefficients of the four 6S basic components for RH = 50%. The aim was to derive a set of wet aerosol extinction models, where the changes in the size-distribution curves and complex refractive index due to the particle growth by condensation were properly taken into account. For this purpose, we assumed that the aerosol particles were grown for RH increasing from 0% to 50%, as indicated by the average particle growth simulation models defined by Hänel (1976) and Hänel and Bullrich (1978) for aerosol polydispersions of different origins (sea-spray particle samples from the North Atlantic, maritime aerosol samples from the Atlantic and containing Saharan dust, urban aerosol samples from the industrialized area of Mainz (Germany), and background continental aerosol samples from the top of the Hohenpeissenberg mountain in summer). The estimates of the growth radius factor G_r defined by Hänel (1976) and Hänel and Bullrich (1978) were taken into consideration for both increasing and decreasing RH = 50%. They were used to represent with a good approximation the hygroscopic properties of the WS, OC, DL and SO components in the 6S code, together with the values of growth radius factor G_r proposed by Shettle and Fenn (1979) for rural, maritime, urban and free-tropospheric aerosol polydispersions.

Using the above-mentioned growth factor estimates, the average values of G_r given in Table 11.1 were determined for the WS, OC, DL and SO components, relative to RH = 50%. Correspondingly, the particle size-distribution curves of the four 6S basic components were modified with respect to those shown in Fig. 11.1 for dry-air conditions, by (i) assuming new values of the mode radius r_c, calculated by multiplying the mode radii of the dry-air 6S basic polydispersions by the corresponding factor G_r given in Table 11.1, and (ii) assuming the same values of geometric standard deviation σ adopted by Vermote et al. (1997a, 1997b) and given in Table 11.1, for which the log-normal size-distribution curves of wet particles (for RH = 50%) were defined. Calculating the overall particle volumes V of the original 6S log-normal size-distribution curves for dry-air particles and the total grown particle volumes $V_g = V + \Delta V$ of the new log-normal size-distribution curves of grown particles (as given in Table 11.1), the percentage volume ratios $\Delta V/V$ of such growth processes were then determined, as given in Table 11.1, separately for the WS, OC, DL and SO basic components. For these calculations, the volume percentages V_1 and V_2 of dry particulate matter and liquid water fractions were subsequently determined and given in Table 11.1.

According to the above evaluations, appreciable changes in the particle mass density are expected to occur. Their calculations were made for (i) the values of dry particulate matter density ρ obtained by Hänel (1976) and Hänel and Bullrich (1978) for the sea-spray, maritime, continental, and urban particle models, and (ii) the values of ρ proposed by Volz (1972a, 1972b, 1973) for water soluble substances, soot particles, Saharan dust, and coal-fire dust. The values of dry particle density ρ given in Table 11.1 were first determined for the four 6S basic components. Multiplying these values of ρ by the corresponding volume percentages V_1 and V_2 given in Table 11.1, the mass percentages Γ_1 of dry particles and Γ_2 of liquid water droplets were then calculated, from which the values of wet particle mass density ρ_w were obtained for liquid water mass density equal to g cm^{-3}. The results are given in Table 11.1.

For the values of Γ_1 and Γ_2 used as weights, the spectral values of the real part $n(\lambda)$ and imaginary part $k(\lambda)$ of wet particulate matter refractive index were subsequently determined for the four 6S basic components. The weighted values were obtained by assuming the dry-air particulate values of $n(\lambda)$ and $k(\lambda)$ provided by Vermote et al. (1997a) and those of liquid water refractive index calculated according to Irvine and Pollack (1968) and Hale and Querry (1973). The spectral patterns of $n(\lambda)$ and $k(\lambda)$ obtained for the four 6S basic components are shown in Fig. 11.5 for both dry-air (RH $= 0\%$) and wet-air (RH $= 50\%$) conditions. Fig. 11.5 provides evidence that only negligible variations take place in the optical properties of particulate matter for such variations in RH, except for those of $n(\lambda)$ relative to the soot (SO) component and those of $k(\lambda)$ at the 3.75 μm wavelength for the WS and DL components. For the values of $n(\lambda)$ and $k(\lambda)$ shown in Fig. 11.5 and for the size-distribution curves of the four 6S basic components obtained for wet particles (i.e. for RH $= 50\%$), we calculated the spectral values of the single scattering albedo $\omega(\lambda)$ at the 11 wavelengths selected above between 0.30 and 3.75 μm. It was found that: (i) the values of $\omega(\lambda)$ determined for the OC component are all very close to unity, giving a mean value higher than 0.99 at visible and near-infrared wavelengths; (ii) values of the WS component are relatively high at all wavelengths $\lambda \leq 1\,\mu$m (giving a mean value of ~ 0.92), and appreciably lower at wavelengths $\lambda > 1\,\mu$m; (iii) values of the DL component are relatively low at visible wavelengths and gradually increase with λ to assume a mean value of ~ 0.70 at the near-infrared wavelengths; and (iv) values of the SO component are rather low in the visible, and decrease with λ until becoming very small at mid-infrared wavelengths, where a mean value of ~ 0.15 was assumed. Using these evaluations, the mean values of $\omega(\lambda)$ were calculated for wet particles over the 0.30–3.75 μm wavelength range. The results are presented in Table 11.1, showing that the OC and SO wet components yield the extreme values of this optical parameter, which are nearly unit for the OC wet particle component and smaller than 0.15 for the SO one. The above estimates were employed to calculate a set of 14 aerosol models as linear combinations of the four 6S wet particle components and the liquid water (LW) fraction. The average values of ω were calculated in Table 11.1 as simple arithmetic averages made over the 0.30–3.75 μm wavelength range. They were found to decrease gradually from $\sim$1.00 in the first M-type model (pure oceanic aerosol) to about 0.65 in the last M-type (M-14) model (heavy polluted aerosol). Such a low value of ω was obtained for a mixed polydispersion of particles, mainly consisting of sulfate substances, and

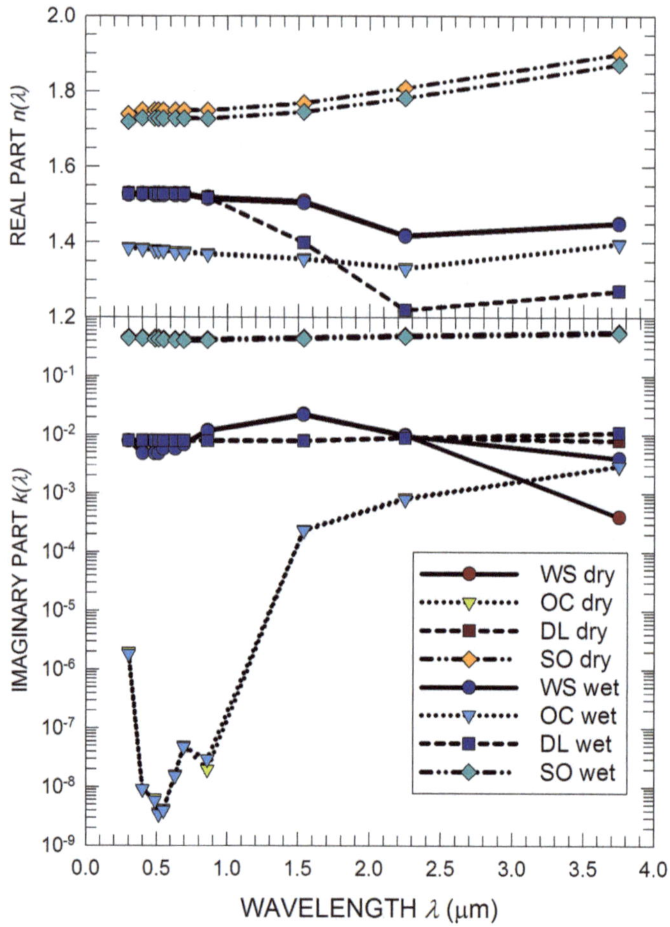

Fig. 11.5. Comparison between the spectral curves of real part $n(\lambda)$ (upper part) and imaginary part $k(\lambda)$ (lower part) obtained for the four 6S basic components of Vermote et al. (1997b) over the 0.30–3.75 μm wavelength range, as calculated for dry-air (RH = 0%) and wet-air (RH = 50%) conditions.

containing moderate concentrations of dust-like (DL) and soot (SO) particulate matter.

Using the single scattering albedo properties of the above aerosol models, it is important to take into account that about 87.8% of the extraterrestrial solar irradiance $I_o(\lambda)$ belongs to the 0.30–1.60 μm wavelength range, presenting the maximum at $\sim$0.48 μm wavelength (Iqbal, 1983). In order to evaluate realistically the single scattering albedo effects induced by aerosol polydispersions on solar radiation, it is therefore of basic importance to bear in mind that the aerosol scattering and absorption effects occurring at visible and near-infrared wavelengths are particularly significant in evaluating the DARF effects. Thus, considering that aerosols are mainly concentrated within the lower part of the troposphere, it was decided to use the weighted average single scattering albedo ω^* as key parameter in the calculations of the DARF effects. It was calculated by weighting the spectral values of $\omega(\lambda)$ in Fig. 11.5 by the model-based function $I^*(\lambda)$, which defines the spectral curve of direct solar irradiance $I(\lambda)$ measured at sea level for relative optical air mass $m = 2$, i.e. for $\theta_o \approx 60°$ (Tomasi et al., 1998), after its passage through the U. S. Standard Atmosphere 1976 (Anderson et al., 1986). The choice of $\theta_o = 60°$ was made bearing in mind that the Sun never sinks below the solar elevation angle of 30° during the middle part of the day throughout the entire year at mid-latitude observation sites. In order to determine a realistic spectral curve of weight function $I^*(\lambda)$, it was assumed that atmospheric particulate extinction of direct solar irradiance was due to the rural aerosol model of Kneizys et al. (1996), normalized to the 23 km visual range at surface-level, thus representing background particulate extinction features of columnar aerosol in a relatively clean-air atmosphere. The values of ω^* obtained following the above procedure are given in Table 11.1, where they vary between ~1.00 (OC component) and 0.152 (SO component). In practice, they can be used as mean spectral evaluations of the broadband single scattering albedo of columnar aerosol over the visible and near-infrared wavelength range.

The results presented in Fig. 11.5 and in Table 11.1 indicate that linear combinations of the four 6S wet-particle basic components and of the LW component accounting for the liquid water uptake can be suitably used to represent the radiative properties of wet aerosol polydispersions of different origin for RH = 50%. These aerosol models present single scattering albedo characteristics that fully cover the variety of radiative properties usually observed in the Earth's atmosphere. For the volume percentages and the mass density values given in Tables 11.1 and 11.4 for the WS, OC, DL, SO and LW components, the four following 6S modified (M-type) aerosol models were determined:

(1) Model M-1, which represents a 'pure oceanic' aerosol polydispersion consisting of the OC component for 81.77% and the LW component for 18.23%, thus yielding a value of $\omega^* = 0.999$. The size-distribution curve of this wet-air aerosol polydispersion is unimodal, with mode radius $r_c \approx 0.32\,\mu$m and, hence, a very high contribution (~ 5%) of coarse particles to the total particle number density, leading to the predominance of their extinction effects over those of fine particles.

(2) Model M-2, which represents a traditional 'maritime' aerosol polydispersion very similar to that of the 6S Maritime particle model. It consists of a volume percentages of 77.61% for the OC component, 4.08% for the WS component (with $r_c \approx 5.1 \times 10^{-3}\,\mu$m) and 18.31% for the LW component, providing values of complex refractive index and volume scattering and absorption coefficients at the 11 above-selected wavelengths giving a value of $\omega^* = 0.987$.

(3) Model M-8, which defines a traditional 'continental' aerosol polydispersion consisting of volume percentages equal to 67.29% for DL (with $r_c \approx 5.04 \times 10^{-1}\,\mu$m), 27.88% for OC, 0.96% for SO (with $r_c \approx 1.9 \times 10^{-2}\,\mu$m) and 3.87% for LW, yielding values of refractive index and volume scattering and absorption coefficients for which $\omega^* = 0.852$ was obtained.

(4) Model M-14, which represents a 'heavy polluted' aerosol polydispersion consisting of 55.60% WS, 18.00% SO, 11.60% DL, and 14.80% LW, yielding a value of $\omega^* = 0.651$. The SO concentration agrees very well with that chosen by Vermote et al. (1997a) for their 6S-U model and is very close to that obtained by

Table 11.4. Volume percentages of the 6S basic wet aerosol components WS (water-soluble), OC (oceanic), DL (dust-like), SO (soot) and of LW (liquid water) used in the present study to determine the radiative parameters of the 14 M-type modified wet aerosol models having different composition (labeled with letter M and increasing numbers from 1 to 14), each normalized to give a total particle number concentration $N_{\text{tot}} = 1000\,\text{cm}^{-3}$. The corresponding values of the mean Ångström's exponent $\alpha(0.40\text{–}0.86\,\mu\text{m})$ (+), mean single scattering albedo ω, weighted average single scattering albedo ω^*, and wet particulate mass density ρ_w measured in $\text{g}\,\text{cm}^{-3}$ are given in the last lines

6S wet	Volume percentages for the 14 M-type modified wet-aerosol models for RH = 50%													
aerosol components	M1 pure oceanic	M2 maritime	M3 mixed maritime-continental	M4 mixed maritime-continental	M5 mixed maritime-continental	M6 mixed maritime-continental	M7 mixed maritime-continental	M8 pure continental	M9 continental-polluted	M10 continental – polluted	M11 continental – polluted	M12 anthropogenic – continental	M13 anthropogenic – continental	M14 heavy polluted
WS	–	0.0408	0.0940	0.1421	0.2128	0.2198	0.2535	0.2788	0.4056	0.4546	0.5075	0.5451	0.5550	0.5560
OC	0.8177	0.7761	0.6151	0.4351	0.2314	0.1863	0.0946	–	–	–	–	–	–	–
DL	–	–	0.1452	0.3108	0.4813	0.5215	0.5912	0.6729	0.5094	0.4267	0.3262	0.2481	0.1908	0.1160
SO	–	–	–	–	–	0.0037	0.0066	0.0096	0.0283	0.0464	0.0725	0.0927	0.1214	0.1800
LW	0.1823	0.1831	0.1457	0.1120	0.0745	0.0687	0.0541	0.0387	0.0567	0.0723	0.0938	0.1141	0.1328	0.1480
Mean Ångström's exponent α (*)	−0.092	0.243	0.544	0.756	0.967	1.008	1.086	1.149	1.215	1.239	1.262	1.277	1.289	1.301
Mean ω	0.995	0.986	0.966	0.946	0.925	0.902	0.878	0.852	0.810	0.771	0.725	0.687	0.652	0.615
Weighted average ω^*	0.999	0.987	0.964	0.943	0.921	0.898	0.877	0.855	0.820	0.787	0.747	0.714	0.684	0.651
Wet particulate mass density ρ_w ($\text{g}\,\text{cm}^{-3}$)	2.022	2.005	2.023	2.089	2.127	2.133	2.144	2.161	2.059	2.000	1.925	1.864	1.812	1.748

(+)Ångström's exponent α was calculated for the values of volume extinction coefficient $\beta_{\text{ext}}(\lambda)$ determined at the 0.400, 0.488, 0.515, 0.550, 0.633, 0.694 and 0.860 μm wavelengths.

Bush and Valero (2002) for aerosol samples collected in the equatorial region of the Indian Ocean during the INDOEX experiment on days characterized by heavy anthropogenic pollution.

The volume percentages of the four 6S basic and LW components were modified step by step to determine the intermediate values of the volume percentages of the five components, in such a way as to determine a further 10 aerosol extinction models, which have been labeled using acronyms from M-3 to M-7 and from M-9 to M-13. The respective volume percentages are given in Table 11.4, while the corresponding values of mass density are reported in Table 11.1, from which the mass percentages of the above-mentioned five components were calculated to define the corresponding spectral series of $n(\lambda)$ and $k(\lambda)$. For these optical parameters and the size-distribution curves given in Fig. 11.6, the values of ω^* were then determined for all 14 M-type wet aerosol models, as given in Table 11.4, showing that the values of ω^* are (i) near unity in models M-1 and M-2 relative to maritime aerosols, where water-soluble particles occupy a relative volume fraction of a few percentage points only; (ii) regularly decreasing from about 0.96 to less than 0.88 in models M-3 to M-7, consisting of increasing volume fractions of both WS and DL components, together with decreasing volume fractions of the OC component and null or slightly increasing percentages of the SO component; (iii) close to 0.85 in the M-8 model, in agreement with the characteristics of the 6S components relative to the WS, DL and SO substances; and (iv) gradually decreasing from 0.82 to about 0.65 in the six remaining models from M-9 to M-14, consisting of increasing volume fractions of both WS and SO components, together with gradually decreasing volume fractions of the DL components and null percentages of the OC component.

Table 11.4 also presents the values of wet particulate mass density ρ_p relative to the 14 M-type aerosol models, which are (i) close to 2 g cm^{-3} for models M-1 and M-2, (ii) slowly increasing from 2.05 to 2.17 g cm^{-3} for models from M-3 to M-8, and (iii) decreasing from 2.09 to 1.84 g cm^{-3}, for models from M-9 to M-14, these variations being mainly due to the gradual increase in the soot particulate mass fraction. Table 11.4 also reports the mean values of the Ångström (1964) exponent $\alpha(0.40–0.86\,\mu\text{m})$, each obtained from the negative slope coefficient of the best-fit line drawn for each spectral series of the natural logarithms of $\beta_{\text{ext}}(\lambda)$ (calculated at the 0.400, 0.488, 0.515, 0.550, 0.633, 0.694 and 0.860 μm wavelengths for each of the 14 M-type wet-air aerosol models) plotted versus the natural logarithm of wavelength. Exponent $\alpha(0.40–0.86\,\mu\text{m})$ assumes the lowest value for model M-1 (pure oceanic particles), and gradually increase for the subsequent M-type models, because (i) the coarse particle mass fraction decreases as the OC component mass fraction diminishes passing from M-1 to M-7 model, and (ii) the fine particle mass fraction increases for the gradually higher contents of the WS and SO components. It can be clearly seen in Fig. 11.6 that the number and volume size-distribution curves of the 14 M-type aerosol models determined for wet air (RH = 50%) conditions, present variable multimodal features, with more pronounced variations within the coarse particle radius range of the aeolian (DL) and anthropogenic (SO) particle components.

The corresponding spectral curves of refractive index parts $n(\lambda)$ and $k(\lambda)$, volume extinction coefficient $\beta_{\text{ext}}(\lambda)$, and single scattering albedo $\omega(\lambda)$ are shown in

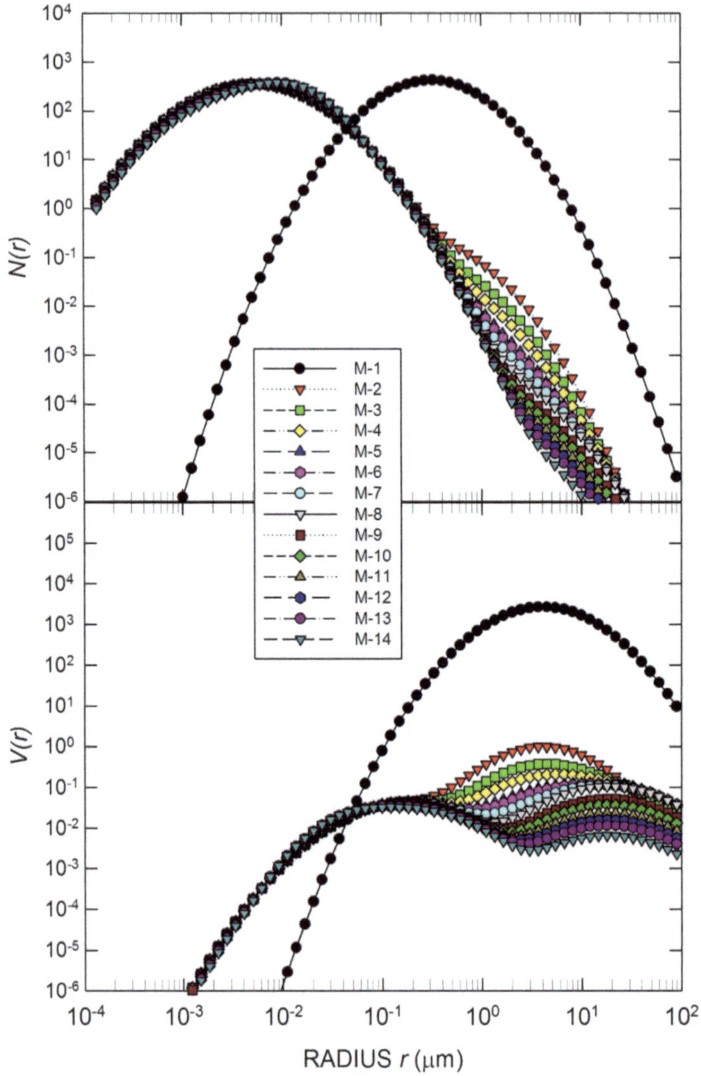

Fig. 11.6. As in Fig. 11.1, for the 14 modified (M-type) aerosol models defined in the present study as linear combinations of the 6S basic aerosol components. Models are labeled with letter M and increasing numbers from 1 to 14, all normalized to give a value of the overall particle number concentration $N_{\mathrm{tot}} = 1000\,\mathrm{cm}^{-3}$.

Fig. 11.7. It presents an exhaustive picture of the gradual and well-spaced variations of the four radiative parameters occurring from one model to another at the visible and infrared wavelengths. In particular, the spectral values of $\beta_{\mathrm{ext}}(\lambda)$ obtained for the M-1 model are greater by about three orders of magnitude than those determined for the other 13 M-type models at all wavelengths, due to the relatively high number concentration of coarse particles ($\sim 50\,\mathrm{cm}^{-3}$, i.e about 5%) compared to that of fine particles ($\sim 950\,\mathrm{cm}^{-3}$). At the same time, the size-distribution curves

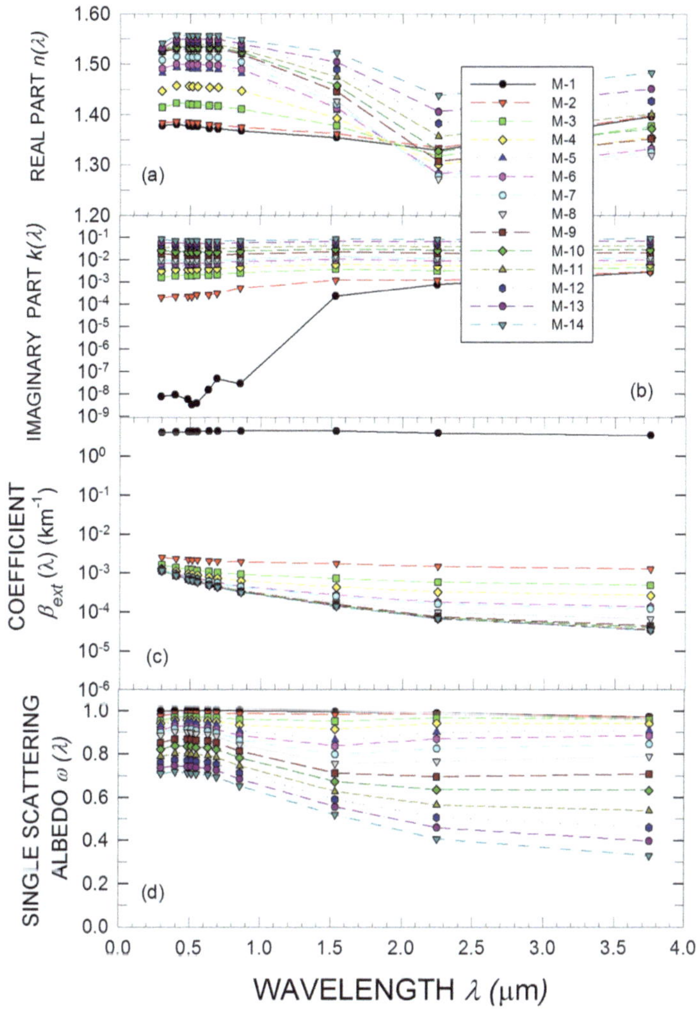

Fig. 11.7. As in Fig. 11.2, for the 14 modified (M-type) aerosol models shown in Fig. 11.6, all normalized to give a value of the overall particle number concentration $N_{\text{tot}} = 1000\,\text{cm}^{-3}$.

of the other M-type models are multimodal, with a more greatly prevailing number concentrations of fine particles and gradually lower number concentrations of coarse particles. In fact, the relative coarse particle number concentration is equal to about $1.5 \times 10^{-3}\%$ in the M-2 model, about $3 \times 10^{-4}\%$ in the M-3 model, and gradually even lower in the subsequent M-type models, as can be clearly verified in Fig. 11.6 by examining the particle number density and volume size-distribution curves of the 14 aerosol M-type models. Fig. 11.7 also shows that the spectral series of $\omega(\lambda)$ exhibit slowly decreasing values with wavelength, passing from the M-1 to the M-14 model, with discrete percentage variations from the near unity values of the M-1 model to the spectral values of the M-14 model, which are approximately equal to 0.70 in the visible.

11.2.4 The OPAC aerosol models

OPAC is a software package containing a set of aerosol models defining the radiative properties of various atmospheric particles (water droplets, aerosol, and ice crystals) over the spectral range including both the solar and terrestrial radiation. Airborne aerosol particles were assumed to consist of mixtures of different components, in which particle sizes vary as a function of RH. The OPAC optical parameters were calculated at 61 wavelengths selected between 0.3 and 40 μm and for 8 values of ambient RH equal to 0%, 50%, 70%, 80%, 90%, 95%, 98%, and 99%, respectively. The conception and the first design of the OPAC models are due to d'Almeida et al. (1991), while the revision of the code was carried out by Hess et al. (1998), who defined 10 aerosol components suitable for use to determine externally mixed compositions and simulate a wide variety of tropospheric aerosol radiative parameters, calculated as linear combinations of those evaluated for the 10 basic components and modeled under the assumption of particle sphericity. Each of the 10 components is attributed to a particular origin and represented using an individual log-normal particle size-distribution and adopting specific spectral features of complex particulate matter refractive index. The 10 basic components of the OPAC aerosol models were defined on the basis of older descriptions (Shettle and Fenn, 1979; Deepak and Gerber, 1983; d'Almeida et al. 1991; Koepke et al. 1997). They are:

(1) The water-insoluble (INS) component, consisting mainly of soil particles with a certain amount of organic substances.

(2) The water-soluble (WAS) component that mainly originates from gas-to-particle conversion and consists of various kinds of sulfates (of anthropogenic origin, with mass density equal to only about half that of the water-soluble component), nitrates, and other (mainly organic) substances mixed together, for which the optical effects of the dimethyl sulfide-related aerosol forming in the oceanic regions were also modeled.

(3) The soot (SOO) component, mainly containing black carbon (BC), which strongly absorbs the solar radiation and is assumed to be unsoluble. Several assumptions were made in defining this component: (i) the particles do not grow with increasing RH; (ii) the density of soot is equal to 1 $\mathrm{g\,cm^{-3}}$, because the soot particles sampled on filters and used to determine aerosol weight per air volume are in general fluffy particles with space inside; (iii) the optical properties were evaluated by neglecting the chainlike character of these particles, while the size distribution contains a significant amount of very small particles, with particulate matter density $\rho = 2.3\,\mathrm{g\,cm^{-3}}$; and (iv) no coagulation of soluble aerosol and soot was considered in the formation of the soot particle component.

(4 and 5) The two sea-salt particle components, both consisting of various kinds of salt contained in sea-water, and presenting the first a sea-salt accumulation particle mode (SAM), and the second a sea-salt coarse particle mode (SCM), as originated by different wind-speed dependent effects on the particle number density in the various size ranges (Koepke et al. 1997).

(6, 7 and 8) The three mineral aerosol components, consisting of mixtures of quartz and clay minerals and modeled using three different monomodal curves for the nucleation (MNM), accumulation (MAM) and coarse (MCM) particle components, which present relative amounts of large particles varying with the atmospheric turbidity conditions.

(9) The mineral particle (MTR) component of desert origin, used to describe the properties of desert dust transported over long distances, and consisting mainly of (i) mineral aerosol particles not growing with increasing RH, and (ii) a reduced amount of large particles.

(10) The sulphate (SDR) component, consisting of 75% H2SO4 and used to describe the sulfate particles found in the Antarctic aerosol and in the stratospheric background aerosol layers consisting mainly of sulphuric acid droplets.

The upper and lower limits of the monomodal particle size-distributions given in Table 11.5 for the 10 OPAC components were taken into account in the Mie calculations of the aerosol radiative parameters. The overall mass concentration M^* of each size-distribution was measured in $\mu g\,cm^{-3}$ per unit number concentration N, and calculated with a cutoff radius equal to $7.5\,\mu m$. For the aerosol particle polydispersions subject to condensation growth, the mode radius and the two radius range limits were assumed to increase with increasing RH (Hänel and Zankl, 1979). Each log-normal size-distribution was determined for RH = 50% by calculating the mode radius increase according to Hänel and Zankl (1979) and assuming that the standard deviation σ given in Table 11.5 remains unchanged as RH varies. The main shape-parameters of the log-normal size-distributions characterizing the 10 OPAC components are also given in Table 11.5, together with the values of wet particulate matter density ρ_w calculated on the basis of the growth factors determined by Hänel and Zankl (1979) for the various components.

Figure 11.8 shows the the size-distribution curves of particle number density $N(r)$ and particle volume $V(r)$, as obtained for an overall number concentration $N_{tot} = 10^3\,cm^{-3}$, highlighting the widely varying values of mode radius r_c used by d'Almeida et al. (1991) and Hess et al. (1998) to define the curve of $N(r)$, which were assumed in Table 11.5 to vary between a minimum of $1.18 \times 10^{-2}\,\mu m$ (for the SOO (soot) component) and a maximum of $1.90\,\mu m$ (for the MCM (mineral dust, coarse mode) component. Fig. 11.8 shows that the mode radius of $V(r)$ varies between a minimum of $5.0 \times 10^{-2}\,\mu m$ (for the SOO (soot) component) and a maximum of $11.00\,\mu m$ (for the MCM (mineral dust, coarse mode) component).

The real and imaginary parts of the refractive index were calculated for wet (RH = 50%) aerosol components, obtaining the spectral values shown in Fig. 11.9, which indicate that the real part $n(\lambda)$ varies between 1.33 and more than 1.70 at visible wavelengths, while the imaginary part $k(\lambda)$ presents particularly marked variations over the 0.30–1.50 μm wavelength range. The spectral patterns of volume extinction coefficient $\beta_{ext}(\lambda)$ and single scattering albedo $\omega(\lambda)$ are presented in Fig. 11.9, showing that $\beta_{ext}(\lambda)$ varies by several orders of magnitude passing from one component to another (for constant particle number concentration), and that $\omega(\lambda)$ varies at visible wavelengths between near unity values (for the WAS, SCM and SDR components) to a value lower than 0.3 (for the SOO component).

The 10 OPAC aerosol models were determined using the above 10 aerosol components and their optical, composition and microphysical characteristics, and as-

Table 11.5. Log-normal size-distribution shape-parameters and microphysical properties of the 10 OPAC wet aerosol components (Hess et al., 1998), as defined for RH = 50%

Aerosol component	Geometric standard deviation σ	Mode radius r_c (μm) of the particle number curve $N(r)$	Mode radius r_c (μm) of the particle volume curve $V(r)$	Lower limit $r_{\min}$(μm) of the particle radius range	Upper limit $r_{\max}$(μm) of the particle radius range	Wet particle mass density ρ_w (g cm^{-3})
(1) Insoluble (INS)	2.51	0.4710	6.00	0.005	20.0	2.0
(2) Water-soluble (WAS)	2.24	0.0212	0.15	0.005	20.0	1.8
(3) Soot (SOO)	2.00	0.0118	0.05	0.005	20.0	1.0
(4) Sea-salt (accumulation mode) (SAM)	2.03	0.2090	0.94	0.005	20.0	2.2
(5) Sea-salt (coarse mode) (SCM)	2.03	1.7500	7.90	0.005	60.0	2.2
(6) Mineral dust (nucleation mode) (MNM)	1.95	0.0700	0.27	0.005	20.0	2.6
(7) Mineral dust (accumulation mode) (MAM)	2.00	0.3900	1.60	0.005	20.0	2.6
(8) Mineral dust (coarse mode) (MCM)	2.15	1.9000	11.00	0.005	60.0	2.6
(9) Mineral-transported (MTR)	2.20	0.5000	3.00	0.020	5.0	2.5
(10) Sulfate droplets (SDR)	2.03	0.0695	0.31	0.005	20.0	1.7

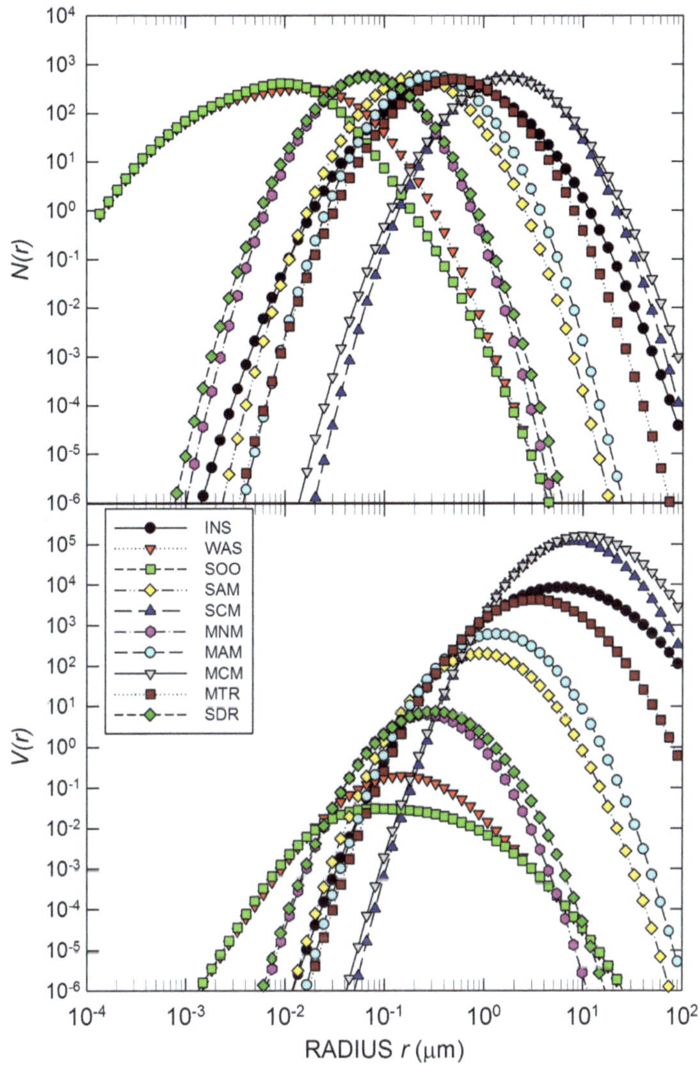

Fig. 11.8. As in Fig. 11.1, for the 10 OPAC aerosol components listed in Table 11.5, calculated for RH = 50% and normalized to give a value of the overall particle number concentration $N_{\mathrm{tot}} = 1000\,\mathrm{cm}^{-3}$.

suming the presence of soot particles (SOO component) in the polluted aerosol models only:

(1) The *Continental clean* (CC) aerosol model represents the aerosol polydispersion monitored in remote continental areas, with very low anthropogenic influences and, consequently, a very low mass concentration of soot substances ($< 0.1\,\mu\mathrm{g\,m}^{-3}$). The composition assumed in Table 11.6 does not contain soot substances, thus constituting a lower benchmark with respect to absorption in the solar spectral range.

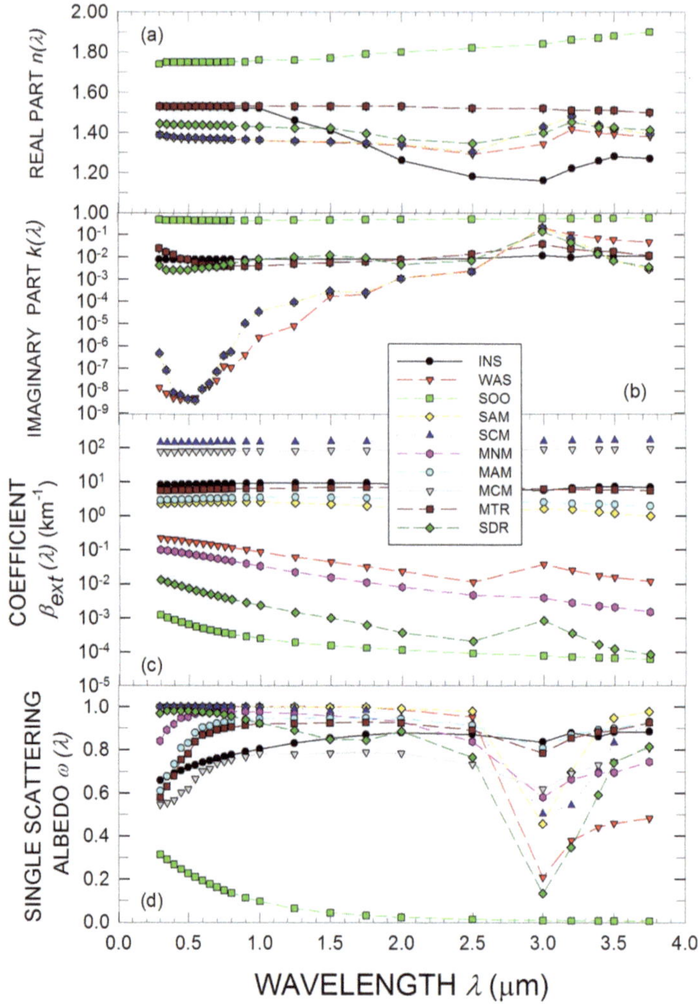

Fig. 11.9. As in Fig. 11.2, for the 10 OPAC aerosol components listed in Table 11.5, calculated for RH = 50% and normalized to give a value of the overall particle number concentration $N_{\text{tot}} = 1000\,\text{cm}^{-3}$.

(2) The *Continental average* (CA) aerosol model is used in cases where anthropogenic particles are present together with continental aerosols, this kind of particulate matter containing soot substances with lower concentrations of insoluble and water-soluble substances.

(3) The *Continental polluted* (CP) aerosol model represents the particle polydispersion usually sampled in highly polluted areas, mainly due to man-made activities, with mass density of soot matter assumed as equal to $2\,\mu\text{g}\,\text{m}^{-3}$, and that of water-soluble substances more than double the mass density usually found in continental average aerosol.

(4) The *Urban* (UR) aerosol model represents cases of strong pollution in urban areas, with the soot mass concentration assumed to be relatively high

Table 11.6. Values of the particle number concentration N_j and mass percentage Γ_j of the basic components giving form to the 10 OPAC wet aerosol models defined in the present study for RH = 50% (Hess et al., 1998), in which the total particle number concentration was assumed to be equal to $N_{tot} = 1000\,\text{cm}^{-3}$

OPAC models with their acronyms (in brackets)	Components	Particle number concentrations N_j (cm^{-3}) of the components	Mass percentages Γ_j of the components
Continental clean (CC)	Water soluble	999.940	0.591
	Insoluble	0.060	0.409
Continental average (CA)	Soot	542.470	0.021
	Water soluble	457.504	0.583
	Insoluble	0.026	0.396
Continental polluted (CP)	Soot	686.000	0.044
	Water soluble	313.999	0.658
	Insoluble	0.001	0.298
Urban (UR)	Soot	822.785	0.079
	Water soluble	177.214	0.563
	Insoluble	0.001	0.358
Desert (DE)	Water soluble	869.511	0.018
	Mineral dust (nucleation)	117.167	0.033
	Mineral dust (accumulation)	13.260	0.747
	Mineral dust (coarse)	0.062	0.202
Maritime clean (MC)	Water soluble	986.840	0.071
	Sea salt (accumulation)	13.158	0.908
	Sea salt (coarse)	0.002	0.021
Maritime polluted (MP)	Soot	575.5556	0.006
	Water soluble	422.222	0.160
	Sea salt (accumulation)	2.222	0.814
	Sea salt (coarse)	0.0004	0.019
Maritime tropical (MT)	Water soluble	983.333	0.058
	Sea salt (accumulation)	16.6668	0.928
	Sea salt (coarse)	0.0002	0.014
Arctic (AR)	Soot	802.798	0.044
	Water soluble	196.913	0.382
	Sea salt (accumulation)	0.2875	0.544
	Insoluble	0.0015	0.029
Antarctic (AN)	Sulphate	998.783	0.910
	Sea salt (accumulation)	1.094	0.045
	Mineral transported	0.123	0.045

($\sim 7.8\,\mu\text{g m}^{-3}$) and the mass concentrations of both water soluble and insoluble substances about twice those assumed in the continental polluted aerosol, as it was often found in aerosol samples collected in central urban areas.

(5) The *Desert* (DE) aerosol model is used to describe aerosol suspended over the desert areas of the world, consisting of the mineral aerosol components in

a combination that is representative for average atmospheric turbidity conditions, together with a certain mass fraction of the water-soluble (WAS) component.

(6) The *Maritime clean* (MC) aerosol model represents particle polydispersions sampled in undisturbed remote maritime areas without anthropogenic influences, with no soot and a limited mass concentration of water-soluble (WAS) aerosol used for representing the non-sea-salt (nss) sulfate particles. In general, maritime aerosol polydispersions contain sea salt particles in amounts depending on the wind speed: for instance, a concentration of 20 sea-salt particles per cubic meter was assumed in this model for a wind speed of $8.9\,\mathrm{m\,s^{-1}}$.

(7) The *Maritime polluted* (MP) aerosol model refers to a maritime environment under anthropogenic influence, with highly variable amounts of soot (SOO) and anthropogenic water-soluble (WAS) particles having mass density equal to 0.3 and $7.6\,\mu\mathrm{g\,m^{-3}}$, respectively, and both sea-salt components kept unchanged compared to clean maritime conditions.

(8) The *Maritime tropical* (MT) aerosol model was assumed to have a very low mass concentration of water-soluble (WAS) substances and was defined for a lower wind speed ($5\,\mathrm{m\,s^{-1}}$) than those assumed in the previous two models and, hence, a lower number concentration of sea-salt particles.

(9) The *Arctic* (AR) aerosol model represents the airborne particles found in the Arctic region at latitudes higher than 70°N, and describes atmospheric turbidity conditions characterized by the presence of a relatively high amount of soot (SOO) particles transported from the mid-latitude continental areas to the Arctic. This model is therefore particularly suitable for representing the aerosol characteristics during springtime, while it is less appropriate for representing the Arctic aerosol radiative properties during the other seasons, when single scattering albedo was found to range on average between about 0.93 and 0.95 (Tomasi et al., 2012).

(10) The *Antarctic* (AN) aerosol model represents the airborne particles found over the Antarctic continent, and consists mostly of sulfate droplets, containing also lower concentrations of mineral and sea-salt particles (typical of the coastal sites) and rather high number concentrations of nss sulfate aerosols (typical of the inner region), these composition features being valid especially for summer conditions, when average values of ω of around 0.96–0.98 were found (Tomasi et al., 2012).

The particle size-distribution curves of $N(r)$ and $V(r)$ determined for the 10 OPAC wet aerosol models are presented in Fig. 11.10, showing that all the curves of $N(r)$, except that of the AN model exhibit similar features over the radius range $r < 10^{-1}\,\mu\mathrm{m}$, while they differ appreciably one model from another over the upper radius range, where the DE (desert) model has the highest content of coarse particles and the UR (urban) model the lowest one. The size-distribution curves of $V(r)$ more clearly show that marked differences exist between the fine particle and coarse particle contents of the 10 aerosol models over the whole radius range, evidencing their multimodal characteristics. Fig. 11.11 shows the spectral curves of parameters $n(\lambda)$, $k(\lambda)$, $\beta_{\mathrm{ext}}(\lambda)$, and $\omega(\lambda)$ determined for the 10 OPAC models, indicating that (i) the values of $n(\lambda)$ range between 1.38 and 1.50 in the visible; (ii) the values of $k(\lambda)$ vary between 10^{-3} and about 0.5 in the visible; (iii) $\beta_{\mathrm{ext}}(\lambda)$

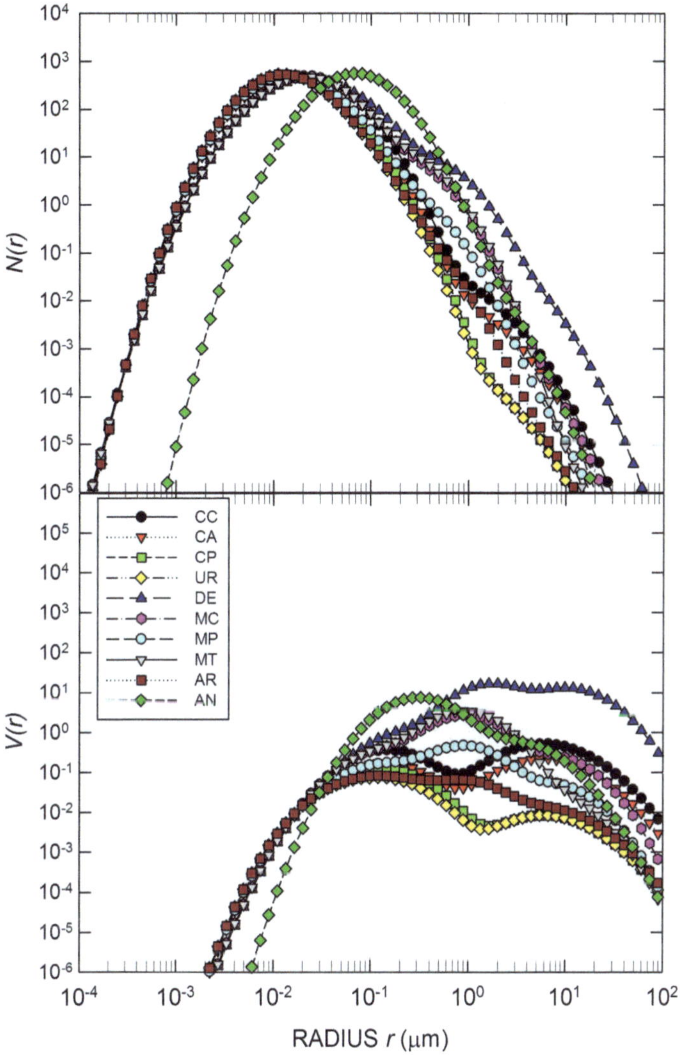

Fig. 11.10. As in Fig. 11.1, for the 10 OPAC aerosol models listed in Table 11.6, calculated for RH = 50% and normalized to give a value of the overall particle number concentration $N_{\mathrm{tot}} = 1000\,\mathrm{cm}^{-3}$.

decreases more or less rapidly as a function of wavelength λ, substantially depending on the variable percentage contents of fine and coarse particles; and (iv) $\omega(\lambda)$ assumes values ranging between 0.74 and near unity in the visible.

Table 11.7 presents the values of the most significant radiative parameters of the 10 OPAC aerosol models calculated for RH = 50%:

- the monochromatic values of volume extinction coefficient $\beta_{\mathrm{ext}}(0.55\,\mu\mathrm{m})$, volume scattering coefficient $\beta_{\mathrm{sca}}(0.55\,\mu\mathrm{m})$, and volume absorption coefficient $\beta_{\mathrm{abs}}(0.55\,\mu\mathrm{m})$, as calculated for the total particle number concentration $N_{\mathrm{tot}} = 10^3\,\mathrm{cm}^{-3}$;

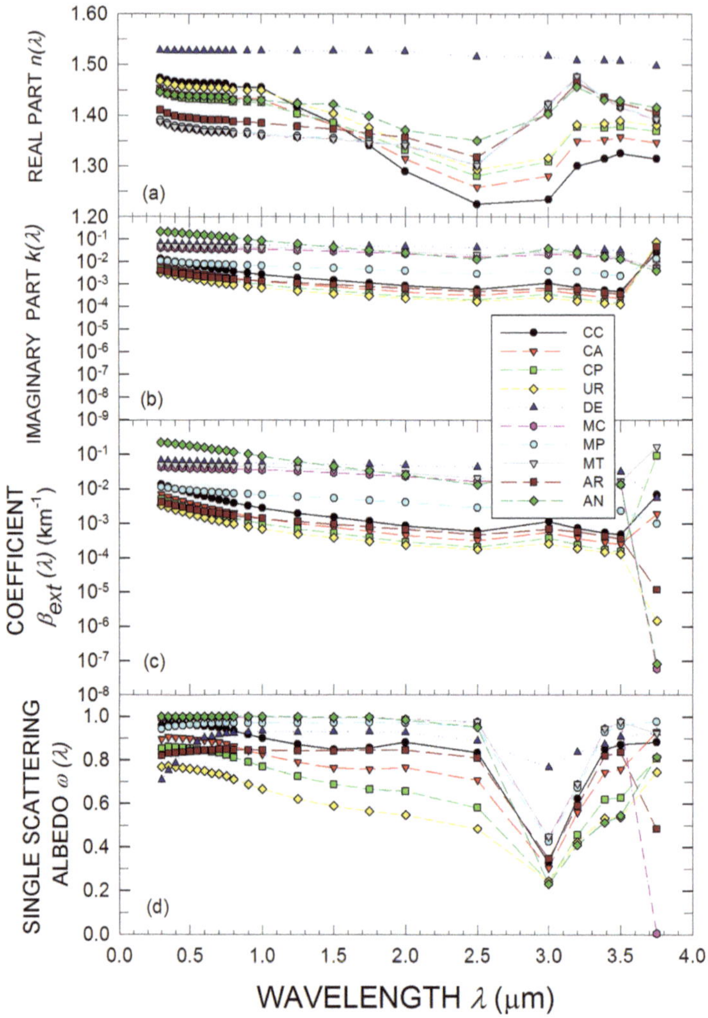

Fig. 11.11. As in Fig. 11.2, for the 10 OPAC aerosol models listed in Table 11.6, calculated for RH = 50% and normalized to give a value of the overall particle number concentration $N_{\text{tot}} = 1000\,\text{cm}^{-3}$.

- the Ångström exponent defined over the 0.40–0.86 μm wavelength range, found to vary between 0.13 (MC model) and ~ 1.43 (CP model);
- the monochromatic single scattering albedo $\omega(0.55\,\mu\text{m})$, estimated to vary between 0.742 (UR model) and 0.997 (MT model);
- the mean single scattering albedo ω calculated over the 0.40–3.70 μm wavelength range and evaluated to vary between 0.686 (UR model) and 0.993 (MC and MT models);
- the weighted average single scattering albedo ω^*, ranging between 0.741 (UR model) and 0.998 (AN model);

Table 11.7. Values of the main radiative and physical parameters calculated for the 10 OPAC wet aerosol models (Hess et al., 1998), as defined in the present study for RH = 50% and relative to total particle number concentration $N_{tot} = 1000\,cm^{-3}$

Radiative and physical parameters	OPAC aerosol models for RH = 50%									
	Contin. clean (CC)	Contin. aver. (CA)	Contin. poll. (CP)	Urban (UR)	Desert (DE)	Maritim. Clean (MC)	Maritim. poll. (MP)	Maritim. trop. (MT)	Arctic (AR)	Antarctic (AN)
Volume extinction coefficient $\beta_{ext}(0.55\,\mu m)$	$6.586\ 10^{-3}$	$3.435\ 10^{-3}$	$2.481\ 10^{-3}$	$3.498\ 10^{-3}$	$6\ 410\ 10^{-2}$	$3.965\ 10^{-2}$	$8.643\ 10^{-2}$	$4.864\ 10^{-2}$	$2.434\ 10^{-3}$	$1.981\ 10^{-1}$
Volume scattering coefficient $\beta_{sca}(0.55\,\mu m)$	$6.574\ 10^{-3}$	$3.069\ 10^{-3}$	$2.106\ 10^{-3}$	$2.594\ 10^{-3}$	$5.305\ 10^{-2}$	$3.905\ 10^{-2}$	$8.328\ 10^{-2}$	$4.831\ 10^{-2}$	$2.050\ 10^{-3}$	$1.931\ 10^{-1}$
Volume absorption coefficient $\beta_{abs}(0.55\,\mu m)$	$2.818\ 10^{-4}$	$3.661\ 10^{-4}$	$3.751\ 10^{-4}$	$9.038\ 10^{-4}$	$7.048\ 10^{-3}$	$1.475\ 10^{-4}$	$3.154\ 10^{-4}$	$1.470\ 10^{-4}$	$3.848\ 10^{-4}$	$4.231\ 10^{-3}$
Ångström exponent $\alpha(0.40–0.86\,\mu m)$	1.386	1.389	1.427	1.397	0.127	0.131	0.410	0.088	0.952	0.826
Single scattering albedo $\omega(0.55\,\mu m)$	0.959	0.893	0.849	0.742	0.883	0.996	0.963	0.997	0.842	0.979
Mean single scattering albedo ω over the 0.40–3.70 μm range	0.926	0.847	0.782	0.686	0.886	0.993	0.964	0.993	0.840	0.950
Weighted average single scattering albedo ω^*	0.952	0.884	0.837	0.741	0.875	0.997	0.964	0.997	0.842	0.998
Asymmetry factor $g(0.55\,\mu m)$	0.681	0.673	0.665	0.745	0.727	0.756	0.737	0.759	0.692	0.764
Wet particulate mass density ρ_w ($g\,cm^{-3}$)	1.61	1.59	1.52	2.56	1.97	1.32	1.31	1.30	1.33	1.77

- the monochromatic asymmetry factor $g(0.55\,\mu m)$, varying between 0.665 (CP model) and 0.764 (AN model); and
- the particulate mass density (for RH = 50%), varying between 1.30 g cm^{-3} (MT model) and 2.56 g cm^{-3} (UR model).

11.2.5 The Shettle and Fenn (1979) aerosol models

Four classical aerosol models (hereinafter referred to as SF) were proposed by Shettle and Fenn (1979) to represent the extinction characteristics of rural, maritime, tropospheric and urban particle polydispersions for different RH values (0%, 50%, 70%, 80%, 95%, 98% and 99%). The size-distribution curves of the four aerosol models were represented by means of monomodal or bimodal log-normal size-distributions to define the particle number density size-distribution function $N(r)$. Each mode has the analytical form defined in Eq. (11.1), for the values of mode radius r_c (μm) and geometric standard deviation σ_i, and the unimodal values of particle number density Ni (cm^{-3}) given in Table 11.8 for RH = 50%. These log-normal curves give form to the Rural (bimodal), Urban (bimodal), Maritime (unimodal) and Tropospheric (unimodal) size-distributions of Shettle and Fenn (1979). Their main characteristics are as follows:

(1) The *Rural* (SF-R) *aerosol model* represents an aerosol polydispersion not directly influenced by urban and/or industrial sources, whose particles are assumed to be composed of a mixture of 70% of water-soluble substances (ammonium and calcium sulfates, with organic compounds) and 30% of dust-like aerosols. The number density and volume size-distribution curves of the SF-R model are presented in Fig. 11.12, as grown for RH = 50%, providing evidence of their bimodal features associated with the small rural and large rural particle modes. The spectral curves of $n(\lambda)$ and $k(\lambda)$ were determined as a function of RH over the 0.25–3.75 μm wavelength range according to Volz (1972a, 1973) and are shown in Fig. 11.13 for RH = 50%. The spectral curves of $\beta_{ext}(\lambda)$ and $\omega(\lambda)$ obtained for this bimodal size-distribution are also shown in Fig. 11.13 over the same wavelength range.

(2) The *Urban* (SF-U) *aerosol model* is given by a linear combination of two size-distribution curves, where the first represents a monomodal polydispersion of background rural aerosols consisting of water-soluble substances only, and the second represents a secondary aerosol polydispersion originating from combustion products and industrial sources. Therefore, such a model consists of a mixture of rural aerosol (80%) and carbonaceous soot-like particles (20%). The two size-distributions of small urban and large urban aerosol components were assumed to have the same values of number density $N(r)$ and geometric standard deviation σ adopted to give form to the bimodal rural SF-R model. The number density and volume size-distribution curves of the SF-U model are presented in Fig. 11.12 for RH = 50%. The spectral values of the soot-like particulate matter refractive index were calculated on the basis of the soot data provided by Twitty and Weinman (1971), while their variations as a function of RH were determined using the growth factors proposed by Hänel (1976) for his urban aerosol model (model 5). The spectral curves of $n(\lambda)$ and $k(\lambda)$ evaluated for RH = 50% are shown in Fig. 11.13 over the 0.25–3.75 μm range, together with those of parameters $\beta_{ext}(\lambda)$ and $\omega(\lambda)$

Table 11.8. Shape-parameters and characteristics of the unimodal log-normal size-distribution curves used by Shettle and Fenn (1979) in the calculations of the shape-parameters defining the four SF aerosol models for RH = 50%. The four models are all normalized to give a total particle number concentration $N_{\text{tot}} = 1000\,\text{cm}^{-3}$

Aerosol model	Modes	Type	Unimodal log-normal size-distribution shape-parameters		
			Particle number densities N_1 (1st mode) and N_2 (2nd mode)	Mode radius r_c (μm)	Geometric standard deviation σ
(SF-R) Rural	1st mode (water-soluble) 2nd mode (dust-like)	Mixture of water-soluble, and dust-like aerosols	999.87 0.125	0.02748 0.43770	2.239 2.500
(SF-U) Urban	1st mode (rural) 2nd mode (soot-like)	Rural aerosol mixture with soot-like aerosols	999.875 0.125	0.02563 0.41130	2.990 2.500
(SF-M) Maritime	1st mode (sea-salt)	Sea-salt solution in water of oceanic origin	1000.00	0.1711	2.500
(SF-T) Tropospheric	1st mode (rural, fine particles)	Rural aerosol mixture	1000.00	0.02748	2.990

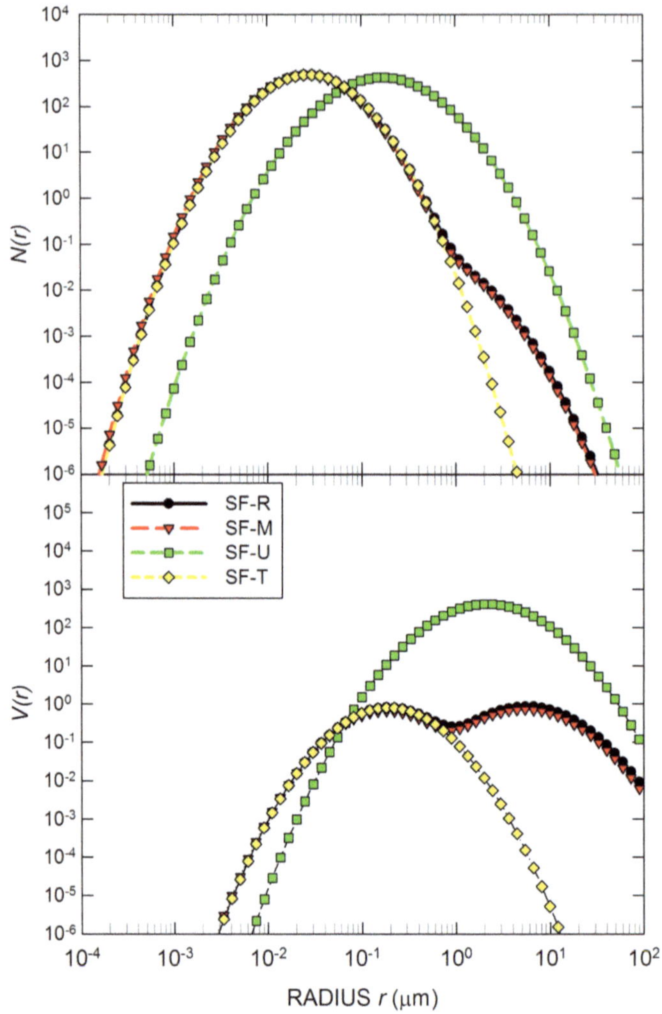

Fig. 11.12. As in Fig. 11.1, for the four aerosol models SF-R (rural), SF-M (maritime), SF-U (urban) and SF-T (tropospheric) defined by Shettle and Fenn (1977) for RH = 50%, The four models are all normalized to give a value of the overall particle number concentration $N_{\text{tot}} = 1000\,\text{cm}^{-3}$.

obtained for this bimodal model (and for RH = 50%), which clearly evidence that the patterns of $\beta_{\text{ext}}(\lambda)$ are very similar to those of the SF-R model, while those of $\omega(\lambda)$ are considerably lower, due to the greater relative content of soot substances.

(3) The *Maritime* (SF-M) *aerosol model* is represented with a monomodal size-distribution curve of particles consisting of sea-salt particles formed through the evaporation of sea-spray droplets that were subsequently grown as a result of water aggregation. The composition of these particles was assumed to be given mainly by (i) a component of oceanic origin, and (ii) a continental component with a higher number concentration than the previous one, added to constitute a background particle polydispersion. The linear combination of these two polydispersions

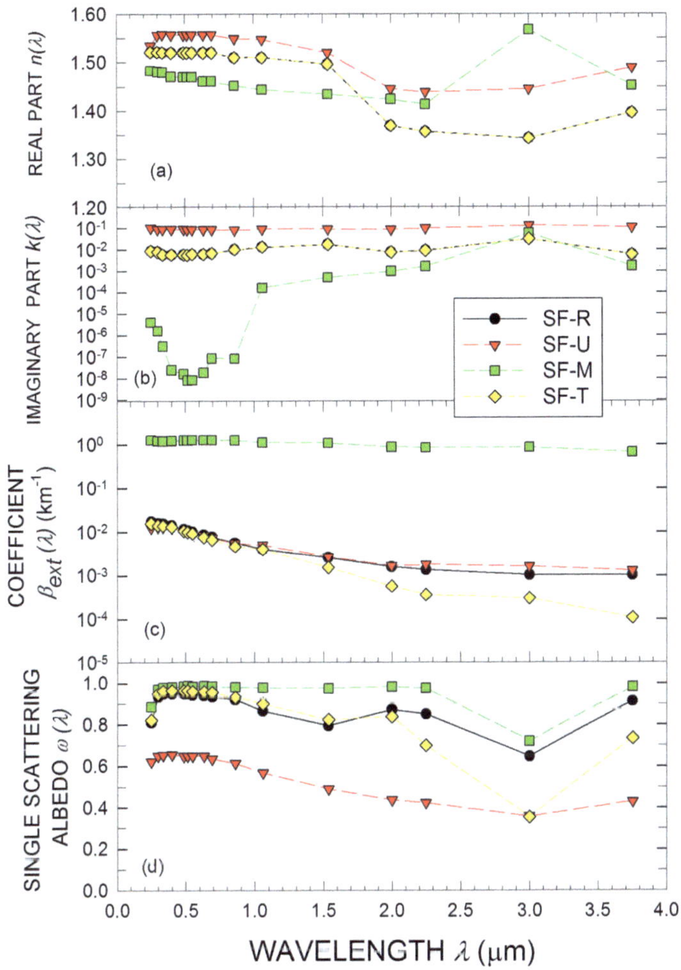

Fig. 11.13. As in Fig. 11.2, for the four aerosol models SF-R (rural), SF-M (maritime), SF-U (urban) and SF-T (tropospheric) defined by Shettle and Fenn (1977) for RH = 50%. The black circles of the SF-R model are in part hidden by the overlapping yellow diamonds used to label the SF-T (tropospheric) model data.

of marine and continental particles constitute a fairly uniform maritime aerosol model, which is representative of the marine aerosol usually sampled within the atmospheric boundary layer of 2–3 km depth over the oceans. The number density and size-distribution shape-parameters of the SF-M aerosol model are given in Table 11.8. The spectral features of the complex refractive index of the SF-M particulate matter were primarily defined taking into account the Volz (1972b) data, while its variations as a function of RH were evaluated using the evaluations of the growth factor made by Hänel (1976) for a sea-spray aerosol polydispersion. Fig. 11.13 shows the spectral curves of the optical parameters $n(\lambda)$, $k(\lambda)$, $\beta_{ext}(\lambda)$ and $\omega(\lambda)$ obtained over the 0.25–3.75 μm wavelength range for such a unimodal aerosol model defined for RH = 50%. The first parameter has values slightly higher than 1.40 in the visible, while the second has very low values close to 10^{-8}, due to the very weak absorption of solar radiation by these particles. The spectral curve of $\beta_{ext}(\lambda)$ exhibits values higher by more than two orders of magnitude than those of the other three SF models, this marked difference being due to the considerably higher relative concentration of coarse particles, as shown in Fig. 11.12. The spectral curve of $\omega(\lambda)$ has values very close to unity throughout the whole wavelength range, due to the very poor absorption properties of these particles.

(4) The *Tropospheric* (SF-T) *aerosol model* represents a particle polydispersion suspended in the troposphere above the boundary layer. These background tropospheric particles were therefore assumed to have the same composition as that of the SF-R aerosol model, forming a dry particle polydispersion consisting of 70% water-soluble and 30% dust-like substances. The size-distribution was assumed to be unimodal. Thus, the SF-T model was obtained by removing the second mode of large particles from the SF-R model and using the shape-parameters presented in Table 11.8 to give form to the unique mode consisting mainly of fine particles. In fact, because of the longer residence time of fine particles above the boundary layer and the consequent differential loss of larger particles, only one log-normal size-distribution consisting of accumulation particles was considered by Shettle and Fenn (1979) in defining this unimodal aerosol model, as can be seen in Fig. 11.12. The dependence of particle sizes on RH is described by the same analytical functions adopted for the small particle component of the SF-R model. For this reason, the spectral curves of $n(\lambda)$ and $k(\lambda)$ shown in Fig. 11.13 for the SF-T model are similar to those determined for the small component of the SF-R model. The spectral variations of parameters $\beta_{ext}(\lambda)$ and $\omega(\lambda)$ are also shown in Fig. 11.13 for RH = 50%, over the 0.25–3.75 μm. Their comparison with the curves of the two parameters determined for the bimodal SF-R model gives evidence of the differences arising from the absence in the SF-T model and the presence in the SF-R model of a second mode of large rural aerosols.

The monochromatic values of volume extinction coefficient $\beta_{ext}(0.55\,\mu m)$, volume scattering coefficient $\beta_{sca}(0.55\,\mu m)$ and volume absorption coefficient $\beta_{abs}(0.55\,\mu m)$ are given in Table 11.9 for the four SF aerosol models, together with those of (i) Ångström's exponent α calculated over the 0.40–0.86 μm wavelength range, (ii) monochromatic single scattering albedo ω(550 nm), (iii) mean single scattering albedo ω determined over the 0.40–3.70 μm wavelength range, (iv) weighted average single scattering albedo ω^*, and (v) monochromatic asymmetry factor $g(0.55\,\mu m)$. The results indicate that $\alpha(0.40\text{–}0.86\,\mu m)$ varies between 0.028 (SF-M) and 1.355 (SF-T), while $\omega(0.55\,\mu m)$ ranges between 0.644 (SF-U) and 1.000 (SF-M), ω between 0.585 (SF-U) and 0.996 (SF-M), ω^* between 0.637 (SF-U) and 0.999 (SF-M), and $g(0.55\,\mu m)$ between 0.640 (SF-T) and 0.753 (SF-M). Table 11.9 also provides the values of wet particulate mass density obtained for decreasing RH = 50%, as determined: (i) for the SF-R and SF-T models, using the particle growth estimates made by Hänel (1976) for his model 6 sampled at Hohenpeissenberg (Germany) in summer 1970; (ii) for the SF-U model, using the evaluations of Hänel (1976) for his model 5 derived from an urban aerosol sample collected at Mainz (Germany) in January 1970; and (iii) for the SF-M model, using the estimates of Hänel (1976) for his model 2 of sea-spray aerosol.

Table 11.9. Values of the main radiative and physical parameters of the four SF aerosol models of Shettle and Fenn (1979), calculated in the present study for RH = 50% and total particle number concentration $N_{tot} = 1000\,cm^{-3}$

Radiative and physical parameters	Shettle and Fenn (1979) aerosol models for RH = 50%			
	SF-R (Rural)	SF-U (Urban)	SF-M (Maritime)	SF-T (Tropospheric)
Volume extinction coefficient $\beta_{ext}(0.55\,\mu m)$	$1.011\ 10^{-2}$	$9.268\ 10^{-3}$	1.262 100	$9.171\ 10^{-3}$
Volume scattering coefficient $\beta_{sca}(0.55\,\mu m)$	$9.528\ 10^{-3}$	$5.982\ 10^{-3}$	1.262 100	$8.809\ 10^{-3}$
Volume absorption coefficient $\beta_{abs}(0.55\,\mu m)$	$5.800\ 10^{-4}$	$3.286\ 10^{-3}$	0.000	$3.620\ 10^{-4}$
Ångström exponent $\alpha(0.40–0.86\,\mu m)$	1.193	1.050	0.028	1.355
Single scattering albedo $\omega(0.55\,\mu m)$	0.943	0.644	1.000	0.960
Mean single scattering albedo ω over the 0.40–3.70 μm range	0.914	0.585	0.996	0.904
Weighted average single scattering albedo ω^*	0.933	0.637	0.999	0.953
Asymmetry factor $g(0.55\,\mu m)$	0.652	0.668	0.753	0.640
Wet particulate mass density ρ_w ($g\,cm^{-3}$)	1.778	2.677	1.878	1.408

11.2.6 The seven additional aerosol models

Seven additional aerosol models were prepared in the present study to represent (i) a pair of multimodal polydispersions defined by examining two Saharan dust samples collected by Tomasi et al. (1979) in Northern Italy; (ii) one pre- and two post-Pinatubo volcanic particle polydispersions determined from samples performed at different stratospheric altitudes by Pueschel et al. (1993); and (iii) two polydispersions of biomass burning smoke particles sampled by Carr (2005) at Jabiru (Australia), the first in the free troposphere and the second within the atmospheric boundary layer:

(1, 2) The two *Saharan Dust* SD-1 and SD-2 models are based on the size-distribution curves defined by Tomasi et al. (1979) analysing a pair of particle samples collected at Sestola (Apennines, Northern Italy) during two transport episodes of desert dust from North Africa. Both size-distribution curves were found to consist of three modes, each represented in terms of the Deirmendjian (1969) modified gamma function. It has the analytical form:

$$dN(r)/dr = Cr^D \exp\left[-\frac{D}{\gamma}(r/R_c)^\gamma\right], \qquad (11.3)$$

for the following values of shape-parameters D, γ and mode radius R_c:

(a) In the SD-1 model: $D = 2$, $\gamma = 0.815$, and $R_c = 0.051\,\mu$m for the 1st mode; $D = 5$, $\gamma = 1.515$, and $R_c = 0.554\,\mu$m for the 2nd mode; and $D = 6$, $\gamma = 2.490$, and $R_c = 1.535\,\mu$m for the 3rd mode.
(b) In the SD-2 model: $D = 4$, $\gamma = 1.533$, and $R_c = 0.120\,\mu$m for the 1st mode; $D = 2$, $\gamma = 1.294$, and $R_c = 0.449\,\mu$m for the 2nd mode; and $D = 6$, $\gamma = 2.959$, and $R_c = 2.018\,\mu$m for the 3rd mode.

To represent the size-distribution models by means of the same analytical log-normal function adopted for the other aerosol models considered in the present study, the original number concentration data recorded by Tomasi et al. (1979) were examined through a best-fit procedure, determining the six corresponding log-normal curves having the analytical form of Eq. (11.1). The best-fit values of the shape-parameters are given in Table 11.10 for the six modes giving form to the SD-1 and SD-2 size-distribution curves, which are normalized to yield an overall particle number concentration $N_{\text{tot}} = 10^3\,\text{cm}^{-3}$. The multimodal size-distributions of $N(r)$ and $V(r)$ of the two SD models determined for dry-air conditions are presented in Fig. 11.14, showing very large differences from one to the other in the coarse and giant particle contents.

The spectral values of $n(\lambda)$ were calculated at 23 selected wavelengths chosen over the 0.25–3.70 μm range using the evaluations proposed by (i) Hänel (1968, 1972) for his aerosol model sampled over the Atlantic in April 1969 and containing Saharan dust, (ii) Volz (1973) for Saharan dust samples collected over the Caribbean region, and (iii) Vermote et al. (1997b) for the 6S DL component. Similarly, the spectral values of $k(\lambda)$ were determined at the same 23 wavelengths according to the estimates of Hänel (1968, 1972), Volz (1973), Patterson (1977) and Patterson et al. (1977) for Saharan dust particles. The mass density was assumed to be equal to $2.60\,\text{g cm}^{-3}$ in both the SD aerosol models, according to Hänel (1968, 1972), this value being in good agreement with that of $2.50\,\text{g cm}^{-3}$ found by Volz (1973) for dry-air conditions.

(3) The *background stratospheric* PV-1 aerosol model was represented using a unimodal log-normal size-distribution of aerosol particles, as observed over the Antarctic continent in 1987 during a long volcanic quiescence period and assumed by Pueschel et al. (1989) to describe realistically the stratospheric turbidity conditions preceding the Pinatubo eruption. The size-distribution of the PV-1 aerosol model assumes the analytical form in Eq. (11.1) for the shape-parameters given in Table 11.10. In defining the radiative properties of this aerosol model, it was taken into account that Pueschel et al. (1989) highlighted the predominance of sulphuric acid in the stratospheric particulate matter, finding that the chemical composition of these particles was given by mass fractions of 72% sulphuric acid, 24% liquid water, and 4% water-soluble (nitrate) substances. Therefore, the spectral values of $n(\lambda)$ and $k(\lambda)$ were first calculated over the 0.36–3.70 μm range for a 75% solution of sulphuric acid, using (i) the values of $n(\lambda)$ defined by Palmer and Williams (1975) over the entire range; (ii) the values of $k(\lambda)$ given by Palmer and Williams (1975) over the 0.70–3.70 μm range; and (iii) the values of $k(\lambda)$ given by Hummel et al. (1988) over the 0.36–0.70 μm range. The values were then appropriately reduced to a 72% solution of sulphuric acid and integrated with mass fractions equal to 24%

Table 11.10. Values of the shape-parameters of the unimodal size-distributions and of the main physical and radiative parameters giving form to the seven additional aerosol models proposed in the present study: i.e. particle mass density ρ ($g\,cm^{-3}$); volume extinction coefficient $\beta_{ext}(0.55\,\mu m)$, volume scattering coefficient $\beta_{sca}(0.55\,\mu m)$ and volume absorption coefficient $\beta_{abs}(0.55\,\mu m)$; mean single scattering albedo ω; average weighted single scattering albedo ω^*; and mean Ångström's exponent α determined over the 0.40–0.86 μm wavelength range

Aerosol models	Shape parameters of the log-normal unimodal size-distributions: First mode N_1 (cm^{-3})	σ	r_c (μm)	Second mode N_2 (cm^{-3})	σ	r_c (μm)	Third mode N_3 (cm^{-3})	σ	r_c (μm)	ρ ($g\,cm^{-3}$)	β_{ext} (0.55 μm) (km^{-1})	β_{sca} (0.55 μm) (km^{-1})	β_{abs} (0.55 μm) (km^{-1})	Monochromatic ω (0.55 μm)	Mean single scattering albedo ω	Weighted average single scattering albedo ω^*	Mean Ångström's exponent α
SD-1 (Saharan dust, trimodal)	940	2.0	0.051	59	1.5	0.554	1	1.4	1.535	2.6 (*)	$4.13\ 10^{-1}$	$3.44\ 10^{-1}$	$6.91\ 10^{-2}$	0.833	0.881	0.837	−0.09
6S-D2 (Saharan dust, trimodal)	733	1.78	0.120	261.5	2.12	0.449	5.5	1.4	2.018	2.6 (*)	1.55 100	1.19 100	$3.62\ 10^{-1}$	0.767	0.819	0.772	−0.05
PV-1 (background stratospheric, unimodal)	1000	1.80	0.070	–	–	–	–	–	–	1.658 (**)	$3.59\ 10^{-2}$	$3.58\ 10^{-2}$	$6\ 10^{-5}$	0.998	0.895	0.996	1.688
PV-2 (Post-Pinatubo 2-month stratospheric, bimodal)	586.207	1.50	0.090	413.83	1.50	0.31	-	-	-	1.65 (+)	$5.27\ 10^{-1}$	$5.27\ 10^{-1}$	0.00	1.00	0.923	0.999	0.186
PV-3 (Post-Pinatubo 9-month stratospheric, trimodal)	757.785	1.40	0.060	214.533	1.50	0.18	27.682	1.20	0.75	1.65 (+)	$1.92\ 10^{-1}$	$1.92\ 10^{-1}$	0.00	1.00	0.913	0.999	0.054
FT model (Biomass burning smoke in free troposphere, unimodal)	1000	1.546	0.022	–	–	–	–	–	–	1.480 (++)	$6.40\ 10^{-5}$	$4.3\ 10^{-5}$	$2.1\ 10^{-5}$	0.675	0.444	0.616	2.425
BL model (Biomass burning smoke in boundary layer, trimodal	404.218	1.533	0.0505	594.886	1.591	0.0655	0.896	1.644	0.605	1.485 (++)	$1.56\ 10^{-2}$	1.48 10-2	$7.8\ 10^{-4}$	0.950	0.942	0.948	1.546

(*) Tomasi et al. (1979); (**)Pueschel et al. (1989); (+) Hofmann and Rosen (1983a); (++) Carr (2005).

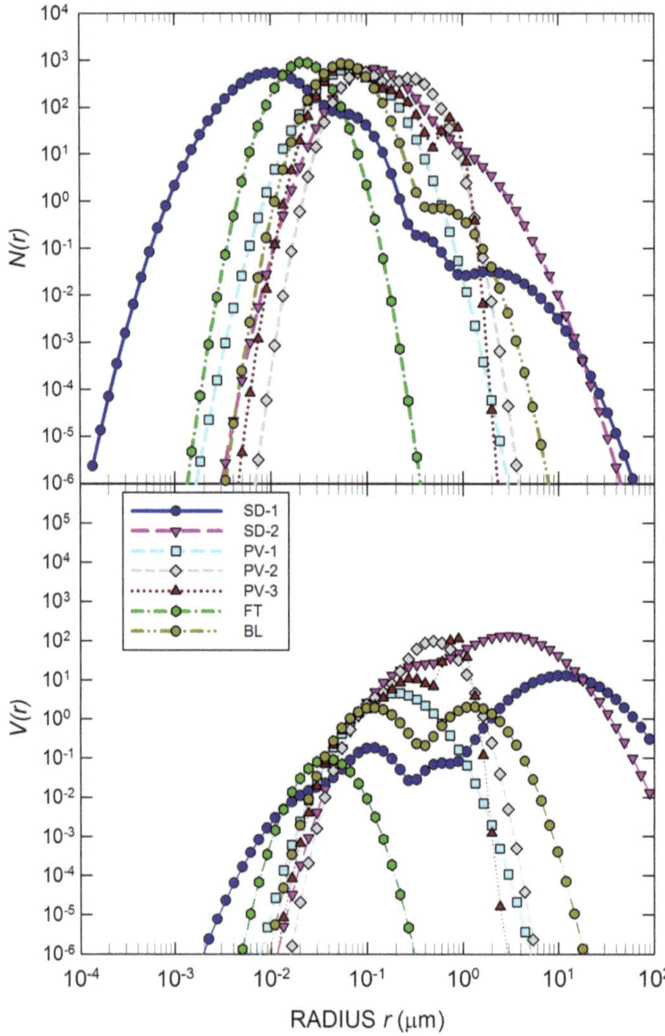

Fig. 11.14. As in Fig. 11.1, for the seven additional aerosol models defined in the present study for Sahara dust particles (SD-1 and SD-2), background stratospheric aerosol during volcanic quiescence periods in Antarctica (VQ), 2-month age Pinatubo volcanic stratospheric particles (PV-1), 9-month aged Pinatubo volcanic stratospheric particles (PV-2), biomass burning smoke particles in free troposphere (FT), and biomass burning smoke particles in the boundary layer (BL). All the additional models are normalized to give a value of the overall particle number concentration $N_{\text{tot}} = 1000\,\text{cm}^{-3}$.

of liquid water (using the Hale and Querry (1973) estimates) and 4% of nitrates, as given by Vermote et al. (1997b) for the 6S WS component.

(4, 5) The *volcanic stratospheric* PV-2 and PV-3 aerosol models were determined by adopting the log-normal multimodal size-distribution curves defined by Pueschel et al. (1993) from *in situ* sampling measurements performed at stratospheric altitudes to represent the post-Pinatubo volcanic particle polydispersions forming 2 and 9 months after the Mt. Pinatubo eruption of June 15, 1991 at the 16.5 km and 12.5 km heights, respectively. The shape-parameters of the log-normal modes of the PV-2 (bimodal) and PV-3 (trimodal) size-distributions, all having the analytical forms in Eq. (11.1), are given in Table 11.10. The complex refractive index of these stratospheric particles was defined according to Pueschel et al. (1993), who assumed that particulate matter consisted of complementary mass fractions of 75% sulphuric acid and 25% liquid water. Therefore, the spectral values of $n(\lambda)$ and $k(\lambda)$ were determined at the selected wavelengths over the 0.36–3.70 μm range for both the PV-2 and PV-3 models, using the data of Palmer and Williams (1975) defined for an aqueous solution of 75% sulphuric acid. Considering that the values of $k(\lambda)$ proposed by Palmer and Williams (1975) are lacking within the 0.36–0.70 μm wavelength range, use was made of the values given by Hummel at al. (1988) within the $0.36 \leq \lambda \leq 0.70\,\mu$m range.

(6, 7) Two *biomass burning smoke* FT and BL aerosol models were derived from the data obtained by Carr (2005) for aerosol samples collected at Jabiru (Australia) in September 2003 within the free troposphere (FT model) and the atmospheric boundary layer (BL model), for nearly dry-air conditions. The first consists of fine particles only, including both Aitken nuclei and accumulation particles, which form a unimodal log-normal size-distribution having the form of Eq. (11.1) for the values of shape-parameters N_o, r_c and σ given in Table 11.10. An average fine particle density of 1.48 g cm^{-3} was estimated by Carr (2005) for the FT model. Correspondingly, the spectral values of $n(\lambda)$ and $k(\lambda)$ were determined by assuming a chemical composition of particulate matter consisting of a mass fraction of water-soluble substances equal to 92%, and complementary percentages equal to 5% of mineral dust containing quartz and silicates, 1.9% of sea-salt and 1.1% of soot substances, for which an average value of $n(\lambda)$ close to 1.552 and an average value of $k(\lambda)$ slightly higher than 0.01 were found in the visible, which agree closely with the average values of $n = 1.558$ and $k = 1.06 \times 10^{-2}$ measured directly by Carr (2005) in the visible.

The BL model was assumed mainly to consist of coarse particles and to have a trimodal size-distribution curve including three log-normal curves containing Aitken nuclei, accumulation particles and coarse particles, respectively, with the shape-parameters N_o, r_c and σ reported in Table 11.10. The average particle density of fine and coarse particles was estimated by Carr (2005) to be equal to 1.48 g cm^{-3} and 1.49 g cm^{-3}, respectively. The spectral values of $n(\lambda)$ were determined by assuming a chemical composition of particulate matter consisting of a 46% mass fraction of mineral dust containing quartz and silicates, 35% water-soluble substances, 15% sea salt, and 4% soot substances, finding an average value of n close to 1.565 in the visible and, hence, only slightly higher than $n = 1.546$ measured by Carr (2005). The spectral values of $k(\lambda)$ were derived from the Carr (2005) measurements, finding an average value of around 4.75×10^{-3} in the visible.

The size-distribution curves of $N(r)$ and $V(r)$ presently defined for the seven additional models are shown in Fig. 11.14. The comparison shows that large differences exist among the various number density and particle volume size-distributions of the polydispersions, especially over the coarse particle radius range. With respect to the other five models, considerably higher values of $N(r)$ and $V(r)$ are presented

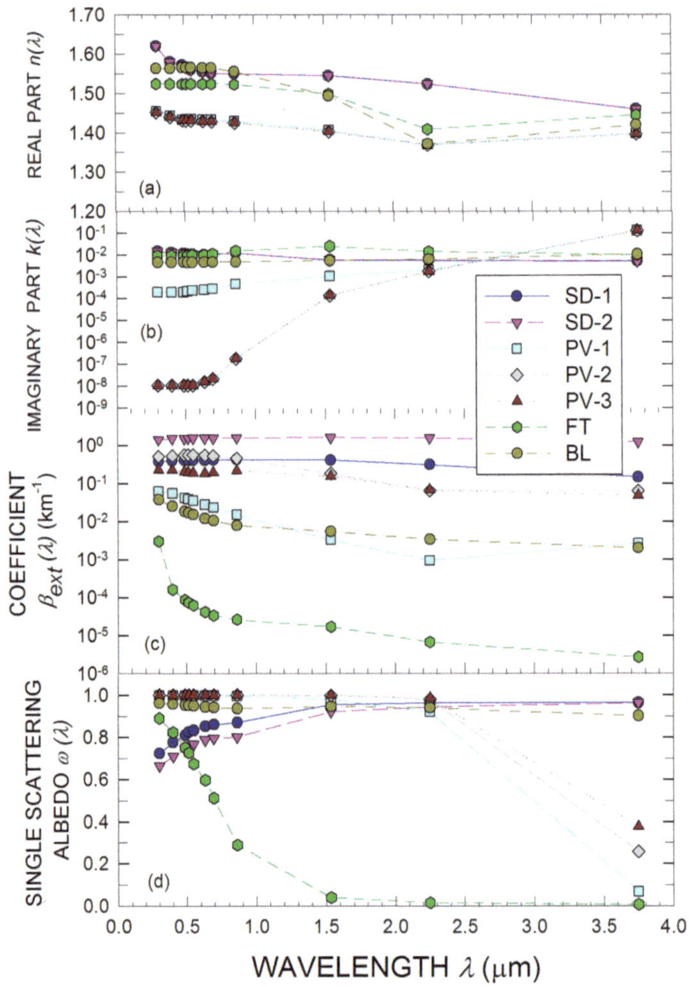

Fig. 11.15. As in Fig. 11.2, for the seven additional aerosol models defined in the present study for Sahara dust particles (SD-1 and SD-2), background stratospheric aerosol during volcanic quiescence periods in Antarctica (PV-1), 2-month aged Pinatubo volcanic stratospheric particles (PV-2), 9-month aged Pinatubo volcanic stratospheric particles (PV-3), biomass burning smoke particles in free troposphere (FT), and biomass burning smoke particles in the boundary layer (BL). All models are normalized to give a value of the overall particle number concentration $N_{\text{tot}} = 1000\,\text{cm}^{-3}$.

by the SD-1 and SD-2 models over the radius range typical of coarse and giant particles with respect to the other five models. The PV-1, PV-2, PV-3 and BL models all exhibit important contents of coarse particles but with radii smaller than 1 μm, while the FT model results to consist predominantly of fine particles. The spectral curves of $n(\lambda)$ and $k(\lambda)$ and those of radiative parameters $\beta_{\text{ext}}(\lambda)$ and $\omega(\lambda)$ are presented in Fig. 11.15 over the 0.30–3.70 μm wavelength range, showing that: (i) $n(\lambda)$ varies mainly between 1.40 and 1.60 at all visible and infrared wavelengths;

(ii) $k(\lambda)$ assumes values varying between about 10^{-8} (PV-2 and PV-3 models) and around 10^{-2} (SD-1, SD-2 and FT models); (iii) $\beta_{ext}(\lambda)$ presents values ranging mainly between $10^{-5}\,km^{-1}$ (FT model) and more than $1\,km^{-1}$ (SD-2 model) at visible wavelengths, such a large variability being due to the different contents of coarse and giant particles characterizing the various models; and (iv) single scattering albedo $\omega(\lambda)$ determined at visible wavelengths for the FT and BL models was evaluated to assume values varying between 0.6 and 0.8, while those of the three PV models were found to be very close to unity and those of the two SD models to vary between 0.8 and 0.9.

11.2.7 Comparison among the radiative properties of the 40 aerosol models

A comparison among the 40 aerosol models of different origin described above is presented in Fig. 11.16, showing the values of weighted average single scattering albedo ω^* as a function of the corresponding Ångström's exponent $\alpha(0.40–0.86\,\mu m)$ calculated over this narrow visible and near-infrared spectral range. Values of ω^* very close to unity characterize the maritime models, which exhibit values of $\alpha(0.40–0.86\,\mu m)$ lower than 0.5, while values of ω^* ranging between 0.88 and 0.94 pertain to the desert dust models. The mixed maritime-continental aerosol models exhibit gradually decreasing values of ω^*, from more than 0.96 to less than 0.88 as the relative content of continental particles increases, while $\alpha(0.40–0.86\,\mu m)$ correspondingly increases from 0.5 to more than 1.0, until assuming spectral characteristics similar to those of the aerosols present in the Arctic and Antarctic regions. The

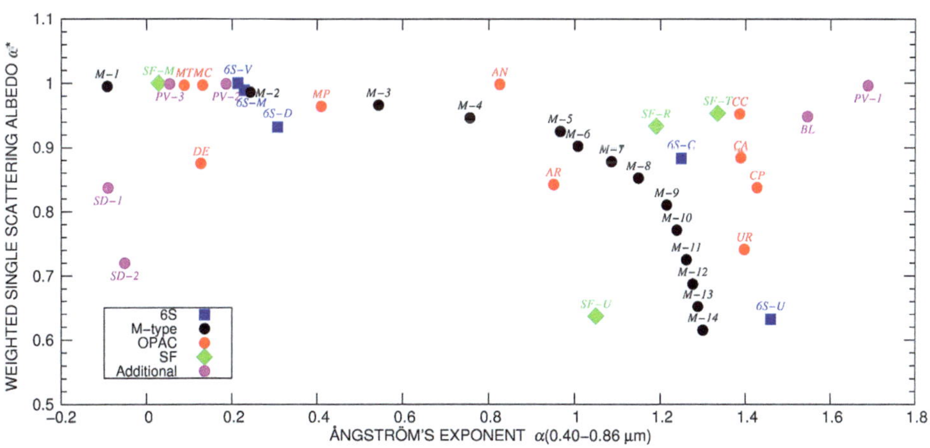

Fig. 11.16. Scatter plots of the weighted average single scattering albedo ω^* versus the Ångström's exponent $\alpha(0.40–0.86\,\mu m)$ defined in Tables 11.3, 11.4, 11.7, 11.9 and 11.10, as obtained for (i) the five 6S dry-air aerosol models (blue circles), (ii) the 14 M-type wet-air (RH = 50%) aerosol models (black circles), (iii) the 10 OPAC wet (RH = 50%) aerosol models (red circles), (iv) the 4 SF (Shettle and Fenn, 1979) wet-air (RH = 50%) aerosol models (green diamonds), and (v) the 7 present additional aerosol models (fuchsia circles). The FT additional model is not shown in the graph, because it yields a value of $\alpha(0.40–0.86\,\mu m) = 2.48$.

continental models obtained for various atmospheric turbidity conditions present gradually lower values of ω^*, from 0.85 to less than 0.60, as the mass fraction of polluted aerosol increases, while the corresponding values of $\alpha(0.40–0.86\,\mu\text{m})$ continue to increase gradually until exceeding the value of 1.50 in cases pertaining to the most marked anthropogenic pollution conditions characterizing the urban areas. The additional FT (Free Troposphere) model derived from samples taken in the part of the atmosphere located above the boundary layer does not appear in the graph since its value of $\alpha(0.40–0.86\,\mu\text{m})$ is equal to 2.48, and hence, much higher than the values defined over the whole atmospheric column for the other 39 models. Fig. 11.16 clearly shows that all such aerosol models cover a large variety of airborne aerosol extinction features observed at various latitudes, as generated by different sources, thus allowing the calculation of a large number of DARF terms induced by aerosol polydispersions in real cases.

11.3 Underlying surface reflectance characteristics

Using the two-stream approximation procedure (Coakley and Chylek, 1975) for simulating the radiative transfer processes in a plane atmosphere containing a layer of aerosol particles with well-defined absorptance and reflectance characteristics, Chylek and Coakley (1974) demonstrated that the aerosol particle layer can induce cooling or warming effects in the atmosphere, depending on the surface albedo conditions for each ratio between absorptance and reflectance of such aerosol particles. This implies that the DARF effects induced at the ToA level by a certain particle load tends to change sign from negative (cooling) to positive (warming) as the aerosol particles under investigation are transported from oceanic areas, typically presenting relatively low surface albedo conditions, to polar regions covered by snow fields and glaciers, e.g. the interior of Greenland and the Antarctic Plateau, which generally present very high surface albedo properties. The Chylek and Coakley (1974) calculations also evidenced that for each surface albedo value, the DARF effect at the ToA level tends to change sign from cooling to warming, as the absorptance properties of particulate matter increase with respect to reflectance, until exceeding a critical value of the absorptance/reflectance ratio. In order to investigate the dependence features of the DARF effects on the surface reflectance properties, as they occur at the ToA and BoA levels as well as in the atmosphere, a set of 16 surface reflectance models were determined in the present analysis, using four classes of bidirectional reflectance distribution functions (BRDF) to improve the representations of the geometrical and spectral characteristics of surface reflectance defined in the 6S code (Vermote et al., 1997a). A bidirectional reflectance distribution function is determined in terms of ratio $f_e = dL\uparrow(\eta,\phi)/dF\downarrow(\theta_o,\phi_o)$ between the radiance $dL\uparrow$ reflected upward by the surface in the direction individuated by the pair of polar zenith and azimuth angles η and ϕ, and the incident irradiance $dF\downarrow$ coming from the direction (θ_o,ϕ_o). It is measured in sr^{-1}, and assumes a constant value equal to $1/\pi$ for an ideal lambertian reflector. An exhaustive analysis of the BRDF function and its derived quantities is available in Nicodemus et al. (1977). The first surface reflectance models were developed in the 80s (Jupp, 2000), when it became of crucial importance to represent the anisotropic features of the reflected solar radiation field for analyzing more correctly the Earth observation data recorded from space-borne platforms (Kriebel, 1978; Hapke 1986). Therefore, the BRDF models were implemented during those years by retrieving the surface reflectance and atmospheric transmittance characteristics from satellite-based data. In addition, in order to determine with improved accuracy the spectral and geometrical dependence features of the BRDF parameters, more realistic approaches were implemented through numerous tests and specific applications (Vermote at al, 1997b; Wanner et al., 1997), which allowed the determination of bidirectional reflectance models suitable for use to perform DARF calculations for different surface reflectance characteristics. For practical purposes, it is useful to insert the concept of bidirectional reflectance factor $R(\lambda, \theta_o, \phi_o, \eta, \phi)$, which is commonly used to represent the ratio between the real BRDF surface reflectance and the BRDF reflectance of an ideal (100%) lambertian reflector. This factor is assumed to depend not only on wavelength λ but also on the four angular coordinates of the Sun-surface–external viewer system. Varying as a function of such angular parameters, factor R provides the ratio between (i) the upwelling irradiance $F\uparrow(\eta, \phi)$ reflected by the surface in a certain upwelling direction (defined by nadir angle η and azimuth angle ϕ) and (ii) the incident flux of a collimated incoming radiation beam with direction defined by solar zenith angle θ_o and azimuth angle ϕ_o. The plane perpendicular to the surface, containing both the Sun and the ground-level reference spot, defines the principal plane of reflection. Function $R(\lambda, \theta_o, \phi_o, \eta, \phi)$ used to represent the surface relectance is commonly assumed to exhibit a cylindrical symmetry with respect to the principal plane of reflection. Thus, its mathematical representation is in general made considering only the difference between the two angles ϕ_o and ϕ, which will be hereinafter named as angular difference ϕ'.

To obtain precise calculations of DARF at the ToA-level for each geometrical configuration of the surface–atmosphere system defined by the angular parameters θ_o, η and ϕ', it was decided to use a set of BRDF models based on rigorous physical concepts, allowing us to perform the calculations of the DARF terms planned in the present study. This choice allowed us to achieve realistic simulations of the deep surface–atmosphere coupling effects induced by the complex radiative transfer processes occurring inside the surface–air interface layer. Various parameterization criteria were adopted for this purpose, based on both hyperspectral (H type) and non-hyperspectral (N-H type) models, where the reflected radiance fields generated by surfaces having various characteristics are represented at each wavelength of the 0.30–4.00 μm spectral range as a function of the above-mentioned angular coordinates, to totally cover the 2π upward solid angle. To represent the surface reflectance properties characterized by various spectral and angular dependence features, the BRDF models were determined in terms of the following functions typical of each surface:

(1) The *spectral directional hemispherical reflectance* (black-sky albedo) $R_{bs}(\lambda, \theta_o)$, obtained through integration of the BRDF function $R(\lambda, \theta_o, \phi_o = 0°, \eta, \phi)$ over the 2π upward solid angle, to represent the spectral curve of surface reflectance as a function of solar zenith angle θ_o (Nicodemus et al., 1977; Román et al., 2010). The function is considered to be valid in the ideal case in which the diffuse component of the global (direct + diffuse) solar radiation field is assumed to be null. It is worth remarking that this spectral function provides the spectral curve of the so-called

black-sky albedo for $\theta_o = 0°$, expressed in terms of the following analytical form,

$$R_{bs}(\lambda, \theta_o = 0°) = \frac{1}{\pi} \int_0^{2\pi} \int_0^{\pi/2} R(\lambda, \theta_o, \phi_o, \eta, \phi) \cos\eta \sin\eta \, d\eta \, d\phi \,. \tag{11.4a}$$

(2) The *spectral bi-hemispherical reflectance* (white-sky albedo) $R_{ws}(\lambda)$, obtained by integrating the function $R_{bs}(\lambda, \theta_o)$ over the 2π upward solid angle, and assuming that the incoming solar radiation field consists only of the diffuse component $D\downarrow(\lambda)$ characterized by isotropic features. Therefore, function $R_{ws}(\lambda)$ in practice represents the spectral curve of surface albedo relative to the diffuse component of incoming solar radiation:

$$R_{ws}(\lambda) = \frac{1}{\pi} \int_0^{2\pi} \int_0^{\pi/2} R_{bs}(\lambda, \theta_o) \cos\theta_o \sin\theta_o \, d\theta_o \, d\phi_o \,. \tag{11.4b}$$

Bearing in mind that the white-sky albedo is given in Eq. (11.4b) by the double integral of the black-sky albedo $R_{bs}(\lambda, \theta_o)$ over the entire intervals of the two downwelling polar angles, and that $R_{bs}(\lambda, \theta_o)$ is obtained in Eq. (11.4a) through the double hemispherical integration of the birectional reflectance factor R over the entire ranges of the two upwelling polar angles, it is evident that $R_{ws}(\lambda)$ does not depend on the geometrical configuration of the Sun-surface–external viewer system. In cases of isotropic surface reflectance conditions, white sky albedo can be assumed to be that of an equivalent lambertian reflector. Its features can be better defined employing satellite data derived from the observations of the Sun-synchronous multispectral sensors mounted on the Terra polar platform, such as MISR (Diner et al., 1998; Bothwell et al., 2002) and MODIS (Christopher and Zhang, 2002).

(3) The spectral curve of *surface albedo* $R_L(\lambda, \theta_o)$, which is obtained as the weighted average of the spectral surface reflectance contributions, associated with the black-sky and white-sky albedo conditions of the solar radiation field, respectively (Lewis and Barnsley, 1994; Lucht et al., 2000). This approximate function can be calculated in terms of the analytical form of the Lewis (1995) function defined by the following equation,

$$R_L(\lambda, \vartheta_o) = R_{bs}(\theta_o)\,[1 - D\downarrow(\lambda)] + R_{ws}(\theta_o) D\downarrow(\lambda) \,, \tag{11.4c}$$

where $D\downarrow(\lambda)$ is the spectral curve of the diffuse fraction of downwelling (global) solar radiation $I\downarrow(\lambda)$ reaching the surface, which can be calculated as a function of solar zenith angle θ_o using the 6S code (Vermote et al., 1997b) for any atmospheric content of aerosol particles.

In order to obtain realistic evaluations of the DARF effects occurring inside the surface–atmosphere system, the BRDF models need to (A) be completely defined over the whole spectral range of the incoming solar radiation from about 0.30 to 4.0 μm, and (B) represent exhaustively the variety of surface albedo conditions most commonly observed in the various regions of the Earth, presenting average values of surface albedo ranging from less than 0.1 over the oceans to more than 0.8 over the ice-covered polar regions.

- With regard to the first point (A), it is worth noting that the BRDF surface reflectance models commonly used in the literature need to be defined with a good spectral accuracy over the 0.40–2.50 μm wavelength range. This range includes 88.8% of incoming extra-terrestrial solar radiation, while only limited percentages pertain to the wavelength intervals below 0.40 μm (8.0%) and beyond 2.50 μm (3.2%) (Iqbal, 1983). On this matter, it can be noted that: (i) surface reflectance usually causes an overall reflected irradiance flux within the range $\lambda > 2.50\,\mu$m, which is so small as to be totally negligible for practical purposes; (ii) aerosol effects are in general very weak at wavelengths $\lambda > 2.50\,\mu$m; and (iii) calculation errors made in evaluating the differences between the radiation fluxes passing through turbid and pristine atmospheres are often of opposite sign and tend to annul each other. Therefore, it was decided to limit the present surface reflectance calculations to the wavelength range from 0.40 to 2.50 μm.

- With regard to the second point (B), different self-adapted BRDF models were considered in the present study (Morel, 1988; Rahman et al., 1993; Kuusk, 1994) over the entire spectral range chosen above, to realistically represent the surface reflectance characteristics of different surfaces, such as those typical of various sea-water areas, vegetation-covered and agricultural land regions, bare soil and arid areas, and snow- and ice-covered polar regions.

11.3.1 The non-lambertian surface reflectance models

A set of BRDF non-lambertian surface reflectance models were defined to constitute a finite lattice of spectral surface albedo values, increasing gradually from less than 0.05 to more than 0.9. For the present analysis, four classes of surface reflectance models were determined pertaining to the ocean surfaces (OS class), vegetated surfaces (VS class), bare soils and arid terrain (BS class), and polar surfaces (PS class). The spectral and angular characteristics of these reflectance models are described for each class as follows:

(1) *OS class* (Ocean Surfaces), which consists of four BRDF surface reflectance models representing the typical oceanic surface reflectance conditions described by the OCEAN hyperspectral model (Morel, 1988) and developed using the OCEAN subroutine given in the 6S code (Vermote et al., 1997b). The four models also take into account the characterictics defined by the whitecaps model of Koepke (1984) (improved modeling features of spectral reflectance of whitecaps were most recently proposed by Kokhanovsky (2004)), including the sun glint reflectance effects (Cox and Munk, 1954) and effects due to Fresnel's reflection (Born and Wolf, 1975). The OCEAN subroutine was used to calculate the BRDF function curves for all the triplets of angular coordinates θ_o, η and ϕ', as a function of the wind speed V_w, and for other constant pre-fixed values of the following supplemental parameters: (i) the wind direction D_w, assumed to lie on the vertical plane $\phi_o = 0°$ for all the OS models; (ii) the sea-water pigment concentration C_p, equal to 0.1 mg/m^3, this assumption being made considering that variations in C_p of more than four orders of magnitude can cause only relatively small changes in the surface reflectance leading to relative variations in the reflectance that do not exceed 10%; and (iii) sea-water salt concentration C_s equal to 34.3 ppt, this value being assumed taking into account that an increase in C_s from 0 to 48 ppt is estimated to induce surface reflectance changes much smaller than 1%. For the above characteristics,

the following four OS models were obtained: (i) OS1 for $V_w = 2\,\mathrm{m\,s^{-1}}$; (ii) OS2 for $V_w = 5\,\mathrm{m\,s^{-1}}$; (iii) OS3 for $V_w = 10\,\mathrm{m\,s^{-1}}$; and (iv) OS4 for $V_w = 20\,\mathrm{m\,s^{-1}}$.

The corresponding spectral curves of surface albedo $R_L(\lambda, \theta_o = 60°)$ are shown in Fig. 11.17, while the monochromatic values of this parameter are reported in Table 11.11 for 21 selected wavelengths over the spectral range from 0.40 to 2.50 μm, as obtained for the four OS models, for their use in application studies of the surface albedo characteristics. Table 11.12 provides the values of reflectance parameters $R_{bs}(\theta_o = 0°)$ and R_{ws}, and those of broadband albedo $A(\theta_o)$ calculated for 9 increasing values of θ_o taken in steps of 10° over the 0°–80° range for each of the four OS models. The values of $A(\theta_o)$ were calculated through integration of $R_L(\lambda, \theta_o)$ over the 0.40–2.50 μm wavelength range. They are reported in Table 11.12, as determined for the M-8 continental aerosol model and aerosol optical thickness $\tau_a = 0.10$ in the visible. Due to the fact that $A(\theta_o)$ is calculated in terms of the following equation,

$$A(\theta_o) = \frac{\int_{0.4\,\mu\mathrm{m}}^{2.5\,\mu\mathrm{m}} R_L(\lambda, \theta_o) I\!\downarrow(\lambda, \theta_o)\, d\lambda}{\int_{0.4\,\mu\mathrm{m}}^{2.5\,\mu\mathrm{m}} I\!\downarrow(\lambda, \theta_o)\, d\lambda}, \tag{11.4d}$$

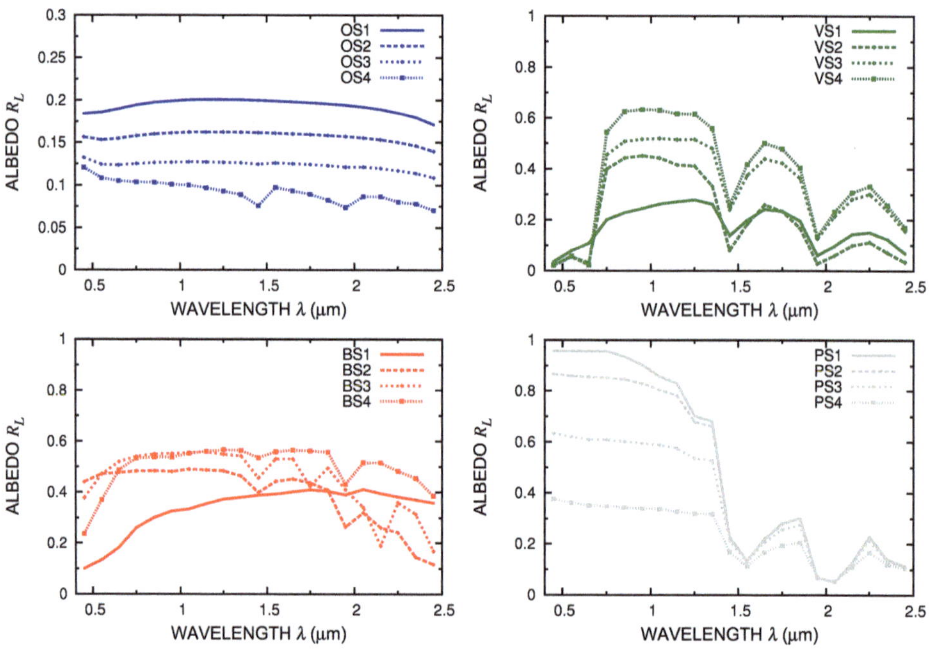

Fig. 11.17. Spectral curves of surface albedo $R_L(\lambda)$ (Lewis and Barnsley, 1994), as defined in Eq. (11.4c) over the 0.40–2.50 μm wavelength range for the 16 BRDF surface reflectance models considered in the present study. All the reflectance models were determined for the radiance field features defined for (i) the optical characteristics of the US62 atmosphere model (Dubin et al., 1966), (ii) the scattering and absortion properties of the M-8 aerosol model (as described in Figs. 11.6 and 11.7), consisting of pure continental particles (see Table 11.4), (iii) aerosol optical depth $\tau_a(0.55\,\mu\mathrm{m}) = 0.10$, and (iv) solar zenith angle $\theta_o = 60°$. Note that the range of R_L is from 0 to 1 for the VS, BS and PS models, and from 0 to 0.3 for the OS models.

Table 11.11. Monochromatic values of surface albedo $R_L(\theta_o, \lambda)$ defined in Eq. (11.4c) and shown in Fig. 11.17, as obtained at 21 wavelengths selected over the 0.40–2.50 μm spectral range for the 16 BRDF surface reflectance models subdivided into the OS, VS, BS and PS classes

Wavelength λ (μm)	OS1	OS2	OS3	OS4	VS1	VS2	VS3	VS4	BS1	BS2	BS3	BS4	PS1	PS2	PS3	PS4
0.45	0.184	0.156	0.132	0.120	0.034	0.023	0.019	0.021	0.099	0.439	0.378	0.237	0.957	0.866	0.633	0.377
0.55	0.186	0.153	0.124	0.108	0.077	0.054	0.053	0.056	0.133	0.471	0.470	0.370	0.956	0.859	0.620	0.362
0.65	0.190	0.155	0.123	0.105	0.108	0.029	0.022	0.021	0.182	0.476	0.519	0.486	0.955	0.855	0.609	0.350
0.75	0.194	0.158	0.125	0.103	0.200	0.399	0.455	0.543	0.260	0.481	0.538	0.531	0.956	0.852	0.607	0.348
0.85	0.197	0.160	0.126	0.103	0.228	0.442	0.507	0.624	0.300	0.482	0.548	0.536	0.935	0.845	0.601	0.343
0.95	0.199	0.161	0.127	0.101	0.244	0.451	0.516	0.632	0.325	0.480	0.550	0.536	0.903	0.830	0.594	0.338
1.05	0.200	0.162	0.127	0.100	0.262	0.442	0.519	0.629	0.333	0.489	0.554	0.551	0.857	0.803	0.588	0.338
1.15	0.201	0.162	0.127	0.097	0.271	0.415	0.514	0.615	0.354	0.486	0.556	0.559	0.828	0.780	0.574	0.327
1.25	0.201	0.162	0.127	0.093	0.278	0.411	0.515	0.614	0.372	0.483	0.547	0.565	0.702	0.678	0.535	0.319
1.35	0.200	0.162	0.126	0.088	0.262	0.332	0.479	0.558	0.379	0.462	0.540	0.562	0.680	0.660	0.527	0.318
1.45	0.200	0.161	0.124	0.075	0.138	0.079	0.239	0.258	0.387	0.398	0.454	0.534	0.227	0.226	0.211	0.167
1.55	0.199	0.161	0.126	0.097	0.199	0.184	0.374	0.418	0.392	0.442	0.530	0.558	0.134	0.134	0.129	0.112
1.65	0.198	0.160	0.125	0.093	0.241	0.258	0.439	0.500	0.401	0.451	0.530	0.564	0.220	0.219	0.206	0.167
1.75	0.197	0.159	0.124	0.089	0.233	0.233	0.422	0.477	0.409	0.436	0.422	0.561	0.280	0.278	0.255	0.193
1.85	0.196	0.158	0.123	0.082	0.195	0.167	0.363	0.403	0.403	0.408	0.494	0.556	0.302	0.299	0.273	0.204
1.95	0.194	0.157	0.121	0.073	0.059	0.027	0.125	0.135	0.390	0.264	0.410	0.431	0.068	0.068	0.068	0.068
2.05	0.192	0.155	0.121	0.086	0.096	0.059	0.213	0.232	0.410	0.319	0.340	0.515	0.050	0.050	0.051	0.054
2.15	0.189	0.153	0.119	0.086	0.143	0.098	0.281	0.307	0.395	0.261	0.188	0.514	0.126	0.125	0.121	0.109
2.25	0.185	0.150	0.117	0.080	0.150	0.111	0.301	0.331	0.381	0.240	0.359	0.482	0.229	0.227	0.211	0.167
2.35	0.179	0.146	0.114	0.078	0.122	0.070	0.237	0.257	0.370	0.144	0.313	0.454	0.139	0.138	0.134	0.119
2.45	0.171	0.140	0.109	0.070	0.067	0.033	0.156	0.169	0.358	0.116	0.168	0.385	0.113	0.113	0.111	0.105

Table 11.12. Values of directional-hemispherical reflectance $R_{bs}(0°)$ (black-sky albedo) at null zenith illumination angle, bi-hemispherical reflectance R_{ws} (white-sky albedo, which does not depend on solar zenith angle θ_o) and broadband albedo $A(\theta_o)$ calculated in terms of Eq. (11.4d) for nine values of solar zenith angle θ_o (increasing from 0° to 80° in steps of 10°), as obtained for the 16 surface reflectance models defined in the present study and the spectral characteristics of the US62 Standard Atmosphere model (Dubin et al., 1966) and the M8 (pure continental) aerosol model defined in Table 11.4. The values of relative optical air mass m given in brackets for the various values of $A(\theta_o)$ were evaluated using the algorithms of Kasten and Young (1989) and Tomasi et al. (1998). The values of $A(80°)$ highlighted with the background light green colour are affected by high uncertainties, due to the fact that the four OS (Oceanic Surface) models are quite failing for $\theta_o > 75°$

Model	$R_{bs}(0°)$	R_{ws}	$A(0°)$ (1.0000)	$A(10°)$ (1.0148)	$A(20°)$ (1.0634)	$A(30°)$ (1.1536)	$A(40°)$ (1.3037)	$A(50°)$ (1.5526)	$A(60°)$ (1.9928)	$A(70°)$ (2.8999)	$A(80°)$ (5.5803)
OS1	0.026	0.070	0.030	0.030	0.030	0.035	0.048	0.083	0.193	0.601	2.098
OS2	0.026	0.069	0.030	0.030	0.030	0.033	0.043	0.071	0.158	0.425	1.202
OS3	0.028	0.069	0.031	0.031	0.032	0.034	0.042	0.065	0.127	0.284	0.714
OS4	0.044	0.081	0.048	0.048	0.049	0.051	0.057	0.072	0.105	0.180	0.385
VS1	0.134	0.153	0.135	0.134	0.133	0.139	0.141	0.149	0.155	0.175	0.202
VS2	0.170	0.203	0.172	0.450	0.450	0.178	0.450	0.450	0.210	0.450	0.450
VS3	0.201	0.243	0.203	0.202	0.204	0.213	0.223	0.239	0.258	0.287	0.340
VS4	0.250	0.292	0.253	0.253	0.256	0.264	0.274	0.289	0.306	0.333	0.386
BS1	0.225	0.237	0.226	0.225	0.224	0.228	0.229	0.235	0.240	0.256	0.278
BS2	0.431	0.457	0.434	0.431	0.429	0.438	0.441	0.453	0.461	0.487	0.510
BS3	0.456	0.484	0.459	0.456	0.455	0.463	0.467	0.480	0.489	0.519	0.549
BS4	0.424	0.450	0.426	0.424	0.422	0.430	0.433	0.446	0.455	0.485	0.521
PS1	0.824	0.847	0.827	0.827	0.830	0.834	0.840	0.847	0.854	0.862	0.865
PS2	0.720	0.761	0.726	0.727	0.732	0.739	0.749	0.761	0.775	0.789	0.796
PS3	0.461	0.536	0.472	0.475	0.483	0.496	0.514	0.537	0.564	0.591	0.608
PS4	0.214	0.296	0.223	0.226	0.234	0.249	0.269	0.296	0.329	0.365	0.390

by integrating the spectral quantities $R_L(\lambda, \theta_o) \times I\!\downarrow(\lambda, \theta_o)$ and $I\!\downarrow(\lambda, \theta_o)$ over the 0.40–2.50 μm wavelength range, some minor variations in the broadband albedo $A(\theta_o)$ can arise from the changes affecting the diffuse fraction $D\!\downarrow$ of the incoming solar irradiance (see Eq. (11.4c)) which are associated with variations in the columnar contents of the atmospheric constituents.

It is worth noting in Table 11.12 that the values of $A(\theta_o)$ found for the four OS models (a) slowly increase passing from OS1 to OS4 for low values of θ_o, and (b) decrease appreciably for $\theta_o \geq 30°$, presenting even more rapid variations as one passes from the OS1 model to the OS4 one, as θ_o assumes gradually higher values. This behavior is presumably due to an increasing specular reflectance effect occurring for calm wind conditions ($V_w = 2\,\mathrm{m\,s^{-1}}$).

(2) *VS class* (Vegetated Surfaces), which consists of four BRDF reflectance models relative to vegetated surfaces, as derived from the AK subroutine reported by the 6S code (Vermote et al., 1997b), used to evaluate the reflectance properties of vegetated surfaces in terms of the hyperspectral Kuusk (1994) model simulations. This subroutine utilizes the PROSPECT code (Jacquemoud and Baret, 1990) for simulating the chlorophyll absorption features, and the Nilson and Kuusk (1989) algorithm for representing the reflectance anisotropy characteristics of the one-layer canopy coverage. The angular BRDF values were calculated as a function of (i) various optical parameters characterizing the physical, optical and biological properties of vegetation (i.e. chlorophyll content C_{AB}, leaf water equivalent thickness C_w, effective number N_e of elementary layers inside a leaf, ratio C_n of refractive indices of the leaf surface wax and internal material, and weight S_1 of the first Price (1990) function for the soil reflectance), and (ii) structural parameters (i.e. Leaf Area Index U_L, elliptical eccentricity E_e of the leaf angle distribution, modal inclination Q_m of the leaf distribution, and relative leaf size S_L with respect to the canopy depth). Varying the above parameters, the following four VS models were obtained: (i) VS1, derived from the Kuusk (1994) corn model for $C_{AB} = 100\,\mu\mathrm{g/cm^2}$, $C_w = 0.04$ cm, $N_e = 1.09$, $C_n = 0.9$, $S_1 = 0.213$, $U_L = 0.1$, $E_e = 0.972$, $Q_m = 10.7°$, and $S_L = 0.1$; (ii) VS2, derived from the Kuusk (1994) corn model for $C_{AB} = 100\,\mu\mathrm{g/cm^2}$, $C_w = 0.026$ cm, $N_e = 1.09$, $C_n = 0.9$, $S_1 = 0.213$, $U_L = 1.5$, $E_e = 0.972$, $Q_m = 10.7°$, and $S_L = 0.1$; (iii) VS3, derived from the Kuusk (1994) soybean model for $C_{AB} = 82.2\,\mu\mathrm{g/cm^2}$, $C_w = 0.005$ cm, $N_e = 1.24$, $C_n = 0.9$, $S_1 = 0.225$, $U_L = 2.5$, $E_e = 0.965$, $Q_m = 45.8°$, and $S_L = 0.1$; and (iv) VS4, derived from the Kuusk (1994) soybean model for $C_{AB} = 82.2\,\mu\mathrm{g/cm^2}$, $C_w = 0.005$ cm, $N_e = 1.24$, $C_n = 0.9$, $S_1 = 0.225$, $U_L = 5.0$, $E_e = 0.965$, $Q_m = 45.8°$, and $S_L = 0.1$. The four spectral curves of albedo $R_L(\lambda, \theta_o = 60°)$ obtained for the VS models are shown in Fig. 11.17 over the 0.40–2.50 μm wavelength range, while the values of broadband albedo $A(\theta_o)$ are given in Table 11.12 for each model, as obtained for values of θ_o increasing in steps of 10° from 0° to 80°. The typical spectral signature of a vegetation-covered surface, given by the typical sharp increase of reflectance at around 0.70 μm wavelength, and commonly called 'red edge', is well reproduced by the Kuusk (1994) model. Model VS1 appears to be suitable for representing a canopy clearly affected by drought conditions, for which the soil spectral signature emerges from the background, while the VS2 to VS4 models can be more confidently adopted to represent vegetation-coverages with increasing Leaf Area Index U_L, and the gradually more enhanced features of the 'red edge'. The monochro-

matic values of surface albedo $R_L(\lambda, \theta_o = 60°)$ obtained for the four VS models are reported in Table 11.11 at the 21 selected wavelengths over the 0.40–2.50 μm range, for their use by the readers. As shown in Table 11.12, the white-sky albedo R_{ws} increases from 0.15 to nearly 0.30 passing from the VS1 model to the VS4 one.

(3) *BS class* (Bare Soil), which consists of four bare soil BRDF surface reflectance models derived from the non-hyperspectral model proposed by Rahman et al. (1993) and used as subroutine to implement the 6S code (Vermote et al., 1997a). These models are based on four spectral curves of surface reflectance defined for surfaces covered by (i) dry sand (BS1), as directly taken from the 6S code; (ii) illite (BS2), as represented by a mixing of clay minerals having crystal structures very similar to that of muscovite; (iii) alunite (BS3), as derived by assuming the spectral properties of a mineral consisting of a hydrous potassium aluminum sulfate and presenting massive forms mixed with rhombohedral crystals, and (iv) montmorillonite (BS4), as represented by a soft clay mineral consisting mainly of a hydrous aluminum silicate in which aluminum was exchanged abundantly with magnesium and other bases. The last three curves were taken from the USGS Library (http://speclab.cr.usgs.gov), giving the wavelength-dependence features of surface reflectance $R_o(\lambda, \theta_o = 0°)$ shown in Fig. 11.17 over the 0.40–2.50 μm spectral range. The angular-dependence characteristics of the above BRDF surface reflectance functions were defined by (i) assuming that they are similar to those determined by Rahman et al. (1993), and (ii) using the pair of parameters K and ξ, of which the first serves to define the anisotropy degree of surface reflectance, and the second to evaluate its asymmetry degree, which allows us to regulate the relative intensities of forward and backward scattering. In the four BS models: (a) parameter K was assumed to be equal to 0.648 within the wavelength intervals $\lambda < 0.630\,\mu$m and to 0.668 for $\lambda > 0.915\,\mu$m, respectively, and (b) asymmetry parameter ξ was kept as equal to -0.290 and -0.268 within the two above-defined wavelength intervals, respectively. These assumptions were made in agreement with the results obtained by Rahman et al. (1993) from the data sets recorded by Kimes et al. (1985) for a bare soil surface (ploughed field): at intermediate wavelengths from 0.630 to 0.915 μm, the values of K and ξ were calculated through a linear interpolation procedure in wavelength between the values established above. The spectral patterns of the four BS models are shown in Fig. 11.17, while (a) the monochromatic values of $R_L(\lambda, \theta_o = 60°)$ determined for the four BS models are given in Table 11.11 at the 21 above-chosen wavelengths for application studies, and (b) the corresponding values of broadband albedo $A(\theta_o)$ relative to 9 values of θ_o increasing in steps of 10° from 0° to 80° are given in Table 11.12, as obtained using the above values of parameters K and ξ. The white-sky albedo is approximately equal to 0.24 for the BS1 model (and, hence, very similar to that of the BS3 model) and assumes values equal to 0.45 and 0.48 for the BS2 and BS4 models, respectively, as shown in Table 11.12.

(4) *PS class* (Polar Surfaces), which consists of four surface reflectance models suitable to represent snow and glacier surfaces, as derived from the hyperspectral model describing the surface reflectances relative to (i) the direct component of the spectral directional hemispherical reflectance (black-sky albedo) $R_{bs}(\lambda, \theta_o)$ defined in Eq. (4a), and (ii) the diffuse component of spectral bi-hemispherical reflectance

(white-sky albedo) $R_{ws}(\lambda, \theta_o)$ of incoming solar radiation defined in Eq. (4b). The main parameters of the PS models are the size-distributions of the snow and black carbon grains, represented by means of log-normal size-distribution functions, and the concentration of dust and/or black carbon in the snow surface layer (Wiscombe and Warren, 1980; Warren and Wiscombe, 1980). In modeling such surface reflectance features, it was taken into account that the effects produced on reflectance by a volume concentration of black carbon particles equal to 1 ppb are comparable with those produced by a volume concentration of dust particles equal to 100 ppm. Both types of this particulate matter have been observed to cause a relevant reduction of albedo over the $\lambda < 1\,\mu$m range of solar radiation spectrum, where soot particles produce a rather flat spectral curve of reflectance, and dust particles cause an appreciable increase in albedo at the 0.6–0.7 μm wavelengths. Snow grains with sizes varying between 50 and 500 μm were found to produce only weak variations in the visible part of the solar spectrum. These features agree in general with the recent results found by Kokhanovsky and Breon (2012), who investigated the anisotropic characteristics of the snow BRDF reflectance, defining a semi-empirical spectral model based on detailed representations of the forward scattering maximum dependence on view zenith angle and azimuthal reflectance variations.

Thus, various simulations of surface albedo were made by assuming that different additional concentrations of soot particles are present (having log-normal size-distribution curves centred at radius $r = 0.1\,\mu$m) together with the main component of the surface layer constituted by snow grains having a mean radius of 100 mm. A Mie algorithm was defined to calculate the spectral values of single scattering albedo $\omega(\lambda)$ and asymmetry factor $g(\lambda)$ of such a snow grain size-distribution at 217 selected wavelengths over the 0.40–2.50 μm range (Warren and Wiscombe, 1980). The two parts of the complex refractive index of snow grains were determined according to the Warren (1984) estimates, while those of black carbon were calculated at the same 217 wavelengths by applying an interpolation procedure in wavelength for the 11 monochromatic values provided by the 6S (Vermote et al., 1997b) parameterization for the aerosol soot component. Following this procedure, values of the real part ranging between 1.75 and 1.90 were obtained, together with values of the imaginary part varying between 0.43 and 0.57 over the entire solar spectrum, according to the values of $n(\lambda)$ and $k(\lambda)$ given in Table 11.2 for the SO (soot) component of Vermote et al. (1997b). The direct and diffuse solar radiation components were then calculated following the semi-empirical parameterization method of Wiscombe and Warren (1980), based on the use of the above-calculated values of optical parameters $\omega(\lambda)$ and $g(\lambda)$ of the bi-modal size-distribution consisting of soot particles and snow grains. Four models have been obtained following the above procedure, by assuming gradually increasing values of the soot volume concentration C_s equal to 0.002, 0.04, 0.40 and 2.0 ppm, which cause a gradual decrease in albedo. They are: (1) the PS1 model for almost pure snow ($C_s = 0.002$ ppm), (2) the PS2 model for slightly contaminated snow ($C_s = 0.04$ ppm), (3) the PS3 model for simulating the experimental observations performed at South Pole (Grenfell and Maycut, 1977) for a snow coverage with $C_s = 0.4$ ppm, and (4) the PS4 model for heavy carbon contaminated snow ($C_s = 2$ ppm). The spectral curves of surface albedo $R_L(\lambda, \theta_o = 60°)$ are shown in Fig. 11.17, while the monochromatic values of

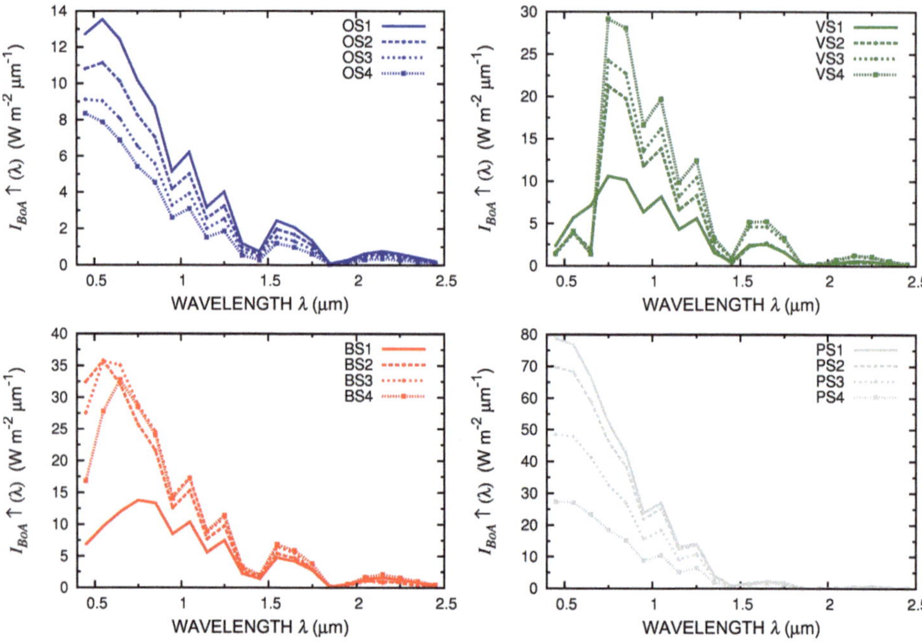

Fig. 11.18. Spectral curves of the upwelling irradiance $I_{\mathrm{BoA}}\uparrow(\lambda)$ at the BoA-level, given by the product of spectral surface albedo $R_L(\lambda, \theta_o = 60°)$ by the incoming global (direct + diffuse) irradiance $I\downarrow$ at the BoA-level, as obtained for the 16 surface reflectance models shown in Fig. 11.17 for $\theta_o = 60°$. All the reflectance models were determined for the radiance field features defined for (i) the optical characteristics of the US62 atmosphere model (Dubin et al., 1966), (ii) the scattering and absorption properties of the M-8 aerosol model (as described in Figs. 11.6 and 11.7), consisting of pure continental particles (see Table 11.4), and (iii) aerosol optical depth $\tau_a(0.55\,\mu\mathrm{m}) = 0.10$. Note the variable range of $I_{\mathrm{BoA}}\uparrow(\lambda)$ is adopted for the four sets of surface reflectance models.

this parameter obtained for the four PS models are given in Table 11.11 at the 21 above-selected wavelengths over the 0.40–2.50 μm spectral range. The corresponding values of the broadband albedo $A(\theta_o)$ determined at 9 values of θ_o from 0° to 80° for the four PS models are reported in Table 11.12. The spectral signature of snow surface is characterized by an appreciable decrease in reflectance as λ increases throughout the middle infrared range beyond 1 μm wavelength, with well-defined features of the water absorption bands labeled with the Greek capital letters Ψ (over the 1.25–1.54 μm spectral range) and Ω (over the 1.69–2.08 μm range), which are well reproduced by the model. Broadband white-sky albedo varies between 0.30 for extremely polluted snow cover (PS4 model) to nearly 0.85 for clean snow cover (PS1 model). In order to calculate the DARF effects, it is necessary to calculate the differences between the irradiance relative to pristine atmospheric transparency conditions and the irradiance relative to a turbid atmosphere. Therefore, it is useful to have a clear picture of the spectral dependence features of the upwelling irradiance $I_{\mathrm{BoA}}\uparrow(\lambda)$ reflected at the surface, as it results from the combined effects of the incoming solar irradiance reaching the surface (characterized by the spectral

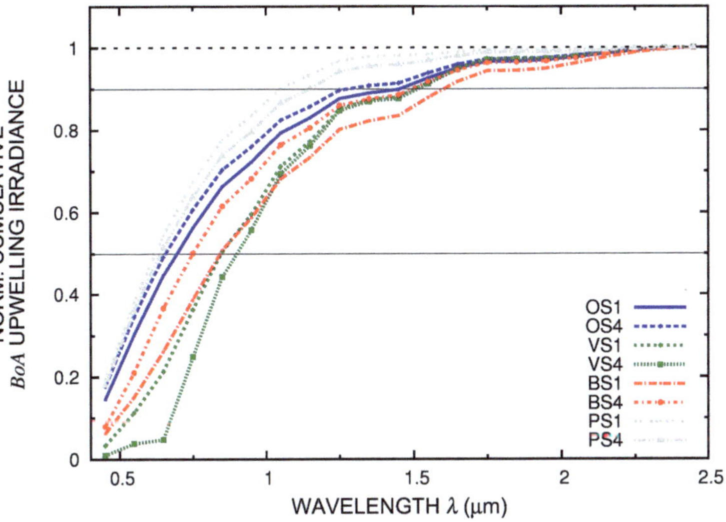

Fig. 11.19. Spectral curves of the normalized cumulative upwelling irradiance at the BoA-level, calculated as the ratio $\int_{0.4}^{\lambda} I_{\text{BoA}} \uparrow (\lambda)\, d\lambda / \int_{0.4}^{2.5} I_{\text{BoA}} \uparrow (\lambda)\, d\lambda$ over the 0.40–2.50 μm wavelength range, as determined for eight of the 16 BRDF non-lambertian surface reflectance models defined in the present study. All the curves are normalized to give a unit value of upwelling irradiance over the entire spectral range.

extinction features of the atmosphere) with those related to the spectral signature characteristics of each surface reflectance model. The spectral curves of the upwelling irradiance $I_{\text{BoA}} \uparrow (\lambda)$ are shown in Fig. 11.18, each obtained in terms of the product of the global (including both direct and diffuse components) downwelling irradiance by spectral surface albedo $R_L(\lambda, \theta_o)$. To integrate such an information, Fig. 11.19 shows the cumulative upwelling irradiance at the BoA-level, normalized to the whole broadband irradiance obtained through the spectral integration of the product $R_L \times I\downarrow$. It is interesting to note that about 50% of such a cumulative curve of upwelling irradiance is already available at nearly 0.60 μm wavelength in the polar surface cases, while this half percentage is reached at around 0.80 μm for the vegetated surfaces (VS models), which present the lowest reflectivity characteristics over the visible spectral range. It is also evident in Fig. 11.19 that about 90% of the cumulative curve of upwelling irradiance is reached at wavelengths $\lambda > 1.0\,\mu$m for the PS reflectance models, and only at wavelengths ranging between 1.4 and 1.6 μm for the VS and BS reflectance models.

Fig. 11.20 shows the spectral curves of white-sky albedo $R_{ws}(\lambda)$, which will be further used for determining the reflectance conditions of the equivalent lambertian reflectance models belonging to the OS, VS, BS and PS classes. The monochromatic values of $R_{ws}(\lambda)$ obtained at 21 wavelengths chosen over the spectral range from 0.40 to 2.50 μm are presented in Table 11.13 for the 16 BRDF surface reflectance models of the OS, VS, BS and PS classes defined above, since these data could be suitable for application studies on the white-sky albedo of various surfaces. Parameter $R_{ws}(\lambda)$ assumes values very close to those of $R_{bs}(\lambda)$ calculated for solar zenith angle $\theta_o = 60°$, as is typical of natural surfaces. In other works, this quantity was

Table 11.13. Monochromatic values of white-sky albedo $R_{ws}(\lambda)$ defined in Eq. (11.4b) and shown in Fig. 11.20, as obtained at 21 wavelengths selected over the 0.40–2.50 μm spectral range for the 16 BRDF surface reflectance models subdivided into the OS, VS, BS and PS classes

Wavelength λ (μm)	OS1	OS2	OS3	OS4	VS1	VS2	VS3	VS4	BS1	BS2	BS3	BS4	PS1	PS2	PS3	PS4
0.45	0.087	0.087	0.087	0.102	0.034	0.023	0.019	0.020	0.099	0.437	0.376	0.236	0.954	0.857	0.612	0.351
0.55	0.071	0.070	0.071	0.086	0.076	0.051	0.050	0.053	0.132	0.468	0.467	0.368	0.953	0.847	0.592	0.330
0.65	0.066	0.065	0.066	0.081	0.106	0.029	0.022	0.020	0.181	0.473	0.515	0.483	0.950	0.840	0.578	0.316
0.75	0.065	0.064	0.065	0.078	0.201	0.389	0.432	0.524	0.259	0.478	0.535	0.528	0.950	0.836	0.574	0.312
0.85	0.065	0.064	0.065	0.077	0.229	0.432	0.483	0.604	0.298	0.480	0.544	0.533	0.928	0.828	0.566	0.305
0.95	0.065	0.064	0.064	0.074	0.244	0.441	0.491	0.611	0.324	0.477	0.547	0.532	0.891	0.810	0.558	0.300
1.05	0.065	0.064	0.064	0.073	0.262	0.432	0.495	0.608	0.332	0.486	0.550	0.547	0.840	0.781	0.551	0.299
1.15	0.064	0.064	0.063	0.070	0.270	0.406	0.490	0.594	0.352	0.483	0.553	0.555	0.809	0.756	0.536	0.289
1.25	0.064	0.063	0.063	0.066	0.277	0.401	0.491	0.593	0.370	0.480	0.544	0.561	0.671	0.646	0.495	0.280
1.35	0.064	0.063	0.062	0.061	0.260	0.323	0.457	0.537	0.377	0.459	0.537	0.559	0.648	0.626	0.487	0.279
1.45	0.064	0.063	0.061	0.049	0.135	0.076	0.225	0.243	0.385	0.396	0.451	0.531	0.193	0.192	0.179	0.140
1.55	0.063	0.062	0.062	0.070	0.197	0.178	0.354	0.398	0.390	0.439	0.526	0.554	0.110	0.110	0.105	0.091
1.65	0.063	0.062	0.062	0.066	0.239	0.251	0.417	0.479	0.399	0.449	0.526	0.561	0.187	0.186	0.175	0.139
1.75	0.062	0.062	0.061	0.063	0.231	0.226	0.400	0.456	0.407	0.434	0.420	0.557	0.242	0.240	0.219	0.163
1.85	0.062	0.061	0.060	0.056	0.192	0.161	0.343	0.383	0.401	0.406	0.491	0.553	0.263	0.260	0.236	0.172
1.95	0.061	0.061	0.059	0.047	0.058	0.026	0.115	0.125	0.387	0.263	0.407	0.428	0.054	0.054	0.054	0.054
2.05	0.061	0.060	0.059	0.060	0.094	0.056	0.198	0.217	0.408	0.317	0.338	0.511	0.039	0.039	0.040	0.043
2.15	0.060	0.059	0.059	0.061	0.141	0.094	0.263	0.289	0.392	0.260	0.187	0.511	0.102	0.102	0.099	0.088
2.25	0.059	0.058	0.057	0.055	0.148	0.106	0.282	0.312	0.379	0.239	0.357	0.479	0.194	0.192	0.178	0.139
2.35	0.058	0.057	0.056	0.053	0.120	0.067	0.221	0.241	0.368	0.144	0.312	0.452	0.113	0.113	0.109	0.097
2.45	0.055	0.055	0.053	0.047	0.066	0.032	0.144	0.157	0.356	0.115	0.167	0.383	0.091	0.091	0.090	0.085

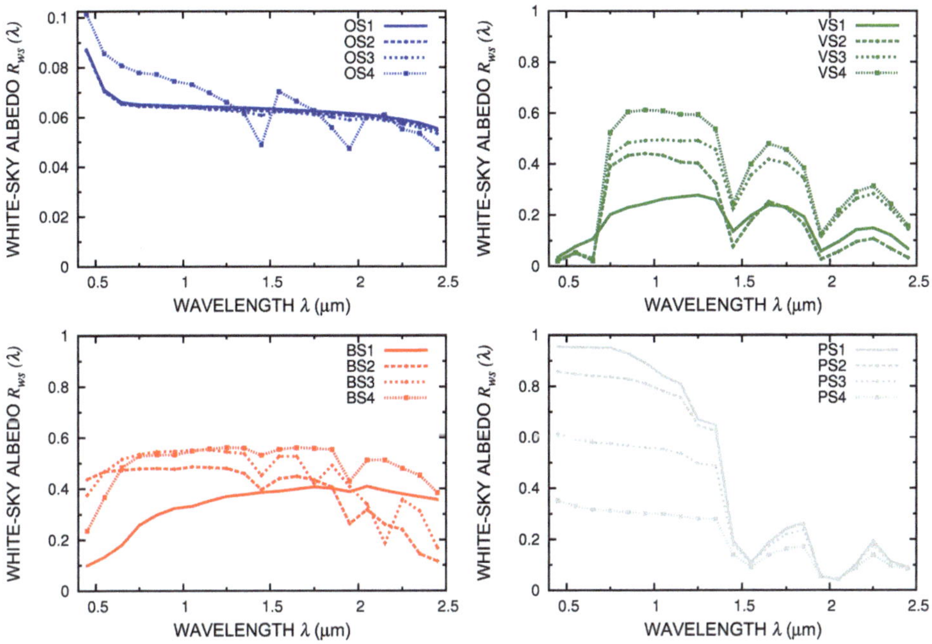

Fig. 11.20. Spectral curves of white-sky albedo $R_{ws}(\lambda)$ defined in Eq. (11.4b) over the 0.40–2.50 μm spectral range for the four OS, VS, BS and PS classes of surface reflectance models. Note that the range of $R_{ws}(\lambda)$ is from 0 to 1 for the VS, BS and PS models, and from 0 to 0.1 for the OS models.

used as a comparison term for evaluating the effects of the lambertian assumption on the DARF calculations (Ricchiazzi et al., 2005). However, the white-sky albedo is here preferably used, considering that (i) it is independent of the geometric configuration of the Sun-surface–external viewer system, and (ii) it can be conveniently determined on the basis of satellite-based products, as mentioned above.

The effects of the assumption of non-lambertian surface reflectance conditions are summarized in Fig. 11.21 for the four OS3, VS3, BS3 and PS3 surface models chosen as examples. The graph shows the spectral curves of surface albedo $R_L(\lambda, \theta_o)$ for solar zenith angles increasing from 0° to 80° in step of 10°, compared with the corresponding white-sky albedo. For each model the curve of R_{ws} was found to lie between the two black-sky albedo curves $R_{bs}(\lambda, \theta_o = 50°)$ and $R_{bs}(\lambda, \theta_o = 60°)$. Thus, in order to weight the single scattering albedo effects on solar radiation, the scattering and absorption effects occurring at visible and near-infrared wavelengths were taken into account, bearing in mind that aerosols are mostly present within the lower part of the troposphere. On this basis, we decided to use the weighted average single scattering albedo ω^* as key parameter for evaluating the relevance of the aerosol radiative effects in inducing the DARF processes. For this purpose, the spectral curve $I^*(\lambda)$ was used as weight function, as defined above in subsection 11.2.3 adopting the spectral distribution curve of direct solar irradiance passing through the U.S. Standard Atmosphere (1976) (Anderson et al., 1986) and reaching the sea-level for $\theta_o = 60°$ (i.e. for relative optical air mass $m = 2$ (Tomasi et al., 1998)).

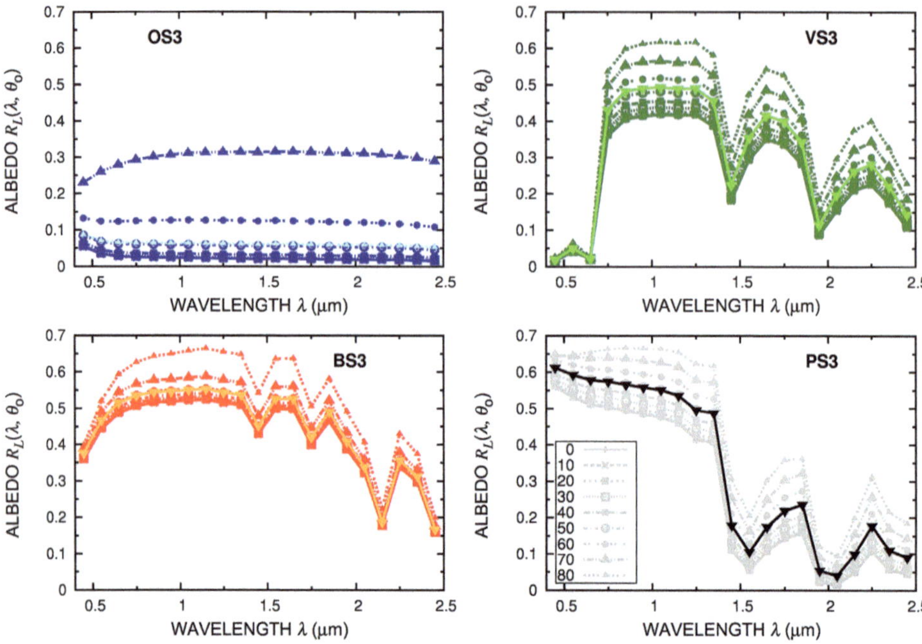

Fig. 11.21. Spectral curves of surface albedo $R_L(\lambda, \theta_o)$ defined in Eq. (11.4c) over the 0.40–2.50 μm spectral range, as obtained for different values of solar zenith angle θ_o (in steps of 10° from 0° to 80°) and for the OS3, VS3, BS3 and PS3 surface reflectance models in a surface–atmosphere system, where the atmosphere is assumed to have (i) the optical characteristics of the US62 atmosphere model (Dubin et al., 1966), (ii) the scattering and absortion properties of the M-8 aerosol model (illustrated in Figs. 11.6 and 11.7), consisting of pure continental particles (see Table 11.4), and (iii) aerosol optical depth $\tau_a(0.55\,\mu\text{m}) = 0.10$. Note that the spectral curve of $R_L(\lambda, \theta_0)$ for the OS3 reflectance model at $\theta_0 = 80°$ assumes values higher than 0.9 for such a very high value of solar zenith angle and, hence, is not reported in the OS3 graph.

11.3.2 The isotropic (lambertian) surface reflectance models

It was assumed in subsection 11.3.1 that the instantaneous DARF terms can be properly evaluated by taking into account the spectral features of the non-lambertian (anisotropic) surface reflectance models. However, the isotropic surface reflectance models are often used to calculate such instantaneous DARF terms, generally obtaining less realistic evaluations in the cases presenting high values of $\tau_a(\lambda)$ in the visible and solar zenith angles $\theta_o > 70°$ (Chyleck and Wong, 1995; Russell et al., 1997). In order to achieve a measure of the relative differences between the evaluations made using isotropic surface reflectance models in place of BRDF non-lambertian models, calculations of the DARF terms were performed by assuming that the reflector is lambertian and, hence, using a set of 16 isotropic surface reflectance models having average values of surface reflectance (over the visible and near-infrared wavelength range) equal to those obtained in subsection 11.3.1 for the 16 BRDF non-lambertian reflectance models. For this purpose, the spectral curve of white-sky albedo R_{ws} defined in Eq. (11.4b) was determined through a bi-hemispherical integration of each of the 16 BRDF non-lambertian surface reflectance models described in subsection 11.3.1.

Applying this procedure to the lambertian surface reflectance models, the white-sky albedo R_{ws} was found to assume average values that are in practice independent of θ_o. The evaluations of R_{ws} obtained for these 16 equivalent lambertian surface reflectance models were used to calculate the aerosol-induced change flux terms for isotropic characteristics of surface reflectance. To achieve this goal, each isotropic (lambertian) spectral reflectance model was defined by using the geometrically integrated reflectance output originated from the BRDF scheme applied to the non-lambertian models in subsection 11.3.1. Following the calculation scheme shown in Fig. 11.22, the variations in the radiation flux density were then calculated for (i) the pristine geometrical Sun-surface–atmosphere (without aerosol) configuration, and (ii) the configuration of the Sun-surface–atmosphere system (with aerosol). Adopting this procedure, the influence of surface isotropic reflectance (assumed in the lambertian models) on the calculations of the DARF terms was evidenced, providing the results presented in Fig. 11.23, as obtained for different pairs of surface models and various wavelengths, and for $\theta_o = 30°$, the 6S-C aerosol model, and $\tau_a(0.55\,\mu\text{m}) = 0.10$. The examined cases were shown in three separate columns pertaining to (i) the VS1 model and $\lambda = 0.55\,\mu\text{m}$, (ii) the VS1 model and $\lambda = 0.75\,\mu\text{m}$, and (iii) the OS2 model and $\lambda = 0.75\,\mu\text{m}$. Each column is subdivided into four panels relative to as many radiation quantities, to highlight their variations

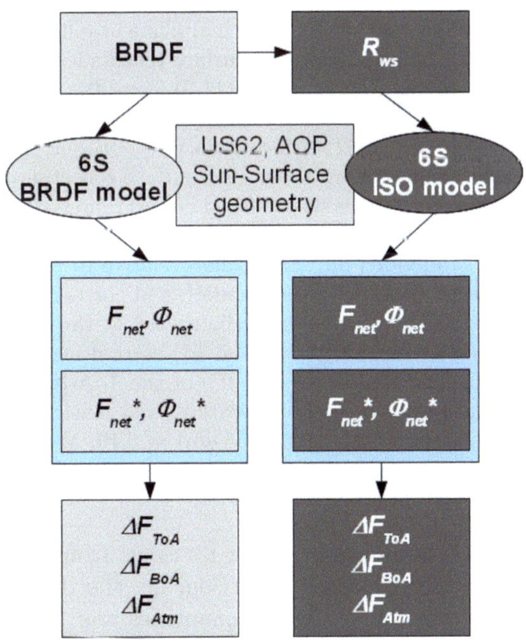

Fig. 11.22. Calculation scheme adopted to evaluate the effects of anisotropic (BRDF, non-lambertian) and isotropic (ISO, lambertian) surface reflectance models on the instantaneous direct aerosol-induced radiative forcing terms ΔF_{ToA}, ΔF_{BoA}, and ΔF_{Atm}, as defined in Eqs. (11.5), (11.9) and (11.13), respectively.

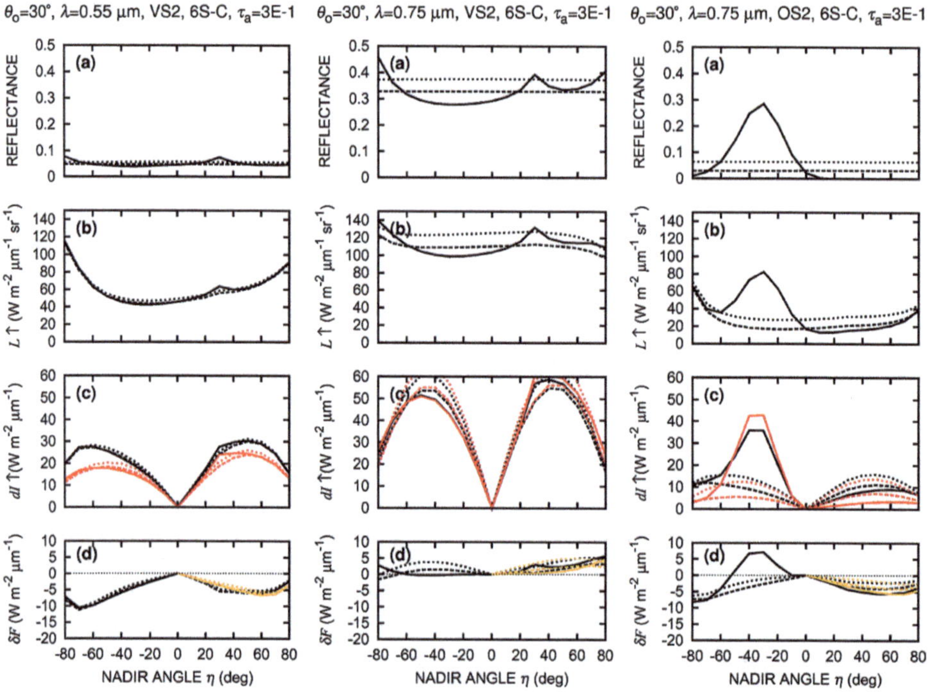

Fig. 11.23. Dependence curves of parameters R, $L\uparrow$, $dI\uparrow$, and $\delta F\uparrow$ on the view nadir angle η ranging from $-80°$ to $+80°$ in the principal plane of reflection ($\phi' = 0°$) for three cases with: (11.1) $\theta_o = 30°$, $\lambda = 0.55\,\mu$m, VS2 surface reflectance model, and 6S-C (continental aerosol) extinction model (left column); (11.2) $\theta_o = 30°$, $\lambda = 0.75\,\mu$m, VS2 surface reflectance model, and 6S-C (continental aerosol) extinction model (middle column); and (11.3) $\theta_o = 30°$, $\lambda = 0.75\,\mu$m, OS2 surface reflectance model, and 6S-C (continental aerosol) extinction model (right column). For each surface reflectance model, solid curves refer to the BRDF non-lambertian version, dashed curves represent $R_{bs}(\theta_o = 30°)$ for the isotropic reflectance version, and dotted curves represent R_{ws} for the isotropic reflectance version. Each column is subdivided into four panels relative to the following parameters: (a) surface reflectance R for non-lambertian models, and corresponding isotropic parameters $R_{bs}(\theta_o = 30°)$ and R_{ws}; (b) upwelling radiance $L\uparrow$ at the ToA-level; (c) upwelling differential irradiance $dI\uparrow$ at the ToA-level; and (d) aerosol-induced differential change $\delta F\uparrow$ in the net outgoing flux $F\uparrow(\lambda, \theta_o, \eta, \phi' = 0°)$ at the ToA-level. Red curves in panel (c) refer to the pristine atmosphere without aerosols, and black curves to the turbid atmosphere with aerosol optical thickness $\tau_a(0.55\,\mu\text{m}) = 0.10$. Yellow curves in panel (d) refer to the perpendicular reflection plane with respect to the Sun-target plane.

as a function of the view nadir angle η over its range from $-80°$ to $+80°$ (whose positive values refer to the backward reflection, i.e. the direction containing the source of illumination). The four radiation quantities are:

(a) the surface bidirectional reflectance factor R, calculated for the BDRF non-lambertian version of the surface reflectance models, and compared with the black-sky albedo $R_{bs}(\theta_o = 30°)$ and white-sky albedo R_{ws}, both determined for the isotropic version of the surface reflectance models;

(b) the upwelling radiance $L\uparrow$, determined at the ToA-level;
(c) the upwelling differential irradiance $dI\uparrow$ at the ToA-level; and
(d) the aerosol-induced differential change $\delta F\uparrow$ in the net outgoing flux $F\uparrow$ $(\lambda, \theta_o, \eta, \phi' = 0°)$ at the ToA-level.

The reflectance panels labeled with letter (a) in the three columns of Fig. 11.23 clearly show the existence of hot-spot effects on vegetated surfaces, reaching a local maximum of bidirectional reflectance R at $\eta = \theta_o$, and give evidence of sun glint features in the sea surface cases presenting a reflectance maximum in the forward direction for $\eta = -\theta_o$. Panels (b) show that the anisotropy features of the surface reflectance are not completely masked by the upwelling radiance passing through the atmosphere. Such an atmospheric effect depends obviously on the magnitude of the total optical thickness τ of the atmosphere, being the anisotropy less evident for higher values of τ. To describe these aerosol-induced effects exhaustively, panels (c) show the angular dependence patterns of upwelling irradiance $dI\uparrow = L\uparrow \cos\eta \sin\eta \, d\eta \, d\phi$, defined for both turbid (with aerosol) and pristine (without aerosol) atmospheric transparency conditions. It was found that the strongest contributions to the upwelling irradiance are given in the directions with $\eta = 45°$, due to the fact that $dI\uparrow$ is proportional to the product $\cos\eta \times \sin\eta$. The three panels (d) of Fig. 11.23 present the angular features of the aerosol-induced change flux differential terms as a function of η, showing that a strong difference can arise from the use of the reflectance scheme shown in Fig. 11.22 to carry out the present calculations. It is also worth noting that the present lambertian approach can cause different marked effects with respect to those obtained using the full BRDF scheme, which in general vary rather considerably as a function of angle η. Consequently, the integration made to obtain the spectral irradiance term generally masks the BRDF effects characterizing the DARF calculations.

In cases of very low reflectance conditions, such as those occurring in the visible for green vegetation (as in the case of the VS2 model shown in Fig. 11.23), the absolute differences arising from the two non-lambertian and isotropic calculation schemes are very low (and, hence, negligible in practice) with respect to the errors made in defining the aerosol and surface parameters (as can be seen, for instance, in panel (d) of Fig. 11.23). In fact, as the wavelength increases from the visible to about $0.75\,\mu$m, the vegetation reflectance increases as well, and the absolute differences become more pronounced. Actually, panel (c) exhibits different values in the upwelling irradiance $dI\uparrow$, which are close to $15\,\mathrm{W\,m^{-2}\,\mu m^{-1}}$, while the value of $50\,\mathrm{W\,m^{-2}\,\mu m^{-1}}$ was obtained for $\theta_o = -40°$ (forward direction). These changes are even more evident in the case of ocean surface reflectance, as clearly shown by the results given in Fig. 11.23 for the OS2 model, where neglecting the sun glint features leads to obtain underestimated values of $dI\uparrow$ equal to nearly $-25\,\mathrm{W\,m^{-2}\,\mu m^{-1}}$ with respect to the absolute value of $40\,\mathrm{W\,m^{-2}\,\mu m^{-1}}$ found for the anisotropy assumption. In this case, the error made in estimating the differential forcing change $\delta F\uparrow$ leads to a positive value of $+5\,\mathrm{W\,m^{-2}\,\mu m^{-1}}$ in place of a slightly negative value (see panel (d) of Fig. 11.23).

For lambertian surface reflectance features, more evident changes were found with respect to the non-lambertian configurations in all cases where pristine atmosphere conditions (red curves) were considered, while less pronounced variations were obtained in cases presenting turbid atmosphere conditions (black curves).

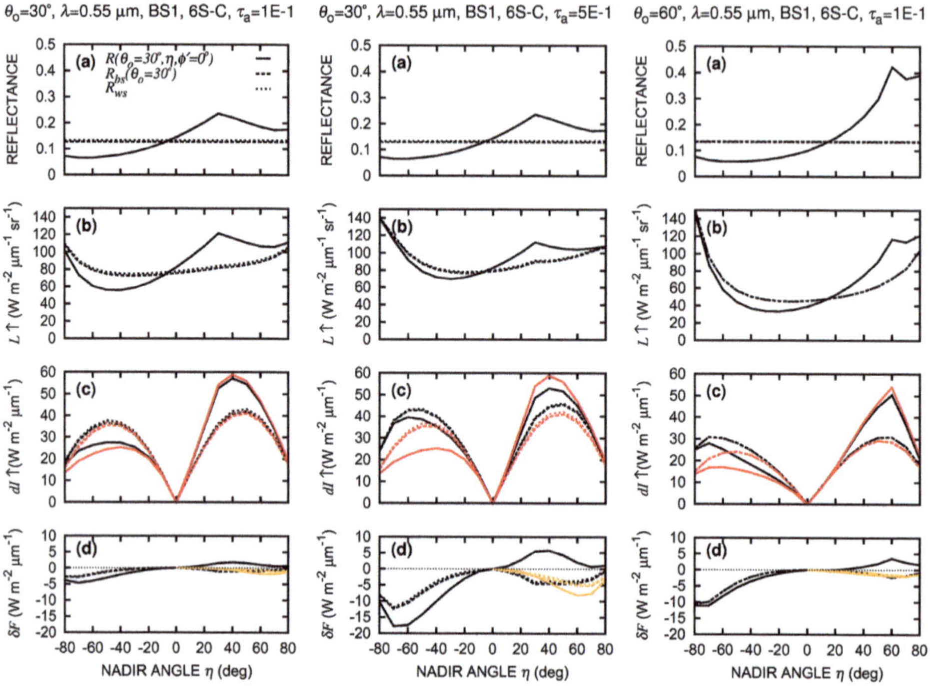

Fig. 11.24. As in Fig. 11.23, for the following three cases: (11.1) $\theta_o = 30°$, $\lambda = 0.55\,\mu$m, BS1 surface reflectance model, and 6S-C (continental aerosol) extinction model defined for aerosol optical thickness $\tau_a(0.55\,\mu\text{m}) = 0.10$ (left column); (11.2) $\theta_o = 30°$, $\lambda = 0.55\,\mu$m, BS1 surface reflectance model, and 6S-C (continental aerosol) extinction model defined for aerosol optical thickness $\tau_a(0.55\,\mu\text{m}) = 0.05$ (middle column); and (11.3) $\theta_o = 60°$, $\lambda = 0.55\,\mu$m, BS1 surface reflectance model, and 6S-C (continental aerosol) extinction model defined for aerosol optical thickness $\tau_a(0.55\,\mu\text{m}) = 0.10$ (right column). For each surface reflectance model, solid curves refer to the BRDF non-lambertian version, dashed curves represent $R_{bs}(\theta_o = 30°)$ for the isotropic reflectance version, and dotted curves represent R_{ws} for the isotropic reflectance version. Each column is subdivided into four panels relative to the following parameters: (a) surface reflectance R for non-lambertian models, and corresponding isotropic parameters R_{bs} evaluated at $\theta_o = 30°$ (left and middle columns) or $\theta_o = 60°$ (right column) and R_{ws}; (b) upwelling radiance $L\uparrow$ at the ToA-level; (c) upwelling differential irradiance $dI\uparrow$ at the ToA-level; and (d) aerosol-induced differential change $\delta F\uparrow$ in the net outgoing flux $F\uparrow(\lambda, \theta_o, \eta, \phi' = 0°)$ at the ToA-level. Red curves in panel (c) refer to the pristine atmosphere without aerosols, and black curves to the turbid atmosphere with aerosol optical thickness $\tau_a(0.55\,\mu\text{m}) = 0.10$. Yellow curves in panel (d) refer to the perpendicular reflection plane with respect to the Sun-target plane.

The changes are associated with atmospheric turbidity conditions, which in general mask the anisotropic features of the surface reflectance, because the product $R_{ws}(\theta_o) \times D\downarrow(\lambda)$ in the second term of Eq. (11.4c) turns out to be enhanced by the increase in the diffuse fraction $D\downarrow(\lambda)$ of downwelling irradiance at the BoA-level originating from the increase in aerosol optical thickness τ_a. These aspects are clearly shown in Fig. 11.24, where the calculations of the radiation flux terms are reported for the BS1 surface reflectance model and the 6S-C continental aerosol model, considering the following configurations of the geometric and atmospheric conditions: (i) $\theta_o = 30°$ and $\tau_a(0.55\,\mu\text{m}) = 0.10$; (ii) $\theta_o = 30°$ and $\tau_a(0.55\,\mu\text{m}) = 0.50$; and (iii) $\theta_o = 60°$ and $\tau_a(0.55\,\mu\text{m}) = 0.10$. The increase in the diffuse fraction $D{\downarrow}(\lambda)$ is thus produced by the increase in τ_a (second column) or the increase in θ_o (third column). In both such cases, the absolute values of the differences between the evaluations of dL and, hence, of $dI{\uparrow}$, both performed for isotropic (ISO) surface reflectance models, were found to be considerably lower than those obtained for BRDF non-lambertian (anisotropic) reflectance models.

11.4 Instantaneous direct aerosol-induced radiative forcing (DARF)

The net flux of short-wave radiation at the ToA-level (or at another level close to the tropopause) is given by the difference between the incoming flux $F{\downarrow}$ and the outgoing flux $F{\uparrow}$, both measured in W m^{-2}. In all cases where this quantity varies as a result of scattering and/or absorption of radiation by an atmospheric constituent, the energy available for 'governing' the Earth's climate is subject to change, and the net flux variation is referred to as the radiative forcing of climate. It implies the concept that climate is consequently subject to a cooling (or warming) trend, depending on the negative (or positive) sign of the net flux change. In fact, the conventional definition of aerosol-induced radiative forcing establishes that it is due to a perturbation in the content and optical properties of columnar particulate matter and is evaluated as the net radiative flux change induced at the tropopause (or at the ToA) level, keeping the concentrations of all the other atmospheric constituents constant (WMO, 1986). In other words, radiative forcing is an imposed change in the net (downwelling minus upwelling) radiation flux at the tropopause altitude (or at ToA-level) (IPCC, 1996; Hansen et al., 1997). Therefore, in the case of atmospheric aerosol, the ToA-level DARF term gives a measure of the energy input provided by such an atmospheric constituent to the climate system. This energy deficit or surplus at the ToA-level depends not only on the magnitude of the radiation flux change at ToA-level but also on the vertical profiles of the aerosol radiative parameters (Hansen et al., 1997).

11.4.1 Definitions

Because of the predominance of the short-wave effects produced by aerosol polydispersions over those occurring throughout the long-wave radiation spectrum, as indicated by the Mie (1908) theory, DARF is commonly evaluated considering only the short-wave (solar) radiation flux change, and neglecting the radiative effects produced by aerosols on the long-wave radiation. These short-wave evaluations are suitable for providing quantitative estimates of the overall change induced by columnar aerosols in the net radiative budget of the atmosphere, with the required precision and accuracy. They are generally evaluated instantaneously for pre-fixed hours of the day, and then calculated as diurnal averages over the 24-hour period

(Bush and Valero, 2003). Instantaneous DARF term ΔF_{ToA} can be simply determined at ToA-level as the difference between (i) the net radiative flux for the turbid atmosphere containing a certain columnar content of aerosol particles, and (ii) the same quantity in a pristine atmosphere without aerosols. Thus, the instantaneous forcing ΔF_{ToA} induced at a certain time by aerosol particles can be represented in terms of the following formula:

$$\Delta F_{\mathrm{ToA}} = F_{\mathrm{net}} - F^{*}_{\mathrm{net}} \,, \tag{11.5}$$

where (i) the ToA-level net flux F_{net} is given by the difference between the short-wave downwelling flux $F\!\downarrow$ and the short-wave upwelling flux $F\!\uparrow$, both determined at the ToA-level for an atmosphere including all its constituents, i.e.:

$$F_{\mathrm{net}} = F\!\downarrow - F\!\uparrow \,, \tag{11.6}$$

where the downwelling flux $F\!\downarrow$ is obviously independent of the aerosol extinction processes taking place in the atmosphere, and (ii) the net flux F^{*}_{net} at ToA-level is given by the difference between the corresponding short-wave downwelling flux $F\!\downarrow^{*}$ and short-wave upwelling flux $F\!\uparrow^{*}$, both calculated in the pristine atmosphere without aerosols, i.e.:

$$F^{*}_{\mathrm{net}} = F\!\downarrow^{*} - F\!\uparrow^{*} \,, \tag{11.7}$$

where the downwelling flux $F\!\downarrow^{*}$ is not altered by atmospheric aerosols.

Therefore, flux $F\downarrow^{*}$ in Eq. (11.7) is equal to flux $F\downarrow$ given in Eq. (11.6). Consequently, the instantaneous term ΔF_{ToA} calculated from Eqs. (11.5), (11.6) and (11.7) is directly given by the difference,

$$\Delta F_{\mathrm{ToA}} = F\!\uparrow^{*} - F\!\uparrow \,, \tag{11.8}$$

which shows that this DARF term can be correctly evaluated by simply subtracting the upward solar radiation flux emerging from the real atmosphere with aerosols from the upwelling solar radiation flux emerging from the pristine atmosphere without aerosols (Hänel et al., 1999). According to the WMO (1986) conventional definition of radiative forcing in the atmosphere, negative values of ΔF_{ToA} indicate that aerosols cause an increase in the upwelling flux of solar radiation and, hence, an increase in the albedo of the surface–atmosphere system, producing direct cooling effects on the climate system. Conversely, positive values of ΔF_{ToA} indicate that lower upwelling solar radiation fluxes are induced by aerosols, leading to a decrease in the overall albedo and, consequently, causing significant atmospheric warming effects (Loeb and Manalo-Smith, 2005; Christopher et al., 2006; Zhao et al., 2008).

The aerosol radiative forcing ΔF_{BoA} at the surface (i.e. BoA-level) gives a measure of the perturbation in the net flux reaching the surface, which is induced by airborne aerosols. Therefore, ΔF_{BoA} can be defined as the difference between the net flux at surface-level in the atmosphere with aerosols and the net flux at surface-level in the same atmosphere assumed without aerosols (Satheesh and Ramanathan, 2000; Bush and Valero, 2002, 2003). It can be expressed at a given time as the difference:

$$\Delta F_{\mathrm{BoA}} = \Phi_{\mathrm{net}} - \Phi^{*}_{\mathrm{net}} \,, \tag{11.9}$$

where, the net flux Φ_{net} at the surface is given by the difference between the downwelling flux $\Phi\downarrow$ and the upwelling flux $\Phi\uparrow$, i.e.

$$\Phi_{\text{net}} = \Phi\downarrow - \Phi\uparrow . \tag{11.10}$$

Assuming that A is the average surface albedo over the short-wave spectral range, the upwelling flux $\Phi\uparrow$ at the surface can be expressed as the product,

$$\Phi\uparrow = A \times \Phi\downarrow . \tag{11.11}$$

Therefore, combining Eqs. (11.10) and (11.11), the net flux Φ_{net} at the surface can be written in the following form:

$$\Phi_{\text{net}} = (1 - A)\Phi\downarrow . \tag{11.12}$$

To evaluate the downwelling flux of solar radiation reaching the Earth's surface after its passage through the pristine atmosphere without aerosols and that passing through the turbid atmosphere (and, hence, to calculate the net flux Φ_{net}), various radiative transfer codes can be used, such as the MODTRAN 4.0 atmospheric model of Kneizys et al. (1996)) or the 6S (Second Simulation of the Satellite Signal in the Solar Spectrum) code of Vermote et al. (1997a,b), applied for instance to the U.S. Standard Atmosphere models defined by Anderson et al. (1986) or to sets of vertical profiles of temperature, pressure, and water vapor partial pressure obtained from local meteorological radiosounding measurements. In such calculations, however, it is important to take into account that the field measurements of aerosol composition, optical parameters and particle size-distribution are affected by not negligible errors, and their use in clear-sky radiative transfer calculations can lead to significant errors in estimating the radiative forcing effects, as pointed out by Valero and Bush (1999).

The occurrences of radiative forcing effects at the ToA- and BoA-levels imply that an aerosol thermodynamic forcing ΔF_{Atm} is produced by aerosols within the atmosphere. It can be evaluated as an instantaneous effect equal to the difference between ΔF_{ToA}, as defined in Eq. (11.8), and ΔF_{BoA}, as defined in Eq. (11.9), according to Ramanathan et al. (2001b):

$$\Delta F_{\text{Atm}} = \Delta F_{\text{ToA}} - \Delta F_{\text{BoA}} . \tag{11.13}$$

Radiative forcing ΔF_{Atm} constitutes a change in the atmospheric energy budget that is not explicitly caused by aerosol-induced radiative effects. Unlike the DARF terms determined at the ToA- and BoA-levels, this DARF term does not modify the net energy budget of the surface–atmosphere system, but rather redistributes it internally and then affects temperature gradients and atmospheric circulation. In fact, the main contribution to ΔF_{Atm} is given by the change in the amount of latent heat released by aerosol-induced changes in clouds and precipitations. Therefore, it can be expressed as a variation in the latent heat flux passing through the atmosphere, which is measured in W m^{-2}, as done for the other DARF terms ΔF_{ToA} and ΔF_{BoA}.

As pointed out above, large uncertainties still exist regarding the role of columnar aerosol loading in causing climate change effects within global circulation models (Hansen et al., 1997, 1998). In particular, large gaps remain in the knowledge of (i) the dependence of the aerosol radiative effects on the microphysical and composition characteristics of the particle polydispersions suspended in the vertical atmospheric column, (ii) the vertical distribution curves of both number and mass concentration and those of the columnar aerosol radiative parameters, often characterized by multi-layered features, and (iii) the spectral and directional characteristics of surface reflectance. Concerning this point, it is of crucial importance to take into account that surface reflectance plays a fundamental role in the DARF calculations, since its effects are combined with those induced by airborne aerosols through interactions varying with surface reflectance and single scattering albedo of aerosols, as clearly shown by Chylek and Coakley (1974). In addition, the lambertian reflection models commonly used to calculate the energy budget of the surface–atmosphere system do not realistically describe the reflectance characteristics of land and ocean surfaces. Thus, their use can lead to biases in the model calculations of aerosol radiative forcing, especially for high values of solar zenith angle θ_o (Ricchiazzi et al., 2005). For this reason, it seems more realistic and appropriate to use the so-called Bidirectional Reflectance Distribution Function (BRDF) models for representing the characteristics of surface reflectance, which are in general non-lambertian in real cases. Using the BRDF models, it is also important to represent the spectral albedo curves of the various surfaces as the sum of two terms, relative to the so-called black-sky and white-sky albedo concepts of Lewis (1995), as defined in Eqs. (11.4a) and (11.4b), respectively. These schematic representations allow us to consider the two distinct albedo contributions of the surface albedo separately, as appropriately evaluated using the spectral percentages of the direct and diffuse components as weight functions in Eq. (11.4c). It must also be taken into account that such two contributions are differently subject to vary as a function of the numerous radiative parameters characterizing the columnar aerosol polydispersions (such as the shape-parameters of the particle number size-distribution and the complex refractive index of columnar particulate matter).

11.4.2 Theory

The upwelling and downwelling solar radiation fluxes $F\uparrow$ and $F\downarrow$ at ToA-level (used to determine the net flux in Eq. (11.6)) are given by the integrals of the monochromatic fluxes over the entire short-wave spectrum. As pointed out above, flux $F\downarrow$ at ToA-level is independent of the atmospheric composition parameters, while flux $F\uparrow$ is given by the upwelling radiation passing once through the atmosphere, and subsequently reflected by the surface to pass again through the atmosphere until reaching outer space. Therefore, flux $F\uparrow$ closely depends on both the surface reflectance characteristics and scattering and absorption properties of the atmosphere, which in turn strongly depend on the airborne aerosol particles. This implies that the instantaneous forcing ΔF_{ToA} at ToA-level defined in Eq. (11.8) can be realistically evaluated at a given time only in cases where the radiative parameters of atmospheric columnar aerosol and the geometrical and spectral characteristics of surface reflectance are known with good accuracy.

The incoming flux $F\downarrow$ of solar radiation at ToA-level presents a spectral distribution very similar to that of a black-body having a temperature close to about 6000 °K, with a maximum centred at about 0.480 μm wavelength. Its integral is commonly referred to as the 'solar constant', which has been estimated to be equal to 1367 W m^{-2} in the '80s (Iqbal, 1983), and found to decrease from 1365.4 $\pm$ 1.3 W m^{-2} in the 1990's to 1360.8 $\pm$ 0.5 W m^{-2} in 2008 (Kopp and Lean, 2011). About 8% of solar radiation belongs to the 0.25–0.40 μm wavelength range, about 81% to the 0.4–1.6 μm range, and only 10% to the 1.6–3.75 μm range, while the remaining 1% pertains to wavelengths $\lambda > 3.75\,\mu$m.

Similarly, the upwelling and downwelling solar radiation fluxes $\Phi\uparrow$ and $\Phi\downarrow$ at the BoA-level, used to determine the net flux at the surface in Eq. (11.10), can be calculated by integrating the spectral distribution curves of these radiation terms over the whole spectral range. Flux $\Phi\downarrow$ mainly depends on the scattering and absorption properties of the atmosphere, including those of the columnar aerosol, and, to a considerably lesser extent, on the surface reflectance effects and the subsequent multiple scattering effects occurring in the lower part of the atmosphere. Thus, flux $\Phi\uparrow$ strongly depends on the surface reflectance characteristics. During its passage through the cloudless atmosphere, solar radiation is attenuated by an intense absorption due to various atmospheric gases (mainly water vapor, and less strongly ozone, carbon dioxide, nitrogen dioxide, and other minor gases), scattering by air molecules (Rayleigh scattering), and scattering and absorption by aerosols. Therefore, the incoming flux $\Phi\downarrow$ of solar radiation reaching the Earth's surface consists of a direct component and a diffuse component. The former can be calculated as the product of extraterrestrial solar irradiance $I_0(\lambda)$ by atmospheric transmissivity $T(\lambda, \theta_o)$, which varies rapidly as a function of wavelength λ over the spectral intervals where numerous and strong absorption bands of water vapor and other weaker bands of the above-mentioned atmospheric gases are active. As a result of these absorption processes, combined with Rayleigh molecular scattering and aerosol extinction, the direct solar irradiance flux at the BoA-level tends to vary quite slowly in time throughout the day, as a function of the so-called relative optical air mass m. This dimensionless quantity is given by the length of the atmospheric path described by the solar radiation parallel beam passing through the atmosphere, and is calculated as the integral of the atmospheric medium density along the sun-path, which is approximately equal to the inverse of the cosine of solar zenith angle θ_o (Kasten and Young, 1989; Tomasi et al., 1998). The diffuse component of solar radiation reaching the Earth's surface for cloudless-sky conditions arises from Rayleigh and aerosol scattering processes. Therefore, it increases as (i) the solar elevation angle equal to $90° - \theta_o$ decreases, and (ii) relative optical air mass m increases. As shown in Eq. (11.11), the upwelling flux $\Phi\uparrow$ of solar radiation at the surface depends strongly on the surface reflectance characteristics. Thus, the angular distribution and spectral characteristics of the direct and diffuse components of flux $\Phi\uparrow$ do not depend only on those of the two components of incoming flux $\Phi\downarrow$, but are also considerably influenced by the spectral characteristics of surface reflectance. After the surface reflection, both direct and diffuse components of $\Phi\uparrow$ again cross the atmosphere to reach outer space. They are even more extinguished during the second passage, because of the same absorption and scattering processes that have attenuated the incoming flux $F\downarrow$ during its first passage through the atmosphere. Consequently, the upwelling flux $F\uparrow$ of solar radiation measured at ToA-level consists of both direct and diffuse components, which are both strongly attenuated not only by the gaseous absorption (mainly within the spectral intervals occupied by the strong absorption bands of water vapor) but also by the Rayleigh-scattering and aerosol extinction processes, both characterized by continuous features over the entire spectrum, presenting intensity features gradually decreasing with wavelength.

As previously mentioned, the simulation of the absorption and scattering processes affecting both the downwelling flux $\Phi\downarrow$ at the surface and the upwelling flux $F\uparrow$ at ToA-level can be realistically made using the 6S code (Vermote et al., 1997a,b). On this matter, Russell et al. (1997) attempted to define a simplified formula for estimating the instantaneous change in the upwelling flux $F\uparrow$ at ToA-level due to an absorbing aerosol layer. They took into account the dependence of aerosol layer transmission and reflection on the aerosol radiative properties, and of surface albedo on solar zenith angle θ_o, obtaining the following equation:

$$\Delta\,\mathrm{Re}_F(\mu_o) = \frac{\tau_a\,\{\omega\,[\beta_a(\mu_o)(1-R_s(\mu_o)) - 2\mu_o B_a R_s(\mu_o)\,(1-A)]\}}{\mu_o\,\{1+[\beta_a(\mu_o)\tau_a/\mu_o]\}} - \frac{\tau_a\,\{(1-\omega)R_s(\mu_o)(1+2\mu_o)\}}{\mu_o\,\{1+[\beta_a(\mu_o)\tau/\mu_o]\}}\,, \tag{11.14}$$

for representing the variation in the ratio $\mathrm{Re}_F(\mu_o) = F\uparrow/F\downarrow$ as a function of $\mu_o = \cos\theta_o$, aerosol optical thickness τ_a, single scattering albedo ω of columnar aerosol, aerosol hemispherical upward scattering fraction $\beta_a(\mu_o)$ (in practice the fraction of incoming solar radiation scattered by the columnar aerosol and directed into the upward hemisphere), and surface reflectance $R_s(\mu_o)$. Eq. (11.14) was obtained after the introduction of some simplifications, using parameter B_a to indicate the average aerosol hemispheric upward scattering fraction, and parameter A adopted in Eq. (11.11) to indicate the average surface albedo over the entire wavelength range of solar radiation. Eq. (11.14) states that the instantaneous DARF term at ToA-level induces cooling effects in all cases where the first term (minuend) prevails over the second term (subtrahend) giving a positive value of $\Delta\,\mathrm{Re}_F(\mu_o)$, while heating effects are produced in the opposite cases where $\Delta\,\mathrm{Re}_F(\mu_o) < 0$.

Although obtained employing averaging simplifications, the formula in Eq. (11.14) appears to be quite complex, even neglecting the multi-layered features of the vertical profile of aerosol number and mass concentrations. Eq. (11.14) provides evidence of the difficulties encountered in parameterizing the instantaneous DARF terms ΔF_{ToA} in Eq. (11.8), ΔF_{BoA} in Eq. (11.9) and ΔF_{Atm} in Eq. (11.13) for the variety of aerosol polydispersions generally found in the atmosphere, because the atmospheric content of airborne aerosol particles is often characterized by very different values of parameters ω and β_a from one layer to another or from one case to another, and for different surface reflectance characteristics over sea and land regions. Therefore, it appears suitable to investigate the dependence features of the DARF terms on the aerosol radiative parameters by considering a certain number of aerosol models presenting different radiative properties together with a set of various surface reflectance models.

Ramanathan et al. (2001a) found that significant variations in ω due to changes in the chemical composition and, hence, in the aerosol radiative parameters of airborne aerosol particles can induce important changes in the radiation fluxes within and below the aerosol layers. They can induce thermodynamic forcing effects that do not substantially modify the net energy budget of the surface–atmosphere system, but rather cause an internal redistribution of the energy surplus, thus altering the amount of latent heat released by aerosol-induced changes in clouds and precipitations. Because of these exchanges, mainly related to aerosol indirect effects, the atmospheric stability conditions may be considerably altered, influencing heating rates, surface temperatures, and cloud formation and persistence, all of which contribute to cause appreciable changes in the local cooling and warming processes occurring in the atmosphere.

The aerosol radiative parameters tend to vary considerably in cases where the chemical composition of particulate matter changes as a result of variations in the transport mechanisms of aerosol particles from different sources. Therefore, it is important to bear in mind that the atmospheric aerosol content maintains in general stable radiative properties only over short periods of a few days, and is then influenced and altered rapidly by the dynamic patterns of the atmosphere over the observation site and by the so-called indirect effects.

In the present study, the calculations of instantaneous upwelling fluxes $F\uparrow^*$ and $F\uparrow$ at ToA-level and fluxes $\Phi\downarrow$, $\Phi\uparrow$, $\Phi\downarrow^*$ and $\Phi\uparrow^*$ at BoA-level were made using the 6S code (Vermote et al., 1997a, 1997b) to obtain reliable simulations of such radiative transfer episodes, in which:

(i) Rayleigh scattering was taken into account using a single scattering approximation for air molecules, which guarantees a relative accuracy of the transmission function better than 0.70% in all cases where the differences between the exact computations of spherical albedo and the 6S code expression are around 0.003 for total atmospheric optical thickness τ equal to 0.35 in the visible, corresponding to the most unfavorable conditions (Vermote et al., 1997a);
(ii) aerosol scattering properties were defined using the Sobolev (1975) approximation for the reflectance, the approach of Zdunkowski et al. (1980) for representing the transmission funtion, and a semi-empirical formula for defining the spherical albedo. For these approximations, aerosol scattering calculations were made by Vermote et al. (1997a) using the 6S code, obtaining for these calculations a relative accuracy better than ± 1% (in reflectance units), especially at large view and Sun angles or for high values of aerosol optical thickness $\tau_a(\lambda)$;
(iii) integrations of downward and upward radiation fluxes were made over the 2π solid angle, in each case determining the radiance at more than 14 scattering angles over the scattering angle range from $-90°$ to $+90°$, and then calculating the overall flux through integration over the whole angular field;
(iv) the calculations of radiative transfer through the atmosphere were performed by defining the optical properties of the atmosphere within several layers of variable geometrical depth along the vertical path, and subdividing the atmosphere into at least 13 layers, in which the scattering properties of the columnar aerosol loading and those of the Rayleigh-scattering system were determined from the surface up to the ToA-level;
(v) scattering–absorption coupling was taken into account, especially the contribution arising from the coupling between water vapor absorption and aerosol scattering;
(vi) the spectral resolution used to define the aerosol radiative parameters and the radiative properties of the atmosphere (Rayleigh scattering, gaseous absorption, atmospheric transmittance, etc.) was generally equal to 2.5 nm over the

entire solar radiation wavelength range from 0.25 to 4.00 μm, directly given by the 6S database or calculated through spectral interpolation procedures; and

(vii) the surface reflectance effects were taken into account in the calculations of upwelling radiance fluxes at the BoA- and ToA-levels by using appropriate surface reflectance models to represent non-lambertian and lambertian (i. e. isotropic) angular configurations in terms of Bidirectional Reflectance Distribution Function (BRDF) models.

The present calculations of the instantaneous DARF effects were made for (a) the 40 aerosol models described in Section 11.2, and (b) the two sub-sets of surface reflectance models defined in Section 11.3. Each sub-set consisted of 16 models, the first representing the BRDF non-lambertian surface reflectance features, and the second defining the less realistic lambertian surface reflectance characteristics. The two sub-sets of reflectance models are described in detail in Section 11.3, each sub-set consisting of four OS, VS, BS and PS surface reflectance models. It is worth noting that the 16 lambertian reflectance models were defined for the same surface characteristics of the corresponding BRDF non-lambertian models, by appropriately adjusting the surface reflectance characteristics to provide the same average surface albedo values over the whole solar spectrum that were obtained for the BRDF models.

11.4.3 Dependence of instantaneous DARF on aerosol properties

The absorptance and reflectance characteristics of a columnar atmospheric load of aerosol particles are calculated as a function of the shape-parameters of the size-distribution curve and radiative parameters of particulate matter. The higher the columnar mass content of aerosol particles is, the more marked the scattering and absorption effects produced by them along the atmospheric path of incoming solar radiation. This implies that the DARF effects occurring at the ToA level are expected to increase as the atmospheric content of aerosol mass becomes more pronounced and aerosol optical thickness assumes in general higher values. On the other hand, taking into account the two-stream approximation evaluations of Chylek and Coakley (1975), it is evident that the DARF effects at the ToA level are strongly influenced by the single scattering albedo characteristics of atmospheric aerosol particles in relation to the local surface albedo conditions. These aspects are analysed in the two following subsections.

11.4.3.1 Aerosol optical thickness influence

Numerous works are available in the literature (Chyleck and Coakley, 1974; Charlson et al., 1990; Chyleck and Wong, 1995; Russell et al., 1997; Remer and Kaufman, 2006), showing that the DARF effects at the ToA-level vary almost linearly (and without modifying the sign) as the aerosol optical thickness τ_a at a chosen visible wavelength increases, in all cases where this parameter is no higher than 0.10 at visible wavelengths. Conversely, nonlinear dependence features are usually observed for atmospheric turbidity conditions presenting values of τ_a considerably greater than 0.1. For such atmospheric turbidity conditions, ΔF_{ToA} is given by the combined radiative effects due to the columnar content of atmospheric aerosol, varying with the optical properties (mainly as a function of the single scattering albedo $\omega(\lambda)$ of columnar aerosol), and surface reflectance effects (substantially varying as a function of the spectral and geometrical characteristics of surface). In the present analysis of the dependence of instantaneous DARF terms on τ_a, the calculations of the radiative forcing parameters were performed for each of the 40 aerosol models defined in Tables 11.1–11.10, and for 5 values of τ_a increasing from 0.1 to 0.9 in steps of 0.2, in addition to its null value used to illustrate the pristine atmosphere conditions. The choice allowed us to evaluate with good accuracy the dependence features of the DARF terms on τ_a, as determined at the ToA- and BoA-levels and within the atmosphere, for various aerosol polydispersions presenting different composition and optical characteristics, and for considerably different surface reflectance models. Figure 11.25a shows three examples of the dependence features of instantaneous ΔF_{ToA} on $\tau_a(0.55\,\mu\mathrm{m})$ and θ_o, made for the VS3 surface model and the three M-1, M-8 and M-14 aerosol models defined in Table 11.4, selected to span the large range of weighted average single scattering albedo ω^* from 1.00 (M-1 model) to 0.65 (M-14 model). The BRDF non-lambertian calculations of ΔF_{ToA} are compared in Fig. 11.25a with those determined for the equivalent isotropic surface reflectance models, showing that important variations in their isopleths arise from the use of the simplified isotropic reflectance models in place of equivalent BRDF non-lambertian models. It is also interesting to note that pronounced discrepancies were evident between the values of ΔF_{ToA} obtained for the M-8 (pure continental) aerosol model and the M-1 (pure oceanic) aerosol model, while the evaluations of ΔF_{ToA} made for the M-14 (heavy polluted) aerosol model and BRDF non-lambertian surface reflectance conditions are less marked than those determined for the combination of the same aerosol model with isotropic surface reflectance conditions, especially within the $\theta_o < 50°$ range and for $\tau_a(0.55\,\mu\mathrm{m}) > 0.40$.

The calculations of the instantaneous DARF terms were made for each pair of the above-selected fixed values of parameters τ_a and θ_o. A database consisting of 34,560 files ($9 \times 6 \times 40 \times 16$) was therefore collected for the BRDF non-lambertian surface models, and an equivalent database consisting of as many files for the corresponding lambertian surface reflectance models. The pair of datasets will be referred to hereinafter as the 'Look Up Table (LUT)' of DARF evaluations, each one containing spectrally integrated data over the 0.4–2.5 μm wavelength range. These computational simulation data are available to the scientific community on the ISAC-CNR Bologna website (http://www.isac.cnr.it/~radiclim/), consulting the AEROCLOUDS folder that will be accessible on request of a password to the corresponding author. Each of the six panels presented in Fig. 11.25a describes an ensemble of 54 files (obtained for 9 values of θ_o and 6 values of τ_a, including the null one), from which the vertical and horizontal cross sections of each LUT representation can be extracted, determining the dependence features of instantaneous ΔF_{ToA} on parameters θ_o and τ_a, respectively.

In order to more deeply investigate the dependence features of the three instantaneous DARF terms ΔF_{ToA}, ΔF_{BoA} and ΔF_{Atm} on aerosol optical thickness $\tau_a(0.55\,\mu\mathrm{m})$, the dependence curves of these three DARF terms are presented in Figs. 11.25b and 11.25c, as obtained for (i) three values of θ_o equal to 0°, 30° and 60°, (ii) a set of aerosol models chosen among the 40 models defined in the present study, and (iii) some selected pairs of BRDF non-lambertian and isotropic sur-

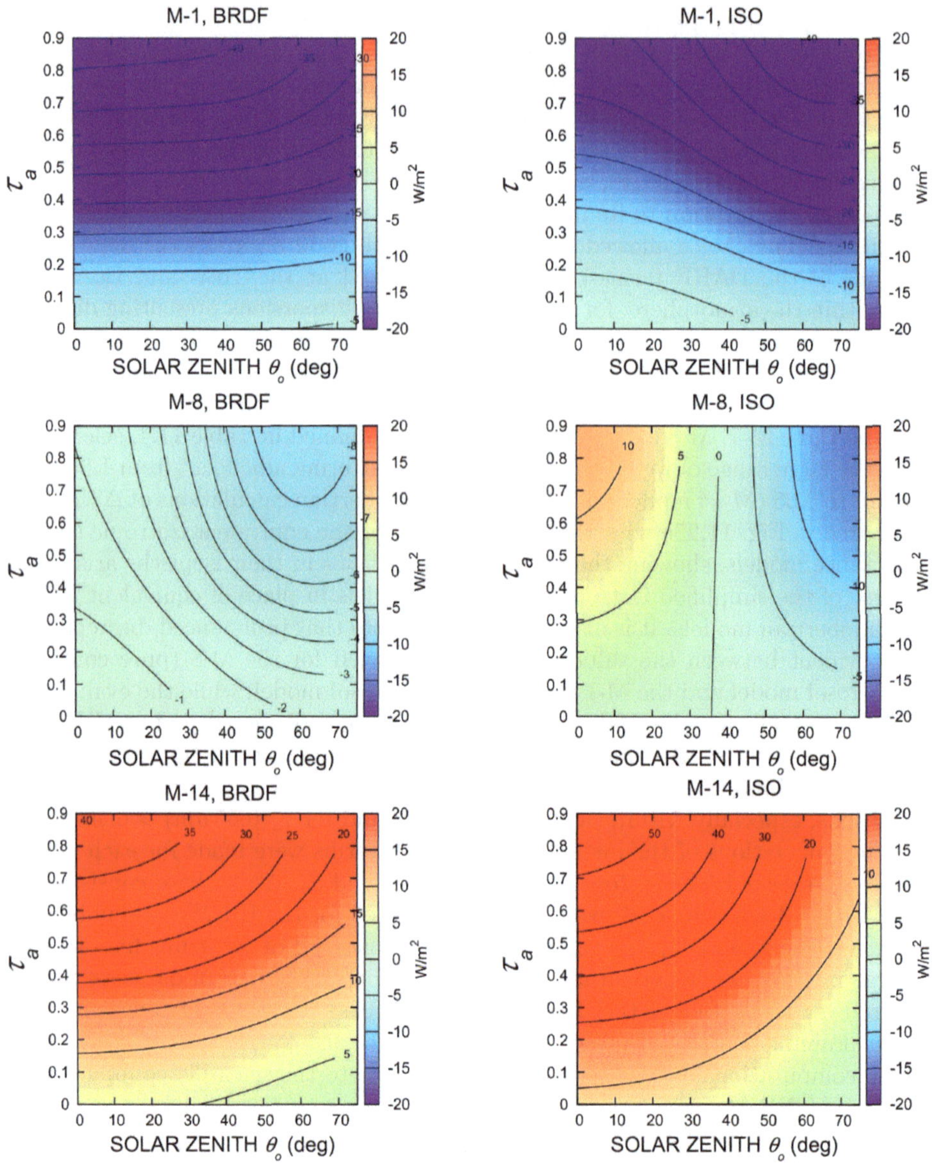

Fig. 11.25a. Mesh plots of the instantaneous short-wave direct aerosol-induced radiative forcing ΔF_{ToA} defined in Eq. (11.5) as a function of aerosol optical thickness $\tau_a(0.55\,\mu\mathrm{m})$ and solar zenith angle θ_o, for the BRDF non-lambertian VS3 surface reflectance model (left) and the equivalent isotropic (ISO, lambertian) VS3 surface reflectance model (right) and the three M-type aerosol models labeled M-1 (pure oceanic aerosol, upper part), M-8 (pure continental aerosol, middle part) and M-14 (heavy polluted aerosol, lower part). The colour scale of ΔF_{ToA} is measured in W m^{-2} and reported on the right site of each graph.

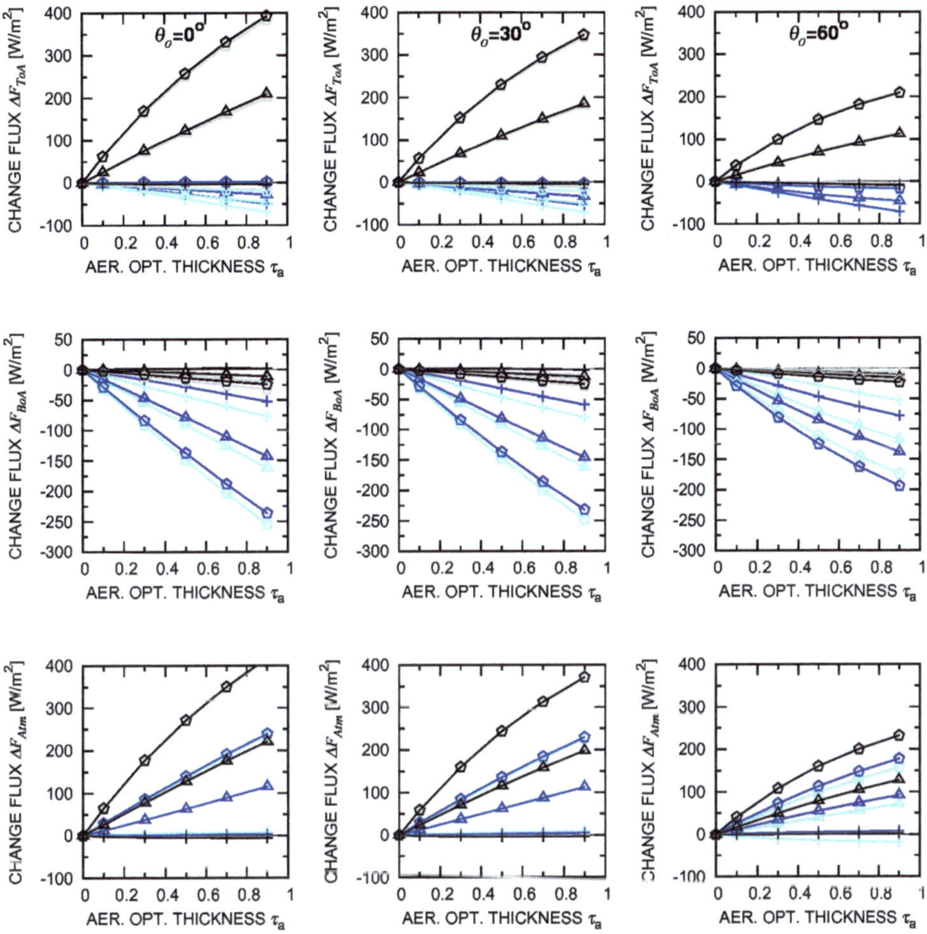

Fig. 11.25b. Dependence curves of instantaneous direct aerosol-induced radiative forcing terms ΔF_{ToA} (upper part), ΔF_{BoA} (middle part) and ΔF_{Atm} (lower part) plotted as a function of aerosol optical thickness $\tau_a(0.55\,\mu\mathrm{m})$, as obtained for (i) the solar zenith angles $\theta_o = 0°$ (left), $\theta_o = 30°$ (middle) and $\theta_o = 60°$ (right), (ii) the M-1 (plus signs), M-8 (up triangles) and M-14 (pentagons) aerosol models defined in Table 11.4, and (iii) the OS3 (blue) and PS1 (gray) models used to represent the non-lambertian surface reflectance characteristics (light colors) and the corresponding equivalent lambertian surface reflectance models (dark colors), as given in Table 11.12.

face reflectance models. Figure 11.25b shows the dependence patterns of the three DARF terms on $\tau_a(0.55\,\mu\mathrm{m})$, as determined for the the M-1 (pure oceanic), M-8 (pure continental) and M-14 (heavy polluted) aerosol models, and the OS3 (sea surface with wind speed of $10\,\mathrm{m\,s^{-1}}$) and PS1 (almost totally pure snow) surface reflectance models. The results show quite linear patterns for all three instantaneous DARF terms derived for (a) the OS3 cases associated with the M-1 (non-absorbing, pure oceanic) and the M-8 (weakly absorbing, continental) aerosol models, and (b) the PS1 case combined with the M-1 aerosol model. At the same time, markedly

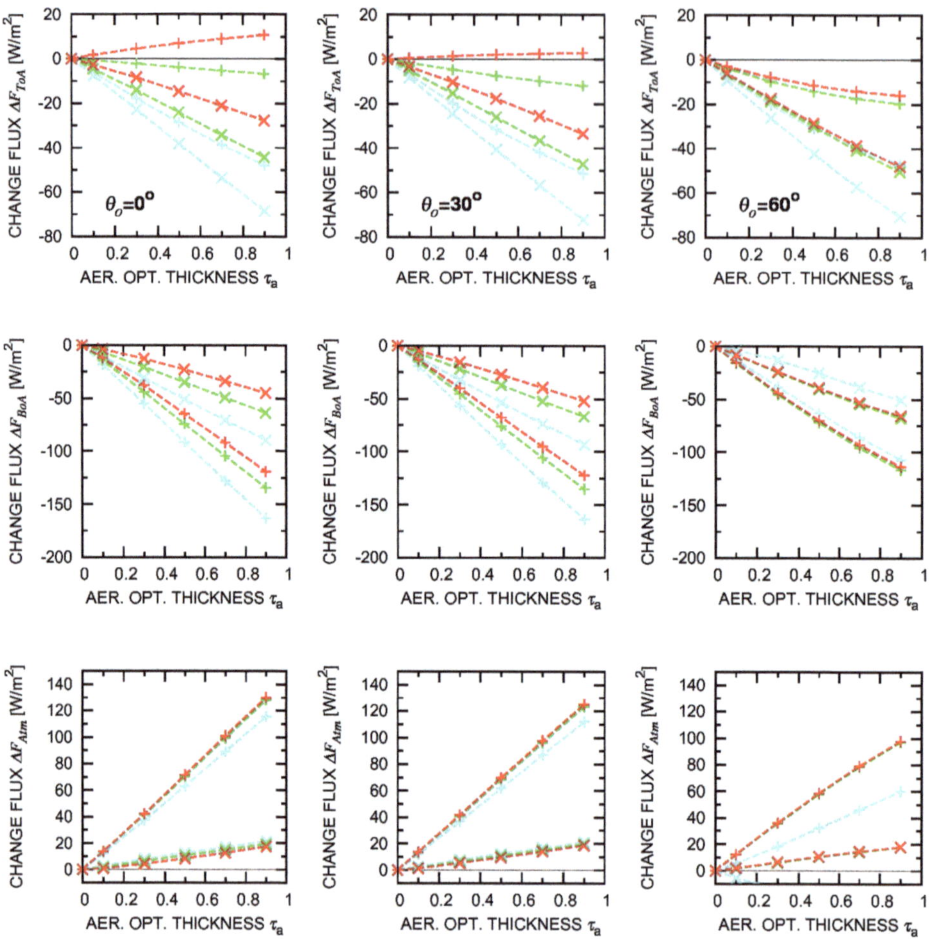

Fig. 11.25c. Dependence curves of instantaneous direct aerosol-induced radiative forcing terms ΔF_{ToA} (upper part), ΔF_{BoA} (middle part) and ΔF_{Atm} (lower part) plotted as a function of aerosol optical thickness $\tau_a(0.55\,\mu\mathrm{m})$, as obtained for (i) the solar zenith angles $\theta_o = 0°$ (left), $\theta_o = 30°$ (middle) and $\theta_o = 60°$ (right), (ii) the 6S-C continental (plus signs) and the 6S-M maritime (ics crosses) aerosol models defined in Table 11.3, and (iii) the OS2 (blue), VS2 (green) and BS1 (red) models used to represent the non-lambertian surface reflectance characteristics, as given in Table 11.12.

convex patterns were described by the ΔF_{BoA} and ΔF_{Atm} determined for the OS3 case associated with the M-14 (strongly absorbing) heavy polluted aerosol model, and the PS1 cases combined with the M-8 and M-14 aerosol models. Correspondingly, the forcing term ΔF_{ToA} was found to assume negative and positive values for the OS3 and PS1 surface reflectance models, respectively, presenting an evident shift to more positive (warming) values, passing from pure oceanic to heavy polluted aerosol models. The average slope coefficients of the various dependence curves give a measure of the ΔF_{ToA} efficiency (i.e. of such DARF term per unit $\tau_a(0.55\,\mu\mathrm{m})$), estimated to vary between $-80\,\mathrm{W\,m^{-2}}$ for the M-1 non-absorbing aerosol polydispersion suspended over the OC3 surface, and $+400\,\mathrm{W\,m^{-2}}$ for the M-14 strongly absorbing aerosol polydispersion over the PS1 snow surface, which is particularly suitable for representing the very high surface reflectance features in the visible and near-infrared that are most frequently observed over the Antarctic Plateau.

More closely linear dependence patterns were obtained in Fig. 11.25c for all the three above-selected values of θ_o, where the 6S-C (continental) and 6S-M (maritime) aerosol models were combined with the OS2 (sea surface, with wind speed of $5\,\mathrm{m\,s^{-1}}$), VS2 (corn field surface) and BS1 (dry sand) surface reflectance models. In these cases, approximately linear dependence patterns of instantaneous ΔF_{ToA} were obtained, with average slope coefficient gradually decreasing from slightly positive to markedly negative values, passing from the 6S-C to the 6S-M model and from the BS1 to the OS2 model. The instantaneous forcing term ΔF_{BoA} was also found to present nearly linear patterns, characterized by gradually more pronounced negative slope coefficients, passing from the 6S-M to 6S-C model and from the BS1 to OS2 model. Increasing patterns of ΔF_{Atm} as a function of $\tau_a(0.55\,\mu\mathrm{m})$ can be noted in Fig. 11.25c for all three surface reflectance models, with a higher slope coefficient found for the 6S-C aerosol model. Therefore, it can be stated that in general the instantaneous DARF terms vary almost linearly as $\tau_a(0.55\,\mu\mathrm{m})$ increases over the lower range of this optical parameter, and for poorly absorbing aerosol particles giving values of single scattering albedo not far from unity. Measurements and modeling evaluations of the radiative flux changes induced by aerosols over the North Atlantic coast of the United States were performed by Russell et al. (1999) during the TARFOX experiment, yielding instantaneous daytime uwpelling flux changes ranging between $+14$ and $+48\,\mathrm{W\,m^{-2}}$ for average values of mid-visible aerosol optical depth (measured over the 0.30–0.70 μm spectral range) varying between 0.20 and 0.55. The changes in the instantaneous upwelling flux were found to be approximately proportional to aerosol optical thickness, while they decreased considerably as $\omega(0.550\,\mu\mathrm{m})$ diminished from 1.00 to 0.86. These results confirmed that the DARF effects may depend on aerosol optical thickness with features varying largely with the single scattering albedo characteristics of atmospheric aerosol loading. Bush and Valero (2002) also applied this concept to the DARF forcing at the BoA-level during the INDOEX (Indian Ocean) Experiment conducted at the Kaashidhoo Climate Observatory (Republic of Maldives), finding a diurnal average value of atmospheric forcing at the surface equal to $-72.2 \pm 5.5\,\mathrm{W\,m^{-2}}$ per unit aerosol optical depth, over the total solar broadband spectrum, subject to vary appreciably as a function of single scattering albedo $\omega(0.50\,\mu\mathrm{m})$, evaluated to be on average equal to 0.874 ± 0.028. The above results clearly demonstrate that the DARF terms depend closely on the single scattering albedo properties of atmospheric particles.

11.4.3.2 Single scattering albedo influence

The influence of single scattering albedo of columnar aerosol polydispersions on the DARF effects occurring at ToA-level has been highlighted in numerous works (Schwarz et al., 1995; Russell et al., 1997, 1999; Takemura et al., 2002). These dependence features are here analysed by evaluating the DARF effects for the surface reflectance models defined in Section 11.3 and the number of aerosol models

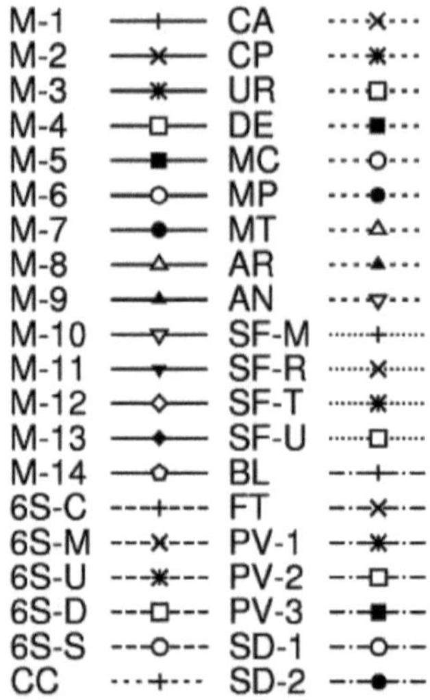

Fig. 11.26. Legend of the 40 aerosol models used in the calculations of the DARF terms presented in Figs. 11.27 to 11.30. Solid lines refer to the series of 14 M-type aerosol models, dashed lines to the 5 6S models, short-dashed lines to the 10 OPAC models, dotted lines to the four SF models, and dot-dashed lines to the seven additional models defined in the present work.

described in Section 11.2. Fig. 11.16 provides evidence that the single scattering albedo of the aerosol models considered in the present study covers with continuous features the range of weighted single scattering albedo ω^* from about 0.6 to 1.0, which is commonly observed in various areas of the Earth. Figure 11.26 shows the legend of the symbols used to represent the 40 aerosol models defined in Section 11.2, for which the DARF effects have been evaluated in the following Figs. 11.27–11.30, here shown to illustrate the dependence patterns of DARF on (i) single scattering albedo by means of aerosol models, (ii) surface reflectance by means of surface reflectance models, and (iii) solar zenith angles taken over the range from 0° to 80°, in steps of 10°.

The examples shown in Fig. 11.16 for maritime, continental and urban aerosol polydispersions indicate that the absorptance/reflectance ratio of columnar particles increases as the absorption properties of particulate matter become gradually stronger (mainly due to the increase in the mass content of soot substances) and, hence, the weighted average single scattering albedo ω^* calculated over the short-wave range gradually tends to decrease. The results obtained by Chylek and Coakley (1974) indicate that aerosol particles causing cooling effects over relatively low-albedo surfaces, such as oceanic areas and vegetated land regions, can generate warming effects in the atmosphere when transported over high-albedo surfaces, such as those of the polar regions covered by snow fields and glaciers, thus causing a change in the sign of radiative forcing ΔF_{ToA}.

To emphasize the crucial importance of these climatic effects dependent on aerosol single scattering albedo, the variations in the instantaneous DARF terms ΔF_{ToA}, ΔF_{BoA} and ΔF_{Atm} occurring as a result of changes in parameter ω^* are presented in Figs. 11.27a to 11.27e over the $0.6 \le \omega^* \le 1.0$ range for (a) solar zenith angles $\theta_o = 30°$ and $\theta_o = 60°$, (b) both BRDF non-lambertian and isotropic configurations of the OS1, VS2, BS3 and PS1 surface reflectance models defined in Table 11.12, and (c) the following five sets of aerosol models, each presented in:

(i) Figure 11.27a, for the five 6S-M, 6S-C, 6S-U, 6S-D and 6S-V 6S aerosol models defined in Table 11.3;
(ii) Figure 11.27b, for the four SF-R, SF-U, SF-M and SF-T aerosol models defined in Tables 8 and 9;
(iii) Figure 11.27c, for the 14 M-type aerosol models defined in Table 11.4;
(iv) Figure 11.27d, for the 10 OPAC models defined in Tables 11.6 and 11.7; and
(v) Figure 11.27e, for the 7 additional aerosol models defined in Table 11.10.

All the graphs present similar results, showing that:

(a) Instantaneous ΔF_{ToA} exhibits decreasing patterns for all the considered aerosol models as ω^* increases from ~ 0.60 to about 1.00, with (i) values mainly ranging between $-10\,\mathrm{W\,m^{-2}}$ and $+175\,\mathrm{W\,m^{-2}}$ for ω^* varying between 0.63 and about 0.75, and (ii) values mainly ranging between $-40\,\mathrm{W\,m^{-2}}$ and no more than $+5\,\mathrm{W\,m^{-2}}$ for ω^* close to 1.00.
(b) Instantaneous ΔF_{BoA} presents negative values for ω^* lower than 0.75, ranging mainly between around $-130\,\mathrm{W\,m^{-2}}$ and $-5\,\mathrm{W\,m^{-2}}$, which were evaluated to gradually increase as ω^* approaches the unity, until reaching values ranging mainly between $-40\,\mathrm{W\,m^{-2}}$ and $+5\,\mathrm{W\,m^{-2}}$ for nearly unit values of ω^*.
(c) Instantaneous ΔF_{Atm} assumes mainly positive values over the total range of ω^*, varying between $+40\,\mathrm{W\,m^{-2}}$ and $+180\,\mathrm{W\,m^{-2}}$ for $\omega^* < 0.75$, and further decreasing as ω^* increases until assuming weakly positive or slightly negative values for $\omega^* \approx 1.00$, varying within the range of $\pm 5\,\mathrm{W\,m^{-2}}$ for $\theta_o = 30°$, and within the range of $\pm 25\,\mathrm{W\,m^{-2}}$ for $\theta_o = 60°$.

The above patterns indicate that the aerosol polydispersions of oceanic or volcanic origins, consisting mainly of weakly absorbing particulate matter (with $\omega^* > 0.95$) generally induce negative (cooling) values of both ΔF_{ToA} and ΔF_{BoA}, thus providing very low values of ΔF_{Atm} with both signs. Conversely, aerosol polydispersions containing strong contents of soot substances and, hence, presenting values of ω^* no higher than 0.75, are evaluated to induce (i) weakly negative values of ΔF_{ToA} leading to only very weak cooling effects over the oceanic surfaces, (ii) moderate warming effects ranging between around + 20 and $+30\,\mathrm{W\,m^{-2}}$ over all the land regions covered by vegetation, (iii) intense warming effects varying between about +70 and $+90\,\mathrm{W\,m^{-2}}$ over the arid regions, and (iv) particularly marked warming effects often exceeding the value of $+100\,\mathrm{W\,m^{-2}}$ over the polar regions covered by snow fields and glaciers. The values of ΔF_{BoA} for $\omega^* \approx 0.63$ are negative in all cases, being equal to around $-15\,\mathrm{W\,m^{-2}}$ over polar areas, $-50\,\mathrm{W\,m^{-2}}$ over bare soils, $-70\,\mathrm{W\,m^{-2}}$ over vegetated surfaces, and $-100\,\mathrm{W\,m^{-2}}$ over oceanic areas. Correspondingly, instantaneous ΔF_{Atm} is estimated to be very weak over all the surfaces for aerosol particles presenting values of ω^* close to unity. Conversely, for $\omega^* \approx 0.63$, average evaluations of ΔF_{Atm} of around $+180\,\mathrm{W\,m^{-2}}$ were obtained over polar snow-covered regions, $+135\,\mathrm{W\,m^{-2}}$ over bare soils and arid terrains, $+110\,\mathrm{W\,m^{-2}}$ over vegetation-covered surfaces, and $+95\,\mathrm{W\,m^{-2}}$ over oceans.

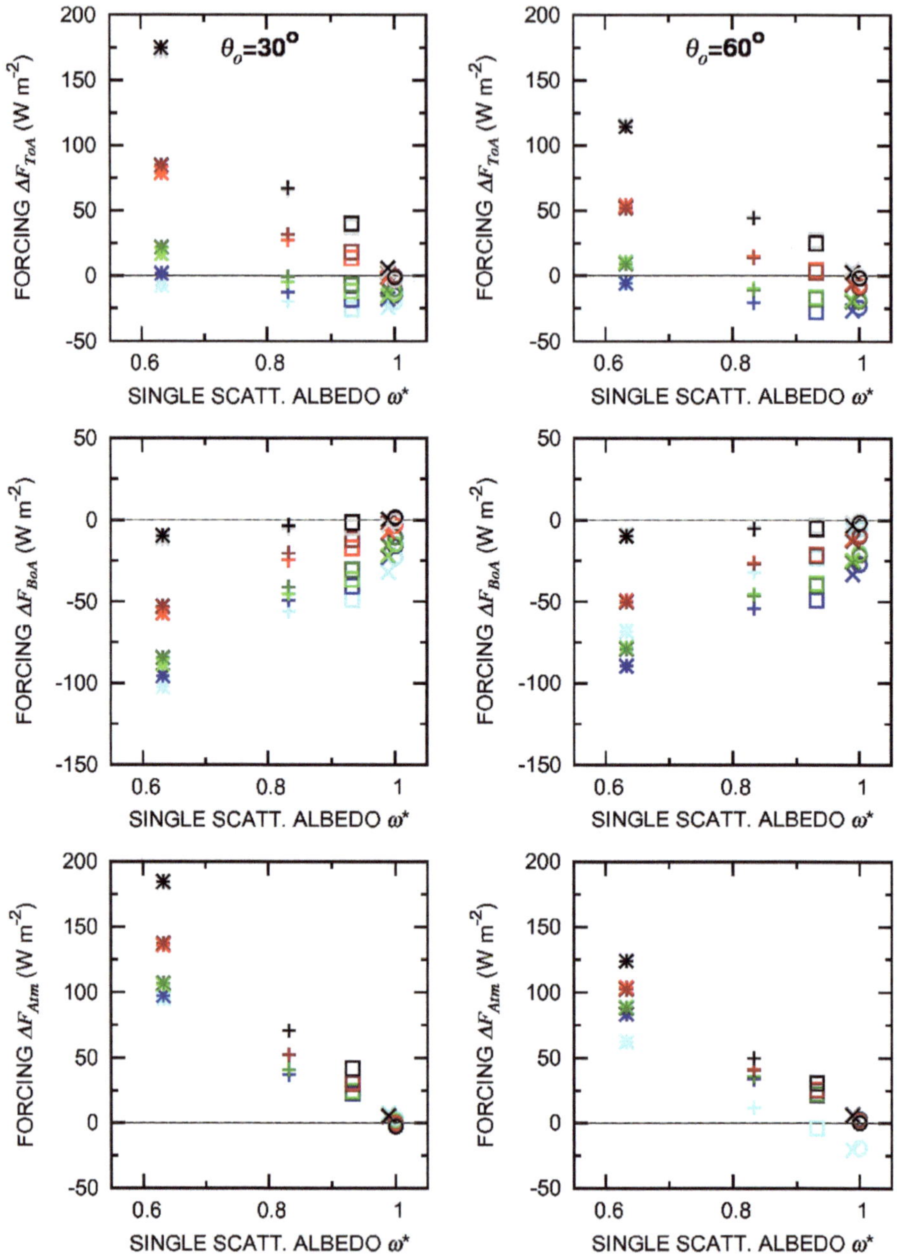

Fig. 11.27a. Instantaneous direct aerosol-induced radiative forcing terms ΔF_{ToA} (upper part), ΔF_{BoA} (middle part) and ΔF_{Atm} (lower part) plotted as a function of weighted average single scattering albedo ω^*, as obtained for (i) the five 6S aerosol models defined in Table 11.3 (labeled using the symbols given in Fig. 11.26), and giving aerosol optical thickness $\tau_a(0.55\,\mu\mathrm{m}) = 0.30$, (ii) solar zenith angles $\theta_o = 30°$ (left) and $\theta_o = 60°$ (right), and (iii) the OS1 (blue), VS2 (green), BS3 (red) and PS1 (gray) BRDF lambertian surface reflectance models (light colors) and the corresponding equivalent lambertian surface reflectance models (dark colors), as given in Table 11.12.

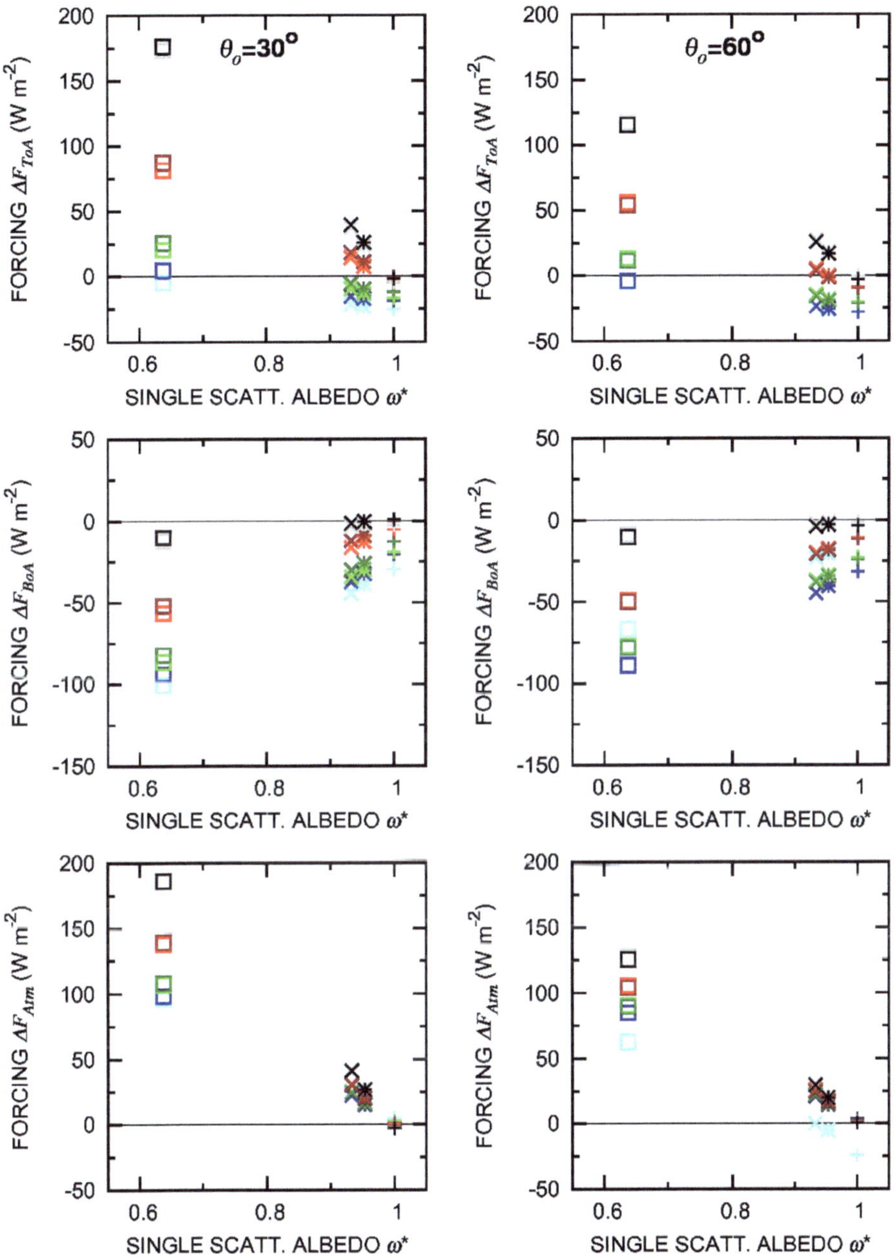

Fig. 11.27b. As in Fig. 11.27a, for (i) the four SF aerosol models defined in Tables 11.8 and 11.9 (labeled using the symbols given in Fig. 11.26), giving aerosol optical thickness $\tau_a(0.55\,\mu\text{m}) = 0.30$, (ii) solar zenith angle $\theta_o = 30°$ (left) and $\theta_o = 60°$ (right), and (iii) the OS1 (blue), VS2 (green), BS3 (red) and PS1 (gray) BRDF lambertian surface reflectance models (light colors) and the corresponding equivalent lambertian surface reflectance models (dark colors), as given in Table 11.12.

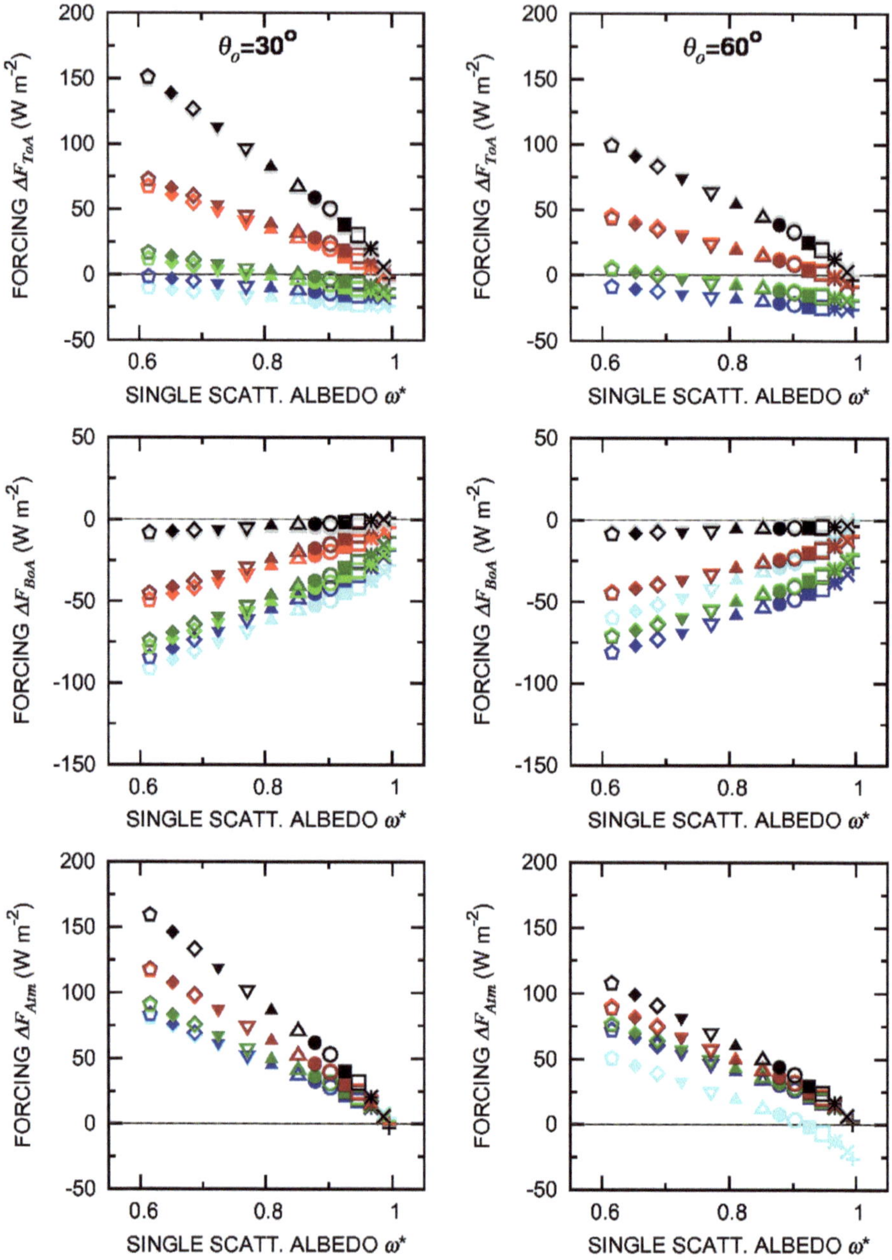

Fig. 11.27c. As in Fig. 11.27a, for (i) the 14 M-type aerosol models defined in Table 11.4 (labeled using the symbols given in Fig. 11.26), giving aerosol optical thickness $\tau_a(0.55\,\mu\mathrm{m}) = 0.30$, (ii) solar zenith angles $\theta_o = 30°$ (left) and $\theta_o = 60°$ (right), and (iii) OS1 (blue), VS2 (green), BS3 (red) and PS1 (gray) BRDF lambertian surface reflectance models (light colors) and the corresponding equivalent lambertian surface reflectance models (dark colors), as given in Table 11.12.

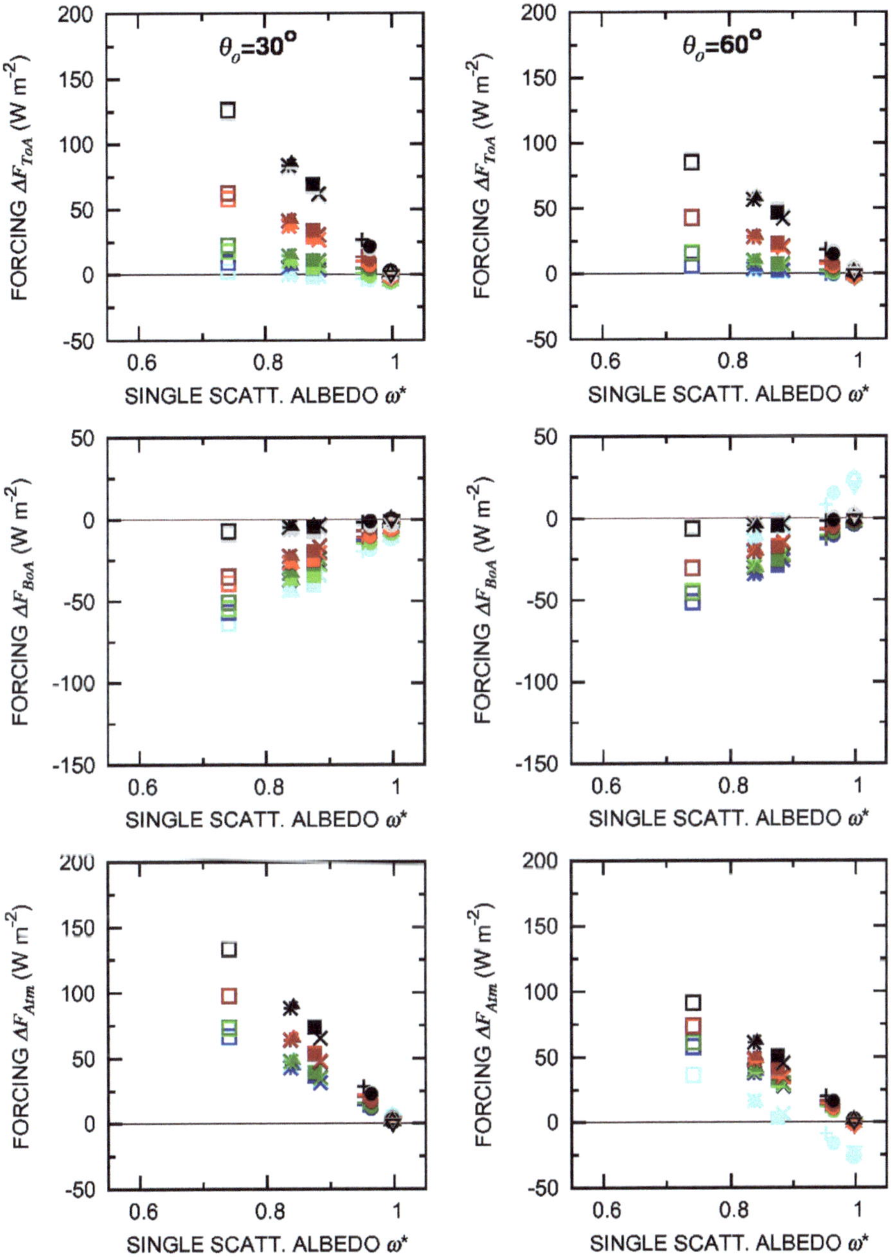

Fig. 11.27d. As in Fig. 11.27a, for (i) the 10 OPAC wet aerosol models defined in Tables 11.6 and 11.7 (labeled using the symbols given in Fig. 11.26), giving aerosol optical thickness $\tau_a(0.55\,\mu\text{m}) = 0.30$, (ii) solar zenith angles $\theta_o = 30°$ (left) and $\theta_o = 60°$ (right), and (iii) the OS1 (blue), VS2 (green), BS3 (red) and PS1 (gray) BRDF lambertian surface reflectance models (light colors) and the corresponding equivalent lambertian surface reflectance models (dark colors), as given in Table 11.12.

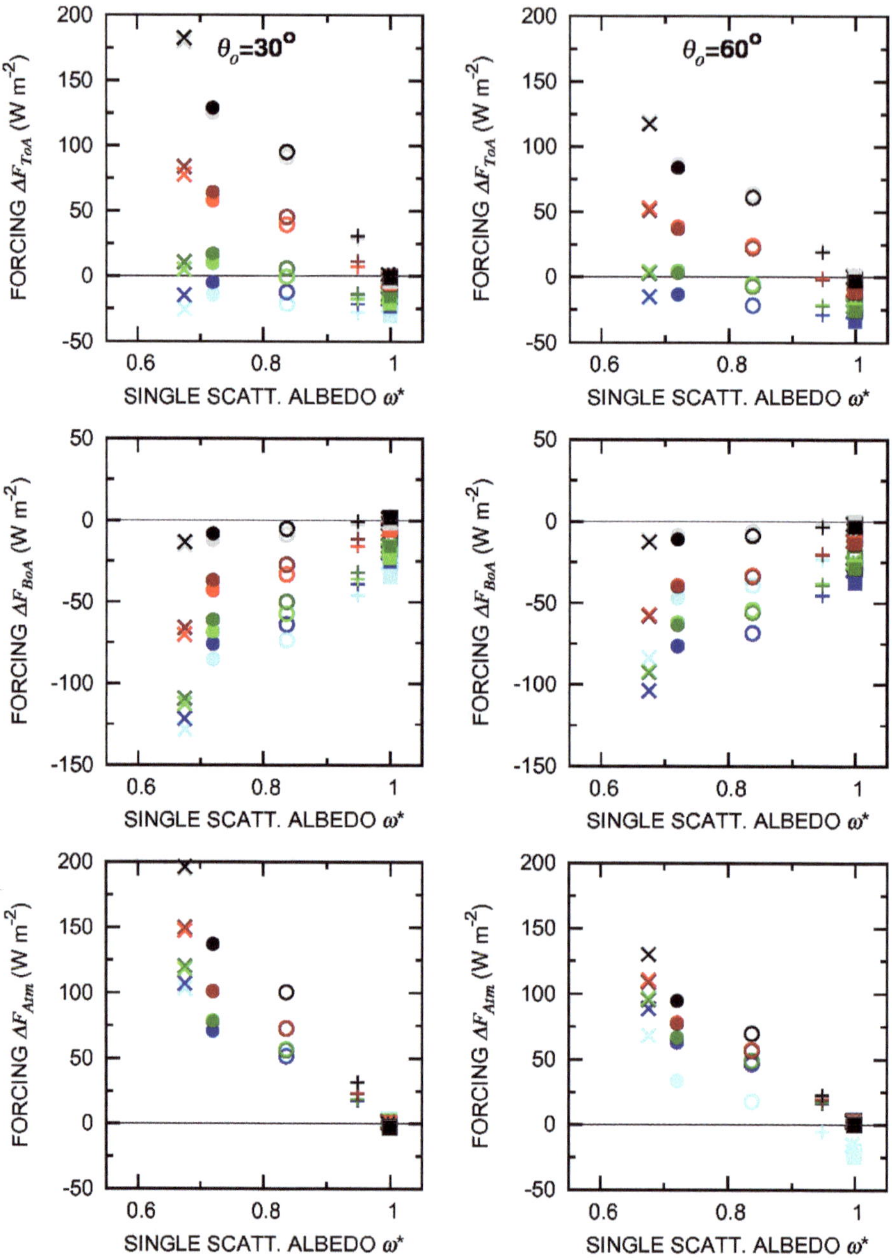

Fig. 11.27e. As in Fig. 11.27a, for (i) the seven additional aerosol models defined in Table 11.10 (labeled using the symbols given in Fig. 11.26), giving aerosol optical thickness $\tau_a(0.55\,\mu\text{m}) = 0.30$, (ii) solar zenith angles $\theta_o = 30°$ (left) and $\theta_o = 60°$ (right), and (iii) the OS1 (blue), VS2 (green), BS3 (red) and PS1 (gray) BRDF lambertian surface reflectance models (light colors) and the corresponding equivalent lambertian surface reflectance models (dark colors), as given in Table 11.12.

To better evaluate the variability of ΔF_{ToA} as a function of ω^*, as described in Figs. 11.27a to 11.27e, the average slope coefficients of the patterns of ΔF_{ToA} plotted versus ω^* were calculated for the two cases $\theta_o = 30°$ and $\theta_o = 60°$. In the first case, the slope coefficient calculated for the set of additional models defined in Table 11.10 varied between $-16.1\,\mathrm{W\,m^{-2}}$ (OS1 model) and $-522.1\,\mathrm{W\,m^{-2}}$ (PS1 model) per unit variation of ω^*. In the second case, this average slope coefficient per unit variation in ω^* was estimated to vary between (i) $-32.9\,\mathrm{W\,m^{-2}}$ (OS1 model) for the set of OPAC aerosol models defined in Tables 11.6 and 11.7, and (ii) $-346.0\,\mathrm{W\,m^{-2}}$ (PS1 model) for the set of additional models defined in Table 11.10. No evidence of large variations in the forcing terms ΔF_{ToA}, ΔF_{BoA} and ΔF_{Atm} arose from the analysis of the data presented in Figs. 11.27a to 11.27e over the range of ω^* determined for the various sets of aerosol models giving $\tau_a(0.55\,\mu\mathrm{m}) = 0.30$, confirming the results presented in Tables 11.14a, 11.14b and 11.14c for the 18 above-chosen aerosol extinction models.

11.4.4 Dependence of instantaneous DARF on underlying surface reflectance

The results presented in Fig. 11.25a offer clear evidence of the appreciably different evaluations of instantaneous forcing terms ΔF_{ToA}, ΔF_{BoA} and ΔF_{Atm} that can be obtained for the same aerosol model characterized by a certain value of single scattering albedo, using different pairs of BRDF non-lambertian and isotropic surface reflectance models. To analyse the dependence of such radiative forcing effects on surface reflectance characteristics more closely, the calculations of the instantaneous DARF terms are shown in Figs. 11.28a to 11.28e as a function of the broadband surface albedo $A(\theta_o)$ defined in Eq. (11.4d), as determined for solar zenith angles $\theta_o = 30°$ and $\theta_o = 60°$. These two values of θ_o were chosen bearing in mind that they are representative of the two distinct ranges of θ_o, within which substantial differences were found to exist between the angular dependence features of instantaneous DARF terms. Figures 11.28a to 11.28e were drawn for different combinations of the most representative aerosol models chosen among the 40 models defined in Tables 11.1 to 11.10, and the BRDF non-lambertian and isotropic surface reflectance models defined in Section 11.3. More precisely, the instantaneous terms ΔF_{ToA}, ΔF_{BoA} and ΔF_{Atm} are presented as follows:

(i) in Fig. 11.28a, for the 6S-M, 6S-C and 6S-U aerosol models defined in Table 11.3 and the non-lambertian and isotropic versions of the OS1, VS1, BS1 and PS1 surface reflectance models defined in Table 11.12;
(ii) in Fig. 11.28b, for the SF-M, SF-R, SF-T and SF-U aerosol models defined in Tables 11.8 and 11.9, and the OS1, VS1, BS1 and PS1 surface reflectance models;
(iii) in Fig. 11.28c, for the M-1, M-8 and M-14 aerosol models defined in Table 11.4 and the OS2, VS2, BS2 and PS2 surface reflectance models defined in Table 11.12;
(iv) in Fig. 11.28d, for the 6S-M, 6S-C and 6S-U aerosol models defined in Table 11.3 and the OS3, VS3, BS3 and PS3 surface reflectance models defined in Table 11.12; and

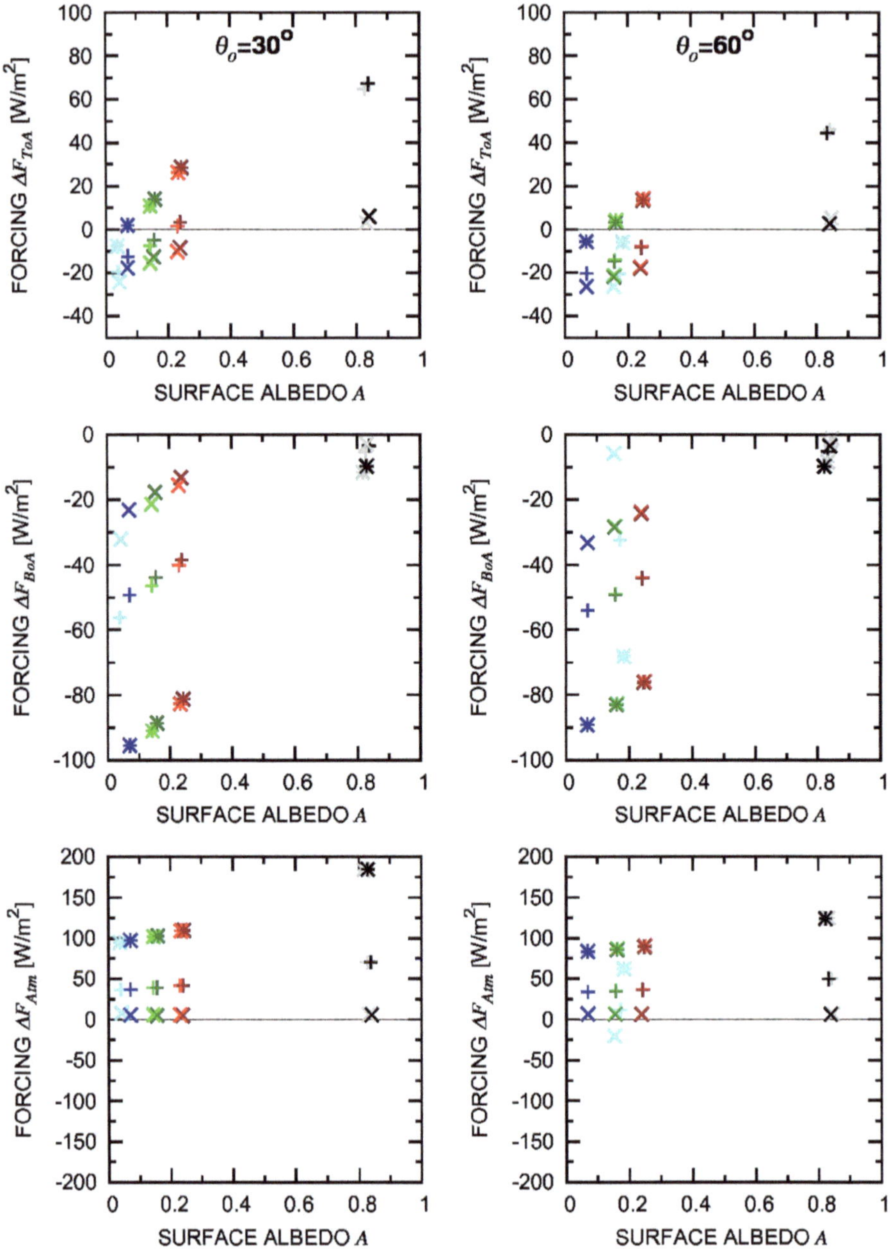

Fig. 11.28a. Instantaneous direct aerosol-induced radiative forcing terms ΔF_{ToA} (upper part), ΔF_{BoA} (middle part) and ΔF_{Atm} (lower part) plotted as a function of broadband surface albedo $A(\theta_o)$, for aerosol optical depth $\tau_a(0.55\,\mu\mathrm{m}) = 0.30$ and solar zenith angles $\theta_o = 30°$ (left) and $\theta_o = 60°$ (right), as calculated for the 6S-M (maritime), 6S-C (continental) and 6S-U (urban) aerosol models defined in Table 11.3 (labeled using the symbols given in Fig. 11.26), and using both the BRDF non-lambertian (light colors) and the equivalent isotropic lambertian (dark colors) versions of the OS1 (blue), VS1 (green), BS1 (red) and PS1 (gray) surface reflectance models defined in Table 11.12.

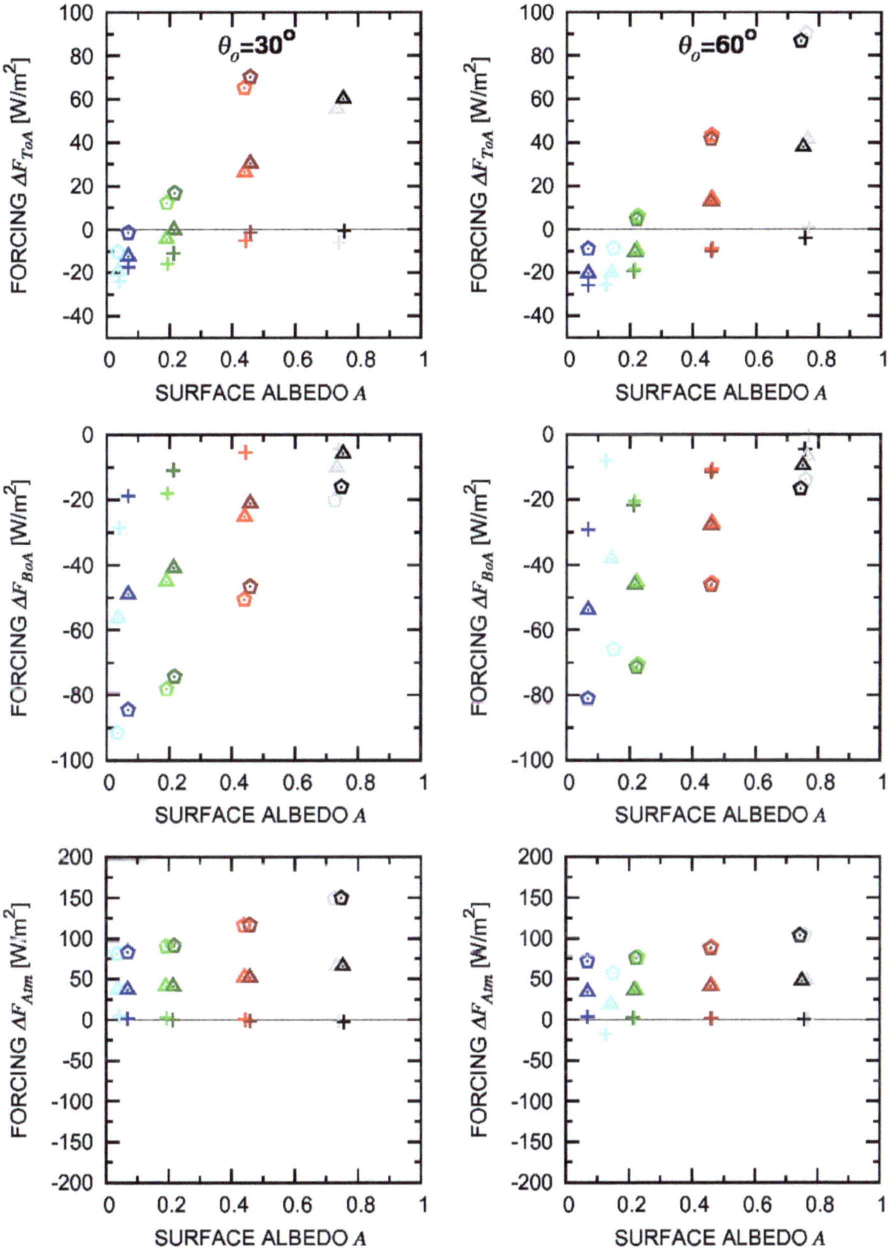

Fig. 11.28b. As in Fig. 11.28a, for aerosol optical depth $\tau_a(0.55\,\mu\text{m}) = 0.30$ and solar zenith angles $\theta_o = 30°$ (left) and $\theta_o = 60°$ (right), as calculated for the SF-M, SF-R, SF-T and SF-U aerosol models defined in Tables 11.8 and 11.9 (labeled using the symbols given in Fig. 11.26), and using both the BRDF non-lambertian (light colors) and the equivalent isotropic lambertian (dark colors) versions of the OS1 (blue), VS1 (green), BS1 (red) and PS1 (gray) surface reflectance models defined in Table 11.12.

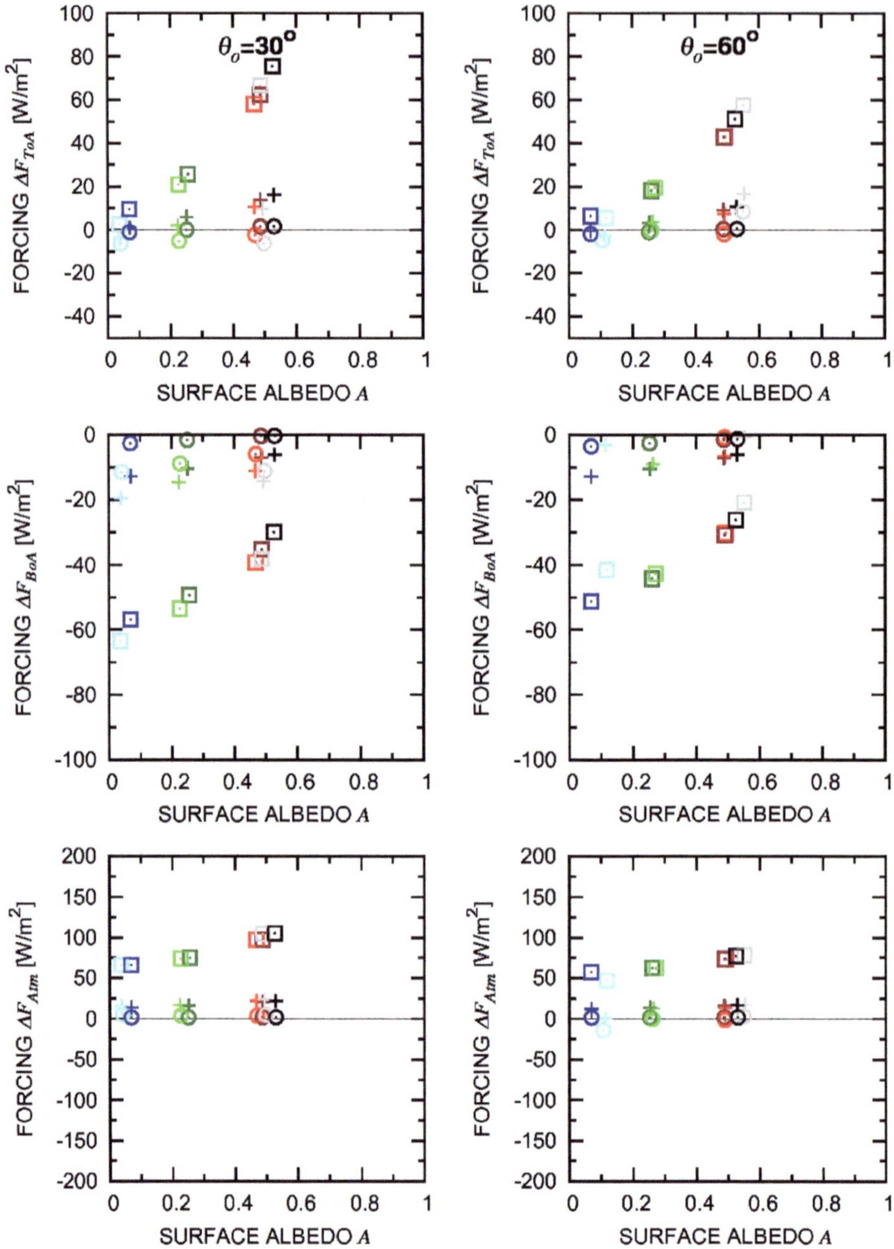

Fig. 11.28c. As in Fig. 11.28a, for aerosol optical depth $\tau_a(0.55\,\mu\text{m}) = 0.30$ and solar zenith angles $\theta_o = 30°$ (left) and $\theta_o = 60°$ (right), as calculated for the M-1, M-8 and M-14 aerosol models defined in Table 11.4 (labeled using the symbols given in Fig. 11.26), and using both the BRDF non-lambertian (light colors) and the equivalent isotropic lambertian (dark colors) versions of the OS2 (blue), VS2 (green), BS2 (red) and PS2 (gray) surface reflectance models defined in Table 11.12.

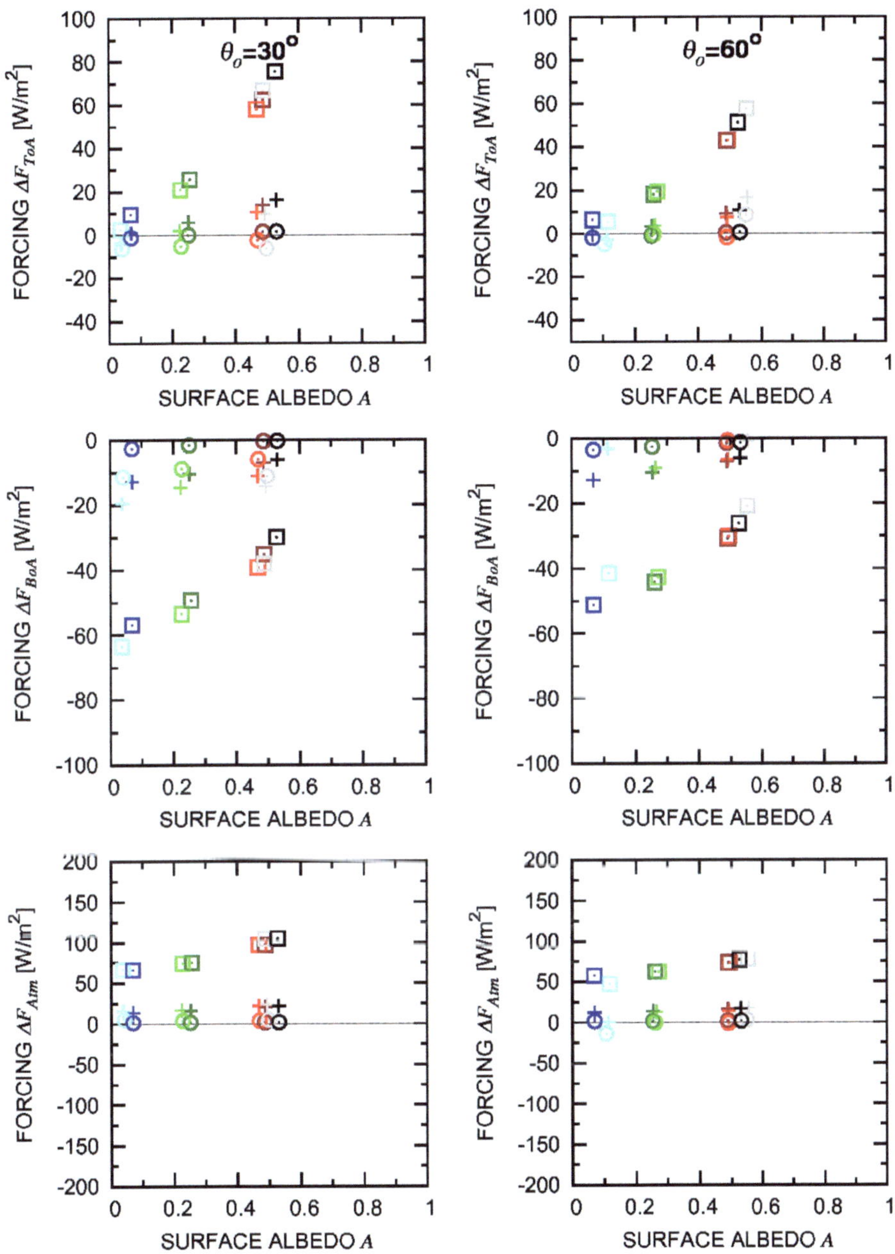

Fig. 11.28d. As in Fig. 11.28a, for aerosol optical depth $\tau_a(0.55\,\mu\text{m}) = 0.30$ and solar zenith angles $\theta_o = 30°$ (left) and $\theta_o = 60°$ (right), as calculated for the MC, CC and UR aerosol models chosen among the 10 OPAC wet aerosol models defined in Tables 11.6 and 11.7 (labeled using the symbols given in Fig. 11.26), and using both the BRDF non-lambertian (light colors) and the equivalent isotropic lambertian (dark colors) versions of the OS3 (blue), VS3 (green), BS3 (red) and PS3 (gray) surface reflectance models defined in Table 11.12.

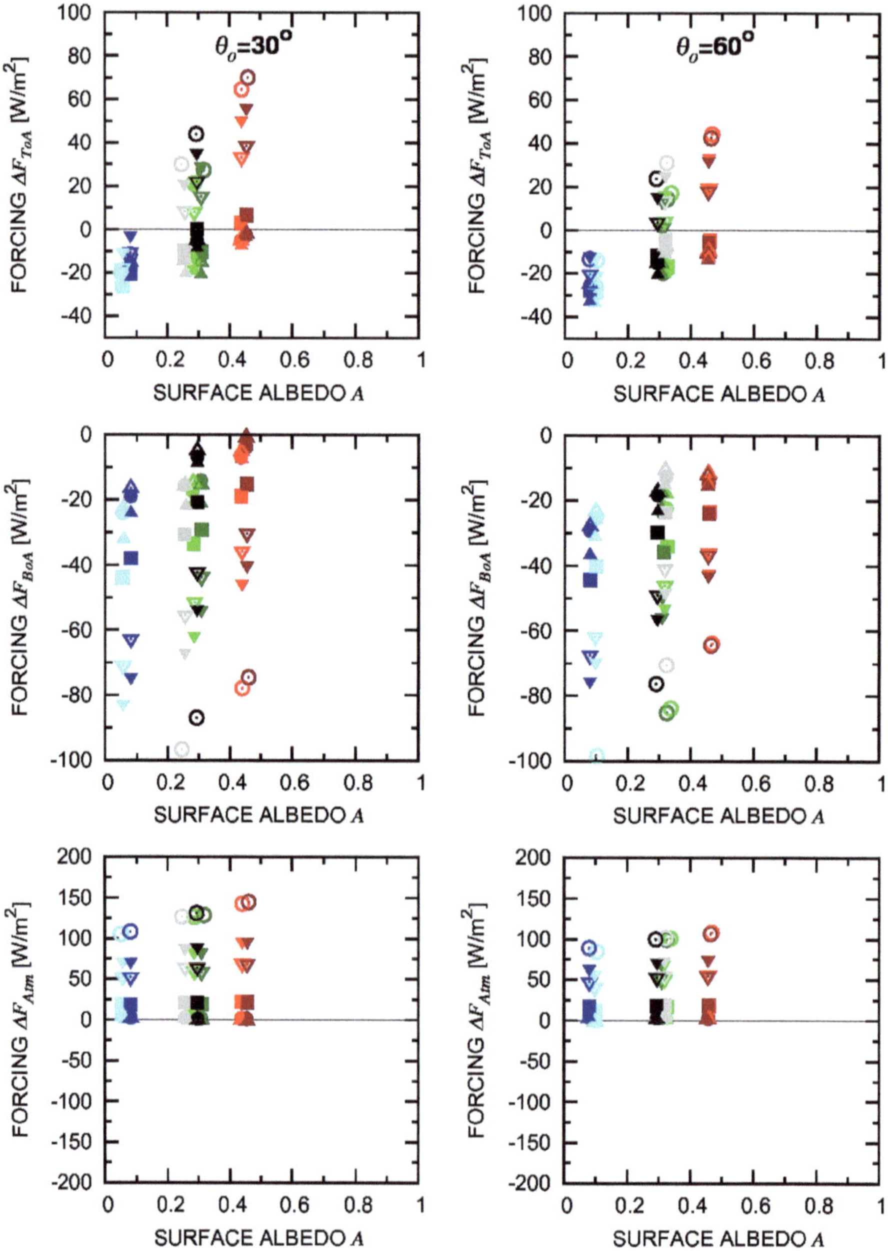

Fig. 11.28e. As in Fig. 11.28a, for aerosol optical depth $\tau_a(0.55\,\mu\text{m}) = 0.30$ and solar zenith angles $\theta_o = 30°$ (left) and $\theta_o = 60°$ (right), as calculated for the seven additional models defined in Table 11.10 (labeled using the symbols given in Fig. 11.26), and using both the BRDF non-lambertian (light colors) and the equivalent isotropic lambertian (dark colors) versions of the OS4 (blue), VS4 (green), BS4 (red) and PS4 (gray) surface reflectance models defined in Table 11.12.

(v) in Fig. 11.28e, for the seven additional aerosol models defined in Table 11.10 and the OS4, VS4, BS4 and PS4 surface reflectance models defined in Table 11.12.

It can be noted in Fig. 11.28a that:

(i) ΔF_{ToA} clearly increases with $A(\theta_o)$ defined for the two above-selected values of θ_o, passing from negative to positive values, where the positive values were mainly obtained for combinations of oceanic and vegetated surfaces with maritime and continental aerosols, implying the occurrence of cooling effects in the surface–atmosphere system, and the positive values were found for bare soil and polar snow surfaces associated with urban aerosol, causing warming effects;
(ii) ΔF_{BoA} assumes negative values, thus inducing cooling effects at the surface, which become gradually more intense as $A(\theta_o)$ increases; the most marked negative values were obtained for aerosol polydispersions that absorb more strongly the solar radiation and, therefore, exhibit rather low values of single scattering albedo, while the nearly null values were obtained for the maritime aerosol polydispersions having single scattering albedo close to unity; and
(iii) ΔF_{Atm} was evaluated to assume mainly positive values, therefore causing more or less marked warming effects in the atmosphere, which slowly increase with $A(\theta_o)$ and, hence, provide the lowest forcing value for maritime aerosols (with $\omega^* = 0.989$) and the highest one for urban aerosols (with $\omega^* = 0.632$).

Similar features to those of Fig. 11.28a were also determined in (a) Fig. 11.28b for the Maritime (SF-M), Tropospheric (SF-T), Rural (SF-R) and Urban (SF-U) aerosol models defined by Shettle and Fenn (1979), giving values of ω^* equal to 0.999, 0.953, 0.933, and 0.697, respectively, and (b) Fig. 11.28c for the M-1 (pure oceanic), M-8 (pure continental) and M-14 (heavy polluted urban) aerosol models, giving values of ω^* equal to 0.999, 0.855 and 0.651, respectively. Similar results were also found in Fig. 11.28d for the three MC, CC and UR OPAC aerosol models (giving values of ω^* equal to 0.997, 0.952 and 0.741, respectively). In these cases, appreciably higher values of ΔF_{ToA} were achieved at both $\theta_o = 30°$ and $\theta_o = 60°$ than those correspondingly obtained in the previous applications for lower values of $A(\theta_o)$. The results suggest that more intense warming effects are induced within the surface–atmosphere system by aerosol polydispersions having comparable values of ω^*, when they are suspended over surfaces characterized by higher reflectance properties. The evaluations of the DARF terms in Fig. 11.28e were made for all the seven additional aerosol models giving values of ω^* ranging between 0.616 (FT model) and 0.999 (PV-2 and PV-3 models). They substantially agree with the evaluations of ΔF_{ToA} shown in Figs. 11.28a to 11.28d, suggesting that an aerosol polydispersion consisting of non-absorbing or weakly absorbing particulate matter, and leading to a negative (cooling) value of this DARF term for low surface reflectance properties, can cause an effect of opposite sign (warming) when it is transported over a surface characterized by very high surface reflectance conditions.

The physical meaning of these trends can be better understood considering that the instantaneous absorptance/reflectance ratio Y of an aerosol polydispersion can vary greatly if the columnar aerosol is associated with gradually higher values of broadband surface albedo $A(\theta_o)$. This concept is illustrated in Fig. 11.29, where ratio Y is plotted versus $A(\theta_o = 30°)$. In the graph, the curve of ratio Y was calculated using the analytical form

$$Y = [(1 - A(\theta_o)2] / 2A(\theta_o) , \tag{11.15a}$$

defined by Chylek and Coakley (1974) (see also Chylek and Wong, 1995) applying the two-stream approximation to the radiative transfer equation in the surface–atmosphere (with aerosol) system for incident isotropically diffused radiation. Thus, ratio Y in Eq. (11.15a) can be defined for a certain aerosol polydispersion in terms of the following equation,

$$Y = (1 - \omega^*)/\omega^* \beta_a(\cos\theta_o) , \tag{11.15b}$$

providing the ratio between (i) the fraction of incoming short-wave radiation absorbed by atmospheric aerosols, equal to the difference $1 - \omega^*$, and (ii) the fraction $\omega^* \beta_a(\cos\theta_o)$ of incoming solar radiation scattered backward by airborne aerosols for solar zenith angle θ_o, where β_a is the fraction of radiation scattered into the backward hemisphere by aerosols (as defined in Eq. (14)), calculated in terms of the asymmetry factor $g(0.55\,\mu\text{m})$ and using the Wiscombe and Grams (1976) formula for mean daily data. The values of $A(\theta_o = 30°)$ relative to 7 of the 16 BRDF non-lambertian models described in Table 11.12 are presented in Fig. 11.29 to cover the whole broadband albedo range from the nearly null value of 0.035 (OS1 model) to 0.834 (PS1 model). The values of ratio Y calculated for three of the OPAC aerosol models described in Table 11.7 are also shown in Fig. 11.29, relative to the Maritime Clean (MC) ($\omega^* = 0.997$), Continental Average (CA) ($\omega^* = 0.884$) and Urban (UR) ($\omega^* = 0.741$) aerosol polydispersions, so as to cover the most commonly observed range of ω^*.

Examining the three examples shown in Fig. 11.29, it can be seen that the MC aerosol polydispersion suspended over an oceanic surface produces marked cooling effects and can induce gradually less pronounced negative values of ΔF_{ToA} when transported over the high-reflectance regions covered by unpolluted glaciers, such as those of the Antarctic Plateau, without causing appreciable warming effects even for the highest albedo conditions of the remote polar regions. The CC aerosol polydispersion situated over the VS1 surface was estimated to cause cooling effects also, although less pronounced than in the first example. Here, more marked cooling effects could be induced in all cases where these aerosol particles were transported over the oceanic regions, while gradually less pronounced cooling effects would be produced if these aerosol polydispersions were transported over surfaces presenting higher albedo characteristics, yielding nearly null radiative effects for $A(\theta_o = 30°) \approx 0.45$, and gradually more intense warming effects over polar surfaces covered by snow fields and glaciers presenting values of $A(\theta_o = 30°) > 0.50$. The UR (urban) aerosol model associated with the BS1 surface (estimated to provide a value of $A(\theta_o = 30°)$ not largely different from those of the most populated urban areas) turns out to induce only moderate warming effects. However, it could cause neutral effects if transported over vegetated surfaces presenting albedo conditions similar to those of the VS1 surface, and marked cooling effects if transported

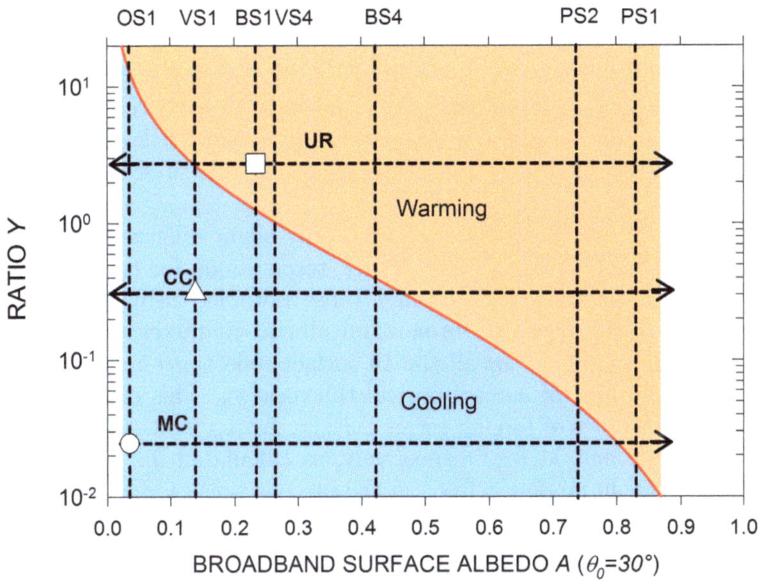

Fig. 11.29. Dependence curve (red colour) of ratio Y (between columnar absorptance and columnar reflectance) as a function of broadband surface albedo $A(\theta_o)$ obtained for solar zenith angle $\theta_o = 30°$ over the entire range of $A(\theta_o)$ usually observed over the Earth's surface, for which no radiative forcing effects at the ToA-level appear to be induced by the columnar aerosol loading. The values of ratio Y were calculated as a function $A(\theta_o)$ in terms of Eq. (15a) derived by Chylek and Coakley (1974) using the two-stream approximation form applied to radiative transfer through the atmosphere. Red curve divides the domain (Y, $A(\theta_o)$) into two sub-domains, where aerosol effects are estimated to induce cooling (light blue background colour) and warming (brownish background colour) effects. Three aerosol models are represented in the domain for values of Y calculated in terms of Eq. (15b) for the three following OPAC wet models: (i) Maritime Clean (MC) aerosol model ($\omega^* = 0.997$) at $A(\theta_o = 30°) = 0.035$ (OS1 model); (ii) Continental Clean (CC) aerosol model ($\omega^* = 0.884$) at $A(\theta_o = 30°) = 0.139$ (VS1 model); and (iii) Urban (UR) aerosol model ($\omega^* = 0.741$) at $A(\theta_o = 30°) = 0.228$ (BS1 model).

over the oceanic regions. In all cases of transport over vegetated surfaces and arid lands presenting higher albedo features or over the polar regions, this polluted aerosol polydispersion is expected to cause marked warming effects that gradually increase for higher surface albedo conditions.

To give a measure of the variability in ΔF_{ToA} shown in Figs. 11.28a to 11.28e, it is important to note that the average slope coefficients of the patterns of ΔF_{ToA} plotted versus $A(\theta_o)$ vary between $+4.2\,\mathrm{W\,m^{-2}}$ (MC model) and $+235.5\,\mathrm{W\,m^{-2}}$ (6S-U) per unit variation of $A(\theta_o = 30°)$, and between $+20.8\,\mathrm{W\,m^{-2}}$ (MC) and $+176.3\,\mathrm{W\,m^{-2}}$ (6S-U) per unit variation of $A(\theta_o = 60°)$, with intercept values varying between $-5.1\,\mathrm{W\,m^{-2}}$ (CC model) and $-32.1\,\mathrm{W\,m^{-2}}$ (PV-3) for $\theta_o = 30°$, and between $-6.9\,\mathrm{W\,m^{-2}}$ (MC) and $-37.6\,\mathrm{W\,m^{-2}}$ (PV-3) for $\theta_o = 60°$. These results arise from the fact that the MC and 6S-U models yield the lowest and the highest absorption features of solar radiation, respectively. The intercept values were determined for null surface albedo conditions, and are therefore indicative of the role of the aerosol polydispersions in scattering the incoming solar radiation toward space during the first passage through the atmosphere, thus contributing to provide an overall negative (cooling) effect. These findings suggest that the most moderate aerosol scattering effects are produced by the CC columnar particles, and the strongest ones by the PV-3 long-age volcanic particles generated by the Pinatubo eruption, which were characterized by very high scattering properties (Valero and Pilewskie, 1992).

Finally, Figs. 11.28a to 11.28e show that only slight differences were obtained between the evaluations of the three DARF terms made for $\theta_o = 30°$ using the isotropic surface reflectance models and those obtained using the BRDF non-lambertian models for the same broadband albedo characteristics, which are in general within a few $\mathrm{W\,m^{-2}}$ for all the 16 surface reflectance models defined here for relatively low values of aerosol optical thickness τ_a. This can be also verified by examining the values of forcing terms ΔF_{ToA}, ΔF_{BoA} and ΔF_{Atm} given in Tables 11.14a, 11.14b and 11.14c, respectively, as calculated for 18 aerosol models selected among the 40 models defined in the present work, for $\tau_a(0.55\,\mu\mathrm{m}) = 0.10$. Figures 11.28a to 11.28e reveal even more limited differences between the DARF terms calculated at $\theta_o = 60°$ for the two sets of isotropic and BRDF non-lambertian models of surface reflectance relative to the VS, BS and PS surface models. By contrast, more marked discrepancies were obtained for the OS surface reflectance models, presumably because of the marked sun glint effects occurring at this rather high solar zenith angle.

11.4.5 Dependence of instantaneous DARF on solar zenith angle

The dependence patterns shown in Figs. 11.25a, 11.25b and 11.25c indicate that the instantaneous DARF terms vary greatly as a function of solar zenith angle θ_o for all the pairs of an aerosol model (chosen among the 40 aerosol models listed in Fig. 11.26) and one of the surface reflectance models defined in the present study. The variability of the instantaneous DARF terms as a function of θ_o is crucial for the calculations of the daily average DARF effects over the 24-hour period, because θ_o varies regularly throughout the day, generally presenting rather low values at noon, but can vary considerably with latitude and season. In order to evaluate how great the variations in DARF are, due to such changes in θ_o, the dependence patterns of instantaneous ΔF_{ToA}, ΔF_{BoA} and ΔF_{Atm} are shown in Figs. 11.30a to 11.30f, as calculated for (i) $\tau_a(0.55\,\mu\mathrm{m}) = 0.30$, (ii) 9 selected values of θ_o taken in steps of 10° from 0° to 80°, (iii) a large number of aerosol models chosen among the above 40 aerosol models defined in Section 11.2, and (iv) various sets of surface reflectance models defined above for both BRDF non-lambertian and isotropic (ISO) surface reflectance characteristics. Figures 11.30a to 11.30f also provide the evaluations of the absolute differences between the estimates of instantaneous DARF terms obtained for ISO and BRDF non-lambertian surface reflectance conditions (and reported as ISO–BDRF differences in the Figs. 11.30a to 11.30f), as performed for the following combinations of aerosol and surface reflectance models:

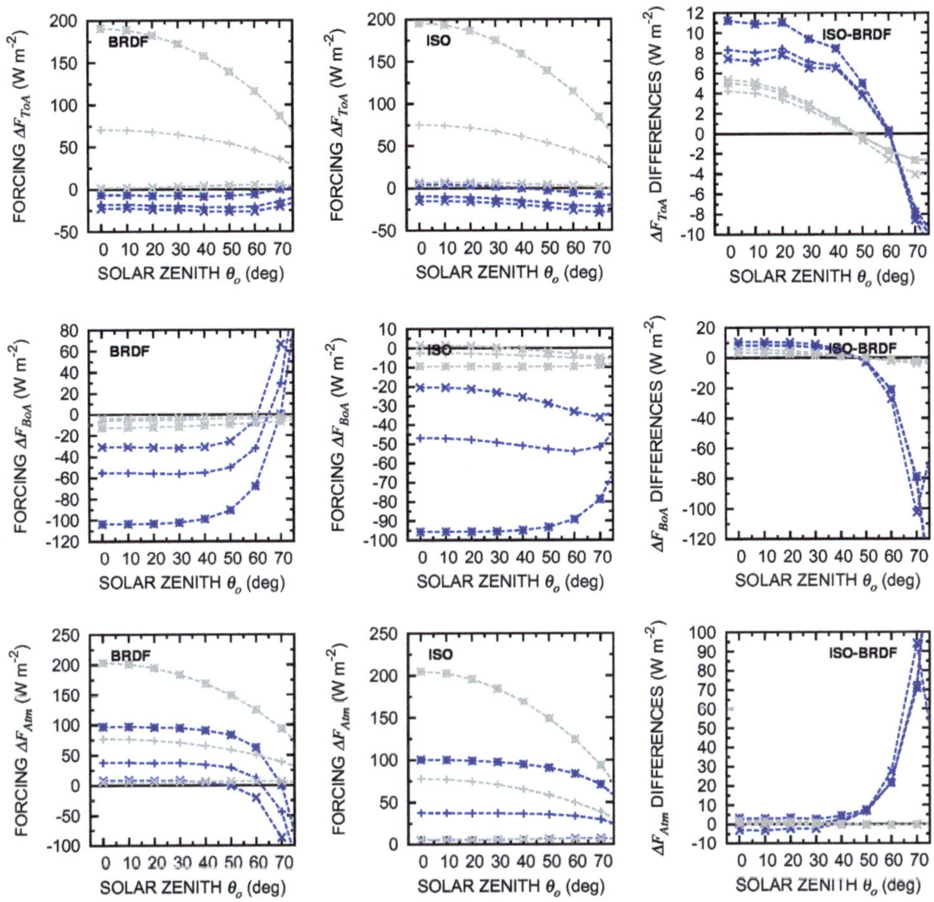

Fig. 11.30a. Instantaneous direct aerosol-induced radiative forcing terms ΔF_{ToA} (upper part), ΔF_{BoA} (middle part) and ΔF_{Atm} (lower part) calculated for aerosol optical depth $\tau_a(0.55\,\mu\mathrm{m}) = 0.30$, and for nine selected values of solar zenith angle θ_o (in steps of 10° from 0° to 80°), as calculated for (i) the 6S-M (maritime), 6S-C (continental) and 6S-U (urban) aerosol models defined in Table 11.3 (labeled using the symbols given in Fig. 11.26), and (ii) the OS1 (blue) and PS1 (gray) models represented for both the non-lambertian (left column) and the corresponding equivalent lambertian (middle column) surface reflectance characteristics, as given in Table 11.12. The absolute differences between the lambertian and non-lambertian evaluations of the DARF terms are shown in the three graphs of the right column.

(i) the 6S-M (maritime), 6S-C (continental) and 6S-U (urban) aerosol models with the OS1 and PS1 models in Fig. 11.30a;

(ii) the 6S-M, 6S-C and 6S-U aerosol models with the VS1 and BS1 models in Fig. 11.30b;

(iii) the SF-R (rural), SF-U (urban), SF-M (maritime) and SF-T (tropospheric) aerosol models with the VS1 and BS1 models in Fig. 11.30c;

(iv) the M-1 (pure oceanic) and M-14 (heavy polluted) aerosol models with the OS2, VS2, BS2, and PS2 models in Fig. 11.30d;

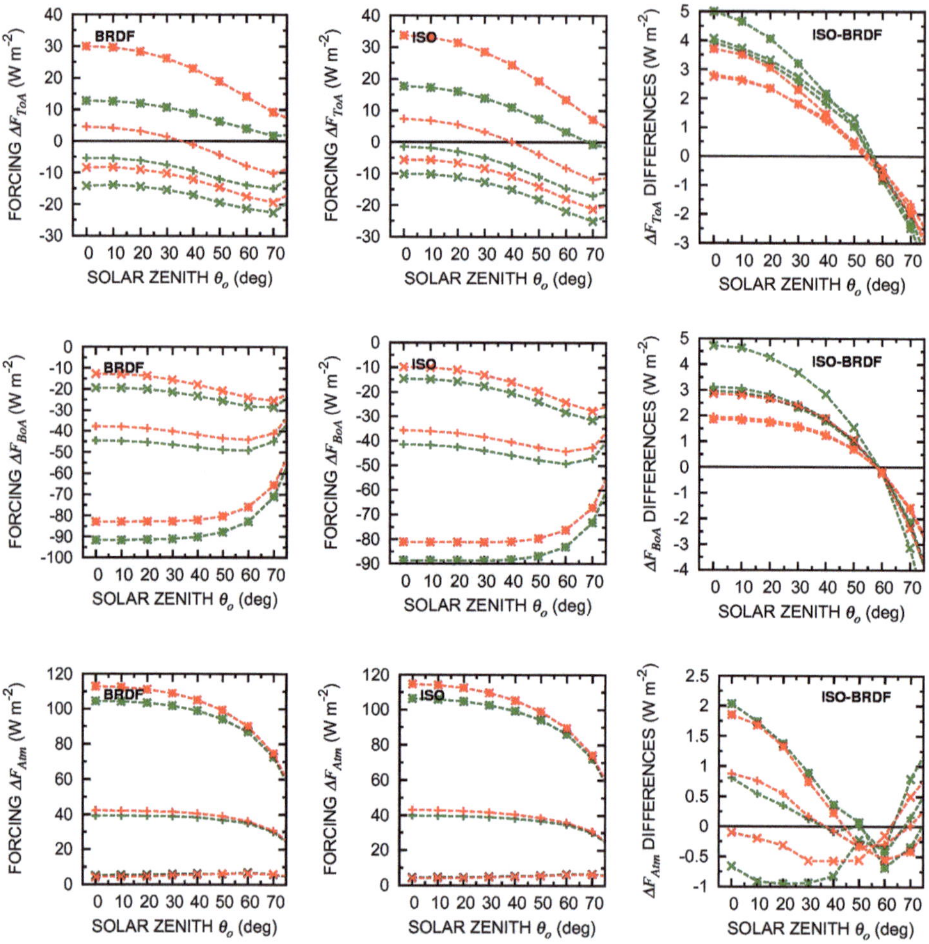

Fig. 11.30b. As in Fig. 11.30a, for $\tau_a(0.55\,\mu\text{m}) = 0.30$, and nine selected values of solar zenith angle θ_o (in steps of 10° from 0° to 80°), as calculated for (i) the 6S-M (maritime), 6S-C (continental) and 6S-U (urban) aerosol models defined in Table 11.3 (labeled using the symbols given in Fig. 11.26a), and (ii) the VS1 (green) and BS1 (red) models represented for both the non-lambertian (left column) and the corresponding equivalent lambertian (middle column) surface reflectance characteristics, as given in Table 11.12. The absolute differences between the lambertian and non-lambertian evaluations of the DARF terms are shown in the three graphs of the right column.

(v) the OPAC UR (urban) and MC (maritime clean) aerosol models with the OS3, VS3, BS3, and PS3 models in Fig. 11.30e; and
(vi) the FT (Free Troposphere) and PV-2 (Post Pinatubo stratospheric particle) aerosol models with the OS4, VS4, BS4, and PS4 models in Fig. 11.30f.

The results presented in Figs. 11.30a to 11.30f were generally obtained for gradually higher reflectance characteristics of the surface. To avoid confusion in the graphs, the results achieved for surface reflectance models belonging to the same class were

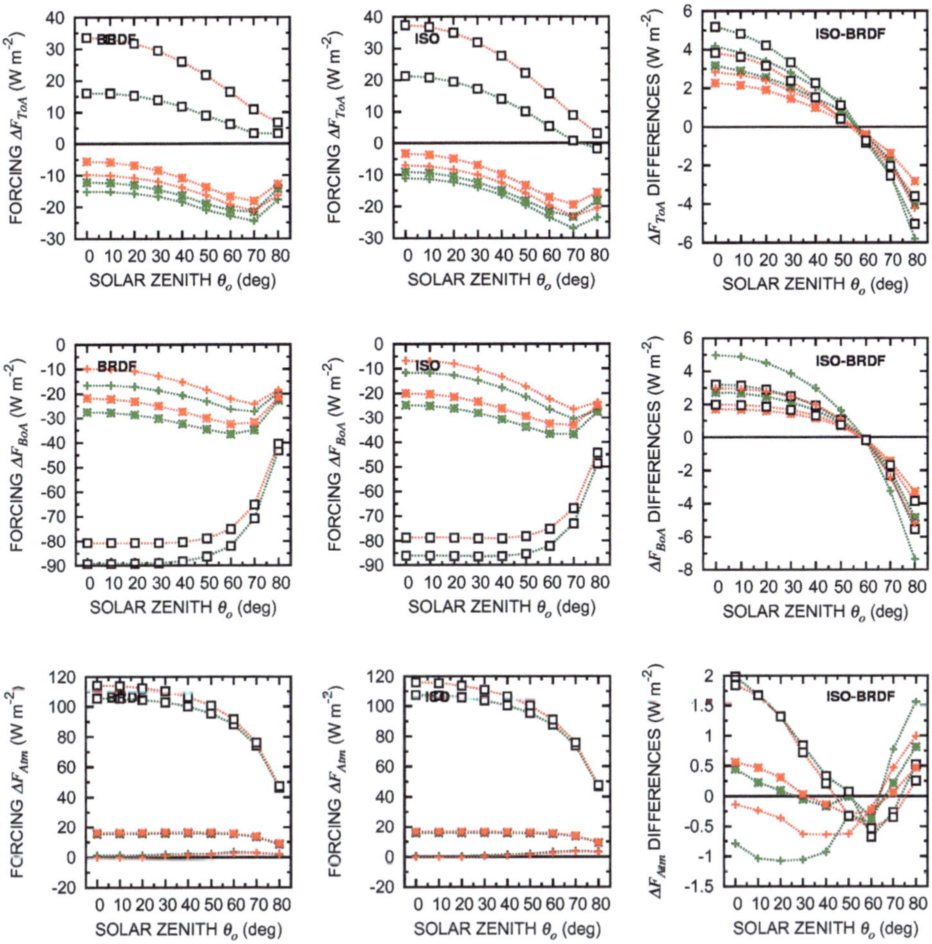

Fig. 11.30c. As in Fig. 11.30a, for $\tau_a(0.55\,\mu\mathrm{m}) = 0.30$ and nine selected values of solar zenith angle θ_o (in steps of 10° from 0° to 80°), as calculated for (i) the four SF aerosol models defined in Tables 11.8 and 11.9 (labeled using the symbols given in Fig. 11.26), and (ii) the VS1 (green) and BS1 (red) models represented for both the non-lambertian (left column) and the corresponding equivalent lambertian (middle column) surface reflectance characteristics, as given in Table 11.12. The absolute differences between the lambertian and non-lambertian evaluations of the DARF terms are shown in the three graphs of the right column.

not compared one with the other in the same graphical representation, in such a way as to identify univocally the various surface model classes (relative to sea surface, vegetation-covered, bare soil and snow- and ice-covered areas) by using clearly different colors. This extensive analysis of the variability of instantaneous DARF terms as a function of θ_o provides a detailed picture of how the DARF effects tend to vary as a function of this key-parameter over the 0°–80° range. Fig. 11.30a reports the results obtained for the three above-mentioned 6S-type aerosol models covering the 0.63–0.99 range of weighted average single scattering albedo ω^*, and for extreme

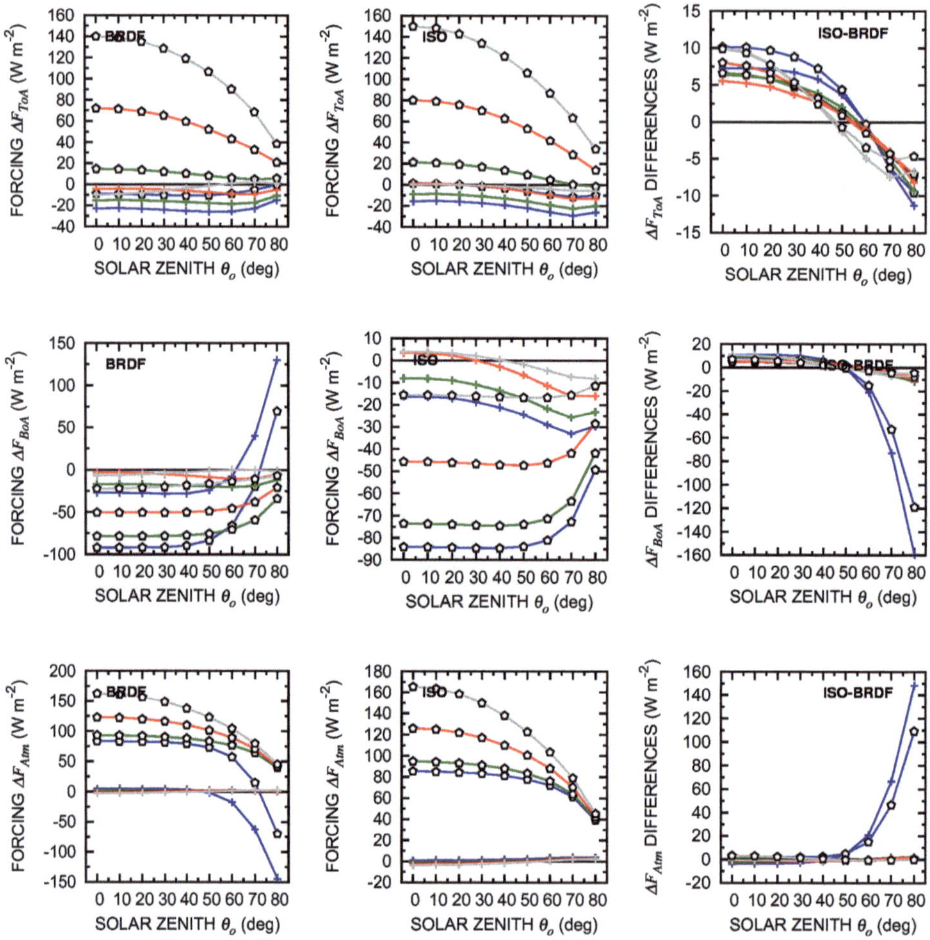

Fig. 11.30d. As in Fig. 11.30a, for $\tau_a(0.55\,\mu\text{m}) = 0.30$ and nine selected values of solar zenith angle θ_o (in steps of 10° from 0° to 80°), as calculated for (i) the pair of M-1 and M-14 aerosol models defined in Table 11.4 (labeled using the symbols given in Fig. 11.26), and (ii) the OS2 (blue), VS2 (green), BS2 (red) and PS2 (gray) models represented for both the non-lambertian (left column) and the corresponding equivalent lambertian (middle column) surface reflectance characteristics, as given in Table 11.12. The absolute differences between the lambertian and non-lambertian evaluations of the DARF terms are shown in the three graphs of the right column.

surface reflectance features represented by the OS1 model (giving $R_{ws} = 0.07$) and the PS1 model ($R_{ws} = 0.85$). The instantaneous forcing term ΔF_{ToA} relative to the OS1 model was found to assume negative values varying between a few W m^{-2} for the 6S-U model and about -25 W m^{-2} for the 6S-M model, while that relative to the PS1 surface assumed positive values varying between a few W m^{-2} (6S-M model) and about $+180$ W m^{-2} (6S-U model) at $\theta_o = 0°$. Forcing term ΔF_{ToA} was estimated to decrease monotonically with increasing θ_o over the polar snow surface, and to exhibit a wide minimum over the ocean surface, at $\theta_o \approx 60°$. It can

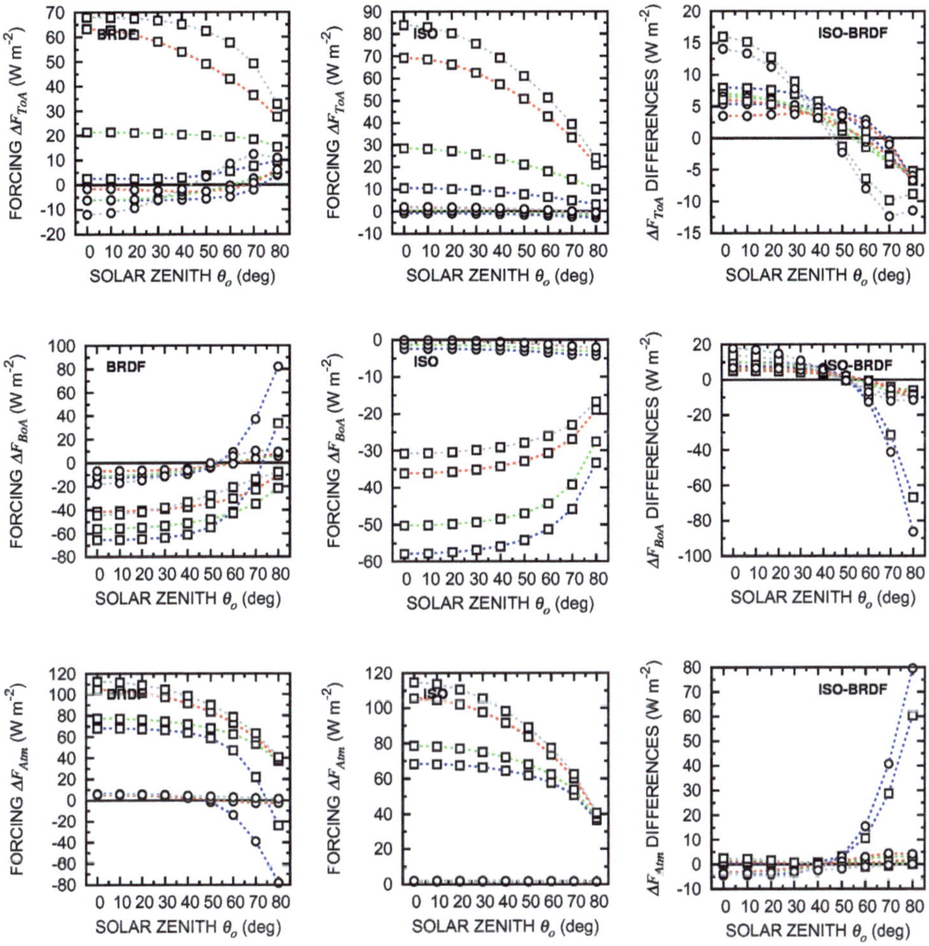

Fig. 11.30e. As in Fig. 11.30a, for $\tau_a(0.55\,\mu\mathrm{m}) = 0.30$ and nine selected values of solar zenith angle θ_o (in steps of 10° from 0° to 80°), as calculated for (i) the UR and MC models chosen among the 10 OPAC wet aerosol models defined in Tables 11.6 and 11.7 (labeled using the symbols given in Fig. 11.26), and (ii) the OS3 (blue), VS3 (green), BS3 (red) and PS3 (gray) models represented for both the non-lambertian (left column) and the corresponding equivalent lambertian (middle column) surface reflectance characteristics, as given in Table 11.12. The absolute differences between the lambertian and non-lambertian evaluations of the DARF terms are shown in the three graphs of the right column.

be also noted in Fig. 11.30a that the isotropic surface reflectance conditions cause in general an overestimation of ΔF_{ToA} over the $0° \leq \theta_o \leq 60°$ range, until yielding values very close to those obtained for BRDF non-lambertian surface reflectance models, and slightly underestimated values at $\theta_o > 60°$. This can be reasonably explained by the fact that white-sky albedo R_{ws} assumes a value comparable with that of albedo R_L over the $50° \leq \theta_o \leq 60°$ range, as can be seen in Fig. 11.21. Because of the different surface reflectance properties, the values of the difference between the isotropic and BRDF non-lambertian values of ΔF_{ToA} were estimated

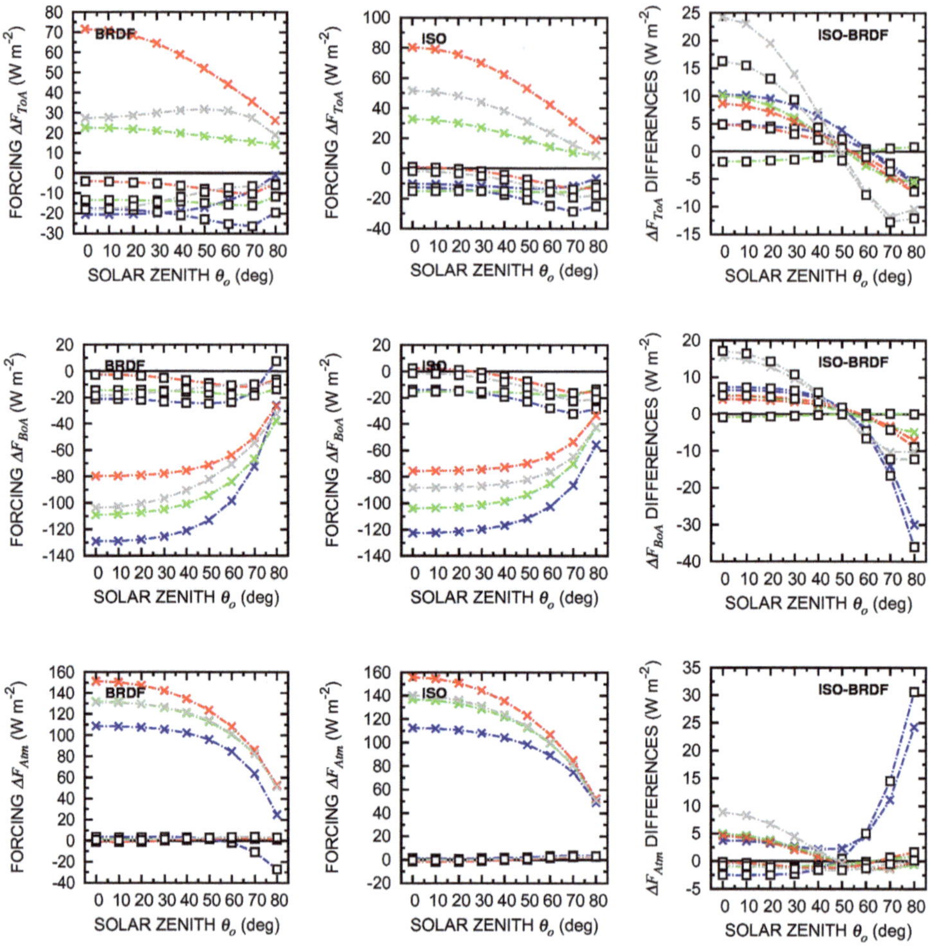

Fig. 11.30f. As in Fig. 11.30a, for $\tau a(0.55\,\mu m) = 0.30$ and 9 selected values of solar zenith angle θ_o (in steps of 10° from 0° to 80°), as calculated for (i) the FT and PV-2 aerosol models chosen among the seven additional aerosol models defined in Table 11.10 for the extreme values of weighted average single scattering albedo ω^* (labeled using the symbols given in Fig. 11.26), and (ii) the OS4 (blue), VS4 (green), BS4 (red) and PS4 (gray) models represented for both the non-lambertian (left column) and the corresponding equivalent lambertian (middle column) surface reflectance characteristics, as given in Table 11.12. The absolute differences between the lambertian and non-lambertian evaluations of the DARF terms are shown in the three graphs of the right column.

to vary between ± 3 and $\pm 12\,\mathrm{W\,m^{-2}}$ in the range of θ_o from 0° to 20°, and then to decrease gradually as θ_o increases, until becoming negative for $\theta_o \geq 60°$. The dependence patterns of ΔF_{ToA} on θ_o showed greatly varying values, which can be of opposite signs depending on both surface reflectance properties and the more or less distant values from unity of albedo ω^* given by the various aerosol models.

Instantaneous forcing ΔF_{BoA} was evaluated to exhibit negative values for both non-lambertian and isotropic surface reflectance models over the range of θ_o from

0° to more than 60°, providing values of the isotropic-minus-BRDF differences that are all positive and varying between 0 and $+10\,\mathrm{W\,m^{-2}}$ over the $0° \leq \theta_o \leq 45°$ range, and negative in the upper range, where they decrease very rapidly to less than $-80\,\mathrm{W\,m^{-2}}$ at $\theta_o = 70°$. Correspondingly, instantaneous ΔF_{Atm} was found to assume positive values ranging between 0 and $+200\,\mathrm{W\,m^{-2}}$ over the $0° \leq \theta_o \leq 50°$ range, followed by mainly negative values for the BRDF non-lambertian models and by appreciably lower positive values for the isotropic models. Consequently, the isotropic-minus-BRDF differences of ΔF_{Atm} turn out to be nearly null for all the aerosol extinction and surface reflectance models over the $0° \leq \theta_o \leq 45°$ range, but become positive over the upper range, increasing gradually to reach a value exceeding $+90\,\mathrm{W\,m^{-2}}$ at $\theta_o = 70°$.

Similar dependence features of instantaneous terms ΔF_{ToA}, ΔF_{BoA} and ΔF_{Atm} on θ_o are shown in Figs. 11.30b to 11.30f for further sets of combinations of aerosol models with surface reflectance models, highlighting the various trends of these DARF terms, which tend often to change significantly throughout the day as a function of θ_o, presenting features that vary with the angular and spectral characteristics of surface reflectance as well as with the scattering and absorption properties of the aerosol polydispersions characterized by different values of ω^*.

Examining the sequence of Figs. 11.30a to 11.30f, without taking into account the variations in the aerosol optical properties, it can be stated in general that the variations in the absolute values of ΔF_{ToA} and ΔF_{BoA} gradually increase as surface reflectance increases, passing from the set of OS1, VS1, BS1 and PS1 models considered in Figs. 11.30a and 11.30b to the set of OS4, VS4, BS4 and PS4 models chosen in Fig. 11.30f. Correspondingly, it can be noticed that the evaluations of ΔF_{Atm} become appreciably higher as the surface reflectance properties increase. In general, the ISO-minus-BRDF differences of ΔF_{ToA} and ΔF_{BoA} were both found to decrease more or less rapidly as θ_o increases, passing from positive values for $\theta_o < 50°$ to negative values for $\theta_o > 60°$, while the ISO-minus-BRDF differences of ΔF_{Atm} were evaluated to vary considerably with θ_o, presenting both negative and positive values for $\theta_o < 50°$ and more frequently positive values for $\theta_o > 55°$. As a result of these variations throughout the range of $\theta_o < 50°$, the ISO-minus-BRDF differences of ΔF_{ToA} generally assumed: (i) values ranging between +2 and $+8\,\mathrm{W\,m^{-2}}$ for relatively low reflectance conditions, and (ii) values varying between +10 and $+25\,\mathrm{W\,m^{-2}}$ for the high-reflectance conditions of the PS models. Conversely, the ISO-minus-BRDF differences of ΔF_{BoA} were found to assume negative values for $\theta_o > 50°$, mainly ranging between −10 and $-100\,\mathrm{W\,m^{-2}}$. It is interesting to note that the most negative values of this quantity were obtained for sea-surface reflectance conditions, leading to predominantly positive values of ΔF_{Atm}. Correspondingly, the ISO-minus-BDRF differences of ΔF_{Atm} were found to assume values close to null for $\theta_o < 50°$, varying mainly between −5 and $+10\,\mathrm{W\,m^{-2}}$.

Therefore, the patterns of the ISO-minus-BRDF differences calculated for the three DARF terms turn out to be quite regular over the $0° \leq \theta_o \leq 50°$ and are subject to vary in sign over the upper range of θ_o. Figures 11.30a to 11.30f show that the instantaneous DARF terms sometimes describe negative or positive trends over the range θ_o, presumably due to opposite effects induced by the diverse influences of surface albedo and aerosol single scattering albedo.

11.5 Concluding remarks

A set of 40 aerosol models determined for different multimodal size-distributions of particles and various radiative properties associated with diverse chemical composition and different origins of both dry and wet particulate matter, was used as basis of an exhaustive analysis of the evaluations of instantaneous DARF terms at the ToA- and BoA-levels and within the atmosphere. For this purpose, use was made of two separate sets of surface reflectance models for oceanic, vegetated land, bare soil (arid) areas and snow and ice-covered regions, which were determined for both BRDF non-lambertian and isotropic reflectance characteristics yielding comparable values of broadband albedo. Thus, the dependence patterns of instantaneous forcing terms ΔF_{ToA}, ΔF_{BoA} and ΔF_{Atm} were carefully investigated to study how they vary as a function of (i) aerosol optical thickness determined for different aerosol models, (ii) single scattering albedo of columnar aerosols, (iii) surface reflectance characteristics and broadband albedo, and (iv) solar zenith angle θ_o for different surface reflectance models. The results clearly indicate that:

(a) the three instantaneous DARF terms increase almost linearly as a function of $\tau a(0.55\,\mu m)$ over the various surfaces in all cases presenting poorly absorbing aerosol polydispersions, and describe convex patterns (particularly marked for the ΔF_{BoA} and ΔF_{Atm} terms), not only when strongly absorbing particles are combined with oceanic surface reflectance models but also in cases where weakly absorbing aerosols are associated with high surface reflectance conditions. For greatly variable aerosol and surface reflectance characteristics, the corresponding DARF efficiency relative to ΔF_{ToA} was evaluated to range between $-80\,\mathrm{W\,m^{-2}}$ per unit variation of aerosol optical thickness $\tau_a(0.55\,\mu\mathrm{m})$ due to non-absorbing maritime aerosol over oceans, and $+400\,\mathrm{W\,m^{-2}}$ for heavy polluted urban aerosol polydispersions suspended over high-reflectance surfaces. In general, it was found that all the three instantaneous DARF terms vary almost linearly for relatively low values of $\tau_a(0.55\,\mu\mathrm{m})$, especially in the presence of non-absorbing maritime or only slightly absorbing continental aerosol particles (presenting values of ω^* close to unity) suspended over oceanic surfaces.

(b) The three instantaneous DARF terms were evaluated to vary almost linearly as a function of weighted average parameter ω^* for all the surface reflectance conditions, evidencing that instantaneous forcing ΔF_{ToA} decreases as ω^* increases from ~ 0.60 to about 1.00, with slope coefficient per unit variation of ω^* that was estimated to range between about $-16\,\mathrm{W\,m^{-2}}$ and $-520\,\mathrm{W\,m^{-2}}$ as one passes from oceanic surfaces to polar ice-covered surfaces. Similarly, instantaneous forcing ΔF_{BoA} was evaluated to increase gradually as ω^* increases until reaching nearly null values for non-absorbing aerosols combined with all the surface reflectance configurations defined in the present study. In contrast, instantaneous forcing ΔF_{Atm} was estimated to vary as a function of ω^* and decrease gradually from values sometimes higher than $+\ 150\,\mathrm{W\,m^{-2}}$ for the most absorbing aerosol polydispersions to around null values for very poorly absorbing aerosols.

(c) The three instantaneous DARF terms were evaluated to vary considerably as a function of surface reflectance characteristics and broadband albedo, with average rates found for BRDF non-lambertian models that differ only slightly from those estimated for equivalent isotropic surface reflectance models. The increasing rate of ΔF_{ToA} per unit variation in broadband albedo $A(\theta_o = 30°)$ was estimated to range between about $+4\,\text{W m}^{-2}$ for maritime clean aerosol and more than $+230\,\text{W m}^{-2}$ for polluted urban aerosol. Similar patterns were also determined for instantaneous ΔF_{BoA} presenting more pronounced rates, while more limited variations were estimated to affect the instantaeous ΔF_{Atm}, as shown in Figs. 11.28a to 11.28e.

(d) The three instantaneous DARF terms are subject to different rates of variation as a function of solar zenith angle θ_o for different surface reflectance characteristics. The dependence patterns of $\Delta FToA$ vary considerably throughout the day, depending on both surface albedo conditions and aerosol single scattering albedo, but do not present relevant differences between the estimates made for BDRF non-lambertian models and those obtained for isotropic surface reflectance models in all cases where θ_o does not appreciably exceed 50°. Instantaneous forcing ΔF_{BoA} was found to exhibit generally negative values for both such surface reflectance configurations at values of θ_o increasing from 0° to around 50°, while ΔF_{Atm} assumed positive values ranging between 0 and $+200\,\text{W m}^{-2}$ over the same range of θ_o.

The calculations made to estimate the instantaneous DARF terms ΔF_{ToA}, ΔF_{BoA} and ΔF_{Atm} given in Tables 11.14a, 11.14b and 11.14c indicate that the discrepancies between the evaluations made for the BRDF non-lambertian surface reflectance models and those for isotropic (lambertian) models are not relevant for solar zenith angle $\theta_o = 30°$. Considering the radiative transfer processes occurring during the middle part of a measurement day at mid-latitude sites far from intense anthropogenic aerosol sources and, hence, for $\tau_a(0.55\,\mu\text{m}) \approx 0.10$, it can be plausibly assumed that the instantaneous DARF evaluations made at this measurement site during the central hours of the day should do not differ appreciably from the evaluations given in Tables 11.14a, 11.14b and 11.14c. For the purpose of evidencing the relevance of the discrepancies between isotropic and BRDF non-lambertian evaluations of the instantaneous DARF terms, a comparison was made in Figs. 11.31a and 11.31b showing the isotropic estimates of these three quantities versus the BRDF non-lambertian evaluations, as obtained for a set of aerosol models giving values of $\tau_a(0.55\,\mu\text{m})$ equal to 0, 0.1, 0.3, 0.5, 0.7 and 0.9, and the values of $\theta_o = 0°$, 30°, 60° and 70°. The results shown in Fig. 11.31a pertain to the M-1 pure oceanic aerosol model and the M-14 heavy polluted aerosol model, and clearly indicate that a close relationship exists between the evaluations of ΔF_{ToA}, ΔF_{BoA} and ΔF_{Atm} made for the isotropic and BRDF non-lambertian configurations of the four OS models and four BS models. Appreciable underestimations of the isotropic evaluations of ΔF_{ToA} (by more than $10\,\text{W m}^{-2}$) were found at $\theta_o = 0°$ and $\theta_o = 30°$ mainly over the negative range of this BRDF DARF term, followed by a closer agreement over the residual positive range. Instantaneous isotropic evaluations of ΔF_{BoA} agree very well with those estimated using the BRDF non-lambertian models at both $\theta_o = 0°$ and $\theta_o = 30°$, but turn out to be much overestimated for all the four OS surface reflectance models to a variable extent decreasing from OS1 to OS4, while a substantial agreement was found for all the four BS surface reflectance models, also at $\theta_o = 60°$ and $\theta_o = 70°$. Similarly a close agreement was found between isotropic and non-lambertian evaluations of instantaneous ΔF_{Atm} at both $\theta_o = 0°$ and $\theta_o = 30°$, while more marked underestimations of isotropic ΔF_{Atm} were found

Table 11.14a. Values of instantaneous DARF term ΔF_{ToA} (measured in W m^{-2}), as calculated using the 16 BRDF non-lambertian surface reflectance models and the corresponding equivalent isotropic surface reflectance models (in brackets) defined in the present study for 18 aerosol extinction models chosen among the 40 models presented in Tables 11.1–11.10 and obtained for solar zenith angle $\theta_o = 30°$ and aerosol optical thickness $\tau_a(0.55\,\mu\mathrm{m}) = 0.10$

Aerosol	Surface reflectance model															
Model	OS1	OS2	OS3	OS4	VS1	VS2	VS3	VS4	BS1	BS2	BS3	BS4	PS1	PS2	PS3	PS4
6S-C	-4.2	-4.2	-4.2	-3.7	-1.6	-0.1	0.8	2.0	1.3	10.7	11.1	9.1	22.6	20.5	13.9	5.4
	(−6.8)	(−6.9)	(−6.7)	(−5.7)	(−2.6)	(−1.6)	(−0.9)	(0.3)	(0.6)	(9.1)	(9.5)	(7.6)	(21.7)	(18.7)	(10.4)	(1.3)
6S-M	-5.8	-5.8	-5.8	-5.5	-4.1	-3.4	-3.0	-5.3	-2.6	0.9	1.0	0.4	1.3	1.8	1.5	-1.2
	(−8.1)	(−8.2)	(−8.0)	(−7.3)	(−5.1)	(−5.1)	(−4.9)	(−4.5)	(−3.3)	(−0.5)	(−0.5)	(−1.1)	(0.2)	(−0.2)	(−2.3)	(−5.4)
M-1	-5.7	-5.7	-5.7	-5.5	-4.0	-3.5	-3.2	-3.0	-2.8	-0.3	-0.3	-0.6	-1.5	-0.7	-0.2	-2.0
	(−7.9)	(−8.0)	(−7.9)	(−7.2)	(−5.1)	(−5.3)	(−5.2)	(−5.0)	(−3.6)	(−1.8)	(−1.8)	(−2.1)	(−2.8)	(−2.9)	(−4.1)	(−6.3)
M-8	-4.2	-4.2	-4.2	-3.6	-1.5	-0.1	0.8	2.1	1.3	10.7	11.1	9.1	22.5	20.4	13.9	5.4
	(−6.8)	(−6.8)	(−6.6)	(−5.6)	(−2.5)	(−1.6)	(−0.9)	(0.4)	(0.6)	(9.1)	(9.5)	(7.6)	(21.6)	(18.6)	(10.3)	(1.3)
M-14	-0.5	-0.5	-0.5	0.4	3.3	5.8	7.3	9.6	8.1	25.9	26.6	22.4	55.6	49.3	33.2	16.2
	(−3.9)	(−3.9)	(−3.7)	(−2.2)	(2.2)	(4.1)	(5.3)	(7.7)	(7.3)	(24.0)	(24.5)	(20.5)	(54.6)	(47.2)	(29.0)	(11.3)
OPAC-UR	3.5	3.4	3.4	4.0	6.1	8.0	9.1	10.9	9.4	22.1	22.7	19.6	46.8	41.2	27.8	14.9
	(0.6)	(0.4)	(0.6)	(1.6)	(5.0)	(6.1)	(7.1)	(9.0)	(8.6)	(20.3)	(20.7)	(17.9)	(46.0)	(39.5)	(24.3)	(11.0)
OPAC MP	0.2	0.2	0.2	0.3	0.7	1.1	1.3	1.6	1.4	3.5	3.6	3.1	7.3	6.5	4.4	2.3
	(−1.9)	(−2.2)	(−2.2)	(−1.8)	(−0.5)	(−1.1)	(−0.8)	(−0.3)	(0.6)	(1.8)	(1.7)	(1.4)	(6.5)	(4.9)	(1.5)	(−1.1)
OPAC AR	2.1	2.1	2.1	2.5	3.8	5.1	5.8	6.9	6.0	14.5	14.8	12.8	30.7	27.0	18.2	9.7
	(−0.5)	(−0.7)	(−0.6)	(0.2)	(2.7)	(3.0)	(3.8)	(5.1)	(5.3)	(12.7)	(12.9)	(11.0)	(29.9)	(25.3)	(14.9)	(6.0)
OPAC AN	-0.3	-0.3	-0.3	-0.3	-0.2	-0.2	-0.2	-0.2	-0.1	0.1	0.1	0.1	0.1	0.1	0.2	0.0
	(−2.0)	(−2.2)	(−2.2)	(−2.0)	(−1.1)	(−1.9)	(−1.8)	(−1.6)	(−0.7)	(−1.3)	(−1.4)	(−1.3)	(−0.5)	(−1.1)	(−2.2)	(−2.7)
SF-R	-5.0	-5.1	-5.1	-4.6	-3.0	-1.8	-1.2	-2.2	-0.8	6.1	6.4	5.0	12.7	11.9	8.2	2.3
	(−7.4)	(−7.5)	(−7.3)	(−6.4)	(−3.8)	(−3.2)	(−2.8)	(−1.9)	(−1.4)	(4.7)	(4.9)	(3.6)	(11.9)	(10.2)	(4.9)	(−1.5)
SF-U	1.7	1.6	1.6	2.6	6.2	9.2	11.0	11.3	11.7	31.5	32.4	27.9	66.2	58.6	39.6	20.2
	(−2.0)	(−2.0)	(−1.8)	(−0.2)	(5.0)	(7.3)	(8.8)	(11.6)	(10.7)	(29.4)	(30.2)	(25.9)	(65.0)	(56.3)	(35.1)	(15.0)
SF-M	-6.1	-6.1	-6.1	-5.9	-4.4	-3.9	-3.5	-3.3	-3.1	-0.2	-0.2	-0.6	-1.2	-0.4	0.0	-2.0
	(−8.4)	(−8.5)	(−8.3)	(−7.6)	(−5.4)	(−5.5)	(−5.4)	(−5.2)	(−3.8)	(−1.6)	(−1.7)	(−2.1)	(−2.4)	(−2.5)	(−3.8)	(−6.3)
SF-T	-5.6	-5.6	-5.6	-5.3	-4.0	-3.2	-2.7	-3.7	-2.1	3.9	4.0	2.8	8.3	8.0	5.6	0.8
	(−7.8)	(−7.8)	(−7.7)	(−6.9)	(−4.7)	(−4.4)	(−4.1)	(−3.4)	(−2.6)	(2.6)	(2.7)	(1.5)	(7.6)	(6.5)	(2.5)	(−2.7)
SD-1	-4.2	-4.3	-4.2	-3.6	-0.2	2.1	3.4	5.1	3.5	15.3	16.0	13.5	32.7	29.3	19.6	8.1
	(−7.7)	(−7.8)	(−7.6)	(−6.3)	(−1.7)	(−0.3)	(0.6)	(2.4)	(2.5)	(13.2)	(13.7)	(11.3)	(31.2)	(26.6)	(14.7)	(2.7)
PV-1	-5.5	-5.5	-5.5	-5.3	-4.5	-4.2	-4.1	-3.9	-3.3	0.2	0.2	-0.6	0.1	0.9	0.9	-1.4
	(−7.3)	(−7.3)	(−7.2)	(−6.6)	(−5.1)	(−5.2)	(−5.1)	(−4.9)	(−3.8)	(−0.9)	(−0.9)	(−1.6)	(−0.4)	(−0.4)	(−1.7)	(−4.4)
PV-2	-5.0	-5.1	-5.1	-4.8	-3.8	-3.4	-3.2	-5.1	-2.7	0.1	0.1	-0.4	-0.9	-0.0	0.5	-1.3
	(−6.9)	(−6.9)	(−6.8)	(−6.2)	(−4.5)	(−4.6)	(−4.6)	(−4.4)	(−3.2)	(−1.0)	(−1.1)	(−1.5)	(−1.7)	(−1.7)	(−2.7)	(−5.0)
PV-3	-7.4	-7.4	-7.4	-7.1	-5.5	-4.9	-4.5	-7.2	-4.0	-0.3	-0.2	-0.9	-1.1	-0.2	0.2	-2.4
	(−9.9)	(−10.0)	(−9.9)	(−9.0)	(−6.6)	(−6.7)	(−6.6)	(−6.3)	(−4.8)	(−1.9)	(−1.9)	(−2.5)	(−2.3)	(−2.3)	(−3.8)	(−6.9)
BL	-7.1	-7.1	-7.1	-6.7	-5.3	-4.5	-4.1	-3.5	-3.2	4.0	4.1	2.4	9.8	9.3	6.1	0.4
	(−9.6)	(−9.7)	(−9.5)	(−8.6)	(−6.2)	(−6.0)	(−5.7)	(−5.0)	(−3.8)	(2.4)	(2.4)	(1.0)	(9.0)	(7.6)	(2.8)	(−3.4)

Table 11.14b. Values of instantaneous DARF term $\Delta F_{\rm BoA}$ (measured in W m^{-2}), as calculated using the 16 BRDF non-lambertian surface reflectance models and the corresponding equivalent isotropic surface reflectance models (in brackets) defined in the present study for 18 aerosol extinction models chosen among the 40 models presented in Tables 1 – 10 and obtained for solar zenith angle $\theta_o = 30°$ and aerosol optical thickness $\tau_a(0.55\,\mu\text{m}) = 0.10$

Aerosol Model	Surface reflectance model															
	OS1	OS2	OS3	OS4	VS1	VS2	VS3	VS4	BS1	BS2	BS3	BS4	PS1	PS2	PS3	PS4
6S-C	-16.3	-16.3	-16.3	-15.9	-14.4	-13.4	-12.9	-12.2	-12.4	-6.4	-6.1	-7.4	-0.9	-1.6	-4.6	-9.7
	(−18.8)	(−18.9)	(−18.8)	(−18.1)	(−15.2)	(−14.9)	(−14.5)	(−13.8)	(−13.0)	(−7.9)	(−7.7)	(−8.7)	(−1.7)	(−3.1)	(−7.7)	(−13.4)
6S-M	-7.4	-7.5	-7.4	-7.2	-5.4	-4.6	-4.1	-6.6	-3.8	-0.1	0.1	-0.5	0.4	0.9	0.4	-2.5
	(−10.7)	(−10.8)	(−10.7)	(−10.0)	(−6.8)	(−7.0)	(−6.8)	(−6.3)	(−4.7)	(−2.0)	(−2.0)	(−2.4)	(−0.8)	(−1.3)	(−3.6)	(−7.1)
M-1	-5.9	-6.0	-6.0	-5.7	-3.8	-3.1	-2.5	-2.2	-2.3	0.8	1.0	0.6	0.5	1.1	1.0	-1.5
	(−9.5)	(−9.6)	(−9.5)	(−8.7)	(−5.4)	(−5.7)	(−5.6)	(−5.1)	(−3.3)	(−1.3)	(−1.2)	(−1.5)	(−0.9)	(−1.3)	(−3.3)	(−6.4)
M-8	-16.2	-16.2	-16.2	-15.8	-14.3	-13.4	-12.8	-12.1	-12.3	-6.3	-6.1	-7.3	-0.9	-1.6	-4.6	-9.7
	(−18.7)	(−18.8)	(−18.8)	(−18.0)	(−15.2)	(−14.8)	(−14.5)	(−13.7)	(−12.9)	(−7.9)	(−7.7)	(−8.7)	(−1.7)	(−3.1)	(−7.7)	(−13.3)
M-14	-28.9	-28.9	-28.9	-28.4	-26.6	-25.3	-24.5	-23.4	-24.1	-15.6	-15.2	-17.1	-2.6	-5.2	-12.3	-20.4
	(−31.3)	(−31.4)	(−31.4)	(−30.5)	(−27.4)	(−26.7)	(−26.0)	(−24.9)	(−24.7)	(−17.1)	(−16.7)	(−18.4)	(−3.3)	(−6.7)	(−15.4)	(−24.0)
OPAC-UR	-19.7	-19.7	-19.7	-19.5	-18.5	-17.6	-17.1	-16.4	-17.2	-12.7	-12.4	-13.4	-2.5	-4.9	-10.6	-15.5
	(−22.1)	(−22.2)	(−22.1)	(−21.6)	(−19.3)	(−19.0)	(−18.6)	(−17.8)	(−17.8)	(−14.2)	(−13.9)	(−14.7)	(−3.1)	(−6.4)	(−13.6)	(−19.1)
OPAC MP	-3.5	-3.5	-3.5	-3.4	-3.1	-2.9	-2.8	-2.7	-2.8	-1.9	-1.8	-2.0	-0.4	-0.7	-1.6	-2.5
	(−6.6)	(−6.6)	(−6.6)	(−6.0)	(−4.4)	(−5.0)	(−5.2)	(−4.9)	(−3.6)	(−3.7)	(−3.8)	(−3.8)	(−1.4)	(−2.7)	(−5.3)	(−6.8)
OPAC AR	-12.9	-12.9	-12.9	-12.8	-12.1	-11.5	-11.2	-10.7	-11.2	-8.2	-8.0	-8.7	-1.6	-3.2	-6.8	-10.1
	(−15.6)	(−15.7)	(−15.6)	(−15.1)	(−13.0)	(−13.2)	(−13.0)	(−12.5)	(−11.9)	(−9.8)	(−9.7)	(−10.2)	(−2.4)	(−4.9)	(−10.2)	(−14.0)
OPAC AN	-0.4	-0.4	-0.4	-0.4	-0.3	-0.3	-0.3	-0.2	-0.2	0.0	0.0	0.0	0.0	0.1	0.1	-0.1
	(−3.0)	(−3.1)	(−3.0)	(−2.6)	(−1.2)	(−1.8)	(−2.0)	(−1.9)	(−0.8)	(−1.5)	(−1.6)	(−1.4)	(−0.7)	(−1.5)	(−3.1)	(−3.8)
SF-R	-12.2	-12.2	-12.2	-11.9	-10.5	-9.7	-9.3	-10.5	-8.7	-3.5	-3.3	-4.4	-0.3	-0.4	-2.2	-6.3
	(−14.7)	(−14.8)	(−14.7)	(−14.0)	(−11.3)	(−11.1)	(−10.9)	(−10.2)	(−9.3)	(−5.0)	(−4.9)	(−5.7)	(−1.0)	(−1.9)	(−5.3)	(−9.9)
SF-U	-32.2	-32.3	-32.2	-31.8	-29.7	-28.1	-27.2	-27.6	-27.1	-18.3	-17.7	-19.6	-3.4	-6.6	-14.8	-23.6
	(−34.8)	(−34.9)	(−34.8)	(−34.0)	(−30.6)	(−29.6)	(−28.9)	(−27.5)	(−27.7)	(−19.8)	(−19.3)	(−21.0)	(−4.1)	(−8.2)	(−18.0)	(−27.3)
SF-M	-6.5	-6.6	-6.6	-6.3	-4.4	-3.6	-3.1	-2.7	-2.8	0.7	0.9	0.4	0.6	1.2	1.1	-1.7
	(−9.9)	(−10.0)	(−9.9)	(−9.1)	(−5.8)	(−6.1)	(−5.9)	(−5.4)	(−3.7)	(−1.3)	(−1.2)	(−1.6)	(−0.7)	(−1.0)	(−3.1)	(−6.4)
SF-T	-10.5	-10.5	-10.5	-10.2	-9.0	-8.4	-8.1	-9.2	-7.4	-2.4	-2.2	-3.3	0.0	0.2	-1.1	-4.9
	(−12.9)	(−13.0)	(−12.9)	(−12.2)	(−9.8)	(−9.7)	(−9.4)	(−9.0)	(−7.9)	(−3.8)	(−3.7)	(−4.5)	(−0.6)	(−1.2)	(−4.0)	(−8.3)
SD-1	-21.2	-21.2	-21.2	-20.8	-17.8	-16.1	-15.0	-14.0	-15.2	-8.7	-8.2	-9.2	-1.3	-2.6	-7.0	-13.4
	(−24.8)	(−24.9)	(−24.8)	(−23.8)	(−19.3)	(−18.8)	(−18.1)	(−16.9)	(−16.2)	(−10.8)	(−10.4)	(−11.3)	(−2.7)	(−4.9)	(−11.3)	(−18.3)
PV-1	-6.3	-6.3	-6.3	-6.0	-5.2	-4.8	-4.7	-4.4	-3.9	0.1	0.1	-0.8	0.5	1.1	0.9	-1.7
	(−8.5)	(−8.6)	(−8.6)	(−7.9)	(−5.9)	(−5.9)	(−5.8)	(−5.5)	(−4.4)	(−1.3)	(−1.3)	(−1.9)	(−0.0)	(−0.2)	(−1.8)	(−5.0)
PV-2	-5.5	-5.5	-5.5	-5.2	-3.9	-3.3	-3.0	-4.9	-2.6	0.8	0.9	0.3	0.7	1.4	1.4	-1.0
	(−8.2)	(−8.3)	(−8.2)	(−7.5)	(−5.0)	(−5.2)	(−5.1)	(−4.7)	(−3.2)	(−0.8)	(−0.8)	(−1.2)	(−0.1)	(−0.3)	(−2.0)	(−5.0)
PV-3	-8.0	-8.0	-8.0	-7.7	-5.6	-4.7	-4.1	-6.9	-3.7	0.7	0.9	0.2	0.9	1.6	1.4	-2.0
	(−11.5)	(−11.6)	(−11.5)	(−10.6)	(−7.1)	(−7.3)	(−7.1)	(−6.5)	(−4.7)	(−1.4)	(−1.3)	(−1.8)	(−0.4)	(−0.7)	(−3.0)	(−7.0)
BL	-12.8	-12.8	-12.8	-12.4	-11.0	-10.3	-9.9	-9.4	-9.1	-3.2	-3.1	-4.4	0.0	0.1	-1.7	-6.2
	(−15.4)	(−15.5)	(−15.4)	(−14.6)	(−11.9)	(−11.8)	(−11.5)	(−10.9)	(−9.7)	(−4.8)	(−4.7)	(−5.7)	(−0.7)	(−1.5)	(−4.9)	(−9.9)

Table 11.14c. Values of the differences between the values of instantaneous DARF term ΔF_{Atm} (measured in $\mathrm{W\,m^{-2}}$), as calculated using the 16 BRDF non-lambertian surface reflectance models and the corresponding equivalent isotropic surface reflectance models (in brackets) defined in the present study for 18 aerosol extinction models chosen among the 40 models presented in Tables 1 – 10 and obtained for solar zenith angle $\theta_o = 30°$ and aerosol optical thickness $\tau_a(0.55\,\mu\mathrm{m}) = 0.10$

Aerosol	Surface reflectance model															
Model	OS1	OS2	OS3	OS4	VS1	VS2	VS3	VS4	BS1	BS2	BS3	BS4	PS1	PS2	PS3	PS4
6S-C	12.1	12.1	12.1	12.2	12.8	13.3	13.7	14.2	13.7	17.1	17.3	16.4	23.5	22.1	18.5	15.1
	(12.0)	(12.0)	(12.1)	(12.4)	(12.7)	(13.3)	(13.6)	(14.1)	(13.6)	(17.0)	(17.2)	(16.3)	(23.4)	(21.8)	(18.1)	(14.7)
6S-M	1.6	1.6	1.6	1.6	1.4	1.2	1.1	1.2	1.2	1.0	0.9	0.9	1.0	0.9	1.0	1.3
	(2.7)	(2.7)	(2.7)	(2.7)	(1.7)	(1.9)	(1.9)	(1.7)	(1.3)	(1.5)	(1.4)	(1.3)	(1.0)	(1.1)	(1.3)	(1.7)
M-1	0.3	0.3	0.3	0.2	-0.2	-0.5	-0.7	-0.9	-0.5	-1.1	-1.2	-1.2	-2.0	-1.8	-1.2	-0.5
	(1.6)	(1.6)	(1.6)	(1.5)	(0.3)	(0.4)	(0.4)	(0.1)	(-0.3)	(-0.5)	(-0.6)	(-0.6)	(-1.8)	(-1.6)	(-0.7)	(0.1)
M-8	12.0	12.0	12.0	12.2	12.7	13.3	13.6	14.2	13.6	17.0	17.2	16.4	23.4	22.0	18.5	15.1
	(12.0)	(12.0)	(12.1)	(12.4)	(12.6)	(13.3)	(13.6)	(14.1)	(13.5)	(17.0)	(17.1)	(16.2)	(23.3)	(21.8)	(18.1)	(14.6)
M-14	28.4	28.3	28.3	28.8	29.9	31.1	31.8	33.0	32.3	41.5	41.8	39.4	58.3	54.5	45.5	36.6
	(27.5)	(27.5)	(27.7)	(28.3)	(29.6)	(30.8)	(31.4)	(32.5)	(32.0)	(41.0)	(41.2)	(38.9)	(57.9)	(54.0)	(44.4)	(35.3)
OPAC-UR	23.1	23.1	23.1	23.5	24.5	25.6	26.2	27.2	26.6	34.8	35.1	33.0	49.3	46.1	38.3	30.5
	(22.7)	(22.7)	(22.7)	(23.2)	(24.2)	(25.1)	(25.7)	(26.8)	(26.4)	(34.5)	(34.6)	(32.6)	(49.1)	(45.9)	(37.9)	(30.0)
OPAC MP	3.7	3.7	3.7	3.7	3.9	4.0	4.1	4.3	4.2	5.5	5.5	5.2	7.7	7.2	6.0	4.8
	(4.6)	(4.4)	(4.4)	(4.2)	(3.9)	(3.9)	(4.4)	(4.6)	(4.2)	(5.5)	(5.5)	(5.2)	(7.9)	(7.6)	(6.8)	(5.7)
OPAC AR	15.0	15.0	15.0	15.3	15.9	16.6	17.0	17.6	17.3	22.7	22.8	21.5	32.3	30.2	25.0	19.8
	(15.2)	(15.0)	(15.1)	(15.3)	(15.7)	(16.2)	(16.8)	(17.5)	(17.1)	(22.5)	(22.6)	(21.2)	(32.3)	(30.2)	(25.1)	(20.0)
OPAC AN	0.1	0.1	0.1	0.1	0.1	0.1	0.1	0.1	0.1	0.1	0.1	0.1	0.0	0.0	0.1	0.1
	(1.1)	(0.9)	(0.8)	(0.6)	(0.1)	(-0.0)	(0.2)	(0.3)	(0.1)	(0.2)	(0.1)	(0.1)	(0.2)	(0.4)	(0.9)	(1.1)
SF-R	7.2	7.2	7.2	7.2	7.5	7.9	8.1	8.3	8.0	9.7	9.8	9.4	13.0	12.3	10.4	8.7
	(7.3)	(7.3)	(7.4)	(7.6)	(7.5)	(7.9)	(8.1)	(8.4)	(7.9)	(9.7)	(9.8)	(9.3)	(12.9)	(12.1)	(10.2)	(8.5)
SF-U	33.9	33.9	33.9	34.4	35.9	37.3	38.2	38.9	38.7	49.8	50.2	47.5	69.5	65.2	54.4	43.8
	(32.8)	(32.9)	(33.1)	(33.8)	(35.5)	(36.9)	(37.7)	(39.2)	(38.4)	(49.2)	(49.5)	(46.8)	(69.2)	(64.5)	(53.1)	(42.3)
SF-M	0.4	0.4	0.4	0.4	-0.0	-0.3	-0.4	-0.6	-0.3	-0.9	-1.1	-1.0	-1.8	-1.6	-1.1	-0.4
	(1.6)	(1.6)	(1.7)	(1.6)	(0.4)	(0.6)	(0.5)	(0.2)	(-0.1)	(-0.4)	(-0.5)	(-0.5)	(-1.7)	(-1.5)	(-0.7)	(0.1)
SF-T	4.9	4.9	4.9	4.9	5.1	5.2	5.3	5.5	5.3	6.2	6.3	6.0	8.2	7.8	6.7	5.7
	(5.1)	(5.1)	(5.2)	(5.3)	(5.0)	(5.3)	(5.4)	(5.5)	(5.3)	(6.4)	(6.4)	(6.0)	(8.2)	(7.7)	(6.5)	(5.6)
SD-1	17.0	17.0	17.0	17.2	17.6	18.2	18.5	19.1	18.7	24.0	24.1	22.7	34.0	31.8	26.5	21.5
	(17.0)	(17.1)	(17.2)	(17.5)	(17.7)	(18.5)	(18.8)	(19.3)	(18.6)	(24.0)	(24.1)	(22.6)	(33.9)	(31.6)	(26.1)	(21.0)
PV-1	0.8	0.8	0.8	0.8	0.7	0.6	0.6	0.5	0.6	0.1	0.1	0.2	-0.3	-0.3	-0.0	0.4
	(1.3)	(1.3)	(1.4)	(1.4)	(0.7)	(0.7)	(0.7)	(0.6)	(0.6)	(0.4)	(0.3)	(0.3)	(-0.4)	(-0.2)	(0.1)	(0.5)
PV-2	0.4	0.4	0.4	0.4	0.1	-0.1	-0.2	-0.2	-0.2	-0.7	-0.8	-0.8	-1.6	-1.4	-0.9	-0.3
	(1.4)	(1.4)	(1.5)	(1.4)	(0.4)	(0.6)	(0.5)	(0.2)	(-0.0)	(-0.2)	(-0.3)	(-0.3)	(-1.6)	(-1.3)	(-0.7)	(0.0)
PV-3	0.6	0.6	0.6	0.5	0.1	-0.2	-0.4	-0.3	-0.2	-1.0	-1.1	-1.0	-2.0	-1.8	-1.1	-0.3
	(1.5)	(1.6)	(1.7)	(1.6)	(0.5)	(0.6)	(0.5)	(0.2)	(-0.1)	(-0.5)	(-0.6)	(-0.6)	(-1.9)	(-1.6)	(-0.9)	(0.0)
BL	5.7	5.7	5.7	5.7	5.7	5.8	5.8	5.9	5.9	7.2	7.2	6.8	9.8	9.2	7.8	6.6
	(5.8)	(5.8)	(5.9)	(6.0)	(5.6)	(5.8)	(5.8)	(5.9)	(5.9)	(7.2)	(7.1)	(6.7)	(9.7)	(9.1)	(7.6)	(6.5)

at $\theta_o = 60°$ over the OS surfaces (by no more than $50\,\mathrm{W\,m^{-2}}$) and at $\theta_o = 70°$, with discrepancies exceeding $150\,\mathrm{W\,m^{-2}}$ over the oceanic surfaces.

Similar results were obtained in Fig. 11.31b, prepared for the same aerosol extinction models associated with the four VS and the four PS surface models (presenting considerably higher surface reflectance properties than in Fig. 11.31a) and the same set of four solar zenith angles. The instantaneous isotropic values of ΔF_{ToA} determined at $\theta_o = 0°$ and $\theta_o = 30°$ were found to be greatly underestimated with respect to the BRDF non-lambertian values over the negative range of such a BRDF DARF term (by more than $60\,\mathrm{W\,m^{-2}}$ in the worst case, relative to a polar surface) as well as over the residual positive range (by more than $40\,\mathrm{W\,m^{-2}}$ for the polar surfaces). Similar although less marked discrepancies were also achieved for the other cases with θ_o equal to 30°, 60° and 70°. The instantaneous isotropic evaluations of ΔF_{BoA} were found to agree closely with those obtained for the BRDF non-lambertian surface reflectance models at both $\theta_o = 60°$ and $\theta_o = 70°$, but exhibit values underestimated by around 10 to $20\,\mathrm{W\,m^{-2}}$ on average for $\theta_o = 30°$ and even more marked discrepancies for $\theta_o = 0°$, especially in the cases pertaining to polar surfaces. A closer agreement was found for the instantaneous forcing term ΔF_{Atm} at all four angles θ_o, with discrepancies no greater than a few $\mathrm{W\,m^{-2}}$ throughout its range from -50 to $+300\,\mathrm{W\,m^{-2}}$.

The present calculation of the three instantaneous forcing terms appears to be useful also for calculating the 24-hour average DARF terms applied to field measurements of aerosol optical thickness performed using multi-wavelength sun-photometer techniques and simultaneous *in situ* measurements of the particle size-distribution curves and aerosol radiative parameters, from which the single scattering albedo of airborne particulate matter can be experimentally derived. In order to define more accurately the discrepancies between isotropic and BRDF non-lambertian evaluations of ΔF_{ToA}, like those that could be used for determining the 24-hour average DARF terms (as planned in a further paper), the scatter plots of isotropic ΔF_{ToA} versus the BDRF non-lambertian ΔF_{ToA} are presented in Fig. 11.32. They were determined for (i) the 18 aerosol models shown in Figs. 11.27a to 11.27e, (ii) the surface reflectance models OS2 (ocean surface), BS2 (bare soil surface), VS2 (vegetated surface) and PS2 (polar surface), (iii) $\tau_a(0.55\,\mu\mathrm{m}) = 0.10$, relative to a background content of columnar aerosols for clean-air atmospheric transparency conditions, giving a visual range of more than 20 km, and (iv) solar zenith angle $\theta_o = 30°$. The data clearly indicate that a close correlation exists between these evaluations of instantaneous isotropic and non-lambertian forcing term ΔF_{ToA} for low atmospheric turbidity conditions, with only weakly underestimated values of isotropic ΔF_{ToA} with respect to the corresponding non-lambertian estimates, graphically evidenced along the bisecting lines of the four graphs. In fact, the root-mean-square standard errors of estimate (SEE) are very small for such low atmospheric turbidity conditions, being equal to $\pm 0.62\,\mathrm{W\,m^{-2}}$ for the OS2 ocean surface, $\pm 0.38\,\mathrm{W\,m^{-2}}$ for the BS2 arid terrain, $\pm 0.40\,\mathrm{W\,m^{-2}}$ for the VS2 vegetated surface, and $\pm 0.44\,\mathrm{W\,m^{-2}}$ for the PS2 polar surface.

However, bearing in mind that field measurements of aerosol optical thickness can vary considerably throughout a field measurement day, due to the transport of large desert aerosol loads suspended over both land and oceanic sites or that of heavy columnar loads of polluted urban aerosols over remote regions, the present re-

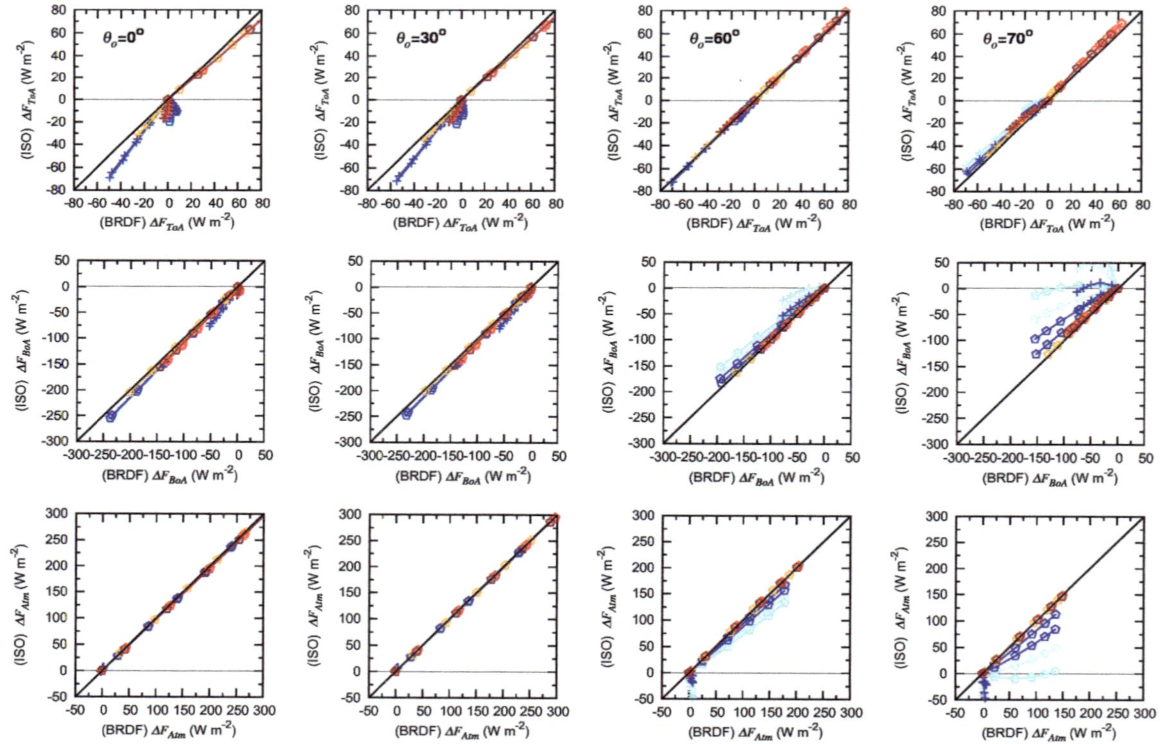

Fig. 11.31a. (a) Scatter plot of the instantaneous direct aerosol-induced radiative forcing terms ΔF_{ToA} (upper part), ΔF_{BoA} (middle part) and ΔF_{Atm} (lower part) calculated using the BRDF non-lambertian (abscissa) and ISO isotropic (ordinates) models in the scheme presented in Fig. 11.22), as obtained for (i) the two M-1 (plus signs) and M-14 (pentagons) aerosol models defined in Table 11.4, (ii) aerosol optical thickness $\tau_a(0.55\,\mu\mathrm{m})$ increasing from 0 to 0.90 in steps of 0.10, and (iii) the four solar zenith angles $\theta_o = 0°$, $\theta_o = 30°$, $\theta_o = 60°$ and $\theta_o = 70°$, using the OS1 (cyan), OS2 (light blue), OS3 (blue), OS4 (dark blue), BS1 (orange), BS2 (light red), BS3 (red) and BS4 (dark red) reflectance models given in Table 11.12.

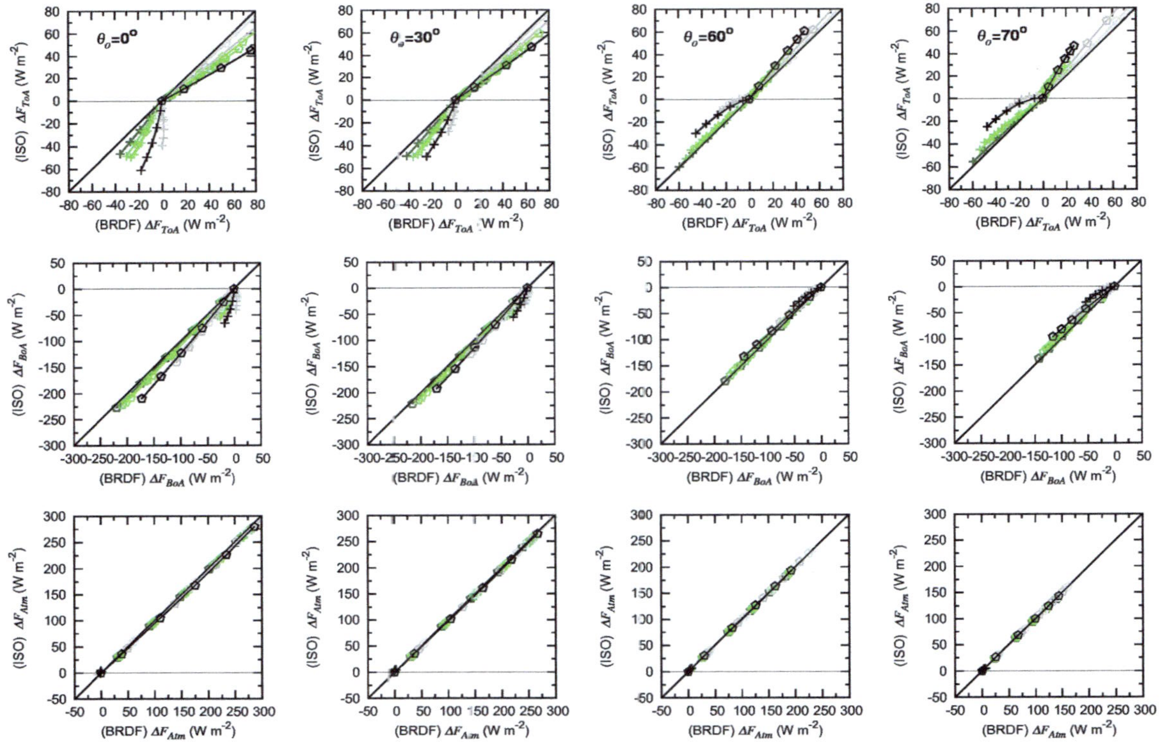

Fig. 11.31b. (b) As in Fig. 11.31a, for the VS1 (light green), VS2 (green), VS3 (less dark green), VS4 (very dark green), PS1 (light gray), PS2 (gray), PS3 (dark gray) and PS4 (black) reflectance models given in Table 11.12.

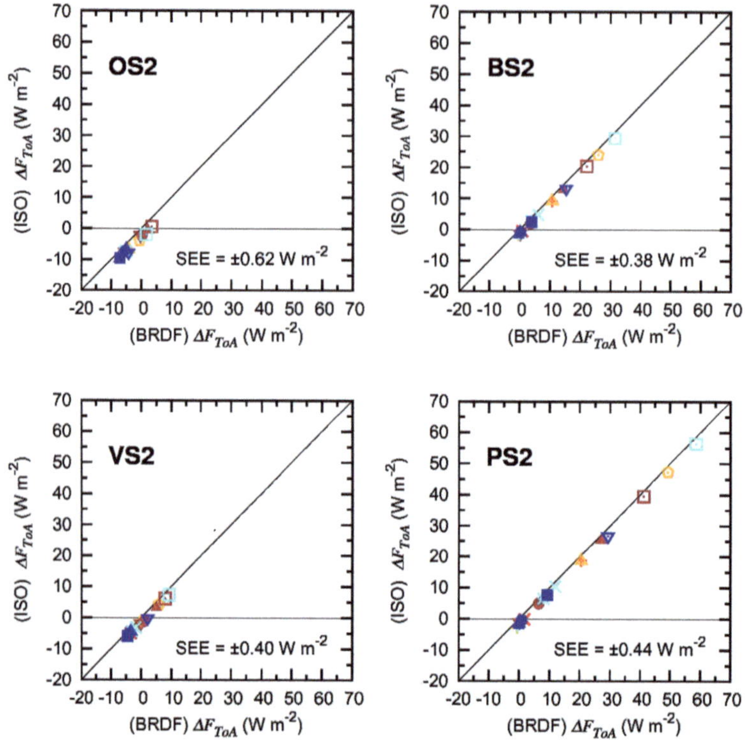

Fig. 11.32. Scatter plots of the values of instantaneous forcing ΔF_{ToA} calculated at solar zenith angle $\theta_o = 30°$ for (i) the 18 aerosol models listed in Tables 11.14a, 11.14b and 11.14c (with $\tau_a(0.55\,\mu\mathrm{m}) = 0.10$), and (ii) the four isotropic models OS2 (ocean surface), BS2 (bare soil surface), VS2 (vegetated surface) and PS2 (polar surface) versus the corresponding values of instantaneous ΔF_{ToA} calculated for the four equivalent BRDF non-lambertian models. The values of the root-mean-square standard errors of estimate (SEE) are also given in each panel.

sults indicate that the most reliable calculations of the instantaneous DARF terms have to be performed using BRDF non-lambertian models of surface reflectance defined with great accuracy, as recommended by Ricchiazzi et al. (2005). Using these non-lambertian surface reflectance models, more suitable evaluations of the direct radiative forcing effects induced by aerosols in the surface–atmosphere system can be derived from the sets of field measurements of aerosol optical thickness and the experimental evaluations of the main radiative parameters of columnar aerosol.

Acknowledgments

The present research was supported by the strategic FISR Programme 'Sustainable Development and Climate Changes' sponsored by the Italian Ministry of the University and Scientific Research (MIUR), and developed in the frame of the cooperative project between CNR and MIUR 'Study of the direct and indirect effects of aerosols and clouds on climate (AEROCLOUDS)'. The authors gratefully acknowledge the colleagues M. Nanni and F. Bedosti of the Institute of Radioastronomy

(IRA), National Institute of Astrophysics (INAF, Bologna, Italy) for the support in the use of Comput-ER GRID computing system to carry out the time-consuming calculations of the DARF terms.

References

Ackerman, A. S., O. B. Toon and P. V. Hobbs (1994), Reassessing the dependence of cloud condensation nucleus concentration on formation rate, *Nature*, **367**, 445–447, doi:10.1038/367445a0.

Ackerman, A. S., O. B. Toon, D. E. Stevens, A. J. Heymsfield, V. Ramanathan and E. J. Welton (2000), Reduction of tropical cloudiness by soot, *Science*, **288**, 1042–1047, doi: 10.1126/science.288.5468.1042.

Ackerman, A. S., M. P. Kirkpatrick, D. E. Stevens and O. B. Toon (2004), The impact of humidity above stratiform clouds on indirect aerosol climate forcing, *Nature*, **432**, 1014–1017, doi: 10.1038/nature03174.

Albrecht, B. (1989), Aerosols, cloud microphysics and fractional cloudiness, *Science*, **245**, 1227–1230, doi: 10.1126/science.245.4923.1227.

Anderson, G. P., S. A. Clough, F. X. Kneizys, J. H. Chetwynd and E. P. Shettle (1986), AFGL Atmospheric Constituent Profiles (0–120 km), *Environ. Res. Papers, No. 954*, AFGL-TR-86-0110, Air Force Geophysics Laboratory, L. G. Hanscom Field, Massachusetts, 43 pp.

Anderson, T. L., R. J. Charlson, N. Bellouin, O. Boucher, M. Chin, S. A. Christopher, J. Haywood, Y. J. Kaufman, S. Kinne, J. A. Ogren, L. A. Remer, T. Takemura, D. Tanré, O. Torres, C. R. Trepte, B. A. Wielicki, D. M. Winker and H. Yu (2005), An 'A-Train' strategy for quantifying direct climate forcing by anthropogenic aerosols, *Bull. Amer. Met. Soc.*, **86**, 1795–1809, doi: 10.1175/BAMS-86-12-1795.

Andreae, M. O. and A. Gelencsér (2006), Black carbon or brown carbon? The nature of light absorbing carbonaceous aerosols, *Atmos. Chem. Phys.*, **6**, 3131–3148, doi: 10.5194/acp-6-3131-2006.

Ångström, A. (1964), The parameters of atmospheric turbidity, *Tellus*, **16**, 1, 64–75.

Bates, T. S., T. L. Anderson, T. Baynard, T. Bond, O. Boucher, G. Carmichael, A. Clarke, C. Erlick, H. Guo, L. Horowitz, S. Howell, S. Kulkarni, H. Maring, A. McComiskey, A. Middlebrook, K. Noone, C. D. O'Dowd, J. Ogren, J. Penner, P. K. Quinn, A. R. Ravishankara, D. L. Savoie, S. E. Schwartz, Y. Shinozuka, Y. Tang, R. J. Weber and Y. Wu (2006), Aerosol direct radiative effects over the northwest Atlantic, northwest Pacific, and North Indian Oceans: estimates based on in-situ chemical and optical measurements and chemical transport modeling, *Atmos. Chem. Phys.*, **6**, 1657–1732, doi:10.5194/acp-6-1657-2006.

Bellouin, N., O. Boucher, D. Tanré and O. Dubovik (2003), Aerosol absorption over the clear-sky oceans deduced from POLDER-1 and AERONET observations, *Geophys. Res. Lett.*, **30**, 1748, doi:10.1029/2003GL017121.

Bellouin, N., O. Boucher, J. Haywood and M. S. Reddy (2005), Global estimate of aerosol direct radiative forcing from satellite measurements, *Nature*, **438**, 1138–1140, doi: 10.1038/nature04348.

Blanchet, J. P. (1989), Toward estimation of climatic effects due to arctic aerosols, *Atmos. Environ.*, **23**, 2609–2625, doi: 10.1016/0004-6981(89)90269-2.

Born, M. and E. Wolf (1975), *Principles of Optics, Electromagnetic Theory of Propagation, Interference and Direction of Light.* Pergamon Press, Oxford, Fifth Edition, 1975.

Bothwell, G. W., E. G. Hansen, R. E. Vargo and K. C. Miller (2002), The Multi-angle Imaging Spectro-Radiometer science data system, its products, tools, and performance, *IEEE Trans. Geosci. Remote Sens.*, **40**, 1467–1476, doi: 10.1109/TGRS.2002.801152.

Burley, J. D. and H. S. Johnston (1992), Ionic mechanisms for heterogeneous stratospheric reactions and ultraviolet photoabsorption cross sections for NO_2^+, HNO_3, and NO_3^- in sulphuric acid, *Geophys. Res. Lett.*, **19**, 13, 1359–1362, doi: 10.1029/92GL01115.

Bush, B. C. and F. P. J. Valero (2002), Spectral aerosol radiative forcing at the surface during the Indian Ocean Experiment (INDOEX), *J. Geophys. Res.*, **107**, D19, 8003, doi: 10.1029/2000JD000020.

Bush, B. C. and F. P. J. Valero (2003), Surface aerosol radiative forcing at Gosan during the ACE-Asia campaign, *J. Geophys. Res.*, **108**, D23, 8660, doi: 10.1029/2002JD003233.

Carlson, T. N. and S. G. Benjamin (1980), Radiative heating rates for Saharan dust, *J. Atmos. Sci.*, **37**, 1, 193–213.

Carlson, T. N. and R. S. Caverly (1977), Radiative characteristics of Saharan dust at solar wavelengths, *J. Geophys. Res.*, **82**, 21, 3141–3152, doi: 10.1029/JC082i021p03141.

Carr, S. B. (2005), The Aerosol Models in MODTRAN: Incorporating Selected Measurements from Northern Australia, Technical Report of the Defence Science and Technology Organisation, No. DSTO-TR-1803, Edinburgh, South Australia (Australia), 67 pp, (see http://www.ewp.rpi.edu/hartford/~brazw/Project/Other/Research/Soot/Carr2005_AerosolModelsInMODTRAN.pdf).

Charlson, R. J., J. Langner and H. Rodhe (1990), Sulphate aerosol and climate, *Nature*, **348**, 22–26, doi:10.1038/348022a0.

Charlson, R. J., J. Langner, H. Rodhe, C. B. Leovy and S. G. Warren (1991), Perturbation of the northern hemisphere radiative balance by backscattering from anthropogenic sulfate aerosols, *Tellus*, **43 B**, 152–163, doi: 10.1034/j.1600-0870.1991.00013.x.

Charlson, R. J., S. E. Schwartz, J. M. Hales, R. D. Cess, J. A. Coakley, jr., J. E. Hansen and D. J. Hofmann (1992), Climate forcing by anthropogenic aerosols, *Science*, **255**, 423–430, doi: 10.1126/science.255.5043.423.

Chin, M., T. Diehl, P. Ginoux and W. Malm (2007), Intercontinental transport of pollution and dust aerosols: implications for regional air quality, *Atmos. Chem. Phys.*, **7**, 5501–5517, doi:10.5194/acp-7-5501-2007.

Christopher, S. A. and J. Zhang (2002), Shortwave aerosol radiative forcing from MODIS and CERES observations over the oceans, *Geophys. Res. Lett.*, **29**, 18, 1859, doi:10.1029/2002GL014803.

Christopher, S. A., J. Zhang, Y. J. Kaufman and L. A. Remer (2006), Satellite-based assessment of top of atmosphere anthropogenic aerosol radiative forcing over cloud-free oceans, *Geophys. Res. Lett.*, **33**, L15816, doi: 10.1029/2005GL025535.

Chung, C. E., V. Ramanathan, D. Kim and I. A. Podgorny (2005), Global anthropogenic aerosol direct forcing derived from satellite and ground-based observations, *J. Geophys. Res.*, **110**, D24207, doi:10.1029/2005JD006356.

Chylek, P. and J. A. Coakley, jr. (1974), Aerosols and climate, *Science*, **183**, 75–77, doi: 10.1126/science.183.4120.75.

Chylek, P. and J. Wong (1995), Effect of absorbing aerosols on global radiation budget. *Geophys. Res. Lett.*, **22**, 929–931, doi: 10.1029/95GL00800.

Coakley, J. A., jr. and P. Chylek (1975), The two-stream approximation in radiative transfer: including the angle of the incident radiation, *J. Atmos. Sci.*, **32**, 409–418.

Coakley, J. A., jr., R. D. Cess and F. B. Yurevich (1983), The effect of tropospheric aerosols on the Earth's radiation budget: A parameterization for climate models, *J. Atmos. Sci.*, **40**, 116–138.

Cook, J., and E. J. Highwood (2004), Climate response to tropospheric absorbing aerosols in an intermediate general-circulation model, *Q. J. R. Meteorol. Soc.*, **130**, 596, 175–191, doi: 10.1256/qj.03.64.

Cox, C. and W. Munk (1954), Statistics of the sea surface derived from sun glitter. *J. Marine Res.*, **13**, 198–227.

d'Almeida, G. A., P. Koepke and E. P. Shettle (Eds.) (1991), *Atmospheric Aerosols: Global Climatology and Radiative Characteristics*, Hampton, Virginia, A. Deepak Publishing, 561 pp.

Deepak, A. and H. E. Gerber (Eds.) (1983), Report of the experts meeting on aerosols and their climatic effects. WCP-55, 107 pp. (Available from World Meteorological Organization, Case Postale No. 5, CH-1211 Geneva, Switzerland).

Deirmendjian, D. (1969), *Electromagnetic Scattering on Spherical Polydispersions*, New York, Elsevier, pp. 75–119.

Diner, D. J., J. C. Beckert, T. H. Reilly, C. J. Bruegge, J. E. Conel, R. A. Kahn, J. V. Martonchik, T. P. Ackerman, R. Davies, S. A. W. Gerstl, H. R. Gordon, J.-P. Muller, R. B. Myneni, P. J. Sellers, B. Pinty and M. M. Verstraete (1998), Multi-angle Imaging Spectro Radiometer (MISR) instrument description and experiment overview, *IEEE Trans. Geosci. Remote Sens.*, **36**, 1072–1087, doi: 10.1109/36.700992.

Dubin, M., N. Sissenwine and S. Teweles (1966), *U. S. Standard Atmosphere Supplements, 1966.* Environmental Science Services Administration, National Aeronautics and Space Administration, United States Air Force, Washington, D. C. 20402, 289 pp.

Dubovik, O. and M. D. King (2000), A flexible inversion algorithm for retrieval of aerosol optical properties from Sun and sky radiance measurements, *J. Geophys. Res.*, **105**, D16, 20673–20696, doi: 10.1029/2000JD900282.

Dubovik, O., A. Smirnov, B. N. Holben, M. D. King, Y. J. Kaufman, T. F. Eck, and I. Slutsker (2000), Accuracy assessments of aerosol optical properties retrieved from AERONET sun and sky radiance measurements, *J. Geophys. Res.*, **105**, D8, 9791–9806, doi: 10.1029/2000JD900040.

Ervens, B., G. Feingold and S. M. Kreidenweis (2005), Influence of water-soluble organic carbon on cloud drop number concentration, *J. Geophys. Res.*, **110**, D18211, doi:10.1029/2004JD005634.

Feichter, J., E. Roeckner, U. Lohmann and B. Liepert (2004), Nonlinear aspects of the climate response to greenhouse gas and aerosol forcing, *J. Climate*, **17**, 2384-2398.

Feingold, G. (2003), Modeling of the first indirect effect: Analysis of measurement requirements, *Geophys. Res. Lett.*, **30**, 1997, doi:10.1029/2003GL017967.

Feingold, G., W. R. Cotton, S. M. Kreidenweis and J. T. Davis (1999), The impact of giant cloud condensation nuclei on drizzle formation in stratocumulus: Implications for cloud radiative properties, *J. Atmos. Sci.*, **56**, 4100–4117.

Forster, P., V. Ramaswamy, P. Artaxo, T. Berntsen, R. Betts, D.W. Fahey, J. Haywood, J. Lean, D.C. Lowe, G. Myhre, J. Nganga, R. Prinn, G. Raga, M. Schulz and R. Van Dorland, (2007), Changes in atmospheric constituents and in radiative forcing. Chapter 2 in: *Climate Change 2007: The Physical Science Basis. Contribution of Working Group I to the Fourth Assessment Report of the Intergovernmental Panel on Climate Change* [Solomon, S., D. Qin, M. Manning, Z. Chen, M. Marquis, K.B. Averyt, M.Tignor and H.L. Miller (eds.)]. Cambridge University Press, Cambridge (United Kingdom) and New York (NY, USA).

Gatebe, C. K., O. Dubovik, M. D. King and A. Sinyuk (2010), Simultaneous retrieval of aerosol and surface optical properties from combined airborne- and ground-based direct and diffuse radiometric measurements, *Atmos. Chem. Phys.*, **10**, 2777–2794, doi:10.5194/acp-10-2777-2010.

Grassl, H. (1973), Aerosol influence on radiative cooling, *Tellus*, **25**, 4, 386–395.

Grassl, H. and M. Newiger (1982), Changes of local planetary albedo by aerosol particles, *Sci. Total Environ.*, **23**, 313–320, doi: 10.1016/0048-9697(82)90148-6.

Grenfell, T. C. and G. A. Maycut (1977), The optical properties of ice and snow in the Arctic Basin, *J. Glaciol.*, **18**, 78, 445–463.

Hale, G. M. and M. R. Querry (1973), Optical constants of water in the 200 nm to 200 μm wavelength region, *Appl. Opt.*, **12**, 555–563, doi: 10.1364/AO.12.000555.

Hänel, G. (1968), The real part of the mean complex refractive index and the mean density of samples of atmospheric aerosol particles, *Tellus*, **20**, 371–379, doi: 10.1111/j.2153-3490.1968.tb00378.x.

Hänel, G. (1972), Computation of the extinction of visible radiation by atmospheric aerosol particles as a function of the relative humidity, based upon measured properties, *J. Aerosol Sci.*, **3**, 377–386, doi:10.1016/0021-8502(72)90092-4.

Hänel, G. (1976), The properties of atmospheric aerosol particles as functions of the relative humidity at thermodynamic equilibrium with the surrounding moist air, *Adv. Geophys.*, **19**, 73–88.

Hänel, G. and K. Bullrich (1978), Physico-chemical property models of tropospheric aerosol particles, *Contr. Atmos. Phys.*, **51**, 129–138.

Hänel, G. and B. Zankl (1979), Aerosol size and relative humidity: Water uptake by mixtures of salts, *Tellus*, **31 B**, 478–486, doi: 10.1111/j.2153-3490.1979.tb00929.x.

Hänel, G., W. Adam, U. Bundke, L. Komguem and U. Leiterer (1999), Optical properties of boundary layer particles, columnar absorption and direct radiative forcing by particles in the solar spectral region, *J. Aerosol Sci.*, **30**, Suppl. 1, S171–S172.

Hansen, J., M. Sato, and R. Ruedy (1997), Radiative forcing and climate response, *J. Geophys. Res.*, **102**, D6, 6831–6864, doi:10.1029/96JD03436.

Hansen, J. E., M. Sato, A. Lacis, R. Ruedy, I. Tegen and E. Matthews (1998), Climate forcings in the industrial era, *Proc. Natl. Acad. Sci. USA*, **95**, 12,753–12,758.

Hapke, B. (1986), Bidirectional reflectance spectroscopy. IV. The extinction coefficient and the opposition effect, *Icarus*, **67**, 264–280, doi:10.1016/0019-1035(86)90108-9.

Haywood, J. M., V. Ramaswamy and B. J. Soden (1999), Tropospheric aerosol climate forcing in clear-sky satellite observations over the oceans, *Science*, **283**, 1299–1303.

Haywood, J. and O. Boucher (2000), Estimates of the direct and indirect radiative forcing due to tropospheric aerosols: A review, *Rev. Geophys.*, **38**, 513-543, doi: 10.1029/1999RG000078.

Heintzenberg, J. (1994), The life cycle of the atmospheric aerosol. In *Topics in Atmospheric and Interstellar Physics and Chemistry* (F. Boutron, Ed.), Les Editions de Physique, Sciences, Les Ulis, France ERCA, Vol. 1, Chapter XII, pp. 251–270.

Hess, M., P. Koepke and I. Schult (1998), Optical properties of aerosols and clouds: The software package OPAC, *Bull. Am. Met. Soc.*, **79**, 5, 831–844.

Hignett, P., J. P. Taylor, P. N. Francis and M. D. Glew (1999), Comparison of observed and modeled direct aerosol forcing during TARFOX, *J. Geophys. Res.*, **104**, D2, 2279–2287, doi: 10.1029/98JD02021.

Hofmann, D. J. and J. M. Rosen (1983a), Stratospheric sulphuric acid fraction and mass estimate for the 1982 volcanic eruption of El Chichón, *Geophys. Res. Lett.*, **10**, 313–316, doi:10.1029/GL010i004p00313.

Hofmann, D. J. and J. M. Rosen (1983b), Sulfuric acid droplet formation and growth in the stratosphere after the 1982 eruption of El Chichon, *Science*, **222**, 325–327, doi: 10.1126/science.222.4621.325.

Holben, B. N., T. F. Eck, I. Slutsker, D. Tanré, J. P. Buis, A. Setzer, E. Vermote, J. A. Reagan, Y. J. Kaufman, T. Nakajima, F. Lavenu, I. Jankowiak and A. Smirnov (1998), AERONET – A federated instrument network and data archive for aerosol characterization, *Remote Sens. Environ.*, **66**, 1–16, doi: S0034-4257(98)00031-5.

Hummel, J. R., E. P. Shettle and D. R. Longtin (1988), A new background stratospheric aerosol model for use in atmospheric radiation models, *Scientific Report No. 8*, AFGL-TR-88-0166, Air Force Geophysics Laboratory, Hanscom Air Force Base, Massachusetts, 01731-5000, 30 July 1988.

IPCC (Intergovernmental Panel on Climate Change) (1996), *Climate Change 1995, The Science of Climate Change* [Houghton, J. T., L. G. Meira Filho, B. A. Callander, N. Harris, A. Kattenberg, and K. Maskell (eds.)], Cambridge University Press, Cambridge (United Kingdom), 572 pp.

IPCC TAR (2001), Third Assessment Report, Climate Change 2001, Working Group I: The Scientific Basis [Houghton, J. T., Y. Ding, D. J. Griggs, M. Noguer, P. J. van der Linden, X. Dai, K. Maskell, and C. A. Johnson (eds.)], Cambridge University Press, Cambridge (United Kingdom) and New York (NY, USA), 881 pp.

Iqbal, M. (1983), *An Introduction to Solar Radiation*, Academic Press, Toronto, pp. 1-58.

Irvine, W. M. and J. B. Pollack (1968), Infrared optical properties of water and ice spheres, *Icarus*, **8**, 342–360, doi: 10.1016/0019-1035(68)90083-3.

Jacobson, M. Z. (2002), Control of fossil-fuel particulate black carbon and organic matter, possibly the most effective method of slowing global warming, *J. Geophys. Res.*, **107**, D19, 4410, doi:10.1029/2001JD001376.

Jacquemoud, S. and F. Baret (1990), PROSPECT: A model of leaf optical properties spectra, *Remote Sens. Environ.*, **34**, 75–91, doi: 10.1016/0034-4257(90)90100-Z.

Jiang, H., H. Xue, A. Teller, G. Feingold and Z. Levin (2006), Aerosol effects on the lifetime of shallow cumulus, *Geophys. Res. Lett.*, **33**, L14806, doi: 10.1029/2006GL026024..

Johnson, B. T., K. P. Shine, and P. M. Forster (2004), The semi-direct aerosol effect: Impact of absorbing aerosols on marine stratocumulus, *Q. J. R. Meteorol. Soc.*, **130**, 1407–1422, doi: 10.1256/qj.03.61.

Jupp, D. L. B. (2000), A compendium of kernel and other (semi-)empirical BDRF models. CSIRO, Office of the Space Science Applications, Earth Observation Centre, pp. 18, available on http://www.cossa.csiro.au/tasks/brdf/k_summ.pdf.

Kasten, F. and A. T. Young (1989), Revised optical air mass tables and approximation formula, *Appl. Opt.*, **28**, 4735–4738, doi: 10.1364/AO.28.004735

Kaufman, Y. J. (1987), Satellite sensing of aerosol absorption, *J. Geophys. Res.*, **92**, D4, 4307–4317, doi: 10.1029/JD092iD04p04307.

Kaufman, Y. J., and R. S. Fraser (1997), The effect of smoke particles on clouds and climate forcing, *Science*, **277**, 1636–1639, doi: 10.1126/science.277.5332.1636.

Kaufman, Y. J., D. Tanré and O. Boucher (2002a), A satellite view of aerosols in the climate system, *Nature*, **419**, 6903, 215–223, doi:10.1038/nature01091.

Kaufman, Y. J. and I. Koren (2006), Smoke and pollution aerosol effect on cloud cover, *Science*, **313**, 5787, 655–658, doi: 10.1126/science.1126232.

Kiehl, J. T. and B. P. Briegleb (1993), The relative roles of sulfate aerosols and greenhouse gases in climate forcing, *Science*, **260**, 5106, 311–314, doi: 10.1126/science.260.5106.311.

Kimes, D., W. W. Newcomb, C. J. Tucker, I. S. Zonneveld, W. Van Wijngaarden, J. De Leeuw and G. F. Epema (1985), Directional reflectance factor distributions for cover types of Northern Africa, *Remote Sens. Environ.*, **18**, 1–19, doi: 10.1016/0034-4257(85)90034-3.

King, M., D. Harshvardhan and A. Arking (1984), A model of the radiative properties of the El Chichon stratospheric aerosol layer, *J. Clim. Appl. Meteor.*, **23**, 7, 1121–1137.

King, M. D., Y. J. Kaufman, D. Tanré and T. Nakajima (1999), Remote sensing of tropospheric aerosols from space: past, present and future, *Bull. Amer. Met. Soc.*, **80**, 11, 2229–2259.

Kneizys, F. X., L. W. Abreu, G. P. Anderson, J. H. Chetwynd, E. P. Shettle, A. Berk, L. S. Bernstein, D. C. Robertson, P. Acharya, L. S. Rothman, J. E. A. Selby, W. O. Gallery and S. A. Clough (1996), The MODTRAN 2/3 Report and LOWTRAN 7 model [Abreu, L. W., and G. P. Anderson (eds.)], Contract F19628-91-C.0132, Phillips Laboratory, Geophysics Directorate, PL/GPOS, Hanscom AFB, Massachusetts, 261 pp.

Koepke, P. (1984), Effective reflectance of oceanic whitecaps, *Appl. Opt.*, **23**, 1816–1824.

Koepke, P., M. Hess, I. Schult and E. P. Shettle (1997), Global Aerosol Data Set. MPI Meteorologie Hamburg Report No. 243, 44 pp.

Kokhanovsky, A. A. (2004), Spectral reflectance of whitecaps, *J. Geophys. Res.*, **109**, C05021, doi:10.1029/2003JC002177.

Kokhanovsky, A. A., and F.-M. Breon (2012), Validation of an analytical snow BRDF model using PARASOL multi-angular and multispectral observations, *IEEE Geosci. Rem. Sens. Letters*, **9**, 928–932, doi: 10.1109/LGRS.2012.2185775.

Kopp, G. and J. L. Lean (2011), A new, lower value of total solar irradiance: Evidence and climate significance, *Geophys. Res. Lett.*, **38**, L01706, doi: 10.10129/2010GL045777.

Koren, I., J. V. Martins, L. A. Remer and H. Afargan (2008), Smoke invigoration versus inhibition of clouds over the Amazon, *Science*, **321**, 5891, doi: 10.1126/science.1159185.

Kriebel, K. T. (1978), Measured spectral bidirectional reflection properties of four vegetated surfaces, *Appl. Opt.*, **17**, 253–259, doi: 10.1364/AO.17.000253.

Krüger, O. and H. Grassl (2002), The indirect aerosol effect over Europe, *Geophys. Res. Lett.*, **29**, 1925, doi: 10.1029/2001GL014081.

Kuusk, A. (1994), A multispectral canopy reflectance model, *Remote Sens. Environ.*, **50**, 75–82, doi: 10.1016/0034-4257(94)90035-3.

Lewis, P. (1995), On the implementation of linear kernel-driven BRDF models, *Proc. Ann. Conf. of Remote Sensing Society* '95, 'Remote Sensing in Action', Southampton, UK, 11–14 Sept., 1995, 333–340.

Lewis, P. and M. J. Barnsley (1994), Influence of the sky radiance distribution on various formulations of the Earth surface albedo, *Proc. Sixth Internat. Symposium on Physical Measurements and Signatures in Remote Sensing*, Val d'Isere (France), January 17–21, 1994, pp. 707–715.

Loeb, N. G. and N. Manalo-Smith (2005), Top-of-Atmosphere direct radiative effect of aerosols over global oceans from merged CERES and MODIS observations. *J. Climate*, **18**, 17, 3506–3526, doi: 10.1175/JCLI3504.1.

Lohmann, U. and J. Feichter (2005), Global indirect aerosol effects: A review, *Atmos. Chem. Phys.*, **5**, 715–737, doi:10.5194/acp-5-715-2005.

Lubin, D., S. K. Satheesh, G. McFarquar and A. J. Heymsfield (2002), Longwave radiative forcing of Indian Ocean tropospheric aerosol, *J. Geophys. Res.*, **107**, D19, 8004, doi: 10.1029/2001JD001183.

Lucht, W., C. B. Schaaf and A. H. Strahler (2000), An algorithm for the retrieval of albedo from space using semiempirical BRDF models, *IEEE Trans. Geosci., Remote Sens.*, **38**, 2, Part 2, 977–998, doi: 10.1109/36.841980.

Marsden, D. and F. P. J. Valero (2004), Observation of water vapor greenhouse absorption over the Gulf of Mexico using aircraft and satellite data, *J. Atmos. Sci.*, **61**, 745-753.

Menon, S., J.-L. Brenguier, O. Boucher, P. Davison, A. D. Del Genio, J. Feichter, S. Ghan, S. Guibert, X. Liu, U. Lohmann, H. Pawlowska, J. E. Penner, J. Quaas, D. L. Roberts, L. Schüller, and J. Snider (2003), Evaluating aerosol/cloud/radiation process parameterizations with single-column models and Second Aerosol Characterization Experiment (ACE-2) cloudy column observations, *J. Geophys. Res.*, **108**, D24, 4762, doi:10.1029/2003JD003902.

McClatchey, R. A., H. J. Bolle and K. Ya. Kondratyev (1980), Report of the IAMAP Radiation Commission working group on a Standard Radiation Atmosphere. WMO/IAMAP, 33 pp (available from AFGL, Hanscom Air Force Base, MA 01731).

McCormick, R. A. and J. H. Ludwig (1967), Climate modification by atmospheric aerosols, *Science*, **156**, 3780, 1358–1359, doi: 10.1126/science.156.3780.1358.

Mie, G. (1908), Beiträge zur Optik trüber Medien, speziell kolloidaler Metallösungen, *Annalen der Physik*, Vierte Folge, **25**, 3, 377–445.

Morel, A. (1988), Optical modeling of the upper ocean in relation to its biogenous matter content (Case I waters), *J. Geophys. Res.*, **93**, C9, 10749–10768, doi: 10.1029/JC093iC09p10749.

Nakajima, T., G. Tonna, R. Rao, P. Boi, Y. Kaufman and B. Holben (1996), Use of sky brightness measurements from ground for remote sensing of particulate polydispersions, *Appl. Opt.*, **35**, 2672–2686, doi: 10.1364/AO.35.002672.

Nakajima, T., M. Sekiguchi, T. Takemura, I. Uno, A. Higurashi, D. Kim, B. J. Sohn, S.-N. Oh, T. Y. Nakajima, S. Ohta, I. Okada, T. Takemura and K Kawamoto (2003), Significance of direct and indirect radiative forcings of aerosols in the East China Sea region, *J. Geophys. Res.*, **108**, D23, 8658, doi:10.1029/2002JD003261.

Nicodemus, F. E., J. C. Richmond, J. J. Hsia, I. W. Ginsberg and T. Limperis (1977), Geometrical considerations and nomenclature for reflectance, *Natl. Bur. Stand. Rep., NBS MN-160*, 52 pp.

Nilson, T. and A. Kuusk (1989), A reflectance model for the homogeneous plant canopy and its inversion, *Remote Sens. Environ.*, **27**, 157–167, doi: 10.1016/0034-4257(89)90015-1.

Palmer, K. F. and D. Williams (1975), Optical constants of sulphuric acid; application to the clouds of Venus?, *Appl. Opt.*, **14**, 208–219, doi: 10.1364/AO.14.000208.

Patterson, E. M. (1977), Atmospheric extinction between 0.55 μm and 10.6 μm due to soil-derived aerosols, *Appl. Opt.*, **16**, 2414–2418, doi: 10.1364/AO.16.002414.

Patterson, E. M. (1981), Optical properties of the crustal aerosol: Relation to chemical and physical characteristics, *J. Geophys. Res.*, **86**, C4, 3236–3246, doi: 10.1029/JC086iC04p03236.

Patterson, E. M., D. A. Gillette and B. H. Stockton (1977), Complex index of refraction between 300 and 700 nm for Saharan aerosols, *J. Geophys. Res.*, **82**, 21, 3153–3160, doi: 10.1029/JC082i021p03153.

Penner, J. E., M. Andreae, H. Annergarn, L. Barrie, J. Feichter, D. Hegg, A. Jayaraman, R. Leaitch, D. Murphy, J. Nganga, and G. Pitari (2001), Aerosols, their direct and indirect effects. In: *Climate Change 2001: The Scientific Basis. Contribution of Working Group I to the Third Assessment Report of the Intergovernmental Panel on Climate Change* [Houghton, J. T., Y. Ding, D. J. Griggs, M. Noguer, P. J. van der Linden, X. Dai, K. Maskell, and C. A. Johnson (eds.)], Cambridge University Press, Cambridge (United Kingdom) and New York (NY, USA), pp. 289–348.

Price, J. C. (1990), On the information content of soil reflectance spectra. *Remote Sens. Environ.*, **33**, 113–121, doi: 10.1016/0034-4257(90)90037-M.

Pueschel, R. F., K. G. Snetsinger, J. K. Goodman, O. B. Toon, G. V. Ferry, V. R. Oberbeck, J. M. Livingston, S. Verma, W. Fong, W. L. Starr and K. R. Chan (1989), Condensed nitrate, sulfate, and chloride in Antarctic stratospheric aerosols, *J. Geophys. Res.*, **94**, D9, 11271–11284, doi: 10.1029/JD094iD09p11271.

Pueschel, R. F., S. A. Kinne, P. B. Russell, K. G. Snetsinger and J. M. Livingston (1993), Effects of the 1991 Pinatubo volcanic eruption on the physical and radiative properties of stratospheric aerosols. In *IRS '92: Current Problems in Atmospheric Radiation, Proceedings of the International Radiation Symposium* [Keevallik, S., and O. Kärner (eds.)], Tallinn (Estonia), 2–8 August 1992, A. Deepak Publishing, Hampton, Virginia (USA), pp. 183–186.

Quaas, J., O. Boucher, N. Bellouin and S. Kinne (2008), Satellite-based estimate of the direct and indirect aerosol climate forcing, *J. Geophys. Res.*, **113**, D05204, doi:10.1029/2007JD008962.

Rahman, H., B. Pinty and M. M. Verstraete (1993), Coupled surface–atmosphere reflectance (CSAR) model 2. Semiempirical surface model usable with NOAA Advanced Very High Resolution Radiometer data, *J. Geophys. Res.*, **98**, D11, 20791–20801, doi: 10.1029/93JD02072.

Ramanathan, V., P. J. Crutzen, J. T. Kiehl and D. Rosenfeld (2001a), Aerosols, climate and the hydrological cycle, *Science*, **294**, 5549, 2119–2124, doi: 10.1126/science.1064034.

Ramanathan, V., P. J. Crutzen, J. Lelieveld, A. P. Mitra, D. Althausen, J. Anderson, M. O. Andreae, W. Cantrell, G. R. Cass, C. E. Chung, A. D. Clarke, J. A. Coakley, W. D. Collins, W. C. Conant, F. Dulac, J. Heintzenberg, A. J. Heymsfield, B. Holben, S. Howell, J. Hudson, A. Jayaraman, J. T. Kiehl, T. N. Krishnamurti, D. Lubin, G. McFarquhar, T. Novakov, J. A. Ogren, I. A. Podgorny, K. Prather, K. Priestley, J. M. Prospero, P. K. Quinn, K. Rajeev, P. Rasch, S. Rupert, R. Sadourny, S. K. Satheesh, G. E. Shaw, P. Sheridan and F. P. J. Valero (2001b), Indian Ocean Experiment: An integrated analysis of the climate forcing and effects of the great Indo-Asian haze. *J. Geophys. Res.*, **106**, D22, 28371–28398, doi: 10.1029/2001JD900133.

Ramaswamy, V., O. Boucher, J. Haigh, D. Hauglustaine, J. Haywood, G. Myhre, T, Nakajima, G. Y. Shi, and S. Solomon (2001), Radiative forcing of climate change. In: *Climate Change 2001: The Scientific Basis. Contribution of Working Group I to the Third Assessment Report of the Intergovernmental Panel on Climate Change* [Houghton, J. T., Y. Ding, D. J. Griggs, M. Noguer, P. J. van der Linden, X. Dai, K. Maskell, and C. A. Johnson (eds.)], Cambridge University Press, Cambridge (United Kingdom) and New York (NY, USA), pp. 349–416.

Remer, L. A. and Y. J. Kaufman (2006), Aerosol direct radiative effect at the top of the atmosphere over cloud free ocean derived from four years of MODIS data, *Atmos. Chem. Phys.*, **6**, 1, 237–253, doi:10.5194/acp-6-237-2006.

Ricchiazzi, P., S. Yang, C. Gautier and D. Sowle (1998), SBDART: A research and teaching software tool for plane-parallel radiative transfer in the Earth's atmosphere, *Bull. Am. Met. Soc.*, **79**, 10, 2101–2114.

Ricchiazzi, P., W. O'Hirok and C. Gautier (2005), The effect of non-lambertian surface reflectance on aerosol radiative forcing, *Fifteenth ARM Science Team Meeting Proceedings*, Daytona Beach, Florida, March 14–18, 2005, 8 pp.

Román, M. O., C. B. Schaaf, P. Lewis, F. Gao, G. P. Anderson, J. L. Privette, A. H. Strahler, C. E. Woodcock and M. Barnsley (2010), Assessing the coupling between surface albedo derived from MODIS and the fraction of diffuse skylight over spatially-characterized landscapes, *Remote Sens. Environ.*, **114**, 4, 738–760, doi: 10.1016/j.rse.2009.11.014.

Russell, P. B., S. A. Kinne and R. W. Bergstrom (1997), Aerosol climate effects: Local radiative forcing and column closure experiments, *J. Geophys. Res.*, **102**, D8, 9397–9407, doi: 10.1029/97JD00112.

Russell, P. B., J. M. Livingston, P. Hignett, S. Kinne, J. Wong, A. Chien, R. Bergstrom, P. Durkee and P. V. Hobbs (1999), Aerosol-induced radiative flux changes off the United States mid-Atlantic coast: Comparison of values calculated from sunphotometer and in situ data with those measured by airborne pyranometer, *J. Geophys. Res.*, **104**, D2, 2289–2307, doi: 10.1029/1998JD200025.

Satheesh, S. K. and V. Ramanathan (2000), Large differences in tropical aerosol forcing at the top of the atmosphere and Earth's surface, *Nature*, **405**, 6782, 60–63, doi: 10.1038/35011039.

Schwartz, S. E. and M. O. Andreae (1996), Uncertainty in climate change caused by aerosols, *Science*, **272**, 5265, 1121–1122, doi: 10.1126/science.272.5265.1121.

Schwartz, S. E., F. Arnold, J.-P. Blanchet, P. A. Durkee, D. J. Hofmann, W. A. Hoppel, M. D. King, A. A. Lacis, T. Nakajima, J. A. Ogren, O. B. Toon and M. Wendisch (1995), Group report: Connections between aerosol properties and forcing of climate. In *Aerosol Forcing of Climate* (R. J. Charlson and J. Heintzenberg, Eds), John Wiley & Sons, New York, pp. 251–280.

Seinfeld, J. H. and S. N. Pandis (2006), *Atmospheric Chemistry and Physics, from Air Pollution to Climate Change* (Second Edition), John Wiley & Sons, New York (USA), 1225 pp.

Shettle, E. P. (1984), Optical and radiative properties of a desert aerosol model. In *IRS '84: Current Problems in Atmospheric Radiation, Proceedings of the International Radiation Symposium* [Fiocco, G. (ed.)], Perugia, Italy, 21–28 August 1984, A. Deepak Publishing, Hampton, Virginia, USA, 74–77.

Shettle, E. P. and R. W. Fenn (1979), Models for the aerosols of the lower atmosphere and the effects of humidity variations on their optical properties, *Environ. Res. Papers*, No. 676, Air Force Geophys. Lab., AFGL-Techn. Rep. 79-0214, Hanscom AFB, MA, 94 pp.

Sobolev, V. V. (1975), *Light Scattering in Planetary Atmospheres*, New York, Pergamon Press, 256 pp.

Stott, P. A., J. F. B. Mitchell, M. R. Allen, T. L. Delworth, J. M. Gregory, G. A. Meehl and B. D. Santer (2006), Observational constraints on past attributable warming and predictions of future global warming, *J. Climate*, **19**, 3055–3069.

Takemura, T., T. Nakajima, O. Dubovik, B. N. Holben and S. Kinne (2002), Single-scattering albedo and radiative forcing of various aerosol species with a global three-dimensional model, *J. Climate*, **15**, 333–352.

Takemura, T., T. Nozawa, S. Emori, T. Y. Nakajima and T. Nakajima (2005), Simulation of climate response to aerosol direct and indirect effects with aerosol transport-radiation model, *J. Geophys. Res.*, **110**, D02202, doi:10.1029/2004JD005029.

Tanré, D., J. Haywood, J. Pelon, J. F. Léon, B. Chatenet, P. Formenti, P. Francis, P. Goloub, E. J. Highwood and G. Myhre (2003), Measurement and modeling of the Saharan dust radiative impact: Overview of the Saharan Dust Experiment (SHADE), *J. Geophys. Res.*, **108**, D18, 8574, doi:10.1029/2002JD003273.

Tomasi, C., F. Prodi and F. Tampieri (1979), Atmospheric turbidity variations caused by layers of Sahara dust particles, *Beitr. Phys. Atmos.*, **52**, 3, 215–228.

Tomasi, C., V. Vitale and L. Tarozzi (1997), Sun-photometric measurements of atmospheric turbidity variations caused by the Pinatubo aerosol cloud in the Himalayan region during the summer periods of 1991 and 1992, *Il Nuovo Cimento*, **20 C**, 61–88.

Tomasi, C., V. Vitale and L. V. De Santis (1998), Relative optical mass functions for air, water vapor, ozone and nitrogen dioxide in atmospheric models presenting different latitudinal and seasonal conditions, *Meteorol. Atmos. Phys.*, **65**, 11–30, doi: 10.1007/BF01030266.

Tomasi, C., V. Vitale, A. Lupi, C. Di Carmine, M. Campanelli, A. Herber, R. Treffeisen, R. S. Stone, E. Andrews, S. Sharma, V. Radionov, W. von Hoyningen-Huene, K. Stebel, G. H. Hansen, C. L. Myhre, C. Wehrli, V. Aaltonen, H. Lihavainen, A. Virkkula, R. Hillamo, J. Ström, C. Toledano, V. Cachorro, P. Ortiz, A. de Frutos, S. Blindheim, M. Frioud, M. Gausa, T. Zielinski, T. Petelski and T. Yamanouchi (2007), Aerosols in polar regions: A historical overview based on optical depth and in situ observations, *J. Geophys. Res.*, **112**, D16205, doi:10.1029/2007JD008432.

Tomasi, C., A. Lupi, M. Mazzola, R. S. Stone, E. G. Dutton, A. Herber, V. F. Radionov, B. N. Holben, M. G. Sorokin, S. M. Sakerin, S. A. Terpugova, P. S. Sobolewski, C. Lanconelli, B. H. Petkov, M. Busetto and V. Vitale (2012), An update on polar aerosol optical properties using POLAR-AOD and other measurements performed during the International Polar Year, *Atmos. Environ.*, **52**, 29–47, doi:10.1016/j.atmosenv.2012.02.055.

Turco, R. P., R. C. Whitten and O. B. Toon (1982), Stratospheric aerosols: observation and theory, *Rev. Geophys. Space Phys.*, **20**, 2, 233–279, doi: 10.1029/RG020i002p00233.

Twitty, J. T. and J. A. Weinman (1971), Radiative properties of carbonaceous aerosols, *J. Appl. Meteorol.*, **10**, 4, 725–731.

Twomey, S. A. (1977), The influence of pollution on the shortwave albedo of clouds, *J. Atmos. Sci.*, **34**, 7, 1149–1152.

Valero, F. P. J. and P. Pilewskie (1992), Latitudinal survey of spectral optical depths of the Pinatubo volcanic cloud-derived particle sizes, columnar mass loadings, and effects on planetary albedo, *Gephys. Res. Lett.*, **19**, 2, 163–166, doi: 10.1029/92GL00074.

Valero, F. P. J. and B. C. Bush (1999), Measured and calculated clear-sky solar radiative fluxes during the Subsonic Aircraft Contrail and Cloud Effects Special Study (SUCCESS), *J. Geophys. Res.*, **104**, D22, 27387–27398, doi: 10.1029/1999JD900947.

Vermote, E. F., D. Tanré, J. L. Deuzé, M. Herman and J.-J. Morcrette (1997a), Second simulation of the satellite signal in the solar spectrum (6S): An overview, *IEEE Trans. Geosci. Remote Sens.*, **35**, 3, 675–686, doi: 10.1109/36.581987.

Vermote, E. F., D. Tanré, J. L. Deuzé, M. Herman and J. J. Morcrette (1997b), *Second Simulation of the Satellite Signal in the Solar Spectrum (6S)*, 6S User Guide Version 2, July 1997. Universitè de Lille, France, 218 pp.

Volz, F. E. (1972a), Infrared absorption by atmospheric aerosol substances, *J. Geophys. Res.*, **77**, 6, 1017–1031, doi: 10.1029/JC077i006p01017.

Volz, F. E. (1972b), Infrared refractive index of atmospheric aerosol substances, *Appl. Opt.*, **11**, 4, 755–759, doi: 10.1364/AO.11.000755.

Volz, F. E. (1973), Infrared optical constants of ammonium sulfate, Sahara dust, volcanic pumice, and flyash, *Appl. Opt.*, **12**, 3, 564–568, doi: 10.1364/AO.12.000564.

Vonder Haar, T. H. and V. E. Suomi (1971), Measurements of the Earth's radiation budget from satellites during a five-year period. Part I: Extended time and space means, *J. Atmos. Sci.*, **28**, 305–314, doi:10.1175/1520-0469(1971)028<0305:MOTERB>2.0.CO;2.

Waliser, D. E. and N. E. Graham (1993), Convective cloud systems and warm-pool sea surface temperatures: Coupled interactions and self-regulation, *J. Geophys. Res.*, **98**, D7, 12,881–12,893, doi: 10.1029/93JD00872.

Wanner, W., A. H. Strahler, B. Hu, P. Lewis, J.-P. Muller, X. Li, C. L. B. Schaaf and M. J. Barnsley (1997), Global retrieval of bidirectional reflectance and albedo over land from EOS MODIS and MISR data: Theory and algorithm, *J. Geophys. Res.*, **102**, D14, 17,143–17,161, doi: 10.1029/96JD03295.

Warren, S. G. (1984), Optical constants of ice from the ultraviolet to the microwave, *Appl. Opt.*, **23**, 1206–1225, doi: 10.1364/AO.23.001206.

Warren, S. G. and W. J. Wiscombe (1980), A model for the spectral albedo of snow. II: Snow containing atmospheric aerosols, *J. Atmos. Sci.*, **37**, 12, 2734–2745.

Wiscombe, W. J. and G. W. Grams (1976), The backscattered fraction in two-stream approximations, *J. Atmos. Sci.*, **33**, 12, 2440–2451.

Wiscombe, W. J. and S. G.. Warren (1980), A model for the spectral albedo of snow. I: Pure snow, *J. Atmos. Sci.*, **37**, 12, 2712–2733.

WMO (1983), World Meteorological Organization (WMO/CAS)/Radiation Commission of IAMAP Meeting of experts on aerosols and their climatic effects, WCP 55, Williamsburg, Virginia, U.S.A., March 28–30, 1983.

WMO (1986), Atmospheric Ozone 1985. WMO Global Ozone Research and Monitoring Project, Report No. 16, Geneva (Switzerland).

Yu, H., S. C. Liu and R. E. Dickinson (2002), Radiative effects of aerosols on the evolution of the atmospheric boundary layer, *J. Geophys. Res.*, **107**, 4142, doi:10.1029/2001JD000754.

Yu, H., R. Dickinson, M. Chin, Y. J. Kaufman, M. Zhou, L. Zhou, Y. Tian, O. Dubovik and B. Holben (2004), Direct radiative effect of aerosols as determined from a combination of MODIS retrievals and GOCART simulations, *J. Geophys. Res.*, **109**, D03206, doi:10.1029/2003JD003914.

Zdunkowski, W. G., R. M. Welch, and G. Korb (1980), An investigation of the structure of typical two-stream-methods for the calculation of solar fluxes and heating rates in clouds, *Beitr. Phys. Atmos.*, **53**, 147–166.

Zhang, J. and S. A. Christopher (2003), Longwave radiative forcing of Saharan dust aerosols estimated from MODIS, MISR, and CERES observations on Terra., *Geophys. Res. Lett.*, **30**, 2188, doi:10.1029/2003GL018479.

Zhao, T. X.-P., H. Yu, I. Laszlo, M. Chin and W. C. Conant (2008), Derivation of component aerosol direct radiative forcing at the top of atmosphere for clear-sky oceans, *J. Quant. Spectrosc. Rad. Transfer*, **109**, 7, 1162–1186, doi: 10.1016/j.jqsrt.2007.10.006.

Index

MIX
Papier aus verantwortungsvollen Quellen
Paper from responsible sources
FSC® C105338

If you have any concerns about our products,
you can contact us on
ProductSafety@springernature.com

In case Publisher is established outside the EU,
the EU authorized representative is:
Springer Nature Customer Service Center GmbH
Europaplatz 3, 69115 Heidelberg, Germany

Printed by Libri Plureos GmbH
in Hamburg, Germany